WebElements: the periodic table on the world-wide web

http://www.webelements.com/

Key:

element name atomic number element symbol 1997 atomic weight (mean relative mass)

1	2		3	4	5	6	7	8	9	10	11	12	13	14	15	16	17	18
hydrogen 1 H 1.00794(7)																		helium 2 He 4.002602(2)
lithium 3 Li 6.941(2)	beryllium 4 Be 9.012182(3)												boron 5 B 10.811(7)	carbon 6 C 12.0107(8)	nitrogen 7 N 14.00674(7)	oxygen 8 O 15.9994(3)	fluorine 9 F 18.9984032(5)	neon 10 Ne 20.1797(6)
sodium 11 Na 22.989770(2)	magnesium 12 Mg 24.3050(6)												aluminium 13 Al 26.981538(2)	silicon 14 Si 28.0855(3)	phosphorus 15 P 30.973761(2)	sulfur 16 S 32.066(6)	chlorine 17 Cl 35.4527(9)	argon 18 Ar 39.948(1)
potassium 19 K 39.0983(1)	calcium 20 Ca 40.078(4)		scandium 21 Sc 44.955910(8)	titanium 22 Ti 47.867(1)	vanadium 23 V 50.9415(1)	chromium 24 Cr 51.9961(6)	manganese 25 Mn 54.938049(9)	iron 26 Fe 55.845(2)	cobalt 27 Co 58.933200(9)	nickel 28 Ni 58.6934(2)	copper 29 Cu 63.546(3)	zinc 30 Zn 65.39(2)	gallium 31 Ga 69.723(1)	germanium 32 Ge 72.61(2)	arsenic 33 As 74.92160(2)	selenium 34 Se 78.96(3)	bromine 35 Br 79.904(1)	krypton 36 Kr 83.80(1)
rubidium 37 Rb 85.4678(3)	strontium 38 Sr 87.62(1)		yttrium 39 Y 88.90585(2)	zirconium 40 Zr 91.224(2)	niobium 41 Nb 92.90638(2)	molybdenum 42 Mo 95.94(1)	technetium 43 Tc [98.9063]	ruthenium 44 Ru 101.07(2)	rhodium 45 Rh 102.90550(2)	palladium 46 Pd 106.42(1)	silver 47 Ag 107.8682(2)	cadmium 48 Cd 112.411(8)	indium 49 In 114.818(3)	tin 50 Sn 118.710(7)	antimony 51 Sb 121.760(1)	tellurium 52 Te 127.60(3)	iodine 53 I 126.90447(3)	xenon 54 Xe 131.29(2)
caesium 55 Cs 132.90545(2)	barium 56 Ba 137.327(7)	57-70 *	lutetium 71 Lu 174.967(1)	hafnium 72 Hf 178.49(2)	tantalum 73 Ta 180.9479(1)	tungsten 74 W 183.84(1)	rhenium 75 Re 186.207(1)	osmium 76 Os 190.23(3)	iridium 77 Ir 192.217(3)	platinum 78 Pt 195.078(2)	gold 79 Au 196.96655(2)	mercury 80 Hg 200.59(2)	thallium 81 Tl 204.3833(2)	lead 82 Pb 207.2(1)	bismuth 83 Bi 208.98038(2)	polonium 84 Po [208.9824]	astatine 85 At [209.9871]	radon 86 Rn [222.0176]
francium 87 Fr [223.0197]	radium 88 Ra [226.0254]	89-102 **	lawrencium 103 Lr [262.110]	rutherfordium 104 Rf [261.1089]	dubnium 105 Db [262.1144]	seaborgium 106 Sg [263.1186]	bohrium 107 Bh [264.12]	hassium 108 Hs [265.1306]	meitnerium 109 Mt [268]	ununnilium 110 Uun [269]	unununium 111 Uuu [272]	ununbium 112 Uub [277]		ununquadium 114 Uuq [289]		ununhexium 116 Uuh [289]		ununoctium 118 Uuo [293]

*lanthanides	lanthanum 57 La 138.9055(2)	cerium 58 Ce 140.116(1)	praseodymium 59 Pr 140.90765(2)	neodymium 60 Nd 144.24(3)	promethium 61 Pm [144.9127]	samarium 62 Sm 150.36(3)	europium 63 Eu 151.964(1)	gadolinium 64 Gd 157.25(3)	terbium 65 Tb 158.92534(2)	dysprosium 66 Dy 162.50(3)	holmium 67 Ho 164.93032(2)	erbium 68 Er 167.26(3)	thulium 69 Tm 168.93421(2)	ytterbium 70 Yb 173.04(3)
**actinides	actinium 89 Ac [227.0277]	thorium 90 Th 232.0381(1)	protactinium 91 Pa 231.03588(2)	uranium 92 U 238.0289(1)	neptunium 93 Np [237.0482]	plutonium 94 Pu [244.0642]	americium 95 Am [243.0614]	curium 96 Cm [247.0703]	berkelium 97 Bk [247.0703]	californium 98 Cf [251.0796]	einsteinium 99 Es [252.0830]	fermium 100 Fm [257.0951]	mendelevium 101 Md [258.0984]	nobelium 102 No [259.1011]

Symbols and names: the symbols of the elements, their names, and their spellings are those recommended by IUPAC. After some controversy, the names of elements 101-109 are now confirmed: see Pure & Appl. Chem., 1997, 69, 2471–2473. Names have not been proposed as yet for the most recently discovered elements 110–112, 114, 116, and 118 so those used here are IUPAC's temporary systematic names: see Pure & Appl. Chem., 1979, 51, 381–384. In the USA and some other countries, the spellings aluminum and cesium are normal while in the UK and elsewhere the usual spelling is sulphur.
Periodic table organisation: for a justification of the positions of the elements La, Ac, Lu, and Lr in the WebElements periodic table see W.B. Jensen, "The positions of lanthanum (actinium) and lutetium (lawrencium) in the periodic table", J. Chem. Ed., 1982, 59, 634–636.
Group labels: the numeric system (1–18) used here is the current IUPAC convention. For a discussion of this and other common systems see: W.C. Fernelius and W.H. Powell, "Confusion in the periodic table of the elements", J. Chem. Ed., 1982, 59, 504–508.
Atomic weights (mean relative masses): see Pure & Appl. Chem., 1996, 68, 2339–2359. These are the IUPAC 1995 values. Elements for which the atomic weight is contained within square brackets have no stable nuclides and are represented by one of the element's more important isotopes. However, the three elements thorium, protactinium, and uranium do have characteristic terrestrial abundances and these are the values quoted. The last significant figure of each value is considered reliable to ±1 except where a larger uncertainty is given in parentheses.
 For updates to this table see http://www.shef.ac.uk/chemistry/web-elements/webelements/support/media/pdf/periodic-table.html. Version date: 13 July 1999.

The Structure of Matter

TOPICS IN PHYSICAL CHEMISTRY

A Series of Advanced Textbooks and Monographs

Series Editor

Donald G. Truhlar, University of Minnesota

Spectra of Atoms and Molecules
P. Bernath

Chemical Reactions in Clusters
E. R. Bernstein

Physical Chemistry, Second Edition
R. S. Berry, S. A. Rice, and J. Ross

The Structure of Matter, Second Edition
R. S. Berry, S. A. Rice, and J. Ross

Matter in Equilibrium, Second Edition
R. S. Berry, S. A. Rice, and J. Ross

Physical and Chemical Kinetics, Second Edition
R. S. Berry, S. A. Rice, and J. Ross

Electronic Structure Calculations on Fullerenes and Their Derivatives
J. Cioslowski

Fractals in Molecular Biophysics
T. G. Dewey

An Introduction to Nonlinear Chemical Dynamics: Oscillations, Waves, Patterns and Chaos
I. R. Epstein and J. A. Pojman

Algebraic Theory of Molecules
F. Iachello and R. D. Levine

An Introduction to Hydrogen Bonding
G. A. Jeffrey

Computer Aided Design of High Temperature Materials
A. Pechenik, R. K. Kalia, and P. Vashishta

Molecular Orbital Calculations for Biological Systems
A.-M. Sapse

Hydrogen Bonding: A Theoretical Perspective
S. Scheiner

Quantum Mechanics in Chemistry
J. Simons and J. Nichols

The Structure of Matter

An Introduction to Quantum Mechanics

SECOND EDITION

R. STEPHEN BERRY
The University of Chicago

STUART A. RICE
The University of Chicago

JOHN ROSS
Stanford University

New York • Oxford
OXFORD UNIVERSITY PRESS
2002

Oxford University Press

Oxford New York
Athens Auckland Bangkok Bogotá Buenos Aires Calcutta
Cape Town Chennai Dar es Salaam Delhi Florence Hong Kong Istanbul
Karachi Kuala Lumpur Madrid Melbourne Mexico City Mumbai
Nairobi Paris São Paulo Shanghai Singapore Taipei Tokyo Toronto Warsaw

and associated companies in
Berlin Ibadan

Published by Oxford University Press, Inc.
198 Madison Avenue, New York, New York 10016
http://www.oup-usa.org

Oxford is a registered trademark of Oxford University Press

ISBN: 0-19-514748-0

Full Libary of Congress Cataloging-in-Publication Data for this title can be
found under Physical chemistry, 2nd ed. / R. Stephen Berry, Stuart A. Rice,
John Ross, ISBN: 0-19-510589-3.

9 8 7 6 5 4 3 2 1
Printed in the United States of America
on acid-free paper

To our families

Contents

Appendices

Preface

Soul of the world, inspired by thee,
The jarring seed of matter did agree,
Thou didst the scatter'd atoms bind,
Which, by the laws of true proportion joined
Made up of various parts one perfect harmony

—Text by Nicholas Brady for the *Ode on St. Cecilia's Day, 1692* by Henry Purcell

The first edition of this book was published twenty years ago. Our goal was to introduce an undergraduate student to physical chemistry in a manner that focused attention on the fundamental principles and conceptual structure of the field, following a logical development from the simple to the more complex descriptions of systems and phenomena. We also emphasized how the content, style, expectation of achievement, and even the nature of what is considered explanation, contribute to the definition of physical chemistry as a field of scholarly scientific endeavor, and keep it a vital contemporary discipline.

The current (second) edition of this book is intended to meet the same goal and to also be useful as a text for a first year graduate course in physical chemistry. Our revision of the text has resulted in the removal of errors and improvements in the quality of the arguments used. We have retained an emphasis on correctness and depth of conceptual argument without recourse to advanced mathematics. The basic concepts and their ramifications are presented in ways that establish both their logical structure and their realms of validity. Points that require close scrutiny in some situations are carefully discussed. The applications of physical chemistry connect with forefront problems in science. The additions in this edition are the following: considerable new material on nuclear magnetic resonance, especially on modern methods; expansion of the treatment of the fundamental principles of statistical mechanics and of the treatment of linear irreversible thermodynamics; updating of the descriptions of the equilibrium properties of liquids and solutions and of nonlinear chemical kinetics and electrode processes. Several vignettes covering forefront areas of research in physical chemistry, by leaders in the fields, have been added. Just as when we finished preparing the first edition of this book, even as we complete this edition, we see new results in the scientific journals that we would have incorporated in the text had they appeared in time.

The first edition of this book found use both as a text and as a reference book in physical chemistry. We trust that the second edition will also be used both ways.

Our approach to the teaching of physical chemistry is based on our experience as researchers and educators. In the Preface to the first edition of this book we described that approach in the following words.

Physical chemistry is an empirical science. A science is a set of constructs, called theories, that link fragments of experience into a consistent description of natural phenomena. The adjective "empirical" refers to the common experiences from which the theories grow, that is, to experiments. Simple working hypotheses are guessed by imaginative insight or intuition or luck, usually from a study of experiments. This repetitive interplay in time leads to the formulation of theories that correlate the accumulated experimental information and that can predict new phenomena with accuracy. As scientists we have, throughout our careers, endeavored to combine both experimental and theoretical work. In this book we try to knit these two inseparable parts of the science into a coherent structure that represents accurately the way they interact in physical chemistry today.

Our goal is the presentation of the three major areas of physical chemistry: molecular structure, the equilibrium properties of systems, and the kinetics of transformations of systems. The theoretical foundations of these subjects are, respectively, quantum mechanics, thermodynamics and equilibrium statistical mechanics, and chemical kinetics and kinetic theory. These theories, firmly based on experimental findings, constitute the structure required for the understanding of past accomplishments and a basis for recognition and development of significant new areas in physical chemistry. The presentation of the theories of physical chemistry requires careful discussions at several levels of exposition. Our approach aims toward depth of understanding of fundamentals more than toward breadth of recognition of the multitude of activities that go on under the name of physical chemistry. The organization of the book, with its three principal sections, should make this clear. The mathematical level begins with elementary calculus and rises to the use of simple properties of partial differential equations and the special functions that enter into their solutions. Our intention is to keep the reader's mind on the science rather than on the mathematics, especially at the beginning. This procedure also corresponds to the pattern, followed by many students, of taking physical chemistry and advanced calculus concurrently. Appendices develop the details of the mathematical tools as they are needed.

The text discussion contains more material than can be covered in the traditional one-year physical chemistry sequence; it is designed to fulfill the dual purpose of providing a clear and incisive treatment of fundamental principles at a level accessible to all students while broadening the perspectives and challenging the minds of the best students. Individual instructors will wish to make their own selections of material for inclusion and exclusion, respectively. We have provided guidance on this matter by suggesting, at the end of this Preface, schema for several different types of physical chemistry courses. These schema suggest both how to use the material discussed in a different order than in the text, and how to omit or downplay classes of topics deemed unsuitable to a particular group of students. For example, we have taught the junior-level physical chemistry course both as organized in this text and with the material of Part II preceding that of Part I. To invert the order of Parts II and I, it is only necessary to ask the student to accept the existence of quantized energy levels and a few specific examples of energy level spectra. This has proved quite easy for students, all of whom have heard of these matters in freshman chemistry courses. Students with strong backgrounds will have seen the material in the first chapter and parts of the second chapter. Chapter 10, on intermolecular forces, and Chapter 11, on the structure of solids, contain large blocks of material that could be passed over in a traditionally oriented one-year course. The thermodynamic description of matter can be emphasized and the statistical molecular description de-emphasized, or vice versa, by selection of the relevant sections of Chapters 21 to 26. Chapter 20, dealing with linear irreversible thermodynamics and negative temperature, can easily be omitted if so wished by the instructor. In Part III, the elements of physical kinetics or transport theory are contained in Sections 27.1 to 28.8 and 29.1; the elements of chemical kinetics are contained in Chapter 30. The specialized topics in the remaining sections may be used for more extensive treatments of these subjects, either in a three-semester physical chemistry sequence or in a senior-graduate course on these topics.

At the other end of the scale there is sufficient material in this book for a first-year graduate course for students whose undergraduate preparation in physical chemistry did not emphasize modern aspects of the subject.

At the end of each chapter we have suggested extra reading for the interested student. The book contains about 750 problems. A few of these problems are designed to acquaint the reader with dimensions, units, and simple manipulations. More are intended to develop intellectual skills, to enable students to master the material of the text discussions by the difficult process of thinking through the kinds of questions encountered in the laboratory. Some of the problems are designed to extend the theoretical analysis of the text to special but interesting situations, and some are designed to require the use of a computer for numerical calculation.

Part I opens with a review of the elementary quantities of the atomic world and how they are measured. Many readers will find some or all of the material in Chapter 1 familiar; they may choose to read quickly through this chapter and start their more intensive study with Chapter 2, where we develop the experimental evidence for the quantum structure of matter at the molecular level. In the process, we examine the harmonic oscillator, the primitive model for many kinds of behavior examined later, and the concept of action.

Chapter 3 is more theoretical and mathematical than its predecessors but begins at a level the well-prepared student will find quite elementary. Waves and wave equations are introduced and are used to describe only very simple situations: the states of particles in boxes of various sorts, and of rigid rotators.

Chapter 4 brings us back to the more realistic problems of the quantum-mechanical oscillator and the simplest atom, hydrogen.

In Chapter 5, we further develop the concepts such as orbitals and transitions between quantum states that are introduced for hydrogen in Chapter 4. Chapter 5 treats atoms with more than two electrons, in particular their electronic states and the interactions among the electrons.

Molecules are first introduced in Chapter 6, which is almost analogous to Chapter 4 in that it too goes in depth into the description and behavior of the simplest molecules, H_2^+ and H_2, just as Chapter 4 examined the H atom. The concepts developed in Chapter 6 are than extended to more complex diatomic molecules in Chapter 7. Chapter 6 deals almost exclusively with electronic states; Chapter 7 introduces molecular vibration and rotation and discusses concepts used to correlate and unify our observations regarding diatomic molecules.

Chapter 8 begins with the primitive species H_3^+ and H_3 and then goes on to ideas that begin to be important with three or more nuclei: hybrid orbitals and delocalized molecular orbitals at the level of electronic states, and normal modes of vibration. Larger molecules are discussed in Chapter 9; here we introduce the concepts of chirality and optical activity, and explore some aspects of ligand field theory and the magnetic properties of molecules.

Chapter 10 is the first in which we go beyond the properties of individual molecules. The discussion of intermolecular forces describes the interactions between charge distributions and how one molecule behaves when it collides with another. The material in this chapter is based wholly on the framework of molecular structure but becomes especially useful to us in Parts II and III, where we study the behavior of matter in the aggregate. Part I concludes with another structural aspect of aggregated matter, the structure of solids. Here, we extend the various concepts of bonding previously developed to include the concept of metallic bonding, to describe the structure and states of periodic condensed phases.

Part II is concerned with the equilibrium properties of bulk matter. Our presentation simultaneously develops the statistical molecular theory and the classical thermodynamic theory in a mutually reinforcing fashion. Despite use of this "mixing" of microscopic and macroscopic points of view, no compromise is made with respect to the rigor of classical thermodynamics, and if desired the two points of view can be separated.

In Chapter 12, we begin with a discussion of the zero-th law of thermodynamics and the concept of temperature. By examining the phenomenological bases for the equation of state and the definition of temperature, together with the elements of the kinetic theory of perfect gases, we establish a first connection between macroscopic and microscopic descriptions. The building of bridges between the two classes of description is a principal theme of succeeding chapters.

Chapters 13 and 14 treat the first law of thermodynamics and some of its many applications. Particular care is devoted to the precise definition of work and heat and to how the nature of these quantities exemplifies the differences between the thermodynamic and mechanical descriptions of matter.

Chapter 15 introduces the concept of entropy by way of the microscopic structure of matter. Given only that every sample of matter has an energy-level spectrum, it is shown that there exists a function of the density of states, the entropy, that behaves like a property only of macroscopic variables of the system, despite its definition in terms of the microscopic energy-level spectrum.

Chapter 16 develops the second law of thermodynamics via the classical Clausius and Kelvin principles, and Chapter 17 is devoted to examples of the use of the second law to solve problems of chemical interest. These chapters contain a careful discussion of the nature of irreversibility and its interpretation in terms of thermodynamic and statistical molecular theories. Chapter 18 introduces, discusses, and gives applications of the third law of thermodynamics.

In Chapter 19 we examine the central problem of describing equilibrium as a function of the external constraints on the system. The thermodynamic theory of open systems is developed and is used to derive the several criteria of equilibrium that are suitable to different external constraints. With this background, the notion of ensemble is introduced, and the classical thermodynamics of equilibrium is related to the development of the grand canonical, canonical, and microcanonical partition functions of statistical mechanics. The theory is illustrated by analyzing the velocity distribution in a perfect gas.

Chapter 20 introduces a new point of view into the analysis—it deals with the description of systems whose properties vary slowly in time, and with the extension of thermodynamics to systems with negative temperature.

Chapters 12 through 19 establish the principles and develop the tools needed to study the bulk properties of matter. This study is carried out systematically in Chapters 21 through 26. Chapter 21 deals with gases, Chapter 22 with solids, Chapter 23 with liquids, Chapter 24 with phase transformations, Chapter 25 with solutions of nonelectrolytes, and Chapter 26 with solutions of electrolytes. In each chapter the thermodynamic theory is developed first, then the statistical molecular theory. Extensive use is made of the details concerning molecular behavior that are provided by computer simulation studies. In addition, the principles developed are illustrated with data from experimental situations wherever that is appropriate.

The simultaneous development of classical and statistical thermodynamics, employed through Part II, is designed to overcome the difficulties associated with the very abstract nature of purely thermodynamic reasoning, and also to illustrate the richness of the phenomena that can arise from molecular interactions. Nevertheless, for any given problem, the generality of the thermodynamic approach is made evident, as are the wealth of detail and dependence on assumed models of the statistical molecular approach.

Part III is concerned with time-dependent processes, especially the approach to equilibrium. The topic of physical kinetics (transport processes) is introduced in Chapter 27 with a discussion of the mechanics of molecular collisions, mostly binary collisions. We present as simply as possible the elements of kinematics and dynamics, including the concept of scattering, which is illustrated with the hard-sphere model.

In Chapter 28 we consider the kinetic theory of gases, beginning with how velocity distribution functions change with time because of collisions. We present an elementary discussion of the time-dependent transport equations and show how they govern a gas's approach to equilibrium. With these results we can discuss fluxes of mass, momentum, and energy, and study the process of effusion and the simple transport properties (diffusion, viscosity, and thermal conduction) in dilute gases. We conclude with a brief treatment of energy exchange processes, and sound propagation and absorption in gases.

The transport properties of dense phases are taken up in Chapter 29. Transport in liquids is approached with a discussion of Brownian motion, leading to the relation of transport coefficients to autocorrelation functions. Brief discussions of transport in solids and of electrical conductivity in electrolyte solutions conclude the chapter.

Chemical kinetics is treated in a manner parallel to physical kinetics, with an elementary development followed by selected advanced applications. We begin in Chapter 30 with a presentation of the mechanics of reactive collisions, including both kinematics and dynamics. The emphasis is again on the simple hard-sphere model. The collision-theory approach is compared with the activated-complex theory, and both theories are used for an analysis of kinetics in gases and solutions. After a brief survey of experimental methods, we discuss complex reactions and provide an elementary discussion of chemical reaction mechanisms.

Chapter 31 is devoted to various advanced topics in chemical kinetics, including the RRKM theory of unimolecular reactions, symmetry rules in chemical reactions, chain reactions, oscillatory reactions, photochemistry, homogeneous and heterogeneous catalysis and the kinetics of electrode reactions.

Our educational and professional associations have obviously influenced the way we wrote this book, as have a few remarkable volumes. In particular, the books by

H. Reiss (*Methods of Thermodynamics,* Blaisdell, N.Y., 1965) and F. Reif (*Statistical Physics,* McGraw-Hill, N.Y., 1965) helped to clarify our independently developed presentations.

The current edition of this book draws heavily on the content of the first edition. We owe a continuing debt to all those who contributed to the first edition, in particular to George P. Flynn and Joseph N. Kushick. It is a pleasure to thank the many colleagues who have helped with critiques of the first edition and made suggestions that, we trust, improve the second edition. Amongst these are Stephen Bernasek, John Brauman, Laurie J. Butler, Claude Diou, John Enderby, Michel Fayer, Harold Friedman, Roald Hoffman, George C. Lie, David Oxtoby, Robert Madix, Peter Radziks, Jan Sengers, and Richard Zare. This edition has been enhanced by the contributions of Christopher Chidsey, Robert Eisenberg, R. A. Marcus, H. Schaefer, D. G. Truhlar, M. Vlad, Peter Wolynes, and Achmed Zewail, whose special topic vignettes are distributed throughout the text; we thank them for these contributions. We also thank Andrea Twiss-Brooks and Fritz Whitcomb for the Appendix describing modern methods of searching the scientific literature, and Ron Friedman and Ron Duchovic who have provided new problems that augment the set of problems carried over from the first edition.

To the readers, we say that we hope that you will have as much delight in using and creating physical chemistry as we have.

R. Stephen Berry
Stuart A. Rice
John Ross

Suggested Topic Selections for Various Courses

The following table illustrates possible choices of subject matter for a variety of physical chemistry courses.

Chapter	One-Year Course (initial emphasis on macroscopic approach)	One-Year Course (initial emphasis on microscopic approach)	Three-Semester Course	Senior or Graduate Course
1	Start with Chapters 12 through 26, with emphasis placed on the thermodynamic analysis. Keep only such elements of the statistical molecular analysis as fits with the instructional goal. Then return to Chapter 2 and follow the outline for the one-year course with initial emphasis on microscopics, going rapidly to Section 3.3	Start with Chapter 1, setting pace as fast as students can absorb Chapters 1 and 2, and Sections 3.1 and 3.2. Touch lightly or omit Section 3.10 and Appendix 3A	All, as introduction	Omit
2			All	Read quickly
3			All	Read Sections 3.1 and 3.2 quickly
4		All	All	All
5		All, but Section 5.5 may be omitted for very well-prepared students	All	Touch lightly on Section 5.5
6		All	All	All
7		Section 7.8 may be dropped	All	All
8		All	All	All
9		Omit Sections 9.6 and 9.7	All	All
10		Omit	All	All
11		Sections 11.1 through 11.5, 11.8, and 11.10	All	All
12		All	All	All
13		All	All	All
14		Omit Section 14.6. Only touch lightly on Sections 14.9, 14.10	All	All
15		Omit optional Section 15.1	Omit optional Section 15.1	All
16		Omit optional Sections and Appendix 16A	Omit Appendix 16A	All
17		All	All	All
18		All	All	All
19		Omit Section 19.3	All	All
20		Omit	Touch lightly on Sections 20.1–20.4, keep 20.5	All
21		Omit Appendices 21A, 21B, 21C	All	All
22		Omit Section 22.5 and Appendix 22A	All	All
23		Omit Section 23.4	All	All
24		Omit Section 24.3. Touch lightly on Section 24.4	All	All
25		Omit latter half of Section 25.8. Touch lightly on Section 25.9	All	All
26		Omit Section 26.6	All	All
27		27.1–4	All	All
28		28.1, 4, 6, 7, 8	28.9, 10 optional	All
29		29.1	29.2–6	All
30		Omit 30.5, 6, 9	All	All
31		As time permits Assign as special topics	31.3, 6–11 and as time permits	All

PART I

The Structure of Matter

CHAPTER 1

The Microscopic World: Atoms and Molecules

We begin our study of the microscopic properties of matter with a review of essential elementary concepts and of the logical inferences that led historically to the atomic theory. Well-prepared readers may wish to go directly to Chapter 2, "Origins of the Quantum Theory of Matter." The experimental evidence for the existence of atoms and molecules and for the magnitudes of the fundamental units of mass and charge form the content of this introduction.

What *are* atoms and molecules? For the moment, let us simply use a working definition: For any of the hundred-odd substances (the chemical elements) listed on the inside cover of this book, an *atom* is the smallest bit of matter that can be identified as that substance; for any other chemical substance, a *molecule* (made up of atoms) is the smallest recognizable bit of the substance. Admittedly, definitions like these are somewhat vague, so we shall begin by reviewing some of the evidence that led to the formulation of these concepts.

Granting the existence of atoms and molecules, what do we want to know about them? In this chapter we deal with some of the simplest properties that can be described in gross terms. When one investigates a macroscopic object, it is natural to ask first about its size and its weight. By the same token, how big and how heavy is an atom or a molecule? On the atomic scale, the electrical charges carried by the various bits of matter are central to their behavior, so we must also consider the charges of the atom's components. We shall describe a number of methods for determining these various atomic magnitudes. Although these are not the only possible methods, nor in general the most precise,[1] they are the most straightforward conceptually, in that the quantities measured are related in very simple ways to the properties desired. (Methods based on properties of gases are the subject of Chapter 12.) Finally, we review some of the basic facts of macroscopic chemistry, the atomic weight scale and the periodic table, which must be related to the properties of atoms.

1.1 Development of the Atomic Theory: Relative Atomic Weights

Chemistry began as a qualitative study of the properties and transformations of the substances found in nature. It evolved into an exact science as quantitative regularities (laws) evolved to link these properties and transformations. Its goal became the construction of theories to correlate and predict the regularities. The most important of these regularities was the *law of conservation of mass,* which states that the total mass (quantity of matter) present is the same before and after a chemical reaction. A small number of substances—elements—could not be broken down into anything simpler, whereas other substances—compounds—could be separated into two or more elements. But it was not a trivial matter to distinguish elements from compounds in the early days of chemistry.

All pure[2] samples of a given compound, regardless of origins, contain the same relative masses of each elemental component. There is no continuous change from pure element A to pure element B through a succession of intermediate compositions, only one or more compounds characterized by fixed ratios of the weights of A and B. This is the *law*

[1] In practice it often turns out that the most direct way of doing something is not the most accurate. For example, the assessment that led to the values of the fundamental constants in Table 1.1 used data collected by many methods, but none of them were methods cited here. The results collected by some of the methods described here were examined very carefully, and the assessors decided to use only results taken from less direct but more accurate methods.

[2] The statement of constant composition is not precisely correct in the sense that samples of a substance may have exactly the same elemental composition yet, if they have different cosmic origins, may have different *isotopic* compositions. Such differences provide a powerful tool for cosmochemistry; e.g., in determining whether planets and comets have the same origins.

of definite composition, which suggests that there are basic units of each element having definite masses. Furthermore, if the elements X, Y, and Z combine in pairs to form three binary compounds, then the weight ratio of Y to Z in the Y–Z compound is related by some simple fraction to the X–Y and X–Z weight ratios; that is, we can write

$$\frac{\text{wt. of Y in Y–Z cpd.}}{\text{wt. of Z in Y–Z cpd.}} = \frac{m(\text{wt. of Y in X–Y cpd.})/(\text{wt. of X in X–Y cpd.})}{n(\text{wt. of Z in X–Z cpd.})/(\text{wt. of X in X–Z cpd.})}, \quad (1.1)$$

where m and n are small integers. This implies that the units involved in chemical combination are the same for various compounds, characteristic of the element rather than the compound. These "units" are the *atoms* of John Dalton's atomic theory (1808).

The strongest argument then available for this theory was based on the empirical *law of multiple proportions.* Suppose that elements A and B form several compounds with each other, say compounds 1, 2, 3, and so on. If we form the ratios of the weights of A and B in the various compounds, such as $w_A(1)/w_B(1)$, in which we read $w_A(1)$ as "weight of A in compound 1," then the ratios of these ratios, such as

$$R_{12} = \frac{w_A(1)/w_B(1)}{w_A(2)/w_B(2)}, \quad (1.2)$$

are all simple fractions. For example, oxygen and sulfur form two binary compounds, with w_O/w_S equal to 1.00 in one and 1.50 in the other, giving a ratio $R_{12} = \frac{2}{3}$ (the compounds are in fact SO_2 and SO_3, respectively).

By what logic does the law of multiple proportions—a relation among weights, not numbers of particles—provide such strong evidence for the atomic theory? If we knew only the *relative* masses of the atoms of A and B, the ratio m_A/m_B, where m is an atomic mass, then we could determine the relative numbers (n) of A and B atoms in a given compound,

$$\frac{n_A}{n_B} = \frac{w_A/m_A}{w_B/m_B} = \frac{w_A}{w_B}\frac{m_B}{m_A}, \quad (1.3)$$

since w_A/m_A and w_B/m_B must be the total numbers of A and B atoms, respectively, in our sample of the compound. In Dalton's time the ratios of atomic masses (or *atomic weights,* to use what is still the conventional term) were not yet known. However, if we form the ratio of ratios of relative masses for compounds 1 and 2, we have

$$R_{12} = \frac{w_A(1)/w_B(1)}{w_A(2)/w_B(2)} = \frac{[w_A(1)/w_B(1)](m_B/m_A)}{[w_A(2)/w_B(2)](m_B/m_A)} = \frac{n_A(1)/n_B(1)}{n_A(2)/n_B(2)}. \quad (1.4)$$

Hence the quantity R_{12} also gives the ratio of the ratios of numbers of atoms in the two compounds. Were there no fundamental atomic structure to matter, the relative amounts of A and B could vary continuously, allowing the ratio R_{12} to take on any value. That R_{12} is a ratio of small integers strongly suggests (but does not prove) that A and B consist of discrete atoms which combine in simple proportions, making the n_A/n_B ratios (and *their* ratio R_{12}) simple fractions.

This argument showed that the existence of atoms was a reasonable hypothesis, but alternative models were still possible; some prominent skeptics remained even in 1900. More convincing evidence came much later, when it became possible to study directly the atomic properties we shall examine in this chapter. Now the most refined, high-resolution varieties of microscopy reveal images of individual atoms, such as that in Fig. 1.1.

The law of multiple proportions does not tell us the relative weights of different atoms. Consider, for example, three compounds of hydrogen and oxygen, water, hydrogen peroxide, and the hydroxyl radical. Analysis of their composition gives the weight ratios $w_H/w_O = \frac{1}{8}$ for water and $w_H/w_O = \frac{1}{16}$ for hydrogen peroxide and hydroxyl. If matter is composed of atoms, hydrogen peroxide and hydroxyl must both contain twice as many oxygen atoms per hydrogen atom as does the water molecule. But this does not tell us whether water should be written with the formula HO (one oxygen atom per hydrogen) and hydrogen peroxide as HO_2, or as H_2O and HO, respectively, and so on; to deal with both hydrogen peroxide and hydroxyl, we need additional information about their relative masses, since their relative compositions are the same. We *can* tell, from the weight ratios of water and peroxide, (approximately) $\frac{1}{8}$ and $\frac{1}{16}$, that

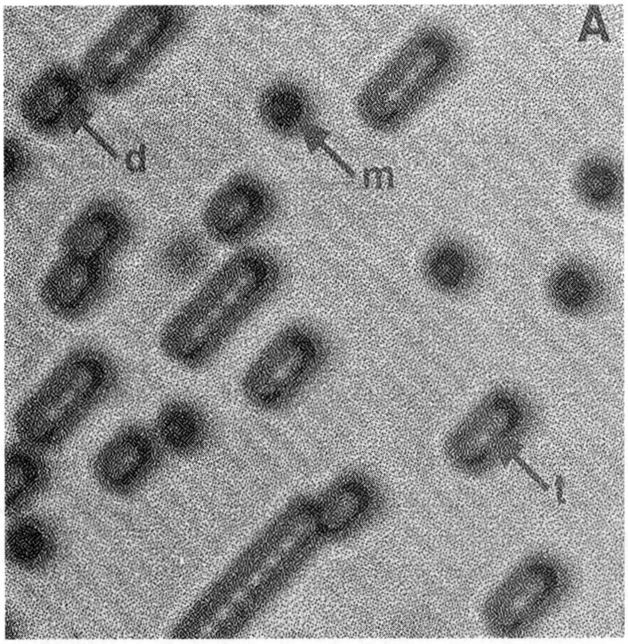

Figure 1.1 The image of individual molecules on a surface, detected by scanning tunneling microscopy (STM). The molecules are carbon monoxide, CO; the surface is copper. [From B. G. Briner, M. Doering, H.-P. Rust and A. M. Bradshaw, *Science 278,* 257–260 (1997).]

an oxygen atom plausibly weighs 4, or 8, or 16, or even 32 times as much as a hydrogen atom (the corresponding formulas for water being HO_2, HO, H_2O, and H_4O). In fact, water was long assumed (on grounds of simplicity) to be HO. The relative combining weights of elements in many compounds led chemists such as Berzelius (**ca.** 1811) to self-consistent sets of relative atomic weights for many elements, long before the more precise microscopic methods available today. But all these ratios of weights were uncertain by various integral factors, as comparison of hydroxyl and hydrogen peroxide show, since they were based on guesswork about the formulas of simple compounds like water.

This difficulty was not overcome until it became possible to obtain relative *molecular weights.* The key came historically from the theory of gases. Experiments showed that when gases reacted, the ratios of their combining volumes were also simple fractions. In 1811 Avogadro proposed that equal volumes of gases at the same temperature and pressure must contain the same numbers of molecules. We shall show in Chapter 12 how this result (*Avogadro's principle*) emerges from microscopic theory. Avogadro's conclusions were long resisted, since popular belief held that elements must occur as single atoms, whereas measurements indicated that the common gases had to be composed of diatomic molecules (hydrogen H_2, oxygen O_2, etc.) for Avogadro's principle to be valid. Finally, in 1860, Cannizzaro showed how one could unify the existing atomic weight data into a consistent system by accepting Avogadro's principle. For example, since two volumes of hydrogen combined with one volume of oxygen to give two volumes of water (as steam), the reaction was correctly described by the equation

$$2H_2 + O_2 \rightarrow 2H_2O,$$

in which the numbers of atoms balance and the numbers of molecules are proportional to the volumes involved.

Other methods of determining molecular weights (especially for substances not available in gaseous form) were developed; we shall mention some of these in later chapters. When these results were combined with the measurements of the combining weights of elements, it became possible to develop a complete set of relative atomic weights. Such a table is given on the inside cover of this book; we shall discuss in Section 1.6 the arbitrary scale to which the numbers are referred.

1.2 Atomic Magnitudes

By the early nineteenth century, many scientists had come to believe that matter is made up of atoms, those of each element having a characteristic mass, and that from macroscopic data alone one can obtain the relative values of these masses. What other things do we want to know about the atom? What are the atomic properties that are ultimately the basis of chemistry, and what experiments or theories will tell us their magnitudes? These magnitudes of course include the absolute values of the atom's mass, its linear dimensions, and, anticipating that atoms have inner structure, the electrical charges of its components and what that inner structure is.

For convenience, we summarize some of the results to be developed in this and subsequent chapters. The atom is composed of a tiny, heavy, positively-charged nucleus around which orbit negatively-charged electrons. The nucleus contains over 99.9% of the atom's mass, but occupies less than 10^{-12} (one-trillionth) of its volume. The nucleus in turn is an aggregation of positively charged protons and uncharged neutrons. (These, in turn, have a substructure that we shall not explore here; the manifestations of that substructure appear, for the most part, only at energies well beyond the range of chemistry.) The simplest atom, hydrogen, contains one proton and one electron. The atom as a whole is normally uncharged but can become ionized by the loss or gain of negative charge in the form of electrons. Experiment shows that the charge on an atom or molecule is always a multiple of a small fixed quantity: Charge as well as mass comes in discrete units. This quantity is the charge of the electron or the proton, which are equal in magnitude but opposite in sign.

Now let us look at some numerical values for the quantities in which we are interested. These numbers cannot be deduced from any known *a priori* theory; they can only be obtained, directly or indirectly, from experiment. Some are given in Table 1.1, and a more extensive list can be found on the inside cover of this book. Constants such as these, which are applicable to all forms of matter and turn up in many fields of science, are referred to as the fundamental constants of nature.

To give meaning to numbers with such extreme exponential factors as those in Table 1.1, let us make comparisons with other, more familiar quantities. For example, the mass of the electron stands in about the same relation to a $2\frac{1}{3}$-g weight as that weight does to the mass of the earth (6×10^{24} kg). From the table, the proton (or the neutron, or the hydrogen atom) weighs about 1800 times as much as the electron, so that on our gram–earth scale the proton corresponds to roughly 4 kg. Atomic nuclei can contain as many as 250 protons and neutrons, so that absolute atomic masses

Table 1.1 Some Atomic Constants[a]

Electron rest mass (m_e)	$9.1093819 \times 10^{-31}$ kg
Proton rest mass (m_p)	$1.6726216 \times 10^{-27}$ kg
Neutron rest mass (m_n)	$1.6749272 \times 10^{-27}$ kg
Elementary electronic charge (e)	$1.60217646 \times 10^{-19}$ C
Planck's Constant (h)	$6.6260687 \times 10^{-34}$ Js
Bohr radius (a_0) (radius of a hydrogen atom)	$5.2917725 \times 10^{-11}$ m

[a] From *Fundamental Physical Constants,* National Institute of Science and Technology, 1999; http://physics.nist.gov

run up to about 4×10^{-25} kg (about a ton on our comparison scale).

Next, how big is the charge of the electron? If the earth had an excess of about 4×10^{15} electrons (weighing less than 4×10^{-15} kg, remember), or for that matter 4×10^{15} protons, then it would have a potential of about 1 V, or two-thirds the voltage of a flashlight battery. Another comparison is this: 1 microampere (μA), a small but readily detectable current, corresponds to a flow of 6×10^{12} (6 trillion) electrons per second past a measuring device. Finally, to visualize the linear dimensions of the atom, consider the Bohr radius, abbreviated a_0. We shall say a good deal about this quantity in the next chapter, but here it can be considered loosely as the radius of the hydrogen atom. A special unit is often used for atomic dimensions, the *Ångstrom* (abbreviated Å), which is 10^{-10} meters (0.1 nm). The Bohr radius is about 0.5 Å, which is to 1 cm roughly as 1 cm is to 2000 km. The largest known atoms have radii of about $4a_0$. As for the atomic nucleus, its radius is less than 10^{-14} m (10^{-4} Å or 10^{-5} nm), or less than one ten-thousandth of the atom's radius.

Let us now begin to consider the experimental determination of these quantities. The mass of the electron can be determined almost directly from the optical spectra of hydrogen and deuterium. (See Problem 8, Chapter 2.) However, most atomic or molecular masses are determined by converting the species of interest into ions and measuring the deflection of these ions in an electric or magnetic field, which reveals their charge-to-mass ratio. The charge is most often identical to the electronic charge *e*, and so far as is known, is always an integral multiple of *e*. This technique was introduced by J. J. Thomson (1897), whose determination of e/m for the electron will be discussed in the next section; in Section 1.5 we shall examine the general application of the technique (mass spectrometry).

To determine masses from a charge-to-mass ratio, one must know the value of *e*, independent of the ratio e/m_e. The charge of the electron was determined by R. A. Millikan (1909, by a refinement of a method of Townsend of 1897) from observation of the motion of charged droplets in an electric field; we discuss this method in Section 1.4. Today a more straightforward method is possible, by measuring the quantity of a current of electrons in two different ways. Of two detectors, one measures the number of individual electrons striking it per unit time, and the other measures the electric current (charge per unit time) that it receives; the ratio of the charge per unit time to the number of electrons per unit time is the charge per electron.

At least two straightforward methods measure atomic dimensions. The more accurate is scattering x-rays of known wavelength from a crystal, and from the pattern produced by the scattered x-rays, deducing the interatomic distances in the crystal. The other is measuring the density of a substance in the bulk, and finding independently the mass of a single atom or molecule, and from the two to compute the volume available to one atom or molecule. We shall postpone further discussion of these methods until our treatment of crystal structure in Chapter 11. One can also determine the distances between atoms in a molecule by various spectroscopic techniques. The "size" of an individual atom is a somewhat vague concept, since atoms have no sharp boundaries; the best one can do is to find the spatial distribution of electrons, as we shall see when we take up the theory of atomic structure.

1.3 The Charge-to-Mass Ratio of the Electron: Thomson's Method

The charge-to-mass[3] ratio of the electron, e/m, was the first atomic parameter to be determined, by J. J. Thomson in 1897. This is also in many ways the most straightforward of the experiments we shall discuss, and our description follows Thomson's method. His experiments were performed on what were then known as cathode rays—rays that we now know consist of streams of electrons. The purpose of these experiments, in fact, was to show that the cathode rays *were* streams of particles. The apparatus used is shown schematically in Fig. 1.2.

In the experiment the tube is evacuated and the cathode *A* is charged negatively with respect to the anode *B*, causing electrons to stream from *A* toward *B*. The anode is constructed with a narrow slit in its center, allowing some of the electrons to pass through; downstream is a second slit *S*. The two slits act to produce a narrow beam of cathode rays (electrons). These travel from *A* through *B* and *S* to reach the luminescent screen *Z*, where they produce a small spot, just an enlarged image of the hole in *S*. If *S* is replaced by a small object that only partially blocks the beam, then one sees on the screen a shadow image of this object. Such a shadow was taken as evidence that the cathode rays did indeed behave like particles. (But those who thought otherwise could point to the fact that the rays passed through thin metal foil.) More significant for us now, however, are the electrical properties of these rays. If plate *X* is charged positively and plate *Y* is charged negatively, then the beam moving from *S* strikes *Z* at some point closer to *X* than to *Y*. In other words, the positive charge on *X* has attracted the beam of cathode rays, and we conclude that the cathode rays consist of something with a negative charge.

We describe the force exerted on the particles traveling between plates *X* and *Y* as the force of an *electric field*. This field of force is a mental construct introduced by Michael Faraday, in which one can imagine literal lines of force, like strings, pulling on a charged particle. The lines of force simply describe the trajectories that particles (initially at rest) would follow under the influence of the field. The total force **F** on the particle is directly proportional to its own charge *q*. The proportionality constant, the electrical force per unit

[3] We shall designate the mass of the electron as *m*, rather than m_e, whenever no ambiguity is involved.

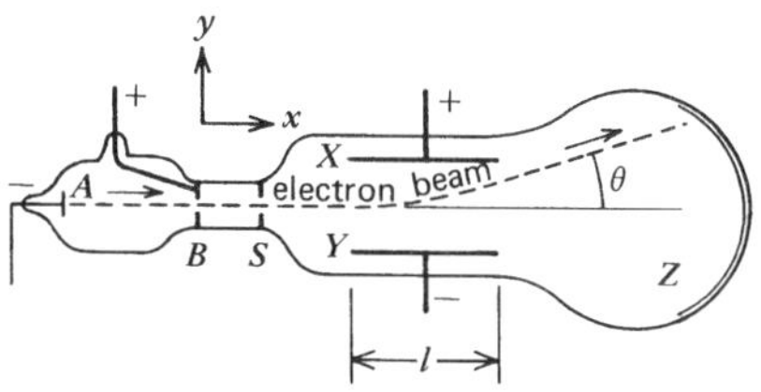

Figure 1.2 Schematic diagram of Thomson's apparatus for determining e/m. (See text for explanation of symbols.)

charge, is called the electric field strength $\mathbf{E}(=\mathbf{F}/q)$. The electric field between X and Y consists essentially of parallel lines of force pointing from X toward Y; that is, a positively charged particle placed between X and Y would be forced from X toward Y. If we neglect the regions near the ends of plates X and Y, where the lines of force curve outward, the field strength between the plates is everywhere the same. We can say that the space between the plates is a region of uniform field, and thus of constant force.

To clarify the notion of a field a bit more, let us consider another familiar force, namely, that exerted between one charged point particle and another in free space. Suppose that the two particles have charges q_1 and q_2. Then the force between them in a vacuum is directed along the line joining them, and by Coulomb's law has the magnitude

$$F = \frac{q_1 q_2}{4\pi\epsilon_0 r^2}, \tag{1.5}$$

where r is the distance between the two charges and ϵ_0 is a constant (the *permittivity of free space*); in SI units[4] (with q in coulombs, C; r in meters, m, and F in newtons, N), we have

$$\epsilon_0 = 8.854187 \times 10^{-12}\ \mathrm{C^2/Nm^2}.$$

(We use the boldface **F** to denote a vector force, but italic F here for the single-radial component of force.)

The electric field strength at particle 2 (defined as $\mathbf{F}/q_2$) thus has a magnitude $q_1/4\pi\epsilon_0 r^2$, which varies with position. Similarly, the electric field in any region of space can be mapped by measuring at various points the force it exerts on some test object of known charge, such as an electron or a charged styrofoam sphere. In general, a field is simply a measurable property, extending through a volume of space, whose magnitude is a function of position.

The force **F** on a charged particle is in fact a *vector* (identified by boldface type), a quantity with both magnitude and direction. It is expressible as a set of three quantities F_x, F_y, F_z, corresponding to the components of force parallel to the x, y, z axes of a Cartesian coordinate system. The magnitude of **F**, denoted by F, is given by the relation

$$|\mathbf{F}| = F = (F_x^2 + F_y^2 + F_z^2)^{1/2}. \tag{1.6}$$

Similarly, any vector **A** is really the set of three quantities A_x, A_y, A_z, and has the magnitude

$$A = (A_x^2 + A_y^2 + A_z^2)^{1/2}. \tag{1.7}$$

We may think of **A** as the generalization in three dimensions of the hypotenuse of a right triangle. In fact, if one of the components A_x, A_y, A_z is zero, the analogy to a right triangle holds exactly. We shall develop other properties of vectors as we need them.

The electric field strength **E** is of course also a vector, since it is defined as the force per unit charge. That is, a test object with charge q will be subjected to a force

$$\mathbf{F} = q\mathbf{E} \tag{1.8}$$

at a point where the field strength is **E**. What this vector equation means, in terms of the components of the vectors, is that $F_x = qE_x$, $F_y = qE_y$, $F_z = qE_z$; thus the magnitudes of the vectors are also proportional, $F = qE$. The electric field strength can be measured in newtons per coulomb, or in the equivalent units of volts per meter.

If a charged particle moves under the influence of a force, then the force has done work on the particle, and thus has changed its energy. When the force is constant and the motion is in a straight line, the work done is the magnitude of the force multiplied by the displacement. Let us now return to the motion of an electron between charged plates, as shown in Fig. 1.3. Since the electron's charge is negative ($q_e = -e$), by Eq. 1.8 the vectors **F** and **E** must have opposite signs, that is, opposite directions; the electron is drawn toward the positive plate, so the electric field vector must point toward the negative plate. If an electron moves from y_1 to y_2 under the influence of a constant field **E**, then the work done is

$$W = F(y_2 - y_1) = eE(y_2 - y_1). \tag{1.9}$$

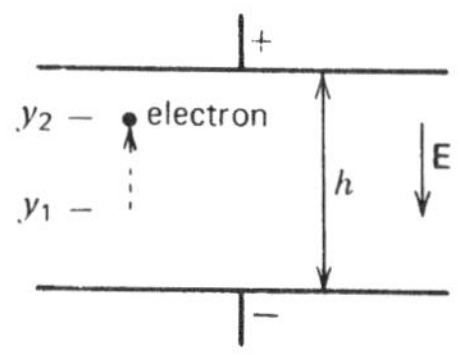

Figure 1.3 Motion of an electron, initially at rest, in a uniform electric field **E**. The electron's position y is measured from the negative toward the positive plate.

[4] Other systems of units deserving mention are the electrostatic and Gaussian systems, in which the constant in Coulomb's law is suppressed: $|\mathbf{F}| = q_1q_2/r^2$, but otherwise cgs (centimeter-gram-second) units are used. The *electrostatic unit* of charge (esu) thus has a value such that there is a force of 1 dyne between two 1-esu charges separated by 1 cm; 1 esu = 3.33564×10^{-10} Coulomb. The esu is also known as the *statcoulomb* when Coulomb's law is taken as its definition, 1 esu = 1 dyn$^{1/2}$ cm, or Franklin (Fr) when taken as an independent unit.

(For force in newtons and distance in meters, W must be expressed in joules.) The electron has moved "downhill" in the electric field, that is, to a point of lower potential energy; the decrease in its potential energy is the same as the work done, $eE(y_2 - y_1)$. Just as we picture the force on a charged particle as the product of charge and electric field strength, so we can describe the particle's potential energy as the product of charge and a quantity we call the *potential.* The potential is the potential energy per unit charge, and is thus measured in joules per coulomb, better known as volts.[5] If the distance between our two plates is h, then the potential energy difference between them is eEh, and the potential (voltage) difference is Eh.

Now we have the vocabulary to give a quantitative description of Thomson's measurement of e/m for the electron. Consider again the diagram of the apparatus in Fig. 1.2. We define a coordinate system with the x axis in the electrons' initial direction of motion and the y axis normal to plates X and Y. Between these plates there is an electric field, which we assume to have the constant value $\mathbf{E}$ everywhere in a region of length l. (That is, we assume that all the lines of force are vertical, neglecting the curvature at the ends.) Let us say that all the electrons enter the field from the left with initial velocity $\mathbf{v}$ in the x direction. At any time when an electron is between the plates, it feels a force in the positive y direction of magnitude eE. This force, by Newton's second law, must be the product of the electron's mass and its acceleration $\mathbf{a}$:

$$\mathbf{F} = -e\mathbf{E} = m\mathbf{a} \qquad (F_y = -eE = ma_y). \tag{1.10}$$

Note that the force and acceleration are exerted in the y direction only, so the x component of the electron's velocity, v_x, is unchanged. Therefore the time the electron spends in the field is

$$t_f = \frac{l}{v_x}. \tag{1.11}$$

During this time the y component of velocity, v_y, changes from zero to $a_y t_f$; the force and thus the acceleration being constant, the velocity is merely the acceleration multiplied by the time it is applied. Consequently, the y component of the velocity reaches the value

$$v_y = a_y t_f = \frac{eE}{m}\frac{l}{v_x}. \tag{1.12}$$

After the electron leaves the region between the plates, it is no longer accelerated, so v_x and v_y are both constant. Over a time t, the electron then moves a distance $v_x t$ in the x direction, and $v_y t$ in the y direction. Then θ, the angle of deflection[6] from the x axis, is given by

$$\tan\theta = \frac{v_y}{v_x}. \tag{1.13}$$

Substituting Eq. 1.12, we obtain

$$\tan\theta = \frac{eE}{m}\frac{l}{v_x^2}. \tag{1.14}$$

Thus if we could measure θ, $\mathbf{E}$, l, and v_x, we could immediately find e/m.

The electric field $\mathbf{E}$ is measurable with a voltmeter and a ruler. It is just the voltage difference between the charged plates divided by their separation. The length l is again a measurement we make when we build the apparatus. The angle θ we find by doing the experiment, for example with a fluorescent screen. The initial speed v_x is a bit more difficult to determine. In principle one can measure it by placing in the path of the beam two shutters that can be opened for brief periods. A small burst of electrons is allowed to pass through the first shutter; one then determines when the second shutter must be opened to permit the electrons to reach a collector. If the separation between the shutters is d, and the time interval that just allows the burst from shutter 1 to pass through shutter 2 is τ, then the speed is simply d/τ. For shutters one could use two wheels, each with a single slot, mounted on a common shaft with variable speed of rotation; this is a method commonly used to determine the velocities of atomic beams,[7] but it is inconvenient for electron beams because electrons move too rapidly. We shall see how fast electrons actually move when we reach the point of evaluating the electron's mass and charge. Fortunately, there are better ways to determine the electron's velocity.

An especially convenient way to deal with v_x is by eliminating it from Eq. 1.14. This method, the one actually used by Thomson, also forms the basis for much of modern mass spectrometry (see Section 1.5). It is based on a second way of deflecting moving charged particles, namely, the use of a *magnetic field.* One could introduce the idea of a magnetic field in a way analogous to Eq. 1.5, by writing an expression for the force between two (hypothetical) magnetic poles. However, the "magnetic pole" is a clumsy and only approximately meaningful concept. It is more logical, as well as more pertinent to our present needs, to define a magnetic field in terms of its interaction with a moving charge.

The magnetic force on a moving charged particle is known as the *Lorentz force.* If a particle with charge q moves

[5] The Gaussian unit of potential is the erg per esu, or *statvolt;* 1 statvolt = 299.793 V.

[6] The angle θ is measured from the center of the region between the plates. Given the uniform field we have defined, the electron will leave the field moving directly away from this point.

[7] Cf. Problem 10 at the end of this chapter.

with velocity **v** through a magnetic field **B** perpendicular to **v**, the Lorentz force **F** has the magnitude

$$F = qvB. \tag{1.15}$$

In SI units, **F** is measured in newtons, q in coulombs, **v** in meters per second, and **B** in volt seconds per meter squared; the latter unit is called the *tesla* (T). This equation could be taken as a definition of the *magnetic induction*[8] (or *magnetic flux density*) **B**, except that we must also establish the direction of the vector **B**. If a coordinate system is defined such that **v** points in the positive x direction and the force on a positively charged particle in the positive z direction, then the magnetic field vector is defined to point in the positive y direction.[9] These directional relationships are shown schematically in Fig. 1.4; they are sometimes called the *right-hand rule,* for reasons indicated in the figure.

All the information given in the previous paragraph is summarized compactly in the vector equation[10]

$$\mathbf{F} = q(\mathbf{v} \times \mathbf{B}). \tag{1.16}$$

The expression $\mathbf{v} \times \mathbf{B}$ is what is known as a *cross product* (or *vector product*). It is itself a vector, with three components related to the components of **v** and **B** by the equations

$$(\mathbf{v} \times \mathbf{B})_x = v_y B_z - v_z B_y, \tag{1.17a}$$
$$(\mathbf{v} \times \mathbf{B})_y = v_z B_x - v_x B_z, \tag{1.17b}$$
$$(\mathbf{v} \times \mathbf{B})_z = v_x B_y - v_y B_x. \tag{1.17c}$$

These equations can readily be extended to the general case. The cross product of any two vectors $\boldsymbol{\alpha}$ and $\boldsymbol{\beta}$ has the magnitude $\alpha\beta\,|\sin\phi|$ where ϕ is the angle between them. Note that $\boldsymbol{\alpha} \times \boldsymbol{\beta}$ and $\boldsymbol{\beta} \times \boldsymbol{\alpha}$ have the same magnitude but opposite directions; that is, their corresponding components have opposite signs. In setting up Fig. 1.3 we simply chose the axes and the orientation of **v** and **B** so as to make the x and y components of force, Eqs. 1.17*a* and 1.17*b*, identically zero. In practice, it is often advantageous to build one's apparatus so that the equipment itself can fulfill simplifying conditions like these.

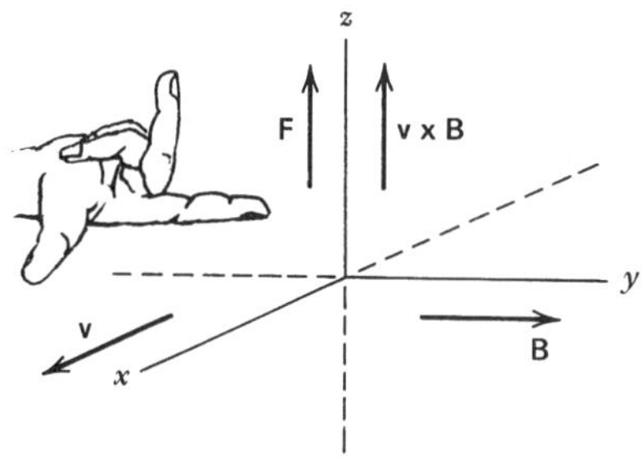

Figure 1.4 Lorentz force on a positively charged particle moving in a uniform magnetic field. The small drawing illustrates the right-hand rule: thumb, index finger, and middle finger point respectively in the directions of **v**, **B**, and **F** (or of the x, y, and z axes, in a right-handed coordinate system).

Now we can derive an expression for the deflection of an electron by a magnetic field, just as we obtained Eq. 1.14 for its deflection by an electric field. We wish to use a set of axes consistent with those in Fig. 1.2, in which the positive z axis can be taken as coming forward out of the paper (right-handed coordinates). We again allow the electron beam to move in the x direction; however, since we have already put an electric field along the (negative) y axis, let us direct the magnetic field along the positive z axis. Then the only nonvanishing term on the right-hand sides of Eqs. 1.17 is the $-v_xB_z$ in Eq. 1.17*b,* so that $\mathbf{v} \times \mathbf{B}$ and thus the Lorentz force must be directed along the y axis. Since for the electron q in Eq. 1.16 is negative, the force will be in the positive y direction, with magnitude

$$F_y = (-e)(-v_x B_z) = e v_x B_z. \tag{1.18}$$

The above is true only initially: Once the force has been applied, v_y is no longer zero, and a component $F_x = -e v_y B_z$ appears, changing v_x in turn. We shall see later how this problem is handled, but for now let us simply assume that the deflection region is short, so that the change in v_x is negligible. Following steps just like those leading from Eq. 1.10 to Eq. 1.14, we then obtain from Eq. 1.18 the result

$$\tan\theta' = \frac{eB_z}{m}\frac{l}{v_x}, \tag{1.19}$$

where θ' is the angle of deflection caused by the magnetic field. We can design the apparatus so the length l is essentially the same for the electric and magnetic deflection regions. For small deflection angles one can approximate $\tan\theta$ by θ and $\tan\theta'$ by θ'; in this approximation we can see that both deflection angles are proportional to field strength but that the electric field deflection θ varies with v_x^{-2} whereas the magnetic field deflection θ' varies with v_x^{-1}.

We were stopped after Eq. 1.14 because we had two unknowns: the ratio e/m, which we want to know, and the electron speed v_x, which we don't particularly care about at this point. But now we have both Eq. 1.14 and Eq. 1.19, two simultaneous equations in these two unknowns, based on

---

[8] It is common to refer to **B** as simply the "magnetic field." Strictly, however, the *magnetic field strength* **H** is a different vector. In a vacuum the two are related by $\mathbf{B} = \mu_0\mathbf{H}$, where $\mu_0 \equiv 4\pi \times 10^{-7}$ N/A^2 or henry/m (the *permeability of free space*); **H** is measured in amperes per meter.

In the Gaussian system μ_0 is suppressed, like the $4\pi\epsilon_0$ in Coulomb's law. The Lorentz force equation then takes the form $F = qvb/c$, where c is the speed of light. With **F** in dynes, q in esu, **v** and c in centimeters per second, the unit of **B** is the gauss (G): 1 G = 10^{-4} T. The Gaussian unit of **H** is the *oersted* (Oe), which has the same dimensions as the gauss. Since in these units **B** and **H** have the same numerical value (in a vacuum), most chemists speak indiscriminately of "magnetic field **H** in gauss."

[9] The direction of **B** is that in which a positive (north) magnetic pole would move under the influence of the field.

[10] In the Gaussian system, $\mathbf{F} = (q/c)(\mathbf{v} \times \mathbf{B})$.

two independent experiments. Combining the two equations to eliminate v_x, and then solving for e/m, we obtain

$$\frac{e}{m} = \frac{\tan^2\theta'}{\tan\theta}\frac{E}{B_z^2 l}. \tag{1.20}$$

All the quantities on the right side of this equation can be determined experimentally, so e/m can now be obtained directly.

Given an apparatus in which the electric and magnetic fields can be applied either simultaneously or separately, there are various ways in which the experiment can be performed. We could, of course, simply pick arbitrary values of E and B_z and measure θ and θ' separately. Or we could choose E and B_z so that the angles θ and θ' are equal (so that $\tan^2\theta'/\tan\theta$ becomes simply $\tan\theta$); that is, we can deflect the electrons with an electric field and find the spot to which the deflected beam moves, then apply a magnetic field just large enough to bring the beam to exactly the same spot. Finally, we can measure θ with the electric field alone, then turn on the magnetic field simultaneously and adjust it so that the beam is again undeflected; in this case we have $\theta + \theta' = 0$, or $\theta' = -\theta$.

Thomson obtained only an approximate value of e/m. Modern measurements, using more accurate techniques, give the currently accepted value

$$e/m = 1.7588201 \times 10^{11}\ \mathrm{C/kg}.$$

This number, the charge-to-mass ratio of the electron, is by itself not very useful to us. We cannot guess from it alone how many electrons there are per gram of carbon, for example, or even what the charge of a single electron is. It becomes useful only if we obtain an independent measure of the charge of the electron itself, a measurement to which the next section is devoted.

1.4 The Charge of the Electron: Millikan's Method

We now turn to the determination of the electronic charge e, by the method R. A. Millikan used in 1909. We should point out once again that, although this method is one of the simplest conceptually and experimentally, and the most important historically, it is no longer among the most precise ways of evaluating e. (It is, nonetheless, one of the ways people have searched for *quarks*, predicted particles with charges of $\frac{1}{3}e$ or $\frac{2}{3}e$.) Later, when we have developed some of the quantum properties of matter, we shall point out other measurements that can yield an accurate value of the charge on the electron. The essence of Millikan's method lies in examining the behavior of small but macroscopic charged particles under the combined influence of gravity, electrical fields, and friction. From a knowledge of the various force laws one can deduce the size of the charge on such a particle. Determinations of many such charges, on particles of different kinds, show that all the particle charges can be expressed as integral multiples of a constant value, identified with the elementary charge e.

In practice, the particles one uses are either droplets of oil or (today) small plastic balls. Millikan tried water first, but water droplets evaporate too fast. (Plastic balls can be made with precisely-known diameters, which simplifies the experiment by removing one unknown.) These droplets or balls manage to pick up negative charges (electrons) spontaneously as they drift in air; one can hasten the process by bringing a bit of radioactive material nearby, so that there are quite a lot of charges in the neighborhood. It is convenient to watch the charged particles with a microscope having calibration lines, so that one can see how long it takes for them to drift downward through a known distance under the influence of gravity. If the region of observation is between two condenser plates to which a charge can be applied, a particle carrying a charge can be forced upward against gravity by an electric field. If one is lucky one can watch the same particle move up and down many times, changing its charge by bringing the radioactive source near. It is convenient to shine light on the particles at right angles to the direction of observation so that the particles shine as bright spots, much like dust motes in a sunbeam. The apparatus is sketched in Fig. 1.5.

We begin by considering the motion of the particle in the absence of the electric field, falling under the influence of gravity alone. Remember that the particle is moving in air, not in vacuum. Since it is very small, its fall is significantly retarded by friction with the surrounding atmosphere. Under this condition, the motion is best described by *Stokes's law*, which we simply state without proof:[11]

$$F = 6\pi\eta r v, \tag{1.21}$$

[11] This equation appears, at first sight, to contradict Newton's second law of motion, which states that acceleration is directly proportional to force; according to Eq. 1.21, the velocity, not the acceleration, is proportional to the force. The contradiction is only apparent, because Stokes's law does follow from the laws of Newtonian mechanics. Stokes's law is a statement of how the *retarding force* of friction in a medium acts on a body moving through that medium. A similar expression determines the maximum velocity of a space craft reentering the atmosphere, or of a skier in a long schuss. The frictional force increases with the velocity, because the moving object collides with more molecules of the impeding medium as its velocity increases, causing it to experience more drag. A falling object reaches a terminal velocity when the gravitational force mg acting on it is balanced by the frictional force. If the frictional force is proportional to velocity, $F_{\mathrm{fric}} = bv$, then the terminal speed must be $v_{\mathrm{term}} = b/mg$. Since the *net* force is zero, there is no further acceleration, and Newton's second law is satisfied. We return to this issue in Chapters 28 and 29.

An object falling at its terminal velocity has a constant kinetic energy and a potential energy that decreases at a constant rate. Where does this energy go? It is transformed into random kinetic and internal energy of the molecules of the medium and the object—into *heat*. If the velocity is high enough, this heat may be sufficient for the object to become incandescent, to melt, to evaporate or to react with the medium. This phenomenon is responsible for the visibility of meteorites and reentering debris from spacecraft. It also required the development of the ablative nose cone for reentry vehicles: Ablation is the process in which surface material is heated up to its softening or melting point, and then, bit by bit, blows away from the vehicle, carrying much of the frictional heat with it.

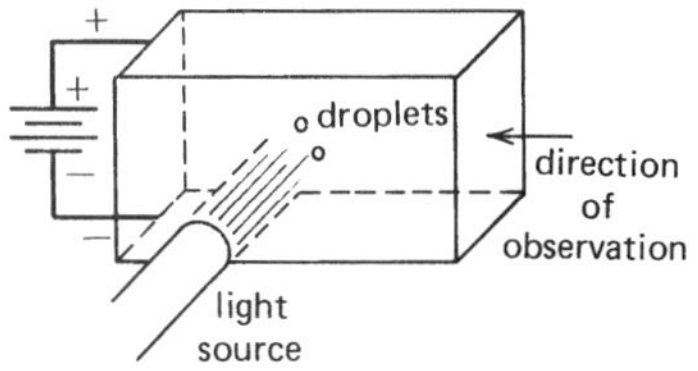

Figure 1.5 Schematic diagram of the Millikan oil-drop experiment. The top and bottom of the chamber are condenser plates charged by the battery at left. The negatively charged droplets will tend to move upward in the electric field but will fall when the battery is disconnected.

where v is the final speed attained by a spherical body of radius r, drawn slowly by a force of magnitude F through a medium of viscosity[12] η. In the Millikan experiment, in the absence of an electric field, F is the net gravitational force *while the sphere is in air*—in other words, the gravitational force in vacuum minus the buoyant force of the air. If ρ_s is the density of the sphere and ρ_{air} is the density of air, then we can write

$$F_{\text{grav}} = (m_s - m_{\text{air}})g = \left(\frac{4\pi}{3}r^3\rho_s - \frac{4\pi}{3}r^3\rho_{\text{air}}\right)g, \tag{1.22}$$

where m_s is the actual mass of the sphere and m_{air} is the mass of an equal volume of air. Substituting in Eq. 1.21 for the gravitational force given by Eq. 1.22, which must equal the retarding frictional force, we obtain

$$(F_{\text{fric}})_1 = 6\pi\eta r v_1 = F_{\text{grav}} = \frac{4\pi r^3}{3}g(\rho_s - \rho_{\text{air}}), \tag{1.23}$$

where v_1 is the limiting (downward) speed under these conditions.

Now suppose that we have an electric field $\mathbf{E}$ exerting an upward force on the particle strong enough to produce a net *upward* speed v_2. In this way we can bring a specific particle back above some calibration mark. Under these conditions the frictional force must equal the net upward force applied to the particle (electrical force minus gravity). If the charge on the particle is ne and $|\mathbf{E}| = E$, we have

$$(F_{\text{fric}})_2 = 6\pi\eta r v_2 = F_{\text{elec}} - F_{\text{grav}} = neE - \frac{4\pi r^3}{3}g(\rho_s - \rho_{\text{air}}). \tag{1.24}$$

This expression contains the unknown elementary charge e multiplied by the also unknown n, the number of such charges that happen to be on the particle observed. Combining Eqs. 1.23 and 1.24 and solving for ne, we obtain

$$ne = \frac{4\pi r^3}{3}(\rho_s - \rho_{\text{air}})\frac{g}{E}\left(\frac{v_2}{v_1} + 1\right); \tag{1.25}$$

v_1 and v_2 must, of course, be measured with the same particle. If we had a way of measuring r independently, then we could use Eq. 1.25 exactly as it is. (Here the advantage of modern plastic spheres of known, uniform radius becomes clear.) The density of the sphere can be determined from the bulk material; the density of air is known accurately; the electric field, as in Thomson's experiment, is the measured voltage divided by the distance between the two metallic plates; and the speeds v_1 and v_2 are just the quantities we measure in the experiment. In practice, however, it is not always possible to measure the radius r; this was the case, for example, in Millikan's original experiments. Fortunately, we can use Eq. 1.23 to determine r from the measured downward speed under gravity alone; rearrangement gives

$$r^2 = \frac{9}{2}\frac{v_1\eta}{g(\rho_s - \rho_{\text{air}})}. \tag{1.26}$$

Substituting this result in Eq. 1.25, we obtain the final expression

$$ne = \frac{4\pi r^3}{3}\left[\frac{9}{2}\frac{v_1\eta}{g(\rho_s - \rho_{\text{air}})}\right]^{3/2}(\rho_s - \rho_{\text{air}})\frac{g}{E}\left(\frac{v_2 + v_1}{v_1}\right) = \frac{36\pi}{E}\left[\frac{\eta^3}{8g(\rho_s - \rho_{\text{air}})}\right]^{1/2}(v_2 + v_1)v_1^{1/2}. \tag{1.27}$$

Now all the quantities on the right are measurable; that is, they were all measurable for Millikan. Some of Millikan's own original data are reproduced in Table 1.2. It can be seen that the measured charges were all almost exactly integral multiples of about 4.9 × 10^{-10} esu (1.63 × 10^{-19} C), which Millikan took to be the value of e.

With plastic spheres of uniform diameter available, one can now do the experiment with Eq. 1.25 as the basis of the calculations, rather than the more cumbersome Eq. 1.27 involving the viscosity of air. Actually, Millikan's original value of e differed from the present value primarily because of an error in the then-accepted value of the viscosity of air. The best value now available for the electronic charge is

$$e = 1.60217646 \times 10^{-19}\ \text{C}.$$

At this point we can combine the value of e with the charge-to-mass ratio e/m of the last section to derive a value for the mass of the electron (strictly speaking, the rest mass, but in this text we shall not consider the effects of relativistic velocities). The value thus found is

$$m_e = 9.1093819 \times 10^{-31}\ \text{kg}.$$

[12] For a discussion of viscosity, see Chapter 28. The viscosity of atmospheric air (as in Millikan's experiment) is about 2 × 10^{-5} N s/m^2.

Table 1.2 Sample Data from Millikan's Report of the Oil-Drop Experiment[a,b]

q (10^{-10} esu)	19.66	24.60	29.62	34.47	39.38	44.42	49.41
n	4	5	6	7	8	9	10
e (10^{-10} esu)	4.915	4.920	4.937	4.923	4.931	4.936	4.941
q (10^{-10} esu)	53.92	59.12	63.68	68.65	78.34	83.22	
n	11	12	13	14	16	17	
e (10^{-10} esu)	4.902	4.927	4.900	4.904	4.897	4.894	

[a] From R. A. Millikan, *Phys. Rev.* **32,** 349 (1911).

[b] These data were all obtained with a single oil droplet; they are not in chronological order, and most values were obtained several times. The first line of the table lists the values obtained for the droplet's charge q at various times, calculated with Eq. 1.27. The integer n is the number of elementary charges assumed to make up q, selected so as to give the best fit for the value of $e(= q/n)$. The average for this droplet was $e = 4.917 \times 10^{-10}$ esu; after hundreds of measurements with many droplets, Millikan obtained an overall average of 4.891×10^{-10} esu (1.631×10^{-19} C).

1.5 Mass Spectrometry

Given the possibility of determining the charge-to-mass ratio of the electron, it is clear that by similar techniques one can just as well (or at least almost as well) determine the charge-to-mass ratio for other charged particles, such as ions—atoms or molecules with charges. Since the charges must always be integral multiples of e, in this way one can measure almost directly the mass of any ion. The mass of one or a few electrons is so small that these masses are essentially identical to the masses of the uncharged atoms or molecules. Such measurements are in fact among the most precise ways used to determine actual atomic weights (as opposed to the relative atomic weights discussed in Section 1.1). One need only be sure that the ions are singly charged, or else that the number of charges can be identified, a relatively simple task in either case. For example, the hydrogen atom yields only one positively charged ion, presumably with a single positive charge (H^+), and with $q/m = 9.57876 \times 10^7$ C/kg; assuming that $q = +e$, we obtain 1.67265×10^{-27} kg or the mass of the proton. The apparatus used for such measurements is usually not precisely of the design used by Thomson; almost invariably, however, the principles and fundamental equations for the process are the same as those described in Section 1.3.

The most common device for measuring precise masses of individual atoms or molecules is called the *mass spectrometer.* Several kinds of instruments go under this name; they have in common their capability for selecting particles with specific values of the charge-to-mass ratio, and thus (for singly charged ions) of the mass. The word "spectrometer" is used because one can select or measure which values actually occur, out of the entire spectrum of possible values of the physical quantity of interest, in this case mass.

Virtually all mass spectrometers have the following features in common:

1. A source region, where neutral atoms or molecules are introduced, converted to ions, and passed on to
2. An analyzing region, where particles of different masses are actually separated from each other in space, and
3. A detecting region, where the ions of a particular mass reach a detector.

The source usually generates ions by either ultraviolet light or electron bombardment. The detector may be an electrical device for counting ions (as particles or as electric current), designed to select only ions of a single mass, or it may be a photographic plate capable of collecting particles of many masses simultaneously at different places. If the information is recorded photographically, on a chart, or directly on a computer, in any form equivalent to the number of ions striking the detector as a function of the mass of the ions, then the instrument is called a mass spectro*graph.* As for the analyzing region, there are several means of separating or *dispersing* the ions according to their masses. We shall now examine two of them, to see how the operations of the instruments are related to the equations of motion and to measurable physical quantities.

One class of instruments, the *magnetic mass spectrometer,* includes Thomson's method and its variations: Devices in this category use a combination of electric and magnetic fields to put moving ions into curved trajectories; the curvature of any ion's trajectory is a function of its charge-to-mass ratio. Commonly, an electric field draws the ions out of the source region and into a region where the trajectories are bent by a magnetic field. Only ions on one specific trajectory can reach the detector; others strike the walls of the chamber. By varying the strength of the magnetic field or the initial electric "drawout" field, one can vary all the trajectories and allow ions of any desired mass to strike the detector. A diagram of such a mass spectrometer is shown in Fig. 1.6. This is the general type of instrument used for accurate measurement of absolute masses; admittedly one now could use a rather fancy version.

The spectrometer of Fig. 1.6 and others of the same general type depend on the Lorentz force given by Eq. 1.16. In such a machine, the acceleration of an ion of mass m and charge q in the magnetic field **B** is

$$\mathbf{a} = \frac{\mathbf{F}}{m} = \frac{q}{m}(\mathbf{v} \times \mathbf{B}). \qquad (1.28)$$

However, we can no longer make the simplifying assumption that the deflecting region is small and v_x essentially constant; rather, we typically have a situation in which

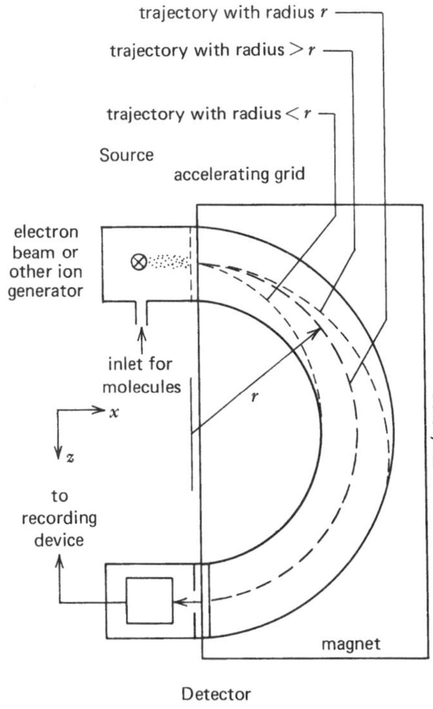

Figure 1.6 Schematic diagram of a magnetic mass spectrometer (180° variety). The trajectory leading to the detector is a semicircle of radius *r*. The magnetic field is directed forward out of the page; this corresponds to the positive *y* direction in the coordinate system of Fig. 1.4.

the deflection is 60 or 180°, with nearly all the trajectory within the deflecting region. We must thus solve the complete equation for the trajectory. Let us use the coordinate system of Fig. 1.4, with the magnetic field in the *y* direction and the ions initially moving in the *x* direction (with initial speed v_0); these coordinates have been indicated in Fig. 1.6. Combining Eq. 1.28 with Eqs. 1.17, we obtain for the components of the acceleration

$$a_x = \frac{d^2x}{dt^2} = -\frac{q}{m} v_z B_y, \tag{1.29a}$$

$$a_y = \frac{d^2y}{dt^2} = 0 \quad \text{(so that } v_y \text{ is always zero)}, \tag{1.29b}$$

$$a_z = \frac{d^2z}{dt^2} = \frac{q}{m} v_x B_y. \tag{1.29c}$$

Notice that ions in the *xz* plane are never accelerated out of this plane by the constant field B_y; rather, the force is always perpendicular to the field itself and to the velocity. The trajectory generated thus lies entirely in the *xz* plane; we shall show that it is a circular arc.

To obtain the actual trajectory, we need to know v_x and v_z as functions of time. Given the initial conditions that $v_x = v_0$ and $v_x = 0$ at the time $t = 0$ (when the ion enters the deflecting field), the simultaneous differential equations 1.29*a* and 1.29*c* have the solution

$$v_x(t) = v_0 \cos\left(\frac{qB_y t}{m}\right),$$

$$v_z(t) = v_0 \sin\left(\frac{qB_y t}{m}\right), \tag{1.30}$$

as can be confirmed by substitution ($a_x = dv_x/dt$, $a_z = dv_z/dt$). These are the equations for the velocity of a body moving in a circle at a constant angular velocity (radians per unit time) of qB_y/m. The speed of such a body is the product of the circle's radius *r* and the angular velocity ω; since the magnitude of the velocity here is always v_0, we can obtain the radius as

$$r = \frac{v_0}{\omega} = \frac{mv_0}{qB_y}. \tag{1.31}$$

The initial speed v_0 is a function of the kinetic energy given to the ion, which must equal the work done by the electric field in the source region. Usually this field is constant; in the coordinates of Fig. 1.6, it is in the *x* direction. If the magnitude of the electric field is E_x, and the distance through which it accelerates the ions is *d*, then the kinetic energy of an ion will be

$$\frac{mv_0^2}{2} = qE_x d, \tag{1.32}$$

so that

$$v_0 = \left(\frac{2qE_x d}{m}\right)^{1/2}. \tag{1.33}$$

In conjunction with Eq. 1.31, this tells us that an ion with mass *m* and charge *q*, accelerated through a distance *d* by an electric field E_x and then moving through a magnetic field B_y, will follow a circular trajectory whose radius *r* is expressible entirely in terms of measurable quantities and the ion's charge-to-mass ratio:

$$r = \frac{m}{qB_y}\left(\frac{2qE_x d}{m}\right)^{1/2} = \frac{(2E_x d)^{1/2}}{B_y}\left(\frac{m}{q}\right)^{1/2}. \tag{1.34}$$

Thus one can fix *r*, and adjust either B_y or E_x to make ions of any desired q/m ratio come around to a fixed detector;[13] or one can fix E_x and B_y, and move the detector to observe ions moving in circles of different radii. Given q/m, one need only estimate the value of *q* to obtain the ionic mass; in practice, singly charged ions ($q = e$) are by far the most common, and atomic weights are known well enough for most other cases to be easily distinguished.

[13] Alternatively, if one has a beam of identical ions with a distribution of velocities, one can use such an instrument as a velocity selector.

This analysis gives the essence of how a magnetic mass spectrometer operates. In practice, certain fixed deflection angles such as 60 or 180° are ordinarily chosen. This is because, with these specific deflection angles, the magnetic field focuses ions that enter the field non-normally (i.e., slightly off the center of the beam) at the same place in space as those (with the same q/m value) that enter along the axis of the beam. Instruments of this type are used for absolute mass determinations because one can determine the quantities r, d, B_y, and E_x with great accuracy, while minimizing other sources of uncertainty such as off-axis trajectories. Such machines are also useful for relative mass measurements, for steady monitoring of the intensities of single-mass beams, and even as sources of beams of ions with a single known mass.

The second type of instrument is called the *time-of-flight mass spectrometer*. Conceptually it is simpler than the magnetic variety. It can scan mass spectra rapidly, and with modern techniques beyond the scope of this discussion, attain resolutions comparable to those of magnetic mass spectrometers. In the simple time-of-flight instrument, shown schematically in Fig. 1.7a, a sudden and short pulse of force accelerates the ions that happen to be in the source chamber when the pulse is applied. This is done by making plate A positive and grid B negative, so that the positively charged ions are accelerated through the grid. The pulse duration is long enough for all the ions in the resulting beam to have passed through the full length d of the electric field, and thus to gain the same kinetic energy. If the initial kinetic energy of the ions is very small compared with the amount added, then their total kinetic energy must be essentially that due to the applied field. Furthermore, if the field is uniform, the ions will all leave the source region going in the same direction (say, the x direction). The ions pass through the grid and into the analyzer, which in this instrument is called the drift tube. Once in the drift tube, they continue moving with the same constant velocity with which they entered the tube, a velocity we shall now show to be dependent on their mass. Ions of different masses thus reach the detector at different times, by measurement of which they can be distinguished.

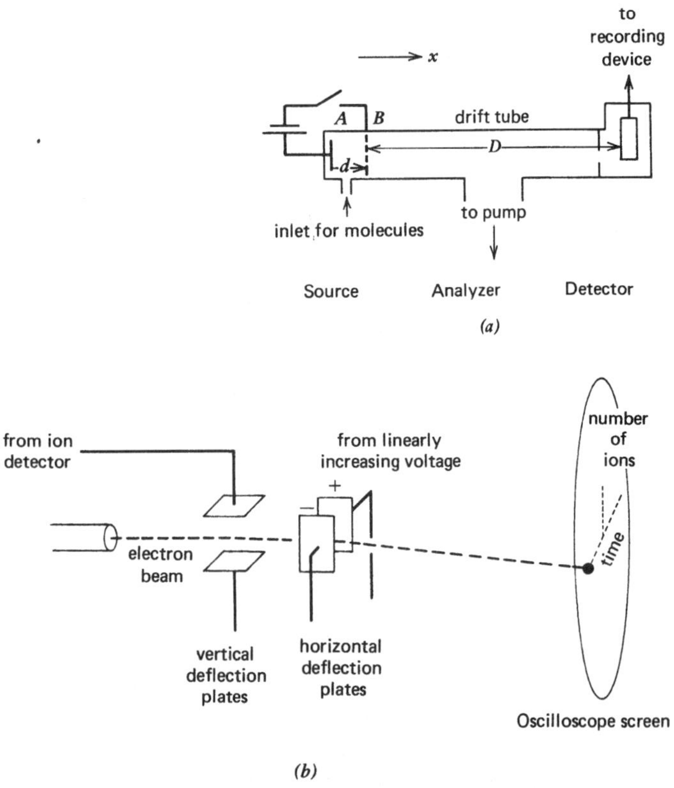

Figure 1.7 (a) Schematic diagram of a time-of-flight mass spectrometer. Molecules are ionized near the plate A; the ions are accelerated through a distance d until they pass through the grid B; they then travel at constant velocity a distance D to the detector. (b) Schematic diagram of the interior of an oscilloscope tube being used to display the mass spectrum from a time-of-flight mass spectrometer. (The tube is normally evacuated.)

Suppose that all the ions are given kinetic energy T by the field. Ions of type j, with mass m_j and charge q_j, will have an x component of velocity determined by the condition

$$T = \frac{1}{2} m_j v_{xj}^2 \quad \text{or} \quad v_{xj} = \left(\frac{2T}{m_j} \right)^{1/2}. \tag{1.35}$$

If the applied field is E_x and the length of the field region is d, the kinetic energy must be $q_j E_x d$. Hence we have

$$v_{xj} = \left(\frac{2 q_j E_x d}{m_j} \right)^{1/2}, \tag{1.36}$$

of the same form as Eq. 1.33. In the time-of-flight instrument, no further fields are required to separate the ions of different masses. Rather, the detector has the capability of distinguishing the pulses of ions reaching it at different times. The time required for an ion of type j to travel the distance D from the source to the detector, after the accelerating pulse has been applied at the source, is then

$$t_j = \frac{D}{v_{xj}} = D \left(\frac{m_j}{2 q_j E_x d} \right)^{1/2}. \tag{1.37}$$

As in the magnetic mass spectrometer, the equations of motion give us a relation from which the ratio q_j/m_j can be obtained, since all other quantities in Eq. 1.37 can be determined experimentally. In practice, one rarely measures the quantities D, d, and E_x precisely for each new set of experiments. Rather, the apparatus is calibrated by measuring t_j for one or more ions of known mass; the times of flight for two singly charged ions of masses m_i and m_j are then in the ratio

$$\frac{t_i}{t_j} = \left(\frac{m_i}{m_j} \right)^{1/2} \tag{1.38}$$

if the other quantities are held constant.

The time of arrival of an ion is measured, typically, as the time interval from the instant when the accelerating pulse is delivered. At that moment, a voltage begins to build at a constant rate. This increasing voltage can be used to displace the spot horizontally on the screen of an oscilloscope, generating a time axis. The arrival of ions of a single mass at the detector produces a pulse of voltage to the oscilloscope that displaces the spot vertically for a moment. Thus one obtains a graph on the oscilloscope screen that may be recorded electronically or on paper: The horizontal axis represents time, and the vertical axis, the appearance of ions. If the voltage of the pulse from the ion detector is proportional to the number of ions of a single mass, then the vertical axis becomes an indicator of the number of ions of that mass. Figure 1.7b shows the display system schematically.

There are still other types of mass spectrometers; some operate with time-varying fields, others use pure electrostatic separation. The two varieties we have described, however, are among the most widely used and are relatively simple to analyze.

1.6 The Atomic Mass Scale and the Mole

In Section 1.1 we discussed how a scale of relative atomic weights can be obtained from macroscopic chemistry, and in subsequent sections we have shown how absolute atomic masses can be measured. It remains for us to obtain the conversion factor between the relative scale and absolute units. We could now abandon the use of the relative scale, of course, but this has not been done—partly because the relative atomic weights were used so long and became so firmly established before this became possible, and partly because they roughly correspond to the numbers of nucleons in the atoms. To explain the latter statement, let us now briefly introduce a few more facts about atomic nuclei.

The nucleus of any atom contains integral numbers of protons and neutrons (the latter usually somewhat in excess, except for light atoms). The electron weighs so little that the nuclear and atomic masses of a given atom do not differ significantly. Since the masses of the proton, the neutron, and the hydrogen atom are almost identical, one might expect all atomic masses to be nearly integral multiples of the mass of the hydrogen atom. This is not quite true, because of the mass defect associated with the nuclear binding energy, which amounts to nearly 1% of the total mass for most nuclei. But it is close enough to true that a scale can be set up on which the masses of all stable atoms are integral within 0.1 unit. This scale corresponds to our relative atomic weight scale.

It is apparent from the table on the inside cover, however, that many of the relative atomic weights obtained are far from integral. This is because a given element may have atoms of two or more different masses, known as *isotopes.* The number of protons in an atom's nucleus equals the number of electrons in the neutral atom and thus determines the atom's chemical behavior, so that this number (the *atomic number, Z*) defines the element. Isotopes of a given element are thus atoms with the same number of protons but different numbers of neutrons. Since isotopes have (nearly) the same chemistry, they are thoroughly mixed in nature,[14] and the observed atomic weight of an element is the average of the isotopic atomic masses weighted by their relative abundances. Isotopic ions have different q/m values, and can thus be separated in a mass spectrometer.

None of these facts about nuclei, of course, were known when the atomic weight scale was first established. Its basis then was simply the convenience of setting the smallest atomic mass approximately equal to unity, yielding near-integral values for most of the more common elements (H, C, O, N, Na, S, etc.). Since nearly all the elements form simple compounds with oxygen, this element was chosen as the standard, the atomic weight of oxygen being defined as exactly 16.0000 . . . ; this gave the *chemical scale* of atomic weights, which was used until 1961. Since natural oxygen consists of three isotopes and can vary slightly in composition, a scale based on a single isotope is inherently more precise (and was needed anyway for mass spectrometric measurements). Physicists therefore introduced the *physical scale,* on which the mass of the isotope[15] of oxygen ^{16}O was defined as exactly 16.0000. . . . The use of two scales often led to confusion, and in 1961 both were replaced by the *carbon-12 scale,* based on the standard ^{12}C = 12.0000. . . . This scale retains the advantage of being based on a single isotope, but it is much closer to the old chemical scale than the physical scale had been.[16] The atomic weights listed on the inside cover are referred to the ^{12}C scale.

The unit in which atomic masses are expressed, the *atomic mass unit* (traditionally amu, now designated d, for *dalton*), is thus one-twelfth of the mass of a ^{12}C atom. By making an absolute measurement of the ^{12}C mass one can obtain the conversion factor 1.660539×10^{-27} kg/amu; its reciprocal is known as *Avogadro's number,* and has the value 6.022142×10^{26} amu/kg (6.022142×10^{23} amu/g). This is a fundamental quantitative relationship between microscopic and macroscopic chemistry.

---

[14] There are small natural variations in isotopic abundances because there is more than one process in which elements are formed in nature, and many natural processes that segregate materials have slightly different rates for different isotopes. Hence, the mean atomic masses from different kinds of samples may differ; for example, isotopic compositions of terrestrial origin may differ from those in meteorites. This is especially the case for radioactive elements and their decay products, which is responsible for the poor precision of the atomic weight of lead and for the possibility of dating archaeological objects according to the amount of radioactive carbon they contain.

[15] By ^{16}O we mean oxygen with *mass number* 16, the nucleus containing 16 nucleons—in this case, 8 protons and 8 neutrons. A more complete notation is $^{16}_{8}O$, which specifies that there are 8 protons in the nucleus.

[16] The numerical values assigned to a given mass M on the three scales are related by the equations

$$M_{chem} = 1.000049 M_{C\text{-}12}, \qquad M_{phys} = 1.00032 M_{C\text{-}12}.$$

Atoms and electrons are such incredibly tiny objects that it would be terribly inconvenient to do chemistry with macroscopic objects while counting particles one at a time. It is much easier to count them many at a time, and to use a general convention defining a basis for the counting system. We thus define a physical quantity called *amount of substance,* the SI unit of which is the *mole* (abbreviated mol). What is a mole? A mole is simply a specific number of things, like a dozen or a ream. However that number defining the mole is the number of atomic mass units (daltons, in SI units) in 1 gram. Hence the mole is exactly the number of carbon atoms in[17] 12.0000 · · · g of ^{12}C. A mole of atomic mass units has a mass of 1 g, and Avogadro's number N_A, the conversion factor between amu and g, gives the number of entities per mole. The terms "gram-atom" and "gram-molecule" were sometimes used for a mole of atoms or molecules, respectively, but the general term "mole" is sufficient as long as the elementary entities are specified. They need not be particles; for example, in "a mole of sodium chloride," the elementary entity is the Na^+–Cl^- ion pair. However, it is important to avoid ambiguity by specifying the species we are counting one mol at a time.

In dealing with physical and chemical properties, we shall find it convenient to use both the macroscopic scale, in which we count particles a mole at a time and energies likewise on a molar basis, and the microscopic scale, in which we count particles one at a time and energies on a per-particle basis. For example, when dealing with the absorption and emission of light by atoms or with collisions between atoms, we shall find it convenient to speak on the scale of single atoms and to define energy units and distances appropriate to this scale. On the other hand, when we deal with bulk chemical phenomena, such as equilibria between phases or the heat evolved in a chemical reaction, we shall consider particles a mole at a time and speak of such quantities as molar volume or energy evolved per mol of reaction. The two methods of counting are entirely equivalent; we need only be careful to realize which we are using.

The values of Avogadro's number and the electronic charge can be related to each other through Faraday's laws of electrolysis. The experiment to establish this relationship is still one of the fundamental methods of determining the physical constants. What Faraday found was that a constant quantity of electricity was required to carry out one equivalent[18] of electrochemical reaction. By the definition of the equivalent, this quantity—now known as the faraday—must equal the total charge on 1 mol of electrons. Precise measurements of the faraday usually involve the electroplating of silver from solution. Chemical and physical evidence shows that silver exists in aqueous solution as Ag^+ ions, so that one electron is required for each atom of silver deposited ($Ag^+ + e^- \rightarrow Ag$) and the equivalent and the mole of silver are the same. From extremely precise mass spectrometric measurements, the atomic weight of silver is known[19] to be 107.868. The faraday is thus the quantity of electricity required to deposit 107.868 g of metallic silver; from precise measurements of weight and charge (current times time), its value is found to be 96484.56 C. *Faraday's constant* ($\mathcal{F}$) is the conversion factor 96484.56 C/mol. Since $\mathcal{F} = N_A e$, dividing this value by Avogadro's number gives the familiar 1.60219×10^{-19} C for the charge of the electron.

1.7 The Periodic Table

So far in this chapter we have sought to develop a feeling for atomic magnitudes, and to introduce the means for dealing with these magnitudes on both microscopic and macroscopic scales. We have emphasized the common properties of atoms, introducing just enough information about the differences between atoms to explain the mass scale. But as we said at the outset, it is these differences that constitute the essential subject matter of chemistry. In subsequent chapters we shall examine the structure of atoms and molecules in detail, eventually relating these properties to macroscopic chemical behavior. At this point let us briefly review the facts we must account for, as summarized in the periodic table of the elements.

At the present time 115 chemical elements are known (although the number may have changed by the time this book is published); about 90 of these exist in appreciable quantities in nature, and 82 have at least one stable isotope. One obvious way of ordering the elements is in the sequence of increasing atomic weights, beginning with hydrogen and extending to the artificial transuranium elements. It became increasingly apparent in the nineteenth century that many properties of the elements vary periodically over this sequence. This is graphically illustrated by the curve of atomic volumes (shown in Fig. 1.8), which Lothar Meyer used in 1870 to establish the principle of periodicity. Dmitri Mendeleev in 1869 had reached the same conclusion on the basis of periodicities in chemical properties, such as valence and metallic character. Mendeleev imposed a further order on the sequence of elements, changing it from a linear array into a sort of spiral on which elements with similar properties were grouped together. This required the brave assumption of some gaps in the sequence, and Mendeleev correctly predicted the properties of the elements later found to occupy those positions. The resulting arrangement became

[17] In the SI it would be more consistent to relate this unit to kilograms rather than grams, but the mole (like the gram itself) was too firmly established for the terminology to be changed.

[18] If an elementary electrochemical process involves the transfer of *n* electrons, then there are *n* equivalents per mole of reaction. Thus in the reaction $Cu^{+2} + 2e^- \rightarrow Cu$ we have $n = 2$, so that in this case an equivalent of Cu is half a mole.

[19] What follows is a circular argument since we need the value of *e* to get the atomic weight. In practice, one combines the results of a wide variety of experiments involving the different physical constants, selecting the set of values that best fits all the data simultaneously.

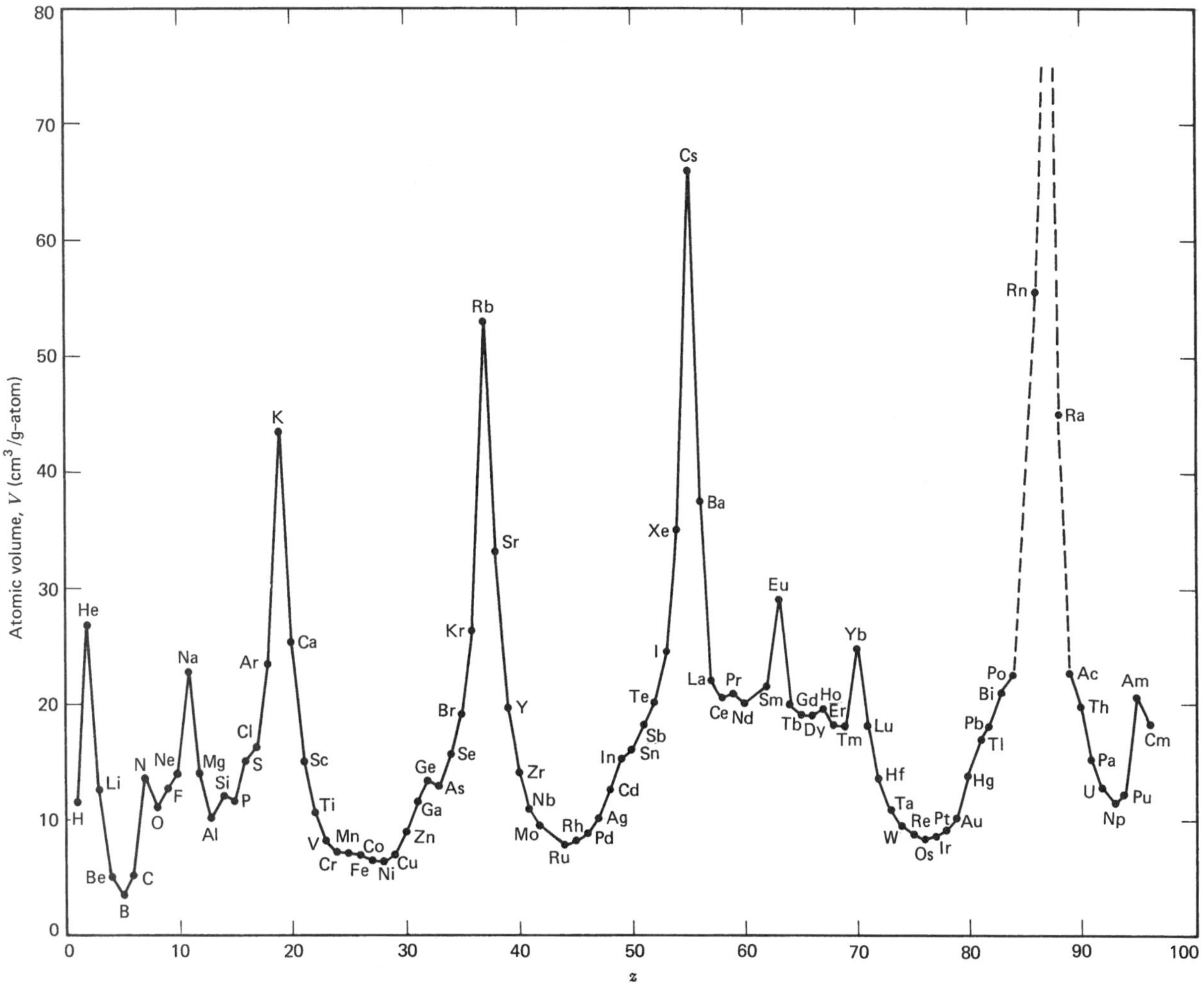

Figure 1.8 Atomic volumes of the elements, from solid-state density data extrapolated to 0 K (room-temperature data used for a few of the less common elements). Note the periodic behavior, especially the alkali metal peaks.

the *periodic table;* a modern version (somewhat refined beyond Mendeleev's) is given as Table 1.3.

As the reader is aware, in the periodic table the groupings of similar elements generally occupy vertical columns. The most clearly defined of such groupings (although unknown to Mendeleev) is that of the "inert" gases, which occupy the last column: The elements helium, neon, argon, krypton, xenon, and radon are all gases at room temperature, and so nonreactive that until a few years ago they were believed to form no chemical compounds. These elements are immediately preceded in the sequence by the halogens (fluorine, chlorine, bromine, iodine, astatine) and immediately followed by the alkali metals (lithium, sodium, potassium, rubidium, cesium, francium), both well defined by their chemical similarities. Other groups, though less clear-cut, similarly fall into place vertically. Several other natural groupings are horizontal or diagonal; one derives from the close resemblance among the elements in sequence from lanthanum to lutetium (the lanthanides) and another is the diagonal relationship among aluminum, germanium, and antimony. The table as finally assembled is an empirical arrangement giving such correlations for all the elements.

To obtain this ordering, it was necessary to take some liberties with the sequence of atomic weights, even after all the vacancies had been filled. For example, tellurium by its properties had to precede the lighter iodine, and similarly potassium and argon had to be interchanged. Elements not found in nature can be assigned the atomic weights of their most long-lived or most stable isotopes. The order actually used is not that of the atomic weights but rather that of the *atomic numbers.* For the moment, we can take the atomic number Z as giving nothing more than the empirical position of an element in the sequence. In fact, however, as mentioned earlier, Z is the number of positive charges in the nucleus and thus the number of electrons in the atom; in Chapter 2 we shall give the evidence for this statement. In subsequent chapters we shall develop the theory of atomic structure, in the light of which we can add more to our interpretation of the patterns of chemistry associated with the periodic table.

Table 1.3 The Periodic Table of Elements

WebElements: the periodic table on the world-wide web

http://www.webelements.com/

Key:

element name atomic number element symbol 1997 atomic weight (mean relative mass)

1	2		3	4	5	6	7	8	9	10	11	12	13	14	15	16	17	18
hydrogen 1 H 1.00794(7)																		helium 2 He 4.002602(2)
lithium 3 Li 6.941(2)	beryllium 4 Be 9.012182(3)												boron 5 B 10.811(7)	carbon 6 C 12.0107(8)	nitrogen 7 N 14.00674(7)	oxygen 8 O 15.9994(3)	fluorine 9 F 18.9984032(5)	neon 10 Ne 20.1797(6)
sodium 11 Na 22.989770(2)	magnesium 12 Mg 24.3050(6)												aluminium 13 Al 26.981538(2)	silicon 14 Si 28.0855(3)	phosphorus 15 P 30.973761(2)	sulfur 16 S 32.066(6)	chlorine 17 Cl 35.4527(9)	argon 18 Ar 39.948(1)
potassium 19 K 39.0983(1)	calcium 20 Ca 40.078(4)		scandium 21 Sc 44.955910(8)	titanium 22 Ti 47.867(1)	vanadium 23 V 50.9415(1)	chromium 24 Cr 51.9961(6)	manganese 25 Mn 54.938049(9)	iron 26 Fe 55.845(2)	cobalt 27 Co 58.933200(9)	nickel 28 Ni 58.6934(2)	copper 29 Cu 63.546(3)	zinc 30 Zn 65.39(2)	gallium 31 Ga 69.723(1)	germanium 32 Ge 72.61(2)	arsenic 33 As 74.92160(2)	selenium 34 Se 78.96(3)	bromine 35 Br 79.904(1)	krypton 36 Kr 83.80(1)
rubidium 37 Rb 85.4678(3)	strontium 38 Sr 87.62(1)		yttrium 39 Y 88.90585(2)	zirconium 40 Zr 91.224(2)	niobium 41 Nb 92.90638(2)	molybdenum 42 Mo 95.94(1)	technetium 43 Tc [98.9063]	ruthenium 44 Ru 101.07(2)	rhodium 45 Rh 102.90550(2)	palladium 46 Pd 106.42(1)	silver 47 Ag 107.8682(2)	cadmium 48 Cd 112.411(8)	indium 49 In 114.818(3)	tin 50 Sn 118.710(7)	antimony 51 Sb 121.760(1)	tellurium 52 Te 127.60(3)	iodine 53 I 126.90447(3)	xenon 54 Xe 131.29(2)
caesium 55 Cs 132.90545(2)	barium 56 Ba 137.327(7)	57-70 *	lutetium 71 Lu 174.967(1)	hafnium 72 Hf 178.49(2)	tantalum 73 Ta 180.9479(1)	tungsten 74 W 183.84(1)	rhenium 75 Re 186.207(1)	osmium 76 Os 190.23(3)	iridium 77 Ir 192.217(3)	platinum 78 Pt 195.078(2)	gold 79 Au 196.96655(2)	mercury 80 Hg 200.59(2)	thallium 81 Tl 204.3833(2)	lead 82 Pb 207.2(1)	bismuth 83 Bi 208.98038(2)	polonium 84 Po [208.9824]	astatine 85 At [209.9871]	radon 86 Rn [222.0176]
francium 87 Fr [223.0197]	radium 88 Ra [226.0254]	89-102 **	lawrencium 103 Lr [262.110]	rutherfordium 104 Rf [261.1089]	dubnium 105 Db [262.1144]	seaborgium 106 Sg [263.1186]	bohrium 107 Bh [264.12]	hassium 108 Hs [265.1306]	meitnerium 109 Mt [268]	ununnilium 110 Uun [269]	unununium 111 Uuu [272]	ununbium 112 Uub [277]		ununquadium 114 Uuq [289]		ununhexium 116 Uuh [289]		ununoctium 118 Uuo [293]

*lanthanides	lanthanum 57 La 138.9055(2)	cerium 58 Ce 140.116(1)	praseodymium 59 Pr 140.90765(2)	neodymium 60 Nd 144.24(3)	promethium 61 Pm [144.9127]	samarium 62 Sm 150.36(3)	europium 63 Eu 151.964(1)	gadolinium 64 Gd 157.25(3)	terbium 65 Tb 158.92534(2)	dysprosium 66 Dy 162.50(3)	holmium 67 Ho 164.93032(2)	erbium 68 Er 167.26(3)	thulium 69 Tm 168.93421(2)	ytterbium 70 Yb 173.04(3)
**actinides	actinium 89 Ac [227.0277]	thorium 90 Th 232.0381(1)	protactinium 91 Pa 231.03588(2)	uranium 92 U 238.0289(1)	neptunium 93 Np [237.0482]	plutonium 94 Pu [244.0642]	americium 95 Am [243.0614]	curium 96 Cm [247.0703]	berkelium 97 Bk [247.0703]	californium 98 Cf [251.0796]	einsteinium 99 Es [252.0830]	fermium 100 Fm [257.0951]	mendelevium 101 Md [258.0984]	nobelium 102 No [259.1011]

Symbols and names: the symbols of the elements, their names, and their spellings are those recommended by IUPAC. After some controversy, the names of elements 101-109 are now confirmed: see Pure & Appl. Chem., 1997, **69**, 2471–2473. Names have not been proposed as yet for the most recently discovered elements 110–112, 114, 116, and 118 so those used here are IUPAC's temporary systematic names: see Pure & Appl. Chem., 1979, **51**, 381–384. In the USA and some other countries, the spellings aluminum and cesium are normal while in the UK and elsewhere the usual spelling is sulphur.
Periodic table organisation: for a justification of the positions of the elements La, Ac, Lu, and Lr in the WebElements periodic table see W.B. Jensen, "The positions of lanthanum (actinium) and lutetium (lawrencium) in the periodic table", J. Chem. Ed., 1982, 59, 634–636.
Group labels: the numeric system (1–18) used here is the current IUPAC convention. For a discussion of this and other common systems see: W.C. Fernelius and W.H. Powell, "Confusion in the periodic table of the elements", J. Chem. Ed., 1982, **59**, 504–508.
Atomic weights (mean relative masses): see Pure & Appl. Chem., 1996, **68**, 2339–2359. These are the IUPAC 1995 values. Elements for which the atomic weight is contained within square brackets have no stable nuclides and are represented by one of the element's more important isotopes. However, the three elements thorium, protactinium, and uranium do have characteristic terrestrial abundances and these are the values quoted. The last significant figure of each value is considered reliable to ±1 except where a larger uncertainty is given in parentheses.
 For updates to this table see http://www.shef.ac.uk/chemistry/web-elements/webelements/support/media/pdf/periodic-table.html. Version date: 13 July 1999.

From WebElements, http://www.webelements.com

• FURTHER READING

D'Abro, A., *The Rise of the New Physics* (Dover Publications, Inc., New York, 1951).

Conn, G. K. T., and Turner, H. D., The *Evolution of the Nuclear Atom* (American Elsevier Publishing Company, Inc., New York, 1965).

Kondratyev, V., *The Structure of Atoms and Molecules* (P. Noordhoff N. V., Groningen, The Netherlands), Chapter 1.

Oldenberg, O., and Holladay, W. G., *Introduction to Atomic and Nuclear Physics* (McGraw-Hill Book Co., New York, 1967).

Richtmyer, F., and Kennard, E. H., *Introduction to Modem Physics* (McGraw-Hill Book Co., New York, 1947), Chapters 1 and 2.

Shamos, M., ed., *Great Experiments in Physics* (Holt-Dryden, New York, 1959).

• PROBLEMS

1. Suppose that one were trying to apply the law of multiple proportions when weights could not be measured with modern accuracy and separations were often not quantitative. In particular, suppose that one were trying to use this law to infer molecular formulas of two salts of Mg, S, and O, and the atomic weight of magnesium by comparing the percent Mg in $MgSO_3$ and $MgSO_4$, but that the apparent weight of sulfite is approximately 20% lower than the true value, whereas the apparent weight of sulfate is within 2% of the correct value. What would one take for the formulas of magnesium sulfite and sulfate, and for the atomic weight of Mg with this error in the data? Assume atomic weights of 16 and 32 for oxygen and sulfur, respectively.

2. Estimate each of the following: the volume of 1 mol of pinheads; the time it takes 1 mol of electrons to pass a monitoring point if the current is 0.1 A; the radius of a hydrogen atom if the nucleus has a radius of 0.1 mm; the mass of all the electrons striking a target in 1 s, from a current of 0.1 A.

3. Design an experiment, by giving a diagram and suitable dimensions, voltages, and magnetic field strengths, to determine the ratio *e*/*m* by Thomson's method for a muon. Assume you know that the charge is approximately that of the electron and that the mass falls between 200 and 500 times that of the electron. Assume the muons can be injected into the apparatus with energies of 100 ± 0.1 eV. Finally, assume that the detector has an area of 1 mm^2.

4. Design an experiment to search for quarks, with charge $+\frac{1}{3}e$, by Millikan's method but with uniform plastic spheres whose density is 0.6 g/cm^3 or 6×10^5 g/m^3. Would you choose a radius $r = 0.1$ mm or 0.01 mm for the spheres? The quark might have a mass as large as that of the proton. Would this affect the experiment?

5. Compute the following: the force on an electron in a uniform electric field of 10^5 V/m in the z direction and the corresponding acceleration; the force on an electron moving 10^5 m/s in the x, y plane at exactly 45° to a magnetic field whose magnetic field strength **H** lies along the y axis and has a magnitude of 10^4 A/m, and the acceleration of that electron; the force on an electron moving at 10^5 m/s along the x axis simultaneously in an electric field of 10^4 V/m along the y axis and a magnetic field in the z direction whose magnetic flux density is 0.1 T, and the acceleration of that electron. Be sure to indicate the direction of each force.

6. Suppose that ions with masses between 1 and 200 amu are to be analyzed by a magnetic mass spectrometer of the type described in Section 1.5. The radius r has been selected to be 0.015 m. It is impractical to use electric fields E_x below about 2×10^5 V/m because, with lower fields, ions emerging from the source in which they are generated leave so slowly that they inhibit the formation and extraction of more ions. This condition is called a "space-charge-limited" condition. It is also impractical to use electric fields much above 3×10^6 V/m because of the problem of electric discharges; avoiding these discharges becomes expensive. Assume that the mass spectrometer will operate with a fixed bending angle of 60° and that the paths of the ions will be determined by the electric accelerating field and the magnetic deflecting field. Select suitable ranges for the electric and magnetic fields, and specify what fields would be used to focus masses of 1, 50, and 200 amu on the detector.

7. Sometimes unstable ions are formed in the source of a mass spectrometer; these break into fragments before they pass through the entire mass analyzing region. Suppose that an ion of mass 150 is produced in the source of a magnetic mass spectrometer, passes through the accelerating field, and then decomposes into fragments with masses 135 and 15. The light fragments are nearly all neutral and furthermore are lost, but the heavy fragments that carry the charges continue to move with approximately their initial velocity as they travel into the region of magnetic deflection. At what apparent mass will these fragments appear at the detector?

8. In an experiment designed by Classen, electrons are emitted from the hot filament cathode labeled C in Fig. 1.9. The electrons are then accelerated toward an anode A by a known potential difference V. A known magnetic field **H** acting at right angles to the plane of the drawing deflects the electrons along a semicircular path EL_1 onto photographic plate P_1. On reversal of the direction of **H** the electrons follow EL_2 and strike plate P_2. Show that *e*/*m* can be calculated if V, **H**, and the radii of curvature of the orbits EL_1 and EL_2 are measured.

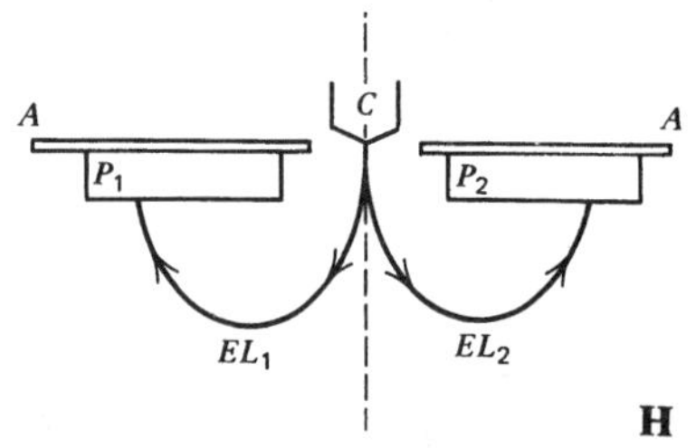

Figure 1.9 Trajectories of electrons in two oppositely directed magnetic fields (Problem 8).

9. Consider the apparatus sketched in Fig. 1.10. The electric field E_y is along the y axis only. A stream of Li^+ ions enters the slit with velocity in the x direction of $v_x = 5 \times 10^7$ cm/s. What must E_y be if, on the photographic plate, the lines corresponding to ^6Li and ^7Li are to be separated by 1 cm? Assume the slit is 10 cm from the photographic plate.

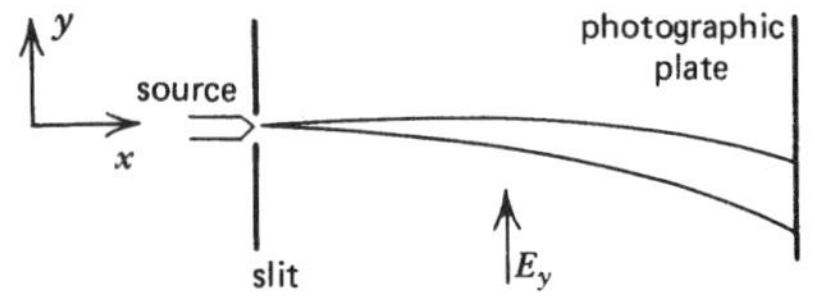

Figure 1.10 Trajectories of ions deflected on their way to a photographic plate (Problem 9).

10. A velocity selector for molecules is made of two slotted disks on a common rotatable shaft. The speed of the shaft is variable. The disks are 0.1 m in diameter and have rim slots 2 mm wide and 0.01 m (10 mm) high. They are 0.1 m apart. It is hoped that the selector can be used to choose velocities in the range 5×10^2 to 5×10^3 m/s. How fast must the selector rotate to achieve such selection? Plot the rotation speed (revolutions/s) against the transmitted velocity, for angles ϕ of 10, 45, and 90° between the slots, when they are viewed along the rotational axis of the shaft. If the angle is 45° and the velocity of rotation is set to pass particles traveling at 10^3 m/s, what are the upper and lower limits of the velocities that also pass, due to the finite widths of the slots, if we assume that velocities are all parallel to the axis of the shaft?

CHAPTER 2

Origins of the Quantum Theory of Matter

In Chapter 1 we dealt with the gross properties of atoms; now we begin to study their internal structure. The broad purpose of this chapter is to develop the experimental evidence leading to what is known as the *quantum theory of matter.* There are two principal ways by which an atom can interact with the rest of the universe: by colliding with other particles, and by absorbing or emitting radiation; either process may change the atom's energy. We shall examine experiments of both kinds, all leading to the conclusion that energy associated with atoms is *quantized.* We find that atoms can *accept or give up* energy only in discrete amounts known as *quanta,* that energy is *stored* in quantized units, and that the energy *in light* is also found only in quantized units. We shall examine the experimental basis for each of these aspects of the quantum theory.

Although quantization is the main theme of the chapter, several other important topics are introduced. These include the evidence for the nuclear structure of the atom; the mechanical concept of action (in terms of which quantization can most easily be formulated); the remarkably useful model of the harmonic oscillator; and phenomena such as spectra and heat capacities, which will later be examined in detail. To close the chapter, we use much of this material in formulating the first and simplest model of atomic structure based on quantization, the planetary model of Niels Bohr.

2.1 The Franck–Hertz Experiment

Some of the clearest evidence for our interpretation of the structure of matter comes from the classic experiment performed in 1914 by James Franck and Gustav Hertz. Although historically this came after much of the other evidence we shall discuss, it is a particularly straightforward example, both conceptually and practically, of a quantum process. In brief, one bombards a chosen gas with electrons of known energy; the electrons collide with the atoms of the gas, losing energy to them; the excited atoms emit light upon returning to their original state. One studies both the light emitted and the energy lost by the electrons. (If an electron collides with an *excited* atom, it is possible, of course, for that excitation energy to transfer to the electron or for the electron to excite the atom still more; energy can flow in either direction.)

A schematic diagram of the Franck–Hertz apparatus is shown in Fig. 2.1. Electrons leave the hot filament both because of its high temperature and because the grid is positively charged relative to the filament. We define the grid-filament and grid-plate potential differences as V_1 and V_2,

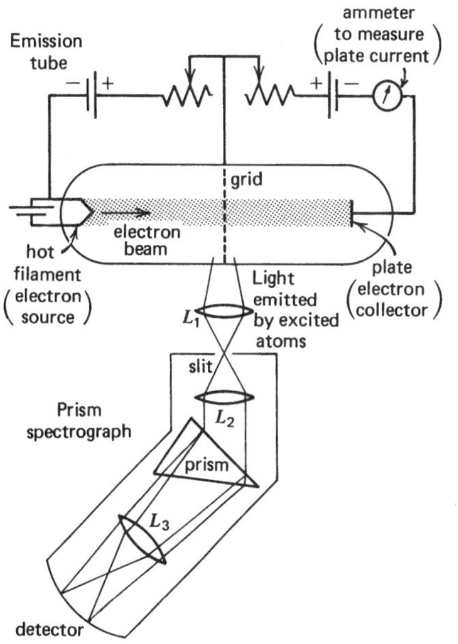

Figure 2.1 Schematic diagram of Franck–Hertz apparatus. The emission tube is filled with a gas at low pressure. The batteries and resistances at the top are adjusted to give the desired electrostatic potentials (cf. Fig. 2.2). The light emitted by excited atoms near the grid is collected by the lens L_1, the focus of which is at the spectrograph slit. The lens L_2 collimates (makes parallel) the light rays, the prism separates the light of different wavelengths, and the lens L_3 focuses the light of each wavelength at a different point on the detector, usually a photographic film.

respectively, as illustrated in Fig. 2.2. The electric field between filament and grid accelerates the electrons to a maximum kinetic energy[1] of

$$T_0 = \tfrac{1}{2} m_e v^2 = eV_1 ; \tag{2.1}$$

V_1 and thus T_0 can be measured fairly accurately. (In Section 1.3 we saw that electrostatic potential is just the potential energy per unit charge.) If the plate were more positive than the grid ($V_2 < 0$), all the electrons passing through the grid would reach the plate and be counted as part of the plate current. However, making the plate more negative than the grid ($V_2 > 0$) allows the apparatus to act as an analyzer of electron energies. The grid-to-plate region is then one of steadily increasing potential energy, analogous to a hill that the electrons must climb. If the plate is more positive than the filament (as in Fig. 2.2), the electrons will have acquired enough kinetic energy in their "downhill" slide to the grid to travel back up the lower hill to the plate—unless they have lost energy along the way. It is, of course, precisely the energy lost along the way in which we are interested. For given values of the potential differences, only electrons that lose an amount of energy less than ΔT will reach the plate, where $\Delta T = e(V_1 - V_2)$. By measuring the plate current as a function of ΔT, one can determine what fraction of the electrons have lost any given amount of energy.

In practice, one usually holds V_2 constant and varies V_1. Figure 2.3 illustrates typical results. Whenever the maximum electron energy eV_1 is less than some threshold value E_{thr}, the plate current rises continuously with increasing V_1, as more electrons are drawn from the filament. After the threshold energy is reached, the current drops sharply, and thereafter shows successive, similar spikes, corresponding to energy losses in fixed amounts. We interpret these current measurements as follows: As long as the electrons cannot excite the internal structure of the atom, the total kinetic energy must be conserved; because electron-atom collisions exchange very little momentum as a consequence of the very large difference in their masses, the kinetic energy of the electron hardly changes and the current rises as V_1 rises. But when an electron has enough energy to excite an internal state or states of the atom, it can lose a fixed large amount of energy equal to the atomic excitation energy. At the threshold this excitation requires virtually all of the electron's energy, so that electrons undergoing collisions of this type cannot reach the plate, and the plate current drops. The potential at which such a drop begins is known as a *resonance potential*. As V_1 continues to increase, the electrons have enough kinetic energy

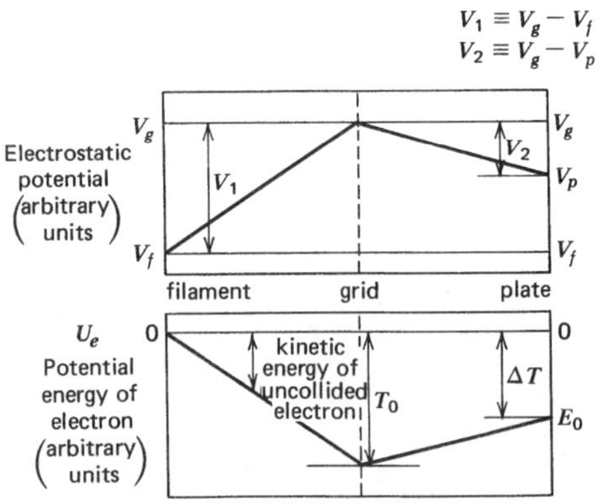

Figure 2.2 Electrostatic potential (V) and potential energy of an electron (U_e) in the Franck–Hertz emission tube. If U_e is set equal to zero at the filament, the two are related by $U_e = -e(V - V_f)$, and the kinetic energy of an electron is $-U_e$ if no energy has been lost by collision. Near the grid the electron that has not collided should thus have kinetic energy $T_0 = eV_1$. The total energy of the electron is initially zero on this scale, but may be lowered by collisions with gas atoms. It must be at least E_0 for the electron to reach the plate, where

$$E_0 = -\Delta T = -e(V_p - V_f) = -e(V_1 - V_2);$$

ΔT is thus the maximum allowable energy loss.

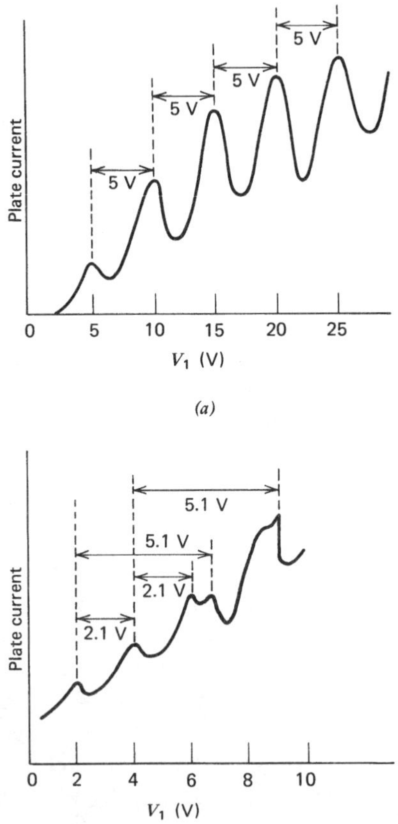

Figure 2.3 Observed plate currents as functions of grid voltage in the Franck–Hertz experiment: (*a*) Hg vapor, resonance potential 4.9 ± 0.1 V; (*b*) Na vapor, first resonance potential 2.1 V.

[1] In measurements of this kind it is most convenient to measure the energy in *electron volts* (eV), where 1 eV is the energy of one electron accelerated through a potential of 1 V. We thus have

$$1 \text{ eV} = (e)(1 \text{ V}) = (1.60219 \times 10^{-19} \text{ C})\left(\frac{\text{J}}{\text{V C}}\right) = 1.602176 \times 10^{-19} \text{ J}.$$

left over to reach the plate once more, and the current resumes its rise. The thresholds after the first correspond to successive excitations by the same electrons, or to the appearance of different kinds of excitation.

In addition to measuring the energy lost by the electrons, one also may look at light emitted by the atoms with which they collide. The simplest observable properties of light are its wavelength and its intensity. The light of different wavelengths can be separated with a *spectrometer* or *spectrograph,* much as the mass spectrometers of Section 1.5 spatially separate ions of different mass. In an optical spectrometer the light is usually dispersed with a prism (as in Fig. 2.1) or a diffraction grating, either of closely spaced parallel grooves or of a holographic image of such grooves. Auxiliary lenses or mirrors then focus the light at the detector, which may be a photographic film, a photoelectric cell, a photosensitive array of semiconductor detectors, a thermocouple (for infrared light), or any other device sensitive to incident radiation. The intensity of the light can be determined from the strength of the signal observed (darkening of film, photoelectric current, etc.). In the Franck–Hertz experiment, one studies how the wavelength and intensity of the emitted light vary with the grid potential; the most important conclusions come from a comparison of these data with the electron energy measurements.

The essential results of the Franck–Hertz experiment are these:

1. Every gas introduced into the tube emits light with its own characteristic spectrum of wavelengths; such an emission spectrum may occur as a set of lines of very sharply defined wavelength, as a broad, continuous spectrum, or as a combination of the two.
2. Each part of a substance's emission spectrum has its own threshold energy for excitation. Each spectral line or band of continuous emission appears only if the energy of the bombarding electrons is equal to or greater than some minimum value characteristic of that line. More than one line may have the same energy threshold.
3. The occurrence and intensity of each spectral line are correlated with one or more of the electron-current resonance potentials. The threshold energy for the appearance of the line is the same as the threshold energy for one of the sharp drops in current (or the difference between two such energies). The intensity of the line is proportional to the number of electrons that have lost just this amount of energy as indicated by the plate current.
4. The shorter the wavelength of a spectral line, the higher is the threshold electron energy E_{thr} required to excite the line.
5. In a large number of cases, the wavelength λ_{min} of the line of shortest wavelength associated with a given E_{thr} is inversely proportional to the threshold energy; that is, frequently we have

$$\lambda_{min}(E_{thr}) \propto (E_{thr})^{-1}\,. \qquad (2.2a)$$

The wavelength λ and the frequency ν of light are inversely proportional to each other, $\nu \times \lambda = c$, the speed of propagation of the wave fronts. Hence a minimum wavelength λ_{min} corresponds to a maximum frequency ν_{max}, and we can rewrite Eq. 2.2*a* as the direct proportionality

$$\nu_{max}(E_{thr}) \propto E_{thr}\,. \qquad (2.2b)$$

The foregoing information tells us that atoms are capable of accepting energy from collisions with electrons and of returning this energy (or at least part of it) as light. The light may be returned as spectral lines or bands characteristic of the substance, and not necessarily as a simple continuous spectrum. Result 3 suggests that each spectral line associated with a single threshold energy is also associated with the acquisition by the atom of the same well-defined amount of energy from the electrons; it is the amount of energy absorbed, rather than the total energy of the electrons, that determines if a particular line appears. In short, the results suggest that any given atom can accept energy only in well-defined (quantized) amounts.

We can use an analogy to interpret this result and emphasize how much it differs from expectations based on classical physical laws. If the electrons in atoms could move about as freely as classical laws would allow, we would expect an atom to be capable of having any energy at all. Instead, we observe that atoms can possess only discrete amounts of energy. A classical analogue of this model is a rock on a hill. The gravitational potential energy of the rock is analogous to the total internal energy of the atom. With no knowledge of the shape of the hill, we expect the rock to roll continuously down the hill. However, a rock behaving like an atom would be found only on one of a set of well-defined ledges. Figure 2.4 illustrates the contrast between the two views. Even as late as 1925, we had no basis except the observation of discrete frequencies of light on which to postulate "ledges" for electrons. Interpretation of the Franck–Hertz experiment, and the photoelectric effect and the discrete lines of atomic spectra as well, required an audacious assumption that defied classical physical ideas.

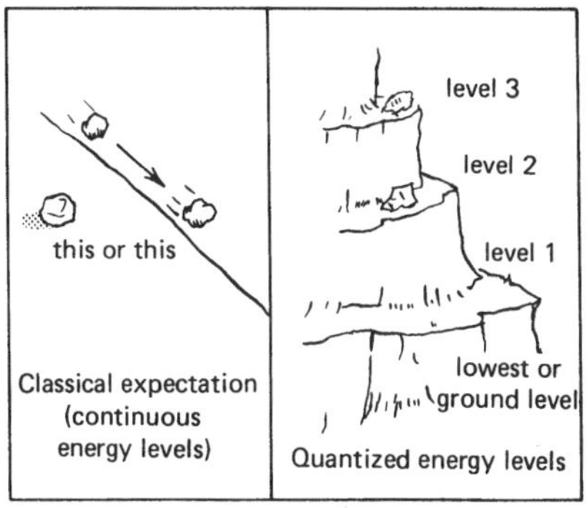

Figure 2.4 Analogue of the quantized energy levels suggested by the Franck–Hertz experiment.

Let us extend our analogy further. If the atom is initially in its "ground state," the threshold energy of the Franck–Hertz experiment corresponds to the potential energy needed to lift a rock to a higher ledge. The emission of light then corresponds to the rock's falling to a lower ledge or the ground. Gravitational potential energy, of course, is proportional to height ($V_{grav} = mgz$). Result 5 above suggests that the frequency ν_{max} corresponds to the height of the first ledge above the ground, whereas the lower frequencies associated with the same E_{thr} give its height above various intermediate ledges. In other words, the frequency of any emission line should be proportional to the energy difference between the excited state (reached by adding E_{thr}) and the atom's state after emission. Thus, if we number the levels as in Fig. 2.4, we would expect an atom excited to level 2, by an amount of energy we can call E_{thr} (2), to emit two spectral lines—one with a frequency (ν_{max}) proportional to E_{thr} (2), the other with a frequency proportional to E_{thr} (2) – E_{thr} (1).

The test of this hypothesis is the proportionality between frequency and energy difference. If we introduce into Eq. 2.2*b* a proportionality constant *h*, so that for the *i*th threshold of a given substance we have

$$E_{thr}(i) = h\nu_{max}(i), \tag{2.3a}$$

then we would expect the other observed frequencies to be given by some combination

$$E_{thr}(i) - E_{thr}(j) = h\nu_{ij}, \tag{2.3b}$$

with the same constant *h*. This is indeed what one finds. Not all *i*–*j* combinations give an observed spectral line (some lines are inherently weak, for reasons we shall discuss later), but the lines that are observed satisfy Eq. 2.3*b*. In fact, not only does this equation hold for a given substance, but the same proportionality constant *h* is found for *all* substances.

We shall have much to say later about the constant *h* (known as *Planck's constant*). It is another fundamental constant like those in Table 1.1, with the value

$$h = 6.626068 \times 10^{-34} \text{ J s.}$$

This value can be used to calculate the threshold energy corresponding to a given spectral line; Table 2.1 gives some early data illustrating this point.

What, then, can we conclude from the Franck–Hertz experiment? It seems that the internal structure of atoms allows for the absorption and emission of energy only in well-defined amounts. For the emission of light by gaseous atoms, at least, the energy given off is related to the frequency of the light by $\Delta E = h\nu$, where *h* is a universal constant. This relationship will turn up again and again, eventually leading us to a general quantum theory of matter. Quantized behavior of this sort occurs not only with visible light, but in all parts of the electromagnetic spectrum; to illustrate this, in the next two sections we consider the properties of ultraviolet light and x-rays.

Table 2.1 Wavelengths and Excitation Potentials for Selected Lines of the Mercury Spectrum[a]

Measured Accelerating Potential at Threshold (V)	Wavelength of Spectral Line (Å)	Theoretical Potential[b] (V)
4.68	2656.5	4.67
4.9	2537.0	4.89
5.47	2270.6	5.46
6.73	1849.6	6.70
7.73	1603.9	7.73
9.37	2656.5	9.33 = 2 × 4.67 (excitation of two atoms)

[a] From J. Franck and E. Einsporn, *Z. Physik* **2,** 18 (1920).

[b] Setting the energies in Eqs. 2.1 and 2.3*a* equal to each other, we have

$$V_1 = \frac{h\nu}{e} = \frac{hc}{e\lambda}$$

for the theoretical threshold potential corresponding to wavelength λ. (The values in the third column are calculated with our current values of the physical constants; Franck and Einsporn themselves used the values $e = 1.592 \times 10^{-19}$ C, $h = 6.545 \times 10^{-34}$ J s, and thus obtained results about 0.6% lower.)

2.2 The Photoelectric Effect

One of the first major phenomena accounted for by the hypothesis of quantization was the *photoelectric effect.* This consists of the emission of electrons when light shines onto a material. The interpretation of this process generated the idea that light contains energy in discrete quantized units.

Photoelectrons are produced when light of frequency higher than some threshold value strikes any substance. (Historically, the effect was first studied for solids.) The threshold frequency is a characteristic property of the substance, and sometimes of the condition of its surface. The alkali metals yield photoelectrons with visible light, but for most substances the threshold lies in the ultraviolet.

We determine the rate of electron emission by measuring the current through a collector plate, as in the Franck–Hertz experiment; we find the electrons' kinetic energy by determining the voltage required to prevent them from reaching the collector, as in Eq. 2.1.

How do the number of electrons and their energies vary with the frequency and intensity of the incident light? Below the threshold frequency ν_0, no electrons at all are emitted. Above ν_0, electrons with a variety of energies escape the surface; their energy distribution is independent of the light intensity, but varies with frequency. In particular, the *maximum* electron energy for a frequency ν obeys the equation

$$E_{max} = h\nu - h\nu_0, \tag{2.4}$$

where *h* is again Planck's constant; $h\nu_0$ is called the *work function.* For a given frequency, the rate of electron emission is proportional to the light intensity.

These results were puzzling in terms of the wave theory of light. Since the energy delivered by a wave increases with the wave's amplitude, one would expect the energy of the liberated electrons to increase with the light intensity. Furthermore, if the energy were distributed uniformly over the wavefront, it would take a rather long time for any one electron to obtain the observed energy, but no such time lag is observed. The answer to the problem was to assume that light can behave as a stream of particles as well as a wave—that is, that light is an entity with both particle and wave aspects. This explanation, like that of the heat capacity of solids, was first given in 1905 by Einstein, one of the very few scientists who used but never really accepted the quantum theory.

Einstein's model, then, amounted to the quantization of light. He assumed that the light striking a photoelectric surface (and, by extension, all light) consists of discrete particles or quanta, each with a definite energy proportional to the frequency of the light:

$$E(\text{per quantum}) = h\nu. \tag{2.5}$$

Each electron emitted from the surface then corresponds to the absorption of a single quantum of light.[2] It must take a certain minimum energy, the work function $h\nu_0$, to remove an electron from a solid surface;[3] quanta with less energy (lower frequency) have no effect. If the quantum absorbed has more than the requisite amount of energy, then the extra energy is retained by the newly freed electron as kinetic energy. The maximum energy given by Eq. 2.4 is that of electrons that needed just $h\nu_0$ for liberation; the photoelectrons with less energy are those that experienced one or more collisions in which they lost some part of $h\nu - h\nu_0$ on their way to free flight. The more quanta of a given frequency strike the surface, the more electrons of the corresponding energy distribution are emitted, giving the observed dependence on light intensity. Einstein's model thus explains all the experimental facts about the photoelectric effect in a clear, simple, and vivid way.

The *Compton effect* (1923) provided even more conclusive evidence of the particle nature of light. When x-rays strike a target, some of the scattered radiation is of lower frequency (and thus energy) than the incident beam; at the same time electrons are emitted by the target. What happens is that the incident *photon* (light particle) gives up some but not all its energy to an electron with which it collides. This interpretation is confirmed by the observation that both energy and momentum[4] are conserved in the process.

[2] According to modern quantum mechanics, this is not absolutely true. But the probability of a single quantum's removing two electrons, or of two quanta together exciting a single electron, is so small that it can be disregarded here.

[3] The same energy is found to be required for *thermionic* emission from a hot filament.

[4] See Section 3.1 for the definition of a photon's momentum.

The idea that light consists of quanta of energy $h\nu$ also explains many other phenomena. Both the Franck–Hertz experiment and x-ray emission, described in Section 2.3, can be regarded as examples of the *inverse* photoelectric effect: Electrons strike a target, giving up energy which is released as quanta of radiation. In the ordinary absorption and emission of light, Section 2.4, electrons *within* atoms gain or lose energy by means of quanta. In all these cases the energy transferred is related to the frequency of the radiation by $E = h\nu$; the particular frequencies involved are determined by the energy levels of the atoms.

The idea that light behaves like a stream of particles seemed at first like a resurrection of the old quarrel between the rival seventeenth- and eighteenth-century schools that supported the corpuscular and wave theories of light. The wave theory had been thought to be proved by such phenomena as interference and diffraction, but now new evidence supported the particle interpretation. The split between these two camps now seems to be only a historical accident. As we shall see in more detail in the next chapter, the particlelike and wavelike properties are merely complementary aspects of the same phenomenon.

2.3 X-rays and Matter

X-rays are electromagnetic waves, similar to light and radio waves but with much shorter wavelengths, of the order of 10^{-10} m. X-rays are produced when electrons, accelerated to high velocities from a cathode, bombard a solid target. The process is similar to that in the Franck–Hertz experiment, except that the electrons have higher energies so that higher-frequency radiation is emitted, and the target is a solid rather than a gas. Figure 2.5 shows a schematic diagram of such an x-ray source.

X-rays enable us to study atomic and molecular properties in several ways. In this section we shall examine the emission spectra produced by an x-ray source (which are precisely analogous to the emission spectra produced in the Franck–Hertz experiment) and the ways in which matter may attenuate a beam of x-rays. These experiments give us further indications of the quantization of atomic energies, and suggest the physical interpretation of the atomic number *Z*. In later chapters we shall see how x-rays can be used to study the structure of solids and liquids.

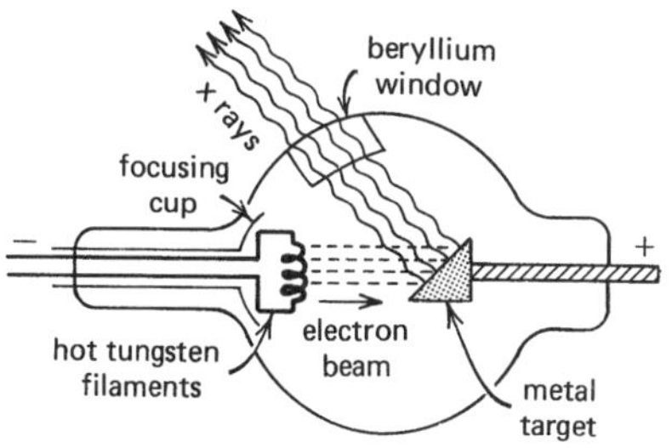

Figure 2.5 Schematic diagram of x-ray tube.

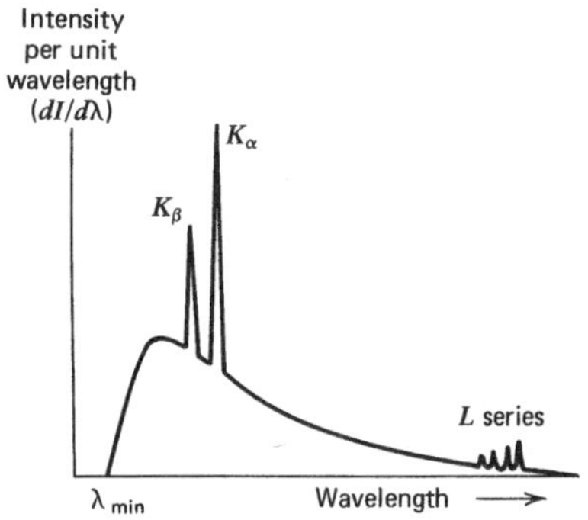

Figure 2.6 A typical x-ray emission spectrum.

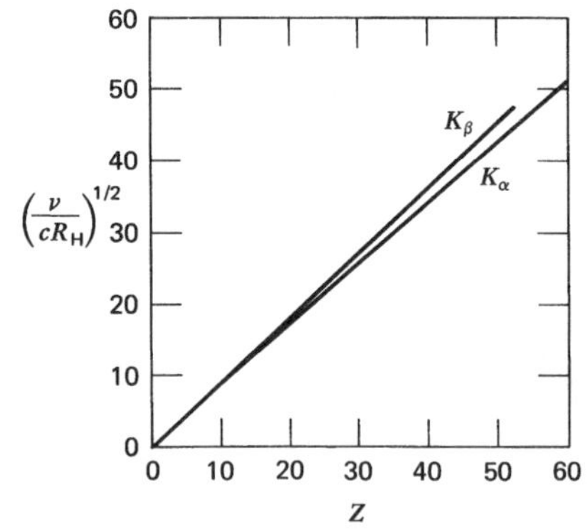

Figure 2.7 Dependence of x-ray emission frequencies on atomic number for the K series. Note that the square root of frequency, $\nu^{1/2}$ is very nearly linear in Z. Each of the curves shown is actually made up of subsets (e.g., K_{α_1} and K_{α_2}) too close together to be distinguished on this scale.

The most precise and useful relationship between the intrinsic structural properties of atoms and their interaction with x-rays involves the x-ray emission spectra of the elements. These spectra are the distributions of intensities, over wavelengths, of the x-rays emitted when a target made of a particular element is bombarded with electrons. X-rays are dispersed by wavelength with a crystal in the role of a diffraction grating, as we shall see in Chapter 11. In general, the x-ray emission spectra of the elements resemble that shown in Fig. 2.6, consisting of a continuous emission on which are superimposed a number of characteristic lines. The shape of the continuous spectrum is independent of target, depending only on the voltage applied to the x-ray tube, that is, on the kinetic energy of the electrons that hit the target. The line emissions, on the other hand, are different for each target, and thus presumably arise from excitation of the target atoms, just as in the Franck–Hertz experiment.

H. G. J. Moseley, between 1912 and 1914, made a very systematic and dramatic series of observations of the characteristic x-ray emission lines produced by various elements. For each element the lines were known to fall into several closely bunched groups, which are called the K, L, M . . . series (proceeding toward greater wavelengths); the K and L series were found for all the elements studied, the later series only for the heavier elements. In each series Moseley found a rather good correlation between the frequency of the x-rays and the atomic weight of the element. However, the correlation with atomic *number* was so remarkable as to prove the real physical significance of Z. Figure 2.7 displays this correlation for the K series.

In every series the relationship between the frequency ν and the atomic number Z was found to be of the form

$$\nu = A(Z-S)^2, \tag{2.6}$$

where A and S are (very nearly) constants. For the *screening constant S*, Moseley found a value close to unity for the K series so that $\nu^{1/2}$ is nearly proportional to Z. Later measurements gave S of about 3 for the K series and 10 to 20 for the L series. As for the constant A, its value for the various series is approximately given by

$$A = cR_H\left(\frac{1}{n_1^2}-\frac{1}{n_2^2}\right), \tag{2.7}$$

where c is the speed of light, R_H is the Rydberg constant (see the next section), and n_1 and n_2 are integers. For the K series n_1 is 1, whereas n_2 is 2 for the K_α (lowest-frequency) lines, 3 for K_β, and so on; in the L series, n_1 is 2 and n_2 = 3, 4, ... We shall see in the next section that the form of Eq. 2.7 is common to many types of spectra, clearly illustrating the quantized nature of atomic energy levels.

What did the appearance of Z in Eq. 2.6 prove? Remember that the atomic number originally gave only the element's place in the empirical periodic table. As we shall see, several lines of evidence already suggested that Z was equal to the number of electrons in the atom, and Bohr's theory of the atom (1912) was based on this assumption. Not until Moseley's work, however, was an exact relationship involving Z shown to apply to an extensive sequence of elements. A firm basis was thus given to the periodic table, eliminating the anomalies that appeared when the sequence was based on atomic weights.

We have mentioned the similarity between x-ray emission and the Franck–Hertz experiment. This similarity extends to the appearance of threshold energies. To produce any given x-ray emission line from a particular target material, the bombarding electrons must be accelerated to a minimum energy characteristic of the particular line. As the electron energy increases, lines of higher and higher frequency are produced, with the threshold energies obeying Eqs. 2.3. Note also that for a given electron energy E continuous emission occurs only with wavelengths greater than some wavelength λ_{min}, which is such that $E = Ve = hc/\lambda_{min} = h\nu_{max}$; this is the point at which *all* of the electron's energy is transferred to the emitted x-ray. All this evidence reinforces the hypothesis that radiation of frequency ν is associated with an energy change $h\nu$ in the material absorbing or emitting the radiation.

We shall see that threshold energies also exist for the absorption of x-rays. First, however, we must discuss the general laws governing the attenuation of x-rays by matter.

If one passes a beam of x-rays through a sample of some material, the intensity I of the beam will fall off with the thickness of the sample. The fractional loss of intensity, dI/I, usually obeys the equation

$$\frac{dI}{I} = -\sigma n\, dx, \tag{2.8}$$

where dI is the change in intensity over a small distance dx, n is the *number density* (in units such as atoms per cubic meter), and σ is a proportionality constant characteristic of the substance. Note that the dimensions of σ are those of area; in fact, σ is called the *attenuation cross section;* the product σn is the *attenuation coefficient.* Integrating over the total thickness x of the sample, we find the total loss of intensity to be

$$\ln I - \ln I_0 = -\sigma n x \quad \text{or} \quad I = I_0 e^{-\sigma n x}, \tag{2.9}$$

where I_0 is the intensity of the originally impinging beam. In other words, the intensity of the beam decreases exponentially with distance as the radiation penetrates the material. This result, usually referred to as *Beer's law,*[5] is a very powerful and general equation, which applies to many kinds of radiation and even to beams of particles.

What causes a beam of radiation to be attenuated? We are concerned here primarily with the process of *absorption,* a term implying that the energy of the beam goes into producing some kind of internal excitation in the atoms (or molecules) of the material. We saw one effect of absorption with the photoelectric effect. However, some of the beam can also be deflected or *scattered* in different directions and thus fail to reach the detector.[6] We shall have a great deal to say later about the scattering of particles. It is not always easy to distinguish between scattering and absorption; in some cases (*inelastic scattering*), the incident radiation is scattered but some of its energy is absorbed. Usually, however, one can analyze the various effects and obtain separate absorption and scattering coefficients, or the corresponding cross sections. For x-rays the absorption effect is much greater at the energies with which we are concerned, and attenuation measurements effectively give the absorption cross section.[7]

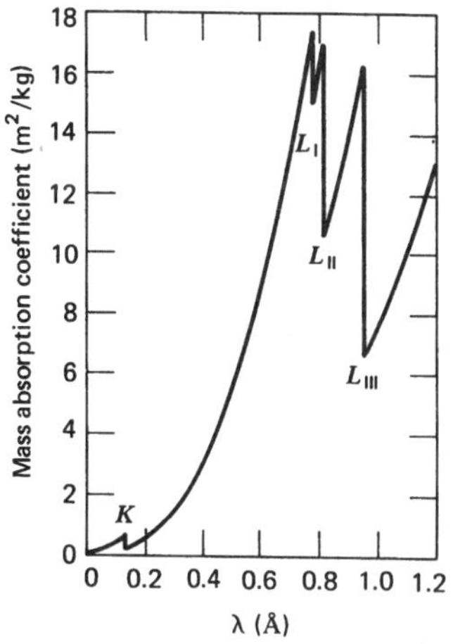

Figure 2.8 X-ray absorption of lead as a function of wavelength. The "mass absorption coefficient" is the absorption coefficient divided by the density. Note the sharp breaks (absorption edges) labeled *K*, L_{I}, L_{II}, and L_{III}; additional sets of absorption edges (*M, N, O,* . . . series) appear at longer wavelengths.

If one measures the absorption of x-rays by a given material as a function of wavelength, the results resemble those shown in Fig. 2.8. In general, the absorption coefficient rises steeply with increasing wavelength. At a series of points known as *absorption edges,* however, it drops precipitously to a lower value before rising again to a new peak. In the direction of higher frequencies (and thus higher energies), at each of these points the absorption increases. In short, each of the absorption edges corresponds to the threshold energy for a new type of absorption. We thus have threshold energies, indicating well-defined energy levels within the atom, for both emission and absorption. The absorption edges, like the emission lines, can be assigned to various series (again labeled *K, L,* . . .). It turns out that the absorption edge frequencies for different elements vary with atomic number in just the same way as do the emission frequencies, that is, in accordance with Eq. 2.6. Moreover, when one follows a series of x-ray absorption lines obeying Eq. 2.7 with n_1 fixed and n_2 ($>n_1$) increasing, one finds that when $n_2 \to \infty$, electrons are emitted. In short, the photoelectric effect sets in.

[5] Or the *Beer–Lambert law;* Lambert discovered the distance effect and Beer (1852) the concentration (density) dependence for the absorption of light by liquid solutions.

[6] Beer's law is strictly valid only in the limit of a very thin sample, and any system for which it does hold is thus called *optically thin.* One reason is the possibility of *double* scattering (out of the beam and back in), which increases with thickness. Reemission in the direction of the detector can also distort the results, as can the heating of the sample caused by energy absorption.

[7] X-ray scattering also had some relevance to the development of our theory. The scattering cross section can be seen to be proportional to the scattering effect of a single atom. According to J. J. Thomson's model of the atom (see Section 2.5), the scattering was done by individual electrons moving nearly freely within the atom, so that σ should be proportional to the number of such electrons per atom. The experiments of Barkla (1909) gave a number equal to about half the atomic weight—and thus approximately equal to the atomic number. This result was largely fortuitous (Thomson's model was wrong, and Barkla's results would have been quite different with x-rays of longer wavelength), but nevertheless it was the first indication that Z gives the number of electrons in the atom.

All that we have said about x-rays and matter thus reinforces two main conclusions: that the energies of atoms are quantized in some way (involving the constant h), and that the atomic number Z is of fundamental significance to atomic structure. But these conclusions were well accepted before either Moseley or Franck and Hertz, even if the evidence was perhaps more circumstantial. In the next section we examine what was historically the first evidence of quantization, the nature of atomic spectra.

2.4 The Emission Spectra of Atoms

The atoms of any element will emit light when sufficiently excited, whether by bombardments as in the Franck–Hertz experiment or simply by heating to a high temperature. As with x-rays, the light emitted consists of a continuous spectrum (black-body radiation, which we shall discuss in Section 2.6) and a spectrum of lines characteristic of the element. We are concerned here with the line spectra; these are actually of the same nature as the x-ray emission line spectra, but at much longer wavelengths—in the ultraviolet, visible, and infrared regions. They can thus be excited by relatively low energies, for example, by an electric discharge in a gas or vapor at low pressure. We shall discuss mainly the spectrum of atomic hydrogen, which is by far the simplest. (Molecules can also be made to emit light, but we shall reserve that discussion for later chapters.)

The hydrogen spectrum consists of several series of lines, with those in each series apparently closely related to one another. The series first discovered (by J. J. Balmer in 1885) lies primarily in the visible region, a similar series is in the far ultraviolet, and several more are in the infrared. Balmer found that the nine lines he observed in the visible series (now known as the *Balmer series*) have wavelengths given very precisely by the formula

$$\lambda = b\frac{n^2}{n^2-4} \quad (n=3,\ 4,\ 5,\ \ldots), \tag{2.10a}$$

Balmer's value for the constant b being 3645.6 Å. This equation is now usually written in the inverted form

$$\frac{1}{\lambda} = R_{\mathrm{H}}\left(\frac{1}{4}-\frac{1}{n^2}\right) \quad (n=3,\ 4,\ 5,\ \ldots); \tag{2.10b}$$

the inverse of the wavelength, $1/\lambda$, is known as the *wave number* (nearly always expressed in cm^{-1})—the number of wavelengths per linear unit. The constant R_{H}, which we mentioned in the last section, is known as the *Rydberg constant* (for hydrogen), and has a value of 109678 cm^{-1}. Each value of n corresponds to one spectral line. The ultraviolet series, called the *Lyman series,* fits a formula resembling Eq. 2.10*b*, but with the fraction $\frac{1}{4}$ replaced by the integer 1, and the values of n beginning with 2. Similarly, each successive series[8] in the infrared region can be fitted to such an equation with the first constant in brackets replaced by $\frac{1}{9}$, $\frac{1}{16}$, and so on. In general, one can write

$$\frac{1}{\lambda} = R_{\mathrm{H}}\left(\frac{1}{n_1^2}-\frac{1}{n_2^2}\right) \quad (n=1,\ 2,\ 3,\ \ldots),$$
$$(n_2 = n_1+1,\ n_1+2,\ \ldots). \tag{2.11}$$

The value of n_1 fixes the series to which the line belongs, the value of n_2 fixes which particular line it is. Note that each series converges to a low-wavelength limit R_{H}/n_1^2 as $n_2 \to \infty$; beyond this limit a continuous emission spectrum appears. Eq. 2.11 is a remarkably accurate and succinct expression for all the lines in the spectrum of atomic hydrogen.

What can we deduce from the form of Eq. 2.11? The very existence of line spectra, of course, is suggestive of quantized energy levels. If the energy of light is proportional to its frequency, $E = h\nu = hc/\lambda$ (as we concluded from the Franck–Hertz experiment), then Eq. 2.11 can be satisfied if the hydrogen atom has quantized states with energies given by $E_n = -hcR_{\mathrm{H}}/n^2$ (n_2 = initial state, n_1 = final state). The problem, however, was how to formulate a model of atomic structure that would have such energy levels. As we shall see in Section 2.11, this was the achievement of Niels Bohr.

Expressions similar to Eq. 2.11 also describe the emission spectra of vapors of the alkali metals (Li, Na, K, Rb, Cs). The alkali spectra are more complex than that of hydrogen, however, in that they contain several series of rather weak lines in addition to the strong and well-defined series analogous to hydrogen. Furthermore, the integers n_1 and n_2 of Eq. 2.11 are replaced by numbers n_j^*, which are usually nearly but not quite integers:

$$n_j^* = n_j - \delta \quad (j=1,2). \tag{2.12}$$

For higher members of any one series in an element's spectrum, the *quantum defect* δ is essentially a constant. Also, the constant R_{H} of Eq. 2.11 is replaced in each of the corresponding alkali formulas by another constant differing from it only very slightly (by less than 0.1%, in fact). From the similarity of their spectra, we might assume that hydrogen and the alkalis share some very striking common property. We therefore expect that, given a theoretical explanation of the spectrum and structure of the hydrogen atom, it should be relatively easy to extend this explanation to the alkali atoms.

When we go on to the spectra of other elements, the situation immediately becomes far more complex, but not completely intractable. Some regularities were recognized long before the spectra could be fully interpreted. These were largely in the form of repeating sets of lines having the

[8] The series with n_1 = 3, 4, 5 are known as the *Paschen, Brackett,* and *Pfund series,* respectively.

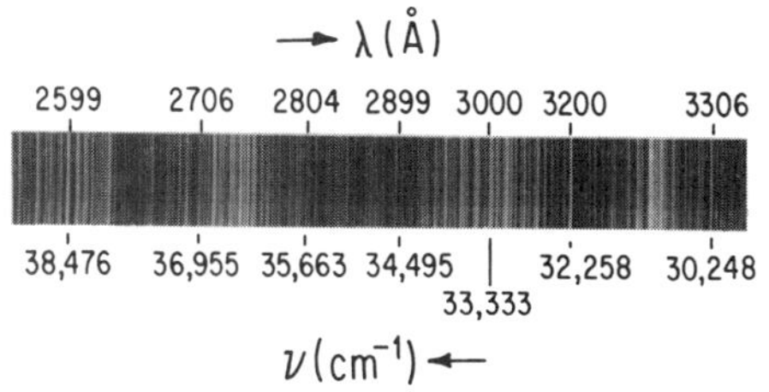

Figure 2.9 A portion of the emission spectrum of iron, in the region between approximately 2600 and 3300 Å, or 260 and 330 nm.

same frequency intervals. These repeating sets, for example, triplets or quartets of lines, can sometimes be recognized even in very complex patterns like those in Fig. 2.9. It is not surprising that each element's spectrum is unique and easily distinguishable from any other element's, so that one can use these spectra to recognize the presence of specific elements, even in the atmospheres of stars.

One can generate emission spectra in several ways. Electric discharges in gases at low pressures are one way, as we already mentioned. Electric arcs also excite a number of lines of most elements, and it is easy to introduce solid or liquid samples into such arcs (struck, for example, between two pieces of high-purity carbon). Many more lines are excited by a spark discharge, in which the electric current is subjected to a much higher accelerating voltage than in the arc. In the spark, because of the higher electric field, the accelerated ions and electrons reach much higher energies than in an arc, and one can recognize lines characteristic not only of neutral atoms, but of the positive ions of many elements, even of those with several electrons removed. One can achieve more precise control of the excitation by using a beam of electrons of known, controlled energy to collide with the target species and excite them. Given still higher excitation energies, one can obtain emission lines at shorter and shorter wavelengths, eventually reaching the x-ray spectra of the last section; it should be clear from the similarity of Eqs. 2.7 and 2.11 that the processes involved are related.[9]

So far we have spoken only of emission, but of course each element also has a characteristic absorption spectrum. The two spectra are in fact the same: Any substance that absorbs light of a given wavelength can be stimulated to emit light of that wavelength. We conclude that absorption is simply the inverse process to emission, with atoms going from lower to higher energy levels. The absorption spectrum can be observed by passing continuous radiation through a sample and determining which wavelengths are removed; the amount of absorption usually obeys Beer's law, with the absorption coefficient a function of wavelength. The dark (Fraunhofer) lines in the spectra of the sun and other stars result from absorption by gases in the outer layers of their atmospheres. Light—and more generally electromagnetic radiation throughout the spectrum—is also capable of exciting emission spectra, but unless it is extremely intense, it must be of a frequency corresponding to some absorption line.

2.5 The Nuclear Atom

We have now introduced most of the experimental facts needed to formulate a theory of atomic structure. One key piece of information remains: the evidence for the nuclear structure of the atom. Let us review what we already know about the nature of atoms, corresponding to what was known before 1910. We know approximately the sizes and weights of individual atoms. We know that they can be ionized into particles with positive and negative charges, and that, because atoms are electrically neutral, the total amounts of positive and negative charge must be the same in an intact atom. We know from measurements of charge-to-mass ratios that the negatively charged electrons weigh far less than the rest of the atom. Various measurements we have not discussed suggest that the total number of electrons (and thus of positive charges) in an atom is of the order of its atomic number Z. The final link in this chain is the evidence, which we consider in this section, that the atom has a nuclear structure—that the positive charge and its associated large mass are concentrated in a very small volume at the atom's center, the *nucleus,* whereas the light electrons occupy the remainder of the atom's volume. In the simplest formulation, we say that the atom is like a miniature solar system, with the electrons orbiting the nucleus as the planets orbit the sun.

What alternatives are there to the nuclear model of atomic structure? For example, could it be that the positive charges are in orbit around the lighter negative charges? Probably not; such a model is physically unreasonable. If the atom does consist of particles moving about one another, some very light and some very heavy, then obviously the lighter particles must move about the heavier ones. The light particles must have much higher velocities than the heavy particles and must be much further from the center of mass. Indeed, if the positive charge is concentrated in a small volume, any volume smaller than that of the atom as a whole, then the model with the positive particles at the nucleus must surely be the correct one. The other principal alternative is one in which the positive charge and the mass associated with it are spread in some moderately uniform way throughout the entire volume occupied by the atom. This model, sometimes called the "plum pudding model," was proposed by J. J. Thomson, who favored it until experiment gave incontrovertible proof that the nuclear model was the correct one.

---

[9] In fact, the first line of the Lyman series ($n_1 = 1$, $n_2 = 2$) is actually the hydrogen K_α line, shifted into the ultraviolet by the low value of Z (the screening constant is zero for hydrogen).

Just how does one establish that one model is right and another wrong? Each model can be used to predict the results of various experiments. One finds an experiment for which the two predictions are different, then carries out the experiment. One hopes that the results will be clearly inconsistent with one of the hypotheses, which can then be rejected. This does not *prove* the alternative hypothesis, but merely demonstrates that it can be accepted for want of other equally valid models. (This is the ideal case, of course; sometimes no experiment can be performed to distinguish between two models—and sometimes *all* the models turn out to be wrong!) It is almost always in this way that new scientific theories replace old ones. The crucial experiment that established the nuclear theory of atomic structure involved the scattering of particles by atoms; it was first performed in 1909 by H. Geiger and E. Marsden, and the results were soon explained by Ernest Rutherford.

Specifically, the experiment consisted of the scattering of a beam of α particles ($^4He^{2+}$ ions) by a thin metal foil, for example, gold foil about 2 μm thick. Fig. 2.10 illustrates the type of apparatus used in measurements of this sort. (The detectors shown in the figure are more sophisticated than Geiger and Marsden's. They measured the actual scattered intensity with a ZnS screen that showed a flash of light when an α particle struck it. The flashes were counted by observing the screen through a microscope.) The basic quantity measured is the number of particles scattered in various directions, that is, as a function of the *scattering angle* θ; from this measurement can be obtained what we shall call the *differential cross section.*

First, however, let us consider the *total cross section,* which corresponds to the quantity σ introduced in Eq. 2.8. Particle scattering and x-ray attenuation are so different that it is worthwhile to look at the general expression again, with a slightly different terminology. We refer to Fig. 2.10. Suppose that the beam of projectile particles emerging from B has a *flux density* of f_0 particles per unit area per unit time. Suppose also that in the scattering region C there are n target particles per unit volume, so that a thin layer of thickness dx contains $n\,dx$ particles per unit area. Let us assume that each of these targets has an effective cross-sectional area σ for scattering; that is, whenever a projectile strikes the area σ around a target particle, it will scatter away from the beam and thus not reach the detector D. Then the fraction of the beam flux scattered out of the beam will be just the fraction of the target area effectively covered by target particles. The fractional change in the beam's intensity in the layer of thickness dx will thus be

$$\begin{aligned}\frac{df}{f} &= -(\text{area per particle})(\text{particles per unit area})\\ &= -\sigma n\,dx \end{aligned} \tag{2.13}$$

This is the same expression that we stated empirically as Eq. 2.8, but now we can see why σ is called a cross section. As before, we can integrate to obtain

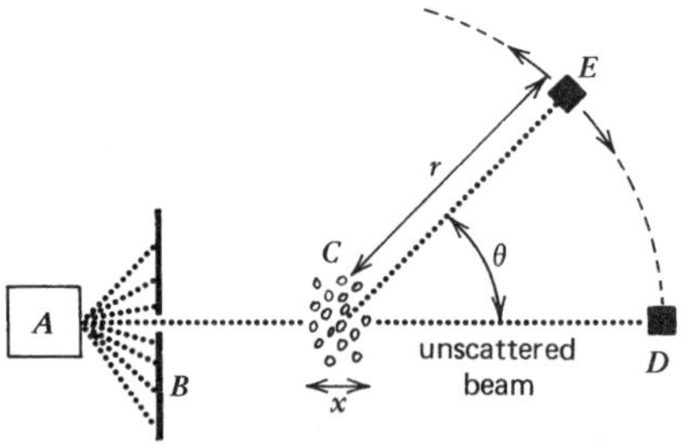

Figure 2.10 Schematic diagram of a scattering apparatus. The source *A,* which may be a hot oven, radioactive material, and so on, emits particles (black dots) in random directions. The collimating aperture *B* forms a reasonably well-defined beam of these projectile particles. The projectiles strike target particles (open circles) in the scattering region *C,* with thickness *x;* this may be a confined chamber, a solid film or foil, or just the intersection of two beams (projectiles and targets). The unscattered projectiles continue on to the detector *D,* which measures attenuation and total cross section. Other projectiles are scattered in all directions, one trajectory being chosen for illustration. The movable detector *E* is used to determine the number of particles scattered in a given direction; usually the scattering is symmetric around *D,* and only the scattering angle θ need be varied. To determine the differential cross section as a function of θ, one must move the detector *E* over the full range of angles into which a significant amount of scattering takes place.

$$f_D = f_0 e^{-\sigma n x} \tag{2.14}$$

for the flux density of the beam emerging from the scattering region and reaching the detector D. In the x-ray case the beam attenuation was due primarily to absorption; here it is due entirely to scattering, and we refer to σ as the *total scattering cross section.*

The detector D measures the loss of intensity of the direct beam, the number of particles scattered *out* of the incident direction; one can then use the detector E to measure the number of particles scattered *into* a particular scattering angle. Experimentally, θ is the angle between E and the original beam, measured from the center of the scattering region. Of course, the angle θ defines not a single trajectory but a whole cone of trajectories, symmetric around the unscattered beam; ordinarily, however, the same number of particles will be scattered into all these directions, so that the orientation of E and D does not matter.[10]

Since the detector and the scattering region both have nonzero areas, the particles entering the detector at any given position have actually been scattered through a range of angles (Fig. 2.11); if the two are far enough apart, however, the average scattering angle will still be θ. We can thus make our calculations as if all the particles radiated from the

[10] To obtain scattering that is not symmetric around the original beam, one must use target particles with nonrandom orientation. An example of this is the scattering of x-rays by crystals (Chapter 11).

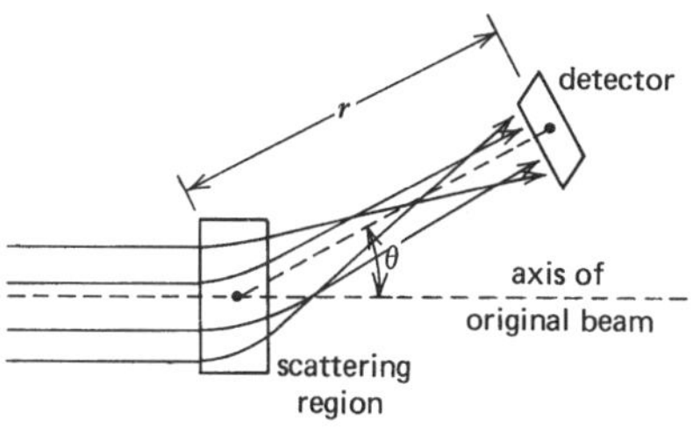

Figure 2.11 Uncertainty in the scattering angle. The angle θ is defined by the centers of the scattering region and the detector, but the trajectories in the figure show that some particles scattered through angles greater or less than θ will also be detected. The greater the distance r, the smaller will be the range of angles.

center of the scattering region, relative to which the detector E subtends the solid angle

$$\Omega_E = \frac{A_E}{r^2}, \tag{2.15}$$

where A_E is the detector's area. Given the flux of particles reaching the detector and the solid angle subtended by the detector, one can obtain[11] $df(\theta)/d\Omega$, the flux density scattered into θ per unit solid angle—that is, the number of particles per unit area of the incident beam that are scattered per unit time into unit solid angle in the vicinity of the angle θ. The *fraction* of the incident beam flux scattered into θ per unit solid angle by the layer dx is then given by an expression analogous to Eq. 2.13,

$$\frac{1}{f} d\left[\frac{df(\theta)}{d\Omega}\right] = \sigma(\theta) n\, dx\,; \tag{2.16}$$

as before, f is the flux density of the beam incident on the layer dx, and n is the number density of targets. This equation defines the quantity $\sigma(\theta)$, the *differential scattering* cross section per unit solid angle, which is obviously a function of the scattering angle. The differential and total cross sections are related by the equation[12]

$$\sigma = \int \sigma(\theta)\, d\Omega = 2\pi \int_0^{\pi} \sigma(\theta) \sin\theta\, d\theta, \tag{2.17}$$

so that $\sigma(\theta) = (d\sigma/d\Omega)_\theta$.

Both σ and $\sigma(\theta)$ are numerical quantities that must be determined either from experiment or from microscopic information about how the projectiles and targets interact. They are parameters of the system, and cannot be derived from any line of deductive reasoning or purely mathematical construction; some real physical information about the forces of interaction is needed. Equations 2.13 and 2.16 are precise mathematical statements of some very general relationships, but they tell us nothing about any specific system. For example, Eq. 2.16 says that the fraction of the incident particles scattered into our detector is proportional to the number of scatterers (targets) available, and that there exists a function $\sigma(\theta)$ that gives this proportionality. We can identify $\sigma(\theta)$ with some fraction of an effective cross section σ for purposes of guiding our ideas, but to say that either is a specific and recognizable physical area may often be misleading. Equation 2.16 is in fact valid under a wide variety of circumstances in which $\sigma(\theta)$ can be treated as no more than a proportionality factor. The graphic interpretations suggested by the names "cross section" and "differential cross section" are strictly valid only for a few classical systems. Equations like 2.13 and 2.16 are what we call *phenomenological* relationships. They express physical phenomena in very general and precise mathematical terms, but at the same time leave the characteristic behavior of individual and specific systems contained in a small number of parameters such as the cross section. These parameters must be derived from some entirely different theory or from experiment. We shall frequently make use of phenomenological relationships throughout this text. We shall also encounter and make use of the contrasting approach, often called a microscopic theory—the kind of approach in which one derives properties such as the value of a cross section, or its dependence on the scattering angle, from the structures of the colliding particles.

All this discussion of cross sections has told us nothing about atomic structure, but it has given us a vocabulary with which we can examine Geiger and Marsden's results for α-particle scattering. Remember that there were two fundamentally different microscopic pictures of atomic structure, namely, the plum pudding model and the nuclear model. For each of these models one can calculate the dynamics of atomic collisions and obtain the dependence of the differential cross section $\sigma(\theta)$ on the scattering angle θ. If the density and thickness of the scattering foil are known, one thus obtains from Eq. 2.16 a prediction of the fraction of a beam that will be scattered into an angle θ, the quantity measured directly in the experiment. According to *both* hypothetical models, the heavy positively charged part of the atom must deflect the heavy α particles; the light electrons, rather than deflecting the α particles, must themselves be deflected from their positions or trajectories as the α particles pass near them. The plum pudding model leads to the conclusion that the scattered intensity should fall off with the angle θ according to a Gaussian distribution, that is,

$$\sigma(\theta) \propto e^{-\theta^2/\theta_m^2}, \tag{2.18}$$

[11] If the detector records n_E particles per unit time, we have

$$\frac{df(\theta)}{d\Omega} = \frac{n_E}{A_0 \Omega_E},$$

where A_0 is the cross-sectional area of the original beam. Integration over all solid angles should give

$$\int \frac{df(\theta)}{d\Omega} d\Omega = f_0 - f_D$$

for the total scattered flux density.

[12] Experimentally, the integration of Eq. 2.17 must begin, not at $\theta = 0$, but at some small angle just outside the unscattered beam.

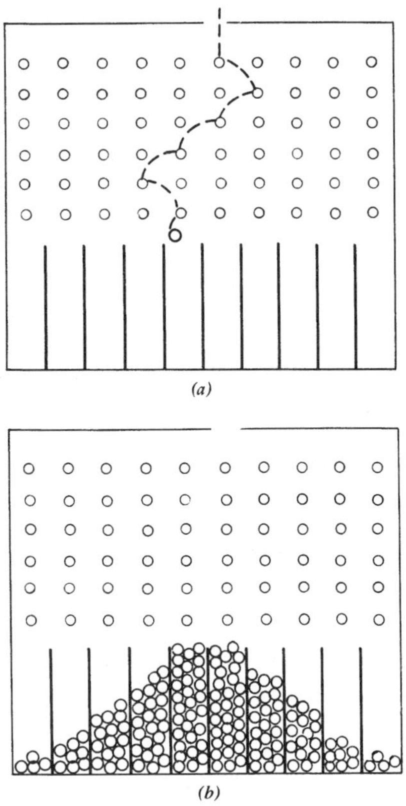

Figure 2.12 A ball-and-pegboard demonstration of the effect of multiple scattering and how it gives rise to a Gaussian distribution: (*a*) A single ball and its trajectory, (*b*) The result of dropping many balls through the board.

where θ_m is the mean deflection of the particles passing through the scattering region. The mean angle θ_m is proportional to the mean deviation due to a single atom but includes the effects of scattering by more than one target atom. The form of expression 2.18 is characteristic of the accumulation of many small effects—here, many small deviations in trajectory, with an average of θ_m for each—and is often called a random distribution. A ball cascading through a pegboard is an example of how a single α particle would behave if this model were correct. With many balls cascading from a single slot into many receptacles, one gets a distribution like 2.18 (Fig. 2.12). If one can determine θ_m experimentally, then one can predict quantitatively the intensity at any other angle θ relative to that at θ_m. The other model, the nuclear model, assumes the nuclei to be so small that more than one encounter is very unlikely, so that double or multiple scattering events can be disregarded. On the other hand, any individual encounter in the nuclear model is likely to be a much more violent experience for a projectile α particle than would be the total effect of passage through several plum puddings. Specifically, according to the nuclear model the differential cross section should obey the relation

$$\sigma(\theta) \propto \frac{1}{\sin^4(\theta/2)}; \qquad (2.19)$$

note that this quantity is divergent, becoming infinite as $\theta \to 0$. The cross section for the nuclear model is also predicted to be proportional to Z^2, the square of the atomic number, and inversely proportional to the square of the initial kinetic energy of the α particles. Rutherford's derivation of expression 2.19 is given in Appendix 2A.

Expression 2.19 predicts that the differential cross section in the nuclear model should drop sharply as θ increases from zero ($\theta = 0$ being the forward direction of the beam) and tend to level off as θ approaches its maximum of π. The infinite cross section for the forward direction reflects the very long range of the Coulomb potential, the longest-range potential we know; in crude terms, this means that every particle in the beam feels some effect from at least one single scatterer. The flatness of $\sigma(\theta)$ at large θ reflects the very strong interaction associated with head-on collisions. These are the collisions that scatter particles backward—into the region for which $\theta > \pi/2$—and they can exist only if there is a concentrated center of force responsible for the scattering. The plum pudding model, without concentrated scatterers, predicts much weaker backward scattering. For gold foil 2 μm thick, θ_m is inferred from experiment to be about 1°. According to expression 2.18, for the plum pudding model a vanishingly small number of α particles could be deflected to all angles very much larger than 1°; one certainly would not expect to see particles deflected through 45° or more. The nuclear model and expression 2.19, however, open the possibility of seeing large-angle deflections.

Which model is correct? The original experiments of Geiger and Marsden showed that about one α particle in 8000 was deflected through an angle greater than 90°. This in itself is strongly suggestive,[13] but it is not conclusive evidence that the nuclear atom is the correct model. We must have something firmer and more quantitative if we are going to be convinced. The convincing proof came from further measurements by Geiger and Marsden. The intensity scattered into unit area, or better still into unit solid angle, can be plotted against $1/\sin^4(\theta/2)$. This is a little inconvenient, however, because both quantities vary over many orders of magnitude within the range of the experimental data. It is more convenient to plot the logarithm of one against the logarithm of the other. If expression 2.19 is correct, then this log–log plot must give a straight line with a slope of unity. Figure 2.13 shows clearly that this is the case. The experiment thus demonstrated clearly that the nuclear model must be the preferable one.

An additional result of Geiger and Marsden's work was to confirm that the atomic number Z gives the positive charge on the atomic nucleus and thus the number of electrons in the neutral atom, since this was an essential assumption in Rutherford's derivation (Appendix 2A). This was

[13] Rutherford, who had thought the plum pudding model to be correct, later said that it was "as if you fired a 15-inch shell at a piece of tissue paper and it came back and hit you." It was in fact only *after* this experiment that he began to devise the nuclear model.

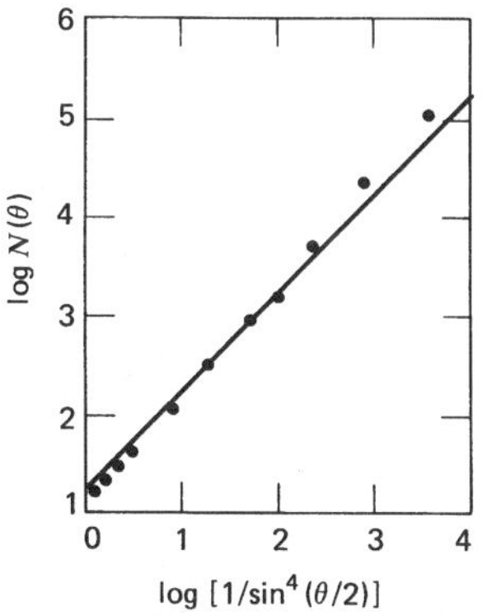

Figure 2.13 Geiger and Marsden's data for the scattering of α particles by silver foil [replotted from intensities reported in *Phil. Mag.* **25,** 604 (1913)]. $N(\theta)$ is the number of particles scattered per minute into unit area of the detector screen at a scattering angle θ. The data can be seen to give a reasonably good fit to the 45° straight line predicted by expression 2.19.

established by using foils of various materials as targets: As predicted, the scattering cross section was proportional to Z^2, where Z was the element's place in the periodic table. Further confirmation of the significance of Z was soon provided by Moseley's x-ray measurements.

2.6 The Problem of Black-Body Radiation

One stone remains to be set in the foundation of our theory. We have identified the basic components of the atom—a heavy nucleus surrounded by electrons—and we must still explain the way in which they are put together. We began this chapter by describing several types of evidence suggesting that atoms can gain or lose energy only in certain well-defined amounts. The key concept explaining these phenomena is that of the *quantum of action,* an idea foreign to classical physics. In the next several sections we shall develop its meaning and show how it constrains the energy and angular momentum in microscopic systems.

The quantum concept dominated the development of atomic and molecular physics in the early years of this century. A dramatic change occurred when three very different phenomena, all inexplicable within the framework of classical physics, yielded to rather thorough interpretation within less than 12 years; all three interpretations were intimately grounded in the concept of the quantum of action. One is the photoelectric effect, discussed in Section 2.2; another, the heat capacity of solids, will be considered in Section 2.9. We now turn to the problem of what is best known as *black-body radiation.*

Imagine a solid object with a hollow interior. Now suppose that the exterior of this object is completely insulated from the outside world, except for some heating wires and some telephone wires to bring us information such as the object's temperature. If the object is heated to, say, 700 or 800°C, it should glow bright orange. This is a property we know well from common experience, from looking at "red hot" solids. But the object of our new and peculiar experiment must glow only in its own interior; after all, we have insulated its exterior, so it can lose no energy to the outside world. Although we cannot see the glowing surface, we know that the inside walls are glowing, emitting bright orange light (and also infrared radiation) into the empty volume or *cavity* in the object's interior. If our object is really *very* well insulated, it will soon reach a state of thermal equilibrium, when it will require essentially no additional energy from the heating wires to maintain a constant temperature. But those inside walls are still radiating. Will the solid object lose all its energy to the cavity simply by emission of radiation? Clearly it will not. Some of that radiation must be reabsorbed by the walls. As soon as our perfectly insulated object draws no more energy from the heating wires, that is, as soon as it has reached a steady temperature, then no *net* energy enters the cavity. Every bit of energy that appears as radiation must be paid for by an equal amount of radiation energy absorbed by the walls. The object is called a *black body,* and is characterized by its radiation, which is in equilibrium with the container at a precise temperature.

The classical problem, the very first problem to involve the concept of the quantum, was this: How much energy is present *as radiation* in a given volume of "empty" space within such a simple hot object held at any definite temperature T? A subsidiary question is this: What is the distribution of the radiation energy in the cavity, as a function of the wavelength? For example, is the radiation all of the same frequency? This is a conceivable answer, but a simple experiment tells us that it is not correct. All one has to do is steal just a bit of radiation from the cavity (through a small hole) and examine it with a spectrometer. As shown in Fig. 2.14, the radiation intensity gives a broad, smooth distribution that varies with the temperature of the black body. This is quite similar to the radiation given off by any hot object; the importance of the ideal (perfectly insulated) black body, whose radiation is independent of the material from which the walls are made, is that it can be treated by relatively simple theories that do not depend on the material constituting the walls. The curves in the figure are primarily in the infrared, but with increasing temperature the peak shifts to lower wavelengths; the light given off by the sun (ca. 6000°C) or an incandescent lamp has its peak in the visible region.

What one wants to know, naturally, is why the energy distribution of black-body radiation has this particular form. The answer to this question lies primarily in the domain of thermodynamics, the subject that comprises Part II of this book. Despite the division that is sometimes drawn between thermodynamics and atomic and molecular structure, the beginnings of all our modern theories of atomic and molecular structure lay in Max Planck's exposition of the thermodynamic problem of black-body radiation in 1901. Because its analysis would take us prematurely into thermodynamics, we shall not go into Planck's theory in detail. We merely indicate why there was a problem and what key assumptions Planck made to solve it.

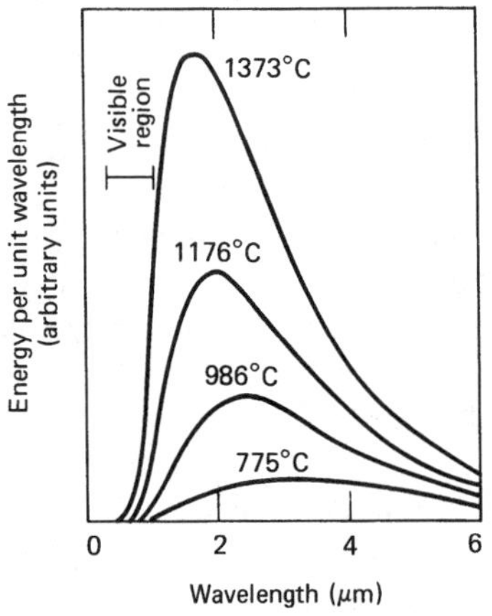

Figure 2.14 Energy distribution of black-body radiation at several temperatures. At the lower temperatures the energy in the visible region is negligible on this scale, but still enough to be observed as a glow.

The work of James Clerk Maxwell and Heinrich Hertz during the latter part of the nineteenth century made it clear that light and many other electromagnetic phenomena behave as waves. What this means in detail we shall examine in Chapter 3. At the moment we can rely simply on our intuitive concepts of waves (in strings or on oceans) to obtain a graphic feeling for the processes involved. Rayleigh and Jeans made a great stride toward the solution of the black-body radiation problem when they suggested that one think of the radiation within the black body as standing waves, forced to remain in the cavity by the condition that there be no wave displacement in any direction at the boundary walls. These waves are analogous to the vibrations in a jump rope or a violin string whose ends are held tightly so that they cannot be displaced. This picture creates a model on which the physicist or chemist can proceed. Rayleigh and Jeans computed how many modes of wave motion[14] can be contained in a given cavity and satisfy the stated boundary condition; they found that the number of modes should increase steadily with frequency, becoming infinite in the short-wavelength limit. According to the classical theory applied by Rayleigh and Jeans, the radiation energy should be distributed evenly over all these modes, and thus an infinite amount of energy should be stored in the short-wavelength modes. By contrast, the experimental data show that the energy distribution drops sharply at short wavelengths. The theory did give a satisfactory description of the long-wavelength end of the spectrum, but was patently wrong in the short-wavelength region. This discrepancy was known as the *ultraviolet catastrophe,* and could not be explained by classical means.

Planck resolved the problem by restricting the way in which the radiation energy could be distributed over the modes of wave motion in the cavity. Classically, one assumed that this distribution could be carried out by a method like that used to define a derivative or an integral in elementary calculus: One begins by chopping a system into small but discrete bits, then imagines that the bits get smaller and smaller as their number gets greater and greater, all the way to the limit of infinitely small bits. Planck found that to explain the observed black-body spectrum, he had to assume that the radiation field could not be subdivided into infinitely small bits—that it was made up of some kind of indivisible elements. He called the modes of wave motion *oscillators* (for reasons that we shall see later) and said that their properties were *quantized.* By making this assumption, Planck was able to calculate a theoretical energy distribution in excellent agreement with the experimental data for a black body, giving curves like those in Fig. 2.14. To see why Planck's assumption made so great a difference, we must begin by describing the properties of his oscillators. We can carry out this description on two levels.

In the simpler formulation, Planck assumed that the possible energies of the oscillators must be separated by finite amounts, or *quanta.* The energy associated with any oscillatory or wave motion is proportional to the frequency; Planck proposed that each mode of the radiation field could have only specific and precise energies, and not any energy between these characteristic values. In the classical model the increment between any two allowed energies could be arbitrarily small, but Planck fixed the increment at a constant small value, proportional to the oscillator frequency ν. If we call the proportionality constant h (*Planck's constant*), then the increment, or quantum of energy, between allowed energies of a given oscillator is given by

$$\Delta E = h\nu. \qquad (2.20)$$

The constant h is the same as that we introduced in Eqs. 2.3, with the value of 6.626068×10^{-34} J s. On the basis of then existing data, Planck estimated a value of 6.55×10^{-34} J s. This was the first formulation of the idea of quantization of energy.

Now we can understand the black-body energy distribution. The number of possible oscillators (modes of wave motion) still becomes infinite as $\nu \to \infty$ ($\lambda \to 0$), but so does the minimum nonzero energy that each oscillator can have. Classically, even if each of the high-frequency oscillators had only a tiny amount of energy, their great number meant that most of the energy would be at that end of the spectrum. In Planck's model, however, an oscillator of frequency ν that has any energy at all must have at least the amount $h\nu$. As ν increases, $h\nu$ becomes a greater and greater fraction of the available energy, and it becomes more and more likely for the same energy to be divided among many oscillators of lower ν. This is why the energy density falls off sharply at high ν (low λ). As the temperature increases, more total energy is available and the fall-off occurs at higher frequencies.

[14] That is, how many possible standing waves with different frequencies and/or directions of propagation. This number is limited, since the waves must have an integral number of wavelengths in the distance between opposite walls.

We said that the quantization of energy was one of two ways to interpret Planck's quantization process. The other and perhaps more illuminating interpretation is in terms of the quantity called *action,* the quantum of action being h itself. To understand what this means, we must discuss the concept of action at some length.

2.7 The Concept of Action

Action is a quantity with the same dimensions as Planck's constant. These dimensions can be expressed as energy × time or as momentum × distance, and are thus the same as those of angular momentum. Action is a standard variable of classical mechanics; it is related to momentum in the same way that work is related to force. To express these relationships in their most general forms, we need to review more information about vectors.

A vector, it will be recalled from Section 1.3, can be expressed as a set of three quantities, the x, y, and z-components. In Eqs. 1.17 we introduced the cross product of two vectors, $\boldsymbol{\alpha} \times \boldsymbol{\beta}$, which is itself a kind of vector. A second way of multiplying vectors gives a different quantity, the *dot product* (or *scalar product*) $\boldsymbol{\alpha} \cdot \boldsymbol{\beta}$. This is not a vector, but a scalar (pure number) given in terms of the components of the vectors $\boldsymbol{\alpha}$ and $\boldsymbol{\beta}$ by

$$\boldsymbol{\alpha} \cdot \boldsymbol{\beta} = \alpha_x \beta_x + \alpha_y \beta_y + \alpha_z \beta_z . \qquad (2.21)$$

Alternatively, the dot product can be expressed as

$$\boldsymbol{\alpha} \cdot \boldsymbol{\beta} = \alpha\beta \cos \phi, \qquad (2.22)$$

where α, β are the magnitudes of the vectors $\boldsymbol{\alpha}$, $\boldsymbol{\beta}$, and ϕ is the angle between them; it is thus the magnitude of $\boldsymbol{\alpha}$ times the component of $\boldsymbol{\beta}$ in the direction of $\boldsymbol{\alpha}$ (or vice versa, since $\boldsymbol{\alpha} \cdot \boldsymbol{\beta} = \boldsymbol{\beta} \cdot \boldsymbol{\alpha}$).

Now how does this apply to work and action? Suppose that an object moves an infinitesimal distance; since this displacement has both magnitude and direction, it is expressed as a vector $d\mathbf{s}$ with components dx, dy, dz. If this motion is produced by a force $\mathbf{F}$, the infinitesimal increment of work performed by the force[15] is defined as

$$dW = \mathbf{F} \cdot d\mathbf{s}, \qquad (2.23)$$

or the displacement times the component of force in the direction of the displacement. (The elementary formulation "work = force × distance," as in Eq. 1.9, applies only when force and displacement are parallel.) The total work done when an object moves from an initial point 1 to a final point 2 is thus the sum of such infinitesimal increments,

$$W_{12} = \int_1^2 dW = \int_1^2 \mathbf{F} \cdot d\mathbf{s}. \qquad (2.24)$$

An expression of the form $\int_1^2 \mathbf{F} \cdot d\mathbf{s}$ is known as a *line integral,* since the variable of integration is the displacement along some path. The value of the integral in general depends on the route chosen between points 1 and 2, and this path must be specified before one can calculate it.

We can now define the action, in which the momentum $\mathbf{p}$ plays the role that the force $\mathbf{F}$ plays in defining work. Specifically, the infinitesimal increment of action is

$$dA = \mathbf{p} \cdot d\mathbf{s}, \qquad (2.25)$$

and the total action associated with motion from point 1 to point 2 is the line integral

$$A_{12} = \int_1^2 dA = \int_1^2 \mathbf{p} \cdot d\mathbf{s}. \qquad (2.26)$$

It is all very well to define action by its formal analogy to work, but we need a better understanding of its physical meaning. Let us examine a specific case where the action becomes identical with a more familiar quantity. The momentum of a particle, of course, is the product of its mass and velocity,

$$\mathbf{p} = m\mathbf{v}. \qquad (2.27)$$

Suppose that we compute the action given by Eq. 2.26 for a particle moving at constant speed on a circle of radius r. The displacement along the circle is given by

$$ds = r \, d\theta, \qquad (2.28)$$

where θ is the angular displacement in radians. Since the displacement and the momentum at any given time are both in the same direction, the $\cos \phi$ of Eq. 2.22 is always unity and we have simply

$$dA = \mathbf{p} \cdot d\mathbf{s} = m\upsilon r \, d\theta. \qquad (2.29)$$

To obtain the total action for one complete orbit around the circle, we integrate to obtain

$$A_{\mathrm{orbit}} = \int_0^{2\pi} m\upsilon r \, d\theta = m\upsilon r \int_0^{2\pi} d\theta = 2\pi m\upsilon r. \qquad (2.30)$$

The action is therefore simply 2π times the magnitude of the angular momentum,

$$L = m\upsilon r. \qquad (2.31)$$

---

[15] Here W is the work performed on the object *by* the force, and is thus positive when force and displacement are in the same direction. This is not important here, but later we shall have to be careful in defining the sign of work.

This illustrates the point made earlier that the dimensions of action are the same as those of angular momentum.

2.8 The Harmonic Oscillator

We shall calculate the action explicitly for one other classical example, that of the *harmonic oscillator.* First, however, we must examine at some length the properties of the ubiquitous harmonic oscillator, which is far and away the most powerful and most widely used single model in all of physical science. It underlies our theories of solids and, to some degree, of liquids; it is the basis of much of our understanding of the behavior of molecules and of electromagnetic radiation. The reason for this generality is that a great variety of physical systems are described by equations that are mathematically equivalent to (or can be transformed into) those of the harmonic oscillator. In fact, the oscillator model is an obvious starting point for describing any system that remains near but not at some position or state of equilibrium.

The entire physical description of the harmonic oscillator is contained in its force law, which is simpler than those we have encountered earlier. To understand the nature of this law, consider a spring rigidly fixed at one end and constrained to move along a straight line (Fig. 2.15*a*). The spring has some equilibrium length l_0: If stretched, it will spontaneously contract; if compressed, it will expand again. The spring's length l will then oscillate back and forth around its equilibrium value (until friction stops the process). For small deformations in either direction, the restoring force is proportional to the deformation (Hooke's law): $F \propto (l - l_0)$. The harmonic oscillator is an idealization of this model in which a similar force law always holds.

The ideal one-dimensional harmonic oscillator, then, is a mass point moving frictionlessly along a straight line; it is subject to a force that always acts to return it toward the same specific position; and the force is proportional to the particle's displacement from this equilibrium position. If we define the equilibrium position as $x = 0$, the displacement is x (corresponding to $l - l_0$ in the spring model), and the oscillator's force law can be written as

$$F = -kx, \tag{2.32}$$

where F is the force in the positive x direction (cf. Fig. 2.15*b*). When x is positive, the force is negative, meaning that it tends to push the particle in the direction of $-x$, that is, back toward the origin. The proportionality constant k is called the *force constant* or spring constant.

We wish to find an equation for the oscillator's displacement x as a function of time. Using Newton's second law, we can rewrite Eq. 2.32 as

$$F = ma = m\frac{dv}{dt} = m\frac{d^2x}{dt^2} = -kx, \tag{2.33}$$

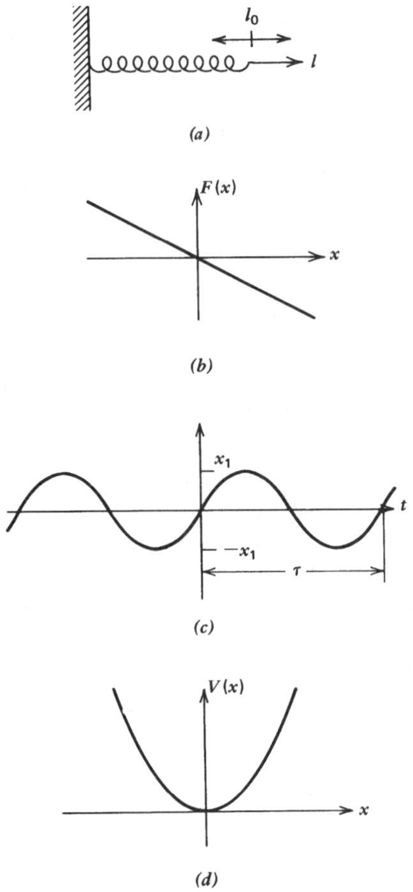

Figure 2.15 The one-dimensional harmonic oscillator. (*a*) The spring model: The spring's length l oscillates around its equilibrium value l_0, (*b*) The force law, Eq. 2.32, where F is the force in the positive x direction, (*c*) The displacement as a function of time, Eq. 2.35, assuming that $t = 0$ at $x = 0$, (*d*) The potential energy, Eq. 2.48.

where m is the mass of the particle. Equating the last two terms, we have a differential equation in x and t to solve. To obtain such a solution, we must add some more information: For example, we can give the particle's position at the arbitrary time $t = 0$ and state its maximum displacement. (Pieces of information of this sort are called *boundary conditions,* about which we shall have much to say in the next chapter; we mentioned another example in our discussion of the black-body problem.) For convenience, we set $t = 0$ when the particle is at its equilibrium position, $x = 0$; and we let x_1 be the particle's maximum displacement from equilibrium (the *amplitude* of the oscillation). The equation of motion and its boundary conditions can then be written compactly together as

$$\frac{d^2x(t)}{dt^2} = -\frac{k}{m}x(t), \quad \text{subject to} \begin{cases} x(0) = 0, \\ |x|_{\max} = x_1. \end{cases} \tag{2.34}$$

There are formal methods for solving such an equation. Two integrations are required, with the two boundary conditions giving the integration constants; however, we proceed in a more direct and intuitive way. We need to know what function $x(t)$ has a second derivative equal to itself times a negative constant. (Since k and m are both positive by definition, $-k/m$ must be negative.) Pulling a rabbit out of the hat, we recognize that the sine or cosine of t, or of any multiple of t, is such a function. Moreover, sin t would be equal to zero at $t = 0$. The maximum displacement is x_1, so $x_1 \sin t$ would satisfy the equation completely except for the constant k/m. If the argument of the sine is not t itself, but some multiple[16] αt, then each differentiation of x with respect to t introduces a multiplicative factor α. Our equation contains two differentiations, so we must have $\alpha^2 = k/m$. We conclude that the displacement as a function of time is given by

$$x(t) = x_1 \sin\left(\frac{k}{m}\right)^{1/2} t, \tag{2.35}$$

with x oscillating continually between positive and negative values (Fig. 2.15*c*). (In a real macroscopic oscillator, of course, friction would steadily decrease the amplitude of the oscillation.)

The displacement of the oscillator is given by Eq. 2.35 as a sine function, whose argument is naturally an angle. This angle,

$$\theta = \left(\frac{k}{m}\right)^{1/2} t, \tag{2.36}$$

is called the *phase* of the oscillator. Since $(k/m)^{1/2}\, t$ is dimensionless, proper cancellation of the units should give θ in radians; θ is thus equivalent to the time measured in units of $(m/k)^{1/2}$. The motion of the oscillator is *periodic*—that is, it repeats the same motions over and over, one such repetition being called a *cycle*. In the cycle beginning at $t = 0$, the sine function goes from 0 to 1 to 0 to –1 and back to 0, as its argument θ goes from 0 to 2π. We thus have

$$2\pi = \left(\frac{k}{m}\right)^{1/2} \tau \quad \text{or} \quad \tau = 2\pi\left(\frac{m}{k}\right)^{1/2}, \tag{2.37}$$

where τ is the *period* of the oscillator, the time it takes to go through one complete cycle. The inverse of the time per cycle is the number of cycles per unit time, which we call the *frequency*,

$$\nu = \frac{1}{\tau} = \frac{1}{2\pi}\left(\frac{k}{m}\right)^{1/2}; \tag{2.38}$$

if t is measured in seconds, the units of ν are s^{-1}, also written as cycles per second or hertz (Hz). Sometimes it is more convenient to describe a periodic motion in terms of the *angular frequency* ω, the number of radians per second; since there are 2π radians to a cycle, we have

$$\omega = 2\pi\nu = \left(\frac{k}{m}\right)^{1/2}. \tag{2.39}$$

The terminology introduced here is applicable to any periodic phenomenon. However, the value of ω or ν is a property of the particular system; for the oscillator it depends on the values of k and m, and cannot be derived from any of the physical laws we have so far written down. We shall see later that frequencies can be estimated for particular physical models, but to do this one must always have additional information, including the masses of the particles and something about the strength of the forces binding them.

We can now draw a parallel between the harmonic oscillator and the last section's example of a particle moving in a circular orbit. Suppose that the center of the circle is the origin of a Cartesian coordinate system. Eq. 2.35 then gives exactly the variation of x with time for a circular orbit of radius x_1 and constant angular velocity $(k/m)^{1/2}$; $y(t)$ is the corresponding cosine function. All the quantities introduced in the last paragraph then have the same values for the circular orbit, for which the "angular" terms can be taken in their literal sense. However, the two systems are not equivalent in all respects, as we can show by calculating the action for the harmonic oscillator.

As with the particle moving in a circle, we wish to know the total action for a complete cycle. Combining Eqs. 2.35 and 2.39, we have

$$x(t) = x_1 \sin \omega t; \tag{2.40}$$

the velocity of the particle at any given time is thus

$$v(t) = \frac{dx(t)}{dt} = \omega x_1 \cos \omega t = v_0 \cos \omega t \quad (v_0 \equiv \omega x_1), \tag{2.41}$$

where v_0 is the maximum velocity. When the displacement is zero, that is, when $\sin \omega t = 0$, then $\cos \omega t$ is a maximum in one direction or the other; in other words, the particle reaches its maximum speed when it passes its equilibrium position. Similarly, since the two quantities are 90° out of phase with each other, the velocity is zero when the displacement is at its maximum. By Eq. 2.40, the increment of the particle's action is

$$dA = m v\, dx, \tag{2.42}$$

since the momentum $m\mathbf{v}$ is always in the same direction as the displacement. The "chain rule" allows us to write $dx = (dx/dt)dt = v\,dt$, and substitute into Eq. 2.41 to obtain

$$dA = m v^2\, dt = m v_0^2 \cos^2 \omega t\, dt. \tag{2.43}$$

---

[16] Since the sine and cosine functions can have only dimensionless arguments, t *must* have some multiplier to appear in such an argument; the multiplier must have the dimensions of frequency, with units such as s^{-1}.

Note that the increment of action is twice the kinetic energy times the time increment dt—again, energy × time. Integrating over the time from 0 to τ, we obtain[17] the total action associated with one cycle of the oscillator,

$$
\begin{aligned}
A_{\text{cycle}} &= \int_0^{\tau} m v_0^2 \cos^2 \omega t \, dt \\
&= \frac{m v_0^2}{\omega} \int_0^{2\pi} \cos^2 \theta \, d\theta \qquad (\theta \equiv \omega t) \\
&= \frac{m v_0^2 \pi}{\omega}.
\end{aligned} \tag{2.44}
$$

Since $v_0 = \omega x_1$, we have $A_{\text{cycle}} = \pi m v_0 x_1$, which is only half as much as Eq. 2.30 gives for the circular orbit of radius x_1. We can rewrite our expression for the action in several equivalent ways:

$$A_{\text{cycle}} = \frac{m v_0^2 \pi}{\omega} = \frac{m v_0^2}{2} \frac{1}{\nu} = \frac{m v_0^2 \tau}{2}, \tag{2.45a}$$

$$A_{\text{cycle}} = \frac{k x_1^2 \pi}{\omega} = \frac{k x_1^2}{2} \frac{1}{\nu} = \frac{k x_1^2 \tau}{2}. \tag{2.45b}$$

We can interpret these expressions by recognizing the physical meaning of certain factors that appear in them. The harmonic oscillator, like any mechanical system with no frictional losses, must conserve its energy. Its kinetic energy is instantaneously $\frac{1}{2} m v^2$, with a maximum value of $\frac{1}{2} m v_0^2$, which we find in Eq. 2.45*a*. At the position of maximum displacement the kinetic energy is zero, and all the oscillator's energy is stored as potential energy. To calculate this we need a definition of potential energy.

The potential energy of a particle is a function of its position and the forces acting on it. It is strictly defined only for *conservative* systems, those with no time-varying, or frictional or other dissipative forces. If in such a system a particle moves from point 1 to point 2 under the influence of a total force **F**, the change in the particle's potential energy is the negative of the work done on the particle,

$$V_2 - V_1 = -W_{12} = -\int_1^2 \mathbf{F} \cdot d\mathbf{s}. \tag{2.46}$$

(Note that in a conservative system the work done is independent of path.) The zero of potential energy can be chosen arbitrarily, since only *differences* in potential energy are physically meaningful.

Let us apply this definition to the harmonic oscillator. Since the force on the article and its displacement both lie along the x axis, $\mathbf{F} \cdot d\mathbf{s}$ reduces to $F_x \, dx$, where F_x is simply the F of Eq. 2.32. The change in potential energy as the particle goes from $x = a$ to $x = b$ is thus

$$
\begin{aligned}
V(b) - V(a) &= -\int_a^b F(x) dx = \int_a^b kx \, dx \\
&= \tfrac{1}{2} k b^2 - \tfrac{1}{2} k a^2.
\end{aligned} \tag{2.47}
$$

It is most convenient to set the potential energy equal to zero at the equilibrium point $x = 0$; if $V(0) = 0$, the potential energy of the oscillator at any other point x is simply

$$V(x) = \tfrac{1}{2} k x^2. \tag{2.48}$$

This is the equation of a parabola, as illustrated in Fig. 2.15*d*. The maximum potential energy is $\frac{1}{2} k x_1^2$, the value at the point of maximum displacement, where the kinetic energy vanishes. The total energy E of the oscillator is the sum of the kinetic and the potential energy;[18] since each of these vanishes when the other has its maximum, each of the two maxima must be equal to the total energy:

$$E = \tfrac{1}{2} m v^2 + \tfrac{1}{2} k x^2 = \tfrac{1}{2} m v_0^2 = \tfrac{1}{2} k x_1^2. \tag{2.49}$$

The information in Eqs. 2.45 can therefore be summarized in the single equation

$$A_{\text{cycle}} = \frac{E}{\nu} = E\tau, \tag{2.50}$$

that is, the action per cycle is the energy per unit frequency or the energy times the period of oscillation. The total energy E is a constant of the oscillator's motion.

There is one further way in which we can visualize the action, with a geometric interpretation that contrasts with the more mechanical picture considered thus far. The geometric concept must be drawn by analogy. Consider a particle moving in a plane, in which we define a Cartesian coordinate system. If the only constraint on its motion is that it remain within the region $0 < x < x_1$, $0 < y < y_1$ (with x and y independent), its position can be anywhere in the rectangle of area $x_1 y_1$ shown in Fig. 2.16*a*. If there were an additional constraint that the sum $(x/x_1)^2 + (y/y_1)^2$ always be less than unity, then the particle would be restricted to the quadrant of the ellipse shown in Fig. 2.16*b*. One can imagine innumerable other restrictions. Basically, however, if there are two variables describing the behavior of a particle, perhaps with some constraints, then one can describe an available region in a plane within which the particle can move.

Now we can construct our analogy for the harmonic oscillator. We choose as one of the two variables the displacement x itself. Then, instead of displacement in another spatial direction (since none exists for our model), we

[17] The last line of Eq. 2.44 makes use of the definite integral

$$\int_0^{2\pi} \cos^2 \theta \, d\theta = \pi.$$

[18] Note that E, like V, has an arbitrary zero.

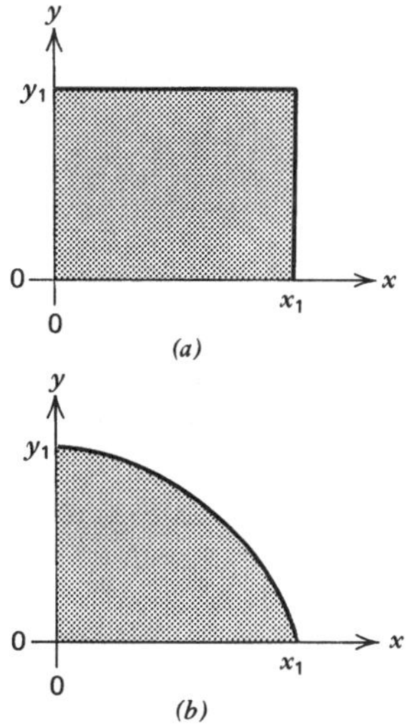

Figure 2.16 Areas available to particles moving in a real plane, constrained to stay within limits $0 < x < x_1$, $0 < y < y_1$. (*a*) *x, y* independent, (*b*) With added constraint $(x/x_1)^2 + (y/y_1)^2 < 1$.

choose as our second variable the momentum p. For any object restricted to finite velocities and to a finite region of space, that is, for anything we can conceivably observe for an arbitrarily long time, each component of both position and momentum is a quantity bounded above and below—it has a maximum and a minimum. Each component of the motion (and our one-dimensional oscillator has only one such component) can thus be thought of as defining an area in a plane. Such a plane—a two-dimensional space—is the simplest example of what is called a *phase space,* a space whose dimensions are spatial coordinates and the corresponding momentum components. The "area" in this plane has the dimensions of momentum × distance, which it will be recalled are the same as those of action. In other words, we can associate with every bounded one-dimensional system an arealike quantity with the dimensions of action which describes the ranges of momentum and displacement available to the system.

Now that we have defined the "action plane," let us use it to describe the motion of the harmonic oscillator. We need an equation for the motion of the harmonic oscillator. We need an equation for the motion in terms of x and p $(= mv)$. We have just such an equation available in Eq. 2.49, which we can rearrange to read

$$\frac{p^2}{(mv_0)^2} + \frac{x^2}{x_1^2} = 1 \qquad (2.51)$$

(where we have divided through by $mv_0^2 = kx_1^2/2$). But from analytic geometry we recognize immediately that this is the equation of an ellipse in the px plane, with its center at the origin and semiaxes of mv_0 and x_1. This ellipse is illustrated in Fig. 2.17. The displacement and momentum of the oscillator are represented by a point moving around the ellipse in the clockwise direction: One trip around the ellipse represents one cycle of the oscillator. The motion of the point is along the curve itself rather than within it (as in Fig. 2.16*b*); nevertheless, the area within the curve is significant. If we combine the definition of the action as $\int p\, dx$ with the familiar interpretation of the definite integral as an area, we realize immediately that the area within the ellipse must equal the action over a cycle. (There is nothing remarkable about this: Any integral over a cycle equals the area inside the closed curve of the integrand.) The area of an ellipse is πab, where a and b are the two semiaxes; the action over a cycle of the harmonic oscillator is thus

$$A_{\text{cycle}} = \pi(mv_0)x_1 = \frac{\pi m v_0^2}{\omega}, \qquad (2.52)$$

in agreement with Eq. 2.44. That is, the geometric interpretation of the action gives the same value as the direct calculation.

2.9 Action Quantized: The Heat Capacity of Solids

After our long digression to clarify the nature of action, we can now return to our main line of argument, broken off in Section 2.6. We spoke then of three phenomena that were explained in terms of the quantum of action; now we can complete our treatment of black-body radiation and proceed to the heat capacities of solids.

Remember that Planck's model for the black-body radiation field was an array of "oscillators," by which he meant harmonic oscillators such as we have just been describing. To the extent that this model is valid, we should be able to extract some useful insight from what we have learned about oscillators. Sure enough, we have Eq. 2.50, which says that an oscillator with energy E and frequency ν has an action per cycle of E/ν. But for Planck's oscillators the allowed values of E are separated by increments of $h\nu$; this means that their action per cycle can only have values separated by increments of $\Delta E/\nu = h$. In terms of Fig. 2.17, the allowable ellipses are only those with areas of exactly h, $2h$, $3h$, ... One can now see why h is called the "quantum of action."

Our second example of the quantization of action involves the classic problem of the heat capacity of a crystal lattice. Again, because a detailed analysis would take us too far afield, we shall limit ourselves here to a simple qualitative description. (See Chapters 11, 14, and 22.) A crystal is composed of an orderly array (lattice) of atoms or ions. The equilibrium position of each atom is called a *lattice site.* Because of the repulsion between adjacent atoms, the structure with each atom occupying its lattice site is one of minimum potential energy. If the atoms have any kinetic energy (as they do at any temperature above absolute zero), then they do not rest at their lattice sites but move about them, striking each other and rebounding. To a good

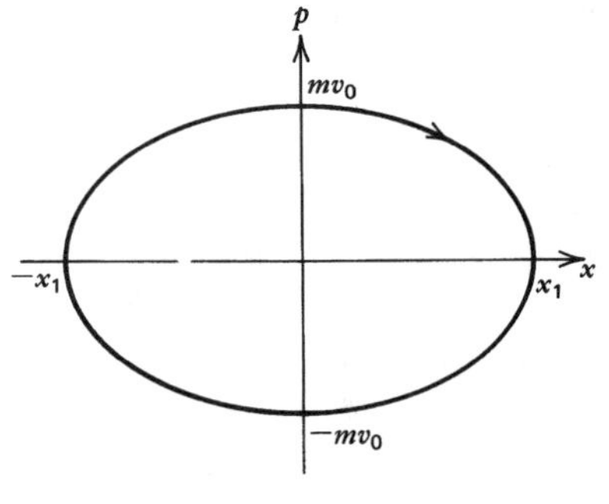

Figure 2.17 Ellipse in the action plane describing the motion of the one-dimensional harmonic oscillator. The two semiaxes are related to each other and the energy by

$$mv_0 = (km)^{1/2}x_1 = (2mE)^{1/2}.$$

The area within the ellipse is the action per cycle.

approximation one can describe this vibration by considering the atoms as harmonic oscillators. (Here they are again!) To be precise, since each atom has three independent directions of motion, one represents it by three one-dimensional oscillators, all oscillating around the lattice site. One can calculate the energy of such an array of oscillators as a function of temperature, and this is where the trouble arose.

The problem is best described in terms of the physical quantity called the *heat capacity* (see Chapter 14). Simply defined, the heat capacity of a given quantity of matter is the amount of energy (in the form of heat) required to increase its temperature by 1°C; that is, it is the temperature derivative of the substance's energy. In microscopic terms, the heat capacity must thus be directly related to the capacity of the atomic oscillators for storing energy. According to classical mechanics the heat capacity of a crystal should be independent of temperature, with a constant value of approximately 25 J/mol°C. This result agrees with the empirical *law of Dulong and Petit* (1819), which holds fairly well for many solids (especially metals) at room temperature, and can thus be used to estimate molecular weights from specific heats. However, a number of solid substances, particularly nonmetals, have heat capacities much lower than 25 J/mol°C at room temperature; furthermore, all heat capacities appear to fall off toward zero at sufficiently low temperatures. The classical theory thus appears to be correct only as a limiting case. The key to the correct theory was found by Albert Einstein in 1907, and of course involved the quantization of action.

The problem of the vibrational energy of solids is cleared up by essentially the same assumption as that used for the problem of black-body radiation. Indeed, the same sweeping hypothesis can be applied to any bounded system undergoing periodic motion. It consists of one simple but very severe constraint: The action integral for a single cycle of the system's motion must equal an integral multiple of Planck's constant h. That is, we must have

$$A_{\text{cycle}} = \int_{\text{cycle}} \mathbf{p} \cdot d\mathbf{s} = nh, \tag{2.53}$$

where n is an integer, both for the system as a whole and for each individual oscillator. In the geometric interpretation of Fig. 2.17, this equation again means that the ellipse in the action associated with an individual oscillator must have an area that is an integral multiple of h. Classically, any value was permissible for the ellipse's area. Referring back to Eq. 2.50, we recognize immediately that the energy of a single oscillator of frequency ν can only be

$$E_{\text{osc}} = \nu A_{\text{cycle}} = nh\nu. \tag{2.54}$$

The energy of an oscillator is thus restricted to integral multiples of the product $h\nu$ and the oscillator can only increase its energy in discrete steps of magnitude $h\nu$. Using this assumption, as we shall see in Chapter 22, Einstein was able to derive a qualitatively correct temperature dependence for the heat capacity of crystals.

We can give here an outline of how quantization of the oscillator energies affects the macroscopic properties of crystals. Energy as heat is transferred to a solid by means that need not be specified. That energy is distributed among the oscillators of the solid. Classical mechanics assumed that the oscillators could exchange energy in any amount. Now, however, Eq. 2.54 tells us that the oscillators are allowed to increase their energy only in jumps of size $h\nu$. If an oscillator's $h\nu$ is larger than the average energy per oscillator, then that oscillator must be rather unaccommodating as a storehouse for energy: It hardly ever gets enough energy all at once to provide an entire quantum. When the temperature is low, the average energy per oscillator is low, so most of the oscillators cannot contribute to the capacity of the solid to store energy. Hence the heat capacity is low at low temperatures. The amount of available energy of course increases with temperature, and at a high enough temperature the heat capacity c approaches the classical (Dulong–Petit) value of 25 J/mol°C. The temperature required for this to occur is proportional to the value of ν, which can thus be estimated from the temperature dependence of c. The highest values of ν are found in hard substances like diamond, for which temperatures of about 2000°C are required for c to approach the classical limit. For most solids, however, the value of ν is low enough to give classical behavior at room temperature, resulting in the Dulong–Petit law. At still lower temperatures the available thermal energy per atom becomes smaller and smaller relative to $h\nu$, eventually becoming so small that the heat capacity of all solids decreases to zero.

Our equations for the energy of light quanta and for the energy of an oscillator, Eqs. 2.20 and 2.54, look very similar. Their content is also similar, but not quite identical. Each quantum has a specific and well-defined energy $h\nu$; the oscillator has a definite frequency ν, but can absorb any number of quanta, changing its state with each energy transfer.[19] What about Planck's oscillators in the black-body radiation field? The space within the black body is filled with a vast number of light quanta with frequencies all over the spectrum. The number of quanta of any given frequency ν must always be some integer n, and the total energy of these quanta is $nh\nu$, just as in Eq. 2.54. The "oscillator" of frequency ν is thus just a convenient way of summing up the quanta of frequency ν; the "oscillator" gains or loses energy when one quantum at a time leaves or enters the walls, yielding Eq. 2.20.

2.10 Some Orders of Magnitude

We have now introduced the fundamental concepts of the quantum theory—quantum levels of energy, the quantization of action, and the particulate or quantum character of light. We are ready to go on to study the structure of the atom. However, as we did in Section 1.2, let us first examine some numerical magnitudes, in order to put some of the esoteric quantities we have been discussing into context. We begin by recalling in familiar units the magnitudes usually associated with energy and with action. The mechanical unit of energy, the joule (1 J = 1 kg m^2/s^2), is not a very large quantity: 1 J is just about the energy required to lift a 100-g weight 1 m; it takes about 4.2 J (1 cal) to heat a gram of water by 1°C. As for action (the unit of which has no special name of its own), for that 100-g weight to drop 1 m would involve an action of about 0.3 J s; to take a cyclic phenomenon, at 30 mph an automobile tire might have an action of 200 J s per revolution.

What about microscopic quantities? To begin with, what do the phenomena just mentioned amount to on a microscopic scale? Some things we can calculate easily enough: A mole of water weighs 18 g and contains Avogadro's number of H_2O molecules, so it takes about

$$\frac{4.2\ \text{J}}{\text{g}} \times \frac{18\ \text{g}}{\text{mol}} \times \frac{\text{mol}}{6\times10^{23}\ \text{molecules}} \approx \frac{1.3\times10^{-22}\ \text{J}}{\text{molecule}}$$

to heat water 1°C. It is a little more complicated to find the action per atom in the automobile tire. Say that the tire weighs 10 kg. What about the atomic weight? Most of the tire is made of rubber (or some synthetic equivalent), which is essentially a hydrocarbon polymer; the rayon or cotton cords are basically cellulose $(C_6H_{10}O_5)_x$. Hydrogen's atomic weight is about 1, carbon's 12, and oxygen's 16; a little playing with formulas shows that the average atomic weight in the tire should not be far from 6, which will make the arithmetic easy. Some parts of the tire are moving faster than others, but the average action per atom for the whole tire will suffice for our purposes. This is easily calculated as

$$\frac{200\ \text{J s}}{10\ \text{kg}} \times \frac{\text{kg}}{1000\ \text{g}} \times \frac{\text{g}}{6\times10^{23}\ \text{amu}} \times \frac{6\ \text{amu}}{\text{atom}} \approx \frac{2\times10^{-25}\ \text{J s}}{\text{molecule}}$$

per turn of the tire.

Something should be said about these calculations, which are typical of the way one often approaches a scientific problem. That is, we don't care what the exact action per atom really is; we just want to know roughly how much it is. To do this we can make rough guesses of the sizes of the numbers. It would be an incredible waste of time to get a tire, measure its exact weight, carry out a chemical analysis, make a detailed calculation of its moment of inertia,[20] and so on. All we want is a simple *order-of-magnitude* estimate. One cannot overemphasize the importance of order-of-magnitude calculations in science. They are our sounding lines, the guides that tell us whether it is reasonable or nonsensical to continue in a given direction. They are the guideposts we use to give us a feeling for unfamiliar quantities, in just the way we are using them now. They are the quick checks that tell us if we have made a gross error in calculation, that give us an on-the-job diagnosis of whether or not an experiment is likely to be working properly. With a little practice, and a few natural constants and conversion factors in one's head, one develops a facility to make these estimates quickly and painlessly at the slightest provocation.

All right, how do the quantities we have calculated compare with the quanta of energy and action? Planck's constant h, the quantum of action, is only about 6.6×10^{-34} J s. This means that one revolution of our automobile tire involves about 300 million quanta of action *per atom!* Obviously h is a small quantity indeed, and it is clear why quantum effects are not observed in macroscopic phenomena. What about the energy of a light quantum? Yellow light, with a wavelength of about 5000 Å, has a frequency of

$$\nu = \frac{c}{\lambda} \approx \frac{(3\times10^{8}\ \text{m/s})}{5\times10^{-7}\ \text{m}} = 6\times10^{14}\ \text{s}^{-1};$$

multiplying this frequency by Planck's constant, we obtain about 4×10^{-19} J for the energy of a single quantum of yellow light. This is not so small: It is 3000 times our figure for heating one water molecule 1°C, and a mole of these quanta[21] would amount to 240 kJ, enough energy to heat 57 liters of water 1°C. But it is well to keep in mind the vast range of the electromagnetic spectrum when one thinks of the energy associated with radiation. X-ray quanta are

[19] The heating of a solid involves the absorption of quanta in the form of vibrational energy (*phonons*) rather than quanta of electromagnetic radiation (photons), but the principle is the same.

[20] To determine the total action. We got our figure by guessing an average radius and applying Eq. 2.30, another order-of-magnitude calculation.

[21] Incidentally, photochemists call a mole of light quanta an *einstein*, for obvious reasons.

usually more than a thousand times as energetic as visible-light quanta, and gamma rays are more energetic still. On the other hand, if we proceed to lower frequencies, in the radio region we commonly find frequencies in megahertz, that is, mere millions of cycles per second; this corresponds to energies of the order of 10^{-27} J per quantum, a very small number even by microscopic standards.

2.11 Bohr's Model of the Atom

This chapter concludes with a discussion of the simplest model of the atom containing the essence of quantization, the model developed by Niels Bohr in 1913. It is not really correct in all its details, or even in some of its most fundamental ideas; however, its utility as a simple device for computing atomic magnitudes is unsurpassed. There is no other computational tool known to which we can turn so quickly to estimate whether a new effect will be observable or beyond our detectable range. We shall make no attempt to explore the model in detail, but even by examining its simplest form we can derive all the atomic properties and magnitudes we have been introducing throughout this chapter. In particular, we wish to establish that atoms are stable entities (i.e., that electrons do not fall into nuclei), that they exhibit sharp spectral lines associated with definite excitation energies, and that the intervals between these spectral lines are remarkably regular.

Our derivation of Bohr's model will not be carried out in quite the way Bohr himself did it, but by a somewhat shorter and, for our purposes, more efficient route. (You can usually improve a derivation *after* you know how it comes out.) We begin with a few simple assumptions about the energy of the system, its mechanical stability, and, above all, the quantization of its action. From this simple model we can derive the characteristic lengths, velocities, frequencies, and times associated with the motion of electrons in atoms. We shall also see, in a surprisingly quantitative way, how the general character of atomic and molecular spectra has its origins in the structure of the atom, and we shall be able to infer from the model many of the general characteristics of more complex atoms and molecules.

Bohr's model was derived for the hydrogen atom, or for any other system in which a single electron orbits a positively charged nucleus. It thus took for granted (and immediately followed) Rutherford's model of the nuclear atom. In most of our derivation we shall make the simplifying assumption that the nucleus is so heavy that it can be considered stationary, with only the electron moving; strictly speaking, both particles orbit about the center of mass.[22] We follow Bohr's original model in assuming that the electron moves in a circular orbit (elliptical orbits were also introduced to account for the fine structure of spectra). We consider the general case of a nucleus with charge $+Ze$; the hydrogen atom corresponds to $Z = 1$.

We begin by defining the energy of the system. The total energy is the sum of the electron's kinetic energy and its potential energy in the field of the nucleus. Let the electron's mass be m, its velocity v, and its distance from the nucleus r. The kinetic energy is simply $\frac{1}{2}mv^2$, while the potential energy is some function of r. We can evaluate Eq. 2.46 for the potential energy between any two charges, using Eq. 1.5 for the force. Since the force is directed radially, we have $\mathbf{F} \cdot d\mathbf{s} = F(r)\,dr$ for a radial displacement and the potential energy is given by[23]

$$V(r) - V(\infty) = -\int_r^\infty F(r)\,dr = \frac{q_1 q_2}{4\pi\epsilon_0}\int_r^\infty \frac{dr}{r^2} = \frac{q_1 q_2}{4\pi\epsilon_0 r}. \tag{2.55}$$

The zero of potential energy can as usual be set arbitrarily, and it is clearly most convenient to set $V(\infty) = 0$. For our atomic system we have $q_1 = +Ze$, $q_2 = -e$, $V(r) = -Ze^2/(4\pi\epsilon_0 r)$, and the total energy is

$$E = \frac{mv^2}{2} - \frac{Ze^2}{4\pi\epsilon_0 r}. \tag{2.56}$$

Our choice of the zero of energy corresponds to setting $E = 0$ for an electron at rest ($v = 0$) at an infinite distance from the nucleus ($r = \infty$). The advantage of this choice is that all negative values of E then correspond to bound states of the electron, states in which it cannot escape from the nucleus. This is so because the kinetic energy $mv^2/2$ is necessarily a positive quantity; for E to be negative we must have $Ze^2/4\pi\epsilon_0 r > mv^2/2$, which sets an upper bound on the value of r. By contrast, a positive value of E corresponds to a free electron, which can have as large a value of r as one wishes; at sufficiently great distances, the potential energy becomes negligible.

The next step in the derivation comes from the assumption that the electron's orbit is stable. By Coulomb's law, the inward attractive force on the electron is $Ze^2/4\pi\epsilon_0 r^2$. The centrifugal force on a particle moving in a circular orbit is mv^2/r. For the electron to continue moving in a stable circular orbit, these two forces must be equal. Consequently, we have

$$\frac{mv^2}{r} = \frac{Ze^2}{4\pi\epsilon_0 r^2}. \tag{2.57}$$

[22] The model can in fact easily be generalized to apply to any two oppositely charged particles. Of particular interest is the case where two particles have equal mass, and orbit about each other like a binary star. There are real physical examples of such species, for example, *positronium,* composed of a positron (positively charged electron) and an ordinary negative electron.

[23] This and all the equations that follow assume the use of SI units. Many texts give the corresponding expressions in electrostatic or Gaussian units; these expressions can be generated by substituting $(4\pi)^{-1}$ for ϵ_0.

The assumption of stability may appear trivial, but this was one of the most revolutionary aspects of Bohr's model. According to classical electromagnetic theory, an orbiting charge would continuously emit radiation, thereby losing energy and spiraling inward to the nucleus. By contrast, the quantum theory assumes not only that the orbit is stable, but that the electron can gain or lose energy only in quantized amounts.

The introduction of this quantization constitutes our final major assumption. As in previous sections, it is the action that must be quantized: We require that the action integral over each cycle be an integral multiple of h. Introducing our earlier result for the action in a circular orbit, Eq. 2.30, we have

$$A_{orbit} = 2\pi m \upsilon r = nh. \tag{2.58}$$

The product $m\upsilon r$ is the electron's angular momentum, which must be an integral multiple of $h/2\pi$. The latter quantity appears so often in subsequent expressions that it is convenient to introduce a special symbol for it,

$$\hbar \equiv \frac{h}{2\pi} = 1.0545716 \times 10^{-34} \text{ J s}. \tag{2.59}$$

Such shorthand ways of combining symbols, especially those for natural constants, make our equations less cumbersome and thus tend to prevent mistakes.

We are now ready to combine our assumptions. Multiplying both sides of Eq. 2.57 by $r/m\upsilon$, we obtain an expression for the velocity of the electron,

$$\upsilon = \frac{Ze^2}{4\pi\epsilon_0 m\upsilon r}. \tag{2.60}$$

But the $m\upsilon r$ in the denominator of this equation must, by Eq. 2.58, be an integral multiple of $\hbar$. We immediately have an explicit expression for the allowed values of the velocity,

$$\upsilon = \frac{Ze^2}{4\pi\epsilon_0 n\hbar} = \frac{Ze^2}{2\epsilon_0 nh}. \tag{2.61}$$

This equation tells us that the velocity of the electron is directly proportional to the nuclear charge Ze and inversely proportional to the integer n, which we call the *principal quantum number*. The essence of Bohr's theory is that only those orbits corresponding to integral values of n are physically allowed. Later we shall obtain numerical values for υ and other atomic quantities, but not until we have completed the formal structure of the theory.

Another quantity whose value we can easily derive is the radius of the electron's orbit. From Eq. 2.58 we immediately have

$$r = \frac{n\hbar}{m\upsilon}, \tag{2.62}$$

and substitution of the value of υ from Eq. 2.61 then gives

$$r = \frac{4\pi\epsilon_0 n^2 \hbar^2}{Ze^2 m} = \frac{\epsilon_0 n^2 h^2}{\pi Ze^2 m} \tag{2.63}$$

in terms of fundamental constants. The radius depends inversely on the velocity and thus on the nuclear charge; a nucleus with a larger positive charge will hold the electron in a tighter orbit, but with a greater orbital velocity. Note also that the radius depends explicitly on the mass m; a charged particle heavier than the electron would orbit closer to the nucleus.[24]

Next we compute the period of the orbit, the time it takes the electron to go around the nucleus once. This time is simply the distance (the circumference) divided by the velocity, or from Eqs. 2.61 and 2.63,

$$\tau = \frac{2\pi r}{\upsilon} = \frac{(4\pi\epsilon_0)^2 2\pi n^3 \hbar^3}{Z^2 e^4 m} = \frac{4\epsilon_0^2 n^3 h^3}{Z^2 e^4 m}. \tag{2.64}$$

The period is inversely proportional to the mass and to the square of the nuclear charge.

It can now be seen that our model of the atom exhibits a series of discrete states, each corresponding to a different integral value of the quantum number n. The values of n begin with 1 and have no upper limit. (A value $n = 0$ would correspond to a particle moving with infinite velocity at zero radius; negative values would merely correspond to reversing the direction of rotation.) Each successive larger value of n defines a new stable orbit, with lower velocity, a larger radius, and a much longer period than the preceding orbit. The orbit for which $n = 1$ must be the closest to the nucleus and, as we shall now see, corresponds to the most tightly bound state.

By "most tightly bound," of course, we mean the state with the lowest energy, which thus requires the greatest amount of work to remove the electron from the atom. Let us then consider the energy of our system. Equation 2.57 can be rewritten as

$$\frac{m\upsilon^2}{2} = \frac{1}{2}\left(\frac{Ze^2}{4\pi\epsilon_0 r}\right); \tag{2.65}$$

what this says is that, for any stable circular orbit, the kinetic energy is half the negative of the potential energy (relative to the energy zero previously defined). Substituting this relation into Eq. 2.56, we immediately obtain a simple expression for the total energy,

$$E = \frac{Ze^2}{4\pi\epsilon_0 r}\left(\frac{1}{2} - 1\right) = -\frac{Ze^2}{8\pi\epsilon_0 r}. \tag{2.66}$$

---

[24] This property can be used to probe the structure of nuclei. Short-lived "atoms" can be made in which negatively charged mesons move in orbits about nuclei. For a meson whose mass is 200 times that of the electron, the stable radii are about 1/200 of those for electron orbits. Such an orbit is close enough to the nucleus to be affected by its detailed internal structure.

Note that this is still a classical expression. We obtain the quantized energy levels by substituting the value of r from Eq. 2.63:

$$E = -\frac{Z^2 e^4 m}{2(4\pi\epsilon_0)^2 n^2 \hbar^2} = -\frac{Z^2 e^4 m}{8\epsilon_0^2 n^2 h^2}. \tag{2.67}$$

The energy of any stable orbit is of course negative, approaching the limit $E = 0$ as n (and thus r) becomes infinite. The absolute value of the energy is proportional to the mass and to the square of the nuclear charge.

The above expression for the energy is rather cumbersome; it is time to introduce another simplification. Designating the energy of the nth quantum state as E_n, we can write Eq. 2.67 in the form

$$E_n = -hcR_\infty \frac{Z^2}{n^2}, \tag{2.68}$$

where hcR_∞ is the energy defined by

$$hcR_\infty \equiv \frac{e^4 m}{8\epsilon_0^2 h^2} = 2.179907 \times 10^{-18}\ \text{J}$$
$$= 13.605692\ \text{eV}. \tag{2.69}$$

Since the joule is an inconveniently large quantity for atomic energies, we also give the value in electron volts (cf. footnote 1 on page 22). Spectroscopists often express energies in terms of the corresponding wave numbers ($1/\lambda$); since for a transition between two energy levels (see below) we have $|\Delta E| = h\nu = hc/\lambda$, the constant corresponding to hcR_∞ is

$$R_\infty \equiv \frac{e^4 m}{8\epsilon_0^2 h^3 c} = 10973731.568\ \text{m}^{-1}. \tag{2.70}$$

Note that R_∞ itself is known more accurately than the other physical constants in the equation. Like the R_H of Eq. 2.10, R_∞ is known as the *Rydberg constant* (the reason for the subscript "∞" will be explained later); the energy hcR_∞ is sometimes called one *rydberg.*

We can now determine the magnitudes of the energies involved in actual interactions. Quanta of visible light, with wavelengths between 4000 and 8000 Å, have energies in the range 2–4 eV. In the hydrogen atom ($Z = 1$), the lowest energy level is $E_1 = -hcR_\infty$, so that about 13.6 eV is required to remove the electron from the atom and leave it at rest at a great distance; this energy corresponds to a wavelength of 911 Å, far in the ultraviolet. If the electron is initially in the state with $n = 2$, only one-fourth as much energy is required to remove it, corresponding to light of 3645 Å, just over the edge into the ultraviolet; and so forth for the higher energy levels. (The energy levels of the hydrogen atom are shown in Fig. 2.18). Suppose that an electron in the nth state absorbs a quantum with energy greater than $-E_n$: Then the electron is not only freed from the atom but given a kinetic

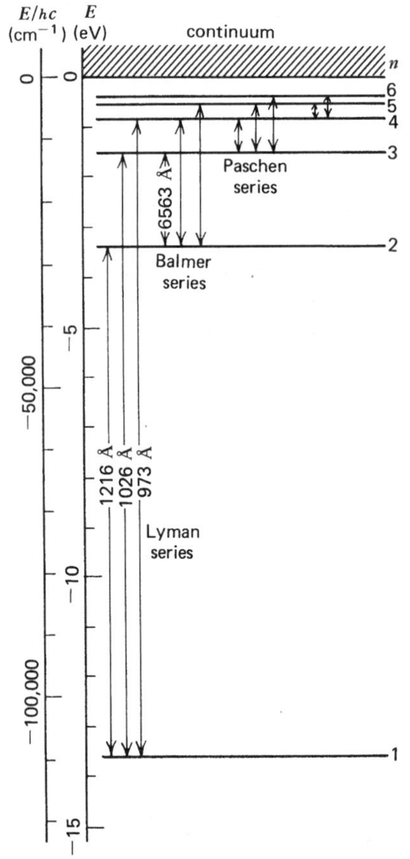

Figure 2.18 Energy levels of the hydrogen atom, with some of the principal spectral lines. The levels are almost exactly given by

$$E_n = \frac{-hcR_H}{n^2},$$

where $R_H = 109678\ \text{cm}^{-1}$ and $hcR_H = 13.60$ eV. The wavelengths of the spectral lines are related to the energy levels by

$$\lambda = \frac{hc}{\Delta E}.$$

energy equal to the difference between the actual energy of the quantum and $-E_n$. Note that at positive values of E the energy is no longer quantized, so this kinetic energy can have any value. It should be obvious that the process we have been describing is simply the photoelectric effect for free atoms.

Of course, an electron can also absorb a quantum of light without being set completely free. For example, an electron in a state with $n = 1$ can go to a state with $n = 2$ if it absorbs a quantum with just the right amount of energy. A transition between any other two states can be brought about by a quantum whose energy equals the energy difference between them. An atomic electron can thus absorb a quantum with any of a set of very specific energies, but none with energies different from these. If continuous radiation passes through a sample of matter, the specific wavelengths corresponding to these energies (characteristic of the sample) will be selectively absorbed, giving a spectrum of sharply defined absorption lines. Quanta with enough energy to liberate an electron completely give a continuous absorption spectrum beyond the limit of the line spectrum.

Electrons can also fall back from higher to lower energy levels, emitting quanta in the process. The energy of the quantum emitted again just equals the energy difference between the two states. This is why any element's emission spectrum exactly corresponds to its absorption spectrum. Since the ground state is most stable, the emission spectrum will be observed only when an atom has first been excited to some higher state. This excitation may be due to earlier absorption of a quantum of light, to electron bombardment (as in the Franck–Hertz experiment), or simply to high temperatures (which increase the rate and energy of collisions between atoms).

Note that in all the processes just described we mention only the initial and final states (the bound states corresponding to quantized energies are called *stationary states*). We say that the electron goes from one to the other by absorbing or emitting a quantum, but do not discuss the mechanism by which this occurs. In fact, Bohr simply assumed an instantaneous "jump," and it took a more advanced theory to explain how the process could occur (cf. Section 4.5). But it is only the energy differences between the various states that concern us now.

We can now interpret the spectrum of atomic hydrogen in terms of quantum states of the hydrogen atom. Let us apply Eq. 2.68 (setting $Z = 1$) to the case of a transition between two states. If the electron in a hydrogen atom goes from an initial state n_2 to a final state n_1, its change in energy is

$$\Delta E = E_{\text{final}} - E_{\text{initial}} = E_{n_2} = -hcR_\infty\left(\frac{1}{n_1^2} - \frac{1}{n_1^2}\right). \tag{2.71}$$

If the atom loses energy in this process ($\Delta E < 0$, $n_2 > n_1$), then this energy must reappear somewhere; if it is not removed by a collision of some kind, it must appear as emitted radiation, that is, as light. The quantum of this radiation then has an energy exactly equal to the energy lost by the atom. This fact enables us to relate the energy change to the wavelength of the emitted light,

$$\Delta E_{\text{atom}} = -E_{\text{emitted light}} = -h\nu - \frac{hc}{\lambda}. \tag{2.72}$$

Combining Eqs. 2.70 and 2.71, then, we have for emission

$$\frac{1}{\lambda} = -\frac{\Delta E}{hc} = R_\infty\left(\frac{1}{n_1^2} - \frac{1}{n_2^2}\right). \tag{2.73}$$

Comparing this result with the experimental Eq. 2.11, we see that the two are identical except for the tiny difference (about 0.05%) between R_H and R_∞, which we explain below. The absorption spectrum of hydrogen is described by the same equations with the roles of n_1 and n_2 reversed ($\Delta E > 0$, $n_1 > n_2$). Figure 2.18 shows the major series of hydrogen spectral lines in relation to the energy levels of the atom.

We now account for the discrepancy between the two Rydberg constants: the experimental $R_H = 109678\ \text{cm}^{-1}$ and the theoretical $R_\infty = 109737\ \text{cm}^{-1}$. This discrepancy is not real, but due merely to a simplification introduced into our derivation for the sake of conciseness. This simplification was the assumption that the nucleus is stationary and only the electron moves. In fact, both particles must revolve about their center of mass. Classical mechanics shows how the motion of two bodies around their center of mass can be expressed in terms of the motion of one body around a force center, the one body having a mass equal to the *reduced mass* of the two bodies. For individual masses m_1 and m_2, the reduced mass is defined by

$$\mu \equiv \frac{m_1 m_2}{m_1 + m_2} \quad \text{or} \quad \frac{1}{\mu} = \frac{1}{m_1} + \frac{1}{m_2} \tag{2.74}$$

(μ is called the harmonic mean of m_1 and m_2). Suppose now that the two particles are the electron and the proton; for the hydrogen atom

$$\mu_H = \frac{m_e m_p}{m_e + m_p} = \frac{m_e(1836.15 m_e)}{m_e(1 + 1836.15)} = 0.9994557\, m_e. \tag{2.75}$$

If we carry through our derivation of the Bohr theory with μ_H replacing the electron mass m (i.e., m_e), we find that the constant R_∞ in Eq. 2.73 is replaced by

$$\frac{e^2 \mu_H}{8\epsilon_0^2 h^3 c} = 0.9994557\, R_\infty = R_H; \tag{2.76}$$

the agreement with the experimental value of R_H is exact. For an atom in which an electron revolves about a heavier nucleus, the value of μ would be still closer to m_e and the corresponding Rydberg constant closer to R_∞. In the limit of a nucleus of infinite mass, the constant would be exactly R_∞, which is therefore more properly called the *Rydberg constant for infinite mass.*

The Bohr theory is thus able to reproduce exactly the spectrum of hydrogen as we described it in Section 2.4.[25] From an extremely simple (although *ad hoc*) model we have easily and directly derived the remarkable Rydberg equation, which elegantly describes hundreds of individual spectral lines. (In principle the number is infinite, since n can be as large as one likes.) One stands in awe at the beauty, clarity and power of Bohr's simple model.

[25] Greater resolution reveals a "fine structure," with each spectral line actually consisting of two or more very closely spaced lines. Much of this can be accounted for within the Bohr theory by assuming elliptical orbits, but modern quantum mechanics is needed for a full explanation.

Of course, this magnificent model *is* only a model. It is quite wrong in representing electrons as ordinary physical objects orbiting around nuclei, even if this assumption is sufficient to explain the spectra. More seriously, it ceases to be accurate when we try to apply it to atoms with more than one electron. This breakdown is due to the fact that each electron moves in the field of all the other electrons as well as the nucleus. In the alkali metals one electron is much farther from the nucleus than all the others, and thus sees the rest of the atom as a net charge of +1 spread over the atom's interior; the Bohr model then applies approximately, with n replaced by the effective quantum number n^* of Eq. 2.12. In general, however, the simplicity of the theory was lost in attempts to explain complex atoms; as a result a still more revolutionary theory had to be devised, as we shall see in the next few chapters.

Nevertheless, the Bohr model is still useful. It gives us expressions from which we can estimate the magnitudes of most atomic quantities, and exhibits in simple form the dependence of these quantities on nuclear charge and on the principal quantum number. Some of the most important of these quantities are given in Table 2.2 for our simple model of one-electron atoms with circular orbits. The energy we have already discussed. We can immediately see that atoms must be of the order of 10^{-10} m (1 Å) in radius. The radius of the first orbit for $Z = 1$, 0.52918 Å, is sometimes called the *Bohr radius,* or 1 *bohr;* it was this value that we listed in Table 1.1 The nuclear charge Z, of course, increases with atomic mass, but so does the value of n for the outermost electron; these effects largely cancel, so that all atoms are of much the same size. (To the very rough extent that this is true, the densities of solid materials should be proportional to their atomic weights.) We also see that a bound electron moves along its orbit at a speed of the order of 1000 km/s; by comparison, a rifle bullet moves at less than 1 km/s, a satellite orbiting the earth at up to 8 km/s. Finally, the period for a single revolution of the electron around the nucleus is of the order of 10^{-16} s; events occurring much faster than this can be considered crudely as if the electron were not moving, whereas much slower events will occur as if the electron were a charge distribution smeared over the entire orbit.

Table 2.2 Characteristic Properties of Circular Orbits[a]

Property	Expression	Value
Energy	$E_n = \dfrac{e^4 m_e}{8\epsilon_0^2 h^2}\dfrac{Z^2}{n^2}$	$\left.\begin{matrix}2.180\times10^{-18}\text{ J}\\13.61\text{ eV}\end{matrix}\right\}\times\dfrac{Z^2}{n^2}$
Radius	$r_n = \dfrac{\epsilon_0 h^2}{\pi e^2 m_e}\dfrac{n^2}{Z}$	$\left.\begin{matrix}5.292\times10^{-11}\text{ m}\\0.5292\text{ Å}\end{matrix}\right\}\times\dfrac{n^2}{Z}$
Velocity	$v_n = \dfrac{e^2}{2\epsilon_0 h}\dfrac{Z}{n}$	$2.188\times10^6\ \dfrac{\text{m}}{\text{s}}\times\dfrac{Z}{n}$
Period	$\tau_n = \dfrac{4\epsilon_0{}^2 h^3}{e^4 m_e}\dfrac{n^3}{Z^2}$	$1.520\times10^{-16}\text{ s}\times\dfrac{n^3}{Z^2}$

[a] The expressions given are for nuclei of infinite mass. For the hydrogen atom, m_e should be replaced by $\mu_H = 0.999456 m_e$.

At this point we have completed our overview of the origins of the microscopic theory of matter. We have examined a range of experimental evidence, from which one can infer the basic structure of the atom and the quantized nature of its energy, and we have rough estimates for the characteristic magnitudes involved. To go further we must modify our ideas of the fundamental nature of matter itself, a nature that must be regarded as having wavelike properties. In the next chapter we continue to develop the quantum theory by introducing the concept of matter waves.

APPENDIX 2A

Rutherford Scattering

The names *Rutherford scattering* and *Coulomb scattering* are used to describe the classical scattering of one particle by another when the force between the particles varies as the inverse square of the distance between them. Hence both electrostatic and gravitational interactions lead to Rutherford scattering, although we consider here only the electrostatic case. This appendix is a brief derivation of the expression for the flux of scattered particles as a function of the scattering angle. (For the terminology of scattering, see Section 2.5, or the more extensive discussion in Chapter 27.)

For simplicity, let us assume that one particle is sufficiently heavy, compared with its collision partner, to be considered at rest; this is the particle shown at the point K in Fig. 2A.1. We call this heavy particle the *scatterer,* and assume that it has a charge Ze. The other particle, the *projectile,* has a charge Q, an initial speed v, and a mass[26] m. The distance between projectile and scatterer at any given time is r; the force of electrostatic interaction between them is then

$$F(r) = \frac{ZeQ}{4\pi\epsilon_0 r^2}, \tag{2A.1}$$

and the corresponding potential energy is

$$V(r) = \frac{ZeQ}{4\pi\epsilon_0 r}, \tag{2A.2}$$

from Eq. 2.55.

If there were no force of interaction, the projectile would pass K in a straight line; the distance of its closest approach to the point K along this hypothetical "unscattered" trajectory is b, called the *impact parameter.* The actual trajectory, however, is deflected through the scattering angle θ, as shown in Fig. 2A.1. For a given trajectory, θ is uniquely determined by the impact parameter and the initial speed of the projectile. One can fairly easily obtain a beam of projectiles with essentially the same speed, but any beam of nonzero width must have a range of values of b. We must therefore determine what fraction of the entire beam will be

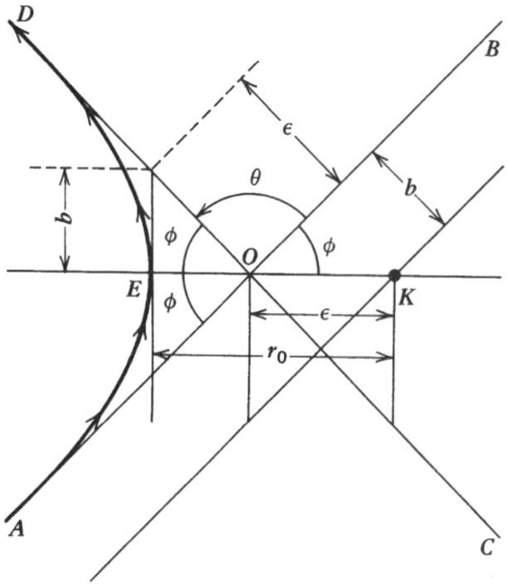

Figure 2A.1 Geometry of Rutherford scattering. The heavy scatterer is at point K; the projectile's trajectory is the curve with arrows. This trajectory is initially along the line AOB, which it would follow if there were no interaction; after scattering, the trajectory becomes asymptotic to the line COD. The scattering angle θ is the angle between these two lines, and the impact parameter b is the distance from K to the line AOB. See the text for definitions of the other quantities shown in the figure. The trajectory is a hyperbola with center at O, focus at K, and vertex at E.

scattered into a given angle θ; this fraction will be expressed in terms of the differential scattering cross section $\sigma(\theta)$, defined in Eq. 2.16.

We need not go through the derivation of the actual trajectory (which can be found in most texts on classical mechanics). It is sufficient to know that the trajectory is a hyperbola with the scatterer at one focus. This is drawn in Fig. 2A.1, on the assumption that the scattering is repulsive, which is the case when both charges are of the same sign (as in Rutherford's measurements). As indicated above, θ is the scattering angle; we also define the angle

$$\phi = \tfrac{1}{2}(\pi - \theta). \tag{2A.3}$$

We need to know the actual distance of closest approach, called r_0, which is the distance from the hyperbola's vertex E to its focus K. From the properties of a hyperbola, this is

$$r_0 = \epsilon + \epsilon \cos\phi, \tag{2A.4}$$

[26] Strictly, we should use the system's reduced mass μ and take the center of mass as point K; cf. Eq. 2.74 and the accompanying discussion. The angle θ that appears in our equations is then the scattering angle in relative (center-of-mass) coordinates, for which see Section 27.1

where ϵ, the distance between focus and center, is given by

$$\epsilon = \frac{b}{\sin\phi}; \tag{2A.5}$$

we thus have

$$r_0 = \frac{b}{\sin\phi}(1+\cos\phi). \tag{2A.6}$$

We can now obtain θ as a function of b and v, using only the principles of conservation of energy and angular momentum.

Since at great distances the potential energy becomes negligible, the total energy of the system is equal to the initial kinetic energy, $\frac{1}{2}mv^2$. The total energy is conserved, so, at any point on the trajectory,

$$\frac{1}{2}mv^2 = \frac{1}{2}mu^2 + \frac{ZeQ}{4\pi\epsilon_0 r}, \tag{2A.7}$$

where the instantaneous speed u is a function of r. In particular, at the point of closest approach, ($r = r_0$), the velocity has an extremum value u_0 (maximum for attraction, minimum for repulsion), we can then write

$$\frac{1}{2}mv^2 = \frac{1}{2}mu_0^2 + \frac{ZeQ}{4\pi\epsilon_0 r_0}. \tag{2A.8}$$

This can be rearranged, with the help of Eq. 2A.6, to give

$$\frac{u_0^2}{v^2} = 1 - \frac{ZeQ}{2\pi\epsilon_0 mv^2 r_0} = 1 - \frac{ZeQ\sin\phi}{2\pi\epsilon_0 mv^2 b(1+\cos\phi)}. \tag{2A.9}$$

Next we consider the angular momentum around point K, which at any given instant is the linear momentum multiplied by the "lever arm."[27] Conservation of angular momentum requires that its initial value equal the value at the point of closest approach:

$$mvb = mu_0 r_0. \tag{2A.10}$$

We therefore have

$$\frac{u_0}{v} = \frac{b}{r_0} = \frac{\sin\phi}{1+\cos\phi}, \tag{2A.11}$$

and squaring this result gives

$$\frac{u_0^2}{v^2} = \frac{\sin^2\phi}{(1+\cos^2\phi)} = \frac{1-\cos^2\phi}{(1+\cos\phi)^2} = \frac{1-\cos\phi}{1+\cos\phi}. \tag{2A.12}$$

Combining Eqs. 2A.9 and 2A.12 to eliminate u_0^2/v^2, and multiplying through by $(1 + \cos\phi)$, we have

$$1-\cos\phi = 1+\cos\phi - \frac{ZeQ}{2\pi\epsilon_0 mv^2 b}\sin\phi, \tag{2A.13}$$

which can be further rearranged to

$$\frac{4\pi\epsilon_0 mv^2 b}{ZeQ} = \frac{2\sin\phi}{2\cos\phi} = \tan\phi. \tag{2A.14}$$

We want our result in terms of the scattering angle θ, of course. From Eq. 2A.3 we have

$$\tan\phi = \tan\left(\frac{\pi}{2} - \frac{\theta}{2}\right). \tag{2A.15}$$

Using the standard trigonometric relation

$$\tan(x-y) = \frac{\sin x\cos y - \cos x\sin y}{\sin x\sin y + \cos x\cos y}, \tag{2A.16}$$

with $x = \pi/2$, $y = \theta/2$, we obtain

$$\tan\phi = \frac{\cos(\theta/2)}{\sin(\theta/2)} = \cot\left(\frac{\theta}{2}\right). \tag{2A.17}$$

Substituting this result in Eq. 2A.14, we have a direct relationship between b and θ for a given trajectory,

$$b = \frac{ZeQ}{4\pi\epsilon_0 mv^2}\cot\left(\frac{\theta}{2}\right). \tag{2A.18}$$

We also need the differential of this equation,

$$\begin{aligned} db &= \frac{ZeQ}{4\pi\epsilon_0 mv^2} d\left(\cot\left(\frac{\theta}{2}\right)\right) \\ &= -\frac{ZeQ}{8\pi mv^2}\frac{d\theta}{\sin^2(\theta/2)}. \end{aligned} \tag{2A.19}$$

To obtain the scattering cross section, we must know what fraction of particles will be scattered into a given range of scattering angles. Consider a plane normal to the incident beam, at a point so far upstream that no appreciable scattering has yet occurred. If the flux density of incident particles across this plane is everywhere uniform, the flux[28] across a

[27] That is, by the shortest distance between point K and a line drawn tangent to the trajectory at that instant. More formally, the angular momentum is defined as

$$\mathbf{L} \equiv \mathbf{r} \times m\mathbf{u},$$

where $\mathbf{r}$ is the radius vector (here drawn from point K) and $\mathbf{u}$ is the instantaneous velocity.

[28] In our terminology, *flux* means the number of particles per unit time, whereas *flux density* is the flux per unit area; however, many authors use "flux" with the latter meaning.

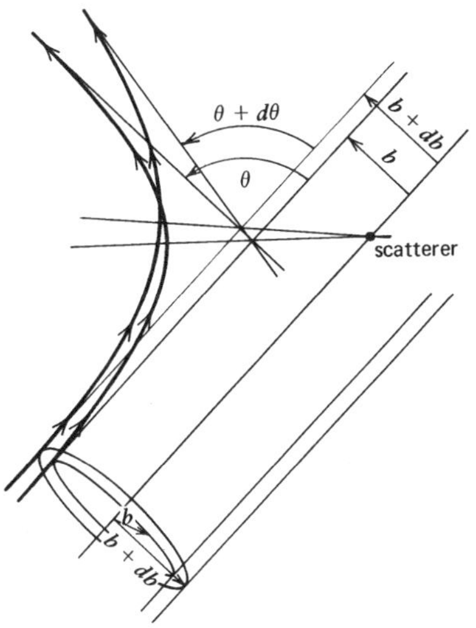

Figure 2A.2 Trajectories of particles with impact parameters between b and $b + db$ and scattering angles between θ and $d\theta$. Note that θ decreases as b increases, so that db and $d\theta$ must have opposite signs (i.e., $db/d\theta$ is negative). The ring shown at lower left is in a plane normal to the incident beam, and has an area $2\pi b\, db$.

given area must be proportional to that area. If f is the incident flux density, and $dn(b)$ the flux of particles with impact parameters between b and $b + db$, then the ratio $dn(b)/f$ must equal the area of the ring shown in Fig. 2A.2. This differential area is simply the radius increment multiplied by the circumference, so that we have

$$\frac{dn(b)}{f} = 2\pi b\, db. \tag{2A.20}$$

Since b and θ are directly related, the range of impact parameters from b to $b + db$ corresponds to a range of scattering angles from θ to $\theta + d\theta$ (cf. figure). Substituting Eqs. 2A.18 and 2A.19, we find

$$\begin{aligned}\frac{dn(\theta)}{f} &= \frac{dn(b)}{f}\\ &= 2\pi\left(\frac{ZeQ}{4\pi\epsilon_0 mv^2}\right)\cot\frac{\theta}{2}\left(\frac{ZeQ}{8\pi\epsilon_0 mv^2}\right)\frac{d\theta}{\sin^2(\theta/2)}\\ &= \pi\left(\frac{ZeQ}{4\pi\epsilon_0 mv^2}\right)^2\frac{\cos(\theta/2)}{\sin^3(\theta/2)}\,d\theta,\end{aligned} \tag{2A.21}$$

where $dn(\theta)$ is the number of particles scattered per unit time into the angles between θ and $\theta + d\theta$; we omit the minus sign of Eq. 2A.19, since we want only the absolute value of $dn(\theta)$.

We can now introduce the cross section, by analogy with Eq. 2.16. In that equation, $n\, dx$ was the number of scatterers per unit area (of a plane normal to the incident beam) in thickness dx^; here we are considering only one scatterer, so we multiply both sides by unit area and eliminate the differential in x. We can therefore write

$$\begin{aligned}\sigma(\theta) &= \frac{dn(\theta)/f}{2\pi\sin\theta\, d\theta}\\ &= \pi\left(\frac{ZeQ}{4\pi\epsilon_0 mv^2}\right)^2\frac{\cos(\theta/2)}{\sin^3(\theta/2)}\frac{1}{2\pi\sin\theta}\\ &= \left(\frac{ZeQ}{8\pi\epsilon_0 mv^2}\right)^2\frac{1}{\sin^4(\theta/2)},\end{aligned} \tag{2A.22}$$

where we have used the fact that

$$\sin\theta = 2\sin\frac{\theta}{2}\cos\frac{\theta}{2}. \tag{2A.23}$$

Eq. 2A.22 is the complete Rutherford scattering formula that we summarized in Eq. 2.19. In combination with the experimental data of Fig. 2.13, it confirms the nuclear model of atomic structure. Our heavy scatterer is the concentrated mass of the atomic nucleus, with a charge of $+Ze$ (where Z is the atomic number). The projectiles in Geiger and Marsden's experiments were α particles, with charge $Q = +2e$; since all had virtually the same energy, the initial speed v could be taken as constant.

• FURTHER READING

D'Abro, A., *The Rise of the New Physics,* Vol. 2 (Dover Publications, Inc., New York, 1951), Chapters 16–18.

Oldenberg, O., and Holladay, W., *Introduction to Atomic and Nuclear Physics* (McGraw-Hill Book Company, Inc., New York, 1967).

Richtmyer, F. K., and Kennard, E. H., *Introduction to Modern Physics,* 6th Ed. (McGraw-Hill Book Company, Inc., New York, 1969).

• PROBLEMS

1. Explain why, in the Franck–Hertz experiment, one never ordinarily observes spectral lines with frequencies greater than E_{thr}/h. Find a rationalization for the fact that sometimes the maximum frequency of the emission lines associated with a given E_{thr} is less than E_{thr}/h.

2. When a Franck–Hertz experiment is conducted with atomic sodium vapor and the energy of the electrons is increased from zero, the first visible spectral emission occurs when the energy of the electrons is 2.103 eV. What is the minimum wavelength at which one might observe atomic spectral lines in this experiment? Sodium vapor lamps in fact emit radiation in lines whose wavelengths are 589.18 and 588.99 nm. The first spectral lines of atomic lithium to appear in such an

experiment have wavelengths of 670.791 and 670.761 nm. What is the minimum voltage of the electrons at which these lines will appear?

3. The lowest electron voltage at which neon exhibits visible emission in a Franck–Hertz experiment is 18.72 V. The emission occurs at 585.24 nm. On this basis, predict the wavelengths of *two* other spectral lines that one might observe for neon.

4. The Franck–Hertz experiment is based on the collision of electrons with atoms and molecules. In some ranges of collision energies, the quantities that best characterize the probability of excitation are the duration of the interaction or, alternatively, the velocity of the electron as it passes a target atom. Presumably the transient electric field of the passing electron is the source of the coupling that passes energy from this electron to others in the target. If so, then passing protons having the same *velocity* as the passing electrons should cause similar excitation, and they often do. Compute the voltage through which a proton must be accelerated to have the same velocity as a 300-V electron. Give a general formula for the voltage V' required to give the proton the same velocity as that of a proton accelerated through a voltage V.

5. Compute the energy emitted by one atom emitting a spectral line with frequency 10^{15} s^{-1}; what is the energy of the light emitted by 1 *mole* of such atoms? Compare this with the kinetic energy of an average person walking (60 kg, 6 km/h), and with the energy of a 60-W light bulb burning 1 h. (1 W is 1 J/s.)

6. If σ of Eq. 2.8 is approximately the same as the cross section one derives from the packing of spherical atoms in a crystal, and the number density n is that of the same spheres in contact, estimate the distance required to attenuate the intensity I of an x-ray beam by 10%. Do this evaluation both by supposing $dI \approx \Delta I$, $dx \approx \Delta x$, and by carrying out the integration and using the integrated form of Eq. 2.9.

7. In an experiment of the type of Geiger and Marsden's, with α particles scattering from gold foil 2 μm thick, what fraction of the α particles would be expected to be found at angles greater than $\pi/2$, if θ_m is 1° and if Eq. 2.18 described the scattering?

8. The correction for the fact that the electron and nucleus revolve around their center of mass is made very simply; one carries through all the equations 2.47–2.65 with the mass m taken as the reduced mass. The *reduced mass* is defined, for two particles of mass m_1 and m_2, by Eq. 2.74.

Calculate the wavelengths of the spectral lines corresponding to the $n = 3 \rightarrow n = 2$ and $n = 4 \rightarrow n = 2$ transitions in the atoms of hydrogen (1.007825 amu) and of deuterium (2.01410 amu), which both have nuclear charges of $+e$. Note that these masses include the electron mass. (See Table 1.1.) Show that a comparison of the positions of atomic spectral lines can give a value for the electron mass.

9. Mercury atoms may be excited by absorbing light of wavelength 2537 Å. When excited Hg atoms are allowed to collide with N_2 molecules, light is emitted with a wavelength of 4047 Å, and no other wavelengths appear in the spectrum. Assume that the kinetic energies of excited Hg and of N_2 may be neglected prior to a collision. What is the total kinetic energy of the Hg and N_2 after a collision in which 4047 Å light is emitted?

10. Assume that the one-electron ion C^{5+} behaves like a hydrogen atom with a nuclear charge of +6. Compute the ionization energy and the wavelength of the first lines in the Lyman and Balmer series for this atom. In what spectral regions do they lie? What is the quantum number n of the lowest state of a transition in which visible light is emitted in a process $n_2 \rightarrow n_1$?

11. The two outer electrons of the magnesium atom are in the same shell. The first ionization potential of Mg is 7.644 eV and the second is 15.031 eV. Using these two values, estimate an approximate average distance of the two outermost electrons from the nucleus in the ground state of this atom.

12. Transitions to the ground state from several excited states of the sodium atom produce spectral lines that form a regular series with wavelengths 589.18, 330.26, 285.28, 268.04, and 259.39 nm. If one replaces the integral quantum number n with $n - \delta$, where δ lies between 0 and 1, this series can be described by a formula analogous to the Balmer formula. Evaluate δ for the lines given; how much variation in δ is necessary to fit these spectral lines within the accuracy quoted? (Note that excited atoms become more hydrogen-like as n increases; the Bohr model may not fit the long-wavelength lines.)

13. Suppose that a detector such as E in Fig. 2.10 subtends a solid angle of 10^{-3} steradians and is used in a scattering experiment in which a beam of fast light particles scatters from a stationary collection of heavy gaseous molecules. The system has cylindrical symmetry as in Fig. 2.10. Suppose it is believed that the scattering cross section σ for this experiment behave as $A \cos(\theta/2)$, and at $\theta = 5°$ or 0.87 radians, the differential cross section is measured to be 2.584×10^{-20} cm^2/10^{-3} sr or $2.584 \times$

10^{-17} cm^2/sr. Compute A and the total cross section from this measured value.

14. Estimate the action per cycle of
(a) A 33-rpm phonograph record;
(b) A piston in an automobile engine oscillating (approximately) harmonically at 1000 rpm;
(c) A 1-g ball bouncing perfectly elastically under the action of gravity, reaching a maximum height of 1 m on every rebound.
What is the action associated with a free hydrogen atom moving through free space for 1 s at 10^6 cm/s?

15. A laser emits 1 W of infrared radiation whose wavelength is 10.7 μm (10.7×10^{-6} m). How many quanta per second are being emitted in this beam? How long would it take to raise the temperature of 200 g of water from room temperature to its normal boiling point if all these quanta were absorbed?

16. Suppose that in another universe, $h = 10$ J s, $m_e = 1$ g, and $e = 0.1$ C. What would be the approximate diameter of a hydrogen atom in this universe? (Assume that all other masses scale as does the electron mass.)

17. Positronium is a short-lived species composed of an electron and its positive counterpart, the positron, whose mass is equal to that of the electron, and whose charge is equal in magnitude but opposite in sign. Use the Bohr model to compute the frequency of the first Lyman line for positronium. Compute the ionization energy of positronium also.

18. Ionization potentials of several atoms and ions are

Li: 5.363 eV	Na: 5.12 eV
Be^{+}: 18.12 eV	Mg^{+}: 14.96 eV
B^{2+}: 37.75 eV	Al^{2+}: 28.31 eV
C^{3+}: 64.22 eV	Si^{3+}: 44.93 eV
N^{4+}: 97.4 eV	p^{4+}: 64.70 eV

Plot the square roots of these energies against the corresponding nuclear charges, and explain the relationship as well as you can in terms of the Bohr model.

19. What must be the force constant of a harmonic oscillator if its mass is 10 amu and its frequency is 10^{14} s^{-1}? What is the natural frequency of an oscillator whose mass is 10^{-6} kg (1 mg) and whose force constant is 0.1 N/m (100 dyn/cm)?

20. A spherically symmetric harmonic oscillator is defined by the equivalent conditions

$$F_x = -kx, \qquad F_y = -ky, \qquad F_z = -kz$$

and/or

$$V(x, y, z) = \frac{k}{2}(x^2 + y^2 + z^2).$$

Calculate the action for this oscillator. Assuming that the action is quantized, what are the possible energy levels of the oscillator?

21. Use the Bohr–Sommerfeld–Wilson quantization rule, that action occurs only in integral multiples of h, to show that a particle moving along a line of length L (repulsive barriers at the ends of this segment) has only a discrete set of energy levels. Show that the possible energies are proportional to n^2 where n is an integer.

22. Use the Bohr model to find the spectrum of a single electron bound to a nucleus of neon. Calculate the wavelengths of the lowest five transitions connecting the ground state with excited states of this species.

23. When the velocity of a particle approaches the speed of light, classical mechanics no longer describes the particle accurately and one must use relativistic mechanics. As the nuclear charge Z increases, the velocity of an electron in a Bohr atom increases. At what value of Z is the velocity of an electron in a Bohr atom 1% of the speed of light? At what Z is the first excitation energy equal to 1% of the rest mass, m_ec^2, of the electron?

24. Suppose that one were to use the Bohr model to describe the motion of the earth around the sun. What would be the principal quantum number of the earth if the sun's mass is assumed infinite?

25. The harmonic oscillator is a very useful model but often we need to use a slightly less idealized way to represent a real system. For example, a real diatomic molecule dissociates, which means that, like a weight on a spring, with enough stretching, the system comes apart. The atoms dissociate, or the spring breaks, and there is no restoring force left at all. For large separations, the forces in these systems are weaker—or softer—than the harmonic forces that dominate when the systems are close to their equilibrium geometries. On the other hand, if a triatomic molecule such as H_2O or CO_2 bends, because the atoms have stiff cores that repel each other strongly, the forces are stronger—or stiffer—than the harmonic forces near equilibrium. We approximate such behavior by adding to the potential energy term $kx^2/2$ a cubic or quartic term, $k'x^3/3$ or $k''x^4/4$, or both. We call such systems "anharmonic oscillators." First, show why a model based on the addition of only a cubic term cannot be physically plausible for all distances x. Then, using a computer-based mathematical package such as MAPLE, MATHEMATICA, MACSYMA, or

DERIVE, find the trajectories for the periodic motions of the following anharmonic oscillators; that is, find the extreme values of the displacement x and the period τ of the oscillation for each of these: (a) mass = 10 dalton, force constant k = 600 newtons/meter, no cubic term, and no quartic term, and an energy of 6×10^{-20} J; how does the energy affect your result? (b) mass = 10 dalton, force constant k = 600 newtons/meter, no cubic term, and a quartic term with k'' of 4×10^{24} N/m^3, and an energy of 6×10^{-20} J; (the seemingly-large numerical value of k'' is the consequence of the displacement being only of subatomic dimensions, so that $k''x^4/4$ is not large compared with the harmonic contribution); (c) the same parameters as in (b) but with three times the energy.

26. The Bohr model is very successful indeed for reproducing many of the properties of the hydrogen atom. Niels Bohr and many of his contemporaries went on to try to use the same kind of approach to describe the helium atom, with a nuclear charge of $+2e$ and two electrons. These attempts were all unsuccessful until 1980, when Leopold and Percival found a way to construct a sort of Bohr model for this system. One of the most interesting approaches was that proposed by Irving Langmuir in 1921. This model has the two electrons of a helium atom on opposite sides of and at the same distance from the nucleus; the electrons vibrate by moving along arcs of a common circle, staying at a fixed distance from the nucleus, like two pendulums moving in synchrony. First, find the radius of the lowest orbit of a single electron attached to a helium nucleus, with $Z = 2$. Then imagine—and sketch—a Langmuir model of the helium atom with the electrons 180° apart, on opposite sides of the nucleus, both at the equilibrium distance for a single electron bound in its lowest state to He^+. Next, using a computer-based mathematical package such as MAPLE, MATHEMATICA, MACSYMA, or DERIVE, find the trajectories, i.e. the extremal angular displacements from 180° and the periods τ (and frequencies ν) for the periodic pendulum-like oscillatory motion of the electrons undergoing this motion. Remember that the interaction potential between the two electrons is their Coulomb repulsion, $e^2/4\pi\epsilon_0 r$, where r is the distance between the electrons. Carry this out for several low energies, enough to estimate an approximate harmonic frequency ν for this system. Finally, estimate the energies and displacements for states of this system if their energies are $h\nu/2$, $3h\nu/2$, and $5h\nu/2$.

CHAPTER 3

Matter Waves in Simple Systems

We have now examined most of the evidence that led to the modern quantum theory of matter. In this chapter we begin to develop that theory itself, the theory known as *quantum mechanics* or (in the form presented here) *wave mechanics.*

The fundamental hypothesis of this theory is that matter, as well as light, has both wavelike and particlelike properties. The whole of classical mechanics dealt with matter as made of particles, whereas nineteenth-century physics generally assumed light to be simply a wave phenomenon. In the last chapter we introduced the evidence that light has particlelike, quantized properties; now we must describe the wavelike properties of matter. We first develop the nature and properties of waves in general, and then formulate the equation that describes "matter waves"—the famous Schrödinger equation. As we shall see, this equation can be rationalized but it cannot be derived; it must be postulated. Its justification is the agreement of its predictions with experiment.

In principle (but not in practice), it should be possible to derive all the properties of matter from the Schrödinger equation. Even for all but the simplest atoms, however, the calculations involved are formidable. In the present chapter we consider only some idealized systems simple enough to treat in detail, that is, systems whose Schrödinger equations can be solved exactly: the free particle in space, and particles constrained by "boxes" of various simple shapes. These solutions introduce many of the principles of quantum mechanics and illustrate the kinds of calculations required to describe matter waves. In particular, we show how the quantization of energy is related directly to the boundary conditions on a system.

3.1 The de Broglie Hypothesis

We have already seen some strong reasons for assuming that light has particlelike as well as wavelike properties—in particular, the photoelectric effect and the concept of the quantum of radiation. In 1923, in an elegant and daring extrapolation, Louis de Broglie proposed that matter should exhibit a parallel to this dual character of light. Just as light can show both particlelike and wavelike properties, matter, he suggested, might well exhibit wavelike properties. de Broglie proposed this hypothesis despite the fact that only the particlelike properties of matter were actually known at the time. Indeed, the theory of quantum mechanics was well developed before experimental proof of "matter waves" was obtained.

This proof was found in 1927 by Davisson and Germer, who showed that electrons impinging on a crystal exhibit a property that had always been associated with waves: *diffraction.* Diffraction appears as a pattern of light and dark areas or areas of high and low intensity, often regular, produced when light waves or other waves encounter some obstacle as they travel toward a screen or detector. (An easy way to observe diffraction is this: With one eye closed, look at a lamp some distance away. Then, holding your hands just beyond your nose, bring the tips of your forefingers *almost* together in front of your open eye. When the blurred edges of your out-of-focus fingertips seem to merge, you will see rather sharply defined lines of light and dark, roughly parallel to the silhouetted edges of the fingertips. This light and dark pattern is a diffraction pattern.) As can be seen in Fig. 3.1, electrons are diffracted by crystals in much the same way as are x-rays; the dimensions of the patterns are related to crystal structure, as we shall explain in Chapter 11. Similar effects can be obtained with much heavier particles, and neutron diffraction is particularly useful in structural analysis.

But let us return to the actual formulation of de Broglie's hypothesis. If there are waves associated with particles of matter, what are their wavelengths? de Broglie drew an analogy to the particles of radiation (photons). By Einstein's relation $E = mc^2$, the mass associated with a photon of frequency ν must be $E/c^2 = h\nu/c^2$; the momentum of the photon is thus mass × velocity, or $h\nu/c = h\lambda$. What de Broglie

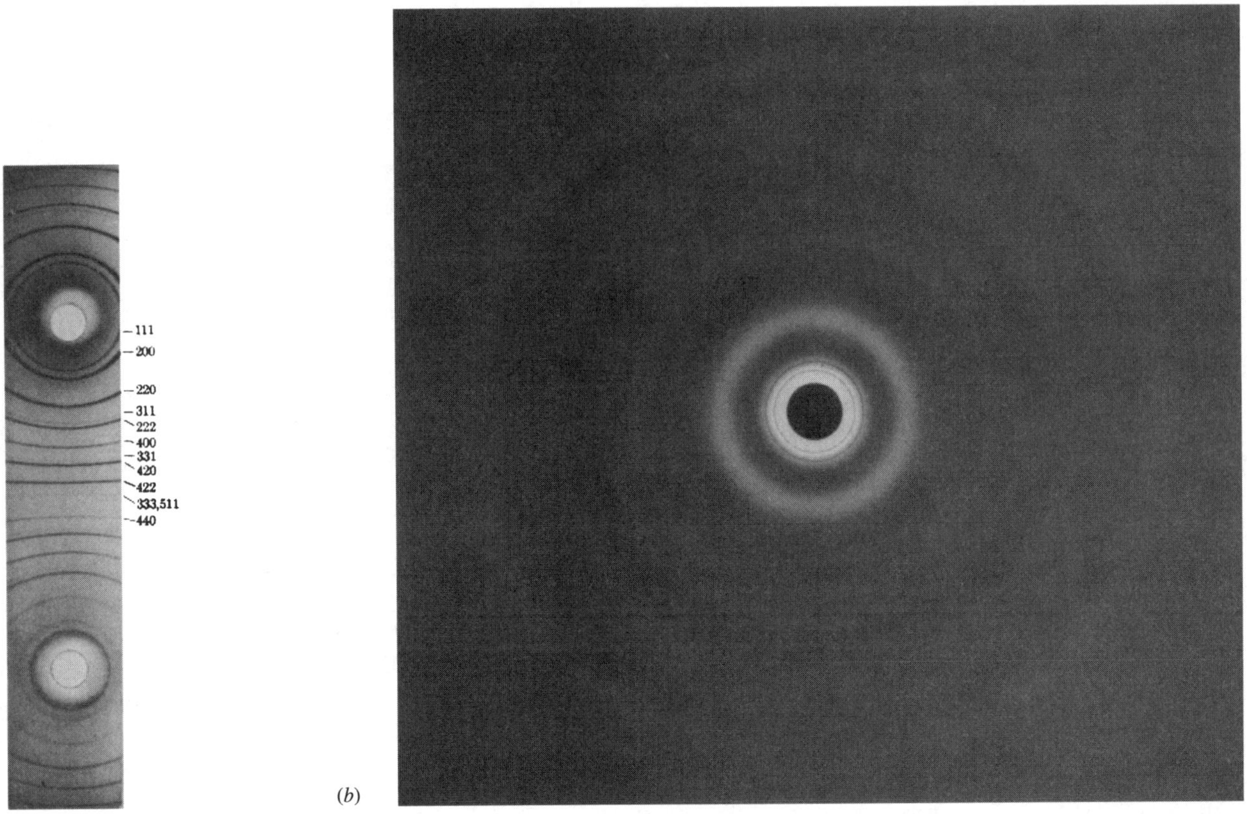

(*a*) (*b*)

Figure 3.1 Diffraction patterns for (*a*) x-rays, (*b*) electrons. The x-ray photograph is a print of a strip of film from a cylindrical film holder. The x-ray beam enters through the hole at the bottom and leaves through the hole at the top. The sample is sodium chloride. The electron diffraction photograph of chlorobenzene is a print from a plate large enough to show the full circles of the diffraction pattern; most of these circles are cut off in the x-ray picture. (x-ray picture courtesy of Prof. S. Bailey; electron diffraction picture courtesy of Prof. Denis Kohl.)

proposed was that the same equation applied to material particles. The *de Broglie wavelength* of a particle is thus simply Planck's constant divided by the particle's momentum *p:*

$$\lambda = \frac{h}{p}. \tag{3.1}$$

Note that it is the wavelength, not the frequency of the wave, that we relate explicitly to the momentum; we must return to this after discussing waves in general, for it is a very important point, which forms part of the basis of our physical picture of atoms.

What sort of wavelengths do material particles have? In gaseous hydrogen at room temperature an average molecule moves at about 1.77 km/s; since the H_2 molecule weighs 3.35×10^{-27} kg, its wavelength is

$$\begin{aligned} \lambda &= \frac{h}{mv} \\ &= \frac{6.63\times10^{-34}\ \mathrm{J\,s}}{(3.35\times10^{-27}\ \mathrm{kg})(1.77\times10^{3}\ \mathrm{m/s})} \times 1\frac{\mathrm{kg\,m/s^2}}{\mathrm{J}} \\ &= 1.12\times10^{-10}\ \mathrm{m}, \end{aligned}$$

or about 1.1 Å, the wavelength of a "hard" x-ray. High-speed electrons also have wavelengths in the x-ray range, accounting for the similarity of x-ray and electron diffraction patterns. The slower or lighter a particle, the longer is its wavelength; the heavier or faster the particle, the shorter is its wavelength. Macroscopic objects thus have extremely small wavelengths; for example, a thrown baseball would have $\lambda \approx 2 \times 10^{-26}$ m, compared with which even an atomic nucleus is huge.

By analogy with the familiar properties of light, we should expect to observe wavelike phenomena, such as diffraction, when matter waves are forced to interact with something having dimensions comparable to their wavelength. A beam of light will be strongly diffracted by a slit if the width of the slit is roughly equal to the wavelength of the light. Naturally, diffraction does occur even with much wider slits, or even with a single sharp edge. However, if a slit is significantly wider than the wavelength, then the light streams through in effectively straight rays with negligible deflection. In the same way, as long as all the dimensions of an apparatus are much greater than the wavelength of the particles under examination, we can treat the particles as though they obeyed the laws of classical mechanics. The estimates above tell us that material "particles" of atomic size will show wavelike properties only when they interact with other particles of roughly the same size, whereas macroscopic objects will have no detectible wavelike properties at all. By contrast, visible light, with wavelengths of several thousand angstroms, can be made to exhibit wavelike properties with such simple pieces of apparatus as a slit made from a pair of razor blades (or even two fingertips!).

Before we can explore how de Broglie's hypothesis of matter waves leads us to the structure of the atom, we must turn aside to study the properties of waves in general—what they are, how they are represented, and how they interact.

3.2 The Nature of Waves

The first images that come to mind for most of us with the word *waves* are water waves. Unfortunately, despite our familiarity with them, water waves are rather complicated, and are not easy models to use. Nevertheless, one of their characteristics is common to most types of waves: The wave is a moving *disturbance* but not a gross motion of the medium. A small bit of water at the surface of a pond does not travel across the pond as a wave passes by; rather, it moves up, down, and a bit backward and forward. It is easy to recognize this when one thinks of the motion of a small stick floating on the surface of the pond. The stick bobs with the waves, but it does not travel any distance.

Now let us speak in more general terms. In the broadest definition, a wave is a disturbance that varies regularly with time. We can immediately think of a classification into two types of waves, *traveling waves* and *standing waves.* The ripples that spread on a pond when a stone is thrown in are traveling waves: The waves themselves propagate outward in concentric circles, regardless of the motion of the floating stick. A simple standing wave almost as familiar as water waves is the wave in a moving jump rope held at both ends and swung in a loop, as shown in Fig. 3.2*a.* As the rope moves around, any given point on the rope describes a circle about the axis of rotation, always maintaining a fixed distance from the axis. Ordinarily a jump rope has only its end points on the axis of rotation. However, most people have noticed that it is possible to set the rope in motion so that one, two, or more points on the rope are fixed on the axis. In other words, one can have the rope rotating in two parts, as in Fig. 3.2*b,* or even in three or more parts. The point or points that remain fixed on the axis, that is, that have no displacement from the axis, are called *nodes,* and the points of maximum displacement from the axis, *antinodes.* In a standing wave the nodes remain fixed in space, and the antinodes at most oscillate (in this case circularly) around fixed positions.

The standing-wave motion just described happens to be circular motion: As a given point on the rope moves about the axis, its behavior is completely described by a statement giving its angle about the axis as a function of time. However, motion need not be circular to be standing-wave motion. For example, an elastic cord with its ends fixed could be stretched into one of the shapes of Fig. 3.2; on being released, the cord would oscillate up and down, but only within the plane of the figure. The characteristic that makes this a standing-wave motion is that the nodes always

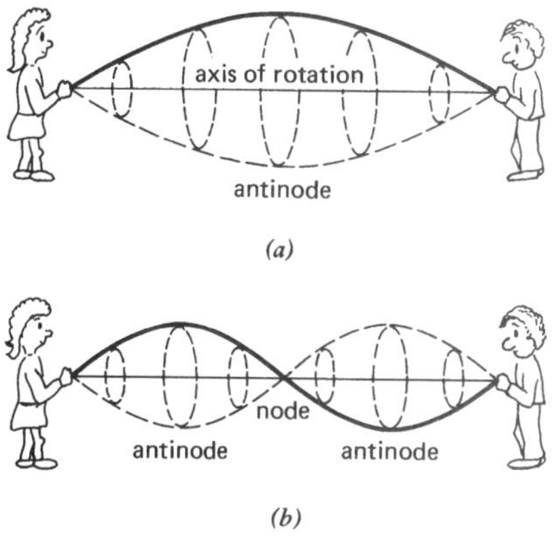

Figure 3.2 Standing waves in a jump rope (*a*) with no nodes except at the ends; (*b*) with one node in the middle.

remain nodes, the antinodes always remain antinodes, and all the sections between them always retain their own positions of displacement relative to neighboring sections. The vibrations of the strings in a violin or other stringed instrument are of this type.

One can also easily set up a traveling wave in a rope. If one end is free and the other is given a shake or twirl, a wave will move down the length of the rope. If one shakes the end repeatedly in a regular way, a steady flow of waves down the rope can be established. A traveling wave also has its nodes and antinodes, but they move steadily in some direction rather than remaining fixed.

One other important variety of waves occurs in material media, best exemplified by sound waves. These arise from the motion of the atoms or molecules of the medium, and give rise to the sensations we call sounds when they strike our ears. In a sound wave the molecules are displaced primarily back and forth along the direction in which the wave moves forward (propagates). Such a wave is called a *longitudinal* wave. The waves in a jump rope are examples of *transverse* waves, in which the individual particles are displaced in a direction perpendicular to the axis of propagation. Water waves are partially transverse and partially longitudinal, one aspect that makes them complicated.

Thus far all our examples have been waves in a medium: a rope, water, or air. But there are other waves that we can describe without reference to a medium. The example that comes to mind first is, of course, a light wave. What we speak of as a wave of light or other electromagnetic radiation is in fact a field of force that propagates as a wave. One can detect such an electromagnetic wave by observing the oscillations of suitable test charges in the space where the wave is supposed to be. For example, we use the electrons in a metal aerial to probe for the electromagnetic waves that we call radio waves. Both radio and light waves propagate perfectly well through the best vacuum known, namely, outer space, where there is no question of a material medium. One of the great scientific challenges at the turn of the century was to discover if there was in fact an all-permeating medium, the so-called luminiferous ether, which was supposed to act as the carrier of light waves. However, experiments[1] showed that there was no detectable medium associated with electromagnetic waves, so the idea of the luminiferous ether has gone the way of phlogiston and the *élan vital.*

Roughly speaking, then, a traveling wave is a disturbance that propagates in some direction or directions, leaving the medium (if there is one) essentially unchanged. A standing wave is one that repeats itself in time with no net forward motion. How do we describe these more precisely in mathematical terms? For a traveling wave that maintains its shape as it moves, this is very simple indeed. Let us choose a wave moving in a single direction for simplicity. Any function of the variable $x - \upsilon t$ describes a traveling wave, where x is the distance, t the time, and υ the speed with which the wave moves forward. Figure 3.3 illustrates such a disturbance moving a distance $x_2 - x_1$ in the time interval $t_2 - t_1 = (x_2 - x_1)/\upsilon$.

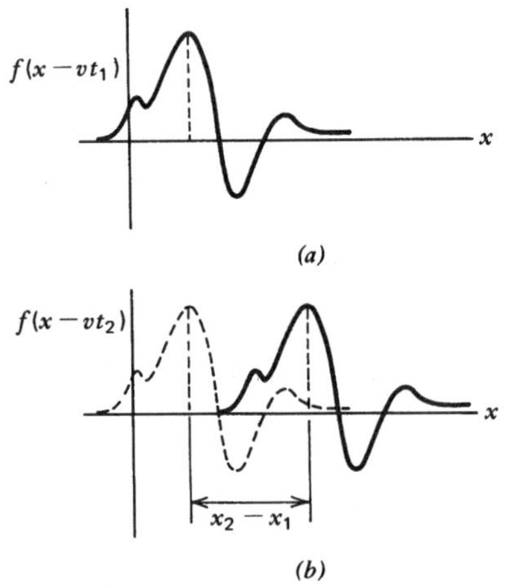

Figure 3.3 A wave whose amplitude is described by the function $f(x - \upsilon t)$, shown at two times: (*a*) at time t_1; (*b*) at time t_2, when each point of the wave has moved a distance $x_2 - x_1 = \upsilon(t_2 - t_1)$, as illustrated for the highest peak.

We have expressed the wave in terms of the variable $x - \upsilon t$, with dimension of length. It is usually more convenient to multiply this by a constant with the dimension of $(\text{length})^{-1}$ to obtain a dimensionless variable with two characteristic scale factors, one for distance and one for time. The representation commonly used has the form

$$\frac{1}{\lambda}(x - \upsilon t) = \frac{x}{\lambda} - \nu t \qquad \left(\nu \equiv \frac{\upsilon}{\lambda}\right), \tag{3.2}$$

in which the scale of length is defined by the *wavelength* λ and the scale of time by the *frequency* ν. If λ is expressed in meters and υ in meters per second, then ν has the units s^{-1} (hertz). The reciprocal of the wavelength is the *wave number* $\tilde{\nu} \equiv 1/\lambda$, usually expressed in cm^{-1}. It is often convenient (cf. our jump-rope example) to describe a wave in terms of some equivalent circular motion. For this purpose we can represent the wave as a function of the variable

$$k(x - \upsilon t) = kx - \omega t \begin{cases} \left(k \equiv \dfrac{2\pi}{\lambda} = 2\pi\tilde{\nu}\right), \\ (\omega \equiv 2\pi\nu = k\upsilon) \end{cases} \tag{3.3}$$

in which k is the *circular wave number* and ω is the *angular frequency,* with SI units of radians per meter and radians per second, respectively.

The words *wavelength* and *frequency* naturally suggest a *periodic* wave, one that repeats itself in space or time. Until now, however, we have made no assumption of periodicity for traveling waves, and many waves—such as tsunamis—are not periodic. We have considered only a disturbance that maintains its shape as it moves forward, and that may or may not be periodic. If it is *not* periodic, of course, the quantities λ, ν, and so on, are simply arbitrary scale factors with no particular physical meaning (except that one must always have $\upsilon = \lambda\nu$). Nevertheless, any such disturbance can be described mathematically in terms of a single variable, which can be $x - \upsilon t$ or the forms of Eqs. 3.2 and 3.3. In the circular representation the single variable is most conveniently expressed as

$$\varphi \equiv kx - \omega t, \tag{3.4}$$

called the *phase* of the wave. Any disturbance described by a function $f(x, t)$ that is also a unique function of some phase,

$$f(x,t) = g(\varphi), \tag{3.5}$$

is a wave, no matter what the form of $f(x, t)$.

What about a wave that *is* periodic—as are those which will concern us here? All the symbols in Eqs. 3.2 and 3.3 are then physically meaningful. The wavelength is the distance over which the wave repeats itself, and the frequency is the number of such replicas that pass a given point per unit time. The time required for a single wavelength to pass—that is, the time between two successive wave crests—is the inverse of the frequency and is called the *period,* $\tau \equiv 1/\nu$. In the circular representation, the period of course corresponds to a single trip around the circle, at the angular velocity ω; the phase φ is an angle and thus varies through 2π in the course of a period.

All this can be recognized as the same language that we used to describe the harmonic oscillator in Section 2.8. The oscillator's motion is given by a sine function, Eq. 2.40. One can show that *any* periodic motion—including waves—can be analyzed into a sum of such sine (or cosine) functions; this is why the harmonic oscillator is such a useful model. A wave of the simple form

$$f(x,t) = A\sin(kx - \omega t + \delta), \tag{3.6}$$

[1] The key experiment was that of Michelson and Morley (1887), showing that the earth's motion relative to the hypothetical ether did not affect the observed velocity of light. This result was an important foundation for Einstein's theory of relativity.

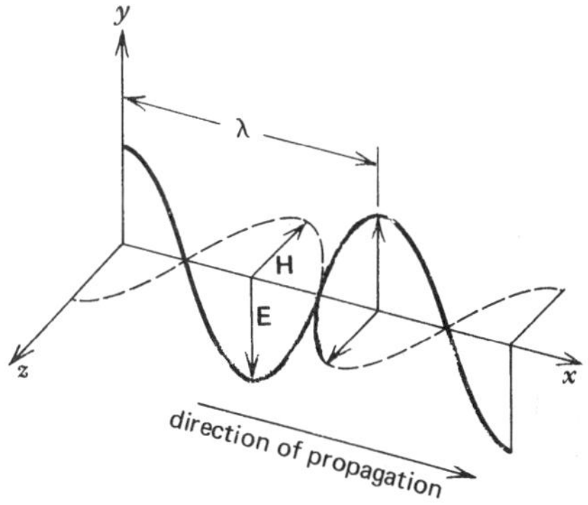

Figure 3.4 An electromagnetic wave. The wave shown is plane-polarized, with the electric-field vector **E** restricted to the xy plane and the magnetic-field vector **H** restricted to the xz plane. (In general, both vectors will rotate about the x axis, but remain at right angles and in phase.) Both **E** and **H** are harmonic (sine) waves of wavelength λ.

where A and δ are constants, is thus called a *harmonic* wave (or sine wave). Electromagnetic waves are of this form, with the electric and magnetic fields in phase but at right angles to each other, as shown in Fig. 3.4.

Standing waves can also be periodic. In our example of a rope fixed at the ends, the waves at any time have the shape of a sine curve. The wavelength can thus be defined as twice the distance between nodes. Since there are nodes at the ends, the rope must contain an integral number of half-wavelengths; this gives a hint of how quantization will enter our theory. We shall see in Section 3.5 that a standing wave is simply the resultant of two identical traveling waves going in opposite directions.

3.3 Dispersion Relations and Wave Equations: The Free Particle

The functions $f(x, t)$ that satisfy Eq. 3.5 are characterized primarily by their values of the physical parameters k and ω—that is, by their wavelengths and frequencies. The wavelength and frequency are not independent, of course, but are connected by the relation

$$\nu\lambda = \frac{\omega}{k} = \upsilon, \tag{3.7}$$

where υ is the speed at which the wave propagates. For light waves in a vacuum this speed is the universal constant $c = 2.99792458 \times 10^8$ m/s, and the relation between λ and ν is straightforward. But what is the velocity of propagation of a matter wave?[2] Fortunately, we do not need this, but can infer it. Whenever a phenomenon can be described in terms of waves, there is always a relation connecting the wavelengths and the frequencies, a *dispersion relation,* that is the key to the behavior of those waves.

What, then, is the dispersion relation for matter waves? Let us first examine a free particle in motion. According to de Broglie's hypothesis, the wavelength of a matter wave must be $\lambda = h/p$, where p is the momentum of the particle. Suppose we also assume that the frequency of this matter wave is related to the energy of the particle by the same law that governs the energy of a photon, the Planck-Einstein relation $E = h\nu$. The energy of a free particle of mass m and velocity υ is simply its kinetic energy,

$$E = \frac{m\upsilon^2}{2} = \frac{p^2}{2m} \tag{3.8}$$

(since $p = m\upsilon$). Combining these three conditions, we obtain

$$h\nu = \frac{p^2}{2m} = \frac{1}{2m}\left(\frac{h}{\lambda}\right)^2, \tag{3.9}$$

which we can rearrange to give our desired dispersion relation in the form

$$\nu = \frac{h}{2m\lambda^2} \tag{3.10}$$

or (with $\hbar \equiv h/2\pi$)

$$\omega = \frac{\hbar k^2}{2m}. \tag{3.11}$$

In the more general case, if a particle is subject to external forces that can be expressed in terms of a potential energy $V(x, y, z)$—for example, an electron in the field of a nucleus—then we can generalize Eq. 3.8 to

$$E = \frac{p^2}{2m} + V(x, y, z), \tag{3.12}$$

from which we obtain the slightly more complicated dispersion relation

$$\omega = \frac{\hbar k^2}{2m} + \frac{V(x,y,z)}{\hbar}. \tag{3.13}$$

Do not get the idea that we have "derived" these equations from fundamental principles. Let us review just what we have done. Experiment shows that matter has wavelike properties; we interpret this as meaning that *something* associated with matter can be described as a wave. We do not know what that "something" is, and we are only guessing at the properties of the wave. The guesses are plausible—that matter waves obey the equations $\lambda = h/p$ and $E = h\nu$ that

[2] Note that the velocity of the waves that constitute a piece of matter (*phase velocity*) is in general not the same as the velocity at which the matter as a whole moves (*group velocity*); υ stands for phase velocity in Eq. 3.7, and for group velocity in $\lambda = h/m\upsilon$.

apply to light waves—but are still unproven. In fact, they are *postulates,* which can be justified only by the conclusions drawn from them. Eqs. 3.11–3.13 are consequences of these postulates, which now must be tested. This is a point that we shall emphasize again and again.

Of what use is a formal dispersion relation when we want to obtain physical information about a system? That is, how do we make use of a relation like Eq. 3.13 to find out how an electron behaves in an atom? We must discover the answer to this question in stages. To begin, having decided to describe matter as a wave phenomenon, we must find a way to determine the amplitude of a matter wave at an arbitrary point in space and time. Therefore we must find some way to obtain an expression for this wave amplitude, or *wave function,* which is usually designated by the Greek letter Ψ (psi). The amplitude must be a function of position and time; we need an equation, a differential equation, whose solution is this function. Again, we do not derive this wave equation; we postulate it, requiring it to be consistent with the dispersion relation.

Why do we not work directly with the algebraic equations, Eq. 3.11 or Eq. 3.13, rather than use the more complex differential equation? The answer is that dispersion equations make no reference to the amplitude of a wave, the quantity we are trying to find. Nevertheless, we can use them to determine the form of a differential equation[3] whose solution will give the wave amplitude we seek. Specifically, given any dispersion relation, one can find an equation involving derivatives of the wave amplitude Ψ from which the dispersion relation can be derived. In general, there may be more than one wave equation consistent with the dispersion relation we want to satisfy; if this happens, we try to pick the simplest one.

Let us extract a wave equation for the simple case of a free particle in motion. For simplicity, we restrict the particle to motion in only one dimension, along the x axis. We then need an equation in terms of a wave amplitude $\Psi(x, t)$, describing a wave that obeys the dispersion relation of Eq. 3.11. We must try to find the simplest equation that relates a space derivative of Ψ, if possible only the first or second derivative, to a corresponding time derivative of Ψ. The equation should have a solution that is physically meaningful and consistent with Eq. 3.11.

This assertion, that the wave equation should involve only simple derivatives, may seem arbitrary. Remember that the validity of our theories rests only upon our obtaining results in agreement with experiment. One reason for trying to stay with the simplest equation is the principle known as Occam's razor; always choose the simplest way to achieve the task. However in this case, we can also show why the simplest choice is the most plausible. One obvious reason why it is desirable to restrict ourselves to first and second derivatives of Ψ is that it makes the problem reasonably simple mathematically. More significant is the reason that in classical mechanics, one needs to specify only two conditions, for example, the initial position and velocity, to determine the entire future history of a particle. This is because Newton's second law, $F = ma$, is an equation involving a *second* derivative, so that two integration constants must be supplied to fix the solution. We would like our quantum mechanics to be economical and as nearly parallel to classical mechanics as possible, because we know that classical mechanics does apply to a vast part of our experience. Invoking third or higher derivatives would be unpleasant on both counts. We shall quickly see that it is possible to restrict ourselves to first and second derivatives.

How, then, must the derivatives enter into a differential equation? Let us first consider the forms of the derivatives themselves. Let us assume that the wave amplitude can be written as an explicit function of the phase,

$$\Psi(x,t) = \Psi(\varphi) = \Psi(kx - \omega t); \tag{3.14}$$

as mentioned before, such an equation describes any traveling wave that maintains its shape as it propagates. (Note that the sign of k determines the direction in which the wave moves.) We can thus apply the simple "chain rule" of differentiation to take partial derivatives[4] of Ψ with respect to x or t. If we want to differentiate Ψ with respect to x at constant t, we take the derivative with respect to the phase φ and multiply this by the partial derivative of φ with respect to x; that is,

$$\left(\frac{\partial \Psi}{\partial x}\right)_t = \frac{d\Psi}{d\varphi}\left(\frac{\partial \varphi}{\partial x}\right)_t. \tag{3.15}$$

Similarly, the derivative with respect to t at constant x can be written as

$$\left(\frac{\partial \Psi}{\partial t}\right)_x = \frac{d\Psi}{d\varphi}\left(\frac{\partial \varphi}{\partial t}\right)_x. \tag{3.16}$$

But the partial derivatives of φ are very simple; differentiation of Eq. 3.4 gives

$$\left(\frac{\partial \Psi}{\partial x}\right)_t = k \quad \text{and} \quad \left(\frac{\partial \varphi}{\partial t}\right)_x = -\omega. \tag{3.17}$$

Equations 3.15 and 3.16 therefore become

$$\left(\frac{\partial \Psi}{\partial x}\right)_t = k\frac{d\Psi}{d\varphi} \quad \text{and} \quad \left(\frac{\partial \varphi}{\partial t}\right)_x = -\omega\frac{d\Psi}{d\varphi}. \tag{3.18}$$

[3] One can also formulate quantum mechanics in terms of integral equations, but the differential equation approach is simpler and was developed first. Ever since Newton, most physical laws had been expressed as differential equations and thought of in terms of changes in variables, and a great deal was known about how to solve such equations.

[4] If you are not familiar with partial derivatives, see Appendix II at the back of the book.

Notice that, since Ψ is a function only of φ, we can use ordinary derivative notation (d rather than ∂) for the derivative of Ψ with respect to φ.

How do we combine these derivatives in a wave equation? We know from the dispersion relation, Eq. 3.11, that ω must be proportional to k^2. Our differential equation must therefore involve the first time derivative of Ψ, to introduce a factor of ω, and the second space derivative of Ψ, to introduce a factor of k^2. The simplest relationship that will give the right proportionality is

$$\left(\frac{\partial \Psi}{\partial x}\right)_x \propto \left(\frac{\partial^2 \Psi}{\partial x^2}\right)_t, \tag{3.19}$$

which by Eq. 3.18 is equivalent to

$$-\omega \frac{d\Psi}{d\varphi} \propto k^2 \frac{d^2\Psi}{d\varphi^2}. \tag{3.20}$$

This is not quite a wave equation, but is very close to it.

We still need to know how Ψ depends on φ. Since no other variable appears in Eq. 3.11, $d\Psi/d\varphi$ and $d^2\Psi/d\varphi^2$ must cancel out (except perhaps for a constant factor). What kind of function $\Psi(\varphi)$ has essentially the same form when it is differentiated once as when it is differentiated twice? The only such function is the exponential: The first and second derivatives of e^{φ} with respect to φ are both just e^{φ} itself. Could Ψ then be just e^{φ} or $e^{-\varphi}$? Formally this is possible, but physically it would be nonsense. The reason is that the function e^{φ} becomes infinite as x goes to $+\infty$, and $e^{-\varphi}$ as x goes to $-\infty$, regardless of the value of t. This would mean that the wave amplitude Ψ would be larger at infinity than at any point in accessible space. We anticipate by saying that the physical significance of Ψ must somehow be related to the chance of finding the particle at a particular point in space. If Ψ were larger at infinity than at any accessible x, then the particle would be found more probably at infinity than at any other place; that is, one could not find it at all. To avoid this problem, we require for Ψ a functional form that remains *bounded*—a functional form that prohibits $|\Psi|$ from exceeding some maximum value.

Fortunately, there does exist a simple function that is bounded yet still exponential. Remember that the complex exponential $e^{i\varphi}$ is equivalent to $\cos \varphi + i \sin \varphi$, in which both terms are clearly bounded. The function $\Psi(\varphi)$ can thus have the form $e^{i\varphi}$ or $e^{-i\varphi}$ and satisfy our mathematical conditions. But what is the physical meaning of a wave with a complex amplitude? On this point let us wait and see what kind of results we can obtain. We therefore write

$$\Psi \propto e^{\pm i\varphi}. \tag{3.21}$$

In the equations that follow, the upper sign (in $\pm$ or $\mp$) corresponds to the choice of $e^{+i\varphi}$ in Eq. 3.21. Given this form for $\Psi(\varphi)$, we obtain from Eqs. 3.18 the partial derivatives

$$\left(\frac{\partial \Psi}{\partial t}\right)_x = -\omega(\pm i\Psi) = \mp i\omega\Psi \tag{3.22}$$

and

$$\left(\frac{\partial^2 \Psi}{\partial x^2}\right)_t = k^2 \frac{d^2\Psi}{d\varphi^2} = -k^2\Psi. \tag{3.23}$$

Substituting for ω in terms of k from Eq. 3.11, we obtain the wave equation[5]

$$\left(\frac{\partial \Psi}{\partial t}\right)_x = \mp i\Psi \frac{\hbar k^2}{2m} = \pm \frac{i\hbar}{2m}\left(\frac{\partial^2 \Psi}{\partial x^2}\right)_t. \tag{3.24}$$

Changing the sign of $(\partial\Psi/\partial t)$, corresponds to reversing the direction of propagation of the wave. Since we can describe this by simply allowing k to have both positive and negative values, the lower signs in the above equations are redundant and can be omitted. Thus our final wave equation for a free particle moving in one dimension is

$$\left(\frac{\partial \Psi(x,t)}{\partial t}\right)_x = \frac{i\hbar}{2m}\left(\frac{\partial^2 \Psi(x,t)}{\partial x^2}\right)_t. \tag{3.25}$$

Remember that we still have no proof that this wave equation is correct. We have made one assumption after another. First we postulated that matter waves obeyed the equations $\lambda = h/p$ and $E = h\nu$, on the basis of which we obtained the dispersion relation of Eq. 3.11. Then we assumed that the waves were described by a simple differential equation. The simplest form that *could* give Eq. 3.11 was that of Eq. 3.19, and $e^{\pm i\varphi}$ is its simplest solution that *does* give Eq. 3.11. All this is plausible, but all it proves is the *possibility* that our result is correct. The real proof of its correctness is that it predicts no results inconsistent with experiment.

Let us now consider the wave function itself. It has the form

$$\Psi(x,t) = Ae^{i\varphi} = Ae^{i(kx-\omega t)}, \tag{3.26}$$

direct substitution of which in Eq. 3.25 gives us back Eq. 3.11 directly. The constant A is the scale factor or normalization factor for the function Ψ, and its value is of no concern to us as yet; the functional form of Ψ *is* of great interest to us. It is very important to recognize that it takes a *complex* function, a function with an imaginary as well as a real part, to yield the dispersion relation that we inferred

[5] Since $(\partial^2\Psi/\partial t^2)_x = -\omega^2\Psi$ we can also write

$$\left(\frac{\partial^2 \Psi}{\partial x^2}\right)_t = \frac{k^2}{\omega^2}\left(\frac{\partial^2 \Psi}{\partial t^2}\right)_x = \frac{1}{v^2}\left(\frac{\partial^2 \Psi}{\partial t^2}\right)_x,$$

which is the general equation obeyed by any kind of wave Ψ moving along the x axis at a speed $v = \omega/k$. This equation is the basis of our statement that any unique function of $\varphi(\equiv kx - \omega t)$ is a wave.

from the Planck-Einstein and de Broglie relationships. We can see this by writing Ψ in the form

$$\Psi(\varphi) = A(\cos\varphi + i\sin\varphi) \tag{3.27}$$

Here $A\cos\varphi$ is the real part and $A\sin\varphi$ is the imaginary part, the part multiplied by i; each of these terms is a harmonic or sine wave like Eq. 3.6. But Ψ itself is not just a simple sine wave. It is composed of two pieces, a cosine term and a sine term, which have the same shape but are 90° out of phase with each other. The cosine term is real and the sine term imaginary. (Strictly speaking, the constant A can also be complex—that is, it can have both real and imaginary parts. But we are at liberty to choose A as real, and we do so.) Thus, at a particular time t_1, the real part of Ψ (Re Ψ) has a spatial variation proportional to $\cos(kx - \omega t_1)$, whereas the imaginary part of Ψ (Im Ψ) varies as the sine of the same argument, $\sin(kx - \omega t_1)$. Conversely, at a fixed point in space x_1, Re Ψ varies with time as $\cos(k_x - \omega t)$, whereas Im Ψ varies with time as $\sin(kx_1 - \omega t)$.

To summarize, we have constructed a wave equation, Eq. 3.25, describing the wave associated with a free particle moving in the x direction. We have found a solution of this equation, $\Psi(x, t)$, given by Eq. 3.26. Equation 3.25 is a form of what is usually called the *time-dependent Schrödinger equation,* after Erwin Schrödinger, who in 1926 first developed it from the stimulus provided by de Broglie's relation. For a particle moving freely in a space of three dimensions, the Schrödinger equation would have the form

$$\left(\frac{\partial\Psi}{\partial t}\right)_{x,y,z} = \frac{i\hbar}{2m}\left[\left(\frac{\partial^2\Psi}{\partial x^2}\right)_{y,z,t} + \left(\frac{\partial^2\Psi}{\partial y^2}\right)_{x,z,t} + \left(\frac{\partial^2\Psi}{\partial z^2}\right)_{x,y,t}\right], \tag{3.28}$$

in which the partial derivatives of Ψ with respect to x, y, and z correspond to the kinetic energy associated with the three components of momentum. The wave function Ψ must then be a function of x, y, z, and t. We shall examine the three-dimensional problem in more detail in connection with confined particles.

3.4 Operators

Let us now approach the wave equation from a somewhat different direction. We want to develop a complete theory of matter, including particlelike as well as wavelike properties; indeed, the theory must yield the laws of classical mechanics in the macroscopic limit. We must therefore relate the wave equation to the mechanical properties of particles. For simplicity we continue to consider the free particle moving in one direction, but the principles we shall introduce have quite general validity.

We obtained the wave equation (Eq. 3.25) by requiring that it satisfy the dispersion relation 3.11, which was written in terms of the wavelike properties described by k and ω (wavelength and frequency). However, we could as well have emphasized the particlelike properties of energy and momentum, since our fundamental assumption is that a "particle" has both sets of properties. From the Planck-Einstein and de Broglie relations, we have for the energy and momentum

$$E = h\nu = \hbar\omega \quad \text{and} \quad p = \frac{h}{\lambda} = \hbar k. \tag{3.29}$$

Substituting for ω and k in Eqs. 3.22 (upper sign only) and 3.23, we obtain

$$i\hbar\left(\frac{\partial\Psi}{\partial t}\right)_x = E\Psi \tag{3.30}$$

and

$$-\hbar^2\left(\frac{\partial^2\Psi}{\partial x^2}\right)_t = p^2\Psi. \tag{3.31}$$

Now we go one step beyond the dispersion relation. Let us *identify* the operations on the left side of Eq. 3.30, time differentiation of Ψ followed by multiplication by $i\hbar$, with the operation on the right side, multiplication of Ψ by the energy E. To generalize, let the symbol E represent the set of operations on Ψ that are equivalent to multiplying Ψ by the energy. Thus we define

$$\mathsf{E} \equiv i\hbar\frac{\partial}{\partial t}, \tag{3.32}$$

with the partial derivative understood to be taken with all spatial variables held constant. Here E is an example of a symbolic representation of a set of instructions to be carried out by operation on some understood thing, in this case the function Ψ representing the wave amplitude. As such, it is an example of an *operator.*[6] The concept of operators is fundamental to quantum mechanics. First we identify operators with various classical quantities like E, then we can rewrite the equations of classical mechanics in terms of the corresponding operators. We shall use sans serif type to denote operators in general; thus E is the operator counterpart of the total energy E, and V will be the operator counterpart of the potential energy V.

First let us say a little about the algebra of operators. When more than one operator is applied in succession, their symbols can be written together as in multiplication: The

[6] More familiar examples of operators are "+" or "÷," denoting the operations of addition and division. Addition and division happen to be defined as operators acting on real or complex numbers; addition is also defined as an operation on vectors, but division is not. Both addition and division are also defined for a function such as Ψ, whose value must be a real or complex number. For multiplication we have the operators "·" and "×," which have the same meaning for numbers but different meanings for vectors: cf. Eqs. 1.17 and 2.21.

expression $\mathsf{AB}f$ means that first the operator B is applied to a function f, then the operator A to the result. One must be cautious in such manipulations, since in general the order of operations cannot be interchanged without affecting the result. For example, if A means "multiply by 2" and B means "add 2," then $\mathsf{AB}f = 2f + 4$, whereas $\mathsf{BA}f = 2f + 2$. If AB is equivalent to BA,

$$\mathsf{AB}f = \mathsf{BA}f \quad \text{or} \quad (\mathsf{AB} - \mathsf{BA})f = 0, \tag{3.33}$$

we say that A and B *commute*. It is generally true, however, that

$$\mathsf{A}f + \mathsf{B}f = \mathsf{B}f + \mathsf{A}f. \tag{3.34}$$

We call A a *linear* operator if

$$\mathsf{A}(f + g) = \mathsf{A}f + \mathsf{A}g \tag{3.35}$$

for any functions f and g; we shall deal only with linear operators. Any sequence of operations can be combined into a single operator, as we combined differentiation and multiplication to define E; similarly, the second-derivative operator d^2/dx^2 is equivalent to one differentiation followed by another.

The basic variables of classical mechanics are positions, momenta, and time; what are the corresponding operators? Since we express Ψ as a function of x and t, we do not want to tamper with those variables; thus the operator corresponding to x (or any Cartesian coordinate) or t is simply multiplication by that variable itself. The same applies to any quantity (such as potential energy) that is a function of only these variables. Momentum is another matter. Just as we identified the energy with $i\hbar(\partial/\partial t)$ on the basis of Eq. 3.30, we can obtain an operation equivalent to multiplication by the momentum p from Eq. 3.31. If we define the vector operator (in this case, still in one dimension),

$$\mathsf{p} \equiv -i\hbar\frac{\partial}{\partial x}, \tag{3.36}$$

we find that

$$\mathsf{p}^2 \equiv -\hbar^2\frac{\partial^2}{\partial x^2}, \tag{3.37}$$

consistent with Eq. 3.31. We know that the *quantities E* and p are related by $E = p^2/2m$, where E is the classical kinetic energy of the free particle. Multiplication of both sides by the wave function Ψ gives

$$E\Psi = \frac{p^2\Psi}{2m}, \tag{3.38}$$

which is a simple algebraic equation. But we have implied that operators can be identified with their corresponding quantities, which suggests that E can be replaced by E and p by p. If we make this replacement in Eq. 3.38, using Eqs. 3.32 and 3.37 to express E and p^2, we obtain

$$i\hbar\frac{\partial\Psi(x,t)}{\partial t} = -\frac{\hbar^2}{2m}\frac{\partial^2\Psi(x,t)}{\partial x^2}, \tag{3.39}$$

which is, of course, the same as Eq. 3.25.

In classical mechanics the energy expressed as a function of coordinates and momenta—in this case, as $p^2/2m$—is known as the *Hamiltonian* (after Sir William Rowan Hamilton, the nineteenth-century Irish mathematician). The corresponding quantum mechanical operator, with p replacing p, is called the *Hamiltonian operator* H, which for the free particle in one dimension is given by

$$\mathsf{H} = \frac{\mathsf{p}^2}{2m} = -\frac{\hbar^2}{2m}\frac{\partial^2}{\partial x^2}. \tag{3.40a}$$

The corresponding Schrödinger equation can therefore be written in operator notation as

$$\mathsf{H}\Psi = \mathsf{E}\Psi = -i\hbar\frac{\partial\Psi}{\partial t}. \tag{3.41}$$

To anticipate, when a classical particle in one dimension is subject to a potential $V(x)$, then H becomes

$$\mathsf{H} = \frac{\mathsf{p}^2}{2m} + \mathsf{V}(x) = -\frac{\hbar^2}{2m}\frac{\partial^2}{\partial x^2} + \mathsf{V}(x) \tag{3.40b}$$

and this new Hamiltonian must be used in Eq. 3.41. Equation 3.41 is the fundamental equation of quantum mechanics. We now postulate that it applies not just to the free particle in one dimension, but (with the appropriate Hamiltonian) to *any* physical system. In other words, for any system described by a wave function Ψ, the Hamiltonian operator H acts on Ψ to give $-i\hbar(\partial\Psi/\partial t)$. Since H (in terms of p) corresponds formally to the energy (in terms of p), it is H and not E that is commonly called the "energy operator."

Our derivation of Eq. 3.39 may have seemed trivial, since we defined the operators E and p in just such a way as to give the Schrödinger equation; furthermore, our "identification" of E and p with E and p was quite arbitrary. This is true: Rather than "deriving" Eq. 3.39, we have merely showed its plausibility in a new way. The real point is the formal expression 3.41, which is still a postulate—but now a general postulate, of which Eq. 3.39 is one special case. We make a very far-reaching claim for this equation: If we write the energy of any classical system in Hamiltonian form and obtain the corresponding operator H, then the solution of Eq. 3.41 with this H is the wave function Ψ that describes the same system in quantum mechanics. If this statement is valid, we can transcribe the whole of classical mechanics into quantum mechanical form. *Is* it valid? Apparently so. How do we know? Once again, because the results obtained in this way agree with experiment.

This question—what it means to say that a theory is valid—deserves some further comment. We should always try to distinguish our assumptions about the nature of matter from mere mathematical manipulations or deductive reasoning. The real substance of any physical theory lies in the choice of variables and in the relations between them that we *assume* to be valid. In classical Newtonian mechanics, the choice of variables—force, acceleration, and momentum—leads us to think in terms of particles and rigid bodies, and to express our assumptions in the form of differential equations. These assumptions are the three laws of motion: the law of inertia, stating that a free body moves with a constant velocity; the law relating externally applied forces to velocity changes, stated as $F = ma$; and the law of reaction. In quantum mechanics, the basic assumptions can be taken as the relations $E = h\nu$ and $\lambda = h/p$, and the validity of classical Newtonian physics in the limit of macroscopic phenomena. These assumptions, together with a series of shrewd guesses, led us to the Schrödinger equation, but in no way "proved" it. In fact, there are other wave equations that could be used to describe matter waves, in that they satisfy the same assumptions. The Schrödinger equation was the first one tried because it is the simplest and therefore the most economical; so far as it works, there is no need to go any further. We have assumed, then, that economy or simplicity is a valid criterion for scientific choice, in the absence of any contradictory evidence. Whether or not this assumption is justified is a question of epistemology, metaphysics, or possibly aesthetics.

3.5 Eigenfunctions and Eigenvalues

So we have a method for converting classical equations into quantum mechanical operator equations. Once we have such an equation, what do we do with it? The answer, of course, is to make predictions about the behavior of physical systems. The rules for making these predictions involve still more postulates; to understand what these are, we must again consider the nature of the wave function.

We continue with the free particle in one dimension. The wave function we have obtained, Eq. 3.26, is the amplitude of a traveling wave. Can we construct a *standing wave* capable of describing a free particle? The answer to this question, strangely enough, is "no," so long as we require the particle to have a precise momentum. The function with which we have been dealing is associated with a definite momentum p, with a value given by Eq. 3.29. This momentum has a definite sign, that is, a definite direction: A positive value of p corresponds to motion in the positive x direction. Suppose we now give up the requirement that the particle have definite momentum, and relax our conditions to finding any function that remains bounded and satisfies Eq. 3.25. Can we find a standing wave to meet these conditions? The answer is *yes,* as we now see.

To obtain such a standing wave, we must somehow separate the space-dependent and time-dependent parts of the wave function. Our previous mathematical arguments are still valid: The function $Ae^{i(kx-\omega t)}$ is a solution of Eq. 3.25 for any value of k and ω; each such value gives a different solution. But since differentiation is a linear operation, any sum or difference of these solutions must also satisfy Eq. 3.25. This *principle of superposition* applies to all linear wave phenomena. Let us then consider an arbitrary wave function for which k has the value k' (which we can assume without lack of generality to be positive); we can identify this wave function as $\Psi_{k'}$. There is, of course, another wave function, $\Psi_{-k'}$, for which k has the value $-k'$; the dispersion relation guarantees that these two functions correspond to the same value of ω, so they correspond to identical traveling waves moving in opposite directions with equal energies and opposite momenta. We can now construct two new wave functions, still satisfying Eq. 3.25, by adding and subtracting $\Psi_{k'}$ and $\Psi_{-k'}$:

$$\begin{aligned}\Psi' &= \frac{\Psi_{k'} + \Psi_{-k'}}{2} = \frac{A}{2}(e^{ik'x} + e^{-ik'x})e^{-i\omega t} \\ &= A\cos k'xe^{-i\omega t}, \\ \Psi'' &= \frac{\Psi_{k'} - \Psi_{-k'}}{2i} = \frac{A}{2i}(e^{ik'x} - e^{-ik'x})e^{i\omega t} \\ &= A\sin k'xe^{-i\omega t}. \end{aligned} \tag{3.42}$$

The factors 2 and $2i$ are included to give the cosine and sine functions, from the trigonometric equalities

$$\cos\theta = \frac{e^{i\theta} + e^{-i\theta}}{2} \quad \text{and} \quad \sin\theta = \frac{e^{i\theta} - e^{-i\theta}}{2i}. \tag{3.43}$$

We have thus achieved our goal. By combining the amplitudes of two traveling waves moving in opposite directions—that is, by superposing the two waves—we have obtained a pair of standing waves, each of which oscillates in time with the frequency ω. More precisely, Ψ' and Ψ'' both have real and imaginary parts, each of which is a standing wave varying as $\cos\omega t$ and $\sin\omega t$, respectively. Direct differentiation of Ψ' and Ψ'' shows that both satisfy Eq. 3.25. Nevertheless, neither Ψ' nor Ψ'' is associated with a specific value of the momentum. To see this, let us compare how the momentum operator p, defined in Eq. 3.26, acts on our original Ψ and on Ψ'. With Ψ we obtain

$$\mathsf{p}\Psi = -i\hbar\frac{\partial\Psi}{\partial x} = \hbar kAe^{i(kx-\omega t)} = \hbar k\Psi = p\Psi, \tag{3.44}$$

Ψ itself multiplied by the momentum $\hbar k$; but with Ψ' we obtain

$$\mathsf{p}\Psi' = -i\hbar\frac{\partial\Psi'}{\partial x} = \frac{h}{i}(-Ak'\sin k'xe^{-i\omega t}) = i\hbar k'\Psi''. \tag{3.45}$$

In other words, applying p to Ψ' not only does not give us Ψ' multiplied by the value of the momentum, it gives us an altogether different function, proportional to Ψ''. A system described by Ψ can be assigned a value ($\hbar k$) equivalent to the classical momentum, but this cannot be done for a system described by Ψ'. Let us now investigate the difference between the two kinds of wave functions.

Actually, for an operation to change one function into some other is the usual situation, not the exception. A case like Eq. 3.44, in which an operation leaves a function unchanged except perhaps for some multiplicative factor, is the unusual situation in mathematics, physics, and chemistry. However unusual it may be, this situation is an exceedingly important one. We have already seen other examples of such behavior, for example, in Eqs. 3.30 and 3.31. This special property, the ability of an operator to change a function only by a numerical factor, is peculiar to specific combinations of operations and functions. The simplest case is perhaps the differentiation of an exponential: $d(e^{ax})/dx = ae^{ax}$. The operator d/dx leaves the function e^{ax} unchanged except to multiply it by the constant a. By contrast, differentiation changes a sine or a cosine into a different function. A function left unchanged except for multiplication by a constant when it is acted on by a particular operator is said to be an *eigenfunction* (*characteristic function*) of that operator. Thus by the equations of Section 3.4 we know that $\Psi = Ae^{i(kx-\omega t)}$ is an eigenfunction of the operators of momentum and energy. We have shown that the functions Ψ' and Ψ'' of Eqs. 3.42 are not eigenfunctions of momentum; however, they are eigenfunctions of the energy ($\mathsf{H}\Psi' = E'\Psi'$, with $E' = \hbar^2k'^2/2m$), since a sine or cosine does become a multiple of itself when it is differentiated twice.

What physical interpretation can we give to the concepts of eigenfunctions and their corresponding operators? Suppose, for example, that a particular wave function is an eigenfunction of the energy operator. This means that if we allow the energy operator (H or E) to operate on the wave function, then the result will be the wave function multiplied by a constant. This result contains one very specific piece of physical information, the value of the characteristic constant. This value is the energy corresponding to that particular eigenfunction, and is thus known as an *eigenvalue* (*characteristic value*) of the energy. If instead of H we used the momentum operator p, and the wave function were an eigenfunction of momentum, then we would obtain the wave function multiplied by the corresponding momentum eigenvalue. Whenever we wish to describe a system in a state having a definite value of some variable, we must use a wave function that is an eigenfunction of the corresponding operator,[7] that particular eigenfunction which corresponds to the given value; only such a function can correctly describe the system as we have assumed it to exist. Thus if we wish to specify that a free particle has an energy E_0, its wave function must be that eigenfunction of the energy operator which has the eigenvalue E_0, that is, the function $Ae^{i(kx-\omega t)}$ for which k and ω have the values given by

$$E = \hbar\omega = \frac{\hbar^2k^2}{2m} = E_0\,. \tag{3.46}$$

On the other hand, any wave function that is not an eigenfunction of a given operator (as with Ψ' and the momentum operator) cannot be assigned a definite value of the variable corresponding to that operator.

By this time the reader may be confused. Are we saying that the wave function of a system depends on how we choose to describe it? In a sense this is true, but "how we describe it" has a real physical meaning. To say that a system has a definite energy, one must carry out some measuring process that gives a particular value of the energy; similarly, any statement specifying the properties of a system implies some kind of measurement. But one of the fundamental principles of quantum mechanics is that the process of measurement itself affects the state of a system—basically by imposing new boundary conditions. This is reflected in what is known as the *indeterminacy* or *uncertainty principle,* which we shall consider in detail in Section 3.7. All that needs to be understood at this point is that specifying (i.e., measuring) different properties does in general cause the system to have different wave functions. It is useless to ask what the state of the system would be in the absence of any measurement, since there is no way for us to know this.

We have perhaps implied that the specification of a single quantity, such as the energy or the momentum, is enough to define the wave function of a system. In general this is not true. There are ordinarily several characteristic quantities with precisely defined values, usually the same quantities that are conserved in classical mechanics: energy, linear momentum, and angular momentum. These properties are constants of the motion, remaining unchanged in the absence of external interference, and are associated with some general symmetry of the physical system itself.

For example, the angular momentum about a particular axis is associated with cylindrical symmetry about that axis. This does not mean that the system itself need be cylindrical, but that its orientation about the axis makes no physically significant difference in its description. Thus, if a particle moves in an elliptical orbit about an attractive center of force, the properties of the orbit (its shape, the angular momentum, the total energy) would be the same for any orientation of the ellipse; if there is no external torque or other dissipative mechanism, the angular momentum and energy will remain constant. The symmetry associated with the conservation of linear momentum reflects our freedom to locate the origin of our coordinate system anywhere in space without affecting the value of the momentum. The three Cartesian components of momentum are conserved separately, so that an external force applied in one direction will change the momentum in that direction but leave the other

---

[7] Note that the eigenvalues of the Cartesian coordinates are continuous: The operator corresponding to x is itself x, so we have $\mathbf{x}\Psi = x\Psi$ for any value of x.

components unaffected. As for the energy, note that it is the total energy that is conserved; the kinetic energy and the potential energy separately can vary, but their sum remains constant. Although the conservation theorems are ordinarily derived from Newton's laws, they can also be shown to follow from the principles of quantum mechanics.

Let us enumerate the quantities that we are sure will be conserved. For the free particle in one dimension, these are just the energy and the momentum. Does specifying these give enough information for a complete determination of the particle's wave function? The answer is almost, but not quite enough. We know that the wave function $\Psi = Ae^{i(kx-\omega t)}$ is a satisfactory solution to the wave equation 3.25, inasmuch as it has well-defined values of the energy and momentum, satisfies the dispersion relation 3.11, and has the aesthetic simplicity of being a very straightforward and uncomplicated kind of wave. But this wave function contains one other very important piece of information to which we have not yet turned our attention. We refer to the constant *A*, which gives the maximum amplitude of the wave. The real part of Ψ is restricted to values between $+A$ and $-A$, and the imaginary part between $+iA$ and $-iA$, no matter how large or small *x* or *t* may become. It will be recalled that we deliberately chose the form of Ψ to be such that the amplitude would be everywhere finite. We can now expand on the reasons for this restriction, in connection with a physical interpretation of Ψ itself.

We have said that the value of Ψ must be somehow related to the probability of finding the particle at a given position. By the "probability" of a given value we mean the fraction of all measurements that should yield that value. (See Section 15.2 for a more extensive discussion.) This probability is analogous to the intensity of a light wave (which is proportional to the density of light quanta). But the intensity of a light wave is proportional not to the amplitude of the wave itself, but to its square; the same is true of the energy carried by a water wave. We can thus interpret the intensity of our matter wave, the probability of finding a particle, as proportional to the square of Ψ. This is another postulate, of course. But the value of Ψ is a complex number, whereas the probability (like any other measurable quantity) must be a real number; to obtain a real result, we need simply take the square of the absolute value of Ψ. This is why it does not matter that Ψ is complex: We never observe Ψ directly, and it is only for mathematical convenience that Ψ rather than $|\Psi|^2$ is the "wave function."

The intensity of the matter wave is thus given by

$$\mathcal{I}(x,t) = |\Psi(x,t)|^2 = \Psi^*(x,t)\Psi(x,t), \qquad (3.47)$$

where Ψ^* is the complex conjugate of Ψ, the function identical to Ψ except that every i $(\equiv \sqrt{-1})$ is replaced by $-i$:

$$\Psi^*(x,t) = A^* e^{-i(kx-\omega t)}, \qquad (3.48)$$

in which we write A^* to allow for the possibility that *A* is also complex. The last part of Eq. 3.47 is justified by the fact that $|z|^2 = z^* z$ for any complex number.[8] But when we multiply Ψ^* by Ψ the exponential factors cancel, and we obtain

$$\mathcal{I}(x,t) = A^* A = |A|^2; \qquad (3.49)$$

that is, the intensity of the wave is simply given by the square of the absolute value of *A*.

The physical interpretation of the matter wave is thus clear. For the free particle moving in one dimension, the probability of finding the particle at a given place and time is proportional to the intensity of the wave at that place and time. Consider an infinitesimal region of length *dx* around the point *x;* the probability of finding the particle in this region at time *t* is given by

$$\mathcal{P}(x,t)\,dx \propto \mathcal{I}(x,t)\,dx = \Psi^*\Psi\,dx = |A|^2\,dx, \qquad (3.50)$$

where $\mathcal{P}(x, t)$ is what we call a *probability density.* The most striking fact about this result is that the probability density is a constant, and not a function of *x* or *t* at all. For the particular case of the free particle, all points are equally probable; like Kipling's "cat that walked by himself," all places are alike for it. This result should not be surprising: If all we know about a particle is that it is moving along a line with a given energy and momentum, we can say nothing about *where* on the line it is; this can be determined by a measurement, of course, but the performance of such a measurement will change the wave function.

We wrote Eq. 3.50 as a proportionality, but we might as well define *A* so as to make it an equality. Given this assumption, what is the absolute value of *A*? To answer this question we must carry out what is called a *normalization.* By the definition of probability (see Section 15.2), the total probability over the entire available range must be unity. Since the range of a completely free particle is unlimited, we have

$$\int_{-\infty}^{\infty} \mathcal{P}(x,t)\,dx = \int_{-\infty}^{\infty} \Psi^*\Psi\,dx = \int_{-\infty}^{\infty} |A|^2\,dx = 1, \qquad (3.51)$$

a result that can be true only if $|A| = 0$. The physical meaning of this is also quite simple: If the particle can be *anywhere,* the probability of its being at any particular position is vanishingly small. By contrast, if a measurement of the particle's position localized it within some small range Δx, integration over that range would give a finite, nonzero value $|A|^2 = 1/\Delta x$. The normalization process tells us nothing new here, but we shall see that it is quite useful for bounded systems. Equation 3.51 does illustrate more clearly why Ψ must remain bounded at infinity, the reason for which we rejected forms like $e^{\pm\varphi}$.

[8] Any complex number has the form $z = a + ib$; its complex conjugate is thus $z^* = a - ib$, and we have

$$z^* z = (a-ib)(a+ib) = a^2 - i^2b^2 = a^2 + b^2 = |z|^2,$$

since $|z|$ is $(a^2 + b^2)^{1/2}$ by definition.

Before ending our consideration of the free particle, let us briefly take up the case of the free particle in three dimensions. We have already given the wave equation that must be satisfied, Eq. 3.28. In three dimensions the constant k must be replaced by three constants k_x, k_y, k_z, corresponding to the three components of the momentum. The wave function thus has the form

$$\Psi(x,y,z,t) = Ae^{i(k_x x + k_y y + k_z z - \omega t)} = Ae^{i(\mathbf{k}\cdot\mathbf{r} - \omega t)}. \quad (3.52)$$

This still has the form $Ae^{i\varphi}$, since $\mathbf{k} \cdot \mathbf{r} - \omega t$ is in the form of a phase. The function 3.52 has real and imaginary oscillations in all three directions, x, y, and z. It is still an eigenfunction of the energy operator, so $\mathsf{H}\Psi = \mathsf{E}\Psi = E\Psi$ and the corresponding energy eigenvalue can easily be shown to be

$$E = \hbar\omega = \frac{\hbar^2}{2m}\left(k_x^2 + k_y^2 + k_z^2\right) = \frac{\hbar^2 k^2}{2m}, \quad (3.53)$$

where $k^2 \equiv k_x^2 + k_y^2 + k_z^2$. The momentum $\mathbf{p}$, a vector with components $\hbar k_x$, $\hbar k_y$, $\hbar k_z$, is an eigenvalue of the three-dimensional momentum operator[9]

$$\mathsf{p} = -i\hbar\left(\mathbf{i}\frac{\partial}{\partial x} + \mathbf{j}\frac{\partial}{\partial y} + \mathbf{k}\frac{\partial}{\partial z}\right), \quad (3.54)$$

in which $\mathbf{i}$, $\mathbf{j}$, $\mathbf{k}$ are unit vectors in the x, y, z directions, respectively. The arguments of the last paragraphs still apply, so that the particle is equally likely to be found anywhere in space. We must add some physical restriction to get a system for which $|\Psi|^2$ exhibits some spatial dependence; in the next section we begin to consider such restrictions.

3.6 The Particle in a One-Dimensional Box

The simplest way to generate spatial dependence for $|\Psi|^2$, to produce a varying probability of finding a particle at different points in space, is to leave the particle as free as possible, but within the confines of a box from which it cannot escape. Physically, this means that within the box the particle's potential energy is independent of its position, but that no matter how much kinetic energy it has it cannot get beyond the walls.

We again deal first with a one-dimensional system for simplicity. As before, we consider a particle of mass m moving along the x axis. Let us suppose that this particle is constrained to be between the points $x = 0$ and $x = a$. For convenience we assume that the potential energy between these two limits is identically zero, so that the total energy inside the box is pure kinetic energy. The potential energy outside the box must be so high that the particle cannot escape, no matter how large the kinetic energy may be; in short, it must be infinite. A sketch of such a box is given in Fig. 3.5.

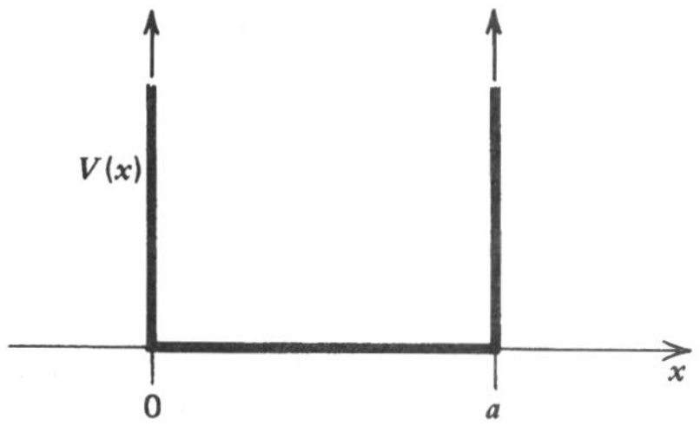

Figure 3.5 The potential energy of the one-dimensional box.

The physical problem before us is clearly to determine the allowable values that the energy of the confined particle may have, and to find the wave function (and thus the spatial probability distribution) associated with each of these energy eigenvalues. The corresponding mathematical problem requires that we solve the Schrödinger equation subject to the specific *boundary conditions* appropriate to this particular problem. These are that the probability of finding the particle, and therefore the value of Ψ, must be identically zero for all values of x beyond the boundaries of the box. This means we require that

$$\Psi(x,t) = 0 \quad \text{for all } x < 0 \text{ and } x > a. \quad (3.55)$$

We must also assume that the wave function is *continuous* through the walls (and everywhere else in space). Like the assumptions that Ψ be finite and single-valued, this is necessary if the results are to make physical sense; a function that satisfies these three conditions is called "well-behaved." In general, the first derivatives of Ψ must also be continuous, but this rule does not apply at an infinite potential jump such as we have here. Given the assumption of continuity, we must have

$$\Psi(0,t) = 0 \quad \text{and} \quad \Psi(a,t) = 0, \quad (3.56)$$

whether we approach the wall from the inside or the outside. This is our first example of an explicit boundary condition.

Our treatment in previous sections assumed that the potential energy was zero everywhere, with the dispersion relation 3.11 yielding the wave equation 3.39. For the general case of nonzero potential energy, we start instead with Eq. 3.13, from which we easily obtain

$$i\hbar\left[\frac{\partial\Psi(x,t)}{\partial t}\right]_x = -\frac{\hbar^2}{2m}\left[\frac{\partial^2\Psi(x,t)}{\partial x^2}\right]_t + \mathsf{V}(x)\Psi(x,t) \quad (3.57)$$

for the one-dimensional time-dependent Schrödinger equation. However, since we have assumed the potential energy

[9] For any scalar function f, the vector whose components are $\partial f/\partial x$, $\partial f/\partial y$, and $\partial f/\partial z$ is called the *gradient* of f, written **grad** f or ∇ f (the operator "∇" is read "del"). Thus we can write $\mathsf{p} \equiv -i\hbar\nabla$.

V(x) to be zero inside the box, we can still use Eq. 3.39 to describe the behavior of the wave function there. It is up to us to find the solution of this equation that satisfies the boundary conditions of Eq. 3.56.

The best way to begin is by separating the spatial dependence of the wave function from its time dependence. We wish our wave function to be an eigenfunction of the energy operator, which means that it must satisfy the equation $\mathsf{E}\Psi = E\Psi$, or

$$i\hbar\left(\frac{\partial \Psi}{\partial t}\right)_x = \hbar\omega\Psi, \tag{3.58}$$

with the usual substitution $E = \hbar\omega$. But this equation can hold only if Ψ has the form

$$\Psi(x,t) = \psi(x)e^{-i\omega t}, \tag{3.59}$$

in which $\psi(x)$ is a function of only the spatial coordinate x. Our solution for the free particle was also of this form, with $\psi(x) = Ae^{ikx}$. In fact, the argument is valid for *any* wave function that is an eigenfunction of the energy, and these are what we usually want to obtain. Substituting Eqs. 3.58 and 3.59 into the Schrödinger equation 3.57, we obtain

$$\hbar\omega\psi(x)e^{-i\omega t} = -\frac{\hbar^2}{2m}\frac{d^2\psi(x)}{dx^2}e^{-i\omega t} + \mathsf{V}(x)\psi(x)e^{-i\omega t}. \tag{3.60}$$

Cancelling $e^{-i\omega t}$ from both sides, again replacing $\hbar\omega$ by E, and rearranging, we have

$$-\frac{\hbar^2}{2m}\frac{d^2\psi(x)}{dx^2} = [E - \mathsf{V}(x)]\psi(x), \tag{3.61}$$

the *time-independent Schödinger equation,* involving only the spatial variable x. This equation is usually written in the form

$$\mathsf{H}\psi = E\psi, \tag{3.62}$$

where the Hamiltonian operator H, representing the energy in terms of coordinates and momenta, is given by

$$\mathsf{H} = \frac{\mathsf{p}^2}{2m} + \mathsf{V}(x) = -\frac{\hbar^2}{2m}\frac{\partial^2}{\partial x^2} + \mathsf{V}(x). \tag{3.63}$$

Equation 3.62, like Eq. 3.41, is quite general: With the appropriate Hamiltonian, it applies to the spatial part of any energy eigenfunction. Since $\mathsf{H}\psi$ must be a function of spatial variables only, the Hamiltonian cannot be an explicit function of time: If V(x) also varies with time, there is no $\psi(x)$ (and thus no E) that can satisfy Eq. 3.61.

We included V(x) in Eq. 3.61 for generality, but it can immediately be dropped again for the region inside the box. We must thus solve the equation

$$\frac{d^2\psi(x)}{dx^2} = -\frac{2mE}{\hbar^2}\psi(x) \tag{3.64}$$

to obtain the spatial part of the wave function. As we have done several times before, we can simplify the equation by removing the dimensional dependence of the variable. We thus make the transformation

$$\xi \equiv \left(\frac{2mE}{\hbar^2}\right)^{1/2} x, \tag{3.65}$$

which allows us to rewrite Eq. 3.64 as

$$\frac{d^2\psi(\xi)}{d\xi^2} = -\psi(\xi). \tag{3.66}$$

The only simple functions that are the negatives of their own second derivatives are the complex exponential, which is never zero, and the sine and cosine functions. Both sine and cosine do of course pass through zero, at $n\pi$ and $\left(n+\frac{1}{2}\right)\pi$, respectively, and can thus satisfy the boundary conditions. A general form for $\psi(\xi)$ should then be

$$\psi(\xi) = A\sin\xi + B\cos\xi. \tag{3.67}$$

Although this wave function is an eigenfunction of the energy, it clearly cannot be an eigenfunction of momentum: Since ψ is real, the equation $-i\hbar(d\psi/dx) = p\psi$ can be satisfied for no real value of p. One can see a "reason" for this in simple terms. The presence of a wall means that the particle must in effect bounce, thereby changing its momentum; hence the momentum cannot be a constant of the motion. The real reason is complex but very important, and will be discussed in the next section.

The general procedure for solving differential equations such as Eq. 3.66 consists of writing down the formal solution, as we have done in Eq. 3.67, and then using the boundary conditions, in our case those of Eq. 3.56, to determine the integration constants, here A and B. The formal solution 3.67 is by no means *the* solution to the problem. A differential equation by itself does not contain all the information required to specify a solution. Each time we carry out an integration, we necessarily introduce an integration constant; fixing the value of these constants is just as much a part of finding a solution as is finding a general form like Eq. 3.67. It is the boundary conditions that fix the integration constants, and changing them sometimes completely changes the entire character of the problem, either in the form of the solution or in the difficulty of obtaining it. We can see this explicitly by comparing the solution for the particle in a box with that obtained for the free particle. After all, in these two cases the Hamiltonians and therefore the wave equations are exactly the same; the problems differ *only* in their boundary conditions.

Now let us turn to obtaining this solution. The boundary conditions tell us that $\Psi = 0$ at $\xi = 0$ ($x = 0$) and $\xi = (2mE/\hbar^2)^{1/2}\, a$ ($x = a$). Where $\Psi = 0$ for all values of t, we must also have $\psi = 0$. We can require that $\psi(0)$ be zero only by setting the constant B in Eq. 3.67 equal to zero. Therefore the solution has the form

$$\psi(\xi) = A \sin \xi \qquad (3.68)$$

or

$$\psi(x) = A \sin\left(\frac{2mE}{\hbar^2}\right)^{1/2} x. \qquad (3.69)$$

This brings us to the crux of the problem of quantization. How can we satisfy the boundary condition at the point $x = a$? With the function 3.69 there is only one way to accomplish this goal: We must require the argument of the sine to be an integral multiple of π; that is, the condition

$$\left(\frac{2mE}{\hbar^2}\right)^{1/2} a = n\pi \qquad (n = 1, 2, \ldots) \qquad (3.70)$$

must be satisfied. (The boundary condition could also be met by setting $n = 0$—and thus $E = 0$—or even by letting both A and B be zero, but in either case the wave function would be identically zero everywhere in the box; this would be of no help to us, because it would correspond to an empty box. Since $-n$ and $+n$ correspond to the same value of E, negative values of n can also be discarded.) Of the quantities in Eq. 3.70, only E and n are not fixed beforehand. The equation thus relates the allowed values of the energy E to the integral values of n. To each n there corresponds an energy E_n, which we find by rearrangement to be

$$E_n = \frac{n^2\pi^2\hbar^2}{2ma^2} \qquad (n = 1, 2, \ldots). \qquad (3.71)$$

The boundary conditions at the two ends of the box—and note that it takes two boundaries, not one—have forced us to accept only a discrete set of possible values for the energy E. These energies vary inversely with the square of the box length, and directly with the square of the quantum number n. The separation between consecutive energy levels, $E_{n+1} - E_n$, thus also increases directly with n, as illustrated in Fig. 3.6*a*.

How do the energy levels vary with the size of the box? If we squeeze the length of the box to $a/2$, we raise the energy of the nth level to four times its original value. If we stretch the box, we correspondingly lower the energy levels. Squeezing the box by a factor of two can be interpreted in another very straightforward way, as equivalent to placing a boundary at the point $x = a/2$; we can then apply the same kind of boundary conditions at $x = a/2$ that we originally applied at $x = a$. The method of solution is exactly the same as before, so the only acceptable wave functions are those that manage to vanish at $x = 0$ and $x = a/2$. In other words, we have in the new box the condition

$$\left(\frac{2mE}{\hbar^2}\right)^{1/2} \frac{a}{2} = n\pi \qquad (n = 1, 2, \ldots), \qquad (3.72)$$

which is exactly equivalent to taking every other energy level of the set found for the original box, those levels corresponding to even values of n in the box of length a.

Now let us examine just how the boundary conditions lead to quantization of the energy levels. Equation 3.66 tells us that the curvature of the wave function ψ is equal to the negative of the function itself. (Recall that the first derivative of a function is its slope, whereas the second derivative is the rate of change of the slope, which we may loosely refer to as the function's "curvature." Strictly, the curvature is $(d^2\psi/dx^2)[1 + (d\psi/dx)^2]^{-3/2}$.) If ψ is positive, then its curvature must be negative, so that the function must be curving back downward toward the x axis. Correspondingly, wherever ψ is negative, the curvature must be positive and the function curving upward toward the x axis. These are conditions that produce a sine- or cosinelike wave. The function 3.69 has zero displacement at the origin. According to Eq. 3.64, the curvature of ψ is proportional to the energy of the state with which it is associated. If the energy E has a value E' near zero, the curvature of ψ is so slight that the function (which must

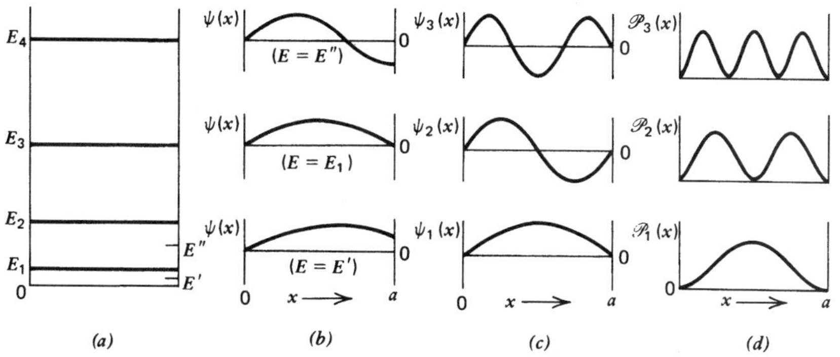

Figure 3.6 Energy levels and wave functions for a particle in a one-dimensional box. (*a*) The first four energy eigenvalues; the energies E' and E'' that are not eigenvalues correspond to the top and bottom wave functions shown in part (*b*). (*b*) Illustration of how only the eigenvalues give wave functions satisfying the boundary condition $\psi(a) = 0$. (*c*) Eigenfunctions corresponding to the first three eigenvalues. (*d*) The corresponding probability densities, $\mathcal{P}_n(x) = |\psi_n(x)|^2$ for $n = 1$, 2, and 3.

be positive near the origin) never returns to the x axis in the interval between 0 and a. If the value of E is somewhat higher, say E'', then ψ increases with increasing x, passes through a maximum, and curves downward, crossing the x axis before the point $x = a$ is reached. After it crosses the axis, ψ must start curving upward again, but let us assume that E'' is not high enough for it to reach the axis a second time before $x = a$. For exactly one value of E between the two arbitrary values E' and E'', the function ψ will just reach the axis at the point $x = a$. This value E_1 is the first permissible value that the energy can assume, according to the boundary condition we have imposed. The behavior just described is illustrated in Fig. 3.6*b*. As E increases beyond E'', eventually an energy will be reached at which ψ just reaches the axis a second time at $x = a$; this is the second allowed energy value, E_2, corresponding to $n = 2$ in Eq. 3.71. Similarly, there are specific values of E for all the higher values of n. Each value of E that satisfies the boundary conditions is an eigenvalue of the energy operator, and the wave function

$$\psi_n(x) = A \sin\left(\frac{2mE_n}{\hbar^2}\right)^{1/2} x \tag{3.73}$$

is the spatial eigenfunction corresponding to E_n. The first three eigenfunctions are plotted in Fig. 3.6*c*.

Using Eq. 3.71, we can rewrite the wave function in the form

$$\psi_n(x) = A \sin\frac{n\pi x}{a}. \tag{3.74}$$

This expression is identical to that for the possible standing waves in a string fastened at both ends (as in Fig. 3.2). This identity should not be surprising, since the two systems are mathematically equivalent: Both wave amplitudes must satisfy a wave equation, with the boundary condition that the amplitude must vanish at the ends. Only the *meanings* of the wave amplitudes are different, and this does not affect their mathematical form.

We still have to evaluate the constant A, which fixes the maximum amplitude of the wave function. This can be done quite simply, as soon as we recognize that there is one more physical condition to be satisfied. We refer to the normalization process introduced in the preceding section; it did us no good there, but here it can give us the value of A. The probability of finding the particle in a region of length dx around x is

$$\begin{aligned}\mathcal{P}(x,t)\,dx &= \Psi^*\Psi\,dx\\ &= \psi^*(x)e^{i\omega t}\psi(x)e^{-i\omega t}\,dx = |\psi(x)|^2\,dx.\end{aligned} \tag{3.75}$$

Note that the probability density is a function of x only, as is true whenever Eq. 3.59 holds. For our eigenfunctions we have

$$\mathcal{P}_n(x) = |\psi_n(x)|^2 = |A|^2 \sin^2\frac{n\pi x}{a}; \tag{3.76}$$

this probability distribution is plotted in Fig. 3.6*d*. The total probability of the particle's being anywhere in its range is again unity. Since we have specified that there is zero probability of finding the particle outside the box, we need integrate only over the length of the box, to obtain

$$\int_0^a \mathcal{P}_n(x)\,dx = |A|^2\int_0^a \sin^2\frac{n\pi x}{a}\,dx = 1. \tag{3.77}$$

Again making the substitution

$$\xi = \left(\frac{2mE}{\hbar^2}\right)^{1/2} x = \frac{n\pi x}{a}, \tag{3.78}$$

we can convert Eq. 3.77 to[10]

$$\begin{aligned}|A|^2\left(\frac{a}{n\pi}\right)\int_0^{n\pi}\sin^2\xi\,d\xi &= |A|^2\left(\frac{a}{n\pi}\right)\left(\frac{n\pi}{2}\right)\\ &= |A|^2\frac{a}{2} = 1,\end{aligned} \tag{3.79}$$

which immediately tells us that

$$|A| = \left(\frac{2}{a}\right)^{1/2}. \tag{3.80}$$

There is no reason not to assume A real, so we set $A = |A|$ and finally obtain

$$\psi_n(x) = \left(\frac{2}{a}\right)^{1/2}\sin\frac{n\pi x}{a} \tag{3.81}$$

for the spatial part of the nth eigenfunction.

Note that the nth eigenfunction, $\psi_n(x)$, has $n - 1$ nodes, not counting the walls. The energy of the system thus increases with the number of nodes in its wave function. We shall see that this relationship between energy and nodes is generally true. As $n \rightarrow \infty$, the nodes become closer and closer together; eventually the oscillations of the wave function blur together, and the probability density over any

---

[10] The integral

$$\int_0^{\pi}\sin^2\theta\,d\theta = \frac{\pi}{2}$$

can be found in any table of definite integrals. Here the range is from 0 to $n\pi$, but the integral over each range of π is of course the same, so that

$$\int_0^{n\pi}\sin^2\theta\,d\theta = \frac{n\pi}{2}.$$

measurable distance averages out[11] to the constant value $\langle \mathcal{P}_n(x) \rangle = 1/a$. This is the *classical limit:* A classical particle would be equally likely to be anywhere in the box. In the same limit E_n becomes so large that the fractional difference between successive energy levels is negligible; the energy is thus effectively continuous. This behavior illustrates the general rule that quantum mechanical systems behave classically in the limit $n \to \infty$ (the *correspondence principle*).

The form of Eq. 3.81 also tells us that the wave $\psi_n(x)$ has n half-waves in a length a, or that $\psi_n(x)$ has a wavelength λ of $2a/n$. The de Broglie condition $p\lambda = h$ implies that the momentum p of the particle in a box is h/λ or $nh/2a$ or $\pi n\hbar/a$. (There is a subtle point being glossed over here: The momentum may be positive or negative for a free particle; by putting up a box, we allow only the absolute value of the momentum to be meaningful.) The kinetic energy, $p^2/2m$, is therefore $n^2\pi^2\hbar^2/2ma^2$, in agreement with Eq. 3.71. In short, our development of the particle in a box is indeed consistent with the de Broglie condition, as it should be. The wavelengths go down with n, the (absolute values of) momenta go up with n, and the energies go up with n^2.

What is the separation between energy levels for the particle in a box? We wish eventually to develop a theory of atomic structure, and we can make the present model a very crude approximation to an atom if we consider an electron ($m = m_e$) in a box of atomic dimensions. From the results in Section 2.11 we can guess in what range the energy levels will lie, but let us make our estimate anew. Let us say that $\pi^2 \approx 10$, $h \approx 10^{-34}$ J s, $m_e \approx 10^{-30}$ kg, and $a \approx 10^{-10}$ m (1 Å). Substituting these numbers in Eq. 3.71, we have

$$E_n \approx \frac{n^2 \times 10 \times (10^{-34}\text{ J s})^2}{2 \times 10^{-30}\text{ kg} \times (10^{-10}\text{ m})^2} \times \frac{1\text{ kg m/s}^2}{\text{J}}$$
$$= 5n^2 \times 10^{-18}\text{ J},$$

equivalent to about $30n^2$ eV (since 1 eV $\approx 1.6 \times 10^{-19}$ J). As we expected, for small values of n these numbers are indeed of the same order of magnitude as those given in Table 2.2 for the Bohr atom. Another example, for which the particle-in-a-box model is more realistic, is that of a gas molecule in a real macroscopic container. (Extending the problem to three dimensions does not change the orders of magnitude involved.) If we set $m \approx 3.3 \times 10^{-27}$ kg (a molecule of H_2) and $a \approx 1$ cm, we obtain $E_n \approx 3n^2 \times 10^{-37}$. This is such a small spacing, even by atomic dimensions, that the energies can be regarded as continuous for most purposes; this is why nearly all the macroscopic properties of gases can be explained in terms of classical mechanics (see Part II).

3.7 The Indeterminacy or Uncertainty Principle

We have already mentioned the indeterminacy or[12] uncertainty principle in Section 3.5, where we pointed out that the wave function of a system depends on which of its properties one measures. Now we can understand the reason for this: Any measuring process imposes a new boundary condition on the system, and thus affects the wave function. The time has now come to examine this principle in detail. The uncertainty principle, derived by Werner Heisenberg in 1927, is perhaps the most general and fundamental point on which the conclusions of quantum mechanics diverge from those of classical mechanics. In classical mechanics it is assumed that one can simultaneously know as many properties of a system as one wishes, to any desired degree of accuracy. It is already apparent that this is not true in quantum mechanics; the most obvious example is that the wave function gives only the probability of a particle's various possible locations, rather than specifying a single location. The uncertainty principle gives the limitations on the accuracy with which any quantity can be known.

The nature of these limitations can best be understood by considering an example. Suppose that we wish to know simultaneously the position and momentum of a particle in motion. We can specify the momentum initially by launching the particle with a known kinetic energy. Its wave function is then an eigenfunction of energy and momentum, which we have shown to have the form $Ae^{i(kx-\omega t)}$; however, this function gives the same probability for any value of x, and thus tells us nothing about the particle's position. We can in principle measure the position by dividing the trajectory into short segments and applying some kind of test to each to determine which contains the particle. (Don't worry about the practicality of this; the difficulty is more fundamental.) But such a test is tantamount to looking for the particle in a highly localized state, that is, confining it to a very narrow box. We must thus apply boundary conditions rather like those of the last section. But confining a wave to a box forces the wave to oscillate through at least a half-cycle within the box. The narrower the box, the more rapid is the oscillation of ψ, the higher the average value of $|\partial\psi/\partial x|$,

[11] The average of a function $f(x)$ over a range Δx is

$$\langle f(x) \rangle = \frac{1}{\Delta x}\int_{\Delta x} f(x)\, dx.$$

Since $\mathcal{P}_n(x)$ is a periodic function with period a/n, its average is simply

$$\langle \mathcal{P}_n(x) \rangle = \frac{n}{a}\int_0^{a/n} \mathcal{P}_n(x)\, dx = \left(\frac{n}{a}\right)\left(\frac{2}{a}\right)\left(\frac{a}{n\pi}\right)\int_0^{\pi} \sin^2 \xi\, d\xi = \frac{1}{a}.$$

[12] Strictly, this principle should be called the "indeterminacy principle," because it is really *indeterminacy* that the principle states—that there is a limit on our capability to determine simultaneous values of pairs of complementary variables, such as position and momentum, along each independent direction. However, in common usage, it is most often called the "uncertainty principle," so, with this caution, we shall generally follow the common practice.

and thus (since $\partial\psi/\partial x$ is proportional to $\mathsf{p}\psi$) the higher the average momentum of the particle. Another way to express the same idea is in terms of Eq. 3.71: Decreasing the box length increases the energy and thus the momentum. In other words, the more accurately we try to specify the particle's position, the more we increase (and thus make uncertain) its momentum. The best we can do is to balance off the uncertainties in position and momentum at some level.

We can make a rough estimate of this level. Suppose that a particle is at rest, and we wish to find its exact position by looking through a microscope. To see the particle we must of course bounce one or more photons off it. If the light we use has a wavelength λ, we cannot expect to determine the position within a distance much shorter than λ because of diffraction; so let us say that λ is the uncertainty in the position measurement. But a photon of wavelength λ has a momentum h/λ, and in the collision it may transfer roughly this much momentum to the particle. The particle is then no longer at rest, and its momentum is uncertain to within h/λ. The product of the position and momentum uncertainties is thus of the order of $\lambda(h/\lambda) = h$, with Planck's constant again making a significant appearance. We shall state this principle more accurately, but first we must define our terms.

Position and momentum are what is called a *conjugate* pair of variables. This is a concept from classical mechanics that we shall not consider in detail. Other examples of conjugate pairs include energy and time, and angular momentum and angle. In each case the pair consists of two variables whose product has the dimensions of action (cf. Section 2.7), and that are associated with the same coordinate or degree of freedom. What this means is that position on the x axis and momentum in the x direction (x and p_x) are conjugate, as are y and p_y, z and p_z, but that x and p_y are not conjugate. When one member of the pair is used as a coordinate, the other is called the *generalized momentum* conjugate to that coordinate even though it may not have the dimensions of momentum. It can be shown by reasoning like that above that the two members of any conjugate pair of variables cannot be simultaneously measured accurately.[13] The uncertainty principle is a quantitative statement of this conclusion.

[13] An equivalent (and more general) formulation is that two variables cannot simultaneously be determined exactly if their corresponding operators do not commute with each other. For example, we have

$$\mathsf{x}\mathsf{p}_x\psi = -i\hbar x\frac{\partial\psi}{\partial x} = -i\hbar\mathsf{x}\frac{\partial\psi}{\partial x}$$

but

$$\mathsf{p}_x\mathsf{x}\psi = -i\hbar\frac{\partial}{\partial x}(x\psi) = -i\hbar\left(x\frac{\partial\psi}{\partial x} + \psi\right),$$

so that $\mathsf{x}\mathsf{p}_x \neq \mathsf{p}_x\mathsf{x}$, and the operators x and p_x do not commute. On the other hand, x does commute with p_y. For any conjugate position and momentum one obtains

$$(\mathsf{q}_i\mathsf{p}_i - \mathsf{p}_i\mathsf{q}_i)\psi = i\hbar\psi;$$

$\mathsf{q}_i\mathsf{p}_i - \mathsf{p}_i\mathsf{q}_i$ is called the commutator of q_i and p_i.

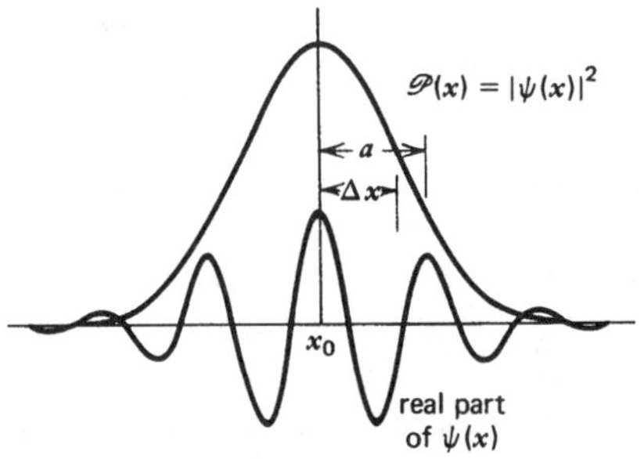

Figure 3.7 Wave packet representing a particle localized in space. The figure is drawn for a wave function

$$Ae^{-(x-x_0)^2/2a^2}e^{ik(x-x_0)}$$

giving the Gaussian probability distribution

$$\mathcal{P}(x) = |\psi(x)|^2 = A^2e^{-(x-x_0)^2/a^2},$$

where x_0 is the average value of x. The scale factor a, defined by $\mathcal{P}(x_0 \pm a) = \mathcal{P}(x_0)/e$, is equal to $\sqrt{2}$ times to root-mean-square deviation Δx.

We must also define just what we mean by the "uncertainty" in a variable. The closest quantum mechanical approximation to a classical particle localized at a point is a sharply peaked probability distribution—a wave function something like that for a particle in a narrow box, but trailing off gradually at the sides (since there are no sharp boundary conditions). Such a distribution, or *wave packet,* is shown in Fig. 3.7. There are various ways in which Δx, the uncertainty in x, can be defined; we have chosen the root-mean-square (rms) deviation, defined by

$$(\Delta x)^2 \equiv \langle (x-x_0)^2 \rangle = \int_{-\infty}^{\infty} (x-x_0)^2\,\mathcal{P}(x)\,dx, \quad (3.82)$$

where x_0 is the average value of x for the wave packet. The probability distribution for any other variable, when plotted against that variable, will resemble that in Fig. 3.7. We can thus define the momentum uncertainty Δp by

$$(\Delta p)^2 \equiv \langle (p-p_0)^2 \rangle = \int_{-\infty}^{\infty} (p-p_0)^2\,\mathcal{P}(p)\,dp, \quad (3.83)$$

and other uncertainties in the same way.

We can now formulate the uncertainty principle. (It can be derived from what we already know about quantum mechanics, but we shall simply give it without proof.) Suppose that p and q are a conjugate pair of variables, with uncertainties Δp and Δq as defined above. The uncertainty principle states that

$$(\Delta p)(\Delta q) \geq \frac{\hbar}{2}. \quad (3.84)$$

In other words, our simultaneous knowledge of any two conjugate variables is limited by the condition that the product of their uncertainties must be greater than or at best equal to

$\hbar/2$. The better we know p, that is, the smaller we make Δp, the larger must Δq be, and vice versa.

Let us apply the uncertainty principle to the particle in a box. The average position is, of course, the center of the box, and the uncertainty in position must be something less than half the box's width. In fact, a detailed calculation with Eq. 3.82 shows that

$$\Delta x = \left[\int_{-\infty}^{\infty} (x - x_0)^2\, \mathcal{P}(x)\, dx\right]^{1/2}$$
$$= \left[\int_0^a \left(x - \frac{a}{2}\right)^2 \left(\frac{2}{a}\right) \sin^2\left(\frac{n\pi x}{a}\right) dx\right]^{1/2} = \frac{a}{2\sqrt{3}} \quad (3.82a)$$

for all values of n. (That $x_0 = a/2$ is proved in the next section.) We can obtain the momentum uncertainty much more simply. We know that the particle's kinetic energy in the nth eigenstate is $E_n = n^2\pi^2\hbar^2/2ma^2$; since $E = p^2/2m$, the absolute value of the momentum must be $|p_x| = n\pi\hbar/a$. Since we have no information on which way the particle is going, the actual value of the momentum may be either $+|p_x|$ or $-|p_x|$; the two values are equally likely, so the average momentum p_0 is zero (as we shall shortly prove analytically). The uncertainty in p_x is the rms deviation from the average, which, from Eq. 3.83, becomes

$$\Delta p_x = \left[\int_{-\infty}^{\infty} (p - 0)^2\, \mathcal{P}(p)\, dp\right]^{1/2}$$
$$= [(+p_x - 0)^2 \times \tfrac{1}{2} + (-p_x - 0)^2 \times \tfrac{1}{2}]^{1/2}$$
$$= |p_x| = \frac{n\pi\hbar}{a}. \quad (3.85)$$

The smallest possible momentum uncertainty is therefore in the lowest state, for which $n = 1$ and $\Delta p_x = \pi\hbar/a$. The product of the uncertainties is thus

$$\Delta p_x \Delta x = \left(\frac{\pi\hbar}{a}\right)\left(\frac{a}{2\sqrt{3}}\right) = 0.907\hbar, \quad (3.85a)$$

which, as Eq. 3.84 predicts, is greater than $\hbar/2$.

Suppose that we did not know the wave function for the particle in the box. We could still say that Δx must be less than $a/2$, so by the uncertainty principle Δp_x must be greater than $\hbar/a$. In an energy eigenstate $|p_x|$ has a definite value, which must be at least as much as Δp_x. Even the lowest energy level must thus satisfy the inequality

$$E = \frac{|p_x|^2}{2m} \geq \frac{(\Delta p_x)^2}{2m} > \frac{\hbar^2}{2ma^2}. \quad (3.86)$$

In short, assuming only that the system is restricted to a region of length a, we conclude that the kinetic energy cannot be less than some minimum value. This illustrates an important general principle, that all confined systems have a positive *zero-point energy* in their lowest energy level; we shall discuss other examples in Chapter 4. The existence of zero-point energy is a direct consequence of the uncertainty principle. One important, if elementary, implication of the uncertainty principle and the existence of zero-point states is the impossibility of electrons in atoms falling into their nuclei. Only by becoming extremely localized within the nuclear volume, and thereby taking on extremely high momenta and kinetic energies, could this happen—in the absence of any other process, such as a reaction of the electron within the nucleus. Without a nuclear reaction, any electron localized briefly within a nucleus would have sufficient kinetic energy and momentum to escape the nucleus. In fact, reactions called K capture processes do occur and electrons do enter nuclei but lose their identity as electrons when those processes occur.

Since energy and time are conjugate variables, we can write the uncertainty principle in the form

$$\Delta E\, \Delta t \geq \frac{\hbar}{2}, \quad (3.87)$$

where Δt is of the order of the minimum time required to measure the energy with an accuracy of ΔE. This means that to determine the energy exactly would require an infinite time of observation ($\Delta t = \infty$). How does this affect our derivation of wave functions corresponding to exact energy eigenvalues? A system in an eigenstate with energy E must obey Eq. 3.62, $\mathsf{H}\psi = E\psi$, with the Hamiltonian H including all the forces acting on the system. But since ψ is a function of spatial coordinates only, this can be true only if H does not explicitly include the time—that is, if all the forces on the system, including those we use to observe it, are independent of time. Equation 3.87 thus expresses the fact that any measurement taking a nonzero time requires the exertion of some outside force on the system. The time-independent results are limiting values that cannot be exactly obtained in any actual measurement.

3.8 Expectation Values; Summary of Postulates

Although one cannot determine a particle's position and momentum exactly, it is possible to determine their average values in a given system. By "average value" we mean the value that we would expect to find as the average of a large number of measurements. We thus call this average the *expectation value*. To find the expectation value of a given variable, one must sum or integrate over all possible values of the variable, weighting each value by the probability of its being observed. The expectation value of Q is thus defined as

$$\langle Q \rangle \equiv \sum_i Q_i\, \mathcal{P}(Q_i), \quad (3.88)$$

where $\mathcal{P}(Q_i)$ is the probability of obtaining the ith value. For example, in the throwing of two dice, Q_i may be 2, 3, 4, 5, 6, 7, 8, 9, 10, 11, or 12. The total number of possible throws is $6 \times 6 = 36$. The probabilities are $\mathcal{P}(2) = \mathcal{P}(12) = \frac{1}{36}$; $\mathcal{P}(3) = \mathcal{P}(11) = \frac{2}{36}$; $\mathcal{P}(4) = \mathcal{P}(10) = \frac{3}{36}$; $\mathcal{P}(5) = \mathcal{P}(9) = \frac{4}{36}$; $\mathcal{P}(6) = \mathcal{P}(8) = \frac{5}{36}$, and $\mathcal{P}(7) = \frac{6}{36}$. A few moments with a pencil or a calculator will show that $\langle Q \rangle = 7$. If the possible values of Q are continuous, we write

$$\langle Q \rangle \equiv \int Q\mathcal{P}(Q)\, dQ, \tag{3.89}$$

where $\mathcal{P}(Q)$ is the probability density within the range dQ. (All this will be discussed in greater detail in Sections 15.1 and 15.2.) Equations 3.88 and 3.89 are true whether the system is classical or quantum mechanical. Where quantum mechanics enters, of course, is in determining the probabilities.

For simplicity, we still consider only a one-dimensional system. Suppose that we wish to know the expectation value of x. If $\mathcal{P}(x)\, dx$ is the probability of finding a particle within the region dx around x, then the expectation value is

$$\langle x \rangle \equiv \int x\mathcal{P}(x)\, dx, \tag{3.90}$$

with the integral taken over the entire possible range of x. But for a quantum system we know that the probability density in the coordinates is given by the absolute square of the normalized wave function,

$$\mathcal{P}(x,t) = \Psi^*(x,t)\Psi(x,t) = |\Psi(x,t)|^2. \tag{3.91}$$

The quantum mechanical expectation value of x at time t is thus

$$\langle x(t) \rangle = \int x|\Psi(x,t)|^2\, dx. \tag{3.92}$$

We are interested at this point primarily in time-independent systems; if the Hamiltonian is time-independent, then $\Psi(x, t)$ is an energy eigenfunction of the form $\psi(x)e^{-i\omega t}$, and $\mathcal{P}(x)$ and $\langle x \rangle$ are also time-independent:

$$\langle x \rangle = \int x|\psi(x)|^2\, dx. \tag{3.93}$$

The expectation value of x is obtained directly from our interpretation of $|\psi|^2$; for other variables we must introduce a new postulate. It turns out that the expectation value of any other variable Q is given by an equation similar to Eq. 3.93, an integral weighted by the square of the wave function. However, one must integrate not the variable Q itself, but the quantum mechanical operator Q that represents it (for x these are the same thing). A complication arises when Q is not a simple multiplicative operator, since the order of the factors in the integral then makes a difference. The correct way to write the integral must be taken as part of the postulate: The expectation value of Q is

$$\langle Q \rangle = \int \Psi^* \mathsf{Q} \Psi\, dx. \tag{3.94}$$

Why this way? Again, because it has to be done this way to make the answers physically reasonable. We are still assuming that Ψ is normalized; if it were not normalized, the above integral would have to be divided by $\int \Psi^*\Psi\, dx$. If the system has more than one dimension, the integral is taken over all the coordinates.

Just what is the meaning of a quantum mechanical expectation value? All the physically admissible information about a system is contained in its wave function Ψ. If Ψ is an eigenfunction of the operator Q, then the system has a definite value Q' of the variable Q such that $\mathsf{Q}\Psi = Q'\Psi$, and a measurement of Q can give only the value Q'; Eq. 3.94 then reduces to $\langle Q \rangle = Q'\int\Psi^*\Psi\, dx = Q'$. If Ψ is not an eigenfunction of Q, however, all we can say about Q is that it has a certain probability distribution. Any single measurement must give one of the eigenvalues of Q, but a series of measurements on identical systems (systems with the same Ψ) will give results distributed among these values.[14] As the number of such measurements increases, the average value of Q should approach the limiting value given by Eq. 3.94.

Let us apply the averaging process to the particle in a box. Using the normalized eigenfunctions of Eq. 3.81, we have[15]

$$\begin{aligned}\langle x \rangle &= \int_0^a x|\psi(x)|^2\, dx = \left(\frac{2}{a}\right)\int_0^a x \sin^2\frac{n\pi x}{a}\, dx \\ &= \left(\frac{2}{a}\right)\left(\frac{a^2}{4}\right) = \frac{a}{2};\end{aligned} \tag{3.95}$$

just as we expected, the average position of the particle is in the middle of the box. For the momentum we must use Eq. 3.94, which yields[16]

$$\begin{aligned}\langle p_x \rangle &= \int_0^a \psi^*(x)\mathsf{p}_x\psi(x)\, dx \\ &= \left(\frac{2}{a}\right)\int_0^a \sin\frac{n\pi x}{a}\left(-i\hbar\frac{d}{dx}\right)\sin\frac{n\pi x}{a}\, dx \\ &= -i\hbar\left(\frac{2}{a}\right)\left(\frac{n\pi}{a}\right)\int_0^a \sin\frac{n\pi x}{a}\cos\frac{n\pi x}{a}\, dx = 0.\end{aligned} \tag{3.96}$$

---

[14] The probability of obtaining a value Q' (that is, the fraction of measurements that should give Q') can be shown to be

$$\mathcal{P}(Q') = \left|\int \psi_{Q'}{}^*\psi\, dx\right|^2,$$

where $\psi_{Q'}$ is the eigenfunction corresponding to Q', and ψ is the system's actual wave function; when $\psi = \psi_{Q'}$, of course, we have $\mathcal{P}(Q') = 1$.

[15] From a table of integrals (or integrating by parts) we can obtain

$$\int x \sin^2 \alpha x\, dx = \frac{x^2}{4} - \frac{x \sin 2\alpha x}{4\alpha} - \frac{\cos 2\alpha x}{8\alpha^2}.$$

We have $\alpha = n\pi/a$ and x varying from 0 to a; thus $2\alpha x$ runs from 0 to $2n\pi$, and the sine and cosine terms drop out.

[16] Since $\int_0^\pi \sin\theta\cos\theta\, d\theta = 0$.

This is what we had deduced from the fact that the particle is equally likely to be traveling in either direction. On the other hand, $\langle p_x^2\rangle$ is simply $2mE_n$, so that $\langle p_x^2\rangle \neq \langle p_x\rangle^2$; this illustrates that one must be careful in manipulating averages. We shall show in Section 15.2 that one always has

$$\left\langle(\Delta Q)^2\right\rangle \equiv \left\langle(Q-\langle Q\rangle)^2\right\rangle = \left\langle Q^2\right\rangle - \left\langle Q\right\rangle^2, \qquad (3.97)$$

where $\langle(\Delta Q)^2\rangle$ corresponds to the mean-square deviation introduced in Eq. 3.82. For the particle in a box we immediately obtain $(\Delta p_x)^2 = 2mE_n$, in agreement with Eq. 3.85.

We have now introduced all the major principles of quantum mechanics that we shall be using. However, the introduction has been piecemeal, and most of the principles have been formulated for the one-dimensional case only. This is a good point to summarize our results with a formal statement of our postulates in general form. The set of postulates we give here is not complete, but covers everything needed in this text.

POSTULATE I. Every physical system is completely described by a wave function $\Psi(q_1, \ldots, q_N, t)$, where the q_i are the coordinates[17] that define the system. The function Ψ and its first derivatives must be everywhere finite, continuous, and single-valued (except that the derivative may be discontinuous at an infinitely high potential barrier); Ψ may be real or complex.

POSTULATE II. The probability that the coordinates of the system are in the range dq_1 around $q_1, \ldots, dq_N$ around q_N is given by

$$\begin{aligned}\mathcal{P}(q_1,\ldots,q_N,t)\, dq_1 \cdots dq_N \\ = \Psi^*(q_1,\ldots,q_N,t)\Psi(q_1,\ldots,q_N,t)\, dq_1 \cdots dq_N.\end{aligned} \qquad (3.98)$$

For a single particle moving in three dimensions, for example, this probability becomes $\Psi^*\Psi\, dx\, dy\, dz$. This definition assumes a normalized wave function, one such that $\int \Psi^*\Psi dq_1 \cdots dq_N$, over all possible values of the coordinates is unity; the integral must in any case be finite.

POSTULATE III. To every variable Q of classical mechanics there corresponds a linear operator Q. These operators are constructed by the following rules: (1) If Q is one of the coordinates q_i, or the time, the operator is simply multiplication by Q. (2) If Q is the momentum p_i conjugate to the coordinate q_i, the operator is

$$\mathsf{p}_i \equiv -i\hbar\frac{\partial}{\partial q_i}. \qquad (3.99)$$

(3) If Q is some function of coordinates and momenta, the operator Q is obtained by substituting Eq. 3.99 for all the momenta.[18]

We must call special attention to the Hamiltonian operator H. As mentioned earlier, the classical Hamiltonian is the total energy expressed as an explicit function of only coordinates, momenta, and (in general) time. Since kinetic energy is given by $p^2/2m$, a single particle moving in a potential field $\mathsf{V}(x, y, z)$ has the Hamiltonian operator[19]

$$\begin{aligned}\mathsf{H} &= \frac{\mathbf{p}\cdot\mathbf{p}}{2m} + \mathsf{V}(x,y,z) = \frac{\mathsf{p}_x^2+\mathsf{p}_y^2+\mathsf{p}_z^2}{2m} + \mathsf{V}(x,y,z) \\ &= -\frac{\hbar^2}{2m}\left(\frac{\partial^2}{\partial x^2}+\frac{\partial^2}{\partial y^2}+\frac{\partial^2}{\partial z^2}\right) + \mathsf{V}(x,y,z),\end{aligned} \qquad (3.100)$$

where $\mathbf{p}$ is the vector whose components are the three momentum operators. This Hamiltonian is written in terms of Cartesian coordinates, but other coordinate systems are often more convenient; we shall discuss later how to express the Hamiltonian in those coordinates.

POSTULATE IV. Any possible measurement of the variable Q for a single atom or molecule can only yield one of the eigenvalues of the corresponding operator Q, that is, a value Q_n such that

$$\mathsf{Q}\Psi_n = Q_n\Psi_n, \qquad (3.101)$$

where Ψ_n, is the eigenfunction of Q corresponding to Q_n. To put it the other way around, any system whose wave function is an eigenfunction of Q has a definite value of Q.

POSTULATE V. All wave functions must satisfy the time-dependent Schrödinger equation,

$$\mathsf{H}\Psi(q_1,\ldots,q_N,t) = i\hbar\frac{\partial\Psi(q_1,\ldots,q_N,t)}{\partial t}. \qquad (3.102)$$

If Ψ is an eigenfunction of the energy, and thus of $i\hbar(\partial/\partial t)$, by Eq. 3.101 it must have the form

$$\Psi(q_1,\ldots,q_N,t) = \psi(q_1,\ldots,q_N)e^{-iEt/\hbar}, \qquad (3.103)$$

[17] One can formulate quantum mechanics with other quantities, for example momenta, as the independent variables. However, we shall restrict our treatment to the formulation in terms of coordinates.

[18] There may be some ambiguity in this process; for example, xp_x, and $p_x x$ are algebraically the same, but for a general function f the operations $-i\hbar x(\partial f/\partial x)$ and $-i\hbar[\partial(xf)/\partial x]$ will not give the same result. We thus need an additional rule: (4) The order of factors must be such that the resulting operator is *Hermitian*. A Hermitian operator is one for which

$$\int\cdots\int \Phi^*(\mathsf{Q}\Psi)dq_1\cdots dq_N = \int\cdots\int \Psi(\mathsf{Q}^*\Phi^*)dq_1\cdots dq_N$$

for any two wave functions Φ and Ψ. This condition will create a problem for us only when we deal with non-Cartesian coordinates.

[19] The operator

$$\frac{\partial^2}{\partial x^2}+\frac{\partial^2}{\partial y^2}+\frac{\partial^2}{\partial z^2}$$

is called the *Laplacian*, and is usually abbreviated as ∇^2 ("del squared" or "nabla squared").

where E is the energy; the spatial part of Ψ then satisfies the time-independent Schrödinger equation,

$$\mathsf{H}\psi(q_1,\ldots,q_N)=E\psi(q_1,\ldots,q_N). \quad (3.104)$$

Remember that this is possible only when H does not explicitly contain the time.

POSTULATE VI. The expectation value of the variable Q (in a system described by the normalized wave function Ψ) is

$$\langle Q\rangle=\int\cdots\int\Psi^*\mathsf{Q}\Psi\,dq_1\cdots dq_N, \quad (3.105)$$

with the integral taken over all possible values of the coordinates. From this rule and Eq. 3.97 one can derive the uncertainty principle.

We have devoted a great deal of effort to showing the plausibility of these postulates, for purely pedagogical reasons. Logically, we could have just set them down and then gone on from there; advanced texts do just that. The whole elaborate structure of quantum mechanics is justified only by its agreement with experiment; the postulates are simply the most concise statement of that structure's building blocks.

3.9 Particles in Two- and Three-Dimensional Boxes

In the remainder of this chapter we deal with particles in boxes of two or three dimensions. We shall have something to say about how one sets up and solves the Schrödinger equation in various systems, but our main concern will be with how the boundary conditions—the shapes of the boxes—affect the solutions. We begin in this section with the relatively simple case of a rectangular two- or three-dimensional box; this is a straightforward extension of our one-dimensional-box theory, but it illustrates some new concepts.

As before, we assume a potential energy that is zero inside the box and infinitely large outside. It is most convenient to treat a box with boundaries perpendicular to the Cartesian axes; we therefore assume a rectangular box extending from $x = 0$ to $x = a$, from $y = 0$ to $y = b$, and (for the three-dimensional case) from $z = 0$ to $z = c$. We shall write most of the following equations in terms of three dimensions, with the understanding that the term or factor involving z (or c) drops out in the two-dimensional case. As before, the wave function must be identically zero outside the box, giving boundary conditions analogous to Eq. 3.56:

$$\begin{aligned}\psi(0,y,z)=\psi(a,y,z)=0 &\quad \text{for all } y,z;\\ \psi(x,0,z)=\psi(x,b,z)=0 &\quad \text{for all } x,z;\\ \psi(x,y,0)=\psi(x,y,c)=0 &\quad \text{for all } x,y. \end{aligned} \quad (3.106)$$

The Hamiltonian operator is given by Eq. 3.100. Inside the box, where $V(x, y, z) = 0$, one can see that H is a sum of three terms, each involving only one of the spatial variables; we can thus write

$$\mathsf{H}=\mathsf{H}_x+\mathsf{H}_y+\mathsf{H}_z. \quad (3.107)$$

Any operator that can be divided into single-variable operators in this way is called *separable.* An operator may be separable in only one coordinate system, or in several; some are not separable at all. Clearly, if the Hamiltonian included a potential energy such as $V(x, y, z) = xyz$, it would not be separable in Cartesian coordinates. The importance of a separable Hamiltonian is that it allows us to separate the wave equation into equations in the individual variables, which are of course much easier to solve. When the Hamiltonian is separable, the energy eigenvalue of Eq. 3.104 can obviously also be written as a sum,

$$E=E_x+E_y+E_z \quad (3.108)$$

(where $\mathsf{H}_x\Psi = E_x\Psi$, etc.). In such a case the spatial part of the wave function can always be expressed as a *product* of single-variable functions,

$$\psi(x,y,z)=f(x)g(y)h(z), \quad (3.109)$$

as we shall now demonstrate.

Given Eqs. 3.107 and 3.109, the time-independent Schrödinger equation can be written as

$$\begin{aligned}&\mathsf{H}_x f(x)g(y)h(z)+\mathsf{H}_y f(x)g(y)h(z)\\ &\quad+\mathsf{H}_z f(x)g(y)h(z)=Ef(x)g(y)h(z).\end{aligned} \quad (3.110)$$

Since the operator H_x acts only on $f(x)$, the functions g and h can be brought outside it as multiplicative factors; treating the H_y and H_z terms similarly and rearranging, we have

$$\begin{aligned}g(y)h(z)\mathsf{H}_x f(x)=&\ Ef(x)g(y)h(z)-f(x)h(z)\mathsf{H}_y g(y)\\ &-f(x)g(y)\mathsf{H}_z h(z).\end{aligned} \quad (3.111)$$

We now divide both sides of the equation by $f(x)g(y)h(z)$:

$$\frac{1}{f(x)}\mathsf{H}_x f(x)=E-\frac{1}{g(y)}\mathsf{H}_y g(y)-\frac{1}{h(z)}\mathsf{H}_z h(z). \quad (3.112)$$

The left side of this equation is a function of x only, whereas the right side is a function of y and z only. Here are two expressions that are functions of entirely different variables, yet are always equal, regardless of the values of the variables. Such a condition can be satisfied only if both are equal to one and the same constant; since this constant is uniquely determined by H_x, $f(x)$, and the boundary conditions on x, it is logical to call it E_x:

$$\frac{1}{f(x)}\mathsf{H}_x f(x)=E_x. \quad (3.113)$$

This is in fact the same E_x as was defined in Eq. 3.108, since $(\mathsf{H}_x f)/f = (\mathsf{H}_x\psi)/\psi$. We see now, however, that $f(x)$ is an eigenfunction of H_x and E_x is the corresponding eigenvalue:

$$\mathsf{H}_x f(x) = E_x f(x). \tag{3.114}$$

The corresponding derivations for y and z are straightforward, yielding

$$\mathsf{H}_y g(y) = E_y g(y) \quad \text{and} \quad \mathsf{H}_z h(z) = E_z h(z); \tag{3.115}$$

E_y and E_z are also the same as in Eq. 3.108, and we have thus proved the validity of Eq. 3.109.

It should be apparent that the above technique is applicable to any eigenvalue equation in which the operator is separable, and we shall have many further occasions to apply it. In each case, an equation in all the variables is replaced by separate equations in the individual variables. Physically, we can say that the behavior of each variable is not affected by any other variable when there are no interactions (i.e., terms in the Hamiltonian) involving both variables. Separability is a property that is particularly useful when we wish to consider systems of several particles, such as many-electron atoms and molecules.

We can now obtain the wave functions and energy levels for a particle in our box. Consider first Eq. 3.114. Since $\mathsf{H}_x = (-\hbar^2/2m)(\partial^2/\partial x^2)$, the form of this equation is identical to that of Eq. 3.64. The boundary conditions in x are also the same, and we can immediately write down the solution

$$\mathsf{H}_x = (-\hbar^2/2m)(\partial^2/\partial x^2) \tag{3.116}$$

corresponding to Eq. 3.81; the corresponding eigenvalues are

$$(E_x)_{n_1} = \frac{n_1^2\pi^2\hbar^2}{2ma^2} \qquad (n_1 = 1, 2, \ldots). \tag{3.117}$$

The solutions for y and z are similar, and substitution in Eq. 3.109 gives

$$\psi_{n_1n_2n_3}(x,y,z) = \left(\frac{8}{abc}\right)^{1/2} \sin\frac{n_1\pi x}{a}\sin\frac{n_2\pi y}{b}\sin\frac{n_3\pi z}{c} \qquad (n_1, n_2, n_3 = 1, 2, \ldots) \tag{3.118}$$

for the complete spatial wave function, with a total energy of

$$E_{n_1n_2n_3} = \frac{\pi^2\hbar^2}{2m}\left(\frac{n_1^2}{a^2} + \frac{n_2^2}{b^2} + \frac{n_3^2}{c^2}\right) \qquad (n_1, n_2, n_3 = 1, 2, \ldots). \tag{3.119}$$

The energy-level spectrum of a particle in a two- or three-dimensional box is naturally more complex than that for a one-dimensional box, because each of the quantum numbers n_1, n_2, n_3 can assume any positive integral value. In a box of arbitrary dimensions, the lengths a, b, c are usually all different; more specifically, they are usually incommensurate with one another, that is, no one length can be expressed as a rational fraction multiplying either of the others. So long as this is the case, every energy level is uniquely defined by a specific set of quantum numbers n_1, n_2, n_3. Change or rearrange any of the numbers, and you are referring to a different energy level. As with the particle in a one-dimensional box, the momentum and its components are not constants of the motion because of the boundaries on the box. But as previously, p_x^2, p_y^2, and p_z^2, and the absolute values of p_x, p_y, and p_z are preserved. Again, we can use the de Broglie condition to relate the wavelength λ of the matter wave, $\lambda = h/p$, to the energy $p^2/2m$, but here, separability of x, y, and z motion allows us to apply our earlier argument for one dimension to each of the components independently.

A very important special situation arises if two or three of the lengths a, b, c become equal. Let us suppose that $a = b$. Then, with a little rearrangement, Eq. 3.119 becomes

$$E - \frac{\pi^2\hbar^2 n_3^2}{2mc^2} = \frac{\pi^2\hbar^2}{2ma^2}(n_1^2 + n_2^2). \tag{3.120}$$

In this situation, for a given value of n_3 there is clearly more than one way of obtaining certain of the energy levels, namely, those for which n_1 is not equal to n_2. Consider the state for which $n_1 = N$ and $n_2 = N'$; the state with $n_1 = N'$ and $n_2 = N$ must have the same energy, since such an interchange leaves the sum $n_1^2 + n_2^2$ in Eq. 3.120 unchanged. If we assumed $a = b = c$, then any states with the same total $n_1^2 + n_2^2 + n_3^2$ would have the same energy; if n_1, n_2, n_3 are all different, there are at least six such states. A situation in which two or more states lie at exactly the same energy is called a *degeneracy.* The n states at the same energy are said to be *degenerate,* and the corresponding energy level to exhibit an n-fold degeneracy. Degeneracy always arises from some natural symmetry of the system. In the first example, we said that the x and y dimensions of the box were indistinguishable. We do not change the physics of the problem if we interchange the names on the x and y axes and thereby interchange n_1 and n_2. A state containing an amount of kinetic energy proportional to n_1^2 in the x direction, and an amount proportional to n_2^2 in the y direction, is physically indistinguishable from the state in which the two kinetic energy terms are interchanged. The energy levels of the particle in a two-dimensional box, for the cases $a \neq b$ and $a = b$, are illustrated in Fig. 3.8.

The eigenfunctions corresponding to degenerate states have the property that we can add or subtract them as we choose and never produce a new combination having an energy different from the original value.[20] This is exactly

---

[20] If $\mathsf{H}\psi_1 = E_1\psi_1$ and $\mathsf{H}\psi_2 = E_1\psi_2$ then for any function ψ_3 given by $\psi_3 = \alpha\psi_1 \pm \beta\psi_2$ we have $\mathsf{H}\psi_3 = E_1\psi_3$.

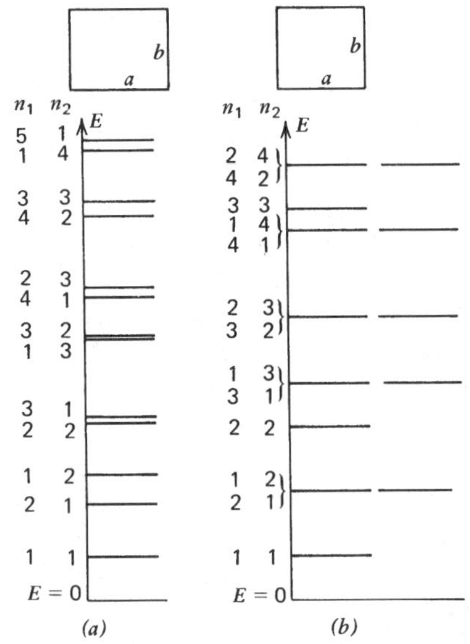

Figure 3.8 Energy levels for a particle in a two dimensional box. (*a*) Unequal sides (levels drawn for $a = 1.25b$), so that all states are uniquely associated with their own energies, (*b*) Equal sides ($a = b$), so that all energy levels with $n_1 \neq n_2$ are degenerate. The energy scales of parts (*a*) and (*b*) are the same if the two boxes have the same area. [The reader may be amused to realize that the example with $a = b$ is richer and more subtle than has been indicated thus far. Additional degeneracies occur with this system because of the existence of "doubly magic" numbers, numbers that can be written in two essentially different ways as the sum of two squares. The states with $n_1 = 7$, $n_2 = 1$ with $n_1 = 1$, $n_2 = 7$, and with $n_1 = n_2 = 5$ are all degenerate, for example. See G. B. Shaw, *J. Phys. A* **7,** 1537 (1974) for a discussion of this problem.]

parallel to what we were able to do for the free particle in one dimension, where a twofold degeneracy is associated with the two directions the momentum can have: We saw that the traveling waves associated with states of definite momentum in the positive and negative x directions could be combined to give a pair of standing waves, Eqs. 3.42, which had the same energy as the traveling waves but were not eigenfunctions of momentum. In the present case, we could combine the two solutions for one of the degenerate energy levels to form a new pair of standing-wave functions.

3.10 Particles in Circular Boxes

A rectangular box of particles is a reasonable model for a macroscopic gas, but hardly for the structure of an atom or molecule. We frequently find in such systems that a particular point forms the natural center of the system, so that distance from the center—say the distance of an electron from a nucleus, or of one atom from another in a molecule—is important but orientation is irrelevant: The electron's energy does not depend on the side of the nucleus where it is found. Any such problem has *circular* or *spherical symmetry,* depending on whether two or three dimensions are involved.

We shall want to treat systems like this, and a good way to approach them is by considering particles in round boxes. This treatment will introduce the coordinate systems used and give some physical intuition for the symmetries and constants of motion. We therefore consider the two-dimensional circular box in this section, the spherical box in the next.

For the two-dimensional system, we wish to examine the effects of circular symmetry. We therefore assume that the potential energy of our particle is a function only of the distance r from some origin, $V = V(r)$, independent of any angular variables giving the orientation about the origin. We must have a circular boundary, of course, which we set at $r = R$. This is to be a box similar to those of the preceding sections, so we define the potential energy as zero for $r < R$, infinite for $r \geq R$. The next step would be to state the Hamiltonian in circular coordinates, but before doing this let us first consider the corresponding classical problem, which can give us some useful insights.

The coordinate system we use is ordinary circular polar coordinates, the radius r from the origin and the angle ϕ measured from the x axis. These coordinates are illustrated in Fig. 3.9, along with their relationships to Cartesian coordinates. The total energy of a particle of mass μ (we shall need m for a quantum number) is simply

$$E = \frac{1}{2\mu}(p_x^2 + p_y^2) + V(x, y), \qquad (3.121)$$

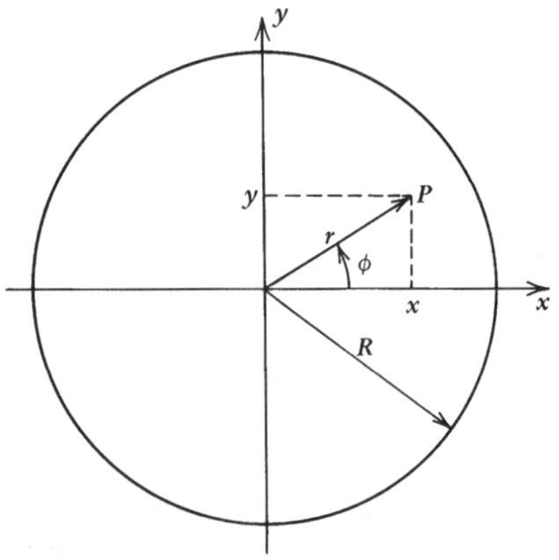

Figure 3.9 Circular polar coordinates and the circular box. The point P can be described by the circular coordinates r, ϕ or the Cartesian coordinates x, y. The two coordinate systems are connected by the conversion equations

$$x = r\cos\phi, \quad y = r\sin\phi$$

or

$$r = (x^2 + y^2)^{1/2}, \quad \phi = \arctan\frac{(y)}{(x)}.$$

The heavy line is the boundary of circular box of radius R.

which we wish to convert to circular coordinates. We have already assumed that $V(x, y)$ becomes simply $V(r)$, but how do we express the momenta? We can obtain these by first putting the transformation equations in differential form. We have

$$dx = d(r\cos\phi) = \cos\phi\, dr - r\sin\phi\, d\phi,$$
$$dy = d(r\sin\phi) = \sin\phi\, dr + r\cos\phi\, d\phi. \quad (3.122)$$

Substituting these into the expressions for the momenta, we obtain

$$p_x = \mu\frac{dx}{dt} = -\mu r\sin\phi\frac{d\phi}{dt} + \mu\cos\phi\frac{dr}{dt} \quad (3.123)$$

and

$$p_y = \mu\frac{dy}{dt} = \mu r\cos\phi\frac{d\phi}{dt} + \mu\sin\phi\frac{dr}{dt}. \quad (3.124)$$

When we write $p_x^2 + p_y^2$ the two cross terms $[\pm\mu r^2 \sin\phi\cos\phi(d\phi/dt) \times (dr/dt)]$ cancel each other, and the energy becomes

$$E = \frac{\mu}{2}\left[r^2(\sin^2\phi + \cos^2\phi)\left(\frac{d\phi}{dt}\right)^2 + (\cos^2\phi + \sin^2\phi)\left(\frac{dr}{dt}\right)^2\right] + V(r)$$
$$= \frac{\mu r^2}{2}\left(\frac{d\phi}{dt}\right)^2 + \frac{\mu}{2}\left(\frac{dr}{dt}\right)^2 + V(r), \quad (3.125)$$

since $\sin^2\phi + \cos^2\phi = 1$. The first term in Eq. 3.125 can be written in several ways:

$$\frac{\mu r^2}{2}\left(\frac{d\phi}{dt}\right)^2 = \frac{\mu}{2}\left(r\frac{d\phi}{dt}\right)^2$$
$$= \frac{\mu}{2}v_\phi^2 = \frac{(\mu v_\phi r)^2}{2\mu r^2} = \frac{p_\phi^2}{2I}. \quad (3.126)$$

We have introduced three definitions in these equations:

$$v_\phi \equiv r\frac{d\phi}{dt} \quad (3.127)$$

is the instantaneous velocity perpendicular to the radial direction;

$$p_\phi \equiv \mu v_\phi r \quad (3.128)$$

is our old friend the angular momentum, written with the subscript ϕ to indicate the angle that changes during the rotational motion; and

$$I \equiv \mu r^2 \quad (3.129)$$

is the *moment of inertia,* of which we shall say more later. In the second term of Eq. 3.125 we simply define

$$p_r \equiv \mu\frac{dr}{dt}, \quad (3.130)$$

the momentum in the radial direction. Substituting these definitions in Eq. 3.125, we obtain

$$E = \frac{p_\phi^2}{2I} + \frac{p_r^2}{2\mu} + V(r). \quad (3.131)$$

The first term in Eq. 3.131 appears to concern only the kinetic energy of angular motion, the second term is the kinetic energy of radial motion, and the third term is the potential energy. So long as V is a function only of r, and not of angle, there is no force acting to affect the angular momentum p_ϕ. Hence, if $V = V(r)$, then p_ϕ must be a constant of the motion. This is a specific example of a completely general and very powerful tool in physics and chemistry, whether in the classical or the quantum mechanical description: *If the energy of a system is independent of some coordinate q_i, then p_i is a constant of the motion of the system.* This principle can be restated in other ways. For example, if the energy is completely independent of some coordinate q, then the uncertainty Δq in this variable is arbitrarily large, and therefore the uncertainty Δp in the conjugate momentum p can be made as small as we choose. For the free particle we can choose the origin $x = 0$ anywhere we wish without affecting the energy, and the momentum is thus a constant. In the present case, energy is independent of the axis from which ϕ is measured. These examples indicate further how the constants of the motion reflect some type of symmetry inherent in the system, as pointed out earlier.

Let us now proceed to the quantum mechanical problem, for which we first need the Hamiltonian operator. One might think that we could simply substitute $-i\hbar(\partial/\partial\phi)$ for p_ϕ and $-i\hbar(\partial/\partial r)$ for p_r in Eq. 3.131, as we have previously done with momenta in Cartesian coordinates. Unfortunately, with non-Cartesian coordinates this method often does not give the correct operators, for mathematical reasons that we need not discuss here.[21] However, we already know the Hamiltonian operator in Cartesian coordinates, in this case

$$\mathsf{H} = -\frac{\hbar^2}{2\mu}\left(\frac{\partial^2}{\partial x^2} + \frac{\partial^2}{\partial y^2}\right) = \mathsf{V}(x, y), \quad (3.132)$$

and this can be transformed directly to the equivalent expression in circular coordinates. The details of this

[21] The basic problem is that the resulting operators may not be Hermitian (see footnote 17 on page 73). A Hermitian operator, Eq. 3A.7, is obtained if one replaces p_r by $\mathsf{r}^{-1}\mathsf{p}_r\mathsf{r}$ and *then* substitutes $-i\hbar(\partial/\partial r)$ for p_r.

transformation are given in Appendix 3A; here we need only the result,

$$\mathsf{H} = -\frac{\hbar^2}{2\mu}\left(\frac{\partial^2}{\partial r^2} + \frac{1}{r}\frac{\partial}{\partial r} + \frac{1}{r^2}\frac{\partial^2}{\partial \phi^2}\right) + \mathsf{V}(r). \qquad (3.133)$$

Inside the circular box we have $\mathsf{V}(r) = 0$, and the time-independent Schrödinger equation becomes

$$-\frac{\hbar^2}{2\mu}\frac{\partial^2 \psi(r,\phi)}{\partial r^2} - \frac{\hbar^2}{2\mu r}\frac{\partial \psi(r,\phi)}{\partial r} - \frac{\hbar^2}{2\mu r^2}\frac{\partial^2 \psi(r,\phi)}{\partial \phi^2} = E\psi(r,\phi), \qquad (3.134)$$

where $\psi(r, \phi)$ is the spatial part of the wave function.

So far, of course, we have done nothing but transform coordinates; the shape of the box does not enter the problem until we introduce the boundary conditions. As in our previous models, the wave function must vanish at the wall of the box where $r = R$, giving the condition

$$\psi(R,\phi) = 0 \qquad \text{for all } \phi. \qquad (3.135)$$

We also need a boundary condition in the angular variable. The wave function must be a single-valued function of position in space, but real space corresponds only to values of ϕ between 0 and 2π. Formally, ϕ can go beyond these limits, but this merely represents going around the same circle again. Two values of ϕ whose difference is exactly 2π (or any integral multiple of 2π) describe the identical physical location. The wave function must therefore be periodic in ϕ, with a period of 2π; otherwise a particular point in real space would correspond to more than one possible value of ψ. Mathematically, this means we must require that

$$\psi(r,\phi) = \psi(r,\phi \pm 2\pi) \quad \text{for all } r,\phi. \qquad (3.136)$$

The Hamiltonian of Eq. 3.133 is not directly separable; however, multiplying through by r^2 gives an operator that *is* separable. We can therefore write the wave function as a product of functions of r and ϕ separately,

$$\psi(r,\phi) = f(r)g(\phi). \qquad (3.137)$$

We could substitute this in the Schrödinger equation and proceed as we did following Eq. 3.109; however, a simpler method is available to evaluate the angular function. We said above that, since ϕ does not appear explicitly in the energy, the angular momentum p_ϕ must be a constant of the motion. In quantum mechanical terms, this means that p_ϕ is an eigenvalue of the angular momentum operator p_ϕ, which is given by

$$\mathsf{p}_\phi \equiv -i\hbar\frac{\partial}{\partial \phi} \qquad (3.138)$$

(see Appendix 3A for proof). It seems logical to assume that $g(\phi)$, the angular part of the wave function, is an eigenfunction of p_ϕ. We can thus write

$$-i\hbar\frac{\partial g(\phi)}{\partial \phi} = p_\phi g(\phi), \qquad (3.139)$$

where p_ϕ is a constant. The solution to this equation is simply

$$g(\phi) = Ae^{ip_\phi \phi/\hbar}, \qquad (3.140)$$

as can be verified by direct differentiation.

Now we can apply the angular boundary condition, which in terms of $g(\phi)$ becomes

$$g(\phi) = g(\phi \pm 2\pi). \qquad (3.141)$$

Since the function e^{ix} has a period of 2π in x, we can satisfy this condition with Eq. 3.140 only if $p_\phi\phi/\hbar$ changes through an integral multiple of 2π while ϕ is changing through 2π. In other words, p_ϕ must be an integral multiple of $\hbar$,

$$p_\phi = m\hbar \quad (m = 0, \pm 1, \pm 2, \ldots), \qquad (3.142)$$

and the angular part of the wave function has the form

$$g(\phi) = Ae^{im\phi} \quad (m = 0, \pm 1, \pm 2, \ldots). \qquad (3.143)$$

(Keep in mind that m is a quantum number, whereas the mass is μ.) The real part of this function behaves as $A \cos m\phi$, and the imaginary part as $A \sin m\phi$; the two have identical shapes, but differ by 90° in phase. Note that the angular part of the probability density, $g^*(\phi)g(\phi)$, is a constant, giving the expected result that all values of ϕ are equally likely.

Equation 3.142 states that the angular momentum is quantized.[22] The allowable values (except for $m = 0$) occur in pairs, corresponding to clockwise and counterclockwise motion with the same absolute value of the angular momentum. Since the angular momentum appears only as p_ϕ^2 in the expression for the energy, $+m$ and $-m$ must correspond to the same energy level; all the levels except that with $m = 0$ are thus doubly degenerate.

We have not yet considered the radial part of the wave function, $f(r)$. The technique used in evaluating $f(r)$ is straightforward but tedious. One substitutes Eq. 3.137 in the Schrödinger equation, reduces to obtain an equation involving r alone, and solves that equation for $f(r)$. Neither the details of this derivation nor the exact form of the solution need be considered here, though some additional informa-

[22] It will be recalled from Section 2.7 that in circular motion the action per cycle is 2π times the angular momentum. Here this would give an action of mh, in agreement with our earlier assumptions on the quantization of action.

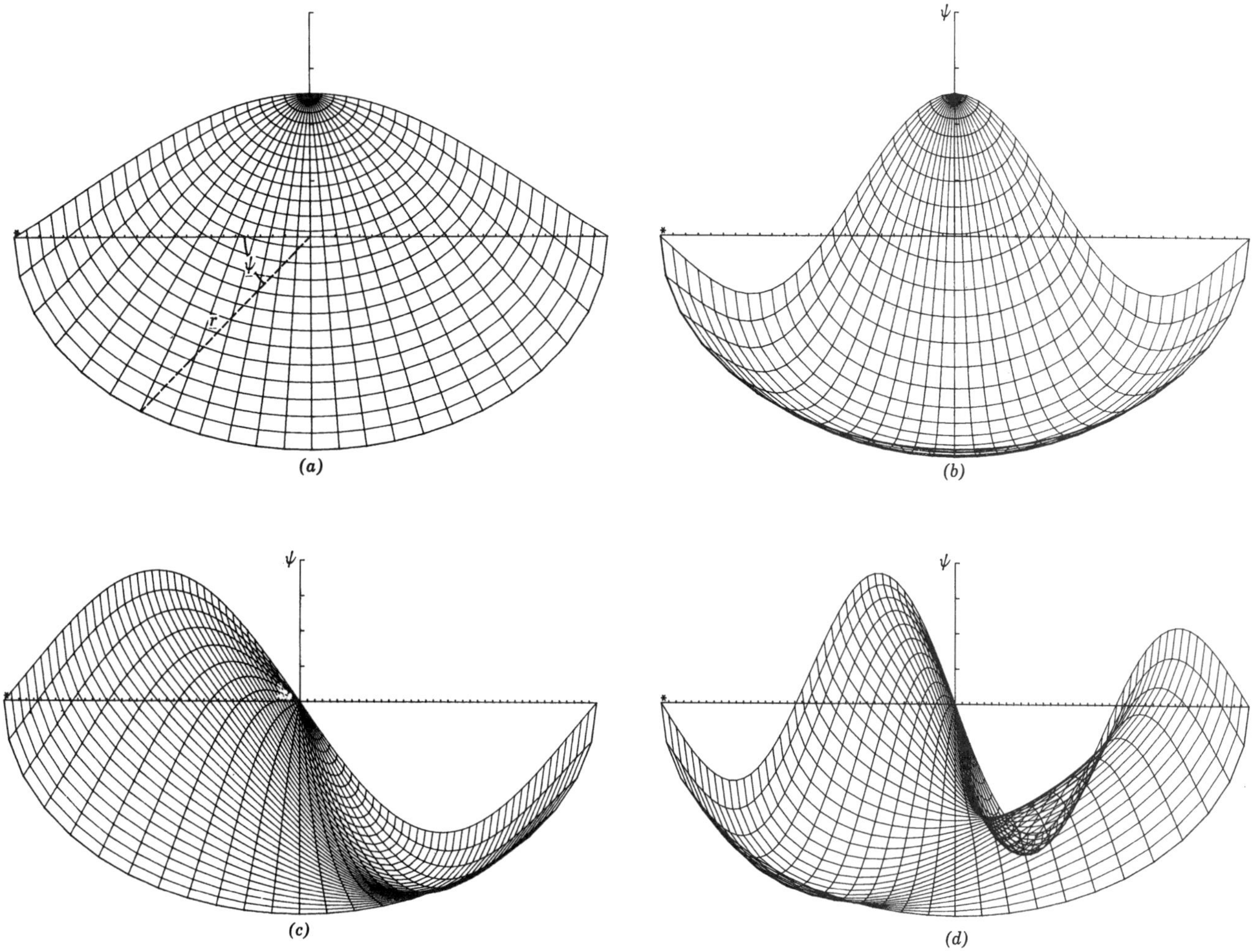

Figure 3.10 Standing waves in a circular box (perspective views). These are the wave functions of Eq. 3.144, with the quantum numbers m and n of: (*a*) 0, 1; (*b*) 0, 2; (*c*) 1, 1; (*d*) 1, 2. Radial nodes are circles, angular nodes are radii in the plane corresponding to $\Psi = 0$. Only half the circle is shown; the semicircle "behind" the plane of the page is omitted.

tion is given in Appendix 3A. The following points about the solution are significant: The function $f(r)$, of course, describes a standing wave in the radial direction. The form of this function depends explicitly on $|m|$, the absolute value of the angular momentum quantum number; and for each value of $|m|$ there is an infinite set of energy eigenvalues that we can enumerate by a second quantum number n, which takes on the values of all the positive integers. The complete spatial wave function can thus be written as[23]

$$\psi_{n,m}(r,\phi) = Af_{n,|m|}(r)e^{im\phi}. \qquad (3.144)$$

[23] The constant A can be evaluated by normalization over the area of the box. The area element in circular coordinates is $r\,dr\,d\phi$, so for one particle in the box we must have

$$\int_{\phi=0}^{2\pi}\int_{r=0}^{R}\left|\psi(r,\phi)\right|^2 r\,dr\,d\phi = 1.$$

As we explain in Appendix 3A, there is no way to express the radial function $f(r)$ directly in terms of simple (algebraic, trigonometric, or exponential) functions, as we have been able to do with other wave functions. We can nevertheless say something about this solution. It turns out that the forms of the functions $\psi(r, \phi)$ described by Eq. 3.118 are exactly the same as the possible forms of a standing wave on a circular drumhead. This parallels our earlier analogy between the wave functions in a one-dimensional box and the vibrations of a string with both ends fixed: In each case, the boundary conditions are mathematically equivalent. Figure 3.10 illustrates some of the simpler standing waves given by Eq. 3.144; for comparison, Fig. 3.11 shows some standing waves for a rectangular box (or drumhead), that is, the functions of Eq. 3.118 in two dimensions. Physical analogies such as those introduced here are common in physics and chemistry: In innumerable cases one finds systems that are very different physically, yet are described by the same mathematical equations. The recognition of these analogies frequently enables us to economize our efforts in solving new

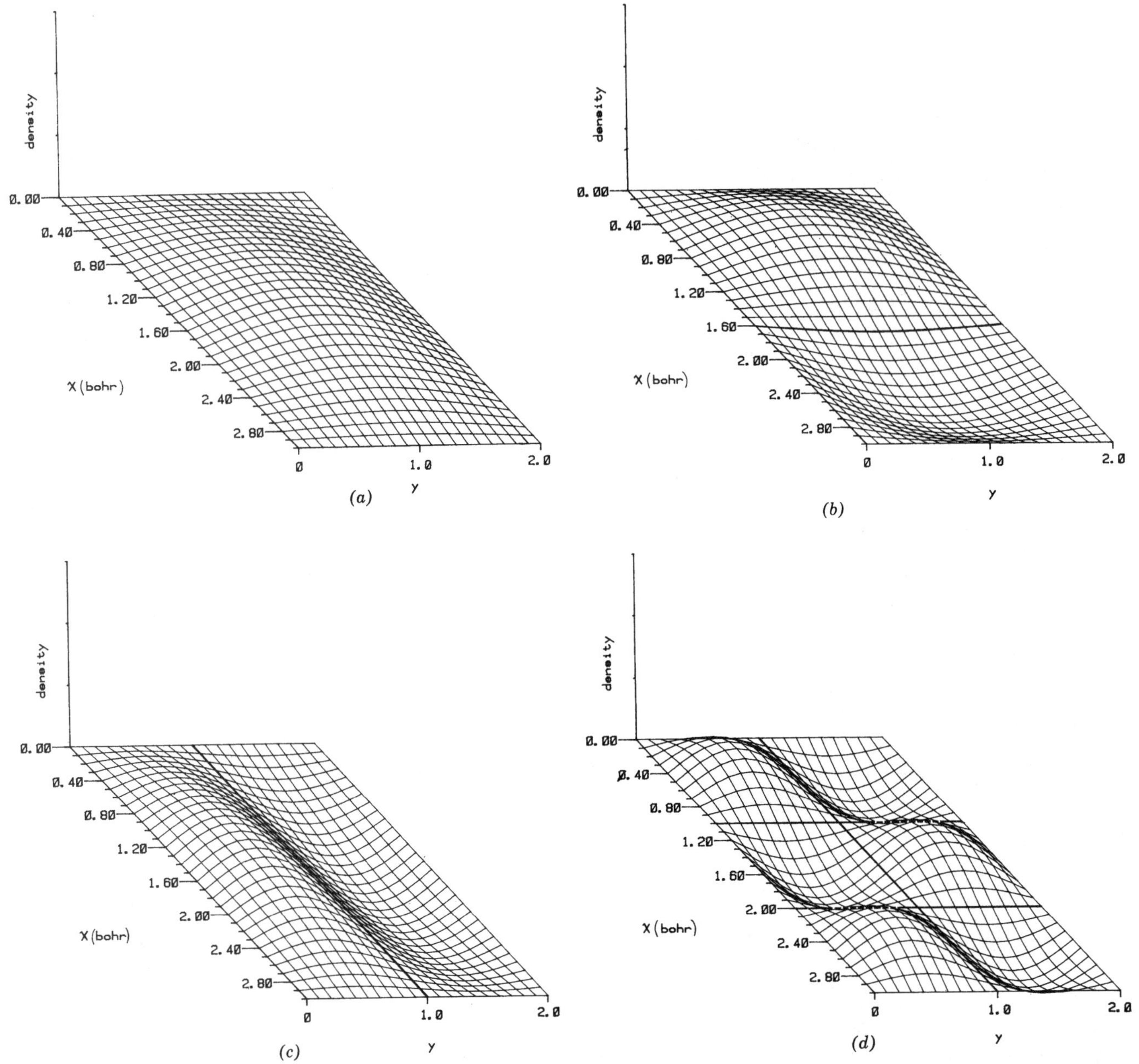

Figure 3.11 Standing waves in a rectangular box (perspective views). The quantum numbers n_1 and n_2 are those used in Eq. 3.118 with (*a*) $n_1 = n_2 = 1$; (*b*) $n_1 = 2$, $n_2 = 1$; (*c*) $n_1 = 1$, $n_2 = 2$; (*d*) $n_1 = 3$, $n_2 = 2$; see Fig. 3.8*a* for the energy levels corresponding to these states. The lines drawn across the rectangle indicate the nodes of the wave function.

problems; in addition, we often find quick and penetrating insights into new systems as we discover exact or nearly exact analogies for them in systems already known to us.

The energy eigenvalues cannot be expressed in simple form either, but some of them are plotted in Fig. 3.12. Each value of $|m|$ has its own set of energy levels, and all those except $|m| = 0$ are doubly degenerate. The quantum number n is an index number that is necessary but not sufficient to specify the state and its energy; n is not directly related to the energy by any simple equation like Eq. 3.71. In general, any two levels with different angular momenta (different $|m|$) also have different energies. The quantity actually plotted in Fig. 3.12 is the dimensionless $\mu R^2 E/\hbar^2$, which is a function of only the quantum numbers; the energy of a given state is thus inversely proportional to the square of the box's radius R.

Notice that by changing our problem ever so slightly, simply by going from a square box to a round box, we have entered a realm where the mathematics begins to look unfamiliar and much more complex. Yet the basic physics of the problem remains very similar. In the square and circular cases, and in any other two-dimensional box with a simple boundary, we have every expectation that the wave function of the lowest energy state will have no nodes in the plane, and that the next higher state will have one node. There may be more than one way of placing this node, and the different

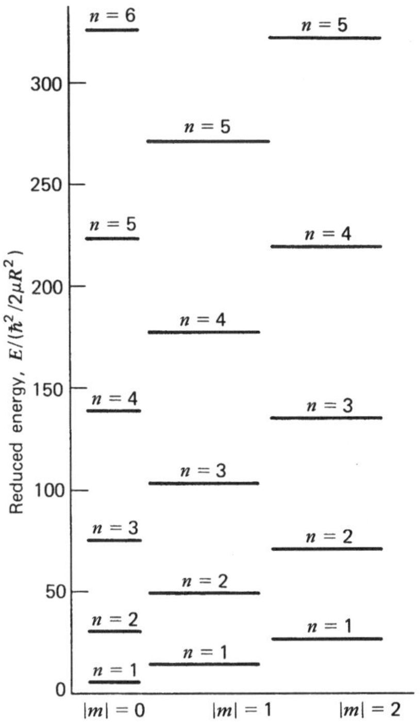

Figure 3.12 Energy levels for a particle in a circular box of radius R. All levels except those with $|m| = 0$ are doubly degenerate. The wave function with quantum numbers n, m has $n + |m| - 1$ nodes: $n - 1$ are radial and m are angular. Note how the energy tends to increase with the number of nodes.

ways of placing the node may correspond to somewhat different energy values. The next functions will have two nodes, and so on.[24] The number of levels will be countably infinite, and in general their spacing grows as the energy gets higher and higher. This similarity is not surprising, since the various boxes form part of a continuous set. Suppose we replace the drumhead of our analogy by a rubber sheet stretched on a flexible rim. As we deform the rim from one shape to another, the standing-wave patterns will vary continuously. In the same way, deformation of our box boundary causes the wave functions and their corresponding energies to vary continuously, but without changing the overall pattern. Degeneracy, of course, occurs only when the box has some kind of symmetry, but this may be subtle; see Fig. 3.8 caption.

Relating the nodal character of a wave function in a circular box to the de Broglie relation is a bit more complicated than it was for a rectangular box. The angular nodes in the real and imaginary parts of $e^{im\phi}$ correspond to the radii, the lines of constant ϕ, marking well-defined wavelengths for angular motion. These wavelengths correspond precisely to the constant angular momentum of the particle in a circular box. However the radial "wavelength" is not a constant; the momentum operator does not give back a multiple of $f_{n,|m|}(r)$ when it operates on this function, and the energy is not simply a sum of independent radial and angular parts. Nevertheless, we can recognize that the more radial nodes $f_{n,|m|}(r)$ has, the more will be the average radial kinetic energy of the particle. More nodes and more waves in a fixed interval along a given direction always mean more curvature, more mean-square average momentum, and more average kinetic energy along that direction.

So far we have considered only problems in which the potential energy is zero (or infinite). Let us briefly examine what happens when this limitation is removed. We again consider a particle in a circular box, but now suppose that the classical potential energy $V(r)$ or its quantum counterpart $\mathsf{V}(r)$ within the box is not constant. We retain the assumption of circular symmetry, so that V depends only on the radial distance r and not on the angle ϕ. We can then still make the separation of Eq. 3.137 and obtain the same angular wave function as before, Eq. 3.143. However, the right side of Eq. 3.134 is replaced by $[E - \mathsf{V}(r)]\psi(r, \phi)$, so that an additional term involving $\mathsf{V}(r)$ will appear in the radial wave equation. The resulting equation will be somewhat harder to solve—how much harder depends on the form of the function $\mathsf{V}(r)$. Nevertheless, the total wave function will still be describable by Eq. 3.144, and the general behavior of the solutions will follow the basic pattern that we have just described. In this case, the analogous vibration problem is that of a vibrating circular drumhead of nonuniform thickness. The nonzero potential energy thus makes no essential difference to the problem (except to complicate the mathematics); here and in general, it is the symmetry and the boundary conditions that determine the basic form of the solutions.

Returning to the zero-potential-energy problem, let us take another look at the form of Eq. 3.144. The spatial wave function obtained is of course complex (except for $m = 0$), because of the angular factor $e^{im\phi}$. However, we can generate real wave functions in much the same way as we obtained Eqs. 3.42, by combining two functions with the same values of $|m|$ and n:

$$\begin{aligned}\psi'_{n,|m|}(r,\phi) &= \frac{\psi_{n,m}(r,\phi)+\psi_{n,-m}(r,\phi)}{2}\\ &= \frac{A}{2} f_{n,|m|}(r)(e^{im\phi}+e^{-im\phi})\\ &= Af_{n,|m|}(r)\cos m\phi;\\ \psi''_{n,|m|}(r,\phi) &= \frac{\psi_{n,m}(r,\phi)-\psi_{n,-m}(r,\phi)}{2i}\\ &= \frac{A}{2i} f_{n,|m|}(r)(e^{im\phi}-e^{-im\phi})\\ &= Af_{n,|m|}(r)\sin m\phi.\end{aligned} \tag{3.145}$$

The functions in 3.145 are not eigenfunctions of angular momentum, just as those in Eqs. 3.42 were not eigenfunctions of linear momentum. Nevertheless, they are still just as good eigenfunctions of *energy* as the original function,

[24] These sets will in general overlap in energy: For example, the state $|m| = 2$, $n = 1$, with two nodes, has a lower energy than $|m| = 0$, $n = 2$, with one node. But for each value of $|m|$ the number of nodes increases directly with energy.

3.144. At the moment, these functions may seem like nothing but mathematical curiosities. However, we shall later face problems in which the angular momentum no longer has a good quantum number, that is, is not conserved for a particular particle. In such a case functions like Eq. 3.145 are more useful than those like Eq. 3.144.

3.11 Particles in Spherical Boxes

The next step to take beyond the circular box is obviously a consideration of its three-dimensional analog, the spherical box. We are interested primarily in what new features are introduced by the added dimension. As just pointed out, the symmetry of a problem governs its solution in a more fundamental way than does the detailed form of the potential energy (although, of course, the potential and the boundary conditions fix the symmetry). The spherical box is therefore of particular interest, since we expect an atom to have spherical symmetry around its nucleus. We shall not give the calculations in as great detail as in the previous section, since they are for the most part similar in form but more complicated.

The system we consider is a particle of mass μ in a spherical box of radius R. The potential energy is $\mathsf{V}(r)$, a function only of the distance from the center of the sphere; we assume $\mathsf{V}(r)$ to equal zero for $r < R$ and to be infinite for $r \geq R$. We shall use the spherical coordinates r, θ, ϕ defined in Appendix 2B, which should be consulted for a diagram. As with the circular box, we begin by outlining the relevant classical relationships.

In three dimensions the kinetic energy of a particle of mass μ is

$$T = \frac{1}{2\mu}(p_x^2 + p_y^2 + p_z^2) = \frac{1}{2\mu}\mathbf{p}\cdot\mathbf{p}, \qquad (3.146)$$

where $p_x = \mu(dx/dt)$, and so on. Using the conversion relations from Appendix 2B, we can transform each of the momenta to spherical coordinates by a method like that of Eqs. 3.96ff. The kinetic energy then becomes

$$T = \frac{1}{2\mu}\left(p_r^2 + \frac{p_\theta^2}{r^2} + \frac{p_\phi^2}{r^2 \sin^2\theta} \right), \qquad (3.147)$$

where

$$p_r \equiv \mu \frac{dr}{dt} \qquad (3.148)$$

is the *classical* radial momentum,

$$p_\theta \equiv \mu r^2 \frac{d\theta}{dt} \qquad (3.149)$$

is the angular momentum corresponding to rotation in the plane formed by the radius vector $\mathbf{r}$ and the z axis, and

$$p_\phi \equiv \mu(r\sin\theta)^2 \frac{d\phi}{dt} \qquad (3.150)$$

is the angular momentum around the z axis (i.e., for rotation in the xy plane, in which $r \sin\theta$ is the projection of $\mathbf{r}$). Equation 3.147 corresponds to Eq. 3.131 for the two-dimensional case; the resemblance is closer if we write μr^2 as I, the moment of inertia.

The total angular momentum around the origin is defined as the vector

$$\mathbf{L} \equiv \mathbf{r} \times \mathbf{p}, \qquad (3.151)$$

in terms of the cross product introduced in Eqs. 1.17. If the motion of the particle is in a plane (as it must be in the absence of external forces), the vector $\mathbf{L}$ is directed perpendicular to that plane. In terms of the spherical coordinates, the square of its magnitude can be shown to be given by

$$L^2 = \mathbf{L}\cdot\mathbf{L} = p_\theta^2 + \frac{p_\phi^2}{\sin^2\theta}. \qquad (3.152)$$

The kinetic energy can thus also be written as

$$T = \frac{p_r^2}{2\mu} + \frac{L^2}{2I} \qquad (I \equiv \mu r^2), \qquad (3.153)$$

i.e., the sum of a part associated with radial momentum and a part associated with the total angular momentum.

Besides the total angular momentum, it is convenient to single out one particular axis and consider the angular momentum around that axis. In our spherical coordinate system the mathematics is simplest if we choose this axis to be the z axis, the angular momentum around which (L_z) is simply p_ϕ. Since the kinetic energy is independent of the angle ϕ, its conjugate momentum p_ϕ must be a constant of the motion. Another such constant is of course the total angular momentum, which is conserved as long as no external torque is applied to the system.

Let us now consider the quantum mechanical problem. The Hamiltonian operator in spherical coordinates turns out to be[25]

$$\mathsf{H} = -\frac{\hbar^2}{2\mu}\left[\frac{1}{r^2}\frac{\partial}{\partial r}\left(r^2 \frac{\partial}{\partial r} \right) + \frac{1}{r^2 \sin\theta}\frac{\partial}{\partial\theta}\left(\sin\theta \frac{\partial}{\partial\theta} \right) + \frac{1}{r^2\sin^2\theta}\frac{\partial^2}{\partial\phi^2} \right] + \mathsf{V}(r), \qquad (3.154)$$

which can be compared with Eq. 3.133 for the circular case. We give this merely for purposes of illustration, since we have no intention of describing how the wave equation is solved. However, it is worthwhile to consider certain general

[25] This can be derived from the Hamiltonian in Cartesian coordinates by a method like that used in Appendix 3A.

characteristics of the solution that depend on the spherical symmetry. Whenever the potential energy is a function of r only, the Schrödinger equation can be split into equations separately involving the radial coordinate r, the angle θ, and the angle ϕ. The wave function is then a product of functions involving the three variables separately,

$$\psi(r,\theta,\phi) = f(r)\Theta(\theta)\Phi(\theta). \quad (3.155)$$

The radial function $f(r)$ will in general depend strongly on the form of the potential energy; even for $V(r) = 0$ it will be a more complicated function than in the circular box, subject to the same boundary condition $f(R) = 0$. We shall say no more about it. The angular functions, however, have several interesting aspects.

Consider first the function involving the angle ϕ. Since ϕ plays exactly the same role here as in the circular coordinate system, we can expect the function $\Phi(\phi)$ to behave in the same way as did our $g(\phi)$ in the circular box. It must thus be an eigenfunction of the operator corresponding to p_ϕ, which is still $-i\hbar(\partial/\partial\phi)$, and the corresponding eigenvalues will be the quantized values of $p_\phi = L_z$, the angular momentum around the z axis. The boundary condition is again a periodicity of 2π in ϕ, and we obtain

$$p_\phi = L_z = m\hbar \qquad (m = 0, \pm 1, \pm 2, \ldots) \quad (3.156)$$

just as before. The term in the kinetic energy involving p_ϕ then becomes

$$\frac{p_\phi^2}{2\mu r^2 \sin^2\theta} = \frac{m^2\hbar^2}{2\mu r^2 \sin^2\theta}. \quad (3.157)$$

We can consider this term as an effective potential energy (corresponding to centrifugal force around the z axis), with a minimum in the xy plane and an infinite maximum all along the z axis; only for $m = 0$ can a particle with finite total energy reach the z axis. This is intuitively reasonable: In Cartesian coordinates we have $p_\phi = xp_y - yp_x$ (cf. Appendix 3A), so that the line $x = y = 0$ can be reached only by achieving infinite values of p_x or p_y (unless p_ϕ, and thus m, is zero).

We also expect the wave function or, more precisely, its angular parts, to be an eigenfunction of the operator corresponding to the total angular momentum, since this quantity is conserved. We might expect the eigenvalues to be defined by an equation similar to Eq. 3.156, for example,

$$L^2 \overset{?}{=} l^2\hbar^2 \qquad (l = 0, 1, 2, \ldots). \quad (3.158)$$

(We are interested only in the magnitude of the angular momentum, which is why we write an equation for L^2 rather than **L**.) We shall see that this is not quite correct, but it is sufficiently accurate for illustrative purposes. If we substitute our quantized equations for L^2 and p_ϕ into Eq. 3.152, we find that the remaining component of angular momentum is given approximately by

$$p_\theta = \left(L^2 - \frac{p_\phi^2}{\sin^2\theta}\right)^{1/2} \approx \left(l^2\hbar^2 - \frac{m^2\hbar^2}{\sin^2\theta}\right)^{1/2}. \quad (3.159)$$

Since p_θ is a measurable quantity, it must have a real value; this means that we require

$$l^2 \sin^2\theta - m^2 \geq 0 \quad (3.160)$$

for all θ, which in turn yields

$$-l \leq m \leq l. \quad (3.161)$$

In short, the length $l\hbar$ of the vector **L** must be at least as great as the length $|m|\hbar$ of its projection on the z axis—a quite logical result.

As we said, Eq. 3.158 is not actually the correct result. Just as we can obtain the eigenvalues of p_ϕ from the wave equation in ϕ, we can expect to obtain the eigenvalues of L^2 from the wave equation in θ. This is a fairly complicated equation, but its solutions can be written in closed form (i.e., in terms of a finite number of simple functions). The boundary condition that must be satisfied is governed by the fact that θ goes only from 0 to π, so that, for example, $\Theta(\pi + \theta)$ must equal $\Theta(\pi - \theta)$; as we shall see, the solutions are all sums of sines and cosines, which automatically satisfy this condition. As with the radial function of Eq. 3.144, the solution depends on the value of $|m|$, each $|m|$ corresponding to an infinite series of solutions numbered with the new quantum number l. It turns out that the eigenvalues of total angular momentum are given by

$$L^2 = l(l+1)\hbar^2 \quad (l = 0, 1, 2, \ldots) \quad (3.162)$$

rather than Eq. 3.158. However, this still yields Eq. 3.161 for the relationship between m and l; for each value of l there are $2l + 1$ possible values of m.

The replacement of l^2 by $l(l + 1)$ is essentially due to the uncertainty principle. Suppose that Eq. 3.158 were correct. Then for the case $|m| = l$, and thus $m^2 = l^2$, Eq. 3.160 would allow only $\sin^2\theta = 1$, $\theta = \pi/2$, which in turn yields $p_\theta = 0$. But by the uncertainty principle we cannot simultaneously have exact knowledge of both θ and p_θ, which are a conjugate pair of variables. This problem is removed when we replace l^2 by $l(l + 1)$, which is always greater than m^2 (unless $l = m = 0$), allowing a range of possible values of θ rather than the single value $\pi/2$. The physical meaning of this result is that the component of angular momentum along a particular axis can never be quite as large as the total (nonzero) angular momentum. (In the case $l = m = 0$, we have total information about all the components of the angular momentum, but the wave function is spherically symmetric, so we have absolutely no information about θ or ϕ.)

Let us consider the term $L^2/2\mu r^2$ in Eq. 3.153, which gives the part of the kinetic energy associated with rotation, the *rotational* or *centrifugal energy* T_{rot}. If the vector of angular momentum, **L**, is conserved, as it must be in the absence of external torque, the centrifugal energy increases as the particle comes closer to the origin, which is the center of force in a potential field $V(r)$. For example, a satellite in an elliptical orbit moves fastest when it is closest to the earth. From Eq. 3.162 we can immediately write the quantum expression for the centrifugal energy,

$$T_{\mathrm{rot}} = \frac{l(l+1)\hbar^2}{2\mu r^2}. \tag{3.163}$$

This would be the total energy for a particle in a circular orbit (r constant). Note that it depends on l but not on m, so that the $2l + 1$ values of m allowable for a given l correspond to a $(2l + 1)$-fold degenerate energy level. The same result is found in the general case, with the energy eigenvalues depending on l and a radial index n, but not on m. Note that the energy levels grow farther and farther apart as l increases, resembling the behavior we have found for other quantum systems.

Let us now take a closer look at the properties of the quantized angular momentum. Remember that the angular momentum is a vector, directed along the axis around which rotation is occurring. Its magnitude is given by

$$L = |\mathbf{r} \times \mathbf{p}| = rp\sin\chi, \tag{3.164}$$

where χ is the angle between the vectors **r** and **p**. The vector **L** forms the positive z axis in a right-handed coordinate system (cf. Fig. 1.3) in which **r** and **p** define the x and y axes, respectively. If the motion is in a circle, then **r** and **p** are always perpendicular ($\sin\chi = 1$), p is constant, and L has the familiar value μvr (or $\mu r^2\omega$, where $\omega = v/r$ is the angular velocity). In an arbitrary coordinate system, **L** will in general have a nonvanishing component (i.e., projection) along any axis chosen at random. Classically, one could specify the components along any such axes, for example, the three Cartesian axes; in quantum mechanics this is no longer true.

We have outlined how the angular parts of the Schrödinger equation give two quantized values of angular momentum. The solution for $\Theta(\theta)$ gives us the magnitude of the total angular momentum, whereas the solution for $\Phi(\phi)$ gives us its component along the z axis. If we were to change the axis about which we measure ϕ, that is, if we were to redefine the z axis, we could obtain the component of angular momentum along the new axis instead. However, no matter how we place the axes, we can determine only one component at a time; the other components must remain unknowable. We cannot simultaneously have precise values for three, or even two, components of the angular momentum unless all are zero, i.e., unless there is no rotation. This can be demonstrated by obtaining the operators (designated by L rather than **L**) corresponding to the total angular momentum and its components: L^2 commutes with L_x, L_y, or L_z, but no two of the components commute with each other. According to the uncertainty principle, then, we can simultaneously measure L and either L_x, L_y, or L_z, but measurement of any one component disturbs the values of the other components.

Let us expand on the consequences of this limitation on our knowledge. In a system with spherical symmetry, all axes are equivalent; thus we can arbitrarily choose any axis as our z axis. We *assume* that this is the axis with respect to which the angular momentum is quantized, that is, we assume that the wave function is an eigenfunction of p_ϕ, with Eq. 3.156 giving the eigenvalues of p_ϕ. We can also obtain the magnitude of the total angular momentum, which is quantized according to Eq. 3.162. It has already been pointed out that a nonzero $|\mathbf{L}|$ must be larger than L_z, so L_x and L_y (which furnish the rest of the length of **L**) cannot both be zero. In fact, we can evaluate

$$L_x^2 + L_y^2 = L^2 - L_z^2 = [l(l+1) - m^2]\hbar^2, \tag{3.165}$$

which is a constant for any given values of m and l. But the uncertainty principle tells us that we cannot know L_x or L_y separately. In fact, just as all values of ϕ are equally likely, so all directions are equally likely for the projection of **L** in the xy plane, the magnitude of which is $(L_x^2 + L_y^2)^{1/2}$. The best we can do is to obtain a probability distribution for the values of L_x and L_y.

We can clarify the quantization of angular momentum by a geometric interpretation, as illustrated in Fig. 3.13. Each allowed value of L^2, corresponding to a value of the quantum number l, fixes the length of the vector **L** and thus defines a sphere about the origin; only one such sphere is shown in the figure. Each allowed value of L_z, corresponding to a value of m, fixes the z component of **L** and defines a plane parallel to the xy plane. When we specify particular values of both l and m, we choose both the sphere and the plane corresponding to $|\mathbf{L}|$ and L_z. The intersection of the

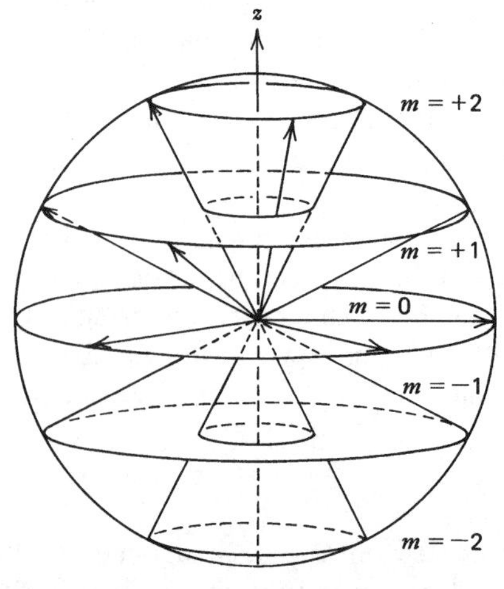

Figure 3.13 A graphic representation of quantized angular momentum. Each cone is the locus of the total angular momentum vector **L** for a particular value of m consistent with $l = 2$. The magnitude of L here is $[l(l + 1)]^{1/2}\,\hbar = \sqrt{6}\hbar$; its z component L_z is $m\hbar$.

sphere and the plane is a circle, which defines the locus of all vectors compatible with the quantized values of $|\mathbf{L}|$ and L_z. If we think of the vector **L** as a movable arrow of fixed length attached by a swivel to the origin, this circle is the base of the cone along whose side **L** may lie.[26] The altitude of the cone is L_z, and the radius of its base is $(L_x^2 + L_y^2)^{1/2}$. All vectors lying on this cone are equally likely representations of **L**. We thus see that not only the magnitude but also the direction of **L** is quantized, with the allowed directions restricted to the cones of Fig. 3.13. For each value of l there are $2l + 1$ such cones allowed, corresponding to the possible values of m from $-l$ to l. This phenomenon is called *space quantization.*

So far we have not looked in any detail at the form of the spherical-box wave functions. The radial functions we can disregard, but the angular functions have a significance much wider than this particular problem. Since the only angular boundary conditions are those imposed by the coordinate system, we must obtain the *same* angular wave functions for any system with spherical symmetry—and all atoms have spherical symmetry.

Until now we have written the angular function as a product of separate functions of θ and ϕ. These are often combined into a single function, the *spherical harmonic*

$$Y_{l,m}(\theta,\phi) = \Theta_{l,|m|}(\theta)\Phi_m(\phi), \tag{3.166}$$

where l and m are the quantum numbers previously introduced. We already know that $\Phi_m(\phi)$ is $e^{im\phi}$, but the derivation of $\Theta_{l,|m|}(\theta)$ is too complicated to justify our including it here. However, since these functions are of such general applicability, it is desirable to have some familiarity with their properties. We therefore include in Table 3.1 a list of all the spherical harmonics with values of l up to 3. The quantum number l specifies the total angular momentum, and a finite number $(2l + 1)$ of m values correspond to each possible value of l; it is thus natural to classify the solutions primarily by the value of l.

We have pointed out earlier that increased momentum is associated with increasing oscillation of the wave function; the same is true of angular momentum and oscillation with angle. We might thus expect that the higher the values of l and m, the more "wiggly" the spherical harmonic will be as a function of θ and ϕ. The nodes of the wave functions for a particle in a spherical box are of three kinds: radial nodes, analogous to the radial nodes in the circular box; nodes in $\Theta(\theta)$, and nodes in the real and imaginary parts of $\Phi(\phi)$. The radial nodes are spherical surfaces; more radial nodes means higher average kinetic energy of motion through the origin, even though the radial momentum or even its square is not constant. The nodes in $\Theta(\theta)$ are cones, surfaces of constant

[26] In addition to $|\mathbf{L}|$ and L_z we also knew L_x and L_y, the circle would be reduced to a single point. But this would be more information than the uncertainty principle allows us to have.

Table 3.1 Lower Spherical Harmonics, $Y_{l,m}(\theta,\phi)$

The functions tabulated are normalized over the surface of a sphere (cf. Appendix 2A):

$$\int_{\phi=0}^{2\pi}\int_{\theta=0}^{\pi} Y_{l,m}(\theta,\phi)\sin\theta\, d\theta\, d\phi = 1.$$

$$l=0:\ Y_{0,0} = \left(\frac{1}{4\pi}\right)^{1/2}$$

$$l=1:\ Y_{1,0} = \left(\frac{3}{4\pi}\right)^{1/2}\cos\theta$$

$$Y_{1,1} = \left(\frac{3}{8\pi}\right)^{1/2}\sin\theta\, e^{i\phi}$$

$$Y_{1,-1} = \left(\frac{3}{8\pi}\right)^{1/2}\sin\theta\, e^{-i\phi}$$

$$l=2:\ Y_{2,0} = \left(\frac{5}{16\pi}\right)^{1/2}(3\cos^2\theta - 1)$$

$$Y_{2,1} = \left(\frac{15}{8\pi}\right)^{1/2}\cos\theta\sin\theta\, e^{i\phi}$$

$$Y_{2,2} = \left(\frac{15}{32\pi}\right)^{1/2}\sin^2\theta\, e^{2i\phi}$$

$$Y_{2,-1} = \left(\frac{15}{8\pi}\right)^{1/2}\cos\theta\sin\theta\, e^{-i\phi}$$

$$Y_{2,-2} = \left(\frac{15}{32\pi}\right)^{1/2}\sin^2\theta\, e^{-2i\phi}$$

$$l=3:\ Y_{3,0} = \left(\frac{63}{16\pi}\right)^{1/2}\left(\frac{5}{3}\cos^3\theta - \cos\theta\right)$$

$$Y_{3,1} = \left(\frac{21}{64\pi}\right)^{1/2}(5\cos^2\theta - 1)\sin\theta\, e^{i\phi}$$

$$Y_{3,2} = \left(\frac{105}{32\pi}\right)^{1/2}\sin^2\theta\cos\theta\, e^{2i\phi}$$

$$Y_{3,3} = \left(\frac{35}{64\pi}\right)^{1/2}\sin^3\theta\, e^{i\phi}$$

$$Y_{3,-1} = \left(\frac{21}{64\pi}\right)^{1/2}(5\cos^2\theta - 1)\sin\theta\, e^{-i\phi}$$

$$Y_{3,-2} = \left(\frac{105}{32\pi}\right)^{1/2}\sin^2\theta\cos\theta\, e^{-2i\phi}$$

$$Y_{3,-3} = \left(\frac{35}{64\pi}\right)^{1/2}\sin^3\theta\, e^{-3i\phi}$$

θ. The greater is l for a given m, the greater is the average angular momentum over the poles. The nodes in the real and imaginary parts of $\Phi(\phi)$ are planes of constant longitude. The greater is m for a given l, the greater is the angular momentum around the polar axis.

In more detail, the real part of $e^{im\phi}$ has m nodes between 0 and π (the m nodes between π and 2π are on the same nodal planes extended through the origin, and are not counted separately). The function $\Theta(\theta)$, a polynomial of order l in $\sin\theta$ and $\cos\theta$, has $l - m$ nodes ($\theta = 0$ and $\theta = \pi$ do not count, since the nodal planes in ϕ all pass through the

z axis). Let us look at specific cases. For the simplest case, $l = m = 0$, the angular momentum is zero; this means that the system has no angular preferences, and the angular part of the wave function is simply a constant. The functions with $l = 1$ have a single angular node: for $\cos\theta$ the node is on the equatorial plane $\theta = \pi/2$; for $\sin\theta e^{\pm i\phi}$ there is a node on the plane defined by $\phi = 0$ and $\phi = \pi$. Similarly, the functions with $l = 2$ have two nodes, and so forth. In general, for a given value of l there are $2l + 1$ different spherical harmonics, each of which has l angular nodes.

We shall see that the angular behavior of the spherical harmonics—especially the *location* of the nodes—is fundamental to the distribution of electrons in atoms, and thus to the shapes of molecules. The electrons in each atom, and thus the bonds they form, are most likely to be at those angles where the wave function has its maxima. We return to this subject in the next chapter, where we consider the wave functions of the hydrogen atom; diagrams of the spherical harmonics will be given at that point.

3.12 The Rigid Rotator

The particle of the previous section was in a spherically symmetric environment, but was free to move both radially and rotationally. One can also envision a simpler system, known as a *rigid rotor,* or sometimes *rigid rotator,* whose only form of motion is rotation. An example would be a point mass rigidly fixed at a constant distance from a center about which it is free to swing. There are no pure rigid rotators in nature, but the rotation of molecules can be described quite well by this model.

We consider first the elementary system just described, a mass μ at a constant distance R from the origin, with no external forces. Since the mass can have no radial motion, the kinetic energy of Eq. 3.153 reduces to $T = L^2/2\mu R^2$. The wave function is a function of θ and ϕ only, and must have exactly the same form as the angular solutions in the spherical box—that is, the spherical harmonics of Table 3.1. Since the total energy consists of only the energy of rotation, its Hamiltonian is a simplified form of Eq. 3.154,

$$\mathsf{H} = \frac{\hbar^2}{2\mu R^2}\left(\frac{1}{\sin\theta}\frac{\partial}{\partial\theta}\sin\theta\frac{\partial}{\partial\theta} + \frac{1}{\sin^2\theta}\frac{\partial^2}{\partial\phi^2}\right), \qquad (3.167)$$

with no radial contribution to the kinetic energy. The Schrödinger equation for the rigid rotator is therefore

$$\mathsf{H}\psi(\theta,\phi) = \frac{\hbar^2}{2\mu R^2}\left(\frac{1}{\sin\theta}\frac{\partial}{\partial\theta}\sin\theta\frac{\partial}{\partial\theta} + \frac{1}{\sin^2\theta}\frac{\partial^2}{\partial\phi^2}\right)\psi(\theta,\phi) = E\psi(\theta,\phi), \qquad (3.168)$$

which simplifies drastically when we use Eqs. 3.153, 3.163, and 3.166:

$$\mathsf{H}\psi(\theta,\phi) = \frac{\mathsf{L}^2}{2\mu R^2}Y_{l,m}(\theta,\phi) = \frac{\hbar^2}{2\mu R^2}l(l+1)Y_{l,m}(\theta,\phi). \qquad (3.169)$$

Consequently, the eigenvalues of energy for the rigid rotator are those given in Eq. 3.163:

$$E = \frac{l(l+1)\hbar^2}{2\mu R^2} \quad (l = 0,1,2,\ldots), \qquad (3.170)$$

with each energy level $(2l + 1)$-fold degenerate.

We know that the spacing between energy levels increases with increasing l, but how does the total number of quantum states vary with energy? By a "state" we mean a solution with a particular set of quantum numbers, so that the $2l + 1$ values of m corresponding to a given l define distinct states. Let us determine the number of such states per unit energy, the *density of states,* in a given energy range. For a given value of l this is $2l + 1$ times the number of energy levels (values of l) per unit energy. In most molecules the spacing between rotational energy levels is very small, and at ordinary temperatures the average value of l is so large that l can be taken as a continuous variable. In this limit ($l >> 1$) we can obtain the density of states by differentiating Eq. 3.170:

$$\text{density of states} = (2l+1)\frac{dl}{dE} = (2l+1)\left(\frac{dE}{dl}\right)^{-1} = (2l+1)\left[\frac{\hbar^2}{2\mu R^2}(2l+1)\right]^{-1} = \frac{2\mu R^2}{\hbar^2}. \qquad (3.171)$$

That is, for the rigid rotator the density of states at high l is a constant, independent of energy or quantum number and depending only on the moment of inertia μR^2. By contrast, for a particle in a three-dimensional box[27] the density of states is proportional to $E^{1/2}$. We shall see in Chapter 15 that the density of states is a very important characteristic of any system, affecting the distribution of particles over the available energy levels and thus the system's thermodynamic properties.

The result of Eq. 3.170 is valid not only for the simple mass-point rotator, but for any system that can be reduced to

[27] For example, in a cubical box the energy is $n^2\pi^2\hbar/2ma^2$ where $n^2 \equiv n_1^2 + n_2^2 + n_3^2$. As n becomes large, the number of states between n and $n + dn$ approaches the limit $4\pi n^2\, dn$. The density of states thus becomes

$$4\pi n^2\left(\frac{dn}{dE}\right) = 4\pi n^2\left(\frac{dE}{dn}\right)^{-1} = 4\pi n^2\left(\frac{n\pi^2\hbar^2}{ma^2}\right)^{-1} = \left(\frac{4ma^2}{\pi\hbar^2}\right)n,$$

proportional to n and thus to $E^{1/2}$.

it. By "reducing" we mean the sort of thing we spoke of in Section 2.11: The motion of the two particles in a hydrogen atom is equivalent to the motion of a single particle around a center of force. Similar reductions can be carried out for more than two particles if their arrangement is sufficiently symmetric. The systems of particles with which we are concerned here are of course molecules, the particles being their constituent atoms.

Let us see what "sufficiently symmetric" means in this context. In general, the rotational energy of a molecule is a very complicated function of the positions and momenta of the individual atoms. However, one can to a fairly good approximation treat a rotating molecule as a rigid body (neglecting vibrations and electronic motion). For any rigid body one can always find a Cartesian coordinate system in terms of which the rotational energy has the simple form

$$E_{\text{rot}} = \frac{L_x^2}{2I_x} + \frac{L_y^2}{2I_y} + \frac{L_z^2}{2I_z}, \tag{3.172}$$

where the L's are the components of the total angular momentum $\mathbf{L} = \Sigma_i \mathbf{r}_i \times \mathbf{p}_i$, summed over all particles, and the I's are constants. The axes of this coordinate system are called the *principal axes,* and have their origin at the center of mass. The *principal moments of inertia* I_x, I_y, I_z then have the form

$$\begin{aligned} I_x &\equiv \sum_i m_i (y_i^2 + z_i^2), \\ I_y &\equiv \sum_i m_i (x_i^2 + z_i^2), \\ I_z &\equiv \sum_i m_i (x_i^2 + y_i^2) \end{aligned} \tag{3.173}$$

(m_i = mass), with each sum taken over all the atoms in the molecule.

Suppose now that all three moments of inertia are equal, $I \equiv I_x = I_y = I_z$; such a molecule is called a *spherical top.* Symmetry of this type is found for all molecules with tetrahedral, octahedral or cubic symmetry, such as CH_4, SF_6 or cubane, C_8H_8, respectively. For a spherical top the rotational energy reduces to simply $L^2/2I$, just as for the single-particle rotator. The rotational energy levels obtained are thus those of Eq. 3.170, with μR^2 replaced by I.

The next case in order of complexity is the *symmetric top,* a molecule in which only two of the moments of inertia are equal: $I_0 \equiv I_x = I_y \neq I_z$. Molecules of this type include the pyramidal NH_3 (ammonia) and CH_3Cl (methyl chloride), the triangular BCl_3, and the hexagonal C_6H_6 (benzene); each has an axis of at least threefold symmetry, which is defined as the z axis. Note that a spherical top can be converted to a symmetric top by a simple distortion that reduces its symmetry. For example, the six bonds in SF_6 are identical; but if we stretch the two bonds along one axis, spinning the molecule about that axis will make it look something like a football rather than a sphere. Alternatively, changing one of the H atoms in methane to Cl lowers the symmetry by shifting the center of mass (the bond length changes, but the added mass has a much greater effect). Let us see what effect such a reduction in symmetry has on the energy levels.

The rotational energy of the symmetric top is of course

$$E_{\text{rot}} = \frac{1}{2I_0}(L_x^2 + L_y^2) + \frac{L_z^2}{2I_z}. \tag{3.174}$$

If we add $L_z^2/2I_0$ to the first term and subtract it from the second, we have the equivalent equation

$$E_{\text{rot}} = \frac{L^2}{2I_0} + \frac{1}{2}\left(\frac{1}{I_z} - \frac{1}{I_0}\right)L_z^2. \tag{3.175}$$

Replacing L^2 and L_z^2 by their equivalent operators, we have for the angular part of the Schrödinger equation

$$\begin{aligned} \mathsf{H}_{\text{rot}} Y(\theta,\phi) &= \frac{1}{2I_0}\mathsf{L}^2 Y(\theta,\phi) + \frac{1}{2}\left(\frac{1}{I_z} - \frac{1}{I_0}\right)\mathsf{L}_z^2 Y(\theta,\phi) \\ &= E_{\text{rot}} Y(\theta,\phi), \end{aligned} \tag{3.176}$$

where $Y(\theta,\phi) = \Theta(\theta)\Phi(\phi)$ is the angular part of the wave function. We would like this function to be of the type already discussed for the case of spherical symmetry, that is, a spherical harmonic. It turns out that this is indeed the case, since the spherical harmonics are specifically defined to be eigenfunctions of both L^2 and L_z (and thus L_z^2). We can thus obtain the eigenvalues of L^2 and L_z separately, using Eqs. 3.162 and 3.156, and substitute to obtain the total rotational energy:

$$E_{\text{rot}} = \frac{\hbar^2}{2}\left[\frac{l(l+1)}{I_0} + m^2\left(\frac{1}{I_z} - \frac{1}{I_0}\right)\right], \quad \left(\begin{matrix} l = 0,1,2,\ldots; \\ m = -l,\ldots,l \end{matrix}\right). \tag{3.177}$$

Compare this result with Eq. 3.170. The most obvious change is that the energy has now become a function of m, that is, of the angular momentum about the z axis. This axis is no longer arbitrary, of course, but the unique symmetry axis of the molecule. The energy levels can now be no more than doubly degenerate (for $+m$ and $-m$, corresponding to clockwise and counterclockwise rotation about the z axis), rather than the $(2l + 1)$-fold degeneracy of the earlier case. It is generally true that a reduction in symmetry tends to result in a removal of degeneracy, a splitting of energy levels (in this case those of a given l). The extent of this splitting depends on how asymmetric the molecule is: If I_z and I_0 are nearly the same, the coefficient of m^2 and thus the splitting are small, but if one moment of inertia is much larger than the other, the energy spectrum will be substantially altered.

The *linear molecule,* with all its atoms aligned on a single axis, is an important special case of the symmetric rotor. This class includes all diatomic molecules, as well as some others, such as CO_2 and C_2H_2 (acetylene). In such a molecule the only rotation about the z axis is that of each atom about its own nucleus. As long as we are considering the atoms as mass points, we have $L_z = 0$; that is, there is no z term in the rotational energy,[28] which reduces to $L^2/2I_0$. This is again of the same form as for the single-particle rotator, giving $l(l+1)\hbar/2I_0$ for the eigenvalues of rotational energy. For a diatomic molecule, if R is the bond length and μ the reduced mass $m_1m_2/(m_1+m_2)$, the moment of inertia is μR^2, as in Eq. 3.170; this demonstrates the equivalence of the one-body and two-body problems, which we stated earlier without proof.

Finally, we have the *asymmetric top,* a molecule in which all three moments of inertia are different. This class includes molecules as simple as H_2O and SO_2. In the asymmetric top there is no axis around which angular momentum is conserved, so the angular wave function can no longer be a spherical harmonic. The solution is very complicated, and the energy levels cannot be expressed by a simple equation; there are still $2l+1$ levels for each value of the total angular momentum, but none of the levels is degenerate.

[28] Of course, in a real molecule the motions of the electrons in general produce a nonzero L_z; but this goes beyond what we can treat in terms of a rigid-rotator model. (See Section 7.2.)

APPENDIX 3A

More on Circular Coordinates and the Circular Box

Circular polar coordinates are introduced in Section 3.10. In this appendix we give supplementary information on the following topics: (1) the expression of the Hamiltonian and angular momentum operators in circular coordinates; (2) the solution of the radial wave equation for the circular box. This material is given to illustrate the types of manipulations required when one uses non-Cartesian coordinates.

As mentioned in the main text, the most straightforward way to obtain differential operators in non-Cartesian coordinates is to obtain the operators in Cartesian form and then transform the coordinates. For a particle moving in a plane, we know that the Hamiltonian operator in Cartesian coordinates is given by Eq. 3.132. Given circular symmetry, we can replace $\mathsf{V}(x, y)$ by $\mathsf{V}(r)$; this leaves the problem of transforming the partial derivatives.

In Appendix I at the end of the book, we discuss the transformation of coordinates in general, with the Cartesian-circular transformation used as an example. There we obtain Eqs. I.16 for the operators $(\partial/\partial y)_y$ and $(\partial/\partial y)_x$ in terms of circular coordinates. The second derivatives that appear in the Hamiltonian can be obtained by applying these equations twice (we omit subscripts to avoid cluttering the equation):

$$\begin{aligned}\frac{\partial^2}{\partial x^2} &= \cos^2\phi\frac{\partial^2}{\partial r^2} - \cos\phi\sin\phi\frac{\partial}{\partial r}\left(\frac{1}{r}\frac{\partial}{\partial\phi}\right)\\ &\quad -\frac{\sin\phi}{r}\frac{\partial}{\partial\phi}\left(\cos\phi\frac{\partial}{\partial r}\right)+\frac{\sin^2\phi}{r^2}\frac{\partial^2}{\partial\phi^2},\\ \frac{\partial^2}{\partial y^2} &= \sin^2\phi\frac{\partial^2}{\partial r^2} + \sin\phi\cos\phi\frac{\partial}{\partial r}\left(\frac{1}{r}\frac{\partial}{\partial\phi}\right)\\ &\quad +\frac{\cos\phi}{r}\frac{\partial}{\partial\phi}\left(\sin\phi\frac{\partial}{\partial r}\right)+\frac{\cos^2\phi}{r^2}\frac{\partial^2}{\partial\phi^2}.\end{aligned} \tag{3A.1}$$

When we add these, the second terms cancel, the first and last terms are simplified by $\sin^2\phi + \cos^2\phi = 1$, and in the third terms we have

$$\begin{aligned}&-\frac{\sin\phi}{r}\left(\cos\phi\frac{\partial^2}{\partial\phi\partial r} - \sin\phi\frac{\partial}{\partial r}\right)\\ &+\frac{\cos\phi}{r}\left(\sin\phi\frac{\partial^2}{\partial\phi\partial r} + \cos\phi\frac{\partial}{\partial r}\right) = \frac{1}{r}(\sin^2\phi+\cos^2\phi)\frac{\partial}{\partial r}\\ &= \frac{1}{r}\frac{\partial}{\partial r}.\end{aligned} \tag{3A.2}$$

Combining these results gives

$$\frac{\partial^2}{\partial x^2}+\frac{\partial^2}{\partial y^2} = \frac{\partial^2}{\partial r^2}+\frac{1}{r}\frac{\partial}{\partial r}+\frac{1}{r^2}\frac{\partial^2}{\partial\phi^2}, \tag{3A.3}$$

yielding Eq. 3.133 for the Hamiltonian operator. Similar procedures can be followed to obtain H in other coordinate systems.

For the angular momentum operator we first need the classical form in terms of Cartesian coordinates. From Eqs. 3.127 and 3.128 and the transformation relations (Fig. 3.9), this is

$$\begin{aligned}p_\phi &= \mu v_\phi r = \mu r^2\frac{d\phi}{dt}\\ &= \mu(x^2+y^2)\frac{x(dy/dt)-y(dx/dt)}{x^2+y^2}\\ &= xp_y - yp_x.\end{aligned} \tag{3A.4}$$

The quantum equivalent of this equation is

$$\mathsf{p}_\phi = -i\hbar\left(x\frac{\partial}{\partial y} - y\frac{\partial}{\partial x}\right). \tag{3A.5}$$

Using Eqs. I.16 again, we transform the partial derivatives to obtain

$$\begin{aligned}\mathsf{p}_\phi &= -i\hbar\left[r\cos\phi\left(\sin\phi\frac{\partial}{\partial r}+\frac{\cos\phi}{r}\frac{\partial}{\partial\phi}\right)\right.\\ &\qquad\left. - r\sin\phi\left(\cos\phi\frac{\partial}{\partial r}-\frac{\sin\phi}{r}\frac{\partial}{\partial\phi}\right)\right]\\ &= -i\hbar\frac{\partial}{\partial\phi},\end{aligned} \tag{3A.6}$$

as might be expected. However, a similar calculation starting with Eq. 3.130 gives for the radial momentum operator not $-i\hbar(\partial/\partial r)$ but

$$\mathsf{p}_r = -i\hbar\left(\frac{\partial}{\partial r}+\frac{1}{r}\right), \tag{3A.7}$$

illustrating the point that the form of an operator depends very much on the choice of coordinates. Substituting the last two equations in Eq. 3.131 is another way to obtain the Hamiltonian operator.

Finally, we consider the solution of the radial wave equation for a particle in a circular box. If we substitute the factored form of the wave function, Eq. 3.137, into Eq. 3.134, we have

$$-\frac{\hbar^2}{2\mu}\left[g(\phi)\frac{d^2 f(r)}{dr^2}+\frac{g(\phi)}{r}\frac{df(r)}{dr}+\frac{f(r)}{r^2}\frac{d^2 g(\phi)}{d\phi^2}\right]=Ef(r)g(\phi). \tag{3A.8}$$

Dividing through by $f(r)g(\phi)$, multiplying by r^2, and rearranging to separate the terms in r and ϕ gives

$$\frac{r^2}{f(f)}\frac{d^2 f(r)}{dr^2}+\frac{r}{f(r)}\frac{df(r)}{dr}+\epsilon r^2=-\frac{1}{g(\phi)}\frac{d^2 g(\phi)}{d\phi^2},\quad\left(\epsilon\equiv\frac{2\mu E}{\hbar^2}\right). \tag{3A.9}$$

As in Eq. 3.112, each side of this equation must equal the same constant; given the form of $g(\phi)$ from Eq. 3.143, we can use the right-hand side to identify the constant as m^2, where m is the angular momentum quantum number. Further rearrangement then gives

$$\frac{d^2 f(r)}{dr^2}+\frac{1}{r}\frac{df(r)}{r}+\left(\epsilon-\frac{m^2}{r^2}\right)f(r)=0 \tag{3A.10}$$

for the radial wave equation.

Equation 3A.10 is a form of *Bessel's equation,* one of the most thoroughly studied equations in all of mathematical physics. It is beyond the scope of this text to develop the methods for its solution, and in any case the solutions cannot be expressed in terms of simple functions (except as infinite series). However, these solutions are of such widespread application that they are defined as functions in their own right, called *Bessel functions;* like the trigonometric or other simple functions, their values can be found in tables or from standard computer programs. There are several different kinds of Bessel functions: Some look rather like sines or cosines, others like exponentials, still others like hyperbolic sines or cosines. Those appropriate to the circular box are the Bessel functions of the first kind, which resemble sine curves with slowly decreasing amplitude (Fig. 3A.1); they are designated as $J_{|m|}(x)$, where the index number m is the same as in Eq. 3A.10.

The boundary condition we must satisfy is given by Eq. 3.135, and means that $f(r)$ must go to zero at $r = R$. The radial wave function is then given by

$$f(r)=J_{|m|}(r\epsilon^{1/2}), \tag{3A.11}$$

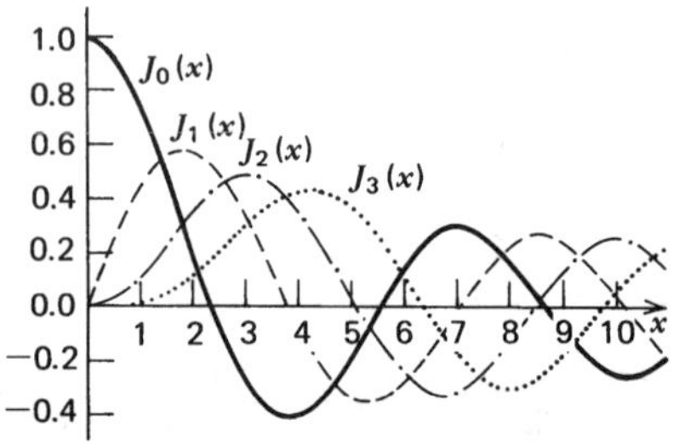

Figure 3A.1 Some Bessel functions of the first kind: $J_n(x)$ for $n = 0$, 1, 2, and 3. Note that these functions are drawn with a common scale for x, not with a scale to make their first zeros appear at the same point on the abscissa, as one would in order to show the radial functions with $n = 1$ for particles in a circular box.

where ϵ (i.e., the energy) must be such that

$$J_{|m|}(R\epsilon^{1/2})=0. \tag{3A.12}$$

For any particular $|m|$, Eq. 3A.12 is satisfied only by a specific discrete (but infinite) set of values of ϵ, which we number by the additional index n. These correspond to the values of x ($= R\epsilon^{1/2}$) for which $J_{|m|}(x) = 0$, given by the points at which the curves of Fig. 3A.1 cross the x axis. The radial wave function $f_{n,|m|}(r)$ thus has the shape of $J_{|m|}(x)$ out to its nth zero, which corresponds to the box wall; for a given n, the two states with $m = \pm|m|$ have different angular functions $g(\phi)$, but the same $f(r)$. Since R is a constant, going to higher values of ϵ (higher n) simply squeezes more waves together in the radial direction; this squeezing increases the average value of $|df(r)/dr|$, and therefore increases the absolute value of the radial momentum and kinetic energy.

This problem parallels that of the one-dimensional box, for which we also have an equation in one variable that is satisfied only by a specific set of energy eigenvalues. Mathematically, the difference between the two problems is the result of the two terms in Eq. 3A.10 that were not present in Eq. 3.64. The resulting solution, Eq. 3A.11, is certainly less familiar to the reader than is the sine function of Eq. 3.69. But the difference is not just one of familiarity. From Eq. 3.69 we were able to derive an *explicit* expression, Eq. 3.71, giving the values of E for which the wave function satisfies the boundary conditions—that is, the energy eigenvalues. We could write down this expression because we know where the sine function has its zeros, at integral multiples of π. By contrast, the zeros of the Bessel function follow no such simple analytic pattern; in general, they must be derived by numerical estimation (or looked up in a table or graph like Fig. 3A.1). Some of the circular-box energy eigenvalues have been shown in Fig. 3.12.

FURTHER READING

Avery, J., *Quantum Theory of Atoms, Molecules and Photons* (McGraw-Hill Book Co., Inc., New York, 1972), Chapters 2–4.

Karplus, M., and Porter, R. N., *Atoms and Molecules* (W. A. Benjamin, Inc., Menlo Park, Calif., 1970), Chapter 2.

Kauzmann, W., *Quantum Chemistry* (Academic Press, Inc., New York, 1957), Chapters 1–5, 6.I., 7, 8.

Kramers, H. A., *Quantum Mechanics* (North-Holland Publishing Company, Amsterdam, 1958; Dover Publications, Inc., New York, 1964).

Merzbacher, E., *Quantum Mechanics,* 2nd Ed. (John Wiley and Sons, Inc., New York, 1970), Chapters 1–6, 8.

Messiah, A., *Quantum Mechanics,* Vol. 1, translated by G. M. Temmer (John Wiley and Sons, Inc., New York, 1958).

Morrison, M. A., Estle, T. L., and Lane, N. F., *Quantum States of Atoms, Molecules and Solids* (Prentice-Hall, Inc., Englewood Cliffs, N.J., 1976), Chapters 1 and 2.

Schiff, L. I., *Quantum Mechanics,* 3rd Ed. (McGraw-Hill Book Co., Inc., New York, 1968), Chapters 2–4.

PROBLEMS

1. Compute the de Broglie wavelengths of
(a) A hydrogen atom with a velocity of 10^3 m/s,
(b) An electron with an energy of 0.05 eV,
(c) An electron with an energy of 5×10^6 eV,
(d) A xenon atom with an energy of 0.05 eV,
(e) A proton with an energy of 200 GeV (2×10^{11} eV).

2. An expression for the amplitude of a transverse wave might be

$$\mathbf{f}(x,y,z) = [A\cos(kz - \omega t), 0, 0].$$

What is the direction of the displacement of this wave? What is its direction of propagation? Is this a traveling wave or a standing wave? What is its wavelength? Write the expression for a longitudinal wave propagating in the same direction as the transverse wave given above.

3. The internuclear spacings of most diatomic molecules are between 1 Å and 3 Å.
(a) Give a crude estimate of the wavelengths of the electrons that bind atoms into molecules.
(b) Using the answer to (a), give an estimate of the kinetic energies of these binding electrons.
(c) Using the answer to (b), and the information that the lowest ionization potentials are in the range 5 eV to 10 eV, give an estimate of the potential energy of the binding electrons.

4. Suppose that the relation connecting the wavelength of a particle with its momentum depended on *c,* the speed of light, and on *m,* the mass of the particle, so that

$$p = \left(\frac{hcm}{\lambda}\right)^{1/2}.$$

What dispersion relation would follow from this? What implications would such a dispersion relation hold for quantum theory?

5. Show that if $\Psi(x,t)$ is a wave function as in Eq. 3.14, then

$$\left(\frac{\partial^n \Psi}{\partial x^n}\right)_t = k^n \frac{d^n \Psi}{d\varphi^n},$$

and

$$\left(\frac{\partial^n \Psi}{\partial t^n}\right)_x = (-\omega)^n \frac{d^n \Psi}{d\varphi^n}.$$

6. Show that the entries in column (a) are eigenfunctions of the operators in column (b). In each case determine the eigenvalue.

	(a)	(b)
(i)	$a\cos(bt + c)$	d^2/dt^2
(ii)	e^{ibt}	d^2/dt^2
(iii)	$ze^{-z^2/2}$	$-\dfrac{d^2}{dz^2} + z^2$
(iv)	ce^{-ax}	d^n/dx^n

7. The function e^{-ax^2} is an eigenfunction of the operator $(d^2/dx^2) - bx^2$ only under certain conditions. What are these conditions, and what is the eigenvalue when they are satisfied?

8. State, in general terms, under what conditions one may expect the diffraction of matter waves to be important in determining the dynamics of a particle, and under what conditions diffraction is unimportant.

9. Using your knowledge of the boundary conditions the wavefunction must satisfy, explain what happens to the energy levels of a particle in a one-dimensional box if the box length is changed from L to L/n (n = 2, 3, . . .).

10. What is the spacing of the energy levels of a particle in a one-dimensional box if
(a) The mass is m_p, the mass of the proton, and the box side is 5 Å.
(b) The mass is that of the proton, m_p, and the box is made much larger, to correspond to the volumes in which gases are customarily confined, for example, 0.1 m on a side.

(c) The mass is m_p and the box is made much smaller, to a length of 10^{-13} cm, roughly the size of atomic nuclei

The second of these will show very quickly how, in large containers, the quantization of energy levels is not observable because the levels are so closely spaced. The third problem, the nuclear problem, will quickly give some feeling for why nuclear physics experiments require high-energy machines for the study of nuclear reactions, although one can use ordinary light to carry out chemical excitation processes.

11. The molecules $H_2C—(CH{=}CH)_n—CH_2$, $n = 2, 3, \ldots,$ can be considered successively longer and longer one-dimensional boxes for the electrons. Suppose that an electron can move freely the whole length of the molecule. If each bond has length $b_0 = 1.5$ Å, and the end CH bonds are neglected, what are the wavelengths of absorption of the lowest transitions when $n = 2, 3,$ and 4?

12. Set up the functions corresponding to the sum and difference combinations based on the solutions of the form of Eq. 3.118 for the case in which $a = b$, but $n_1 \neq n_2$.

13. What degeneracies arise in the three-dimensional box if $a = b = c$ in Eq. 3.119? Plot the lowest five of these levels on an energy scale and give their degeneracies.

14. Consider a particle in a cylindrical box with radius R and length L. The mass of the particle is μ. Derive an expression analogous to Eq. 3.121 for the classical energy of the particle, beginning with the general expression for the energy of a particle in three dimensions,

$$E = (2\mu)^{-1}(p_x^2 + p_y^2 + p_z^2) + V(x, y, z).$$

Show that the problem is separable in the cylindrical coordinate system $r = (x^2 + y^2)^{1/2}$, $\phi = \arctan(y/x)$, and z, when z is chosen along the cylinder axis.

15. Carry out the quantum mechanical counterpart of Problem 14. Find the first several energy eigenvalues and show how they depend on the relationship between R and L. Show that the problem (i.e., the Hamiltonian) is separable into axial and circular parts and that the energy is a sum of independent axial and circular contributions. What relation between R and L makes the energy of the first excited state of circular motion ($n = 1$, $|m| = 1$ in Fig. 3.12) equal to the energy of the first excited state of motion along the cylinder axis? What relation between R and L makes the state $n = 2$, $|m| = 0$ degenerate with (have energy equal to that of) the first excited state of motion along the cylinder axis?

16. Show that if, in Eq. 3.140,

$$\int_0^{2\pi} |g(\phi)|^2 \, d\phi = 1,$$

then

$$|A| = (2\pi)^{-1/2}$$

17. On radial graph paper, plot the real part of g (of Eq. 3.140), for $p = \hbar$, $2\hbar$, and $4\hbar$, or $m = 1, 2,$ and 4.

18. Prove that

$$\mathbf{p}\cdot\mathbf{p} = \frac{1}{r^2}(\mathbf{p}\cdot\mathbf{r})^2 + \frac{1}{r^2}(\mathbf{r}\times\mathbf{p})\cdot(\mathbf{r}\times\mathbf{p})$$

in classical mechanics but not in quantum mechanics.

19. Using tables of Bessel functions and estimates based on the vertical scale of Fig. 3.12, plot on circular graph paper the nodes of the real and imaginary parts of the functions 3A.11 for the lowest five energy levels shown in Fig. 3.12. Note that for m different from zero the functions all have nodal points at the origin. This is a result of the fact that their angular momenta are greater than zero, so that at $r = 0$ the velocity of the particle must become infinite. But if the velocity were infinite, then the slope of the wave function would be infinite, which would make the wave function infinite, which contradicts our requirement for the wave function. The answer to the dilemma is that the particle can never appear at the origin if it has any angular momentum. Therefore the amplitude of its wave function is zero at the origin for those states with $m \neq 0$.

20. For an electron in a circular Bohr orbit with $n = 2$, what is the centrifugal kinetic energy in joules, electron volts, and cm^{-1}?

21. Suppose that the methane molecule is a rigid tetrahedron with the carbon atom located at its center. What is the moment of inertia about any one axis? Hint: Pick an axis along one of the CH bonds. Show that the moment of inertia about either of the other perpendicular axes is equal to that about the first, and therefore that the moment of inertia is completely independent of the choice of axis in this molecule. Such a molecule is called a spherical top, because its rotations are just as simple as those of a sphere. Calculate the energies of the first three rotational levels of this molecule. What are the separations between them? In what part of the electromagnetic spectrum do these differences lie? Is this a region in which one uses optical methods, microwave or radar equipment, transmitters and detectors, or is it in still another region of the spectrum?

22. Calculate the components of moment of inertia and show graphically the energy level spectrum for each of these two examples: (a) sulfur hexafluoride, SF_6, in which four of the six bonds are 1.58 Å and the other two, chosen opposite each other to lie along a single axis, are 1.60 Å; (b) methyl chloride, CH_3Cl, a tetrahedron in which the CH bonds are 1.1 Å and the C—Cl bond is 1.77 Å. Assume that the chlorine atoms are all isotopes of atomic mass number 35. Our examples have been chosen to show the magnitudes associated with molecular rotation and, perhaps more important, the way in which successively larger deviations from an original simple model lead to successively larger deviations in the "results," that is, in the observables we calculate from these models. In this case, where the rigid rotator is the simple model with a simple spectrum and the symmetric rotator represents the deviation, we are fortunate in that Eq. 3.174 is still both relatively simple and essentially correct for many, many molecules.

CHAPTER 4

Particles in Varying Potential Fields; Transitions

Thus far we have dealt only with particles moving in constant potential fields: Both for the free particle and within our various boxes, the potential energy was constant wherever the particles could go. In this chapter we begin to consider situations in which the potential energy varies from point to point. After examining the general behavior of the wave function when the potential energy changes, we treat two basic models of particles bound within "potential wells."

The first of these models is the ubiquitous harmonic oscillator, which we introduced in Chapter 2 to illustrate quantization. We shall now see how the assumptions of Planck and Einstein ($\Delta E = h\nu$) can be derived from the Schrödinger equation. The results obtained here will turn up again when we treat the vibrations of molecules and solids.

The second model is that of the hydrogenlike atom: an electron in the Coulombic field of a positively charged nucleus. Although the physical picture of the atom is very different from that of Bohr's theory, the energy levels obtained are the same. The principles introduced here will later be extended to many-electron atoms, to show that the forms of the atomic wave functions largely govern the shapes of molecules.

We close the chapter with a look into how transitions occur between atomic energy levels, introducing the fundamental problem of how a quantum mechanical system undergoes change.

4.1 Finite Potential Barriers

In this section we discuss some general characteristics of the wave function's dependence on potential energy. For simplicity, we consider only the one-dimensional case, for which the general time-independent Schrödinger equation is Eq. 3.61. We can rewrite this in the form

$$\frac{d^2\psi(x)}{dx^2} = \frac{2m}{\hbar^2}[V(x) - E]\psi(x). \tag{4.1}$$

On the left-hand side of this equation is the second derivative of $\psi(x)$, which for brevity we shall call the "curvature" of $\psi(x)$. On the right-hand side, a positive constant multiplies the energy difference $V(x) - E$ and the function $\psi(x)$ itself. Now let us examine the behavior of $\psi(x)$ in different regions of space. In classical mechanics the total energy E is the sum of kinetic and potential energy; since the classical kinetic energy ($p^2/2m$) is necessarily a positive quantity, E is always greater than the potential energy alone. As we shall see, this restriction no longer applies in quantum mechanics.

If E is larger than V then the multiplier of $\psi(x)$ on the right-hand side of Eq. 4.1 is negative. When $\psi(x)$ is positive, its curvature must be negative; when $\psi(x)$ is negative, its curvature is positive. Any function $f(x)$ whose sign is opposite to the sign of $d^2f(x)/dx^2$ is concave toward the x axis. The type of function with this property that arises in real physical problems is normally one whose slope changes sign, bringing the function back across the axis and therefore changing the direction of its concavity. In short, it must be oscillatory or sinusoidal. Figure 4.1*a* illustrates such a function. The sine and cosine are the simplest examples of functions that are everywhere concave toward the axis. The wave function must be of this type wherever $E > V$.

The other situation, in which a function has the same sign as its curvature, is illustrated in Fig. 4.1*b;* here there are two possible cases. Let us assume, without losing generality, that the function is positive at some point x_0. If the function has a positive slope at x_0, then for $x > x_0$ the slope must be greater than at x_0, so that the function itself must grow steadily larger as x increases. Such a wave function is physically permissible for a region from $-\infty$ to some fixed value of x, but clearly will not do for arbitrarily large values of x. If we allowed $\psi(x)$ to grow infinitely large in this way, the probability density $|\psi(x)|^2$ would accumulate at $x = +\infty$, leaving essentially zero probability of a particle's being found at any real location. The other case is one in which a positive function has negative slope but positive curvature. In this case, the slope grows smaller and smaller as $x \rightarrow +\infty$. Such a wave function can be used for x greater than some

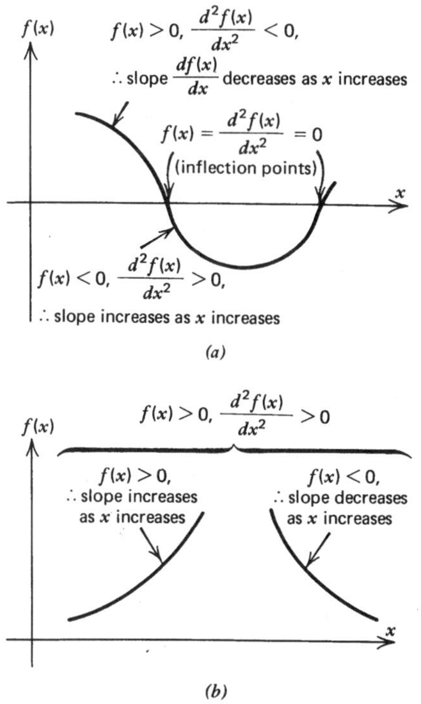

Figure 4.1 Examples of the relationships between functions and their second derivatives: (*a*) The function $f(x)$ and $d^2f(x)$ have opposite signs. (*b*) The function $f(x)$ and $d^2f(x)/dx^2$ have the same sign.

fixed value, but diverges as $x \to -\infty$, with the same catastrophic result as in the first case. It was to eliminate such possibilities that we postulated that the integral $\int \cdots \int \Psi^*\Psi dq_1 \cdots dq_N$ over all allowed values of the coordinates—in this case, $\int_{-\infty}^{\infty} |\psi(x)|^2\, dx$—must be finite. The simplest examples of functions with everywhere the same signs as their curvatures are e^x and e^{-x}, which correspond to our first and second cases, respectively. Wave functions of this type will appear wherever $E < V$.

Until now, however, we have not encountered the possibility that E could be less than V. The free particle has $V = 0$ and $E > 0$ everywhere; the particle in a box is the same where V is finite, and the wave simply does not exist where V is infinite. As we saw in Chapter 3, the wave functions in these systems are indeed sines and cosines. But our various boxes had impenetrable walls, whereas real matter waves (just as light waves) can actually penetrate a little way into a wall, even when the wall acts as an *almost* perfect reflector. For a high (but not infinitely high) potential barrier, a particle with finite energy can have $E < V$ in the region *within* the wall itself. We must therefore examine the consequences of having $E < V$. In this case the multiplier of $\psi(x)$ in Eq. 4.1 is positive, and $d^2\psi(x)/dx^2$ has the same sign as $\psi(x)$ itself. By the arguments of the preceding paragraph, we cannot deal with a wave function whose eigenvalue E is *everywhere* less than $V(x)$; such a function would become infinite in one of the directions $x \to +\infty$ or $x \to -\infty$. On the other hand, there is no objection whatever to having a wave

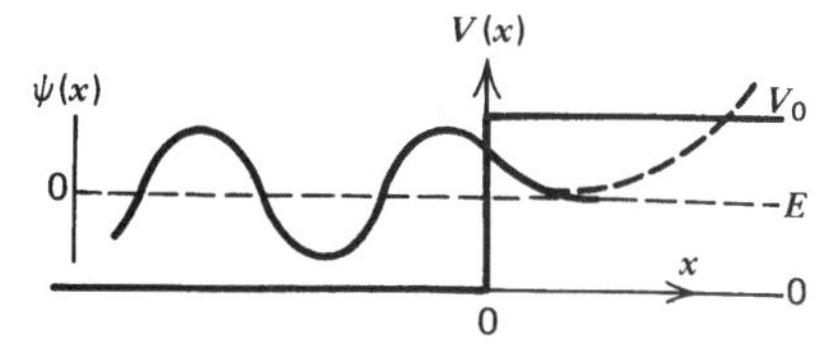

Figure 4.2 The step potential:

$$V(x) = 0 \quad (x < 0),$$
$$V(x) = V_0 \quad (x \geq 0).$$

Superimposed on the potential energy curve we have plotted a possible wave function for energy E, using the line corresponding to energy E as the x axis for $\psi(x)$. The dashed curve (- - -) is an example of the forbidden class of functions that diverge as $x \to \infty$.

function whose eigenvalue E is greater than $V(x)$ for some range of x and less than $V(x)$ for another range of x.

To make this discussion concrete, we have illustrated in Fig. 4.2 a potential having the form of a step. We wish to determine the eigenfunction for a state with energy E less than the barrier height V_0, so that $E > V(x)$ for negative x and $E < V(x)$ for positive x. For $x < 0$ the Schrödinger equation is simply Eq. 3.64, the most general solution of which is

$$\psi(x) = Ae^{i\alpha x} + Be^{-i\alpha x},$$
$$\alpha \equiv \left(\frac{2mE}{\hbar^2}\right)^{1/2} \qquad (x < 0). \tag{4.2}$$

For $x > 0$ we must solve Eq. 4. 1 with $V(x) = V_0$; it is not difficult to show that a solution is

$$\psi(x) = Ce^{\beta x} + De^{-\beta x},$$
$$\beta \equiv \left[\frac{2m(V_0 - E)}{\hbar^2}\right]^{1/2} \qquad (x > 0). \tag{4.3}$$

By the reasoning already described, we must have $C = 0$ if $\psi(x)$ is not to become infinite as $x \to \infty$. (The dashed line in Fig. 4.2 indicates how the wave function might behave without this restriction.) The boundary condition to be satisfied is that both $\psi(x)$ and its first derivative must be continuous across the boundary. That is, one sets $x = 0$ in Eqs. 4.2 and 4.3 and equates the results, and does the same with the two equations obtained by differentiation. The details of this calculation constitute Problem 2 at the end of the chapter. The wave function is in general complex, but qualitatively it must resemble the function plotted in Fig. 4.2: sinusoidal where $E > V(x)$, decaying exponentially as one moves into the potential barrier.

The one-dimensional box of Chapter 3 is simply a region with infinitely high potential barriers on both sides. How does the wave function for a finite-step potential differ from that of the particle in a box? Here the wave function has a nonzero value within the barrier, vanishing only in the limit as $x \to \infty$. For the infinitely high barrier, however, we have $V_0 \to \infty$, $\beta \to \infty$, and $De^{-\beta x} \to 0$ for all $x > 0$; this agrees with our assumption that $\psi(x) = 0$ beyond the box walls. In both cases the wave function itself must be continuous at the boundary; for the finite step we have the additional boundary condition that $d\psi(x)/dx$ be continuous.[1]

The region where $V > E$ is called the classically forbidden region. By Eq. 4.1, the larger the factor $V - E$, the larger is the magnitude of $d^2\psi(x)/dx^2$, and thus the more rapidly $\psi(x)$ decays with distance into the barrier. In a classically allowed region, by contrast, the more E exceeds V, the more rapidly the wave function must oscillate. Whenever E is very close to V whether above or below, the wave function must be relatively flat. In Chapter 3 we interpreted behavior like this in terms of the kinetic energy $(E - V)$: The higher the kinetic energy, the more rapidly the wave function changed. This is still true if we substitute the *magnitude* of the kinetic energy, because in the classically forbidden region we have the remarkable phenomenon of *negative* kinetic energy.

The wave function is nonzero in the classically forbidden region. This means that, whenever a real particle approaches a finite potential barrier, the particle (i.e., the wave that describes it) can penetrate some distance into the classically forbidden region. The larger the mass of the particle (and thus the value of β), the shorter the distance it is able to penetrate; macroscopic objects have masses too great for this effect to be observed. Suppose now that the potential barrier has only a limited extent in space, as shown in Fig. 4.3. In this case, the particle can actually penetrate *through* the barrier, even though it has insufficient energy to go over it classically. Mathematically, this is a consequence of the continuity conditions on the wave function and its derivative. Even though $\psi(x)$ may be a dying exponential coming from the left side of the potential barrier toward the right, it never reaches zero for any finite value of x. Therefore, at the right side of the barrier, this small but still nonzero wave function must join smoothly with an oscillatory function for the region beyond where $E > V$ again. The relative amplitudes of the two sinusoidal waves are related to the transmission coefficient, the probability that a particle will pass through the barrier; for $E << V_0$ this is approximately proportional to $e^{-2\beta a}$, where a is the barrier thickness and β is defined by Eq. 4.3.

What we have described here, of course, is a time-dependent process. How, then, is it described by the time-independent wave function $\psi(x)$? We can think of $\psi(x)$ as describing the steady-state behavior of a continuous stream of particles striking the barrier. At any given time, some particles will be moving toward the barrier from the left and some will be rebounding toward the left; these correspond to the $Ae^{i\alpha x}$ and $Be^{-i\alpha x}$ terms, respectively, in Eq. 4.2. For a barrier of finite thickness, other particles will be moving away from the barrier toward the right; in this region $\psi(x)$ must be of the form $Ge^{i\alpha x}$. The relative values of $|A|^2$, $|B|^2$, and $|G|^2$ give the steady-state fractions of incident, reflected, and transmitted particles. A single particle is described by the same wave function if we know nothing about its position at a given time: $|B|^2/|A|^2$ and $|G|^2/|A|^2$ then represent the probabilities of reflection and transmission. If we wanted to follow the process as a function of time, we would have to solve the time-dependent Schrödinger equation for a wave packet like that in Fig. 3.7, but there is no need to do this when we are interested only in the initial and final states of the system. The same simplification can be made for any collision or scattering process.

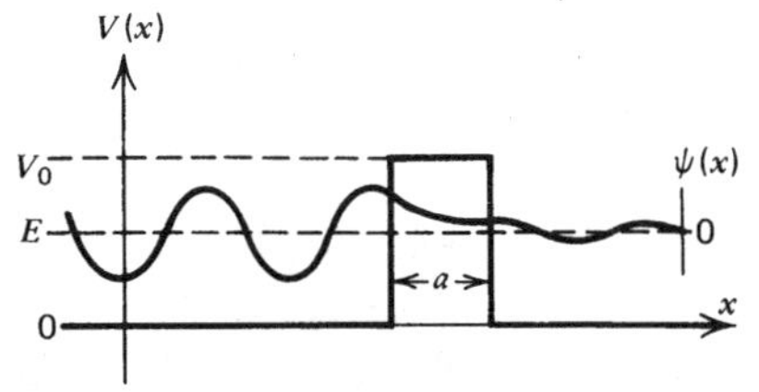

Figure 4.3 Penetration of a potential barrier by a particle wave. As in Fig. 4.2, the wave function is superimposed on the potential energy curve.

The interpretation may be more straightforward if we think in terms of matter waves rather than particles. The fact that the barrier is finite means that it is only an imperfect reflector: Part of the wave is transmitted through the region in which $E < V$, the rest of it is reflected. A simple analogy is the behavior of light striking a thin sheet of metal. Most of the light is reflected, but some is transmitted; if the metal is thin enough, one can even see through it. But the fraction of the light transmitted decreases rapidly with thickness; for most sheets of metal the amplitude of the transmitted wave is indetectably small. The classically forbidden transmission of a particle through a barrier is known as *tunneling*. The situations where this process is important include the escape of α particles from atomic nuclei in radioactive decay, the "umbrella inversion" of pyramidal molecules like NH_3, and the so-called tunnel diode (in which the barrier is the interface between two semiconductors). Tunneling also affects the rates of chemical reactions, as we shall point out in Chapter 30.

So far we have considered two kinds of potential barriers: a potential that rises discontinuously from zero to infinity, and a barrier of finite height. The finite barrier can be of arbitrary shape without affecting our conclusions. In Figs. 4.2 and 4.3 we used potentials with discontinuous jumps only for simplicity; a smoothly rising potential would have given the same qualitative results. One important possibility remains—a potential that goes to infinity gradually; we shall consider this case in the next section.

[1] If $d\psi/dx$ were discontinuous at some boundary, then $d^2\psi/dx^2$ would be either infinite or undefined. By Eq. 4.1, $d^2\psi/dx^2$ can become infinite only if either ψ or V becomes infinite at the boundary. The latter condition is met at an infinitely high potential barrier; for all other types of potentials, ψ itself would have to become infinite. We cannot allow this, so we require the derivative to be continuous everywhere except at infinite potential jumps. (This was included in Postulate I of Section 3.8.)

4.2 The Quantum Mechanical Harmonic Oscillator

In Section 2.8 we introduced the harmonic oscillator, a point mass bound to an equilibrium position by a springlike restoring force whose strength is directly proportional to the particle's displacement. As before, we consider a one-dimensional oscillator. If the equilibrium point is taken as the origin, the force and the classical equation of motion are given by Eq. 2.33 as

$$F(x) = -kx = m\frac{d^2x}{dt^2}. \tag{4.4}$$

If we take our arbitrary zero point of energy to be the energy of the particle at rest at the origin, then by Eq. 2.48 the potential energy is

$$V(x) = \tfrac{1}{2}kx^2. \tag{4.5}$$

To derive the quantum mechanics of the harmonic oscillator, we must solve the Schrödinger equation with this value of $V(x)$.

The one-dimensional potential before us differs in two significant ways from that of our earlier one-dimensional box (Fig. 3.5). First, the oscillator potential is smooth: At no point does it have an abrupt change of value or an infinite derivative, as the box potential does at the walls. Second, it remains finite for all finite values of x. But the oscillator and the one-dimensional box are alike in one very important way: As $x \to \pm\infty$, both potentials approach a positive infinite value, so that no matter how large the energy of a particle, it is always confined within a "potential well."

We recall that Planck assumed the energy levels of an oscillator to be equally spaced integral multiples of $h\nu$, Planck's constant multiplied by the oscillator's fundamental frequency. We certainly expect to find in the wave picture that the energy levels of the oscillator are quantized. Moreover, from what we have already seen, we can make an intuitive statement about how these levels should be spaced. For a particle in a box, the larger the box, the more closely spaced are the levels; if we enlarge the box, each level moves accordingly to a lower energy. The oscillator is a particle in a box that grows wider as the energy increases. We therefore expect that with increasing energy the spacing of the harmonic oscillator energy levels should increase less rapidly than the spacing in a simple square box. In fact, it turns out that the spacing does not increase at all, in agreement with Planck's assumption. To find the energy levels explicitly, we must solve the one-dimensional Schrödinger equation, which for the harmonic oscillator becomes

$$\mathsf{H}\psi(x) = -\frac{\hbar^2}{2m}\frac{d^2\psi(x)}{dx^2} + \frac{1}{2}kx^2\psi(x) = E\psi(x). \tag{4.6}$$

This equation, like those for the particles in circular and spherical boxes, is an extensively studied equation whose eigenvalues and eigenfunctions are quite well known.

Without solving Eq. 4.6 explicitly, what can we deduce about its solutions from the general principles of Section 4.1? Here we have a potential for which $V(x)$ becomes infinite as x approaches $\pm\infty$. Therefore, no matter what a given state's energy eigenvalue E may be, E becomes less than V for sufficiently large $|x|$. In the regions where $E > V$, every wave function must fall off more or less exponentially, as shown in Fig. 4.2, approaching zero as $x \to \pm\infty$. As the wave function for a given state comes in toward the origin, the sign of its second derivative changes at the point where E and V become equal. In the region around the origin, $d^2\psi(x)/dx^2$ must be negative and $\psi(x)$ oscillatory. By analogy with the particle in a box, we expect the lowest state to be one with no nodes, the next state to have one node, and so on as the energy increases. Thus, from purely qualitative considerations, we can infer the general form of all the wave functions of the harmonic oscillator—or indeed of any particle bound in a smooth and infinitely high potential well.

Planck was correct in supposing that the energy spectrum has the remarkable property of equally spaced levels. The levels are in fact given by

$$E_n = (n + \tfrac{1}{2})h\nu \qquad (n = 0, 1, 2, \ldots), \tag{4.7}$$

which differs from Eq. 2.54 only in the extra $\frac{1}{2}h\nu$. Since the energy retains this value even in the lowest state (with $n = 0$), $\frac{1}{2}h\nu$ is called the zero-point energy.

As we pointed out in Section 3.7, the uncertainty principle requires that all confined systems have a zero-point energy. We can rationalize this result for the harmonic oscillator in a different way. If an oscillator had exactly zero energy, it would of course have $E < V$ everywhere except at the origin. This means that the wave function would resemble the curves of Fig. 4.1*b*: If we required $\psi(x)$ to go to zero at both $+\infty$ and $-\infty$, it would become steeper as $x \to 0$ from both sides. There would thus be a discontinuous slope at the origin, a point where $V(x)$ is perfectly smooth. This is simply not an acceptable wave function, because it violates the continuity condition. This argument shows that we just cannot localize a particle at a point in space, even if we try to do it with a smooth potential.

We can generalize from the harmonic oscillator to see how energy-level spacings occur in other potential wells. If a potential is flatter than that of the harmonic oscillator, then the energy levels must come closer together as the energy increases; if a potential is steeper than that of the harmonic oscillator, then the spacing increases as the energy increases. (The particle in a box is an example of the latter case.) The simplest test of "steepness" is the second derivative of the potential energy, which for the harmonic oscillator is a constant. We shall return to this conclusion when we examine the vibrations of molecules.

We shall not carry out a systematic derivation of the solutions of Eq. 4.6, but we can pursue the foregoing qualitative comments a bit further. The relation of Eq. 4.6, stating that the second derivative of $\psi(x)$ is proportional to $x^2\psi(x)$, implies that the argument x appears in an exponent—in fact, that it is x^2, not x, that is in the exponent. The Gaussian, with the form e^{-cx^2}, has just this form. It has the required nodeless behavior of the lowest eigenfunction, it is concave toward the axis from its maximum out to its two points of inflection and it falls off at larger $|x|$, convex toward the x axis. By substituting this form into Eq. 4.6, one can readily show that the Gaussian has the correct form:

$$\frac{d}{dx}e^{-cx^2} = -2cxe^{-cx^2} \qquad (4.6.a)$$

and

$$\frac{d^2}{dx^2}(e^{-cx^2}) = -2ce^{-cx^2} + 4c^2x^2e^{-cx^2}. \qquad (4.6.b)$$

Hence we can identify the constant c with $mE/\hbar^2$ of Eq. 4.6 and $4c^2x^2$ with $mkx^2/\hbar^2$ for the lowest state of the harmonic oscillator. This is adequate to fix the energy of the state in terms of the parameters of the system—its force constant k and mass m—as

$$E = \frac{\hbar}{2}\left(\frac{k}{m}\right)^{1/2}. \qquad (4.7)$$

Recall that $(k/m)^{1/2}$ is the angular frequency ω or $2\pi \times$ the circular frequency ν, so the energy of the lowest state of the harmonic oscillator is $\hbar\omega/2$ or $h\nu/2$, in agreement with Eq. 4.7.

The lowest state of the quantum oscillator is quite different from what one expects for a classical oscillator: The highest amplitude is at the point $x = 0$, where the potential is deepest and therefore where the kinetic energy is greatest. In classical systems, particles spend most of their time in the places where their velocities are low. Quantum mechanical systems with several quanta behave in the same way, but the states that deviate furthest from classical mechanical behavior, the states of lowest energy, do not conform to our classical expectations for particles.

The states above the ground state have wave functions whose forms are products of polynomials and Gaussians. The one-quantum state has a wave function with one node at $x = 0$, and is simply of the form xe^{-cx^2}; the next state has a wave function that is a product of a quadratic and a Gaussian, and thus has two nodes. In general,

$$\psi_n(z) = \left(\frac{m\omega}{\hbar\pi}\right)^{1/4} \frac{e^{-z^2/2}}{(2^n n!)^{1/2}} H_n(z), \qquad (4.8)$$

Table 4.1 Harmonic Oscillator Wave Functions

n	$H_n(z)$	$\psi_n(z)$
0	1	$(m\omega/\hbar\pi)^{1/4}e^{-z^2/2}$
1	$2z$	$(m\omega/\hbar\pi)^{1/4}(2 \cdot 1)^{-1/2}e^{-z^2/2}2z$
2	$4z^2 - 2$	$(m\omega/\hbar\pi)^{1/4}(4 \cdot 2)^{-1/2}e^{-z^2/2}(4z^2 - 2)$
3	$8z^3 - 12z$	$(m\omega/\hbar\pi)^{1/4}(8 \cdot 6)^{-1/2}e^{-z^2/2}(8z^3 - 12z)$
4	$16z^4 - 48z^2 + 12$	$(m\omega/\hbar\pi)^{1/4}(16 \cdot 24)^{-1/2}e^{-z^2/2}(16z^4 - 48z^2 + 12)$

where[2] $z \equiv (m\omega/\hbar)^{1/2}x$, $\omega = 2\pi\nu$, and the *Hermite polynomial* H_n is defined by

$$H_n(z) \equiv (-1)^n e^{z^2} \frac{d^n}{dz^n}(e^{-z^2}). \qquad (4.9)$$

The first five harmonic oscillator wave functions are listed in Table 4.1 and plotted in Fig. 4.4*a* with the energy levels. The correctness of these solutions can be checked by substituting them in Eq. 4.6.

These wave functions have a number of interesting properties. The first thing to note is that the energy levels are nondegenerate: Only one wave function corresponds to each level. The Hermite polynomials are alternately functions of even and odd powers of z (and thus of x): this property makes the functions themselves alternately even and odd with respect to $x = 0$. That is, for $n = 0, 2, 4, \ldots$, the function for $-x$ is exactly the same as that for $+x$, but for $n = 1, 3, 5, \ldots$, the two functions have opposite signs:

$$\begin{aligned} \psi_n(x) &= \psi_n(-x) \qquad (n \text{ even}),\\ \psi_n(x) &= -\psi_n(-x) \qquad (n \text{ odd}). \end{aligned} \qquad (4.10)$$

We say that the even-n and odd-n wave functions have even and odd *parity*, respectively. Finally, as usual, the number of nodes increases with energy, with $\psi_n(x)$ having n nodes.

Let us now look at the oscillator probability distributions, some of which are plotted in Fig. 4.4*b*. Here we can see how much the Gaussian distribution of the lowest state of the quantum oscillator differs from that of the classical oscillator with the same energy. And we can see too how the higher states approach their classical counterparts, whose probability distributions are proportional to $(x_0^2 - x^2)^{-1/2}$. As n increases, however, the probability maxima move outward and $|\psi_n|^2$ is more and more closely approximated by the classical probability, as can be seen in the figure. In the classical limit ($n \to \infty$), the oscillations in $|\psi|^2$ become too close together to distinguish, and the probability over any finite range of x averages out to the classical value.

[2] Note that $(\hbar/m\omega)^{1/2}$ is the classical amplitude x_0 for an oscillator with $E = \frac{1}{2}h\nu = \frac{1}{2}\hbar\omega$ (the zero-point energy); thus, we have $z = x/x_0$.

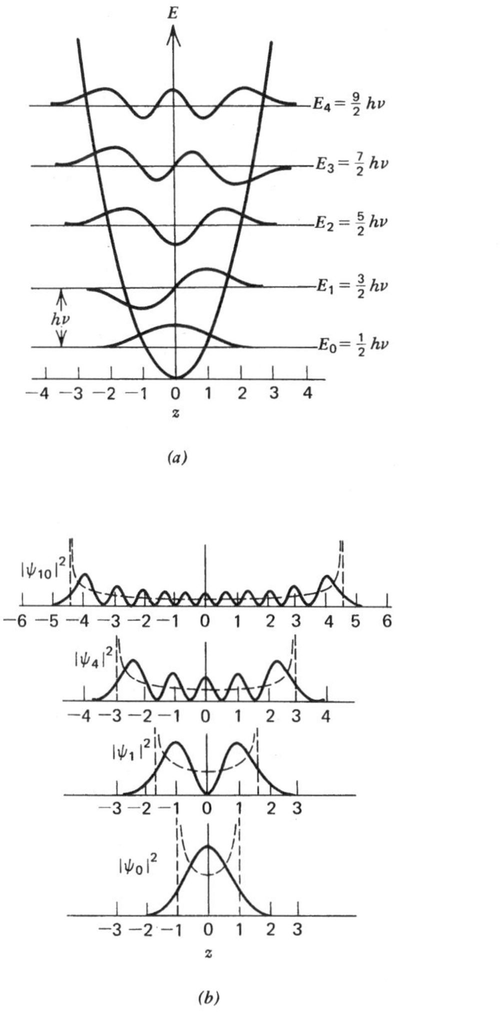

Figure 4.4 The quantum mechanical harmonic oscillator. (*a*) Potential energy curve, energy levels E_n, and superimposed wave functions $\psi_n(z)$, where $z \equiv (m\omega/\hbar)^{1/2}x$. (*b*) Probability densities for $n = 0, 1, 4, 10$, with the dashed lines (- - -) giving the probabilities for classical oscillators of the same energies.

So far we have considered only the one-dimensional harmonic oscillator. The spherically symmetric three-dimensional case, for which

$$\mathbf{F} = -k\mathbf{r} \quad \text{and} \quad V(r) = \tfrac{1}{2}kr^2 = \tfrac{1}{2}k(x^2 + y^2 + z^2) \qquad (4.11)$$

(where $\mathbf{r}$ is the vector giving the displacement from the origin), presents no added difficulties. As in Section 3.9, the Hamiltonian is separable in the three Cartesian coordinates, and the total energy is simply the sum of three terms like Eq. 4.7. In other words, the three-dimensional oscillator is equivalent to three independent one-dimensional oscillators. In general, the motion of any system of oscillating masses can be similarly analyzed into one-dimensional components, provided only that all the vibrations are harmonic. As we shall see in later chapters, crystals and polyatomic molecules are good approximations to such systems.

A comment is in order here about one aspect of Eq. 4.11 and, more generally, of systems in two or three dimensions that have symmetry, such as the spherical symmetry of the potential of Eq. 4.11. This is the property of *degeneracy*, which means that more than one quantum state have the same energy and properties that are identical apart from the choice or naming of axes. A state with one unit of angular momentum of clockwise rotation about the z-axis is indistinguishable, apart from its sense of rotation (and in this case, associated magnetic moment), from the corresponding state with counterclockwise rotation. Degeneracy will play an important role in later discussions.

4.3 The Hydrogen Atom

In Section 2.11 we introduced Bohr's model of the hydrogen atom, which consists of an electron bound to a proton by a Coulomb force. The proton, being much heavier than the electron, is essentially immobile with respect to the center of mass; we therefore take the proton as the origin of our coordinate system.[3] Since the potential energy depends only on the distance between the electron and the proton, that is, on the radial distance of the electron from the origin, we are confronted with a problem of spherical symmetry. Bohr's condition that the action per cycle be quantized led us to conclude that the electron's angular momentum is quantized. Now, knowing what we do about waves in spherical potentials, we come to the same conclusion just from the symmetry of the potential. As soon as we describe the electron by a wave, we find ourselves applying the same periodic boundary conditions that we applied to the particle in a spherical box: In order that the wave function be a single-valued function of θ and ϕ, its angular parts must close on themselves. We conclude immediately that the angular parts of the hydrogen wave functions are exactly the spherical harmonics of Table 3.1.

Let us spell out in more detail the mathematical statement of the last paragraph. The classical expression for the energy of a hydrogen atom is

$$E_{\text{classical}} = T + V(r), \qquad (4.12)$$

where the potential energy is

$$V(r) = -\frac{Ze^2}{4\pi\varepsilon_0 r} \qquad (4.13)$$

[3] Strictly, one takes the center of mass as the origin and replaces the two particles by a single particle of reduced mass μ, as in Eqs. 2.74ff.; for the H atom, $\mu = 0.9994557\, m_e$. Although we shall not go into the details of this refinement, we use μ for mass anyway, to avoid confusion with the quantum number m.

and the kinetic energy is already familiar to us:

$$T = \frac{1}{2\mu}\mathbf{p}\cdot\mathbf{p} \tag{3.146}$$

$$= \frac{1}{2\mu}p_r^2 + \frac{L^2}{2I}. \tag{3.153}$$

The quantum mechanical analog of expression 4.12 is the Hamiltonian operator whose eigenvalues are the allowed energies of this atom (Postulate III):

$$\mathsf{H} = \frac{-\hbar^2}{2\mu}\left[\frac{1}{r^2}\frac{\partial}{\partial r}\left(r^2\frac{\partial}{\partial r}\right) + \frac{1}{r^2\sin\theta}\frac{\partial}{\partial\theta}\left(\sin\theta\frac{\partial}{\partial\theta}\right) + \frac{1}{r^2\sin^2\theta}\frac{\partial^2}{\partial\phi^2}\right] - \frac{Ze^2}{4\pi\varepsilon_0 r}. \tag{3.154}$$

The wave functions $\psi(\mathbf{r})$ for the hydrogen atom, whose existence is supposed in our first postulate, satisfy a time-independent Schrödinger equation if they correspond to eigenstates of energy. According to Postulate V:

$$\mathsf{H}\psi(\mathbf{r}) = E\psi(\mathbf{r}) \qquad (\mathbf{r} \equiv (r,\theta,\phi)). \tag{4.14}$$

The form of Eq. 3.153 tells us that the classical Hamiltonian is the sum of a radial part, $[p_r^2/2\mu + V(r)]$, and an angular part, $L^2/2I$. This in turn assures us that the quantum mechanical operator, Eq. 3.154, is separable into corresponding radial and angular parts, with nothing but the square of the angular momentum operator, L^2 (and a constant scale factor) in the angular part. The wave function $\psi(\mathbf{r})$ can be written as a product $\psi(\mathbf{r}) = R(r)\Theta(\theta)\Phi(\phi)$, and the Schrödinger equation for the hydrogen atom is separable into a radial equation for $R(r)$ (to which we shall return) and an angular equation for $\Theta(\theta)\Phi(\phi)$, involving only the rotational kinetic energy. This latter is identical to the equation discussed in Section 3.11 but never written explicitly there:

$$\mathsf{L}^2\Theta(\theta)\Phi(\phi) = \left(p_\theta^2 + \frac{p_\phi^2}{\sin^2\theta}\right)\Theta(\theta)\Phi(\phi)$$

$$= \langle \mathsf{L}^2 \rangle \Theta(\theta)\Phi(\phi). \tag{4.15}$$

We asserted previously that the eigenvalues of L^2, the constants that may appear in Eq. 4.15, have the form

$$\langle \mathsf{L}^2 \rangle = l(l+1)\hbar^2, \tag{3.162}$$

and the eigenfunctions of Eq. 4.15 are the spherical harmonics $Y_{l,m}(\theta, \phi)$, such as those in Table 3.1.

We thus know the angular behavior of the wave function, and therefore the allowable values of angular momentum, for an electron in a hydrogen atom. But this is still a far cry from knowing either the energy-level spectrum or the spatial distribution of charge in the atom. As we have done before, let us make some general inferences about the relationships between the eigenvalues of angular momentum and energy, and about the general shapes of the wave functions themselves. Then, without going through a derivation, we can examine some of the actual solutions to the hydrogen wave equation, in particular the energy-level spectrum.

The total energy of a hydrogenlike atom is still given by Eq. 2.56, with the zero of energy chosen to correspond to particles at rest infinitely far apart. (For generality, we shall continue to let the nuclear charge be $+Ze$, rather than just $+e$ as in the hydrogen atom itself.) In contrast with the harmonic oscillator or the particle in a box, the potential energy of the Coulomb field given by Eq. 4.13 has no lower bound: It becomes negatively infinite at the origin. Again in contrast with the earlier examples, the Coulomb potential does have an upper bound: It approaches zero as r becomes infinite. This is why our choice for the zero of the energy scale is the most convenient and natural one to deal with the Coulomb potential.

Because the potential has a range $-\infty < V < 0$, any state of a hydrogenlike atom with a negative energy value ($E < 0$) must have $E > V$ near the origin and $E < V$ for large r. (Here, we use V and V interchangeably, because the quantum operator acts simply as a scalar, like the classical potential energy, as we use it.) We say that all states with negative energy must be *bound states.* Following the reasoning of Section 4.1, we immediately recognize that the amplitude of the wave function for any such state must approach zero asymptotically as $r \to \infty$; the particle is effectively confined, and will be found mainly in the region where $E > V$ and the wave function is oscillatory.

By contrast, all states with $E > 0$ have $E > V$ everywhere in space, and are thus *free states.* Their wave functions are oscillatory for all values of r and are therefore analogous to the very first matter waves we examined, those describing a free particle. The positive-energy wave functions of the Coulomb potential are much like those of the free particle, in that they exhibit no discrete quantization and can be represented by either traveling or standing waves. In other ways, however, the two cases differ. We know that the second space derivative of a wave function measures the kinetic energy of the particle to which the wave function corresponds. The free particle feels a constant potential (arbitrarily set equal to zero), and in a state of fixed energy has exactly the same kinetic energy at every point in space. As a result, $\partial^2\psi/\partial x^2$ is constant and the free-particle wave is a simple sinusoidal function. In a Coulomb field, because the potential energy varies smoothly with r, the kinetic energy must also vary with r for a particle with a given total energy. As one follows a particular wave function from large r inward toward the origin, the potential energy drops off toward $-\infty$, the kinetic energy must increase correspondingly, and the magnitude of a $\partial^2\psi/\partial r^2$ must increase. (When the variables are separated, V appears only in the radial wave equation; thus only the radial derivative is affected.) Instead of behaving like simple sinusoidal functions, the waves describing an unbound particle in a Coulomb field thus oscillate more rapidly at small r than at large r. The same property is of course exhibited by the bound states: They too show more rapid oscillations at small *r*. Asymptotically, as *r* grows very large, the distance-dependent wavelength of the wave function increases very slowly toward its limiting value at infinite distance.

We have already seen, from the harmonic oscillator and the particles in various boxes, that the bound states of simple systems show a very regular relationship between their energies and the number of nodes in the wave function. The lowest bound state has no nodes, the next-to-lowest state one node, and so on. For a system with spherical symmetry (like the hydrogen atom), this order holds rigorously for states of the same total angular momentum; however, for an arbitrary spherically symmetric potential, the energy corresponding to a given number of nodes varies with angular momentum (as in Fig. 3.12). If the ground state of the hydrogen atom is to be nodeless, it must be a state with $l = 0$, that is, with no angular momentum (see Table 3.1). The next energy level should correspond to a wave function with one node, which may be either angular or radial. In the former case we have $l = 1$, and the single nodal surface is a plane; in the latter case $l = 0$, and the node is a spherical surface in the radial part of the wave function. Higher energy levels have more nodes; for a state with a given value of *l*, there will be *l* angular nodes, the remainder if any being radial.

What can one say about the relative energies of the two single-node states—or of any states with the same number of nodes? Let us define a quantum number *n* as the number of nodes plus one. In general, as we said above, for a spherically symmetric potential, the energy levels for a given *n* vary with the angular momentum quantum number *l*. However, it is a unique property of the pure Coulomb potential that all bound states with the same *n* have exactly the same energy. Since the energy depends only on *n* and not on the angular momentum or its orientation, we call *n* the *principal quantum number.* Remarkably, this turns out to be the same as the quantum number *n* that appeared in Eq. 2.67 for the energy of the *n*th Bohr orbit. In fact, solution of the Schrödinger equation gives exactly the same expression for the bound-state energy eigenvalues of a hydrogenlike atom:

$$E_n = -\frac{Z^2 e^4 \mu}{8\epsilon_0^2 n^2 h^2} \quad (n = 1, 2, \ldots). \tag{4.16}$$

(We shall not present a derivation of this result. For a derivation, see any of the references cited at the end of this chapter.)

One can now see why the Bohr model is so useful. Unfortunately, things are not really this simple. Despite the similarity of the model's predictions of energy levels to the observed spectrum, precise measurements reveal many small but extremely important differences. A number of effects that we have neglected influence the energy-level structure of Eq. 4.16 and split each level into a group of very closely spaced levels. One speaks of the *fine structure* of the energy spectrum. We shall consider these effects in the next chapter; for our purposes here, we can go on assuming the validity of Eq. 4.16, which is very nearly correct.[4]

Assuming that all states of the same *n* have the same energy, how many such states are there for a given eigenvalue E_n? The wave function corresponding to E_n has a total of $n - 1$ nodes. Since a state with a given value of *l* has *l* angular nodes (and thus $n - l - 1$ radial nodes), it is clear that *l* can never exceed $n - 1$; however, all values of *l* from 0 to $n - 1$ are permitted. Now recall our discussion of space quantization in Section 3.11: For any value of *l* (which gives the total angular momentum), there are $2l + 1$ possible orientations of the angular momentum vector (cf. Fig. 3.13), that is, $2l + 1$ possible values of the quantum number *m*. Incidentally, *l* and *m* are usually called the *azimuthal quantum number* and the *magnetic quantum number* (because the degenerate states with different *m* will split in a magnetic field), respectively. The total number of states with energy E_n is obtained by summing over all the states described above:

$$\begin{aligned} N(E_n) &= \sum_{l=0}^{n-1} (2l+1) = 2\left(\sum_{l=0}^{n-1} l\right) + n \\ &= 2[\tfrac{1}{2} n(n-1)] + n \\ &= n^2 . \end{aligned} \tag{4.17}$$

If we neglect the fine structure, then, the *n*th energy level of a hydrogenlike atom is n^2-fold degenerate. The state with $n = 1$ is nondegenerate, the $n = 2$ state is quadruply degenerate, and so on.

For purely historical reasons, based on the appearance of the spectral lines with which they are associated, the states of specific angular momentum values have come to be known by a set of letters. Illogical as they now seem, these designations are in such common use that we must introduce them here. We refer to states with $l = 0$ as *s* states, those with $l = 1$ as *p* states, those with $l = 2$ as *d* states, and those with $l = 3, 4, 5, 6, \ldots$ as *f, g, h, j, . . .* states (note that *i* is not used). A state with $n = 1$, $l = 0$ is called a 1*s* state, one with $n = 2$, $l = 1$ a 2*p* state, and so on. The letters *s, p, d, f* originally stood for characterizations of spectral lines: "sharp," "principal," "diffuse," and "fundamental." A wave function describing a bound state of a single electron is generally called an *orbital,* a term coined to suggest the connection between the wave function and the classical orbit or trajectory. Small letters, as above, are generally used to denote the quantum numbers of individual orbitals, and capital letters for the total quantum numbers indicate the state of an atom as a whole. For example, an atom with zero total electronic angular momentum has $L = 0$, and is said to be in an *S* state.

[4] If we write Eq. 4.16 for the hydrogen atom as $E_n/hc = -R_H/n^2$, where $R_H = 109677.58\ \text{cm}^{-1}$, the maximum error is 1.18 cm in the ground state; the greatest spread of fine structure levels in hydrogen is $0.36\ \text{cm}^{-1}$ in the $n = 2$ state.

We shall see in the next chapter just what these total quantum numbers are. But in a hydrogenlike atom, with only one electron, the small-letter notation is sufficient.

The energy spectrum of the hydrogenlike atom is clearly very different indeed from that of the particle in a box or the harmonic oscillator, in that its bound states are more and more closely spaced as the energy approaches zero from below. This can be clearly seen in Fig. 2.18. There are, in fact, an infinite number of negative energy levels in a Coulomb potential well. Moreover, since the degeneracy increases with n, the number of bound states at a given energy also increases as $E \to 0$. For $E > 0$ the energy is no longer quantized; that is, the eigenvalues are continuous. This is called the *continuum* region, and corresponds to a free electron passing through the field of a nucleus. Recall that the reason for this is that such a wave is unconfined and hence need not satisfy the finite boundary conditions that induce quantization.

Having discussed the energy levels, let us now see what we can say about the wave functions of the hydrogenlike atom and the distributions implied by Postulate II. If we separate the Schrödinger equation 4.14 according to the variables r, θ, and ϕ, we can immediately write down the radial wave equation as

$$\left[\frac{\mathrm{p}_r^2}{2\mu}+\frac{\hbar^2 l(l+1)}{2\mu r^2}-\frac{Ze^2}{4\pi\epsilon_0 r}\right]R_{nl}(r)=ER_{nl}(r), \qquad (4.18)$$

where $\mathrm{p}_r \equiv i\hbar[(\partial/\partial r)+r^{-1}]$ (see Eq. 3A.7); note that the radial part of the wave function depends on the quantum numbers n and l.

The negative infinity of the Coulomb potential at the origin—or the *singularity* there—places special constraints on the wave function, which must itself remain finite. There are two possible cases, depending on the value of l. When $l > 0$, the kinetic energy includes a contribution from the angular momentum, the centrifugal energy $\hbar^2 l(l+1)/2\mu r^2$. This term, which is a function of the coordinate r only, looks like a contribution to the potential energy, and can be referred to as the *centrifugal potential.* Its derivative is the so-called centrifugal force. The electron thus moves radially as if it were in an effective potential field,

$$V_{\text{eff}}=\frac{L^2}{2\mu r^2}+V(r)=\frac{\hbar^2 l(l+1)}{2\mu r^2}-\frac{Ze^2}{4\pi\epsilon_0 r}. \qquad (4.19)$$

Because the centrifugal term is always positive and increases as r^{-2} when $r \to 0$, it always dominates the effective potential for sufficiently small r. The effective potential thus always becomes *positively* infinite and acts as a repulsive potential in the vicinity of the origin. This means that for $l > 0$ the wave function $R_{nl}(r)$ must go to zero at the origin, representing the fact that an electron with nonzero angular momentum cannot reach the origin.

We can expand upon this conclusion by another route. Without going through the details of the calculation, we can reasonably assume that only the first two terms of Eq. 4.18 are significant at very small r. If this is so, then d^2R_{nl}/dr^2 (part of $\mathrm{p}_r^2 R_{nl}$) will be proportional to $l(l+1)R_{nl}(r)/r^2$, so that $R_{nl}(r)$ itself must vary as either r^{l+1} or r^{-l}. The function r^{-l} becomes infinite when r goes to zero, so this mathematical possibility is physically inadmissable. We conclude, then, that for $l > 0$ the radial wave function must vary as r^{l+1} at small r. This in turn means that, the higher the angular momentum, the lower is the initial slope $[dR(r)/dr]_{r=0}$. In simple terms, centrifugal force pushes the electron away from the nucleus.

We said that there were two possible cases for the behavior of the wave function at the origin, and the second case is that of $l = 0$, that is, of s states. There is then no centrifugal potential and the electron can penetrate to the nucleus. This case is not as simple to analyze without going through the mathematics, but it turns out that $R_{n0}(r)$ rises to a *cusp,* varying as $e^{-r/n}$ as $r \to 0$. The wave function is continuous and finite at the origin, but its slope changes discontinuously along any line through the nucleus; as with the particle in a box, this discontinuity is a result of the potential becoming infinite. Since the wave function is nonzero at the origin, what keeps the electron from collapsing into the nucleus? Here we can appeal to the uncertainty principle: If the electron collapsed to a radius less than some value r, then by Eq. 3.84 the uncertainty in p_r would be greater than about $\hbar/2r$, and the average kinetic energy greater than $\hbar^2/4\mu r^2$. Like the centrifugal term of the earlier argument, this kinetic energy goes to $+\infty$ faster than V goes to $-\infty$, so that no electron with finite total energy can collapse all the way to $r = 0$. In fact, the product $rR_{n0}(r)$, which determines the electron's probability distribution, does vanish as $r \to 0$.

The behavior that we have deduced can be confirmed by looking at the actual radial wave functions, listed in Table 4.2 and plotted in Fig. 4.5*a*. These are the solutions of

Table 4.2 Normalized Radial Wave Functions, $R_{nl}(r)$, for Hydrogenlike Atoms[a]

$$\int_0^\infty |R_{nl}(r)|^2 r^2\,dr=1 \qquad \rho\equiv\frac{Zr}{a_0}$$

Orbital	n	l	
$1s$	1	0	$2\left(\frac{Z}{a_0}\right)^{3/2} e^{-\rho}$
$2s$	2	0	$\frac{1}{2\cdot 2^{1/2}}\left(\frac{Z}{a_0}\right)^{3/2}(2-\rho)e^{-\rho/2}$
$2p$	2	1	$\frac{1}{2\cdot 6^{1/2}}\left(\frac{Z}{a_0}\right)^{3/2}\rho e^{-\rho/2}$
$3s$	3	0	$\frac{2}{81\cdot 3^{1/2}}\left(\frac{Z}{a_0}\right)^{3/2}(27-18\rho+2\rho^2)e^{-\rho/3}$
$3p$	3	1	$\frac{4}{81\cdot 6^{1/2}}\left(\frac{Z}{a_0}\right)^{3/2}(6\rho-\rho^2)e^{-\rho/3}$
$3d$	3	2	$\frac{4}{81\cdot 30^{1/2}}\left(\frac{Z}{a_0}\right)^{3/2}\rho^2 e^{-\rho/3}$

[a] As written, with the Bohr radius a_0 instead of a based on a general reduced mass μ, these expressions describe an atom with an infinitely massive nucleus.

Eq. 4.18. Each $R_{nl}(r)$ consists of an exponential $e^{-\rho/n}$ multiplied by a polynomial of degree n in ρ, where $\rho \equiv Zr/a$ (where a is approximately[5] the Bohr radius a_0, or $4\pi\epsilon_0\hbar^2/e^2m_e$). In agreement with our earlier conclusions, there are $n - l - 1$ radial nodes (not counting the origin, through which pass the angular nodes for $l > 0$).

The next question to be answered is where the electron is most likely to be in any given state. The point of highest probability is, of course, where $|\psi|^2$ has its maximum value. However, we are really more interested in knowing the *radius* at which the electron is most likely to be found. This radius is greater than that of maximum $|\psi|^2$, since a thin spherical shell at large r has a greater volume than an equally thin shell at small r. The probability of finding the electron in a shell of thickness dr is equal to the integral of $|\psi|^2$ over the volume of the shell,

$$\mathcal{P}_{nlm}(r)\,dr = \int_{\text{shell}} |\psi_{nlm}|^2\, dV$$
$$= \int_{\phi=0}^{2\pi}\int_{\theta=0}^{\pi} |\psi_{nlm}|^2\, r^2\, dr \sin\theta\, d\theta\, d\phi, \qquad (4.20)$$

where dV is given by Eq. 2B.2. We separate the variables as usual,

$$\psi_{nlm}(r,\theta,\phi) = R_{nl}(r)\Theta_{lm}(\theta)\Phi_m(\phi)$$
$$= R_{nl}(r)Y_{l,m}(\theta,\phi), \qquad (4.21)$$

and take for the $Y_{l,m}(\theta, \phi)$ the normalized spherical harmonics of Table 3.1. Since the integral of $|Y_{l,m}|^2$ over θ and ϕ is unity, Eq. 4.20 reduces to

$$\mathcal{P}_{nl}(r)\,dr = r^2 |R_{nl}(r)|^2\, dr, \qquad (4.22)$$

with the probability proportional to the square of $rR(r)$ rather than the square of the radial wave function itself. The factor r^2 can be interpreted as being due to the fact that the volume of the spherical shell is $4\pi r^2\, dr$ (remember that $|\psi|^2$ is the probability per unit volume).

Note that in the above equations we have kept explicit the subscripts n, l, m. In fact, the radial wave functions (and thus the probability densities) do depend on both n and l but not on m, and the spherical harmonics depend on l and m, even though the energy is a function of n only. In Fig. 4.5*b* we plot the radial probability densities for the first few states of the hydrogen atom. When the radial probability density has more than one *lobe* (i.e., has one or more nodes), it is always

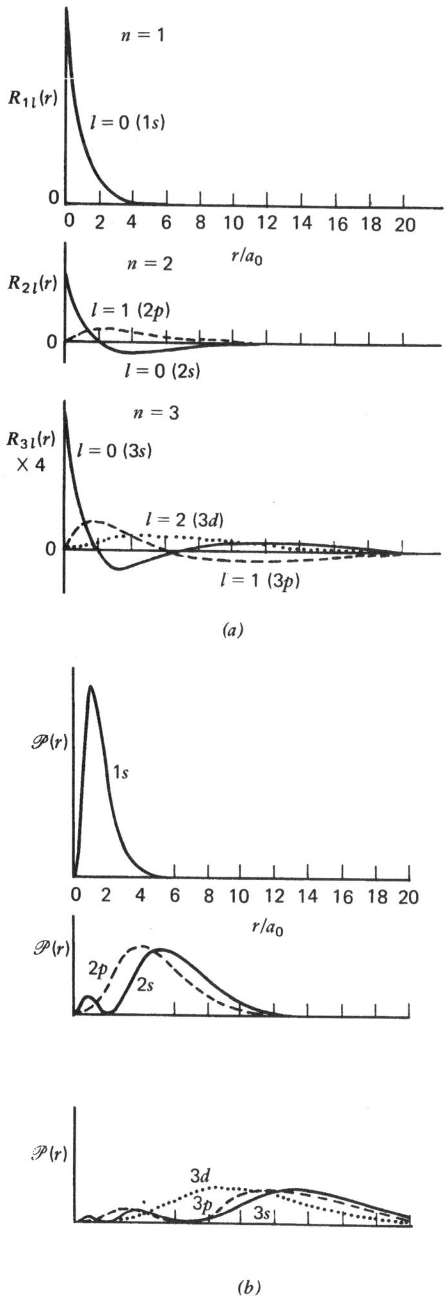

Figure 4.5 Radial wave functions and probabilities for the first few states of the hydrogen atom. (*a*) The radial wave function, $R_{nl}(r)$. (*b*) The radial distribution function, $\mathcal{P}_{nl}(r) = r^2|R_{nl}(r)|^2$. All the graphs are plotted on the same radial scale, r/a_0, where a_0 is the Bohr radius (0.529 Å). Since the wave functions are normalized, the total area under each of the $\mathcal{P}(r)$ curves is the same, corresponding to one electron.

the outermost lobe that has the largest area and the highest amplitude. The $\mathcal{P}_{nl}(r)$ functions with the same n and different l have their outermost maxima in roughly the same region. In particular, the radial function with $l = n - 1$ has only a single maximum,

$$r_{\max}[\mathcal{P}_{n,n-1}(r)] = \frac{4\pi\epsilon_0 n^2\hbar^2}{Ze^2\mu}, \qquad (4.23)$$

---

[5] Strictly, the radius appearing in the solution to Eq. 4.16 is inversely proportional to the reduced mass μ, equal to $m_e m_{\text{nucleus}}/(m_e + m_{\text{nucleus}})$; thus:

$$a \equiv \frac{4\pi\epsilon_0\hbar^2}{e^2\mu},$$

whereas the Bohr radius a_0 is based on a nucleus with infinite mass, so the electron mass m_e appears in a_0, instead of the reduced mass μ.

which is the same as the radius of the nth orbit in the Bohr theory, Eq. 2.63. The most probable radius for the hydrogen $1s$ electron ($Z = 1$, $n = 1$, $l = 0$) is thus the Bohr radius a_0, as can be seen in Fig. 4.5*b*.

Thus, in a rough way, the conclusions based on the wave picture of the hydrogen atom are in agreement with the results of the simple Bohr model. The energy levels are the same (if we neglect the fine structure), and the electron is most likely to be at a radius near the value predicted by Bohr. The physical picture is entirely different, however. In Bohr's model the electron remained in a fixed orbit restricted to a single plane. In the present theory the electron can be anywhere in space except at the nodes of ψ; some places are just more probable than others. We have discussed the radial probability density; in the next section we shall take another look at the angular distribution.

4.4 The Shapes of Orbitals

The potential energy is spherically symmetric for any isolated atom, not just the hydrogen atom. The angular part of all atomic wave functions must thus be given by the spherical harmonics of Table 3.1. As we pointed out earlier, the angular distribution of electrons in atoms governs the shapes of the molecules formed when these atoms join together. We must therefore become acquainted with the actual shapes of the spherical harmonics. Let us consider several ways of representing those shapes.

First, of course, we could simply plot $Y(\theta, \phi)$ as a function of the angles. However, we have the problem that all the $Y_{l,m}$ for $|m| > 0$ are complex, containing the factor $e^{im\phi}$. We can obtain real functions by combining the eigenfunctions for $+m$ and $-m$:

$$Y_{l,\cos m\phi} = \frac{Y_{l,m} + Y_{l,-m}}{\sqrt{2}} = \Theta_{l,m}\left(\frac{e^{im\phi} + e^{-im\phi}}{\sqrt{2}}\right) = \sqrt{2}\Theta_{l,m}\cos m\phi,$$

$$Y_{l,\sin m\phi} = \frac{Y_{l,m} - Y_{l,-m}}{\sqrt{2}i} = \Theta_{l,m}\left(\frac{e^{im\phi} - e^{-im\phi}}{\sqrt{2}i}\right) = \sqrt{2}\Theta_{l,m}\sin m\phi. \qquad (4.24)$$

This is the same method that we used to obtain standing-wave functions for the free particle in Eq. 3.42. (The factor $\sqrt{2}$ here maintains the normalization.) The new functions satisfy the Schrödinger equation as well as the old ones, and thus are valid wave functions of the system. If we look at the spherical-coordinate Hamiltonian, Eq. 3.154, we can see that both must give the same energy eigenvalues for a given m: In both cases $\partial^2 Y/\partial\phi^2 = -m^2 Y$. The functions 4.24 are thus eigenfunctions of the energy and the total angular momentum; unlike the original spherical harmonics, however, they are not eigenfunctions of L_z (cf. Section 3.11).

Using the real forms, we can now plot the angular wave functions in the form of polar graphs, Fig. 4.6. In such a graph, the magnitude of the function at a particular value of the angles is given by the radial distance from the origin. The graphs in the figure are cross sections of the complete three-dimensional graphs, and thus show the behavior of each $Y(\theta, \phi)$ in only a single plane, as a function of either θ or ϕ alone. However, the planes selected for the figure are those that most clearly display the shapes of the individual wave functions.

Let us examine the shapes of these angular wave functions. As expected, the function with $l = 0$ (s orbital) has no dependence on the angles and is thus spherically symmetric. There are three p ($l = 1$) orbitals, each symmetric around one

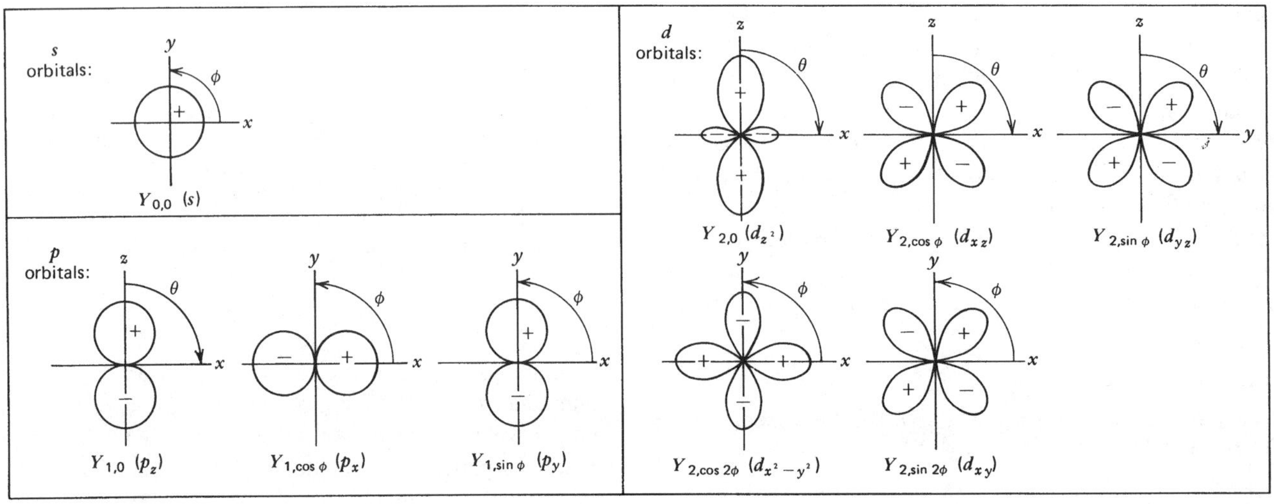

Figure 4.6 Polar graphs of spherical harmonics in real form; see Table 3.1 and Eqs. 4.24 for definitions. The *s, p, d* designations are discussed in the text. The radial distance gives the magnitude of $Y(\theta, \phi)$; the sign of each lobe of the function is indicated by a + or − sign.

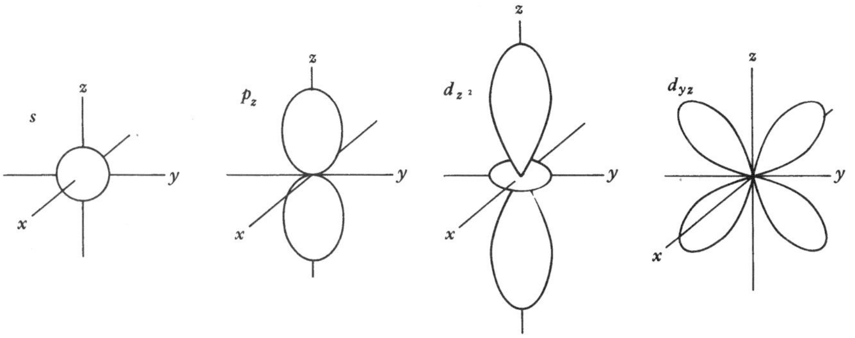

Figure 4.7 Angular probability distributions: perspective views of squared spherical harmonics, $|Y(\theta, \phi)|^2$. Except for orientation, the p_x and p_y orbitals have the same shape as the p_z; similarly, the d_{xy}, d_{xz}, and $d_{x^2-y^2}$ orbitals have the same shape as the d_{yz}.

of the Cartesian axes and thus labeled p_x, p_y, p_z. In a three-dimensional polar graph, each p orbital consists of two spherical lobes tangent at the origin. The function $Y_{2,0}$ is symmetric around the z axis, and thus consists of two elongated axial lobes with a toroidal "collar" around the middle; each of the other four d ($l = 2$) orbitals consists of four lobes whose axes are at right angles in a single plane. The latter four orbitals vary with θ and ϕ in the same way as do the functions xy, xz, yz, and $x^2 - y^2$ over the surface of a unit sphere, and are thus labeled d_{xy}, and so on (see Fig. 4.6); $Y_{2,0}$ varies as $3z^2 - r^2$, and is labeled d_{z^2}. The $f, g, \ldots$ orbitals have even more complicated shapes, but one seldom needs to consider these.

The distribution of electrons is governed not by the wave function itself but by its square. We therefore plot in Fig. 4.7 the squares of the various kinds of real[6] spherical harmonics. This time we give perspective views of the complete three-dimensional polar graphs. It is clear that the squares of the p and d orbitals are even more strongly directional than the functions themselves since squaring weights more heavily those angles for which Y was already large.

Finally, to obtain the electron distribution in an atom, we must multiply the $|Y(\theta, \phi)|^2$ of Fig. 4.7 by the square of the radial wave function $R(r)$. For the hydrogen atom the radial wave functions are the $R_{nl}(r)$ of Fig. 4.5*a*. Combining the two sets of functions, we obtain the hydrogen probability densities[7] shown in Fig. 4.8. The probability densities $|\psi|^2$ here are represented by perspective views; the radial coordinate is now the actual radial distance from the nucleus in the atom. As in Fig. 4.6, Figs. 4.8*a*–*f* show density as functions of r and θ, and can be interpreted as cross sections of the complete three-dimensional density functions.

We now have a comprehensive idea of the electron distribution in the various states of the hydrogen atom. In any other one-electron species (He^+, Li^{2+}, etc.) the distribution will be the same except for the radial scale (with a_0/Z replacing a_0). As we shall see in the next chapter, the individual electrons in more complex atoms can still be considered to have wave functions like those in hydrogen, though distorted by interaction with one another. More significantly, the angular functions $Y(\theta, \phi)$ remain exactly the same, and the basic shapes of $s, p, d, \ldots$ orbitals are thus universal.

4.5 Transitions between Energy Levels

Let us now examine how energy is absorbed or emitted by a hydrogen atom. Any such process must involve a change in the atom's wave function from one standing wave (eigenstate) to another. This change of state must always be associated with emission or absorption of a quantum of energy ΔE; if the energy is in the form of radiation, its frequency must satisfy the Einstein condition $\Delta E = h\nu$. Since the energy levels of the hydrogen atom are correctly given by the Bohr theory, we know that our wave description of the hydrogen atom is consistent with the spectroscopic data on transitions between energy levels, as summarized in Eq. 2.73. Thus we can calculate the exact energy difference or frequency in a transition between any two levels. Up to this point, however, the transition process itself has been just a black box. We have not examined, in either classical or quantum mechanical terms, the mechanism that produces the transition. In this section we shall consider the general nature of the transition process and look at some orders of magnitude for the quantities involved. Here we discuss only the hydrogen atom in any detail, and that on a quite elementary level, but one can use the same ideas in examining more complicated transitions—in molecules, for example.

Let us begin with the crudest sort of model, a classical electron in a circular Bohr orbit in the xy plane. Suppose that a beam of light is incident in the z direction, with its electric field[8] polarized along the x axis. This electric field, which

[6] Note that only the real functions of Eq. 4.24 yield the complete angular variation. If we used the original spherical harmonics containing $e^{im\phi}$, we would have

$$|Y_{l,m}|^2 = Y_{l,m}^* Y_{l,m} = \Theta_{l,m} e^{-im\phi} \Theta_{l,m} e^{im\phi} = \Theta_{l,m}^2,$$

with no ϕ dependence.

[7] The radial distribution function $\mathcal{P}(r)$ of Fig. 4.5*b* is related to the probability density $|\psi|^2$ by Eq. 4.20.

[8] We neglect the magnetic field, which also exerts a force on the moving electron.

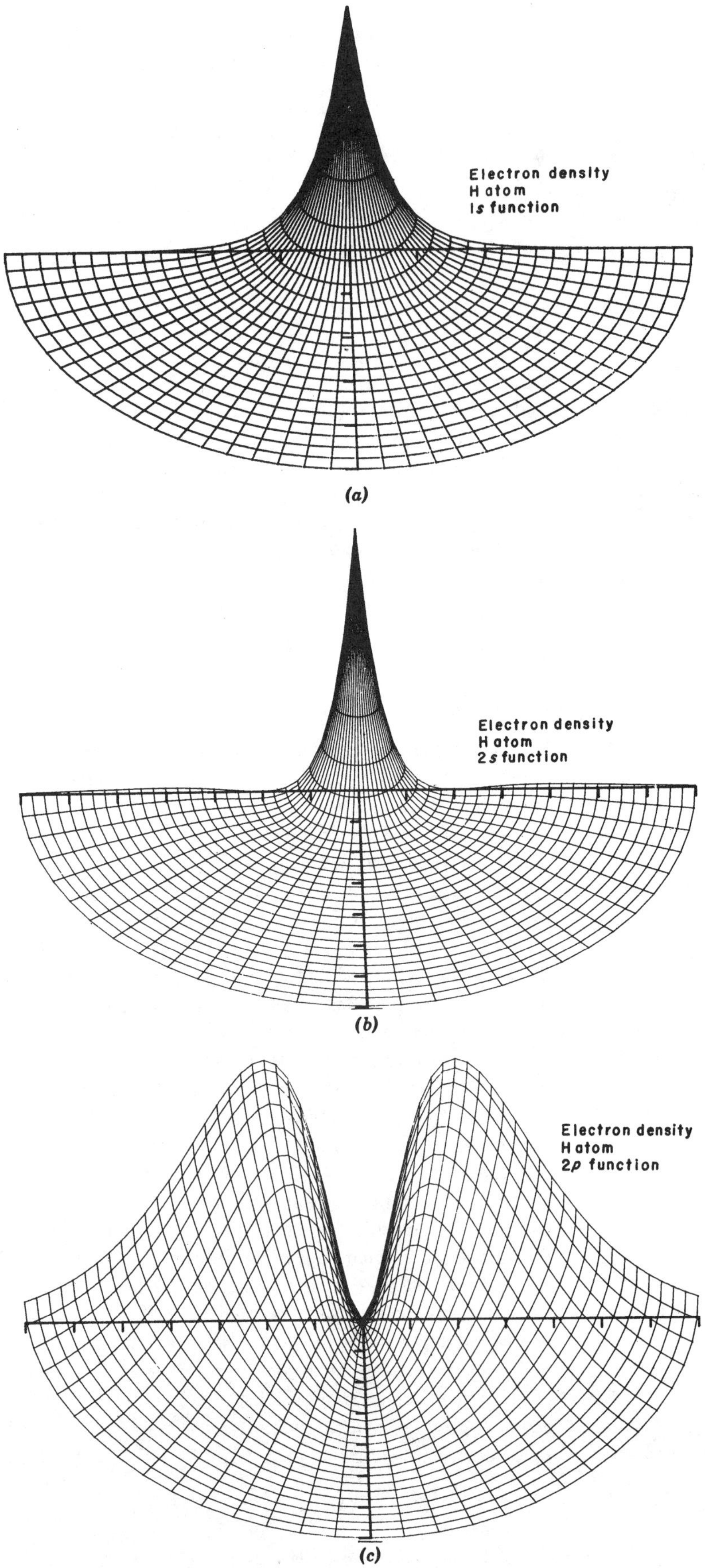

Figure 4.8 Probability densities in hydrogen orbitals. Perspective views of the densities are shown for the (*a*) 1*s*, (*b*) 2*s*, (*c*) 2*p*, (*d*) 3*s*, (*e*) 3*p*, and (*f*) 3*d* orbitals. For the *p* and *d* orbitals, the densities correspond to $m_l = 0$, so they are independent of ϕ. The height in each figure is proportional to

$$|\psi_{nlm}(r)|^2 = |R_{nl}(r)|^2 |Y_{10}(\theta)|^2 .$$

Note how all the *s* orbitals have densities peaking at the origin (the nucleus), the *p* orbital functions have densities rising rapidly from zero at the nucleus, and the 3*d* orbital densities rises slowly from zero at the nucleus. Functions with principal quantum number n exhibit $n - 1$ surfaces of a density. These can be picked out easily except for the 3*s* function, whose odd node is almost indiscernible because it occurs at a value of *r* so large that density is low everywhere in its vicinity. (These figures were supplied through the courtesy of M. E. Dolan.)

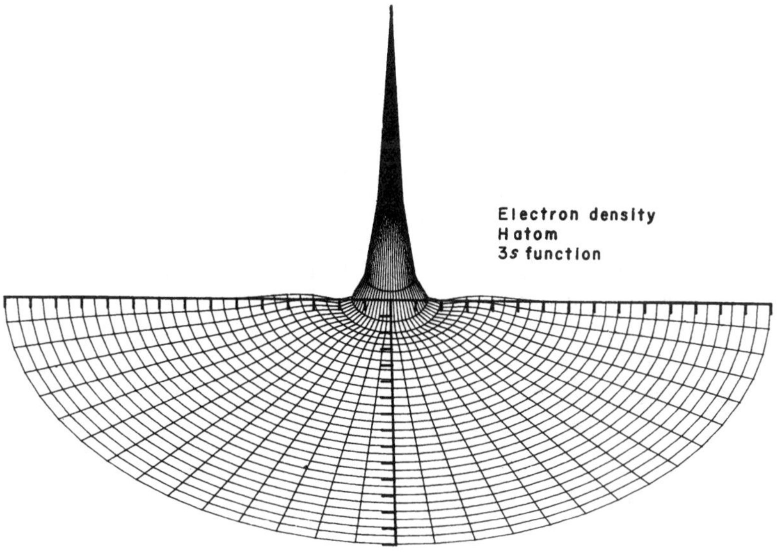
Electron density
H atom
3s function

(d)

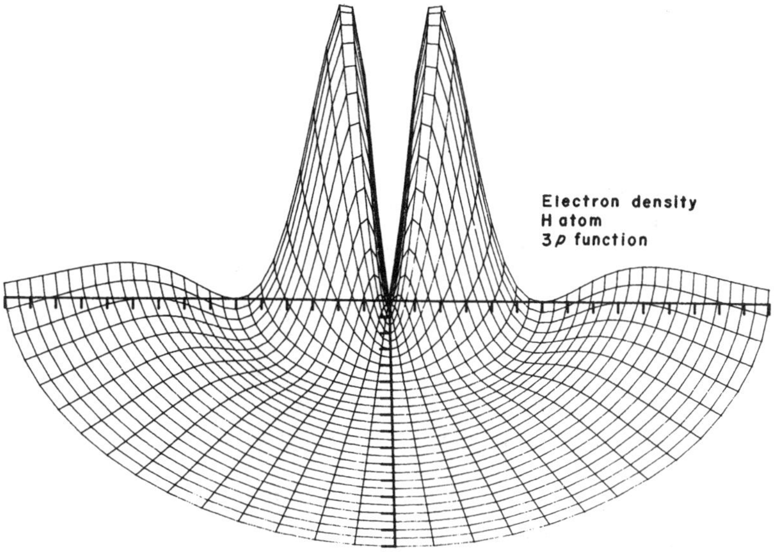
Electron density
H atom
3p function

(e)

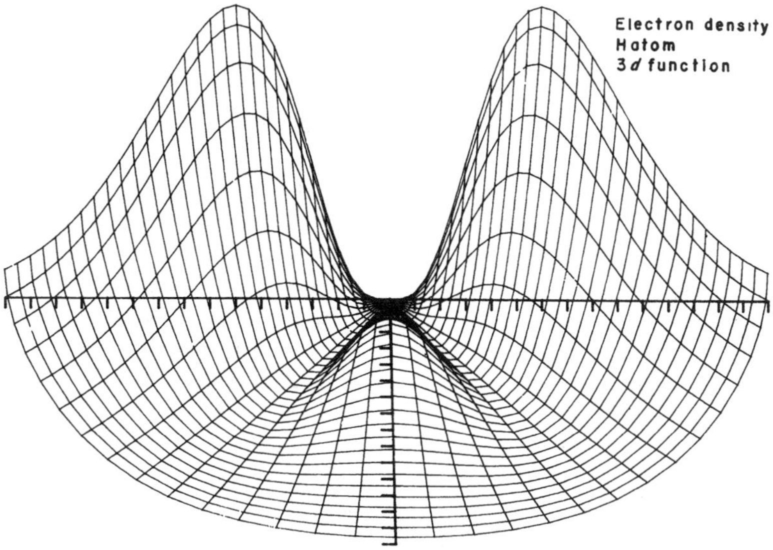
Electron density
H atom
3d function

(f)

exerts a force on the electron, has the form of a wave of frequency ν. Imagine for the moment that ν is very much lower than the frequency of the electron's revolution about the nucleus, so that the electron revolves many times before the force on it changes appreciably. In this case, the force exerted by the electric field has the effect of polarizing the atom, pulling the nucleus in one direction and shifting the *average* position of the electron slightly in the opposite direction, just as a constant electric field would do. The oscillations of the field simply change the direction of polarization slowly from one side to the other. The electron distribution is distorted, and a strong enough field may even pull the electron away from the nucleus altogether (ionize the atom). The net effect of a slowly varying field is to shift the average energy of the atom just so long as the field is applied; no energy is permanently transferred from the field to the atom.

At the other extreme of the frequency scale, suppose that the electric field oscillates much faster than our Bohr electron moves about the nucleus. In this case the atom cannot respond to the fast vibrations of the force field; its inertia is too great. The net effect is to produce an instantaneous polarization varying so rapidly that it averages to zero over a time much shorter than the period of the orbit. The electron can only quiver a bit as it goes round in its orbit, with the orbit itself—and the electron's average energy—remaining unchanged. Thus neither very slow nor very fast electric field oscillations affect the energy of the atom; it is only when the orbital and field frequencies are comparable that energy transfer can occur.

We seek a mechanism for transferring energy from an oscillating field to an atom. The key to this mechanism lies in the concept of *resonance*.[9] Consider a harmonic oscillator of frequency ν, and suppose that it is acted upon by a force field oscillating with the same frequency. For simplicity we assume that the force has its maximum value in a given direction when the oscillator has its greatest displacement in that direction—in other words, that the two oscillations are exactly in phase. This means that on every oscillation the particle receives a push that drives it farther away from the origin. Such a transfer of energy between two oscillations with the same frequency is known as resonance. In principle the process could go on without limit, but no real oscillator is perfectly harmonic; the model must break down in some way. In an analogous (but apocryphal) situation, by establishing a resonance between a sound wave and the natural vibration frequency of a wine glass, Enrico Caruso is said to have covered a good many tables with broken glassware.

Now we cannot apply the same reasoning directly to the hydrogen electron's orbital frequency, which is not fixed but varies with the radius. But we know that a transition between two states can occur when we apply light whose frequency ν (energy $h\nu$) equals the difference between the two orbital frequencies (energies):

$$E = h\nu = h(\nu_2 - \nu_1) = E_2 - E_1 . \tag{4.25}$$

In what sense can the light be in resonance with two states simultaneously?

As in Eq. 4.25, we assume exact resonance, $\nu = \nu_2 - \nu_1$ (with $\nu_2 > \nu_1$). The oscillating electric field has the form

$$\begin{aligned} \mathbf{E} &= \mathbf{E}_0 \sin 2\pi\nu t \\ &= \mathbf{E}_0 \sin(2\pi\nu_2 t - 2\pi\nu_1 t). \end{aligned} \tag{4.26}$$

The second form of Eq. 4.26 leads to what we want to know. Using the trigonometric identity

$$\sin(x-y) = \sin x \cos y - \cos x \sin y, \tag{4.27}$$

we find that the total field can be expressed as the sum of two waves, each constructed from two components with frequencies ν_1 and ν_2:

$$\mathbf{E} = \mathbf{E}_0 (\sin 2\pi\nu_2 t \cos 2\pi\nu_1 t - \cos 2\pi\nu_2 t \sin 2\pi\nu_1 t). \tag{4.28}$$

Thus the field of frequency ν has components that are individually in exact resonance with the oscillatory motion of the electron in states 1 and 2, which can be any two states whose frequency *difference* is ν. Either the first or the second term of Eq. 4.28 will find itself more or less in phase with electrons revolving at frequency ν_1, and will thus tend to excite them to a higher energy level. But the only higher level that is also in resonance with the field (and thus can interact with it) is the one with frequency ν_2. The field pushes the electron out of state 1, and pushes it into state 2. The process can also occur in reverse, with the electron giving up energy to the field and falling from state 2 to state 1. This is why, as we recall from Chapter 2, any system capable of absorbing radiant energy at a particular frequency must be just as capable of emitting radiation of the same frequency.

But we still have not answered the question of what happens to the electron *between* the two states. This is where a classical approach breaks down, and we must appeal to wave mechanical concepts for some understanding. Rather than being a body that undergoes an oscillation, the electron itself *is* an oscillation, that is, a standing wave. What happens is that an oscillation dies out at one place as a new oscillation appears at another place. One can observe something similar with, say, two tuning forks of the same resonant frequency: Set one in motion, and the sound waves from it will start the other vibrating. But the tuning fork continues to exist when it is not oscillating, and the electron does not; its presence is always in the form of *some* wave. Since mass and charge must be conserved, the new oscillation is still an electron, not some other beast, but strictly

[9] The term "resonance" is, of course, derived from the resonance or reverberation of sound waves.

speaking it is meaningless to call it "the same" electron. The actual excitation process, the transfer of energy from one oscillation to the other, requires many cycles of the radiation field. To get a feeling for why this is so, we must examine the magnitude of the interaction energy between the electric field and the atom.

For a first estimate, let us fall back upon the Bohr model, in which the electron and the proton are considered as a pair of point charges. The hydrogen atom thus acts like a rotating *electric dipole,* a pair of equal but opposite charges some distance apart. For charges $+q$ and $-q$, the *dipole moment* is defined as

$$\boldsymbol{\mu} \equiv q\mathbf{r}, \qquad (4.29)$$

where $\mathbf{r}$ is the radius vector from the negative to the positive charge. For a Bohr hydrogen atom in its ground state, the instantaneous dipole moment thus has a magnitude

$$|\boldsymbol{\mu}| = \mu = ea_0 = (1.602\times10^{-19}\,\text{C})(5.292\times10^{-11}\,\text{m}) = 8.478\times10^{-30}\,\text{C m}.$$

Dipole moments are most often reported in *debyes:* 1 debye is the moment of two charges of 10^{-10} esu (about $0.2e$) 1 Å apart, so that

$$1\text{ debye (D)} \equiv 10^{-18}\text{ esu cm} = 3.33564\times10^{-30}\text{ C m}$$

and $ea_0 = 2.542$ D.

The potential energy of interaction between a microscopic dipole and an external field is given by

$$V_{\text{dipole-field}} = -\mathbf{E}\cdot\boldsymbol{\mu} = -E\mu\cos\theta, \qquad (4.30)$$

where θ is the angle between the dipole and field directions; the interaction energy has its greatest magnitude when the dipole is lined up with the field. (We shall derive this equation in Section 10.1.) The electric fields of light waves vary over many orders of magnitude. In a brightly lit room, the electric field strength of the light may be of the order of 25 V/m. On the other hand, a fairly intense monochromatic source like a laser—not by any means the most intense available, but a rather powerful one—can easily produce a beam with a field of 50 million V/m. Thus the instantaneous interaction energy between the field and the atomic dipole should be at most

$$|E\mu| = Eea_0 \approx (25\text{ to }5\times10^7)\,\text{V/m}\times e\times5.3\times10^{-11}\,\text{m} \approx (1.3\times10^{-9}\text{ to }2.6\times10^{-3})\,\text{eV}$$

or about 2×10^{-28} to 4×10^{-22} J. The actual energy must be somewhat less, since $\mathbf{E}$ and $\boldsymbol{\mu}$ will not always be perfectly aligned. However, the total energy that must be absorbed in a single hydrogen atom transition (cf. Fig. 2.18) is of the order of 10 eV, thousands or even billions of times as great. It is hard to conceive in classical terms how such a transfer could occur, unless the field is applied for, say, a million cycles, transferring a millionth of the necessary total energy on each cycle. This reasoning supports our earlier statement that many cycles of the radiation field are required to bring about a transition. This is still not a long time, since for visible light a cycle is only about 2×10^{-15} s.

The calculation just carried out was entirely classical, in so far as it considered the electron as a point charge with a definite position at any given time. How can we reconcile this with the wave model of the electron? We have already interpreted a transition as a replacement of one oscillation by another, analogous to the resonant oscillation of two tuning forks. A different analogy may give some feeling for the mechanism of the process. Imagine that you hold one end of a rope, with the other end attached firmly to a fixed swivel. You can swing the rope so that, although your hand remains practically at rest, the rope takes the shape of a rotating standing wave with no nodes—in other words, the conventional way to swing a jump rope (Fig. 3.2*a*). Now suppose that you want to make the rope oscillate in a standing-wave mode with one node (Fig. 3.2*b*). With a very slight extra wiggle of your hand, you can in a few cycles transform the original motion into the desired new form. What you have done is to apply a weak force in resonance with the rope's motion, thereby changing its mode of oscillation. This is essentially how the rotating electric field transfers energy from one oscillation to another, except that many more cycles are involved.

Thus far, we have considered only transitions resulting from energy exchange between oscillating electric fields and electric charges. This is not the only way that radiation and matter may interact, although it is the strongest form of interaction we encounter in the chemical systems discussed in this book. (Interactions among protons and neutrons and among their constituents are stronger still but lie outside our context.) The most important interactions, after those between electric fields and charges, are those between magnetic fields and atomic and molecular magnets. Electrons with nonzero angular momentum act like currents, generating magnetic fields. The energy of a magnet in an external magnetic field depends on the orientation of the magnet with respect to the field. Our most familiar example is a compass needle—our magnet—in the magnetic field of the earth. Atomic magnets, whether electrons or nuclei, have such levels that, like the electronic energy levels of the hydrogen atom, are discrete and quantized. By applying an oscillatory magnetic field whose frequency matches the spacing between the state of an atomic magnet and another available state, we can induce a transition from the initial state to the new state. The condition for resonance is the same as that for inducing a transition of an electron, that the frequency ν match the energy spacing ΔE; thus, $h\nu = \Delta E$. The spacings between energy levels associated with magnetic interactions are, by and large, small compared with the spacings of electronic energy levels, or, as we shall see, of molecular rotational and vibrational energy levels. Transitions between states associated with magnetic interactions of electrons with external fields occur, with typical magnets, in the microwave region of the frequency spectrum. The study of such transitions is called *electron paramagnetic resonance* (*EPR*) or, if the magnetism is due just to the spin of the electron, *electron spin resonance* (*ESR*). Transitions between states associated with magnetic interactions of nuclei and external fields appear in the radio frequency region. These are the transitions associated with *nuclear magnetic resonance* or *NMR*. We shall discuss magnetic transitions in Chapters 5, 8, and 9.

This section has dealt almost exclusively with analogies and simple models, since the actual quantum mechanical approach to transitions is too complicated to describe at the level of this text. Still, we can outline the basic concepts involved. Consider first an isolated hydrogen atom in its initial state, before any electromagnetic field is applied. This state is time-independent and thus an eigenstate of the energy, which has the value $E_0 = h\nu_0$. Now we apply the oscillating field. Since the system is varying with time, the uncertainty principle (in the form $\Delta E\ \Delta t \geq \hbar/2$) tells us that the energy can no longer be definitely fixed. There is in fact a nonzero probability of the system's being in *any* of its energy eigenstates, and the total time-dependent wave function can be written as a sum over the corresponding eigenfunctions,

$$\Psi(\mathbf{r},t) = \sum_n a_n(t)\Psi_n(\mathbf{r},t) = \sum_n a_n(t)\psi_n(\mathbf{r})e^{-2\pi i\nu_n t} \quad (4.31)$$

where $\nu_n \equiv E_n/h$. This general, time-dependent wave function must, according to Postulate V, be a solution to the general time-dependent Schrödinger equation, whose Hamiltonian operator H includes the kinetic energy T, the time-independent potential interactions V, and the time dependent fields $\mathsf{H}'(t)$ such as that of a light wave:

$$\begin{aligned}\mathsf{H}\Psi(\mathbf{r},t) &= [\mathsf{T}+\mathsf{V}(\mathbf{r})+\mathsf{H}'(\mathbf{r},t)]\Psi(\mathbf{r},t)\\ &= i\hbar\frac{\partial\Psi(\mathbf{r},t)}{\partial t}.\end{aligned} \quad (4.32)$$

If $\Psi(\mathbf{r}, t)$ is expressed as in Eq. 4.31, then Eq. 4.32 can be recast so that the solutions are the expressions for the time-dependent amplitudes $a_n(t)$. We shall not attempt to carry out this calculation. The important result, however, is the same as in the corresponding classical wave problem: $a_n(t)$ is negligibly small except when the field frequency ν satisfies the condition $\nu \approx \pm(\nu_n - \nu_0)$, that is, when the field is in resonance with both the initial state and the nth state. Given the $a_n(t)$, one can go on to calculate the probability and rate of transitions between any two states.

The probability of finding the system in the nth eigenstate at time t is $|a_n(t)|^2$, that is, $|a_n(t)|^2$ is the probability that we find the system with energy E_n and density distribution $|\psi_n(\mathbf{r})|^2$ at time t. The $a_n(t)$'s have the property that the sum of their absolute squares is unity. This is a consequence of the conditions that the eigenstates of energy are mutually exclusive—a system observed to have energy E_n cannot simultaneously be in a state with energy $E_{n'}$—and that every possible eigenstate of energy for the system is included in the set $\psi_n(\mathbf{r})$. For the hydrogen atom, the transition rate is of the same order of magnitude as that we estimated from the rotating-dipole model.

Of course, not all transitions are brought about by an external electromagnetic field. Consider what happens when two atoms or molecules collide with each other. Each molecule is a collection of charged particles in motion and thus generates its own highly localized field, which can induce a transition in the other molecule when the two are close enough. But since the total energy is conserved, this can happen only if both molecules undergo transitions simultaneously. However, one or both of these transitions can be merely between two eigenstates of the molecule's overall motion—in other words, just a change in the molecule's kinetic energy and momentum.[10] Since the energy spectrum for free motion is continuous (or nearly so), a collision between two molecules can induce a wide variety of transitions in their internal states—or even the shuffling of atoms between molecules that we call a chemical reaction.

The relative probability of transitions depends rather sensitively on the system and the two states involved. Some transitions, logically named *forbidden transitions,* almost never occur unless an atom or molecule is so violently disturbed that it virtually loses its identity. These are transitions for which the $a_n(t)$ of Eq. 4.26 are very small even when $\nu = |\nu_n - \nu_0|$, usually because of the symmetry relationships between the initial and final states. For a given atom or molecule, the probability of a transition between states n and n' by light absorption is proportional to $|\mu_{n',n}|^2$, where[11]

$$\mu_{n',n} \equiv \int\cdots\int \psi_{n'}{}^* \mu\psi_n dq_1\cdots dq_N, \quad (4.33)$$

μ being the system's instantaneous dipole moment ($e\mathbf{r}$ for the hydrogen atom); the integral is, as usual, taken over all the coordinates of the system.

For many simple systems the integral of Eq. 4.33 is nonzero only for certain changes in the quantum numbers.

---

[10] A collision in which the total kinetic energy of relative motion is unchanged and both molecules change only their individual kinetic energies and momenta is called *elastic.* We shall treat elastic collisions at length in Part III.

[11] Compare the expectation value of μ in a single stationary state,

$$\langle\mu\rangle_n = \int\cdots\int \psi_n{}^* \mu\psi_n dq_1\cdots dq_N.$$

For the hydrogen atom, for example, the integral over the angular coordinates vanishes unless $\Delta l = \pm 1$ and $\Delta m = 0$ or ± 1. These are examples of what are called *selection rules.* For the harmonic oscillator, with $E_n = (n + \frac{1}{2})h\nu$, the selection rule is $\Delta n = \pm 1$; thus ΔE must equal $\pm h\nu$, and only radiation of the oscillator frequency ν should induce a transition. Similarly, the rigid rotator of Eq. 3.168 also must have $\Delta l = \pm 1$. You may wonder how forbidden transitions manage to occur at all. Actually, the transition probability is given by a series, ordinarily dominated by the $|\mu_{n',n}|^2$ term; the later terms, involving such quantities as magnetic dipoles (cf. Section 5.1) and electric quadrupoles, are still present (though very small) even when $\mu_{n',n}$ vanishes.

In addition to transitions induced by external fields or collisions, it is possible for an atom or molecule to undergo a *spontaneous transition.* That is, rather than changing its energy by $h\nu$ when stimulated by a field of frequency ν, the system itself emits radiation of frequency ν. Of course, such a transition can go only to a state of lower energy. Whereas the rate of induced transitions depends directly on the intensity of the applied field, the rate of spontaneous emission depends only on the initial and final states. The fraction of emissions that occur spontaneously increases rapidly with the difference in energy of those states. For spontaneous radiation of visible light, the rate can be as much as 10^8 or 10^9 s^{-1} (per atom); the inverse of this rate, about $10^{-8} - 10^{-9}$ s, is the average lifetime of the excited state. Forbidden transitions of electrons are usually considered to be those that occur spontaneously more slowly than 10^6 s^{-1}.

If induced transitions are triggered by resonance with the applied field, just what makes a spontaneous transition occur? Of course, for an excited system to give up its energy without inducement is inherently plausible, and in agreement with classical mechanics. (Remember that Bohr had to explain why the orbiting electron didn't radiate *all* its energy.) But the actual wave mechanical mechanism is very complicated. Perhaps the simplest way to describe it is to say that the transition is induced by the zero-point oscillations of the radiation field: If the field is made up of Planck's harmonic oscillators, it must have a zero-point energy, even when no external field is applied. This mechanism accounts for transitions in isolated atoms, but of course no atom is truly isolated from the fields of other atoms or stray radiation.

Consider an excited state from which spontaneous emission is relatively slow. One can often by various means "pump" a large number of atoms into such a state. If the system is then subjected to radiation of the emission frequency ν, a massive amount of induced emission will occur in a short time. The emitted radiation is all of the same frequency ν and in phase with the inducing field; such radiation is called *coherent.* This is the phenomenon of *laser* (or *maser*) action, the name standing for light (or *m*icrowave) *a*mplification by *s*timulated *e*mission of *r*adiation. The typical laser has mirrors at the ends, so that most of the emitted light travels back and forth through the system and induces more and more emission.

To sum all this up, consider a large collection of molecules with two states separated by an energy $E_2 - E_1 = h\nu$, in the presence of radiation of frequency ν. Transitions of all the kinds we have discussed go on simultaneously: transitions induced by the field, transitions induced by collisions, and spontaneous transitions. The first two can go to either the higher or the lower state (the probability per molecule being the same in both directions), whereas spontaneous transitions can go only to the lower state. A dynamic equilibrium is established among all these processes when the total rate of $1 \rightarrow 2$ transitions equals the total rate of $2 \rightarrow 1$ transitions. The ratio of atoms in states 1 and 2 when this occurs is a function of the kinetic energy per molecule, which is proportional to the parameter we call the temperature. As we shall show in Part II, the equilibrium distribution is given by

$$\frac{N_2}{N_1} = \exp\left[\frac{-(E_2 - E_1)}{k_B T}\right] = \exp\left(\frac{-h\nu}{k_B T}\right), \qquad (4.34)$$

where T is the absolute temperature and k_B is a constant. The rates of upward and downward field-induced transitions vary with the intensity of radiation of frequency ν, which for black-body radiation increases with the temperature of the radiation source (cf. Fig. 2.14). Collisions become both more frequent and more likely to produce transitions as the temperature increases (see Part III), but spontaneous transitions are independent of temperature.

Let us make this discussion concrete by considering a gas of sodium atoms, sufficiently dilute that we can ignore collisions. The sodium atom can undergo a transition between two states by absorbing or emitting a quantum of yellow light ($\lambda = 5890$ Å, $\nu = 5.09 \times 10^{14}$ s^{-1}); it is this transition that produces the illumination in sodium-vapor street lights. If the gas is placed in a box whose temperature is about 3000°C, then the rate of induced transitions per atom is about 5000 s^{-1} in each direction. However, the rate of spontaneous emission from the higher state is about 10^7 s^{-1}, so that virtually all the excited atoms will return to the lower state spontaneously rather than by induced emission. Since the downward rate is so much greater than the upward rate, at any time only about 0.05% of the atoms are in the excited state. Now suppose that we increase the intensity of black-body radiation by raising the box temperature from 3000 to 35,000°C. The rate of induced transitions in each direction will then become essentially equal to the rate of spontaneous emission, so that the total emission will be half spontaneous and half induced. With the total downward rate just twice the upward rate, we can expect to find twice as many atoms in the lower state as in the upper state at any time; in other words, at 35,000°C, about one-third of the sodium atoms should be in the excited state.

FURTHER READING

Avery, J., *Quantum Theory of Atoms, Molecules and Photons* (McGraw-Hill Book Co., Inc., New York, 1972), Chapters 3, 4.

Davydov, A. S., *Quantum Mechanics* (Pergamon Press, Oxford, England and Addison-Wesley Publishing Co., Reading, Mass., 1965), Chapters I, II, III, V and VI.

Karplus, M., and Porter, R. N., *Atoms and Molecules* (W. A. Benjamin, Inc., Menlo Park, Calif., 1970), Chapter 3.

Kauzmann, W., *Quantum Chemistry* (Academic Press, Inc., New York, 1957), Chapters 6.11, 9 and 15.

Kramers, H. A., *Quantum Mechanics* (North-Holland Publishing Company, Amsterdam, 1958; Dover Publications, Inc., New York, 1964), Chapters XXX.

Merzbacher, E., *Quantum Mechanics,* 2nd Ed. (John Wiley and Sons, Inc., New York, 1970), Chapters 6, 8, 9.

Messiah, A., *Quantum Mechanics,* Vol. 1, translated by G. M. Temmer (John Wiley and Sons, Inc., New York, 1958).

Morrison, M. A., Estle, T. L., and Lau, N. F., *Quantum States of Atoms, Molecules and Solids* (Prentice-Hall, Inc., Englewood Cliffs, N.J., 1976), Chapter 3.

Schiff, L. I., *Quantum Mechanics,* 3d Ed. (McGraw-Hill Book Co., Inc., New York, 1968), Chapter 4.

PROBLEMS

1. Complete the details of the argument indicated in Section 4.1, that there can be no physically admissible state for which the total energy E is everywhere less than the potential energy V.

2. The penetration of a wave into a steplike barrier as outlined in Eqs. 4.2 and 4.3 is much like the penetration of an electromagnetic wave into a metal. The depth at which the amplitude of the wave is e^{-1} of its amplitude at the surface is called the *skin depth.* Find the general expression for the skin depth of a one-dimensional matter wave for a particle of mass m and energy E_0 striking a barrier of height V, with $0 < E_0 < V$, as shown in the accompanying diagram. Suppose that $m = 10^{-28}$, $E_0 = 1$ eV, and $V = 2$ eV. What is the skin depth? Below what energy is the skin depth greater than 1 Å?

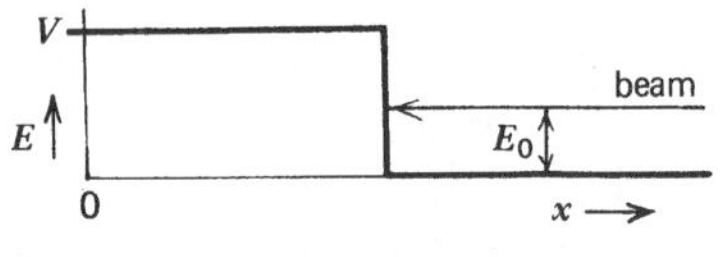

3. Explain why both the wave function and its first derivative must be continuous at the boundaries of a square potential well of finite depth.

4. In Chapter 3, we saw that the energies of the states of a particle in a one-dimensional infinite square well increase as n^2, the square of the quantum number. For the harmonic oscillator, the energies go as n; for the hydrogen atom, as n^{-2}. From these observations, what qualitative statement can one make about the relation between the slope of the walls of a box and the level spacing? What form would you expect for the spacings of the levels if V varied as x^4 instead of x^2? as $x^{3/2}$ instead of x^2?

5. A particle of mass m is bound in a one-dimensional potential $V(x)$ that satisfies

$$\begin{aligned} V &= V_0, & x &< -a; \\ V &= 0, & -a \le x &\le a; \\ V &= V_0, & x &> a; \end{aligned}$$

and $V_0 = 49h^2/128ma^2$.

(a) How many bound states does such a particle have?

(b) What are the energies of all the bound states?

(c) Construct a curve showing each of the bound-state eigenfunctions.

6. A particle of mass m is confined in a two-dimensional box with sides of lengths l_1 and l_2. The potential is zero everywhere inside the box and infinite elsewhere. Calculate the number of energy eigenstates per unit energy interval for such a particle. A particle of mass m is confined inside a three-dimensional box with sides l_1, l_2, and l_3. The potential, like that of the problem above, is zero inside the box and infinite outside. Calculate the number of energy eigenstates per unit energy interval for this particle.

7. Show by substitution that the first three functions of Table 4.1 satisfy the wave equation 4.6, that they are normalized to unity, and that they are orthogonal, that is, that $\int_{-\infty}^{\infty} \psi_i{}^*(z)\psi_i(z)\,dz$ is zero if $i \neq j$.

8. A particle with mass 12 amu is bound in a harmonic potential well with a force constant k of 5×10^5 dyn/cm. What are the zero-point energy and the interval between adjacent energy levels for this system? What happens to these numbers if the force constant is doubled?

9. Graph the five functions of Table 4.1 for a particle with mass of 12 amu and a force constant k of 5×10^5 dyn/cm (500 kg/s^2).

10. In the discussion of the states of the hydrogen atom, with $-\infty < V < 0$, it was pointed out that states with energy $E < 0$ are bound states. What is the form of the wave function for a physically reasonable state whose wave function has an amplitude bounded everywhere, and with $E > 0$?

11. Calculate the radii for which the probability $\mathcal{P}$ is a maximum for $2p$, $4f$, and $6h$ levels of the hydrogen atom. How much energy is required to remove a $5p$ electron from a hydrogen atom and leave it infinitely far away, with no kinetic energy? Li^{2+} has a hydrogenic structure with nuclear charge $Z = 3$. What are the energies of the $1s$ and $2p$ states of Li^{2+}? What are the radii of maximum probability for these two states?

12. Compute the average value of r, the most probable value of r, and the root-mean-square value of r for the $1s$, $2s$, and $2p$ levels of the hydrogen atom. Compare the three kinds of values and explain the origin of their differences.

13. A hydrogenlike atom can be formed from a proton and a negative muon whose mass is approximately 206 times that of the electron. What are the energies and most probable radii for the $1s$ and $2p$ levels of this atom?

14. Find the general expression for the distance r_0 at which the effective potential V_{eff} of the Coulomb Hamiltonian, Eq. 4.19, becomes zero. How do this distance and the slope of V_{eff} at r_0 depend on the quantum number l? What is the distance at which V_{eff} is a minimum?

15. Show by substitution that the first three functions of Table 4.2 satisfy Eq. 4.18.

16. At what value of r do the hydrogenic $3s$, $3p$, and $3d$ orbitals have their outermost maximum amplitudes?

17. The radial equation for the hydrogen atom, Eq. 4.18, can be analyzed in part by finding the form of its solutions near $r = 0$ and as $r \to \infty$.

(a) Transform the equation to eliminate the first derivative. This is done by making explicit the condition that $|\psi(\mathbf{r})|^2$ for any wave function describing a particle in a central potential in three dimensions must fall off as r^{-2}, or that $\psi(\mathbf{r})$ must have the form $r^{-1}f(r)\,Y(\theta, \phi)$.

(b) Transform Eq. 4.18 by substituting $r^{-1}f(r)$ for $R_{nl}(r)$. Then let $r \to 0$ in the Hamiltonian and drop all but the largest term, to show that $f(r)$ behaves as r^l or r^{-l-1} near the origin. Only one of these two possibilities is physically admissable; which one, and for what reason?

(c) Now let $r \to \infty$ and drop all but the largest terms of the differential equation for $f(r)$. Show that $f(r)$ drops off exponentially with r and derive the relationship between the energy of the state and the coefficient α multiplying r in the exponential $e^{-\alpha r}$.

(d) Conclude by writing

$$f(r) = r^l e^{-\alpha r} g(r)$$

and find the differential equation for $g(r)$. Write the function $\psi(\mathbf{r})$ in terms of all the factors you have now derived, with the oscillatory function $g(r)$ left undetermined.

18. Compute the approximate number of cycles of the electric field and the time interval required for an atom to absorb energy equivalent to one quantum, $h\nu$, of light from a laser exerting a field of 5×10^7 V/m, whose light has wavelength 589 nm (5890 Å), if the atom absorbs the maximum interaction energy on every cycle of the light field.

19. Compute all the *transition moments* of the electric dipole operator $e\mathbf{r}$ (really the three components of this vector quantity, corresponding to ex, ey, and ez), where $\psi_0(\mathbf{r})$ and $\psi_1(\mathbf{r})$ are the wave functions for the ground and (threefold degenerate) first excited state of a three-dimensional isotropic harmonic oscillator whose mass is 1 amu and whose force constant is 5×10^5 dyn/cm (500 kg/s^2).

20. A proton passes an atom at a velocity of 5×10^6 cm/s, with a distance of closest approach of 1 nm (10 Å). What is the proton's electric field at the atom when the two are at their closest point? At what distance is the field e^{-1} times as strong as at its maximum? What is the time interval during which the field rises from its first "e^{-1}" point and then falls to its second "e^{-1}" point? To approximately what frequency of oscillation would this correspond, and what wavelength would correspond to a spectral line at this frequency?

21. Refer to the wave functions of the harmonic oscillator of Table 4.1 and Eq. 4.8. Prove the footnote to Eq. 4.8 (footnote 2) and extend it to the state $n = 1$. That is, find the classical turning points for the states with $n = 0$ and 1, the values of z at which the potential energy is exactly equal to the kinetic energy for these states. Show that the curvature of the wave function is zero at a classical turning point for *any* system satisfying the Schrödinger equation.

22. Using Eqs. 4.8 and 4.9, derive the expressions for $\psi_0(z)$, $\psi_1(z)$, and $\psi_2(z)$ of Table 4.1.

23. The peak electric field of a beam of light applied to a hypothetical model system is 25,000 V/m. The system consists of an electron harmonically bound to an infinitely heavy nucleus; the force constant is 10^6 dyn/cm or 10^3 kg/s^2. Calculate the maximum distance of displacement due to this applied electric field.

24. Determine the nodes of the $n = 4$ harmonic oscillator wave function. To find the zeroes of the Hermite polynomial, use the root finder of a computer-based mathematical package such as MAPLE, MATHEMATICA, MACSYMA, or DERIVE.

25. The classical turning points for the harmonic oscillator with $n = 0$ and $n = 1$ were determined in Problem 4.21. Compute, using a computer-based mathematical package such as MAPLE, MATHEMATICA, MACSYMA, or DERIVE, the total probability of finding the harmonic oscillator in the classically forbidden regions for (a) $n = 0$ and (b) $n = 1$.

26. In Problem 4.12, the average value r_{avg}, the most probable value r_{mp}, and the root-mean-square value r_{rms} of the radius for an electron in the $1s$ level of the hydrogen atom were determined. Using a computer-based mathematical package such as MAPLE, MATHEMATICA, MACSYMA, or DERIVE, calculate the probabilities that the 1s electron in the hydrogen atom will be found at a radius greater than or equal to (a) r_{avg}, (b) r_{mp}, and (c) r_{rms}.

CHAPTER 5

The Structure of Atoms

The last chapter gave us a comprehensive picture of the nature of the hydrogen atom, or of any atom containing a single electron. However, most of our world (though not most of the universe) is made up of atoms containing more than one electron. It is the purpose of this chapter to describe these more complex atoms.

How might we construct a many-electron atom? We naturally begin with a nucleus with the desired positive charge $+Ze$. We know from Chapter 4 what is obtained when we add a single electron to the nucleus: a species whose radial wave functions and energy levels differ from those of hydrogen only by scale factors. The natural step is to add a second electron, then a third, and so on until Z electrons have been added to give a neutral atom. But what rules govern the addition of these electrons? Each interacts not only with the nucleus, but with all the other electrons. These interactions involve not only the electrostatic forces but the magnetic properties of electrons as well. An accurate description of a many-electron atom must thus be quite complex. Fortunately, in many cases one can construct a useful approximation by assigning each electron its own set of quantum numbers, its own corresponding constants of motion, and its own wave function, or orbital, analogous to the wave functions of the hydrogen atom. We must determine the sequence of energies of these functions, which dictates the order in which these orbitals are "occupied." This sequence is governed by the Pauli exclusion principle, which says that each orbital has only a limited capacity to contain electrons. In fact, no two electrons can have the same values of the one-electron quantum numbers n, l, m, and m_s, where m_s is a quantum number for a property we shall introduce in the next section, called the electron spin.

Given the basic concepts of atomic structure, we go on to discuss several effects that influence the details of electron configurations and energies in individual atoms. Finally, we begin to explain how chemical behavior can be interpreted in terms of the microscopic structure of atoms.

5.1 Electron Spin; Magnetic Phenomena

Until now we have treated electrons as though they were simple charged mass points. This is an oversimplification. In reality, electrons (and other particles) behave in some ways as though they are spinning about their own axes. We shall examine some of the evidence for this conclusion, but first let us explain what it means by comparing spin and orbital motion.

We designate the spin angular momentum by the vector **S**, just as we use **L** to represent the electron's orbital angular momentum about the atomic nucleus. The spin angular momentum behaves in most ways like other angular momenta. Its magnitude $|\mathbf{S}|$ for a single electron is given by

$$|\mathbf{S}|^2 = s(s+1)\hbar^2, \qquad (5.1)$$

where s is a quantum number; this is completely analogous to the relation $|\mathbf{L}|^2 = l(l+1)\hbar^2$ for the orbital angular momentum (Section 3.11). Thus $|\mathbf{S}|^2$ is an eigenvalue of an operator $\mathbf{S}^2$, which we shall not discuss further. As with the other quantum numbers, we use the lowercase s for a single electron and the capital letter S for the total spin of an atom. Equations 5.1 and 3.162 still apply to the total atom, with S, L replacing s, l. (Note that we must now write the magnitudes of the angular momenta as $|\mathbf{S}|$, $|\mathbf{L}|$, because S, L are reserved for the quantum numbers.) One major difference between the spin and orbital angular momenta is in the values that the quantum numbers may have. For a single electron, s always has the value $\frac{1}{2}$, whereas you will recall that l may be any non-negative integer. When a system contains two or more electrons, the total spin (or orbital) angular momentum is the *vector* sum of the individual electronic angular momenta; we shall return later to this point.

The components of the spin angular momentum are also analogous to those of the orbital angular momentum. Only one component is a constant of the motion, characterized by a quantum number; as usual, we define this as the z component. Its value is given by

$$S_z = m_s \hbar, \qquad (5.2)$$

analogous to Eq. 3.156, with the quantum number m_s (or M_s). Usually m_s is called the *spin quantum number* and s simply the *spin* ("the electron has spin $\frac{1}{2}$"). The magnetic quantum number that we called m in the hydrogen atom can be designated as m_l (or M_L) for uniformity. Just as m_l can range from $+l$ to $-l$ in integral steps, the values allowed for m_s are $s, s-1, \ldots, -s+1, -s$. Hence, for a single electron, the only possible values are $m_s = +\frac{1}{2}$ and $m_s = -\frac{1}{2}$; the corresponding values of S_z are $+\hbar/2$ and $-\hbar/2$.

The total angular momentum is designated as $\mathbf{J} \equiv \mathbf{L} + \mathbf{S}$, with quantum numbers j, m_j (or J, M_J). We postpone a full discussion of **J** until Section 5.7.

Being charged, any spinning or orbiting electron constitutes an electric current moving in a loop; like any such loop current (as in a solenoid), it acts as a bar magnet or magnetic dipole. From this magnetic behavior, the existence of spin or orbital angular momentum can be deduced. In the presence of an external magnetic field **B**, the potential energy of a microscopic magnetic dipole is given by

$$V_{\text{dipole-field}} = -\boldsymbol{\mu}_m \cdot \mathbf{B} = -\mu_m B \cos\theta, \qquad (5.3)$$

where $\boldsymbol{\mu}_m$ is the *magnetic dipole moment,* analogous to the electric dipole moment $\boldsymbol{\mu}_e$ of Eq. 4.30. The potential energy thus has its minimum when $\boldsymbol{\mu}_m$ and **B** are parallel; since $\boldsymbol{\mu}_m$ by definition is directed from the "south" to the "north" (north-seeking) pole of the magnet, this occurs when the north pole points in the direction of the external field.

Both the electron's orbital motion and its spin about its own axis produce magnetic dipole moments. The orbital moment $\boldsymbol{\mu}_l$ in the Bohr model is given by

$$\boldsymbol{\mu}_l = -\frac{e}{2m_e}\mathbf{L}, \qquad (5.4)$$

from the classical equation for a current moving in a circular loop. Let us take our z axis in the direction of **B**. The component of $\boldsymbol{\mu}_l$ in the direction of the magnetic field is then

$$(\boldsymbol{\mu}_l)_z = \mu_l \cos\theta = -\frac{e}{2m_e}L_z = -\frac{e\hbar}{2m_e}m_l. \qquad (5.5)$$

The quantized unit of $(\boldsymbol{\mu}_l)_z$ is thus

$$\mu_B \equiv \frac{e\hbar}{2m_e} = 9.274009 \times 10^{-24} \text{ J/T} \qquad (5.6)$$

(T = tesla), known as the *Bohr magneton.*[1] Equation 5.5 is not exactly correct, because m_e should be replaced by the reduced mass and there is a slight interaction with the spin moment.

The spin magnetic dipole moment $\boldsymbol{\mu}_s$ does not obey the classical Eq. 5.4, but rather, an equation with an additional factor:

$$\boldsymbol{\mu}_s = g_s\left(\frac{-e}{2m_e}\right)\mathbf{S}, \qquad (5.7)$$

where g_s (the *Landé g factor*) is approximately 2. The actual value of g_s for the electron is 2.002319, differing from exactly 2 because of relativistic and radiative effects. By reasoning like that leading to Eq. 5.5, the component of $\boldsymbol{\mu}_s$ in the direction of the external field must equal $-g_s m_s$ Bohr magnetons; since $m_s = \pm\frac{1}{2}$, we have $(\boldsymbol{\mu}_s)_z \approx \pm\mu_B$. If the electron were literally a charged particle spinning about its axis, the spin magnetic dipole moment and spin angular momentum would be related by the equivalent of Eq. 5.4; that is, g_s would have to be 1. The fact that this is not so shows that the electron spin is an essentially quantum mechanical phenomenon. Strictly, one can say only that the electron somehow has an intrinsic angular momentum **S**; the idea of "spin" is meaningful only as a convenient visualization, but the language has stuck.

We can now see why m_l is called the "magnetic" quantum number. In the absence of a magnetic field, all states with the same n and l have the same energy (except for fine-structure splitting). Once we apply an external field **B**, however, the energy of each state changes by the amount given by Eq. 5.3. The total energy of an atom in a given state is thus shifted by an amount

$$\begin{aligned} E - E_0 &= -(\boldsymbol{\mu}_L + \boldsymbol{\mu}_S)\cdot\mathbf{B} = -(\mu_L + \mu_S)_z B \\ &= -\mu_B(M_L + g_s M_S)B; \end{aligned} \qquad (5.8)$$

the splitting of spectral lines due to this process is called the *Zeeman effect.*[2] In the "normal Zeeman effect," for atoms with $S = 0$ (and thus $M_S = 0$), the adjacent energy levels are separated by an amount $\mu_B B$. In a moderately large field of 1 T (10^4 G), this splitting is about 9.3×10^{-24} J (or 5.8×10^{-5} eV), equivalent to less than 0.5 cm^{-1}. For nonzero spin the spectrum is more complicated ("anomalous Zeeman effect"), and spin–orbit interactions produce splitting even with no external field. It was to account for this behavior that George Uhlenbeck and Samuel Goudsmit (1925) proposed the hypothesis of electron spin.

---

[1] In Gaussian units one writes $\boldsymbol{\mu}_l = -(e/2m_e c)\mathbf{L}$, and the Bohr magneton is $e\hbar/2m_e c = 9.27410 \times 10^{-21}$ erg/G (G = gauss). The SI unit can be written as either joules per tesla or ampere meters^2.

[2] An external electric field produces a similar splitting known as the *Stark effect.*

The electron is not the only particle with spin. In particular, both the proton and the neutron also have spin $\frac{1}{2}$. The total nuclear spin I is the vector sum of the individual nucleon spins; I is seldom large, and often is zero. Nuclear magnetic moments are far smaller than those of electrons, being measured in terms of the *nuclear magneton,*

$$\mu_N \equiv \frac{e\hbar}{2m_p} = \mu_B\left(\frac{m_e}{m_p}\right)$$
$$= \frac{\mu_B}{1836.15} = 5.050783\times10^{-27}\ \mathrm{J/T}. \qquad (5.9)$$

The proton and neutron have moments $\mu_p = 2.793\mu_N$ and $\mu_n = 1.913\mu_N$, respectively. The energy-level splitting of these particles in a magnetic field is correspondingly small, about 10^{-3} of that for an electron; heavier nuclei have even smaller splittings.

In spite of the minuscule size of these energy shifts, one can measure them quite accurately by the method of *nuclear magnetic resonance* (*nmr*). One places the system in a strong magnetic field (≈ 1 T), passes a radiofrequency (rf) signal through it ($\nu \approx 100$ MHz), and varies **B** slowly until ΔE for some magnetic transition is equal to $h\nu$; the energy difference is then in *resonance* with the applied radiation, and absorption occurs. In the original approach, still occasionally used, the operator swept the field strength $|\mathbf{B}|$ and recorded the absorption of rf power, thus producing a *magnetic resonance* spectrum. Now, more sophisticated methods use pulses of rf power, singly or in sequences, producing a time response in the system that can, in effect, be "decoded" to yield the spectrum of resonant energy intervals.[3] Nuclear magnetic moments observed this way can act as probes to tell us a good deal about the structure of atoms and molecules. This is possible because a nucleus is not merely acted upon by an external field, but is affected in a delicate and sensitive way by the fields of all the other particles (nuclei and electrons) in its vicinity. For example, a proton in an —OH group is in a slightly different field than a proton in a —CH_3 group, and protons in these two environments have slightly different spacings of their energy levels. Hence they exhibit resonances at slightly different magnetic field strengths in an oscillating field of fixed frequency. The *chemical shifts* of lines in the magnetic resonance spectrum are very small indeed, of the order of 10^{-6} of the total energy-level splitting, but they can be measured accurately. The detailed shape of the absorption spectrum can be interpreted to reveal much information about the environment of the nucleus studied, whether it is a proton or a heavier, more complex nucleus.

One can, of course, observe similar effects in the energy-level splittings due to electron spin. These are studied by the method of *electron spin resonance* (*esr*), which gives information mainly on the electron distribution in molecules. Because of the larger magnetic moment of the electron, the esr transitions occur at higher frequencies, mainly in the microwave region. However, atoms or molecules with net electron spins different from zero are relatively uncommon, whereas nonzero nuclear magnetic moments are ubiquitous.

Energy levels of magnetic systems and magnetic resonance spectra often serve as useful models because of a special simplicity they have: Such systems have only a finite number of states and energy levels. This is quite different from any of the systems we have examined previously; those all have infinite numbers of states and levels. The hydrogen atom, for example, has an infinite number of bound states and then, of course, has all its continuum levels. We shall see in Chapter 16 and especially in Chapter 20, how a particular concept, that of negative absolute temperature, can be developed consistently for systems with only a finite number of states.

The astute reader will have recognized by now that although we have introduced several kinds of magnetic moments into our model of the atom, associated with electron spin, electron orbital motion, and nuclear spin, we have said little about the obvious problem that such magnetic moments must interact with one another. Indeed they do, but their interactions are weak and need not concern us in detail at this point. The interaction of electron spin and orbital moments, called spin–orbit interaction, is important for the energy levels of heavy atoms, and is measurable but relatively unimportant for the chemical properties of light atoms. Spin–orbit interaction is discussed in Section 5.7. The interactions among nuclear moments and between nuclear and electron moments are crucial for the use of nuclear and electron spin magnetic resonance as analytic and diagnostic tools; the identification of lines in nuclear resonance spectra

[3] The method of "decoding" is called *Fourier analysis*. Any smooth function can be constructed (or expressed) by adding, with suitable amplitudes, sines and cosines. If the function lies within a finite interval, then that sum of sines and cosines is a discrete sum. Thus, for example, if we want to describe a function $f(t)$ in the interval from $t = 0$ to $t = 1$, then we may write $f(t) = \sum_{j=0}^{\infty}[a_j \sin(2\pi jt) + b_j \cos(2\pi jt)]$. If the interval is not that from 0 to 1, then we must just rescale the arguments of the sines and cosines by dividing t by the length of the interval and, if necessary, by shifting the origin so that the new variable lies in the interval from 0 to 1. If the variable t may range to ∞, then the sum must be replaced by an integral, and the discrete index j, by a continuous index with the dimension of $[t]^{-1}$; i.e., a frequency ν. Thus, $f(t) = \int_0^{\infty}[a(\nu)\sin(2\pi\nu t) + b(\nu)\cos(2\pi\nu t)]$. To carry out a modern analysis of a magnetic resonance spectrum, one applies a magnetic field as an intense pulse for which the coefficients $a(\nu)$ and $b(\nu)$ are known—that is, a pulse whose time dependence is known. One measures the response, the changes of that applied pulse that produce an outgoing signal. The changes occur only at the frequencies ν_r at which the system can exhibit a resonant response. Thus, the decoding process is one of finding those frequencies for which $a(\nu)$ and $b(\nu)$ change because the system under study responds to the applied pulse. These pulses may have durations as short as milliseconds or microseconds, they may be applied repetitively as a sample is moved, and may be focused to particular locations in space. These qualities have made magnetic resonance, especially nuclear magnetic resonance, a powerful tool for studying objects as complex as the human body, notably in the form known as *magnetic resonance imaging* (*MRI*).

generally depends on knowing how one nuclear magnetic moment interacts with those of its neighbors.

Something should be said here about the origins of *macroscopic* magnetism. All the individual electrons in a material are tiny magnets, but ordinarily they are aligned at random and, in large numbers, cancel one another on average. An external magnetic field, however, tends to align the microscopic magnets; the net effect is to make **B** slightly higher within the material than in a vacuum.[4] This phenomenon is called *paramagnetism.* Both orbital and spin magnetic moments can contribute to paramagnetism. However, because the orbital moments are largely fixed in space by the molecular structure, it is mainly the spin moments that change their alignment in the presence of a field. Paramagnetism is thus rarely found in substances whose total molecular spin is zero, that is, for which the spin magnets cancel out within the molecule; such substances are called *diamagnetic.* In diamagnetic materials **B** is actually a little less than in a vacuum because the field interferes with the orbital motions, but the effect is much smaller than it is in paramagnetism. Paramagnetic substances are weakly attracted, diamagnetic substances weakly repelled, by a magnet. As for the very large permanent magnetism that can be induced in metals like iron (*ferromagnetism*), this results from a cooperative interaction among spins on different atoms: For certain substances the electron spins are thus aligned in parallel over large regions or *domains,* appreciably reducing the total energy of the crystal.

We have introduced electron spin primarily because it is an important property in its own right: The total state of an electron is specified only when its spin is added to all the other properties characterized by quantum numbers. Moreover, it must be introduced to provide a full basis for the exclusion principle, which, in turn, underlies atomic structure. Spin is one of the properties that could not have been predicted from classical mechanics. It is illustrated by one of the most famous and fundamental experiments on which quantum theory rests, the Stern–Gerlach experiment. Appendix 5A describes this experiment.

5.2 The Pauli Exclusion Principle; the Aufbau Principle

The set of quantum numbers *n, l,* m_l, and m_s, and the physical attributes to which they correspond, give us the basis for our first approach to the structure of complex atoms. In this section we shall see how electrons fit together to give each atom a sort of shell-like structure; this will lead us to a systematic interpretation of the periodic table. The model we discuss here is only approximate. In subsequent sections, we shall consider the validity of our approximations and the refinements necessary for a detailed explanation of atomic structure.

Consider first an ion made up of a nucleus with charge $+Ze$ $(Z > 1)$ and a single electron bound to that nucleus. This is a hydrogenlike ion of the type discussed in Chapter 4, where we characterized the electron by the quantum numbers *n, l, m* (or m_l). We must now add the spin quantum number m_s to give a complete description of the state of the electron. Let us denote these four quantum numbers of the first electron by n_1, l_1, m_{l1}, m_{s1}, because we shall add another $Z - 1$ electron, one by one, to neutralize the total nuclear charge.

Suppose that we add a second electron to the ion, and that there are states in which both electrons are bound to the nucleus. To a first approximation, known as the *central field approximation,* the second electron can be treated as though it moves in the field of the nucleus and the *average* field produced by the first electron. In other words, we pretend that the second electron sees only a smeared-out charge cloud equivalent to the average spatial distribution of the first electron. Carrying the approximation a bit further, we also assume that the first electron responds to the *average* field of the second, rather than to its field at each instant.

The central field approximation permits us to assign to the second electron a one-electron wave function with a set of quantum numbers n_2, l_2, m_{l2}, m_{s2}. The energy levels and the shapes of the spatial wave functions or orbitals for the atom or ion with two electrons are somewhat different from those of the one-electron ion, for reasons we shall see very shortly. However, the quantum numbers retain the same basic meaning in both situations. The principal quantum number *n* is one more than the number of radial nodes in the one-electron wave function, and is the first guide to the energy; *l* specifies the electron's orbital angular momentum, and gives the number of angular nodes; m_l specifies the orientation of the electron's orbital angular momentum **L**; and m_s specifies the orientation of its spin angular momentum **S**.

Suppose that the two-electron species, for example, the helium atom with $Z = 2$, is in its state of lowest energy. It would be natural to assume that both electrons then have the same quantum numbers, presumably those of the lowest state in the one-electron atom. This is correct for the quantum numbers *n, l,* and m_l, which for both electrons have the values $n = 1$, $l = 0$, and $m_l = 0$. (Remember that $n = 1, 2, \ldots,$ whereas $0 \le l \le n - 1$ and $|m_l| \le l$.) However, both electrons cannot have the same value of m_s: One has $m_s = +\frac{1}{2}$, the other $m_s = -\frac{1}{2}$, and the total electron spin is zero. Here is a point where our classical intuition fails us, and we must introduce a quantum mechanical postulate to describe the way in which electrons behave. It is found that any two electrons in a single atom with the same values of *n, l,* and m_l

[4] In a vacuum we have $\mathbf{B} = \mu_0(1 + \chi)\mathbf{H}$, where **H** is the magnetic field strength (see footnote 8, p. 12, Chapter 1); **H** is determined only by the external field source and any permanent magnetic movements of the matter in the field. Within a material we write $\mathbf{B} = \mu_0(1 + \chi)\mathbf{H}$, defining the *magnetic susceptibility* χ; χ is positive in paramagnetic substances, negative in diamagnetic substances. One usually tabulates the molar susceptibility $\chi_M \equiv \chi M/\rho$ (where *M* is molecular weight and ρ is density); a typical value for a paramagnetic metal would be about 10^{-10} m^3/mol. We shall discuss magnetic susceptibility in greater detail in Section 9.6.

always have different values of m_s. More generally, in any atom or ion, *no two electrons ever share the same four quantum numbers.* This rule, proposed empirically by Wolfgang Pauli in 1925, is known as the *Pauli exclusion principle.* For the time being we shall take the exclusion principle as a postulate in its own right; in Section 6.7 we shall give a more general rule from which it can be derived.

The exclusion principle imposes strict conditions on the assignment of quantum numbers. We have assigned $n = 1$, $l = 0$, $m_l = 0$ to the first two electrons, which must have $m_s = +\frac{1}{2}$ and $m_s = -\frac{1}{2}$, respectively. Now suppose that $Z \geq 3$, and we add a third electron, as, for example, in Li with $Z = 3$. There is no way to assign this electron to a state with $n = 1$, because all the possible combinations of n, l, m_l, and m_s correspond to states already occupied. We must thus assign the third electron a quantum number n_3 greater than 1. If we set $n_3 = 2$, then we may have $l_3 = 0$ or $l_3 = 1$. If $l_3 = 0$, then m_{l3} is necessarily 0; if $l_3 = 1$, then m_{l3} may be $0, +1$, or -1; whatever values we assign l_3 and m_{l3}, the quantum number m_{s3} may be either $+\frac{1}{2}$ or $-\frac{1}{2}$. Which of all these states will the electron occupy? It *may* occupy any of them, but the most stable state is that of lowest energy, the ground state. To begin with, the energy nearly always increases with the value of n; an electron will thus ordinarily take the lowest available value of n, in this case 2. Among states with the same value of n, the states with $l = 0$ normally have energies lower than those with $l = 1$, and so forth; the energies increase with increasing l. In the absence of external fields, and with internal electron–electron interactions neglected, states with the same n and l but different m_l or m_s differ only in orientation, and therefore must have the same energy.

In the hydrogen atom, whose electron moves in the Coulomb field of a single proton, the energy depends only on the principal quantum number n; it makes no difference energetically what value we assign to l. This is a special property of two particles interacting through a Coulomb field. In more complex atoms, the field felt by each electron is a combination of the field of the nucleus and the fields of the other electrons, which have probability distributions extending over considerable regions of space. The effect of the other electrons is really twofold. First, each of them exerts a Coulomb force of the classical type; second, the mere presence of an electron at a point **r** reduces the probability that another electron of the same spin be in the vicinity of **r**. The latter is a purely quantum mechanical phenomenon, with no classical equivalent; it is a direct result of the exclusion principle. The net effect is as if there were an additional repulsive force (sometimes called *exchange force*) added to the Coulomb repulsion between electrons of the same spin. This is not a real force, represented by a term in the Hamiltonian; rather, we introduce the effect as a constraint on the form of the wave function, which has the effect of changing the charge distribution from that of the unconstrained wave function. As a result of these interactions the field on each electron is significantly different from that of a simple, central Coulomb field. Because of the more complex field in a many-electron atom, the electron's energy depends on its angular momentum; thus the energies of states with the same n but different l are separated (split). In general, as we said previously, for a given n the energy increases with l. The reason should be clear from Fig. 4.5: The higher the value of l, the less likely an electron is to be found near the nucleus.

At this point we must say something about the word "orbital." We introduced this term in Section 4.3 to refer to a single-electron eigenfunction, but it is also loosely used for the corresponding state: One says that an electron is *in* a given orbital. By referring to one-electron states as orbitals, one can maintain a clear distinction between the true state of the entire atom, a concept that need imply no approximations, and the state of an individual electron, a concept associated with a particular approximate description of nature and therefore limited in its precision and applicability.

The complete wave function of an electron must include the spin as well as the spatial coordinates; we might write it as $\psi(r, \theta, \phi; m_s)$. To a very good approximation, however, the spin–orbit interaction can be neglected, and the Hamiltonian is separable into spatial and spin terms. We can then as usual factor the wave function, say, as $\psi(r, \theta, \phi)\alpha(m_s)$. It is the spatial part of the eigenfunction that is ordinarily called an "orbital." Two electrons with the same values of n, l, m_l have the same spatial wave function, and are thus said to occupy the same orbital.[5] Every orbital can thus contain two electrons, which by the exclusion principle must have different values of m_s. An orbital is characterized by its values of n, l, m_l, but usual notation ignores m_l (which, to the extent that the simple orbital picture is valid, does not affect the energy) and uses the *s, p, d, . . .* notation to identify l. The first two electrons in an atom are thus said to be in a 1*s* ($n = 1$, $l = 0$) orbital, and the third should go into a 2*s* ($n = 2$, $l = 0$) orbital, which has a lower energy than a 2*p* ($n = 2$, $l = 1$) orbital.

The conditions imposed by the exclusion principle, together with our knowledge about energy levels, provide us with a rather accurate guide for describing the structure of atoms of any complexity in their ground states. This guide is a way of assigning quantum numbers, and implicity wave functions, to each electron in the atom. In effect, we build up the atom by assigning one electron at a time to the available orbital of lowest energy. This is called the *Aufbau* (building-up) *principle.* The method is rather like adding marbles, one at a time, to a conical cup. The first marble goes to the bottom (the state of lowest energy in the earth's gravitational field), and successive marbles pile up as they can, each in the lowest place where there is room.

---

[5] The wave function including spin is called a *spin orbital,* and this *is* restricted to a single electron. In this text we do not consider the nature of the function $\alpha(m_s)$ or the corresponding operator, for which a relativistic theory must be used.

We have already considered the ground states of the first few atoms. Now we shall extend this analysis to the entire periodic table.

5.3 Electronic Configurations of Atoms

For most purposes one can classify any state of an atom by giving the principal quantum number n and the angular momentum quantum number l for each of the electrons in the atom. It is these quantum numbers that largely determine the atom's energy and chemical behavior; varying the value of m_l or m_s makes only a very slight difference, which in first approximation we neglect altogether. A given assignment of n and l to all the electrons in an atom is called a *configuration.* Although it is not strictly correct to say that any atomic state is identical with a particular configuration, it is accurate to say that almost all known atomic states can be characterized by, and exhibit properties dominated by single configurations.

For an example of a configuration, let us consider the ground state of the nitrogen atom, which can be built up by adding seven electrons, one at a time, to a nucleus of charge $+7e$. As before, the first two electrons go into a 1s orbital. The third electron goes into a 2s orbital, with $n = 2$, $l = 0$, $m_l = 0$, and $m_s = \pm\frac{1}{2}$. The fourth can go into the same orbital, with the same values of n, l, m_l, and whichever value of m_s is not yet taken by the third electron. The next higher energy level is that with $n = 2$, $l = 1$, the 2p level, which can hold six electrons in three orbitals (with $m_l = +1, 0, -1$, respectively). The fifth, sixth, and seventh nitrogen electrons can thus all go into 2p orbitals. Later we shall have to consider *which* 2p orbitals they enter, but we do not need to know this here. Having said this much, we have specified the ground-state configuration of the nitrogen atom: There are two electrons in a 1s orbital, two in a 2s orbital, and three in 2p orbitals. The shorthand representation for this configuration is $1s^2 2s^2 2p^3$. In this notation each large number specifies the value of n, and the letter following it the value of l for a particular electronic state; the superscript gives the number of electrons with those values of n and l.

If one neglects electron–electron interactions and fine-structure effects, all the orbitals with the same values of n and l but different m_l have the same energy (in the absence of external fields). They differ only in the orientation of the electron's angular momentum with respect to an arbitrarily chosen axis. As in the hydrogen atom, their probability densities differ in orientation but not in radial distribution. The energies of the np, nd, . . . orbitals are moderately close to that of the ns orbital, but not equal to it except in the hydrogen atom. The radial distributions of the np, nd, . . . orbitals are also somewhat similar to that of the corresponding ns orbital at fairly large radii (cf. Fig. 4.5). Because of this similarity in both size and energy, one frequently refers to all the orbitals with the same n as a *shell,* and to those with the same n and l as a *subshell;* this is the "shell model" of the atom. The shells with $n = 1, 2, 3, \ldots$ are also referred to as the K, L, M, . . . shells. This terminology derives from x-ray spectroscopy (Section 2.3): The K series of emission lines are produced by electrons dropping into vacancies in the K or innermost shell, and so forth.

When all the quantum numbers available for a given shell or subshell have been assigned to electrons, it is said to be *filled* or *closed.* In the ground-state configuration of the nitrogen atom, the $n = 1$ or K shell (consisting of only the 1s subshell) is filled, the 2s subshell is filled, but the 2p subshell is only half-filled. There are still three vacancies in the 2p subshell, no matter how we assign the quantum numbers m_l and m_s, provided that we satisfy the exclusion principle. These vacancies are filled as we add more electrons, until in the neon atom ($Z = 10$) the shells with $n = 1$ and $n = 2$ are both completely filled.

To proceed any further we must know the sequence of orbital energy levels. The next section describes some of the ways these are derived; for the present, we can examine the levels as they turn out. One might assume that the energy increases in order from one shell to the next, and in the order s, p, d, . . . within each shell. This would give the sequence 1s, 2s, 2p, 3s, 3p, 3d, 4s, 4p, 4d, 4f, 5s, ... This scheme is valid through the 3p subshell, which is filled in the ground state of the argon atom ($Z = 18$). In the potassium atom ($Z = 19$), however, the nineteenth electron goes into a 4s rather than a 3d orbital. What happens is this: As the number of electrons increases, the energy difference between orbitals in the same shell becomes greater than the average energy difference between shells—in other words, the energy ranges spanned by the shells begin to overlap. One thus ought to know just how the energy levels vary as a function of Z, and we shall discuss this in the next section.

Fortunately, this detailed knowledge is not necessary to obtain a good approximation to the electronic configuration. The sequence in which orbitals are filled in the ground states of the elements is given with remarkable accuracy by the simple mnemonic device shown in Fig. 5.1. One need only remember that any subshell with angular momentum quan-

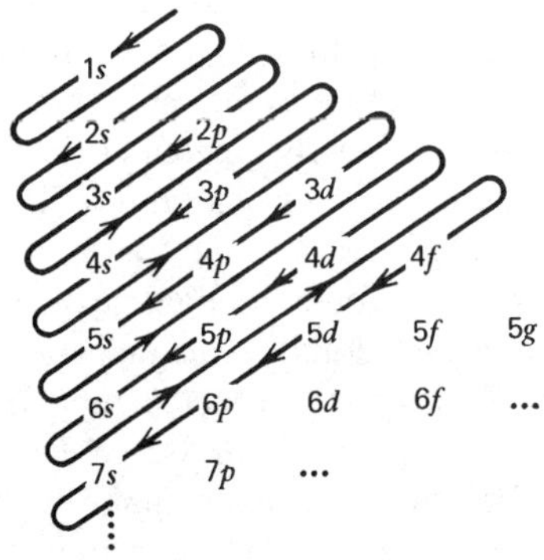

Figure 5.1 Approximate filling order of orbitals for the ground states of the elements.

tum number l can hold up to $2(2l + 1)$ electrons ($2l + 1$ possible values of m_l, each with two possible values of m_s): two in s subshells, six in p subshells, 10 in d subshells, and so forth. Given this information, and filling orbitals in the sequence indicated in Fig. 5.1, one can predict the ground-state configuration for any number of electrons.

The actual ground-state electronic configurations of the elements through Element 106, so far as they are now known, are given in Table 5.1. The configurations are obtained by analyses of electronic spectra, and in some cases (where two or more subshells are very close in energy) they are uncertain. Table 5.1 shows that the sequence predicted by Fig. 5.1

Table 5.1 Electronic Configurations of the Elements in Their Ground States

The configurations that differ from those predicted by Fig. 5.1 are indicated by a star (★). The symbol [(rare gas)] is an abbreviation for the inner electrons in the configuration of the ground state of the particular rare gas atom.

Z	Element	Configuration	Z	Element	Configuration	Z	Element	Configuration
1	H	$1s$	36	Kr	$[Ar]3d^{10}4s^24p^6$	71	Lu	$[Xe]4f^{14}5d6s^2$
2	He	$1s^2$				72	Hf	$[Xe]4f^{14}5d^26s^2$
			37	Rb	$[Kr]5s$	73	Ta	$[Xe]4f^{14}5d^36s^2$
3	Li	$1s^22s$	38	Sr	$[Kr]5s^2$	74	W	$[Xe]4f^{14}5d^46s^2$
4	Be	$1s^22s^2$	39	Y	$[Kr]4d5s^2$	75	Re	$[Xe]4f^{14}5d^56s^2$
5	B	$1s^22s^22p$	40	Zr	$[Kr]4d^25s^2$	76	Os	$[Xe]4f^{14}5d^66s^2$
6	C	$1s^22s^22p^2$	41	Nb	$[Kr]4d^45s\star$	77	Ir	$[Xe]4f^{14}5d^76s^2$
7	N	$1s^22s^22p^3$	42	Mo	$[Kr]4d^55s\star$	78	Pt	$[Xe]4f^{14}5d^96s\star$
8	O	$1s^22s^22p^4$	43	Tc	$[Kr]4d^55s^2$	79	Au	$[Xe]4f^{14}5d^{10}6s\star$
9	F	$1s^22s^22p^5$	44	Ru	$[Kr]4d^75s\star$	80	Hg	$[Xe]4f^{14}5d^{10}6s^2$
10	Ne	$1s^22s^22p^6$	45	Rh	$[Kr]4d^85s\star$	81	TI	$[Xe]4f^{14}5d^{10}6s^26p$
			46	Pd	$[Kr]4d^{10}\star$			
11	Na	$[Ne]3s$	47	Ag	$[Kr]4d^{10}5s\star$	82	Pb	$[Xe]4f^{14}5d^{10}6s^26p^2$
12	Mg	$[Ne]3s^2$	48	Cd	$[Kr]4d^{10}5s^2$	83	Bi	$[Xe]4f^{14}5d^{10}6s^26p^3$
13	Al	$[Ne]3s^23p$	49	In	$[Kr]4d^{10}5s^25p$	84	Po	$[Xe]4f^{14}5d^{10}6s^26p^4$
14	Si	$[Ne]3s^23p^2$	50	Sn	$[Kr]4d^{10}5s^25p^2$	85	At	$[Xe]4f^{14}5d^{10}6s^26p^5$
15	P	$[Ne]3s^23p^3$	51	Sb	$[Kr]4d^{10}5s^25p^3$	86	Rn	$[Xe]4f^{14}5d^{10}6s^26p^6$
16	S	$[Ne]3s^23p^4$	52	Te	$[Kr]4d^{10}5s^25p^4$	87	Fr	$[Rn]7s$
17	Cl	$[Ne]3s^23p^5$	53	I	$[Kr]4d^{10}5s^25p^5$	88	Ra	$[Rn]7s^2$
18	Ar	$[Ne]3s^23p^6$	54	Xe	$[Kr]4d^{10}5s^25p^6$	89	Ac	$[Rn]6d7s^2\star$
						90	Th	$[Rn]6d^27s^2\star$
19	K	$[Ar]4s$	55	Cs	$[Xe]6s$	91	Pa	$[Rn]5f^26d^27s^2\star$
20	Ca	$[Ar]4s^2$	56	Ba	$[Xe]6s^2$	92	U	$[Rn]5f^36d7s^2\star$
21	Sc	$[Ar]3d4s^2$	57	La	$[Xe]5d6s^2\star$	93	Np	$[Rn]5f^46d7s^2\star$
22	Ti	$[Ar]3d^24s^2$	58	Ce	$[Xe]4f5d6s^2\star$ (or $4f^26s^2$)	94	Pu	$[Rn]5f^67s^2$
23	V	$[Ar]3d^34s^2$	59	Pr	$[Xe]4f^36s^2$	95	Am	$[Rn]5f^77s^2$
24	Cr	$[Ar]3d^54s\star$	60	Nd	$[Xe]4f^46s^2$	96	Cm	$[Rn]5f^76d7s^2\star$
25	Mn	$[Ar]3d^54s^2$	61	Pm	$[Xe]4f^56s^2$	97	Bk	$[Rn]5f^86d7s^2\star$ (or $5f^97s^2$)
26	Fe	$[Ar]3d^64s^2$	62	Sm	$[Xe]4f^66s^2$	98	Cf	$[Rn]5f^96d7s^2\star$ (or $5f^{10}7s^2$)
27	Co	$[Ar]3d^74s^2$	63	Eu	$[Xe]4f^76s^2$	99	Es	$[Rn]5f^{10}6d7s^2\star$ (or $5f^{11}7s^2$)
28	Ni	$[Ar]3d^84s^2$ (or $3d^94s\star$)	64	Gd	$[Xe]4f^75d6s^2\star$	100	Fm	$[Rn]5f^{11}6d7s^2\star$ (or $5f^{12}7s^2$)
29	Cu	$[Ar]3d^{10}4s\star$	65	Tb	$[Xe]4f^85d6s^2\star$ (or $4f^96s^2$)	101	Md	$[Rn]5f^{12}6d7s^2\star$ (or $5f^{13}7s^2$)
30	Zn	$[Ar]3d^{10}4s^2$	66	Dy	$[Xe]4f^{10}6s^2$	102	No	$[Rn]5f^{13}6d7s^2\star$ (or $5f^{14}7s^2$)
31	Ga	$[Ar]3d^{10}4s^24p$	67	Ho	$[Xe]4f^{11}6s^2$	103	Lr	$[Rn]5f^{14}6d7s^2$
32	Ge	$[Ar]3d^{10}4s^24p^2$	68	Er	$[Xe]4f^{12}6s^2$	104	Rf	$[Rn]5f^{14}6d^27s^2$
33	As	$[Ar]3d^{10}4s^24p^3$	69	Tm	$[Xe]4f^{13}6s^2$	105	Db	$[Rn]5f^{14}6d^37s^2$
34	Se	$[Ar]3d^{10}4s^24p^4$	70	Yb	$[Xe]4f^{14}6s^2$	106	Sg	$[Rn]5f^{14}6d^47s^2$
35	Cl	$[Ar]3d^{10}4s^24p^5$						

is indeed quite reliable. The exceptions (marked by ★) are those cases in which detailed interactions override the gross features of the shell model; we shall examine these effects later. But already we have essentially explained the form of the periodic table (Table 1.3). The first two columns contain those elements in which *s* subshells are being filled; the last six columns correspond to the filling of *p* subshells; the 10 short columns in the middle (transition metals) to *d* subshells; and the lanthanide and actinide series to *f* subshells.

Thus far we have considered only ground states. Although highly excited states sometimes require more elaborate descriptions, the concept of electronic configuration still gives a satisfactory representation of many excited states. For example, one can obtain a very good representation of the first excited state of helium by assigning one electron to the 1*s* orbital and the other to the 2*s*, the next orbital on the energy ladder. The configuration is then $He^*(1s2s)$, where the asterisk indicates an excited state. At a slightly higher energy one finds the configuration $He^*(1s2p)$. One can proceed all the way up the series of configurations in which one electron is excited to the normally empty orbitals with any values of n and l. Finally, at the limit of the series, sufficient energy removes the electron from the atom altogether; one then has $He^+(1s) + e^-$, a helium ion in its ground state plus a free electron.[6]

To obtain an excited configuration, of course, one need not excite just a single electron. For example, if we excite a helium atom by giving it enough energy to put one electron in the 2*s* orbital and the other electron in the 2*p* orbital, we have the doubly excited configuration $He^*(2s2p)$; or both could be excited to the 2*s* orbital (with opposite spins), giving $He^*(2s^2)$. Such states are actually known experimentally, but their energies are very high indeed compared with states having only one excited electron. We might expect this, particularly for a small atom like helium, because we know that in the hydrogen atom the states with $n = 2$ are three-quarters of the way up from the ground state to the ionization limit. This should be roughly true for helium also, and in fact the first excited state of He is about 80% of the way from the ground state to the energy of $He^+ + e^-$. Therefore the energy required to excite two electrons from the *K* shell to the *L* shell would be over one and one-half times the energy required to remove a single electron from the atom, leaving the other in its lowest state. This is indeed the case: A doubly excited helium atom has more than enough total energy to become a helium positive ion and a free electron. However, much of the time, the energy is divided between the two electrons and is not readily available to just one for purposes of ionization. As a result, when doubly excited helium atoms are produced, it takes some time (corresponding to many Bohr periods of revolution in the $n = 2$ orbit) for an average doubly excited atom to convert itself into a singly charged helium ion and a free electron.

[6] The energy levels of the helium atom are plotted in Fig. 5.10.

Transitions between one electronic configuration and another are responsible for most of the lines in the emission and absorption spectra of atoms. Generally speaking, transitions involving the outer (valence) electrons lie in the visible and adjacent regions. (The highest atomic ionization energy for removal of one outer electron—or *first ionization potential*—is that of helium, corresponding to $\lambda = 504$ Å.) The more tightly bound inner-shell electrons give rise to the x-ray spectrum. Tens of thousands of assignments of atomic spectral lines have been made, meaning that their initial and final states have been determined. Much of modern science and technology could hardly exist without these assignments. (Some examples: The design of lasers requires a knowledge of the specific states of individual atoms and the selection rules governing transitions among them. The industrial analysis of steel uses atomic spectroscopy for the quantitative determination of trace materials.)

5.4 Calculation of Atomic Structures

Now we shall briefly survey the methods used to determine energy levels, quantum states, and wave functions for complex atoms. Unlike those for the hydrogen atom, however, the results obtained are only approximate (though of high accuracy for light atoms).

To describe the state of any quantum mechanical system, one need "simply" write down the Hamiltonian and solve the Schrödinger equation for the wave function. But let us look at such a Hamiltonian. For an atom with N electrons we have the operator

$$\mathsf{H} = \sum_{i=1}^{N}\left(\frac{\mathbf{p}_i^2}{2m} - \frac{Ze^2}{4\pi\epsilon_0 r_i}\right) + \frac{1}{2}\sum_{i=1}^{N}\sum_{j\neq i}^{N}\frac{e^2}{4\pi\epsilon_0 r_{ij}} + \text{magnetic moment interactions,} \tag{5.10}$$

where r_i is the distance of the ith electron from the nucleus and r_{ij} is the distance between the ith and jth electrons. The factor $\frac{1}{2}$ is to prevent counting each interaction twice. Even in classical mechanics one cannot solve exactly the equations of motion for the three-body problem; here we have an $(N + 1)$-body problem, and an exact solution of the Schrödinger equation is clearly impossible. One must therefore resort to approximate methods.

One of the most important ways of obtaining approximate solutions is the *variation method.* Suppose that one guesses a trial wave function ψ, which can be any well-behaved and normalized function of the coordinates. One can show that, for any such ψ,

$$E' \equiv \int \cdots \int \psi^* \mathsf{H} \psi \, dq_1 \cdots dq_N \geq E_0 , \tag{5.11}$$

where H is the Hamiltonian operator of the system and E_0 is its lowest eigenvalue. The equality, of course, holds only when ψ is the true ground-state wave function of the system. The more closely ψ resembles this true wave function, the closer the integral E' will come to the value E_0. If ψ contains one or more adjustable parameters, it is a straightforward calculation to find the values of these parameters that minimize E'. By either trial-and-error or systematic procedures, one can obtain functions that give lower and lower values of E'; when no further improvement can be made, ψ is presumably the best possible approximation to the true wave function within whatever restrictions one has put on one's calculations.[7] There are ways to extend the method to excited states.

To see how the variation method works, let us consider the ground state of the helium atom. If the two electrons did not interact with each other at all, each would independently be in a hydrogenlike $1s$ orbital; since the $1s$ wave function is proportional to e^{-Zr/a_0}, the total wave function for the helium atom would then be of the form

$$\psi^{(1)} = Ae^{-Zr_1/a_0}e^{-Zr_2/a_0}, \tag{5.12}$$

with $Z = 2$. (When the Hamiltonian is separable, the wave function is a product of functions of the individual coordinates.) If $\psi^{(1)}$ is taken as a trial wave function and the operator of Eq. 5.10 as the Hamiltonian (with the magnetic terms omitted), then the integral Eq. 5.11 gives $E' = -74.83$ eV. The actual ground-state energy (E_0) is known from spectroscopic data to be −78.99 eV (and the last figure is, at worst, ±1), so the wave function needs considerable improvement if we want accuracy consistent with our ability to measure this quantity. Since each helium electron is partially screened from the nucleus by the other electron, it is logical to try replacing Z by an adjustable parameter Z':

$$\psi^{(2)} = Ae^{-Z'r_1/a_0}e^{-Z'r_2/a_0}. \tag{5.13}$$

Varying the value of Z', one finds the integral Eq. 5.11 to be minimized when $Z' = \frac{27}{16}$, yielding the value $E' = -77.49$ eV; the error in energy is only one-third as much as before.

[7] The variational principle follows from a generalization of the expansion of a function in Fourier series; i.e., in a series of sines and cosines of the form $\Sigma_n[A_n \sin n\theta + B_n \cos n\theta]$. (This form is appropriate when the range for which the expansion is desired has been scaled to the interval from 0 to 2π.) The generalization is the fact that many sets of functions, notably the complete sets of solutions of any of a very large set of differential equations, including the Schrödinger equation, can also be used to express any arbitrary (not-too-discontinuous) function by using them in a series like the Fourier expansion. If the series expansion is *exact* for the lowest-energy state, ψ_0, then whatever the actual series may be that one uses, we recognize that the series would consist of only one term if the exact solutions of the Schrödinger equation were being used. Operating on ψ_0 with H simply multiplies ψ_0 by the lowest-energy eigenvalue E_0. If we have only achieved an approximation φ_{approx} to the exact solution, as is usually the case, the series expressed in terms of the exact solutions would contain some of the lowest-energy eigenfunction and some of other, higher-energy eigenfunctions. Hence, when we operate on this approximate function $\varphi_{\text{approx}} = a_0\psi_0 + \sum_{j=1} a_j\psi_j$ with the Hamiltonian H and compute the expectation value of the energy, we obtain contributions to that value from not only E_0 but from higher energies as well. Consequently, the approximations to the exact lowest-energy value all lie higher than the exact value, and the exact energy is the greatest lower bound of all such approximations.

The next step carries us beyond the simple configurational wave function, by introducing a term that takes explicit account of the repulsion between the electrons. One way is to use a wave function with terms containing the interelectronic distance r_{12}, for example,

$$\psi^{(3)}A(1+cr_{12})e^{-Z'r_1/a_0}e^{-Z'r_2/a_0}, \tag{5.14}$$

with $c > 0$, which grows larger as the electrons move farther apart; varying both Z' and c, one obtains $E' = -78.66$ eV, only 0.33 eV higher than the experimental value. The function of Eq. 5.14 cannot be labeled by a single configuration as 5.12 or 5.13 can be. Obviously one can carry this process as far as one wants, by using more and more complicated polynomials in the r's. With enough adjustable parameters—and enough computer time—one can get arbitrarily close to E_0. Calculations have in fact been made with ψ's containing over 1000 adjustable parameters, giving an E' that not only agrees with E_0 within experimental accuracy, but is indeed far more precise (and presumably accurate) than experiment. With 26 parameters, one can determine E' to within about 1 part in 10^7. Such results are, for all practical purposes, equivalent to an exact solution. Later in this section we shall give a physical interpretation to the complex wave functions needed to obtain such accuracy.

Because we can compute very accurate wave functions for two-electron atoms, we can construct very reliable pictures of the electron distributions in these atoms. Figure 5.2 shows several such distributions. Note that the atoms and ions in their ground states have very high concentrations of charge density near their nuclei. The total charge at distances far from a nucleus is significant, but in the figures showing the charge density itself (without a weighting factor of r^2), the charge density away from the nucleus is quite small. The total charge at about the distance r is spread out over the entire spherical shell with volume $4\pi r^2\,dr$, so the amount at any point may be quite small although the amount in the entire shell is not.

Executing such elaborate and accurate computations has become feasible, even inexpensive, for the lighter atoms. When one goes beyond the first or second row of the Periodic Table, even the largest computers do not have the capacity to do the job with the same accuracy with which we can compute, for example, the wave function and energy of the helium atom in its ground or even its excited states. But whether one chooses or is limited by available methods, one may go far without achieving spectroscopic accuracy for the total wave function and energy; it is possible to get useful approximations in terms of the central field model and one-electron orbitals.

This method was introduced by D. R. Hartree. The basic assumption is that the wave function can be well represented as a product of one-electron orbitals,

$$\psi = \varphi_1\varphi_2\cdots\varphi_N, \tag{5.15}$$

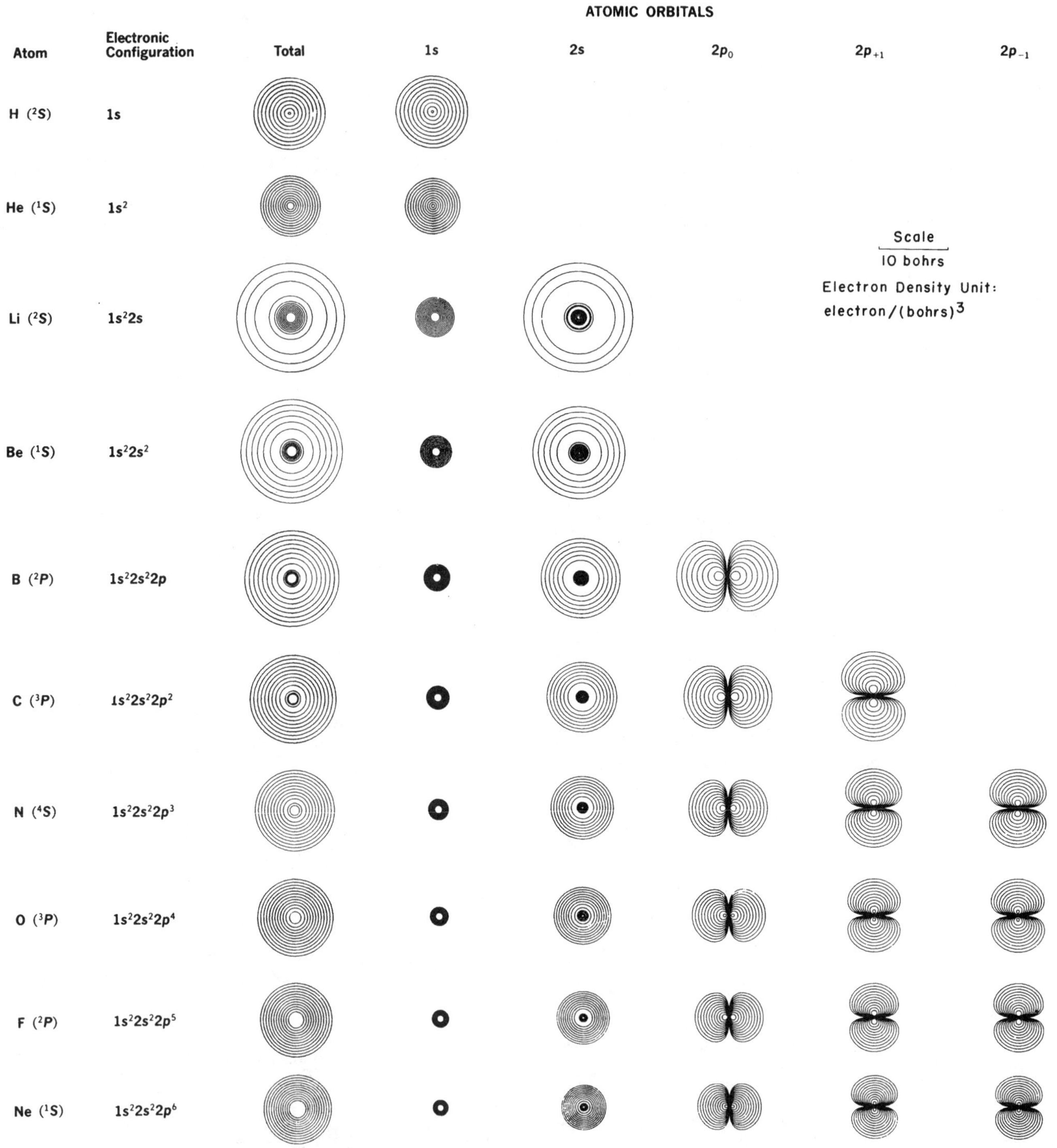

Figure 5.2 Density contours for electronic charge in light atoms. The orbitals are derived from Hartree–Fock orbitals. Plots were made by A. C. Wahl, and published in *Scientific American,* April 1970, p. 55.

where each of the φ_i is a function of the coordinates of the ith electron only. Suppose that this wave function is substituted in Eq. 5.11, using the Hamiltonian 5.10 without the magnetic terms; one can show that the lowest possible value of E' is then obtained when each of the φ_i is a solution of the equation

$$\begin{aligned} \mathsf{H}_i\varphi_i &= \left(\frac{\mathbf{p}_i^2}{2m} - \frac{Ze^2}{4\pi\epsilon_0 r_i} + \sum_{j\neq i} \int |\varphi_i|^2 \frac{e^2}{4\pi\epsilon_0 r_{ij}} dV_j \right) \varphi_i \\ &= \epsilon_i \varphi_i , \end{aligned} \tag{5.16}$$

where $dV_i \equiv r_j^2 \sin \theta_j \, d\theta_j \, d\varphi_j$, and ϵ_i is the orbital energy eigenvalue. Hartree's approximation is thus equivalent to the central field approximation. The instantaneous electron–electron repulsions appearing in Eq. 5.10 are replaced by integrals giving the *average* repulsion over all possible positions of the other electrons (since $|\varphi_j|^2 \, dV_j$ is the probability of finding the jth electron in the volume element dV_j).

It is still necessary to solve simultaneously N equations of the form of Eqs. 5.16; this is done by successive approximation. One starts with a trial set of orbitals $\varphi_i^{(1)}$. These could be hydrogenlike orbitals, for example, although other functions are known that lead to more rapid convergence; the angular part of each orbital, of course, is always taken to be a spherical harmonic. In each of the N Eqs. 5.16 one substitutes the $\varphi_j^{(1)}$ (for $j \neq i$) and solves to obtain a new function $\varphi_i^{(2)}$ (which in general can be expressed only in numerical form). These are again substituted in Eqs. 5.16 and the equations solved to obtain a set of $\varphi_i^{(3)}$. The process is repeated until no further change occurs, that is, until the $\varphi_i^{(n)}$ differ negligibly from the $\varphi_i^{(n-1)}$. In this limit the orbitals consistently reproduce the average field in which each electron moves, and are thus called *self-consistent field* (*SCF*) orbitals.

The initial set of trial orbitals must correspond to the desired electronic configuration of the atom. For example, in the nitrogen atom, two of the φ_i must be $1s$-type functions, two $2s$-type, and three $2p$-type. (What is a "$1s$-type" orbital? As in any other spherically symmetric problem, the wave function with quantum numbers n, l must have l angular nodes and $n - l - 1$ radial nodes.) If one carries out such SCF calculations for a variety of possible configurations, the configuration of the atomic ground state should be the one that gives the lowest total energy. Although the calculations are in fact too approximate to reproduce all the observed sequences of electronic states, they do give the gross features of the orbital-filling sequence.

The Hartree approximation still does not take full account of the interaction between electrons. It includes the average Coulombic repulsion in the $e^2/4\pi\epsilon_0 r_{ij}$ integrals (which we will sometimes abbreviate as e^2/r_{ij}) of Eqs. 5.16, but ignores the *exchange interaction* associated with the exclusion principle. As we shall see in the next chapter, to obtain a wave function that satisfies the exclusion principle, one must replace the simple product of Eq. 5.15 with a determinant, containing all the possible permutations of the electrons among the orbitals. We shall wait until Chapter 6 to go into the details of this approach. One still applies the iterative procedures of the self-consistent field method, leading to what are called *Hartree–Fock* SCF orbitals.

Hartree–Fock SCF calculations have been carried out for all the elements, leading to the results summarized in Fig. 5.3. The quantity plotted is the Hartree–Fock orbital energy ϵ_i, corresponding to the Hartree ϵ_i of Eqs. 5.16. (We need not distinguish one-electron energies, so long as we recognize the Hamiltonian to which they refer.) To the extent that the orbital approximation is valid, ϵ_i should equal the energy required to remove (ionize) one electron from the ith orbital

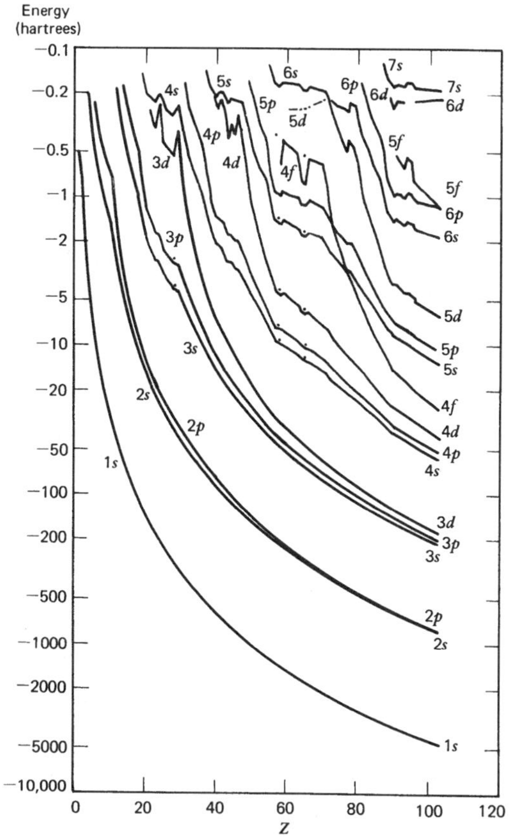

Figure 5.3 Hartree–Fock orbital energies of the elements. The energies plotted are those calculated by F. Herman and R. Skillman, *Atomic Structure Calculations* (Prentice-Hall, Englewood Cliffs, N.J., 1963), for the configurations listed in Table 5.1. Spin–orbit splitting (Section 5.7) is neglected, but relativistic corrections are included. The energy units (hartrees) are defined at the end of Section 5.4.

(*Koopmans' theorem*). Such orbital binding energies can be measured by spectroscopic or electron-scattering techniques. (See especially Section 7.10 for a description of photoelectron spectroscopy, the most direct way to measure orbital binding energies.) The calculated ϵ_i's of Fig. 5.2 generally agree with experimental values to within a few percent. Some typical atomic charge distributions, both for individual orbitals and for total charge distributions, are shown in Fig. 5.3.

The Hartree–Fock calculations also give the approximate orbital-filling sequence, but there are certain anomalies. For example, the $3d$ orbital energy is consistently lower than the $4s$ energy, indicating correctly that a $4s$ electron is more easily ionized. The predicted ground states of Sc and Sc^+ are, respectively, $\cdots 3d4s^2$ and $\cdots 3d4s$. In spite of this, the $4s$ orbital is filled first, because the *total* energy of the atom is lower in the $3d4s^2$ configuration. For Sc again, the configurations $\cdots 3d^24s$ and $\cdots 3d^3$ are, respectively, 2.11 eV and 4.19 eV above the ground state. These anomalies, like those in Table 5.1, show that one cannot completely explain atomic structure in terms of single-orbital energies.

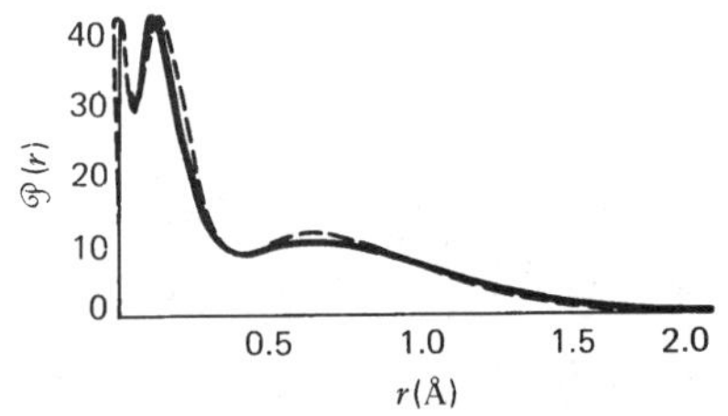

Figure 5.4 Radial distribution of electrons in the argon atom, according to Hartree SCF calculations (- - - -) and electron-diffraction measurements (——).

Another way to check the SCF calculations is in terms of the spatial distribution of electrons. Adding together the squared orbital functions $|\varphi_i|^2$ and integrating over angles, one can obtain the total electron density as a function of r. In Fig. 5.4 the results of such a calculation are compared with an experimental radial distribution function derived from electron-diffraction measurements. The quantity plotted is the probability of finding an electron in the shell between r and $r + dr$. The agreement of the two curves is reasonably good, and in both one can clearly see the shell structure of the atom.

The Hartree–Fock method gives the best possible wave functions describing the motion of each electron in the *mean* potential field of all the other electrons. In mathematical terms, these are the best functions that can be obtained within the central field, one-electron orbital model. But now we must remember that this is only an approximation. At any given instant the field felt by each electron is *not* the spherically symmetrical mean field, but depends on the instantaneous positions of the other electrons. There must be some degree of *correlation* among the positions of the electrons which cannot be simply treated in terms of separate orbitals. Some of this is due to the magnetic moment interactions, which we take up in a later section. The main contribution, however, is from the instantaneous Coulomb and exchange (exclusion-principle) repulsions between electrons; the electrons in a real atom must thus be able to avoid one another more effectively than the central field model would allow. The effect of correlation can be illustrated with our earlier discussion of the helium atom. The trial function 5.13 is a Hartree product of two identical one-electron orbitals; in Eq. 5.14 multiplication by the correlation function $1 + cr_{12}$ gives a better value for the energy, but the wave function can no longer be factored into orbitals.

Figure 5.5 illustrates the effect of correlation in the helium atom, for which it is possible to construct a graphic representation of the phenomenon. We start with the six-dimensional probability density $\mathcal{P}(\mathbf{r}_1, \mathbf{r}_2) = |\psi(\mathbf{r}_1, \mathbf{r}_2)|^2$; one can integrate over three of the variables and obtain a three-dimensional probability density $\rho(r_1, r_2, \theta_{12})$ that depends only on the distances r_1 and r_2 of electrons 1 and 2 from the nucleus, and on θ_{12}, the angle between the vectors $\mathbf{r}_1$ and $\mathbf{r}_2$. From this function of three variables, one can construct the *conditional* probability density for finding electron 2 at r_2 and on θ_{12}, with r_1 at a chosen value r_1'. This conditional probability is the function

$$d(r_2, \theta_{12} | r_1') = \frac{\rho(r_1', r_2, \theta_{12})}{\rho(r_1')}, \tag{5.17}$$

where $\rho(r_1')$ is the probability density for finding r_1 at r_1' whatever r_2 and θ_{12} may be. By constructing graphs of $d(r_2, \theta_{12} | r_1')$ for various values of r_1' and—more important in our present context—for wave functions of different levels of refinement, we can learn how the distribution of probability for one electron is affected by the position of another, and how different wave functions represent the effects of spatial correlation. Figure 5.5*a* shows the conditional probability $d(r_2, \theta_{12} | r_1')$ for the ground state of helium according to a wave function based on a $1s^2$ configuration; Fig. 5.5*c* is the conditional probability for the same atom and for the same value of r_1' but based on a very accurate wave function.

Another way to improve wave functions beyond the level of single-configuration representations is to write them as sums of terms, in which each term corresponds to a different configuration. This method, called configuration interaction, will be discussed in detail in Chapter 6, in connection with the H_2 molecule. For the present, we note only that a wave function composed mostly of a $1s^2$ term, but with a bit of $2s^2$ configuration, is a better representation of the ground state of He than the pure $1s^2$ function, and that addition of some $2p^2$ configuration makes it better still; Fig. 5.5*b* shows the conditional probability for such a function, which is still not as accurate as the 26-term function used for Fig. 5.5*c*. The one rule restricting what functions may be included in such sums is that they all must correspond to the same quantum numbers and constants of motion. For example, to represent the ground state of helium, all the functions in the sum must have total spin $S = 0$ and a total orbital angular momentum quantum number $L = 0$, since S and L are good quantum numbers for the ground state (and all the other states we know) of helium.

The difference between the total Hartree–Fock energy of an atom and the true energy in a given state is defined as the *correlation energy;* the Hartree–Fock method always gives an energy higher than the true value, in accordance with Eq. 5.11. The correlation energy is usually of the order of 1 eV per electron pair; this is of about the same magnitude as chemical bond energies, and thus cannot be neglected if one wishes to proceed accurately from atomic to molecular properties. Various methods exist for estimating correlation energies. To a good approximation, the total correlation energy can be taken as a sum of two-electron correlation energies, which can be calculated by methods like those used in the helium atom; most of the effect seems to be due to the interaction of electron pairs in the same orbital. In Chapter 14 we shall see how this approximation can be used

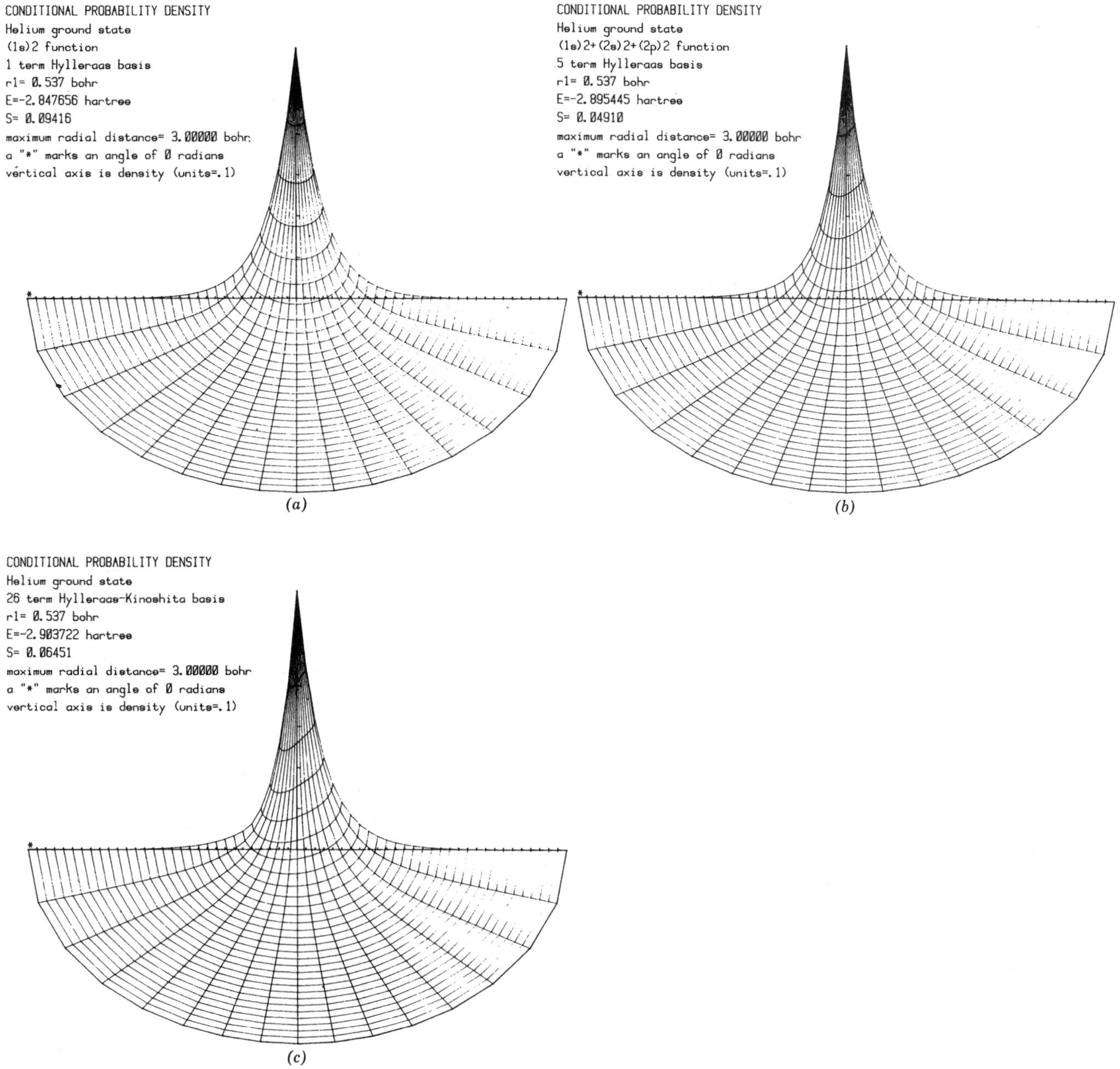

Figure 5.5 Conditional probability densities for the ground state of helium, for three wave functions. (*a*) Probability density for finding electron 2 at any point in space when electron 1 is 0.537 bohr to the left of the nucleus, based on a simple, single-configuration $1s^2$ wave function. (*b*) The corresponding conditional probability density based on a best superposition Of $1s^2$, $2s^2$, and $2p^2$ configurations. (*c*) The corresponding conditional probability density based on an accurate (26-term) wave function. The nucleus is located where the distribution peaks; the distance 0.537 bohr is the most probable distance of electron 1 from the nucleus. Note from the tilt of the contours of constant radius that there is no angular correlation in (*a*), some in (*b*), and more in (*c*). Graphs supplied by Paul Rehmus.

to estimate *molecular* correlation energies from thermochemical data.

An additional shortcoming in the Hartree–Fock method as just described arises from the effects of relativity. In atoms with fairly large *Z*, the inner electrons have such high kinetic energies that their velocities approach the speed of light; their masses thus become significantly larger than the mass of a stationary electron. This results in a shrinkage of the electrons' orbits and an increase in their binding energy. Although Hartree–Fock calculations can include this effect in only an approximate way,[8] the relativistic energy correction is known to be larger than the nonrelativistic correlation

[8] A completely relativistic quantum mechanic requires the revision of the Schrödinger equation itself. In this theory (originated by Dirac), the electron spin appears automatically in the solution of the wave equation; the spin is thus inherently not only a quantum mechanical, but also a relativistic, phenomenon.

energy for heavy atoms. For light atoms this is of little significance. For heavy atoms, the relativistic shrinkage of *s* and *p* shells makes them especially effective screens of the nuclear charge for *d* and *f* shells, which thus enlarge. Closed subshell atoms such as Hg become somewhat inert.[9]

The Hartree or independent-particle approach is not the only way to make a first approximation to the behavior of atomic electrons, particularly valence electrons. It is possible to start with a model in which the electrons move collectively, much as the atoms of small polyatomic molecules do. We shall examine such molecules in Chapters 8 and 9. In many respects, collective motions of valence electrons seem to correspond more closely to conserved quantities than motions of individual electrons. For example, electrons can exchange angular momenta rather readily by collisions within an atom. In terms of practical computations, however, no approach has yet been developed with the power of the Hartree method and its subsequent refinements to approach accurate, if elaborate and complex, representations of real atoms.

In Fig. 5.3 we expressed the energies in *hartrees*. This unit, named for D. R. Hartree, is defined as

$$1 \text{ hartree} \equiv 2 \text{ rydbergs} \equiv 2hcR_\infty \equiv \frac{e^4 m_e}{4\epsilon_0^2 h^2}$$
$$= 4.359743 \times 10^{-18}\text{ J} = 27.21138 \text{ eV}.$$

The ground-state energy of the hydrogen atom is thus –0.5 hartree. Published calculations on atoms and molecules most commonly give energy in hartrees and distance in Bohr radii (a_0 = 0.5292 Å), a combination known as *atomic units.*[10] The utility of such units is obvious: The ubiquitous $e^2/4\pi\epsilon_0 r$ in hartrees is numerically equal to a_0/r. Similarly, m_e is the atomic unit of mass, e of charge, and $\hbar$ of angular momentum. The constants are usually suppressed, leading to equations like $V = -1/r$; this is certainly convenient for making calculations, but is not recommended unless you know just what is missing. In this text we write all such equations out in full.

5.5 Atomic Structure and Periodic Behavior

The physical and chemical properties of the elements vary systematically with atomic number, in a way best summarized by the periodic table. We may now relate this behavior to atomic structure. To do this, at least for light elements, we do not need refinements such as term splitting (Sec. 5.6). The gross features of the periodic table can be explained quite simply in terms of electron configurations and shell structure. In this section we shall look briefly at the chemical behavior of the various families of elements, and at a number of physical properties associated with individual atoms.

The most strikingly similar group of elements is probably the so-called inert gases (rare gases, noble gases): helium, neon, argon, krypton, xenon, and radon. Until 1962 these elements were thought to form no compounds whatsoever; now krypton, xenon, and radon are known to form a limited number of compounds with fluorine and oxygen. Still, they remain by far the most unreactive family of elements. Even the forces between their monatomic molecules are very weak, so that they have very low melting and boiling temperatures. (Radon, the heaviest, boils at –62°C.) Usually one glibly attributes the inertness of these elements to their "closed-shell structure." In fact they have only closed subshells; all but helium have outer shells with the configuration $\cdots ns^2np^6$, and all their occupied inner subshells are filled. But other atoms have closed-subshell structures also, yet are hardly in the same category chemically. Palladium, for example, has the configuration $\cdots 4s^24p^64d^{10}$, and is clearly not inert; even more reactive are the alkaline earth metals, with $\cdots ns^2$ outside an inert-gas configuration, and the family of zinc, cadmium, and mercury. What, then, is really responsible for the inertness?

For a closed-shell atom to form a chemical bond, the outer shell must be "cracked open" in some way. One indication of the difficulty of doing this is the energy of the lowest excited state, in which one electron has been promoted to the lowest empty orbital. In the alkaline earths and the zinc group this excitation is simply $ns \rightarrow np$, between two orbitals that lie quite close together in energy. But for the inert gases the lowest excitation is $np \rightarrow (n+1)s$, all the way to the next higher shell, and the energy required is much greater (cf. Fig. 5.10). To illustrate this point, here are the first excitation energies (in electron volts):

He 19.8	Ne 16.6	Ar 11.5	Kr 9.9	Xe 8.3	Rn 6.8
Be 2.7	Mg 2.7	Ca 1.9	Sr 1.8	Ba 1.1	Ra 1.6
		Zn 4.0	Cd 3.7	Hg 4.7	

(As for palladium, the excitation is $4d \rightarrow 5s$ and takes only 0.8 eV.) The energy gained by forming a chemical bond is usually in the range 2–6 eV, which clearly is enough to "crack the shell" in all but the inert gases.

Another way for an atom to undergo chemical reaction is ionization, in which one electron is removed from the atom altogether. The energy required to remove one electron from a neutral atom is usually called the first *ionization potential*[11]

[9] For relativistic effects, see P. Pyykko and J. P. Desclaux, *Acc. Chem. Res.* **12,** 276 (1979) and K. Pitzer *ibid.,* 282. For a more detailed discussion of atomic orbitals see R. S. Berry, *J. Chem. Educ.* **43,** 293 (1966).

[10] Often written as "a.u." with no further explanation, or not specified at all. This can be confusing, especially since some people use rydbergs instead of hartrees.

[11] Strictly, "potential" implies electric potential or voltage. But for a single electron the ionization potential in volts is numerically the same as the ionization energy in electron volts, and the two terms are often used interchangeably.

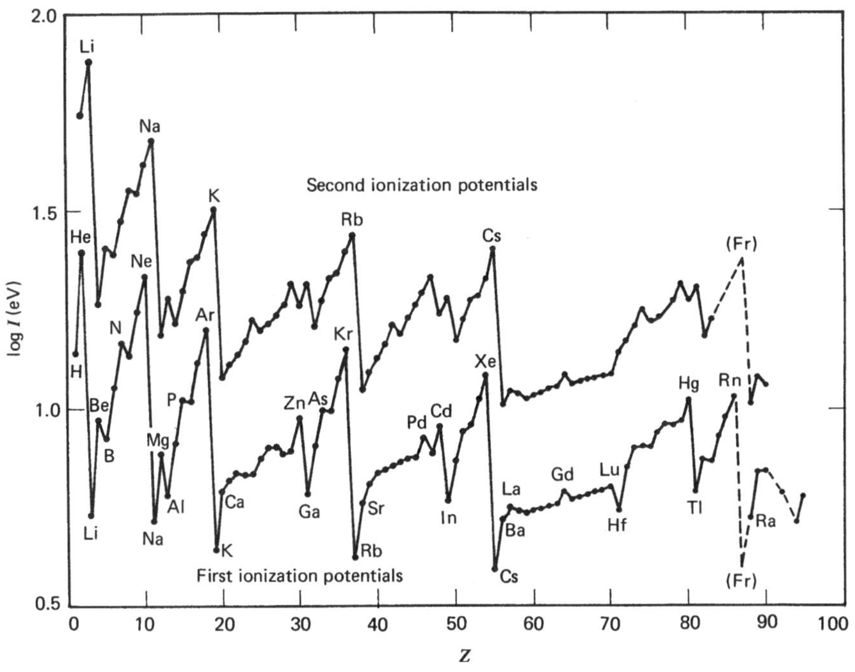

Figure 5.6 First and second ionization potentials of the elements. A number of the more significant peaks in the lower curve have been identified. The broken lines (----) give the probable behavior of the curves where no data are available.

(I or I_1); the second ionization potential (I_2) is the energy of removing a second electron, and so forth. The ionization potential is among the properties that show the most clearly periodic behavior, as can be seen in Fig. 5.6. Note that the second ionization potential is always higher than the first (since there is a greater net positive charge holding the second electron) and has its peaks shifted one unit to higher Z.

Let us examine the structure of the first-ionization-potential curve. The inert gases, as we might expect, have the highest first ionization potentials: It is very hard to remove an electron from a closed p subshell. (There are small peaks for such other closed-shell atoms as the zinc family and palladium.) The lowest values are those for the alkali metals (Li, Na, K, Rb, Cs, Fr), with one loosely bound s electron outside a closed shell; similarly, the next most prominent minima in the curve are for those atoms with a single outer p electron (B, Al, Ga, In, Tl). The small peaks for the lighter $\cdots np^3$ atoms (N, P, As) illustrate the stability of a half-filled subshell. All these details are superimposed on two grosser features, the overall downward trend of the ionization potential with Z and the upward trend within each period; these effects are closely associated with the variation in atomic size, which we shall discuss shortly.

Figure 5.6 can help us to understand a good deal about chemical behavior. After the inert gases, the most clearly defined family of elements is that of the alkali metals. Their ionization potentials are all about 4–5 eV, comparable to the energies normally available in chemical reactions. The result is that these atoms readily lose one electron to form the M^+ ions, and their chemistry is primarily the chemistry of these ions. However, the alkali metals have the *highest* second ionization potentials (ca. 25–75 eV), so that one virtually never sees alkali M^{2+} ions in chemical systems. This is again a matter of closed-shell structure: The alkali M^+ ions are isoelectronic[12] with the inert gas atoms, with the outer electrons even more tightly bound.

The alkaline earth metals (Be, Mg, Ca, Sr, Ba, Ra) have closed-subshell configurations, but we have already shown that this does not lead to inertness. On the contrary, the outer s electrons are relatively weakly bound (the alkaline earths have the lowest second ionization potentials), and the atom readily gives both up to form the M^{2+} ion. Beryllium is an exception: Having the highest ionization potentials of the family, it forms mainly covalent compounds. The elements of the zinc family, with the configurations $\cdots (n-1)d^{10}ns^2$, also generally form M^{2+} ions; however, the ionization potentials are relatively high, and the metals are much less reactive than the alkaline earths.

Since it is easier to remove one electron than two, why do we never see compounds containing ions like Ca^+? The principal reason is the intense Coulomb field of the M^{2+} ion, which is of course twice as strong as that of an M^+ ion at the same distance. This field makes the lattice (bonding) energy of, say, solid $CaCl_2$ much greater than that of CaCl would be—enough greater to outweigh a second ionization potential of more than 10 eV. Similar considerations apply to the hydration energy in solution, where we again find only M^{2+} alkaline-earth ions. We shall consider the energy balance of processes such as these in Chapter 14. One *can* obtain the M^+ ions by applying energy to isolated atoms, as in electrical discharges or very hot gases; in such environments, we

[12] Two atoms or ions are *isoelectronic* if they have the same number of electrons. Isoelectronic systems usually have the same configuration and certain similarities in physical properties. But their chemical behavior is quite different—compare the reactivity of Ar and K^+—because of the dominant effect of the ionic charge and its Coulomb field.

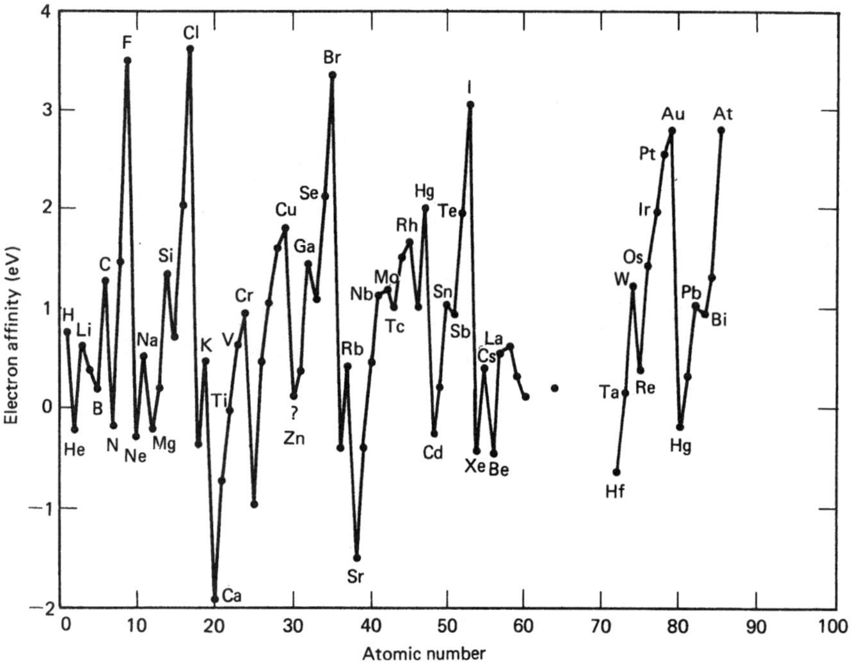

Figure 5.7 Electron affinities of the elements. The values are those selected in the review by H. Hotop and W. C. Lineberger, *J. Phys. Chem. Ref. Data* **4,** 539 (1975); others are those of R. J. Zollweg, *J. Chem. Phys.* **50,** 4251 (1969).

can also observe compounds such as CaCl, which are quite unstable under ordinary, ambient conditions.

Now let us look at the other side of the periodic table, where most of the nonmetallic elements are found. Here the atoms have vacancies in nearly filled *p* subshells, and can attain more stable configurations by *adding* electrons to form negative ions. The *electron affinity* (*A*) of a given atom is the energy gained when one electron is added. The electron affinity of X is thus the same as the first ionization potential of X^-, and a plot of electron affinities should resemble the lower curve in Fig. 5.6 but shifted one unit *lower* in *Z*. Such a plot is given in Fig. 5.7. Note that even the highest atomic electron affinities are less than the lowest atomic ionization potentials. The periodic pattern is clear.

As expected, the highest electron affinities are those of the halogens (F, Cl, Br, I, At), which readily add electrons to form the familiar X^- (halide) ions. Most of the elements near the right side of the periodic table have relatively large electron affinities, and can generally be regarded as electron-accepting species. These elements complement the electron-donating tendency of the alkali and alkaline earth metals, with which they form ionic compounds. The second-highest peaks, surprisingly, are the coinage metals (Cu, Ag, Au), but these form negative ions only in hot gases, in which these materials exist as isolated atoms,[13] and, in the case of Au, in binary, salt-like compounds with the alkali metals, e.g., as NaAu, which behaves like Na^+Au^- in its crystalline form. At the other extreme, many elements actually have *negative* electron affinities—that is, force would be required to make an electron stick to the neutral atom. The dips in the curve are mainly those atoms with exceptionally stable configurations, having either closed subshells (inert gases, alkaline earths, Zn family) or *half*-filled outer subshells $\cdots np^3$: N, P, etc.; $\cdots nd^5$: Mn, Re).

Unlike the case of the ionization potentials, no atoms have positive second electron affinities. The Coulomb repulsion between an X^- ion and an additional electron is so great that no stable X^{2-} ion is known to exist in isolation. Nevertheless, species such as O^{2-} and S^{2-} do appear to exist in ionic crystals, where they are stabilized by the high lattice energy; all these compounds hydrolyze readily in aqueous solution, forming more stable covalently bonded species (OH^-, HS^-, etc.).

Whereas the alkali and alkaline earth metals form almost exclusively ionic compounds, this is not true of the elements at the other side of the periodic table. Besides adding electrons to form negative ions, they can also share their outer electrons to form covalent bonds. In most simple compounds this results in the equivalent of an inert-gas configuration around each atom, as illustrated by the familiar Lewis formulas in which dots represent the outer-shell electrons (e.g., $2H\cdot + \cdot\ddot{\underset{\cdot\cdot}{O}}\cdot = H:\ddot{\underset{\cdot\cdot}{O}}:H$). We shall see in the next few chapters that such electron-sharing can be described in terms of orbitals extending over more than one atom. Covalent bonding occurs primarily between atoms not too different in *electronegativity.* The latter is a rather vague concept, defined by Pauling as "the power of an atom in a molecule to attract electrons to itself." Electronegativity is roughly equivalent to "nonmetallic character," and generally increases toward the upper right of the periodic table. Since atomic ionization potentials and electron affinities both tend to increase in the same direction, one method of defining

[13] However, the bonding in CsAu (and perhaps RbAu) is largely ionic.

numerical electronegativities takes the average of these two quantities. We shall look at this and other methods in Section 7.7, where electronegativities will be found in Table 7.7.

We can now account for the most obvious division of the periodic table, that between metals and nonmetals. Basically, a metallic element is one whose atom has one or more loosely bound valence (outer-shell) electrons; this description fits all the elements whose outermost subshell is *ns* ($n > 1$), as well as many with partially filled *p* subshells. The valence electrons are easily removed to form positive ions, as we have seen in the alkalis and alkaline earths. Metals ordinarily exist as solids in which these electrons are relatively free to move throughout the crystal; the mobility of the electrons accounts for the high electrical and thermal conductivity characteristic of metals. (We shall have more to say about the structure of metals in Chapter 11.) By contrast, the typical nonmetals have nearly full *p* subshells, and can form closed-shell configurations by sharing only a few electrons in small, covalently bonded molecules (Cl_2, O_2, P_4, etc.). The bonding forces within these molecules are strong, but (as in the inert gases) the forces between them are quite weak, so that many of the nonmetals are gases at ordinary temperatures. Along the borderline between metals and nonmetals, the elements often exist in solid forms containing infinite networks of covalent bonds; the best-known examples are the diamond and graphite forms of carbon. The great majority of the elements are metals, since all the transition elements (in which *d* or *f* subshells are being filled) have ns^2 or *ns* outer shells.

The trends in physical and chemical properties across the periodic table can best be understood by considering atomic sizes. The size of an atom cannot have a precise meaning, since at large distances the electron density merely trails off exponentially. However, fairly self-consistent sets of "atomic radii" can be defined for various purposes. In principle, the radii should be such that $r_A + r_B$ gives the actual internuclear distance in a bond between atoms A and B. Unfortunately, there exists no set of radii that can reproduce all bond lengths with experimental accuracy. In Table 5.2 we list *covalent radii,* which should add to give the lengths of single (two-electron) bonds between atoms not too different in electronegativity. These values are most easily obtained by halving the lengths of homonuclear single bonds (e.g., r_{Cl} is half the bond length in Cl_2): In many cases they reproduce other bond lengths within a few percent. A graph of covalent radii would closely resemble the atomic volume graph of Fig. 1.7, except that those elements made up of distinct molecules (most of the nonmetals) have relatively large volumes.

However, there are also many bond lengths quite different from what Table 5.2 predicts; let us digress to consider some of the reasons. A drastic shortening occurs when more than two electrons are shared, that is, in multiple bonds (C—C, 1.54 Å; C=C, 1.34 Å; C≡C, 1.20 Å); we shall see that this results from an increase in bonding energy. There is also a shortening associated with any degree of ionic character (electronegativity difference) of the bond; for example, the C—O single bond is 1.43 Å rather than 1.51 Å ($r_C + r_O$). For crystals essentially composed of ions rather than neutral atoms—the alkali halides, for example—one can derive an altogether different set of *ionic radii,* as we shall see in Chapter 11; the ionic radii depend strongly on charge, positive ions being smaller and negative ions larger than the neutral atoms. Finally, the effective radius of an atom increases with its *coordination number,* the number of atoms to which it is bonded. Since an atom in a solid metal usually has 6 to 12 nearest neighbors, the interatomic distances in metals are typically about 10% more than the naively-expected value of twice the covalent radii. With corrections for all these effects, one can reproduce most bond lengths with reasonable accuracy.[14]

Regardless of all these reservations, Table 5.2 is quite adequate for studying the relative sizes of atoms. To a rough approximation, the atomic radius corresponds to the outermost peak in the electron radial distribution (Fig. 5.4), that is, to the most probable radius of the valence electrons. Since this radius increases with *n,* the atomic size increases with each period in most families of the periodic table.

In an alkali metal atom we have a single outer-shell electron; if it were completely outside the closed-shell core, it would simply feel the Coulomb field of a net +1 charge, its most probable radius would be that of a hydrogen *ns* orbital, $r_n = n^2a_0$, and we could predict the sizes of the alkali atoms from the quantum number *n* of their single valence-shell electrons and the radius of the hydrogen *ns* orbital with the same principal quantum number *n*. The actual increase in radius with *Z* is of course much slower: The 6*s* electron in cesium has its outermost peak at about 2.7 Å, compared with 26 Å for a hydrogen 6*s* electron. The reason is that even the outermost electron has much of its charge density within the electron core (cf. Fig. 4.5), and is thus not perfectly "shielded" from the nucleus. The binding force on each electron thus increases steadily with the nuclear charge *Z,* as we already know from Fig. 5.2. The 6*s* electron in cesium ($Z = 55$) is bound almost as tightly as the 5*s* electron in rubidium ($Z = 37$), and the atomic size is only slightly larger.

We can now understand the trend across each period. One might expect that adding a second *ns* electron would increase the atomic size, because of the repulsion between the two valence electrons. In fact, the increase of *Z* by 1 has a much greater effect, and the alkaline earth atoms are uniformly smaller than their alkali neighbors. Electrons in an open shell generally shield one another from the nucleus quite poorly, since they tend to be in different regions of space (for two electrons, on opposite sides of the nucleus).

[14] For detailed discussions of the various kinds of atomic radii, see L. Pauling, *The Nature of the Chemical Bond,* 3rd ed. (Cornell University Press, Ithaca, N.Y., 1960), and J. C. Slater, *Quantum Theory of Molecules and Solids,* Vol. 2 (McGraw-Hill, New York, 1965).

Table 5.2 Single-Bond Covalent Radii of the Elements (Å)

					Cs 2.29	Fr 2.42
					Ba 1.98	Ra 2.10
					La 1.69	Ac 1.84
					Ce 1.65	Th 1.65
					Pr 1.64	Pa 1.50
					Nd 1.64	U 1.42
					Pm 1.63	Np
			K 1.96	Rb 2.10	Sm 1.62	Pu
			Ca 1.74	Sr 1.92	Eu 1.85	Am
			Sc 1.44	Y 1.62	Gd 1.62	Cm
			Ti 1.32	Zr 1.45	Tb 1.61	Bk
H 0.30	Li 1.34	Na 1.54	V 1.22	Nb 1.34	Dy 1.60	Cf
	Be 0.89	Mg 1.36	Cr 1.18	Mo 1.30	Ho 1.58	Es
	B 0.79	Al 1.25	Mn 1.17	Tc 1.27	Er 1.58	Fm
	C 0.77	Si 1.18	Fe 1.16	Ru 1.25	Tm 1.58	Md
	N 0.74	P 1.10	Co 1.16	Rh 1.25	Yb 1.70	No
	O 0.74	S 1.02	Ni 1.15	Pd 1.28	Lu 1.56	Lr
	F 0.71	Cl 0.99	Cu 1.17	Ag 1.34	Hf 1.44	Rf
He (0.5)	Ne (0.7)	Ar (0.95)	Zn 1.25	Cd 1.41	Ta 1.34	Db
			Ga 1.24	In 1.43	W 1.30	Sg
			Ge 1.22	Sn 1.40	Re 1.28	Bh
			As 1.22	Sb 1.41	Os 1.26	Hs
			Se 1.17	Te 1.37	Ir 1.26	Mt
			Br 1.14	I 1.33	Pt 1.30	
			Kr 1.10	Xe 1.30	Au 1.34	
					Hg 1.44	
					Tl 1.55	
					Pb 1.54	
					Bi 1.52	
					Po 1.53	
					At (1.50)	
					Rn (1.45)	

Only the completion of a closed shell, with its spherically symmetric charge density, gives a significant improvement in shielding. Thus the atomic sizes generally decrease with Z all the way across each period (but slower than Z, since the electron–electron repulsion does increase as the outer shell fills up), and increase again only when a new electron shell is added outside a closed shell.

There are some exceptions to this trend. As we move up the series of the transition metals in the Periodic Table, the $(n - 1)d$ subshell is being filled, whereas the outer shell remains virtually unchanged at ns^2. Since inner-shell electrons shield one another poorly, these atoms are relatively small, and a slight size increase accompanies the resumption of outer-shell filling. The shrinkage due to poor shielding is even more marked for the lanthanides, in which the $4f$ subshell is filled[15] inside a $6s^2$ outer shell. This *lanthanide contraction* is so great that hafnium has virtually the same size as its relative zirconium, and the two elements are extremely difficult to separate by ordinary chemical techniques. A similar contraction appears to occur in the actinide series, in which the $5f$ subshell is filled.

The trends in atomic size are directly correlated with those in ionization potential and electron affinity. Both the latter increase across each period, as the atomic size shrinks and the outer-shell electrons become more tightly bound.

[15] Note the anomalously large sizes of Eu($\cdots 4f^7 6s^2$) and Yb($\cdots 4f^{14} 6s^2$), with their exceptionally stable configurations.

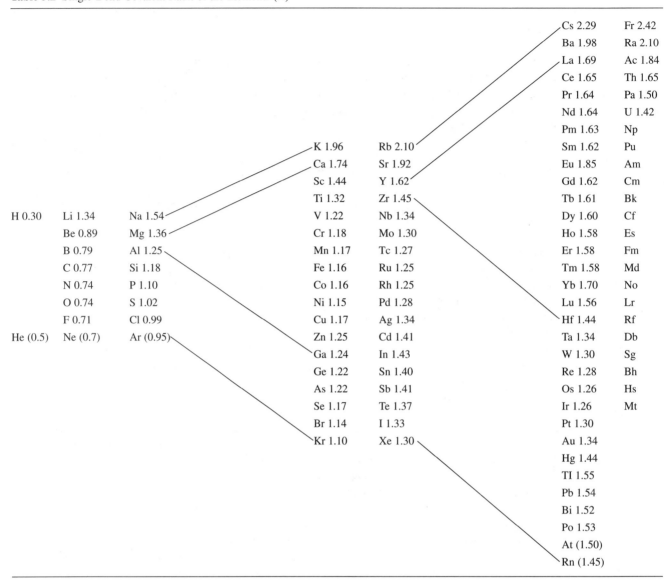

They decrease sharply at the beginning of a new period, as a larger, more weakly bound outer shell is begun. And each period has values lower than the preceding, since the binding energy decreases with distance from the nucleus. This effect too is less than it would be with perfect shielding: Cesium has I_1 = 3.89 eV, compared with the 0.38 eV (= 13.6 eV/n^2) required to ionize a 6*s* hydrogen electron; the effective nuclear charge seen by the valence electron in its ground state is thus $Z^* = (3.89/0.38)^{1/2} = +3.2$, rather than the +1 of perfect shielding.[16]

We have not yet discussed the chemical behavior of the transition metals. Although these elements have only two outer-shell *ns* electrons, the partially filled $(n-1)d$ subshell is quite near the "surface" of the atom and relatively loosely bound. As a result, transition metals ordinarily form compounds in which both the *s* and the *d* electrons are involved in bonding, with a wide variety of possible oxidation states. For example, manganese ($\cdots 3d^5 4s^2$) and oxygen form the species MnO, Mn_2O_3, MnO_2, and MnO_4^{2-}, and MnO_4^-. Even the unoccupied orbitals close to the outer shell can accept electrons from other species (*ligands*), giving rise to the complexes typical of the transition elements, such as $[Fe(H_2O)_6]^{2+}$. These complexes usually have low-lying excited states resulting from the splitting of *d*-orbital levels; since the energy differences often correspond to visible radiation, most such complexes are colored. We shall have much to say about these species in Chapter 9. Unlike the $(n-1)d$ subshells, the lanthanide 4*f* subshell is fairly deep within the atom, and the M^{3+} ions that dominate these elements' chemistry involve the loss of only one 4*f* electron. The actinide 5*f* electrons are not as deeply buried, and the lower actinides (especially uranium) commonly have high oxidation states. Since the binding energies of both *d* and *f* electrons increase with *Z* more rapidly than those of outer-shell electrons (cf. Fig. 5.2), the lower oxidation states again predominate toward the high-*Z* ends of the transition and actinide series.

As was mentioned in Section 5.1, paramagnetism occurs primarily in substances with unpaired electron spins. In most compounds of the nontransition elements, either ionization or the formation of covalent bonds leaves all the electrons paired. This is not true of the transition metals, in which some or all of the *d* electrons often remain unpaired; thus many transition metal compounds are paramagnetic. We shall discuss this behavior (with special reference to complex ions) in Section 9.6. The most intense paramagnetism is found among the lanthanides, with their many unpaired *f* electrons.[17]

We could look at other properties of the elements, nearly all showing some periodic behavior, but our point has been made. A vast amount of "chemistry"—that is, the observed differences between chemical substances—can be well understood in terms of a simple description of atomic structure. We have only scratched the surface of this subject, and further details can be found in advanced texts on inorganic chemistry. We turn in the next two sections to more detailed interactions that determine the states of atoms, in order to complete the groundwork we need to see how atoms are put together to form molecules. Then we shall devote the next four chapters to this latter task.

5.6 Term Splitting and the Vector Model

We have gone far by assuming that the electronic configuration of an atom dominates its energy and other properties. We have examined some aspects of correlation, for which the configuration model is inadequate, but this is nonetheless a very good representation of atoms with closed shells, and a good first approximation for many other real atoms as well. The set of closed-shell species includes the ground states of helium ($1s^2$), the other inert gases ($\cdots ns^2np^6$), and the alkaline earths ($\cdots ns^2$); these configurations are distinguished by having all the occupied subshells filled. Another group that is well represented by single configurations includes hydrogen and the alkalis in their ground states ($\cdots ns$), in which the atom has a single *s* electron outside a closed-shell structure, and the halogens ($\cdots np^5$), with only a single vacancy in an otherwise closed shell. In all these cases the electron configuration corresponds to only a single state or energy level of the atom.

In most atoms, however, a single configuration may describe several different states. These are real, physically distinguishable states, which can differ from one another by as much as several electron volts in energy. One speaks of the splitting of the configuration into *terms;*[18] in this context a "term" is equivalent to a state or group of states. The basic difference between terms of the same configuration lies in the angular momentum of the atom. The configuration defines the magnitudes of the vectors **L** and **S** for each electron, but not their orientations. There is thus more than one possible way to couple (add) these vectors to give the total angular momentum. These different ways of coupling correspond to different assignments of the quantum numbers m_l

[16] For the alkali metal atoms we have $Z^* = n/n^*$, in terms of the effective quantum numbers of Eq. 2.12.

[17] Some typical room-temperature molar susceptibilities (χ_M) in 10^{-12}m^3/mol: nontransition metals, Na 16, Zn −11; transition metals, Ti 153, Mn 529; lanthanides, Pr 5010, Er 44300; nonmetals, S −15, Xe −44. These values are for the elements in their normal states; higher values are often found in compounds (e.g., $MnCl_2$ 14350).

[18] This language derives from spectroscopy: For a transition between states 1 and 2, the frequency of the absorbed or emitted radiation is given by

$$\nu = \left| \frac{E_2}{h} - \frac{E_1}{h} \right|,$$

the difference of two *terms* characteristic of the two states.

and m_s, hence to different spatial distributions of the electrons. Since the spatial relationships of the electrons are different, the terms must differ in energy.

We can describe this term splitting most simply with the *vector model of the atom.* This is a method for organizing the possible states of an atom into terms, by selecting only the most important of the many types of interaction. It also serves as a bookkeeping device for counting and classifying the terms corresponding to a given configuration. The essence of the vector model lies in the ways of adding angular momentum vectors, the topic we postponed discussing in Section 5.1.

In contrast to vector addition in classical physics, the addition of quantum mechanical angular momenta cannot be carried out by determining the *x, y,* and *z* components of the individual vectors and adding these separately. This is because the uncertainty principle forbids us to know more about any angular momentum vector than its magnitude and *one* of its components, usually defined as the *z* component. Knowing only the lengths and the *z* components, however, is sufficient to define the quantum states we need. For any quantized angular momentum vector, say **J**, the two are related by the usual space-quantization rules: The magnitude of **J** is given by an equation of the form of Eq. 5.1,

$$|\mathbf{J}| = [J(J+1)]^{1/2}\,\hbar, \tag{5.18}$$

and the *z* component by an equation of the form of Eq. 5.2,

$$J_z = M_J \hbar \qquad (M_J = J, J-1, \ldots, -J+1, -J), \tag{5.19}$$

with appropriate quantum numbers J and M_J.

We can approach the classification of atomic states by a series of approximations. The first, of course, is the central field approximation, in which the energy of the atom depends only on the configuration—that is, on the values of n and l (and thus $|\mathbf{L}|$) for each of the electrons, but not on the values of m_l and m_s. This would be valid only if each electron moved in a spherically symmetric field; in the real atom, the electron–electron interactions split up each configuration into many states of somewhat different energies. The next approximation rests on the fact that some kinds of interactions are much stronger than others. For the lighter atoms, all states with the same values of the quantum numbers L and S (for the entire atom) have nearly the same energy, whereas states with different values of L or S are appreciably different in energy. One can thus group together all states with the same L, S as a single term; when this is a good approximation, one speaks of *LS coupling.*[19]

Why is the energy so strongly dependent on the values of L and S? The reasons are those we introduced when discussing the exclusion principle: the Coulomb and exchange interactions between electrons. The quantum number L defines the total orbital angular momentum of the atom; as in the hydrogen atom, the value of L governs the symmetry of the atomic wave function. States with different L obviously have different charge distributions, and thus different electron–electron interaction energies. The effect of the spin quantum number S is more indirect. Each value of S corresponds to a particular combination of the electronic m_s's, which by the exclusion principle can be combined only with particular values of m_l (e.g., if two electrons in a given subshell have the same m_s, their values of m_l must differ). Thus a change in S also implies a change in the charge distribution and energy.

The total angular momentum vectors for an atom can be obtained by adding together (*coupling*) the vectors for the individual electrons:

$$\mathbf{L} = \sum_i \mathbf{L}_i \qquad \text{and} \qquad \mathbf{S} = \sum_i \mathbf{S}_i\,. \tag{5.20}$$

The z components of **L** and **S** are then given by

$$L_z = M_L \hbar \quad \left(M_L = \sum_i m_{li} \right) \quad \text{and}$$

$$S_z = M_S \hbar \quad \left(M_S = \sum_i m_{si} \right). \tag{5.21}$$

One can readily show that both **L** and **S** sum to zero over any closed subshell;[20] thus only those electrons outside closed subshells ("open-shell electrons") need to be taken into account. For example, the inert gases and alkaline earths have $\mathbf{L} = \mathbf{S} = 0$, and can thus have only a single term with $L = S = 0$. The alkalis have one *s* electron outside a closed shell and therefore have $L = 0$, $S = \frac{1}{2}$. Group III elements boron, aluminum, etc., with one *p* electron outside a closed s^2 subshell, can have only $L = 1$, $S = \frac{1}{2}$, and also show no term splitting. The same is true for the halogens, with only one vacancy in a subshell: Again we have $L = 1$, $S = \frac{1}{2}$, because m_l and m_s cancel out for all but one of the electrons. It is only in configurations with more than one unpaired electron[21] that things get more complicated and term splitting appears.

The simplest atom for which these complications appear in the ground state is that of carbon, with the configuration $1s^2 2s^2 2p^2$. We need consider only the two open-shell

[19] Also known as *Russell-Saunders coupling.*

[20] From Table 3.1 one can see that $\Sigma_{m_l} \left|Y_{l,m_l}(\theta, \phi)\right|^2$ for any l is independent of θ and ϕ. That is, the total electron distribution for a closed subshell is spherically symmetric, and thus has no orbital angular momentum. It is also clear that the z components of **L** and **S** must vanish, since positive and negative values of m_l and m_s cancel.

[21] That is, in which at least one subshell contains more than one electron *and* more than one vacancy.

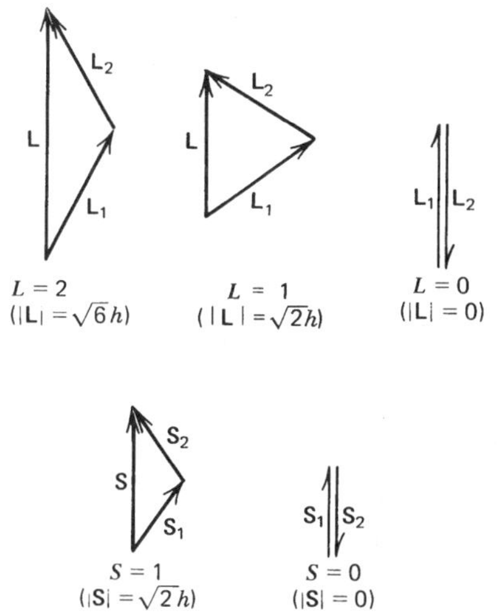

Figure 5.8 Possible ways of coupling angular momentum vectors for two p electrons: $\mathbf{L} = \mathbf{L}_1 + \mathbf{L}_2$, $\mathbf{S} = \mathbf{S}_1 + \mathbf{S}_2$. Each electron has $l = 1$, $s = \frac{1}{2}$, so that

$$|\mathbf{L}_1| = |\mathbf{L}_2| = [l(l+1)]^{1/2}\,\hbar = \sqrt{2}\hbar,$$

$$|\mathbf{S}_1| = |\mathbf{S}_2| = [s(s+1)]^{1/2}\,\hbar = \frac{\sqrt{3}}{2}\hbar.$$

The diagrams show the only ways of adding these vectors consistent with

$$|\mathbf{L}| = [L(L+1)]^{1/2}\,\hbar, \qquad |\mathbf{S}| = [S(S+1)]^{1/2}\,\hbar.$$

The vectors are drawn in their own plane; the z axis is in general not in this plane, but must be at an angle such that $L_{zi} = m_{li}\hbar$, $L_z = M_L\hbar$, $S_{zi} = m_{si}\hbar$, $S_z = M_s\hbar$. There are six possible combinations of L and S, but for equivalent electrons (same n, l) the exclusion principle allows only $L = 1$, $S = 1$; $L = 2$, $S = 0$; and $L = 0$, $S = 0$.

$2p$ electrons. A single $2p$ electron has $n = 2$, $l = 1$, three possible choices of m_l, and two possible choices of m_s, a total of six possible sets of quantum numbers; because of the exclusion principle, only five of these are available to a second electron. There are thus $(6 \times 5)/2 = 15$ possible assignments of the two electrons. (Division by 2 is required because the electrons are indistinguishable, and we cannot tell which is counted first.) Not all of these 15 assignments are physically distinguishable; to see how many separate terms they give rise to, we must determine what values of L and S can correspond to each assignment.

In Fig. 5.8 we show the possible ways in which the angular momentum vectors of two p electrons can be added to give different values of L and S. The value of L for an atom, like that of L for an electron, must be a positive integer or zero ($L = 0, 1, 2, \ldots$); the value of S is integral ($S = 0, 1, 2, \ldots$) for an even number of electrons, half-integral ($S = \frac{1}{2}, \frac{3}{2}, \ldots$) for an odd number. In the LS-coupling approximation, each physically distinct state (term) of the atom corresponds to a specific combination of L and S. These terms can be found in several ways, but the most straightforward method is sufficient for our purposes.

For an electron with given quantum numbers l_i, m_{li}, m_{si} we know both the magnitudes and the z components of the vectors $\mathbf{L}_i$, $\mathbf{S}_i$. Although these vectors can add to give several different atomic values of L or S, there is no ambiguity in the addition of the z components:

$$M_L = \sum_i m_{li}, \qquad M_S = \sum_i m_{si}. \tag{5.22}$$

One can thus make a preliminary classification of atomic states according to the values of M_L and M_S. This is done by tabulating all the possible assignments of m_l and m_s for the open-shell electrons, then obtaining the sums of Eq. 5.22 for each such assignment. In Table 5.3 we have illustrated how this is done for the carbon atom (or any other $\cdots np^2$ configuration). In order to satisfy the exclusion principle, any state in which both electrons have the same m_l must have $m_{s1} \neq m_{s2}$. Since the two electrons are equivalent, two assignments that differ only by interchange of the subscripts "1" and "2" correspond to the same state, which we list only once.[22] We thus obtain the predicted 15 assignments.

Next we decide which values of L and S go with each of these assignments. Since for any state we must have $M_L \leq L$, the existence of states with $M_L = \pm 2$ implies that there must be a term with $L = 2$, that is, a D term. This term must also include states with the other possible orientations of $\mathbf{L}$ relative to the z axis—that is, it includes one each of the states with $M_L = 1, 0, -1$ in Table 5.3. We do not need to specify which states these are, and within the scope of this text one cannot make such a specification. Finally, since the states with $M_L = \pm 2$ have only $M_S = 0$, we conclude that S must be zero for the D term. To sum up, five of the states in Table 5.3 belong to a term with $L = 2$, $S = 0$. We designate any atomic term by a capital letter corresponding to the value of L, with a left superscript equal to $2S + 1$ (called the *multiplicity*); in the present case the term is ^1D, read "singlet D."

That is one term accounted for. There are no other states with $M_L = 2$, and thus no other terms with $L = 2$. Three states each remain unassigned with $M_L = 1, -1$; these must belong to a term (or terms) with $L = 1$, a P term. Each of these groups of three (triplets) is made up of states with $M_S = 1, 0, -1$, which correspond to the possible orientations of $\mathbf{S}$ for $S = 1$. The term containing these states thus has $L = 1$, $S = 1$, with the designation[23] ^3P ("triplet P"). To complete

---

[22] Thus $m_{l1} = 1$, $m_{l2} = 1$, $m_{s1} = \frac{1}{2}$, $m_{s2} = -\frac{1}{2}$ is the same state as $m_{l1} = 1$, $m_{l2} = 1$, $m_{s1} = -\frac{1}{2}$, $m_{s2} = \frac{1}{2}$. But $m_{l1} = 1$, $m_{l2} = 0$, $m_{s1} = \frac{1}{2}$, $m_{s2} = -\frac{1}{2}$ and $m_{l1} = 1$, $m_{l2} = 0$, $m_{s1} = -\frac{1}{2}$, $m_{s2} = \frac{1}{2}$ are different states, since only the m_s's are interchanged. In Chapter 6 we shall designate the spin states of individual electrons by the *Pauli spin functions:* If electron 1 has $m_s = +\frac{1}{2}$, we write $\alpha(1)$ as its spin function; if electron j has $m_s = -\frac{1}{2}$, we write its spin function as $\beta(j)$.

[23] The multiplicity $2S + 1$ gives the number of possible quantized orientations of the vector $\mathbf{S}$. Terms with $2S + 1 = 1, 2, 3, 4, \ldots$ are called *singlets, doublets, triplets, quartets,* etc.

Table 5.3 Addition of z Components of Angular Momenta for Two Equivalent p Electrons (as in the Carbon Atom, $1s^2 2s^2 2p^2$): $l_1 = l_2 = 1$

m_{l1}	m_{l2}	$M_L = \sum_i m_{li}$	m_{s1}	m_{s2}	$M_S = \sum_i m_{si}$	Term
1	1	2	$\frac{1}{2}$	$-\frac{1}{2}$	0	^1D
1	0	1	$\frac{1}{2}$	$\frac{1}{2}$	1	^3P
			$\frac{1}{2}$	$-\frac{1}{2}$	0	^1D, ^3P
			$-\frac{1}{2}$	$\frac{1}{2}$	0	
			$-\frac{1}{2}$	$-\frac{1}{2}$	-1	^3P
1	-1	0	$\frac{1}{2}$	$\frac{1}{2}$	1	^3P
			$-\frac{1}{2}$	$-\frac{1}{2}$	-1	^3P
			$\frac{1}{2}$	$-\frac{1}{2}$	0	^1D, ^3P, ^1S
			$-\frac{1}{2}$	$\frac{1}{2}$	0	
0	0	0	$\frac{1}{2}$	$-\frac{1}{2}$	0	
0	-1	-1	$\frac{1}{2}$	$\frac{1}{2}$	1	^3P
			$\frac{1}{2}$	$-\frac{1}{2}$	0	^1D, ^3P
			$-\frac{1}{2}$	$\frac{1}{2}$	0	
			$-\frac{1}{2}$	$-\frac{1}{2}$	-1	^3P
-1	-1	-2	$\frac{1}{2}$	$-\frac{1}{2}$	0	^1D

NUMBER OF ASSIGNMENTS

M_L \ M_S	1	0	-1
2		1	
1	1	2	1
0	1	3	1
-1	1	2	1
-2		1	

the possible orientations of **L**, the ^3P term must also include a set of states with $M_L = 0$ and $M_S = 1, 0, -1$; these account for three of the as-yet-unassigned states with $M_L = 0$. The ^3P term thus comprises a total of nine states, and only one of the original 15 states remains unassigned, one of the states with $M_L = 0$, $M_S = 0$. This state can only be the single component of a term with $L = 0$, $S = 0$, a ^1S ("singlet S") term.

We have thus shown that the ground-state configuration of the carbon atom contains three terms: ^1D, ^3P, and ^1S. (Note that three of the geometrically possible L, S combinations in Fig. 5.8 do not exist; they are ruled out by the exclusion principle.) Among them these terms account for all 15 possible assignments of quantum numbers: nine states in the ^3P term, five in the ^1D, one in the ^1S. For seven of these 15 states, however, we cannot tell which assignment goes with which term (see Table 5.3); in these states we cannot assign a complete set of quantum numbers to the individual electrons. For example, there are two states with $M_L = 1$, $M_S = 0$, one in the ^1D term and one in the ^3P term; but there is no way to specify in which of these the electron with $m_l = 1$ has $m_s = \frac{1}{2}$. In cases such as this, the orientations of $\mathbf{L}_i$ and $\mathbf{S}_i$ for each electron are not fixed, and the z components of $\mathbf{L}_i$ and $\mathbf{S}_i$ ($m_l\hbar$ and $m_s\hbar$) are not constants of the motion. We say that m_l and m_s are no longer "good quantum numbers." One can interpret this by saying that the electrons interact in such a way as to exert torques on one another, changing the orientation of each electron's angular momentum. This picture explains why we speak of "coupled" angular momenta. What the *LS*-coupling model assumes is that the total angular momenta **L** and **S** *are* conserved, so that L, M_L, S, and M_S are good quantum numbers.

If the electrons do not have definite quantum numbers, how does one set up the wave function for an atom in a given term? In the orbital model the atomic wave function, Eq. 5.15, is a product (or a determinant) of the orbital wave functions φ_i, each of which involves the quantum numbers of a particular electron. What one must do is to obtain such a product, call it $\psi^{(j)}$, for each of the possible assignments in the term; the atomic wave function is then a linear combination of these products, $\Sigma_j c_j \psi^{(j)}$. The problem, which we shall not go into, is to obtain the c_j such that the resulting wave function gives the correct values of L and S, i.e., is an eigenfunction of the operators L^2 and S^2.

The three terms in the carbon atom's ground configuration are physically distinct, each with its own energy, electron distribution, and magnetic properties. The terms ^3P and ^1D, however, are *multiplets,* consisting of nine and five states, respectively. In the *LS*-coupling approximation the states within a given term are degenerate, all having the same energy. In reality there is a further fine-structure splitting, for reasons we shall discuss in the next section.

One can obtain from spectroscopic data the energy levels corresponding to the various terms. In the carbon atom the ^3P term is lowest in energy, the ^1D next, and the ^1S highest (cf. Fig. 5.9). For most ground-state configurations, the order of term energies is conveniently given by *Hund's rules:*

1. Within a given configuration, the term energies increase as S decreases.
2. Among terms with the same S, the term energies increase as L decreases.

This ordering implies that the differences in electron–electron interactions associated with different values of S affect the energy more strongly than the corresponding differences associated with L; this is indeed true when the *LS*-coupling approximation applies. Since the energy is lowest when S has its maximum value, the ground states of open-shell atoms usually have as many electrons as possible with unpaired spins.

It was long believed that the physical basis for Hund's first rule is a simple effect: that the Pauli exclusion principle keeps two electrons of the same spin far enough away from each other to make their repulsive Coulomb interaction

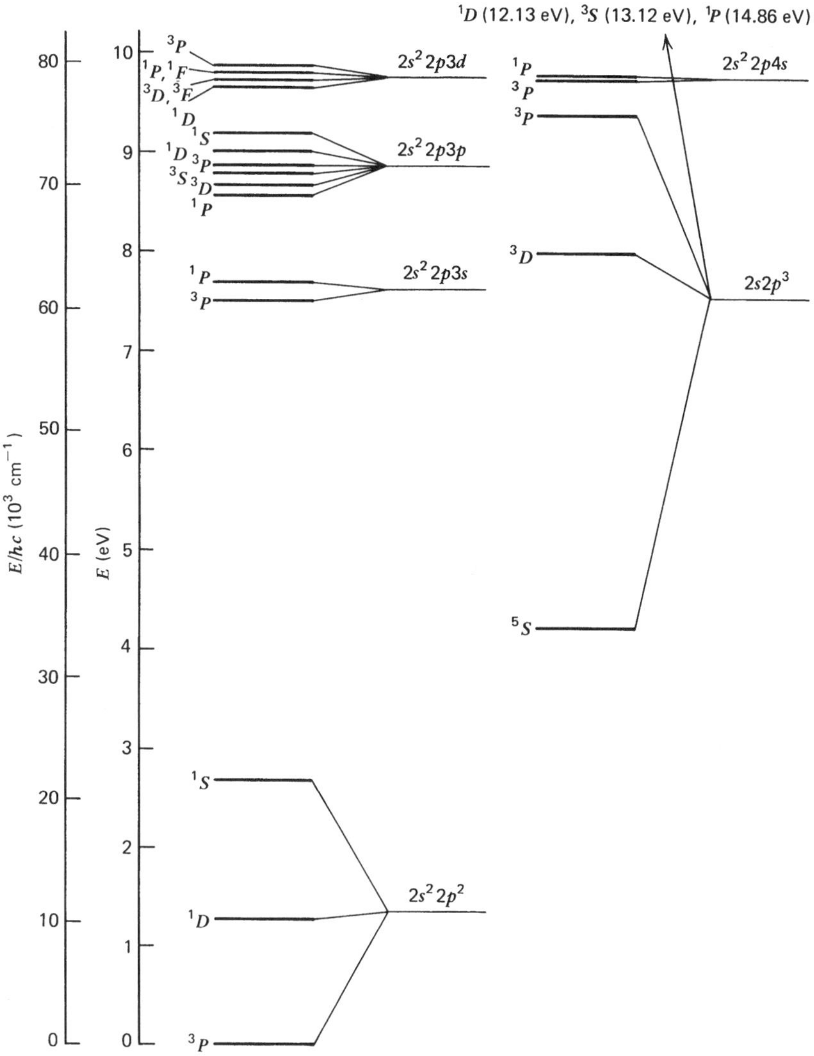

Figure 5.9 Low-lying energy levels of the carbon atom, with term designations. The terms belonging to each configuration have been bracketed together. The energy is measured above the ground state.

smaller than it is in a corresponding singlet, where the exclusion principle does not affect the spatial correlation. Careful studies[24] made it clear that the interelectronic energy of Coulomb repulsion is *larger* in triplet states of simple atoms and molecules than it is in the corresponding singlets based on the same configurations. The reason the energies of the triplets are lower and Hund's first rule holds is that the larger electron–electron repulsion of the triplet is accompanied by an attractive electron–nuclear attraction that more than compensates for the electron–electron repulsion. The electrons in the triplets have larger average values of r_{12}^{-1}, by coming closer, on average, to their nuclei and thus to one another.

The method we have developed for the carbon atom can be applied to other atoms as well. Reviewing the single-term configurations, we see that the closed-shell atoms, with $L = 0$, $S = 0$, have only ^1S terms; the alkalis ($\cdots ns$, $L = 0$, $S = \frac{1}{2}$) only ^2S terms; and the boron-aluminum group ($\cdots np$, $L = 1$, $S = \frac{1}{2}$) only ^2P terms. If a subshell is more than half-filled, it is simpler to work with the vacancies ("holes") in the subshell as if they were electrons; one can show that the values of L and S are the same when calculated either way. For example, the ground state of the oxygen atom, $1s^2 2s^2 2p^4$, with two holes in the $2p$ subshell, can be treated as a case of $2p^2$. The oxygen atom thus has the same three terms as the carbon atom, as do the other $\cdots np^2$ and $\cdots np^4$ elements. Similarly, the halogens ($\cdots np^5$) have only ^2P terms like the boron group. The only nontransition elements remaining are the nitrogen group ($\cdots np^3$), for which the ground state must have $S = \frac{3}{2}$ by Hund's first rule; the rest of the term analysis is left as a problem. (Note that the filling sequence ns^2, ns^2np, . . . , ns^2np^6 has the ground-state multiplicities 1, 2, 3, 4, 3, 2, 1.)

[24] See, for example, E. R. Davidson, *J. Chem. Phys.* **42,** 4199 (1965); J. P. Colpa and R. E. Brown, *Mol. Phys.* **26,** 1453 (1973); E. A. Colbourn, *J. Phys. B* **8,** 1926 (1975). J. Katriel and R. Pauncz, Adv. Quantum Chem. **10,** 145 (1977).

As for the transition elements, one can now see what is implied by most of the anomalies in those electron configurations marked by stars in Table 5.1. Consider the chromium atom ($Z = 24$), which one would expect to be $\cdots 3d^4 4s^2$. By Hund's first rule, the ground state would have $S = 2$, with four parallel spins. But the actual ground-state configuration is $\cdots 3d^5 4s$, which by Hund's rule should have six parallel spins ($S = 3$). Thus the effect of the increase in S on the energy must be greater than the very small difference between the $3d$ and $4s$ orbital energies. In the copper atom ($Z = 29$), the shift from $\cdots 3d^9 4s^2$ to $\cdots 3d^{10} 4s$ does not affect S; however, the exchange interaction is particularly large for a filled (or half-filled) subshell, in which the electrons of each spin have a spherically symmetric charge density, and the resulting term splitting again outweighs the orbital energy difference. Nearly all the anomalous configurations contain (or at least tend toward) filled or half-filled d or f subshells, which are stabilized by these effects.

So far we have discussed only configurations with *equivalent* open-shell electrons, electrons in the same subshell. If the electrons are nonequivalent, there is a slight difference in the bookkeeping. Let us consider the $1s^2 2s^2 2p3p$ excited state of carbon, in which the $2p$ and $3p$ electrons are nonequivalent. The table of quantum number assignments, Table 5.4, is now somewhat longer than in the equivalent-electron case.

Since the two p electrons have different values of n, the exclusion principle no longer limits the possible values of m_s, and all 36 of the possible combinations are distinct assignments. The smaller table shows how these assignments are distributed over the possible values of M_L and M_S. By reasoning like that we used before, it is not difficult to find the terms: These are ^1D (5 states), ^3D (15 states), ^1P (3 states), ^3P (9 states), ^1S (1 state), and ^3S (3 states), the number of states again adding up to 36. This time we have used all possible combinations of L and S (Fig. 5.8).

Other excited configurations can be similarly analyzed into their terms; Fig. 5.9 shows all the energy levels of the carbon atom up to 10 eV above the ground state, each with its configuration and term designation. In general, both the energy differences between configurations and those between terms of the same configuration tend to grow smaller with increasing excitation energy. This is apparent here in the sequence $2p^2$, $2p3s$, $2p3p$, $2p3d$, $2p4s$ (as usual, the $3d$ and $4s$ energies are practically the same): The more a given electron is excited, the greater is its average distance from the nucleus, the less it interacts with the inner electrons, and thus the smaller is the term splitting. In highly excited states, however, the configurations tend to overlap in energy, indicating that the differences between orbital energies in the central field decrease even more rapidly than the term splitting due to deviations from the central field.

For most atoms term assignments have been made for hundreds or even thousands of states, where energy differences are obtained from spectroscopic data. Yet only a fraction of the conceivable transitions between states are observed, many transitions being "forbidden" for reasons outlined in Section 4.5. In the *LS*-coupling approximation the principal selection rules for allowed transitions can be shown to be

$$\Delta l = \pm 1 \quad \text{(for an electron)},$$
$$\Delta L = 0,\ \pm 1, \qquad \Delta S = 0 \quad \text{(for the atom)}; \tag{5.23}$$

these rules hold quite well for the lighter atoms. They are illustrated for the helium atom by Fig. 5.10, which groups the energy levels by their term designations and shows the principal spectral lines. The most striking feature is that transitions between singlet and triplet states are forbidden, in accordance with the rule $\Delta S = 0$. (A line corresponding to $1s^2\,{}^1S - 1s2p\,{}^3P$ does appear, but very weakly.) Both $1s2s$ excited states are metastable: Since they can decay to the ground state only by forbidden transitions, they have relatively long lifetimes. Note also how all the energy levels for a given n approach one another as $n \to \infty$; each series of terms converges to the same limit, the ground state of the He^+ ion.

Table 5.4 Addition of s Components of Angular Momenta for Two Nonequivalent p Electrons (as in $1s^2 2s^2 2p3p$, the Excited State of Carbon): $l_1 = l_2 = 1$

m_{l1}	m_{l2}	$M_L = \sum_i m_{li}$	m_{s1}	m_{s2}	$M_S = \sum_i m_{si}$
1	1	2	$\frac{1}{2},-\frac{1}{2}$	$\frac{1}{2},-\frac{1}{2}$	1,0,0,−1
1	0	1	$\frac{1}{2},-\frac{1}{2}$	$\frac{1}{2},-\frac{1}{2}$	1,0,0,−1
1	−1	0	$\frac{1}{2},-\frac{1}{2}$	$\frac{1}{2},-\frac{1}{2}$	1,0,0,−1
0	1	1	$\frac{1}{2},-\frac{1}{2}$	$\frac{1}{2},-\frac{1}{2}$	1,0,0,−1
0	0	0	$\frac{1}{2},-\frac{1}{2}$	$\frac{1}{2},-\frac{1}{2}$	1,0,0,−1
0	−1	−1	$\frac{1}{2},-\frac{1}{2}$	$\frac{1}{2},-\frac{1}{2}$	1,0,0,−1
−1	1	0	$\frac{1}{2},-\frac{1}{2}$	$\frac{1}{2},-\frac{1}{2}$	1,0,0,−1
−1	0	−1	$\frac{1}{2},-\frac{1}{2}$	$\frac{1}{2},-\frac{1}{2}$	1,0,0,−1
−1	−1	−2	$\frac{1}{2},-\frac{1}{2}$	$\frac{1}{2},-\frac{1}{2}$	1,0,0,−1

NUMBER OF ASSIGNMENTS

M_L \ M_S	1	0	−1
2	1	2	1
1	2	4	2
0	3	6	3
−1	2	4	2
−2	1	2	1

Note: The tabulation of the m_s's is condensed. For each combination of m_{l1} and m_{l2}, there are four possible assignments:

m_{s1}	$\frac{1}{2}$	$\frac{1}{2}$	$-\frac{1}{2}$	$-\frac{1}{2}$
m_{s2}	$\frac{1}{2}$	$-\frac{1}{2}$	$\frac{1}{2}$	$-\frac{1}{2}$
M_S	1	0	0	−1

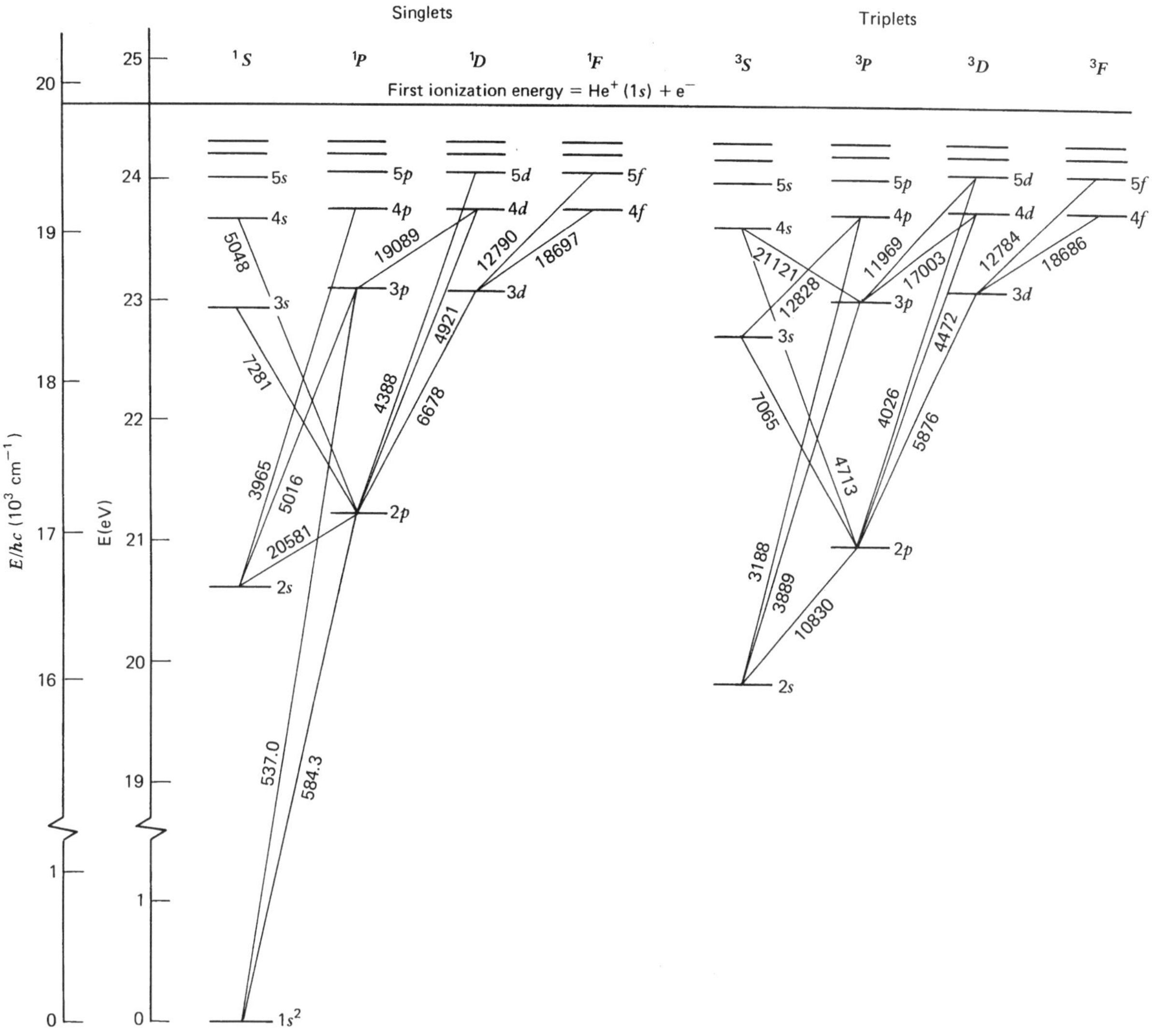

Figure 5.10 Energy levels of the helium atom, arranged by term designations. Except for the ground state, each level is identified by the configuration of only the excited electron. Some of the principal (most intense) spectral lines are shown, each with its wavelength in angstroms.

5.7 Fine Structure and Spin–Orbit Interaction

We still have not come to the end of our analysis of atomic energy levels. Not only do configurations split into terms, but the terms themselves often split into closely spaced levels. This is the fine structure of the energy spectrum, which we have already mentioned on several occasions. Some examples are illustrated in Fig. 5.11.

What is the cause of the fine-structure splitting? The term splitting is due to the nonspherical part of the electrostatic interactions between electrons. Reviewing the Hamiltonian of Eq. 5.10, we see that we still have not considered the magnetic moment interactions. Each electron has both an orbital magnetic moment and a spin magnetic moment, which interact with each other and with the moments of all the other electrons. Since it is precisely in the orientation of **L** and **S** that the states within a given term differ from one another, the electrons in each of these states see a slightly different magnetic field. The states of a multiplet term are thus not really degenerate after all.

The interaction between the spin and orbital magnetic moments is ordinarily far greater than the spin–spin and orbital–orbital interactions. We can thus neglect the latter two and speak only of the *spin–orbit interaction.* The logical extension of the *LS*-coupling model is to assume that this interaction depends only on the total *L* and *S* for an atom. To a fairly high level of approximation, the interaction energy is given by

$$E_{so} = \zeta \mathbf{L}\cdot\mathbf{S} = \langle \zeta \mathsf{L}\cdot\mathsf{S} \rangle, \tag{5.24}$$

where the multiplier ζ is a function of the specific atomic state. The value of ζ is notably difficult to obtain from

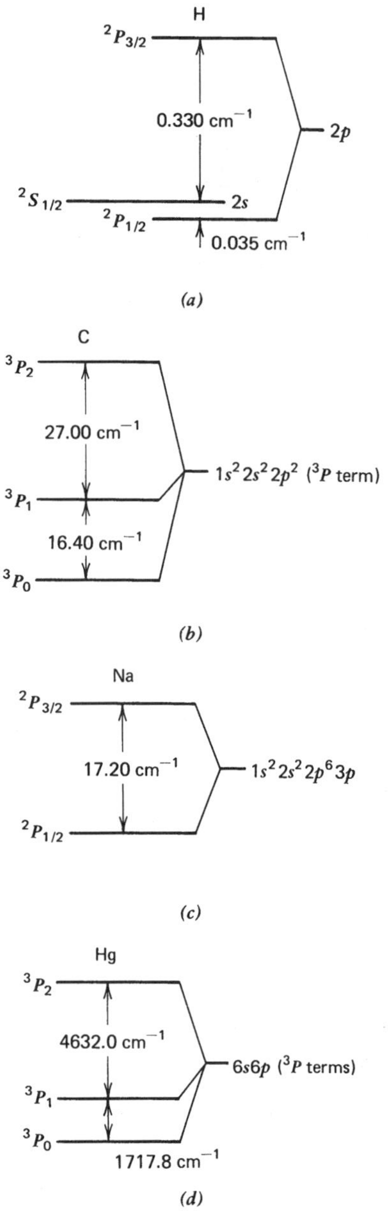

Figure 5.11 Some fine-structure splittings (not drawn to the same scale). Energy differences are expressed as $\Delta E/hc$ in cm^{-1}: 1eV/hc = 8066 cm^{-1}. (*a*) The n = 2 level of hydrogen. (*b*) The ground-state term of carbon (cf. Fig. 5.9). (*c*) The first excited state of sodium: Sodium-vapor lamps emit a doublet line at 5890 and 5896 Å, corresponding to the transitions $2p\ ^2P_{3/2} \rightarrow 2s\ ^2S_{1/2}$ and $2p\ ^2P_{1/2} \rightarrow 2s\ ^2S_{1/2}$. (*d*) The first excited state of mercury, showing the effect of large spin–orbit interaction in heavy atoms.

theory in any general way, and is often treated as an experimental parameter. Equation 5.24 then represents the fact that the interaction of two magnetic dipoles must involve their scalar product, in this case $\boldsymbol{\mu}_l \cdot \boldsymbol{\mu}_s$, which is proportional to $\mathbf{L} \cdot \mathbf{S}$. Leaving ζ as a parameter is often useful when one wishes to make spectral assignments or predict magnetic properties.

The spin–orbit energy can be put in more convenient form in terms of the total angular momentum **J**, which is the vector sum of **L** and **S**:

$$\mathbf{J} \equiv \mathbf{L} + \mathbf{S}. \tag{5.25}$$

Since

$$\mathbf{J}^2 \equiv \mathbf{J} \cdot \mathbf{J} = \mathbf{L}^2 + \mathbf{S}^2 + 2\mathbf{L} \cdot \mathbf{S}, \tag{5.26}$$

substitution in Eq. 5.24 gives

$$E_{so} = \frac{\zeta}{2}(\mathbf{J}^2 - \mathbf{L}^2 - \mathbf{S}^2). \tag{5.27}$$

All three of these squares of angular momenta are capable of being measured simultaneously. Hence it is no surprise that $\mathbf{J}^2$ for an atom is a constant of the motion, with eigenvalues given by

$$\mathbf{J}^2 = J(J+1)\hbar^2; \tag{5.28}$$

J is sometimes called the *inner quantum number.* Like S, J is integral for an even number of electrons, half-integral for an odd number. The z component of **J** is

$$J_z = M_J \hbar \qquad (M_J = J, J-1, \ldots, -J), \tag{5.29}$$

so that for each J there are $2J + 1$ possible space-quantized orientations of J, giving Zeeman-effect splitting in an external field.

Just as in a given configuration the $\mathbf{L}_i$ and $\mathbf{S}_i$ can add to give various values of L and S (cf. Fig. 5.8), so in a given term **L** and **S** can add to give several values of J. The maximum value of the z component for given L, S is clearly

$$\begin{aligned}(J_z)_{\max} &= (L_z + S_z)_{\max} \\ &= (M_L + M_S)_{\max}\, \hbar = (L+S)\hbar,\end{aligned} \tag{5.30}$$

or $(M_J)_{\max} = L + S$; since M_J runs up to J, the maximum value of J is also $L + S$. One can similarly show that the minimum possible value of J is $|L - S|$, so that J can have the values $L + S$, $L + S - 1$, . . . , $|L - S|$. For example, in a ^3D term (L = 2, S = 1) we have J = 3, 2, 1. The term thus splits into states of slightly different energy, designated by a right subscript giving the value of J: $^3D_3$, $^3D_2$, $^3D_1$. Note that no such splitting occurs if either L or S is zero, that is, in all S or singlet terms; this is why the inert gases (^{1}**S**), alkalis (^{2}**S**), and alkaline earths (^{1}**S**) have no fine structure in their ground states. In the *LS*-coupling approximation the selection rules (Eq. 5.23) are supplemented by

$$\Delta J = 0, \pm 1 \tag{5.31}$$

(with $J = 0 \rightarrow J = 0$ forbidden); the various values of J in a given pair of initial and final terms give rise to what is called a spectral multiplet, several closely spaced lines. For light atoms the spin–orbit interactions lead to relatively small differences between the energies of states with the same L, S but different J. In the first row of the periodic table (Li to Ne), these energy differences are at most comparable to mean thermal energies at room temperature (ca. 0.04 eV/molecule). However, the interaction energy increases rapidly with nuclear charge, roughly as Z^4/n^3. The halogen atoms offer a clear illustration of this effect. The ground configuration of a halogen atom is $\cdots np^5$, giving only a ^2P term ($L = 1$, $S = \frac{1}{2}$). This term may have either of two values of J, $\frac{1}{2}$ or $\frac{3}{2}$. The energy separations, $E(^2P_{1/2}) - E(^2P_{3/2})$, for fluorine, chlorine, bromine, and iodine are 0.05, 0.11, 0.46, and 0.94 eV, respectively. This means that both states are effectively part of the ground state in the fluorine atom at ordinary temperatures, but that the higher-energy $^2P_{1/2}$ states must really be considered as excited states for the other halogens. The upper state of iodine was used to store the energy in the first chemical laser, with the radiation due to the transition $^2P_{1/2} \rightarrow {}^2P_{3/2}$. Some other typical spin–orbit splittings are shown in Fig. 5.11;[25] note that the effect occurs even in the hydrogen atom, breaking up the degeneracy of states with the same n.

The *LS*-coupling approximation assumes that all states of the same term have virtually the same energy, regardless of J, with the spin–orbit interaction only a minor perturbation. This clearly breaks down in the heavier atoms, where an approximation known as *jj coupling* is more useful. In this scheme the $\mathbf{L}_i$ and $\mathbf{S}_i$ of each electron are first coupled to form a $\mathbf{J}_i$ (with quantum numbers j_i, m_{ji}), and the $\mathbf{J}_i$ are then coupled to form $\mathbf{J}$ for the atom. Each set of j_i then gives a "j_i term," a group of closely spaced levels with the Coulomb-exchange interaction producing the fine structure. The selection rules for L and S break down and are replaced by $\Delta j = 0, \pm 1$ (for an electron). In the previous section we pointed out how the electronic interactions spoil the quantum numbers m_l and m_s, since each electron's angular momenta no longer have constant orientations. What the spin–orbit interaction does is to make L and S no longer good atomic quantum numbers. Once $\mathbf{L}$ and $\mathbf{S}$ interact with each other, neither is separately conserved. In heavy atoms, the selection rule of Eq. 5.31 is the only one obeyed; violations of the rules of Eq. 5.23 are quite common. One example is the familiar bluish light of a mercury street lamp; the transition responsible for emission of this intense light would be forbidden by the rules of the *LS*-coupling scheme.

Strictly speaking, things are even worse than this. Once the electrons interact at all, the energy and total angular momentum of each electron cannot be conserved, which means that n and l are not really good quantum numbers either. The assignment of atoms to specific configurations (not to mention terms) is thus only an approximation, although for most purposes a very good one. In the most accurate calculations including electron correlation one must take the actual atomic wave function to be a linear combination ("mixture") of the wave functions corresponding to various configurations; this is an extension of the method we described earlier for the wave functions of individual terms. In such a rigorous approach one cannot meaningfully assign any quantum numbers or energies to individual electrons.

Why do we emphasize these limitations on our assumptions when one can get quite good results with the electron configuration model? Unfortunately, there are important quantities for which these results are not good enough, in particular those quantities that depend on small differences between very large numbers. The total electronic binding energy in a large atom may be some thousands of electron volts, but the energy *change* when two such atoms join to form a molecule—the bond energy—is typically of the order of 5 eV. The best Hartree–Fock calculations for atoms are no better than 0.1–0.2% in the energy. In molecules one cannot do even that well, since there is not spherical symmetry and interatomic forces spoil the configuration model still more. It is thus clear that one must use different techniques to obtain bond energies and other quantities of chemical interest for molecules. In the next chapter we shall begin to see how this is done.

The discussion thus far has neglected one other kind of angular momentum and magnetic moment that affects the energy levels of atoms and molecules, albeit on a scale of energies smaller than the spin–orbit and spin–spin interactions of electrons. This is the angular momentum of the nucleus, which produces a magnetic field of order 1/2000 of the field of the electron spin but whose effects are quite detectable in high-resolution spectra. The splittings of energy levels associated with the magnetic electron-nuclear interactions are called *hyperfine* interactions because of the small size of the effects. They have proved to be useful probes of how the electron penetrates the nucleus.

[25] With sufficient resolution one can often observe even finer splittings, known as *hyperfine structure;* the difference between hyperfine levels is usually of the order of a few tenths of a cm^{-1}($\Delta E/hc$). There are two principal causes: (1) each isotope of an element has a different nuclear mass. Thus the reduced masses of the electrons differ slightly among the isotopes, as do their energy levels; cf. Eq. 4.16 and footnote 5 of Chapter 4 for hydrogenlike atoms. (2) The spin magnetic moment of the nucleus interacts with the electron magnetic moments, with an effect similar to that of spin–orbit splitting. In the hydrogen atom both spins are $\frac{1}{2}$, giving rise to two states; the transition between them produces the famous 21 cm line used by radio astronomers to detect hydrogen.

APPENDIX 5A

The Stern–Gerlach Experiment

Among the experiments that illustrate the nature of quantization, perhaps the most vivid is the one first performed by O. Stern and W. Gerlach in 1921. This experiment clearly demonstrated that angular momenta (and the associated magnetic moments) are quantized, and that space quantization makes possible the physical separation of atoms or molecules in different quantum states. It also furnished the evidence from which the existence of electron spin was later deduced. The experiment could be done with atoms having either spin or orbital angular momentum; in fact, Stern and Gerlach worked with silver atoms, which have a spin of $\frac{1}{2}$ (the same as the electron) and zero orbital angular momentum.

In essence, the Stern–Gerlach experiment consists of passing a beam of atoms down a collimating axis and into an *inhomogeneous* magnetic field, whose field strength is greater on one side of the beam than on the other. Such a system is shown in Fig. 5A.1. A uniform field would merely tend to align the atomic magnets; the inhomogeneous field also exerts a net translational force on the atoms. The field as shown in the figure is stronger toward the north pole of the magnet. Any particle entering the region of the field with its own north pole toward the north pole of the field is thus repelled more by the north–north repulsion than by the south–south repulsion, and is driven toward the south pole of the field. Similarly, any particle entering with its own south pole toward the field's north pole is driven toward the north pole of the field. The extent of this deflection depends on the angle between the magnetic moment and the field, and on the field gradient.

If the atomic magnets were classical particles, their moments could initially be directed at any angle to the field. One would then observe a continuous and symmetrical distribution of deflected atoms leaving the region of the inhomogeneous field; some would be deflected toward the field's north pole, some toward the south pole, and some would be undeflected. But what is actually observed is one group of atoms all deflected the same amount toward the north pole, and an equal number deflected the same amount toward the south pole. The distribution is discontinuous. The reason, as we now know, is that the spin moments are not randomly oriented. If the applied field vector defines the z axis, then for $S = \frac{1}{2}$ the z component of $\boldsymbol{\mu}_s$ can only have the values $+\mu_B$ and $-\mu_B$. Atoms in each state are deflected by a characteristic fixed amount, and two distinct beams leave the region of the field. Each of these beams contains only one of the spin states.

Note that the experiment as described assumes no special preparation of the entering beam. One can prepare the beam in ways that affect the relative intensities of the two emerging beams, but one cannot increase the number of emerging beams[26] or produce a continuous distribution of emerging particles. One way of preparing the beam, however, is to select for the entering beam *one* of the beams emerging from an identical apparatus having the same spatial orientation. This choice amounts to selecting only those atoms in a particular quantum state, all with the same value of M_S, and passing only those into the second magnet. Since these atoms all have magnetic moments with the same orientation to the field (which again is directed along the z axis), only a single beam will emerge from the second apparatus. Figure 5A.2*a* illustrates this phenomenon.

The ability to separate particles in different quantum states has made the Stern–Gerlach experiment particularly important in providing a conceptual basis for quantum mechanics. With some kind of trap to remove the other beam

[26] For a given kind of atoms, that is. In general there will be one beam for each possible value of $M_J = M_L + M_S$; for the silver atom in its ground state we have $L = 0$ (S state), $M_L = 0$, $M_J = M_S = \pm\frac{1}{2}$.

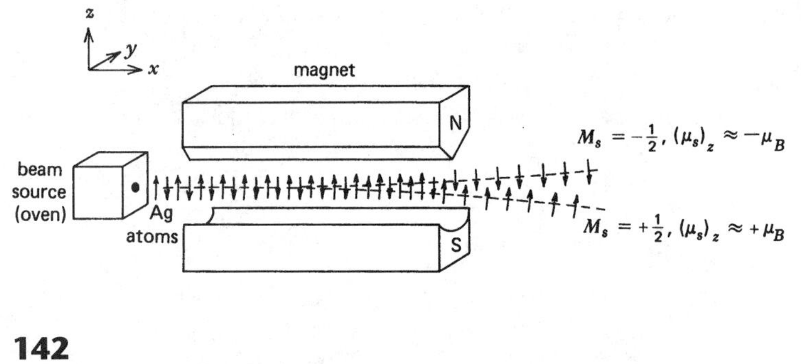

Figure 5A.1 Schematic diagram of the Stern–Gerlach apparatus. The magnetic field is stronger near the pointed north pole than near the smoothly curved south pole. The "atomic magnets" emerging from the oven are driven downward or upward according to whether the z components of the magnetic moments point up or down (as indicated by arrows). The two emerging beams are in distinct quantum states.

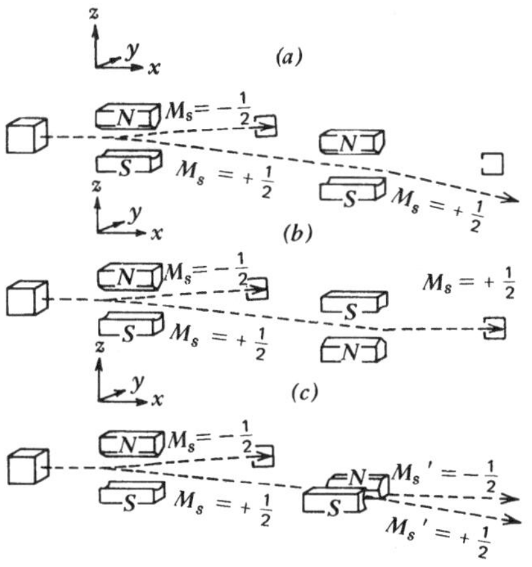

Figure 5A.2 Effects of successive Stern–Gerlach experiments. (*a*) Both apparatuses set to pass only atoms with $M_S = +\frac{1}{2}$: The first "filter" removes half of the original beam, the second "filter" passes all of the residual beam. (*b*) The first apparatus passes only $M_S = +\frac{1}{2}$, the second (with magnetic field reversed) passes only $M_S = -\frac{1}{2}$; none of the beam passes through the second "filter." (*c*) First apparatus selects for z component of spin, passing the beam with $S_z = +\frac{1}{2}\hbar$ (and S_y indeterminate); second apparatus selects for y component of spin, splitting that beam into two beams with $S_y = \pm\frac{1}{2}\hbar$ (and S_z indeterminate).

(or beams), the apparatus is like a filter that passes only those particles in one precisely known quantum state. Once selected, the particles remain in that state so long as they are undisturbed. A second filter that passes the same quantum state leaves the beam unaffected. On the other hand, suppose that the second filter is set to pass a different quantum state—for example, that the first filter passes only the state with $M_S = +\frac{1}{2}$ and the second only the state with $M_S = -\frac{1}{2}$ (both defined with respect to the same z axis). If the filters are so adjusted, then *none* of the beam atoms will get through the second filter. This is shown schematically in Fig. 5A.2*b*. The two possible quantum states, $M_S = +\frac{1}{2}$ and $M_S = -\frac{1}{2}$, are mutually exclusive; an atom in one state cannot be in the other. All this illustrates the distinct identity of quantum mechanical eigenstates, in this case eigenstates of the z component of angular momentum.

The Stern–Gerlach experiment yields a second kind of insight into the quantum nature of matter when one uses a second selector in another way. In our discussion so far, we have spoken about testing for the quantum state transmitted by the first filter, or for another of the quantum states distinguished by the apparatus. As we have described the experiment, these are states with different quantized values of the z component of spin angular momentum, $S_z = M_S\hbar$. The act of measuring S_z by applying an inhomogeneous magnetic field in the z direction forces the atoms to be in one or another of the eigenstates of S_z. But suppose that one uses a second apparatus to test for quantization relative to the x or y axis, perpendicular to the z axis defined by the first apparatus. Recall from our discussion of the quantum properties of angular momentum that one *cannot* simultaneously know the components of angular momentum along two axes. To do so would violate the uncertainty principle. The knowledge that a particular beam emerging from the first apparatus has $M_S = +\frac{1}{2}$ (and thus $S_z = \frac{1}{2}\hbar$) with respect to the z axis is essentially complete knowledge. One might know less, but one can never have any more knowledge than this, about the orientation of the vector **S**.

Very well, suppose that the second apparatus is arranged to separate atoms according to their y components of angular momentum (in the same coordinate system), as shown in Fig. 5A.2*c*. This measurement must force the component S_y to take on one of its eigenvalues; relative to a field in the y direction, S_y has the possible values $M_S'\hbar$, where $M_S' = \pm\frac{1}{2}$, since Eq. 5.2 applies to the component of S in *any one* direction. But now what happens when one puts the beam with $S_z = \frac{1}{2}\hbar$ into this apparatus? For these atoms one has the maximum possible knowledge about the z component of spin angular momentum; one must therefore have the minimum possible knowledge about the x and y components. For a system that can assume only two states, a condition of minimum knowledge can mean only that the two states are equally probable. This in turn means that the beam with $S_z = \hbar/2$ ($M_S = \frac{1}{2}$) must split into two beams of equal intensity, corresponding to $S_y = \hbar/2$ ($M_S' = \frac{1}{2}$) and $S_y = -\hbar/2$ ($M_S' = -\frac{1}{2}$). But what is the value of S_z for these new beams? We cannot know this. The beam entering the second apparatus has a definite value of S_z, and thus is indeterminate in S_y; the outgoing beams have definite values of S_y and are indeterminate in S_z. We have traded information about one kind of quantization for information about another kind of quantization, and the two kinds are mutually exclusive.

The uncertainty principle is particularly well illustrated by the experiment just described. A "complete" description of the system, within the constraints of this principle, consists of the specification of eigenvalues for as many variables as nature allows to be measured simultaneously. The classical notion, that one can simultaneously determine each and every property with unlimited accuracy, is simply invalid. Similar results should be obtained in any other experiment, real or conceptual, in which particles in different quantum states are physically separated.

• FURTHER READING

Bethe, H. A., and Salpeter, E. E., *Quantum Mechanics of One- and Two-Electron Atoms* (Springer-Verlag, Berlin and Academic Press, New York, 1957).

Condon, E. U., and Odabasi, H., *Atomic Structure* (Cambridge University Press, Cambridge, England, 1980).

Condon, E. U., and Shortley, G. H., *The Theory of Atomic Spectra* (Cambridge University Press, Cambridge, England, 1953).

Davydov, A. S., *Quantum Mechanics* (Pergamon Press, Oxford, England and Addison-Wesley Publishing Co., Reading, Mass.), Chapter X.

Hartree, D. R., *The Calculation of Atomic Structures* (John Wiley and Sons, Inc., New York, 1957).
Karplus, M., and Porter, R. N., *Atoms and Molecules* (W. A. Benjamin, Inc., Menlo Park, Calif., 1970), Chapter 4.
Kauzmann, W., *Quantum Chemistry* (Academic Press, Inc., New York, 1957), Chapters 9 and 10.
Kondratyev, V., *The Structure of Atoms and Molecules* (P. Noordhoof N. V., Groningen, The Netherlands, 1964), Chapters 5–8.
March, N. H., *Self-Consistent Fields in Atoms* (Pergamon Press, Oxford and New York, 1975).
Morrison, M. A., Estle, T. L., and Lane, N. F., *Quantum States of Atoms, Molecules and Solids* (Prentice-Hall, Inc., Englewood Cliffs, N. J., 1976), Chapters 5–10.
Schiff, L. I., *Quantum Mechanics,* 3d Ed. (McGraw-Hill Book Co., Inc., New York, 1968), Chapter 12.

• PROBLEMS

1. A magnetic dipole with strength 1.2 Bohr magnetons is in a uniform magnetic field whose strength is 500 oersted. What is the difference in energy between the orientations of the dipole parallel (north–to–north) and antiparallel (north–to–south) to the applied magnetic field? Suppose the dipole has the low-energy orientation; the application of an oscillatory magnetic field at the appropriate frequency can induce absorption of energy and the eventual transition to the high-energy orientation. Sketch the system, indicating the dipole, the external field and the oscillatory field, at three specified times during a single cycle. Be careful to show the *directions* of all vectors. What is the resonant frequency of the dipole in the given field?

2. Using a classical expression analogous to Eq. 5.4, compute the magnetic moment of a 500-g sphere of metal on a 1-m string, whirling at a rate of 1 revolution per second, and carrying an excess negative charge of 10^{-12} $\mathcal{F}$. (One faraday, $\mathcal{F}$, is 1 mol of electrons, or 96,485 C.) What is the energy of interaction of this magnetic dipole with the earth's magnetic dipole if the two are parallel?

3. Compute the frequency of the electromagnetic radiation associated with a transition of an electron between the state with $m_s = +\frac{1}{2}$ and the state with $m_s = -\frac{1}{2}$, if the electron is in a field of 0.5 T (5000 gauss). Compute the corresponding frequency if the particle is a proton instead of an electron.

4. A Stern–Gerlach experiment is conducted with silver atoms passing through a magnetic field whose gradient is 3×10^2 T/m. The length of the region of the field gradient is 0.05 m. The particles travel 0.5 m to their target surface after leaving the field with initial speed 10^5 m/s. How far apart are the points where the two emerging beams strike the target? (Recall the analysis of Thomson's experiment to determine *e*/*m*.)

5. Sodium atoms are put through a Stern–Gerlach apparatus and separated into two beams of equal intensity as shown in Fig. 5A.2*a*. The beam corresponding to $M_S = +\frac{1}{2}$ is then put through a second pair of magnets whose orientation is rotated about the beam axis through an angle θ with respect to the first pair of magnets. Figures 5A.2*b* and 5A.2*c* correspond to special cases with $\theta = 180°$ and $\theta = 90°$, respectively. Calculate the relative intensities of the beams with $M_S' = +\frac{1}{2}$ and $M_S' = -\frac{1}{2}$ for arbitrary θ. Ignore the small nonzero angle at which the beam emerges from the first separating device.

6. Explain, in terms of the symmetry of the forces exerted on an electron, why *l* and *m* are good quantum numbers for each electron in the central field model, but not in real atoms.

7. The most probable value of each electron–nuclear distance in H^- is 1.178 bohr; the average value of the electron–nuclear distance is 2.707 bohr. What qualitative inferences can one make about the shape of the electron probability density from a comparison of these two numbers?

8. Within a series of isoelectronic ions such as Li, Be^+, B^{2+}, C^{3+}, and so on, the relative contribution of the correlation energy to the total energy of the system diminishes with increasing nuclear charge. Explain.

9. Prove that $\frac{27}{16}$ is the value of the "effective nuclear charge" Z' of Eq. 5.13 that minimizes the expectation value E' of energy (Eq. 5.11) when $\psi^{(2)}(r_1,r_2)$ of Eq. 5.13 is used to evaluate E'.

10. Put the entries in the following sets of configurations in order of increasing energy. Then check your intuition by comparing the results with the experimental results from atomic spectroscopy, as given, for example, by C. E. Moore, *Atomic Energy Levels* (National Bureau of Standards Circular 467).

(a) Li, $1s^22s$
$1s2s^2$
$1s2s2p$
$1s^22p$

(b) C, $1s^22s^22p^2$
$1s^22s2p^3$
$1s2s^22p^3$
$1s^22s^22p3s$

11. Using the tables given by C. E. Moore in *Atomic Energy Levels* (National Bureau of Standards Circular 467),

compute the energies of the following atomic configurations by taking the appropriately weighted averages of the atomic state energies in the tables. (The configuration energies can be given relative to the ground state of the atom.)
(a) C, $1s^22s^22p^2$;
(b) N, $1s^22s^22p^3$;
(c) N, $1s^22s^22p^23s$.

12. Explain why the second ionization potential of lithium, for the process $Li^+ \rightarrow Li^{2+} + e^-$, is less than that predicted for a hydrogenlike atom with $Z = 3$.

13. Consider the effect of the nuclear charge on the importance of relativistic effects in atoms, first by estimating from a Bohr model the velocity of a $1s$ electron in an atom of lead. The effective mass of a particle with rest mass m and speed v is $m' = m/\sqrt{1-(v^2/c^2)}$. What is v/c for a $1s$ electron of Pb? What is its effective mass, based on the simple expression above for m'? What velocity v' does one obtain if one carries out the Bohr calculation with the above expression for m', instead of a velocity-independent m?

14. The outer electron of an alkali atom may be treated in an approximate way, as if it were in a hydrogenic orbital. Suppose that one takes the quantum number for the outer electron to be 2, 3, 4, 5, and 6, respectively, for Li, Na, K, Rb, and Cs. What values must Z be given to account for the observed first ionization potentials of these atoms? Explain why they differ from unity.

15. Use the Pauli exclusion principle and Hund's rules to find the number of unpaired electrons and the term of lowest energy for the following atoms:
(a) P
(b) S
(c) Ca
(d) Br
(e) Fe

16. Hydrogen atoms are placed in a strong magnetic field and excited; their emission spectrum is recorded under conditions of moderately high resolution (0.1 cm^{-1} or better). What is the appearance of the portion of the spectrum due to $H(3p) \rightarrow H(2s)$ transitions, and how does it differ from the spectrum of the same transition in the absence of a magnetic field?

17. Show that the lowest configuration of the nitrogen atom, $1s^22s^22p^3$, gives rise to the terms ^2D, ^2P, and ^4S.

18. Derive the terms of the configuration $1s^22s^22p^63s^23p3d$, of the silicon atom.

19. Explain this observation: The energy difference between $1s^22s$, $^2S_{1/2}$ and $1s^22p$, $^2P_{1/2}$ states of Li is 14904 cm^{-1}, whereas that between the $2s$, $^2S_{1/2}$ and $2p^2$, $P_{1/2}$ states of Li^{2+} is only 2.4 cm^{-1}.

20. The first excited state (and higher excited states as well) of He exhibits a positive electron affinity, although the electron affinity of He in its ground state is negative.
(a) What are the configuration and term designations of the first two excited states of the helium atom?
(b) What is the configuration and term designation of the lowest-energy bound state of the negative ion He^-? Note that this cannot be the same as any state built by adding an electron to He in its ground state.
(c) Interpret why atoms in excited configurations generally have positive electron affinities, whereas atoms in closed-shell ground-state configurations seem to have only negative electron affinities.

21. What are the lowest terms for the ions Ti^{2+}, Mn^{2+}, and Fe^{3+}, if the first two electrons to be lost by the corresponding neutrals are the two $4s$ electrons? What are the possible values of the quantum number J for each of these ions?

22. What J values are possible for the ^3P, ^1D, and ^1S states of the silicon atom? Look up the observed splittings of these states in C. E. Moore, *Atomic Energy Levels* (National Bureau of Standards Circular 467) and estimate the spin–orbit splitting parameter ζ for as many of these states as the data permit.

23. Given the following data:

	Li	Na	K	Rb	Cs	Fr
Z	3	11	19	37	55	87
Atomic weight	6.94	23.00	39.10	85.47	132.91	?
Density (g/cm³)	0.53	0.97	0.86	1.53	1.90	?
Atomic volume (cm³)	13.0	23.7	45.5	55.9	70.0	?
Melting point (°C)	186	97.5	62.3	38.5	26	?
Boiling point (°C)	1200	880	760	700	670	?

Predict the numerical values for the atomic weight, density, atomic volume, melting and boiling points of francium, element number 87. What will be its characteristic chemical properties?

24. Show from Eqs. 5.26 and 5.27 that the energy associated with spin–orbit interaction is

$$E_{so} = \tfrac{1}{2}\zeta\hbar^2[J(J+1) - L(L+1) - S(S+1)].$$

25. (Difficult) Substitute $\psi^{(1)}$ of Eq. 5.12 and $\psi^{(2)}$ of Eq. 5.13 into the expression for the expectation value of the energy, Eq. 5.11, and show that the corresponding values given in the text, –74.83 eV and –77.49 eV, respectively, are correct.

26. Electronic structure calculations can be performed using readily available software packages. (One such package, available free of charge from Iowa State University, is GAMESS, Generalized Atomic and Molecular Electronic Structure System—see http://www.msg.ameslab.gov/GAMESS/GAMESS.html.) A popular calculational method is the Hartree–Fock Self-Consistent Field (HF–SCF) calculation with each atomic orbital being represented as a linear combination of Gaussian basis functions of the form e^{-r2}. A variety of basis sets are available.

(a) In a minimal basis STO–*N*G calculation, each atomic orbital is represented by one Slater type orbital of the form e^{-r}, which is itself composed of *N* Gaussians.

(b) In an *m*–*np*G basis set calculation, each inner shell atomic orbital is represented as one linear combination of *m* Gaussians. Each valence shell atomic orbital is represented by two basis functions, a linear combination of *n* Gaussians and a linear combination of *p* Gaussians.

Compute the Hartree–Fock ground electronic state energy of atomic hydrogen using the following basis sets: (a) STO–3G; (b) 3–21G; (c) 6–31G. Note for hydrogen that there are no inner shell orbitals. Compare the computed orbital energy to the exact ionization energy of the hydrogen atom in its electronic ground state.

27. Compute the Hartree–Fock energy of the ground electronic state of the He atom using the (a) STO–3G, (b) 3–21G, and (c) 6–31G basis sets and the HF–SCF method. Compare these calculated energy values to the experimental value of –78.99 eV.

28. Repeat Problem 5.27 for the C atom in its ground electronic state, computing the 1–electron orbital energies (the eigenvalues corresponding to the α and β sets of eigenvectors reported by GAMESS). Compare the least-negative eigenvalue with the experimental first ionization energy (Koopman's Theorem).

29. Repeat Problem 5.27 +2 for the O atom in its ground electronic state.

CHAPTER 6

The Chemical Bond in the Simplest Molecules: H_2^+ and H_2

In this chapter we begin to address the problem of the chemical bond. What makes atoms stick together to form a molecule? Why are some bonds strong and others weak? Can we predict the properties of a molecule—the energy levels, the spectra resulting from transitions among them, the spatial distribution of nuclei and electrons? And how are these properties related to those of the separated atoms? These are typical of the questions we must try to answer.

The simplest molecules, of course, are those with only two atoms. Studying these allows us to examine a single chemical bond in isolation. We begin with a general discussion of the forces between two atoms and how they are responsible for bonding. It will become clear that the key to bonding is how the electron distribution changes when the atoms are brought together. In the rest of the chapter we deal with ways to find this electron distribution, in terms of various computational approaches to obtain the wave function.

The present chapter is concerned explicitly with only the two simplest molecules. Despite their simplicity, they are rich enough to illustrate most of the main principles of bonding and some fundamental ideas that underly our description of molecules. The H_2^+ ion, with only one electron, is the simplest possible molecule; we use it to introduce the most widely employed approximation of bonding theory, the concept of the molecular orbital. From H_2^+ we go on to H_2, in which the presence of two electrons gives us the first complete picture of a "normal" bond. We consider these molecules in both ground and excited states, and in terms of several approximate descriptions. In Chapter 7 we shall extend our theory to diatomic molecules in general.

6.1 Bonding Forces between Atoms

What forces hold a molecule together as an entity? Within any molecule there are attractive forces between electrons and nuclei, and repulsive forces between pairs of nuclei and between pairs of electrons. We must examine how these forces reach a delicate balance that enables molecules to have stable and well-defined structures, structures that neither fall apart at the slightest touch nor collapse into a united atom. We must ask how the attractive forces can just balance the repulsive forces for some particular molecular geometry and make that geometry a stable one. If the molecule is stretched, squeezed, or twisted from its most stable shape, then the forces within the molecule will tend to restore it to that shape. Each molecular structure is thus in true stable equilibrium, corresponding to the bottom of a reasonably deep potential energy well.[1]

But in speaking of molecular structures we are already making an important assumption. The term "structure" implies a rather rigid arrangement of the atomic nuclei, close to the equilibrium positions for which the molecule's energy is minimized. The molecule cannot be totally rigid; it would violate the uncertainty principle for any particle to be at rest at a fixed point. However, the assumption of near-rigid structure is very good indeed. Since nuclei have thousands of times the mass of electrons, Δp for a nucleus can be a quite small fraction of its total momentum and still large enough to give a negligible Δq. To put it another way, the light electrons move so much faster than the heavy nuclei that they have a smooth average distribution over a time during which the nuclei hardly move at all. We can thus obtain a good description of the molecule's electronic structure even if we assume the nuclei to be at rest.

This assumption is part of what is known as the *Born–Oppenheimer approximation* (after Max Born and J. Robert Oppenheimer). What is this approximation? Let us approach it by stages. Consider a diatomic molecule composed of nuclei A and B, with charges of $+Z_Ae$ and $+Z_Be$,

[1] One could imagine a different kind of molecule, in which the energy of the molecule would be quite insensitive to the location of the constituent atoms. Such a molecule would be more like a liquid drop than the conventional ball-and-stick structure. Although this is not a good description of most molecules, many properties of atomic nuclei, especially heavy nuclei, are nicely described by a liquid drop model.

respectively, and some electrons. We define the distance between the two nuclei as R (if there were more atoms we would have to specify R_{AB}), and the distances of the ith electron from the two nuclei as r_{Ai} and r_{Bi}. These coordinates are illustrated in Fig. 6.1. The Hamiltonian operator (neglecting all magnetic interactions) is then

$$\mathsf{H}=\left[\frac{\mathbf{p}_A^2}{2m_A}+\frac{\mathbf{p}_B^2}{2m_B}\right]$$
$$+\left[\sum_i \frac{\mathbf{p}_i^2}{2m_e}+\frac{e^2}{4\pi\epsilon_0}\times\left(\frac{Z_A Z_B}{R}-\sum_i \frac{Z_A}{r_{Ai}}-\sum_i \frac{Z_B}{r_{Bi}}+\frac{1}{2}\sum_i\sum_{j\neq i}\frac{1}{r_{ij}}\right)\right]$$
$$=\mathsf{H}_{\text{nucl}}+\mathsf{H}_{\text{elec}}, \qquad (6.1)$$

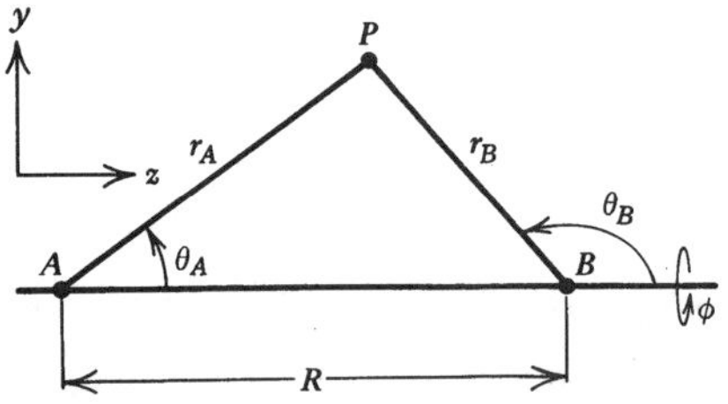

Figure 6.1 Coordinates in a diatomic molecule. The nuclei A and B (charges $+Z_Ae$ and $+Z_Be$) are a distance R apart. Any point P can be described in terms of r_A and r_B, the distances from the two nuclei; θ_A and θ_B, the angles relative to the A–B axis; and an angle ϕ describing rotation about that axis.

where H_{nucl} and H_{elec} are defined by the two sets of square brackets, r_{ij} is as usual the distance between electrons i and j, and the sums are taken over all the electrons; H_{nucl} is the kinetic energy operator for the nuclei, and H_{elec} corresponds to the energy with the nuclei fixed. The molecular wave function ψ is the solution of the equation $\mathsf{H}\psi = E\psi$. Now we make the Born–Oppenheimer approximation, which is simply that the nuclear and electronic motions are separable (cf. Section 3.9), meaning that we can write two Schrödinger equations, one for the electronic energy and wave function, and one for the nuclei. Thus, having collected the kinetic energy of the heavy—and presumably slowly moving—nuclei in the first set of brackets in Eq. 6.1, and all the terms governing the motion of the light, fast electrons in the second set of brackets, we can write

$$\psi(r,R)=\psi_{\text{nucl}}(R)\psi_{\text{elec}}(r,R). \qquad (6.2)$$

Here r stands for all the r_i and r_{ij}. Eq. 6.2 implicitly reveals a key point of the Born–Oppenheimer approximation: The electronic wave function, and therefore the electronic energy, depend on the distance R between the nuclei so that $\psi_{\text{elec}}(r, R)$ must contain R. However $\psi_{\text{elec}}(r, R)$ satisfies the equation

$$\mathsf{H}_{\text{elec}}\psi_{\text{elec}}(r,R)=E(R)\psi_{\text{elec}}(r,R). \qquad (6.3)$$

The Hamiltonian H_{elec} contains R explicitly only in the potential energy of repulsion between the nuclei, but R is implicit in the electron-nuclear attraction terms, $-\sum_i \frac{Z_A}{r_{Ai}}-\sum_i \frac{Z_B}{r_{Bi}}$. The central element of the Born–Oppenheimer approximation is the way R is treated in Eq. 6.3: The internuclear distance is handled as a parameter with a fixed value for each electronic wavefunction $\psi_{\text{elec}}(r, R)$ and each electronic energy we determine. That is, $\psi_{\text{elec}}(r, R)$ is defined as the eigenfunction of the operator H_{elec}, whose eigenvalue $E(R)$ gives the energy of the molecule for a given value of the parameter R. Although ψ_{nucl} must be a function of *all* the nuclear coordinates, only R affects the electronic energy and appears in ψ_{elec}. In other words, $|\psi_{\text{elec}}(r, R)|^2$ and $E(R)$ describe the electron distribution (as a function of r) and energy that the molecule would have if the nuclei were fixed a distance, R, apart. These are the quantities we want to know at this point.

In this chapter we shall address only the solution of the electronic problem expressed in Eq. 6.3, but let us briefly survey the remainder of the problem. To the extent that the Born–Oppenheimer approximation is valid, one can solve Eq. 6.3 to obtain $E(R)$ for any value of R. Combining these values gives a function $E(R)$ from $R = 0$ to $R = \infty$, which for the ground state of a stable diatomic molecule has a form like that in Fig. 6.2. From such a function $E(R)$, we obtain the part of the wave function describing nuclear motion by solving[2]

$$\mathsf{H}_{\text{nucl}}\psi_{\text{nucl}}(R)=[E-E(R)]\psi_{\text{nucl}}(R); \qquad (6.4)$$

in this equation, R is now a quantum-mechanical variable, E is the total energy, $E(R)$ is the effective potential energy in which the nuclei move, and their difference $E - E(R)$ is the kinetic energy of the nuclei associated with changes of R, that is, with their vibration. Since a system in a potential well always has a zero-point energy, the lowest eigenvalue of E must be some distance above the bottom of the well. We shall come back to this problem in Section 7.1.

---

[2] Substituting Eqs. 6.1 and 6.2 into $\mathsf{H}\psi = E\psi$, we have

$$(\mathsf{H}_{\text{nucl}}+\mathsf{H}_{\text{elec}})\psi_{\text{nucl}}(R)\psi_{\text{elec}}(r,R)=E\psi_{\text{nucl}}(R)\psi_{\text{elec}}(r,R)$$

and

$$\psi_{\text{elec}}(r,R)\mathsf{H}_{\text{nucl}}\psi_{\text{nucl}}(R)+\psi_{\text{nucl}}(R)\mathsf{H}_{\text{nucl}}\psi_{\text{elec}}(r,R)+\psi_{\text{nucl}}(R)\mathsf{H}_{\text{elec}}\psi_{\text{elec}}(r,R)=E\psi_{\text{nucl}}(R)\psi_{\text{elec}}(r,R),$$

since H_{nucl} and H_{elec} act only on the nuclear and electronic coordinates, respectively. Substituting Eq. 6.3 in the third term and rearranging gives us

$$\psi_{\text{elec}}(r,R)\mathsf{H}_{\text{nucl}}\psi_{\text{nucl}}(R)+\psi_{\text{nucl}}(R)\mathsf{H}_{\text{nucl}}\psi_{\text{elec}}(r,R)=\psi_{\text{elec}}(r,R)[E-E(R)]\psi_{\text{nucl}}(R).$$

In the second term we find $\mathsf{H}_{\text{nucl}}\,\psi_{\text{elec}}(r, R)$, which contains the usually small interactions between electronic and nuclear motions; the Born–Oppenheimer approximation says that this term can be neglected. If we do this, dividing through by $\psi_{\text{elec}}(r, R)$ yields Eq. 6.4.

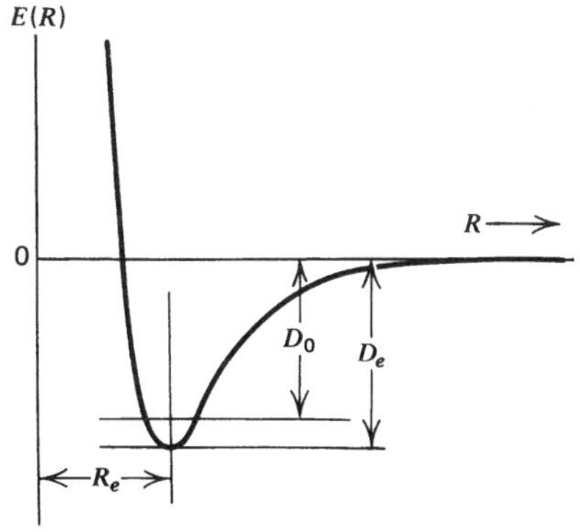

Figure 6.2 Internal energy $E(R)$ of a stable diatomic molecule as a function of the internuclear distance R. The zero of energy is the two separated atoms at rest in the limit $R \to \infty$. The dissociation energy can be defined as D_e, relative to the bottom of the well, at the equilibrium distance R_e, or as D_0, relative to the ground state; $D_e - D_0$ is the zero-point energy of the molecule. In a typical molecule D_e is of the order of 5 eV, and R_e is about 1–2 Å.

When atoms A and B are very far apart, there is essentially no interaction between them; we define the energy of infinitely separated $A + B$ without kinetic energy to be our zero. At the other extreme, when the nuclei are close to each other, the repulsive energy $Z_A Z_B e^2/4\pi\epsilon_0 R$ becomes much larger than all the other terms in Eq. 6.1, and $E(R)$ appears to become infinite as R goes to zero.[3] This repulsion is what keeps molecules from collapsing into single atoms. We are interested in what happens in the region between these limits. A stable molecule can exist only if $E(R)$ has a minimum at some value of R, as in Fig. 6.2. Otherwise the nuclear repulsion would make the atoms fly apart without limit. The depth of this minimum, relative to the energy at infinite separation, is called the *dissociation energy* (D_e) of the molecule, and the value of R at which it occurs is the *equilibrium distance* (R_e). To see how such a minimum can occur, let us examine the balance of forces in a molecule.

In our diatomic molecule with nuclei A and B a distance R apart, the repulsive force between the nuclei is, of course, of magnitude

$$F_{AB} = \frac{Z_A Z_B e^2}{4\pi\epsilon_0 R^2}, \tag{6.5}$$

[3] As earlier discussion indicated, this inference of infinite repulsion is predicated on the assumption that the molecule is not rotating; if it is, then the effective force of centrifugal motion keeps the nuclei apart. However there is another kind of force, largely beyond the scope of this volume, that becomes important when nuclei come close to one another. At *very* small distances—about 10^{-15} m—the strong nuclear forces become dominant. Since these forces are attractive, $E(R)$ goes not to infinity, but to a very high peak bounding a potential well. If we were to push the two nuclei together hard enough, they would cross this barrier and combine into a single nucleus of charge $(Z_A + Z_B)e$; this would be a nuclear *fusion* process, such as occurs in stars and thermonuclear bombs. But since the barrier height is in the MeV range, there is little danger of producing fusion with ordinary chemical energies. As we mentioned in Section 4.1, the emission of α particles by radioactive nuclei corresponds to the reverse process, tunneling through the barrier from inside.

acting directly along the A–B axis. Suppose now that there is a bit of electronic charge, say an amount $-q$, at the arbitrary point P (Fig. 6.1). This charge exerts an attractive force $\mathbf{F}_{AP}$ on nucleus A, in the direction AP, and an attractive force $\mathbf{F}_{BP}$ on nucleus B, in the direction BP. The magnitudes of these forces are

$$F_{AP} = \frac{Z_A eq}{4\pi\epsilon_0 r_A^2} \quad \text{and} \quad F_{BP} = \frac{Z_B eq}{4\pi\epsilon_0 r_B^2}. \tag{6.6}$$

We can define a *bonding* force as one that tends to draw the nuclei together, and an *antibonding* force as one that tends to push them apart. The repulsive force F_{AB} is clearly antibonding; what about the forces exerted by the charge at point P?

Only the force exerted along the A–B axis has any effect on the bonding. We thus break the forces $\mathbf{F}_{AP}$ and $\mathbf{F}_{BP}$ into their components parallel to the A–B axis (our z axis) and perpendicular to it—that is, into the interesting z components and the irrelevant y components. In terms of the coordinates shown in Fig. 6.1, the z components are

$$(\mathbf{F}_{AP})_z = \frac{Z_A eq}{4\pi\epsilon_0 r_A^2}\cos\theta_A \quad \text{and}$$
$$(\mathbf{F}_{BP})_z = \frac{Z_B eq}{4\pi\epsilon_0 r_B^2}\cos\theta_B. \tag{6.7}$$

The bonding force on nucleus A is the force that tends to push it toward B, that is, the component of force in the positive z direction. The bonding force on nucleus B, however, is the force in the *negative* z direction. The net classical bonding force on the nuclei due to the charge $-q$ at point P is therefore the single component of a force vector,

$$\begin{aligned} F_{\text{bonding},P} &= (\mathbf{F}_{AP})_z - (\mathbf{F}_{BP})_z \\ &= \frac{eq}{4\pi\epsilon_0}\left(\frac{Z_A\cos\theta_A}{r_A^2} - \frac{Z_B\cos\theta_B}{r_B^2}\right). \end{aligned} \tag{6.8}$$

The quantum mechanical bonding force operator $\mathsf{F}_{\text{bonding},P}$ looks exactly the same.

To obtain the total bonding force exerted by the electrons, we would have to sum Eq. 6.8 over the entire electron distribution, which we can obtain only by solving the Schrödinger equation. Assume that we have solved Eq. 6.3 to obtain the wave function $\psi_{\text{elec}}(r, R)$ for a given R. The expectation value of the total bonding force is then

$$\begin{aligned} \langle \mathsf{F}_{\text{bonding}}(R)\rangle &= \langle \mathsf{F}_{\text{bonding},P}\rangle - F_{AB} \\ &= \int \psi^*_{\text{elec}} \mathsf{F}_{\text{bonding},P} \psi_{\text{elec}}\, d\tau - \frac{Z_A Z_B}{4\pi\epsilon_0 R} \end{aligned} \tag{6.9}$$

with the integral taken over all possible positions of all the electrons. (The symbol "$d\tau$" is shorthand for what we earlier called "$dq_1 \cdots dq_N$," and is meant to indicate three coordinates for each electron.)

One does not have to evaluate the integral in Eq. 6.9 explicitly to obtain the total bonding force. The operator corresponding to $F_{\text{bonding}}(R)$ is $\partial \mathsf{V}(R, r)/\partial R$, where $\mathsf{V}(R, r)$ is the potential energy of H_{elec}. We thus have

$$\begin{aligned}\langle \mathsf{F}_{\text{bonding}}(R)\rangle &= \left\langle \frac{\partial \mathsf{V}(R,r)}{\partial R}\right\rangle \\ &= \int \psi^*_{\text{elec}} \frac{\partial \mathsf{V}(R,r)}{\partial R} \psi_{\text{elec}}\, d\tau \\ &\quad + \int \frac{\partial \psi^*_{\text{elec}}}{\partial R} E(R)\psi_{\text{elec}}\, d\tau \\ &\quad + \int \psi^* E(R) \frac{\partial \psi_{\text{elec}}}{\partial R}\, d\tau. \end{aligned} \tag{6.10}$$

We shall now see that

$$\begin{aligned}\int \psi^*_{\text{elec}} \frac{\partial \mathsf{H}_{\text{elec}}}{\partial R} \psi_{\text{elec}}\, d\tau &= \frac{\partial}{\partial R}\int \psi^*_{\text{elec}} \mathsf{H}_{\text{elec}} \psi_{\text{elec}}\, d\tau \\ &= \frac{dE(R)}{dR}, \end{aligned} \tag{6.11}$$

which leads to the result that the effective bonding force is the slope of the electronic energy $E(R)$:

$$\langle \mathsf{F}_{\text{bonding}}(R)\rangle = \frac{dE(R)}{dR}, \tag{6.12}$$

as follows. (Note that because the bonding force is defined as a force that *reduces R,* Eq. 6.12 lacks the usual negative sign connecting force with the slope of the potential energy.) We have

$$\begin{aligned}\frac{\partial}{\partial R}\int \psi^*_{\text{elec}} \mathsf{H}_{\text{elec}} \psi_{\text{elec}}\, d\tau &= \int \frac{\partial \psi^*_{\text{elec}}}{\partial R} \mathsf{H}_{\text{elec}} \psi_{\text{elec}}\, d\tau \\ &\quad + \int \psi^*_{\text{elec}} \mathsf{H}_{\text{elec}} \frac{\partial \psi_{\text{elec}}}{\partial R}\, d\tau \\ &\quad + \int \psi^*_{\text{elec}} \frac{\partial \mathsf{H}_{\text{elec}}}{\partial R} \psi_{\text{elec}}\, d\tau. \end{aligned}$$

Since H_{elec} is a real Hermitian operator (cf. Appendix 6B), we can write

$$\int \psi^*_{\text{elec}} \mathsf{H}_{\text{elec}} \frac{\partial \psi_{\text{elec}}}{\partial R}\, d\tau = \int \frac{\partial \psi_{\text{elec}}}{\partial R} \mathsf{H}_{\text{elec}} \psi^*_{\text{elec}}\, d\tau.$$

Substitution of Eq. 6.3 then removes the first two terms on the right-hand side above, yielding

$$\begin{aligned}&\int \frac{\partial \psi^*_{\text{elec}}}{\partial R} E(R)\psi_{\text{elec}}\, d\tau + \int \frac{\partial \psi_{\text{elec}}}{R} E(R)\psi^*_{\text{elec}}\, d\tau \\ &= E(R)\frac{\partial}{\partial R}\int \psi^*_{\text{elec}}\psi_{\text{elec}}\, d\tau = 0, \end{aligned}$$

since $\int \psi^*_{\text{elec}}\psi_{\text{elec}}\, d\tau$ has the constant value unity. The same substitution gives the last term of Eq. 6.11.

Equation 6.12 is known as the *Hellmann–Feynman theorem.* This confirms our earlier statement that $E(R)$ is the effective potential energy for nuclear motion.

Without solving the Schrödinger equation for $\langle \mathsf{F}_{\text{bonding}}(R)\rangle$, we can get a good deal of insight into the nature of the bonding forces simply by looking at the form of Eq. 6.8. Following a method introduced by T. Berlin and K. Fajans, let us see how $F_{\text{bonding},P}$ varies with the location of the point P. First, whenever P lies between A and B ($z_A < z_P < z_B$), we have $\theta_A < \pi/2$ and $\theta_B > \pi/2$; thus $\cos\theta_A$ is positive, $\cos\theta_B$ is negative, and $F_{\text{bonding},P}$ must be positive. This is a very fundamental (if obvious) point, which we must recognize from the outset: Any electronic charge lying between two nuclei is necessarily bonding because it pulls the two nuclei toward each other. In order for charge to be antibonding, it must lie "beyond" the nuclei.

Now suppose we place our test point P to the right of nucleus B ($z_P > z_B$), so that $\cos\theta_A$ and $\cos\theta_B$ are both positive. A negative charge at P then pulls both nuclei to the right, pulling A toward B but B away from A. Which force wins? For simplicity, let us suppose that point P lies on the A–B axis ($\cos\theta_A = \cos\theta_B = 0$). Suppose that the charges Z_A and Z_B are equal. A charge to the right of B must be closer to B than to A ($r_B < r_A$), so the right-hand term in Eq. 6.8 is larger than the left-hand term, and the net force is antibonding. For equal nuclear charges, then, any electronic charge on the axis to the right of B or the left of A is antibonding. This is not true in every case, however. Suppose, for example, that $Z_A > Z_B$, with point P still on the A–B axis to the right of B. For P close to B, the small value of r_B will outweigh the charge difference ($Z_B/r_B^2 > Z_A/r_A^2$) and $F_{\text{bonding},P}$ will be negative. As we move P to the right, the ratio (r_A/r_B) comes closer and closer to unity. Eventually a point is reached at which the larger charge on nucleus A makes the first (bonding) term in Eq. 6.8 exceed the second (antibonding) term, even though the second term has a smaller denominator; for all points on the axis beyond this the charge is bonding. Similar calculations can be made for points not on the A–B axis.

We thus recognize that regions of space in the vicinity of nuclei are inherently bonding or antibonding; whether or not a bond is formed depends on how the electrons are distributed among these regions. To find the positions and shapes of the bonding and antibonding regions, one need merely locate the surfaces that divide them. These boundary surfaces are obtained by solving Eq. 6.8 for those values of r_A, r_B, θ_A, θ_B at which $F_{\text{bonding},P} = 0$. Although we shall not go through the algebra, it is not difficult to simplify this equation to

$$\alpha(1+\rho^2-2\rho\alpha)^{3/2} = \frac{Z_B}{Z_A}(\rho-\alpha), \tag{6.13}$$

where $\rho \equiv R/r_A$ and $\alpha \equiv \cos\theta_A$; Eq. 6.13 contains only one physical parameter (Z_B/Z_A) and two independent variables

(ρ and α). Illustrations of the boundary surfaces for several values of Z_A/Z_B are given in Fig. 6.3. Note that the larger the ratio Z_A/Z_B, the smaller is the antibonding region near the low-Z nucleus B. In a molecule such as HCl, this region includes only that charge very close behind the H nucleus; because of the large ratio Z_{Cl}/Z_H, a charge almost anywhere except beyond the Cl nucleus is in the bonding region.

How much charge is needed to balance the repulsive force between the two nuclei? Suppose that a negative charge $-q$ is at the midpoint of the bond between A and B,

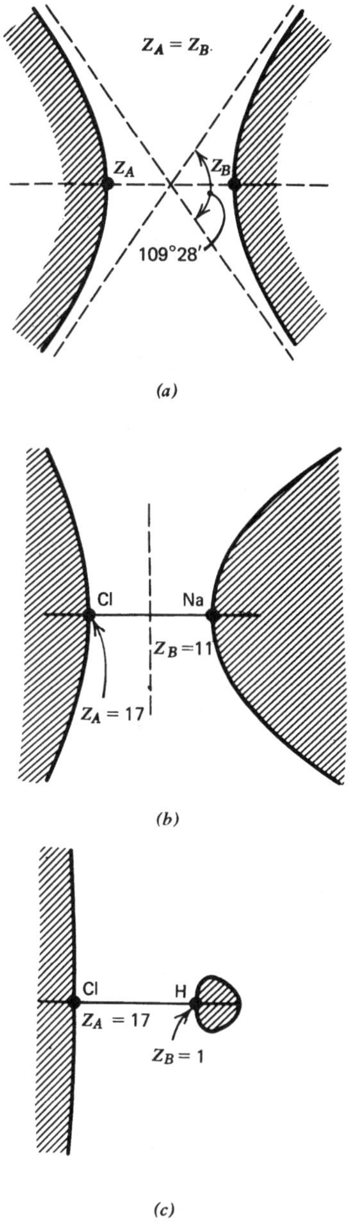

Figure 6.3 Bonding (unshaded) and antibonding (shaded) regions of space for diatomic molecules with various values of Z_A/Z_B. (*a*) Homonuclear molecule (for example, H_2, Cl_2), $Z_A = Z_B$. (*b*) NaCl molecule, $Z_A/Z_B = 17/11 = 1.545$. (*c*) HCl molecule, $Z_A/Z_B = 17$. From Hirschfelder, Curtiss, and Bird, *Molecular Theory of Gases and Liquids* (Wiley, New York, 1954), p. 936.

so that $r_A = r_B = R/2$, $\cos\theta_A = 1$, $\cos\theta_B = -1$. How large must this bit of charge be for the binding force it exerts to equal the repulsive force F_{AB}? Combining Eqs. 6.5 and 6.8, we require that

$$\frac{Z_A eq}{(R/2)^2} + \frac{Z_B eq}{(R/2)^2} = \frac{Z_A Z_B e^2}{R^2}, \tag{6.14}$$

or with a bit of rearrangement,

$$q = \frac{(Z_A + Z_B)e}{4 Z_A Z_B}. \tag{6.15}$$

If the two charges are equal, $Z_A = Z_B = Z$, then q need be only $-Ze/8$ to balance the nuclear repulsion. If Z_A is larger than Z_B, then q must be somewhat larger, but even in the limit $Z_A/Z_B \to \infty$ it need not exceed $Z_Be/4$. From this rough model we see that the amount of negative charge in the bonding region required to stabilize a molecule is very small, at least when the charge lies on the bond axis. The bonding effectiveness of a given charge of course decreases as it moves away from the axis.

Thus, we have a rationalization of how electrons draw nuclei together and counter the mutual repulsion of the nuclei of a diatomic molecule. But closer examination shows that we only have part of the picture. Thus far, the discussion has neglected the electronic kinetic energy, and surely, when an electron with a well-defined total energy passes into a region of low potential energy, its kinetic energy must increase. This implies that if the potential energy of the electron in the region between the nuclei is lowered by the approach of the two nuclei, then the electron must spend less time there. And if that happens, how can we expect charge to accumulate between the nuclei?

This question was examined in detail by Ruedenberg and coworkers, and the results have been discussed by Mulliken and Ermler.[4] The picture that emerges is this: Because the potential between the nuclei shows a maximum, and is relatively flat in that region, the force on an electron there is low and the computed local electronic kinetic energy T_e is low near the midpoint of the internuclear axis. Near the nuclei, however, the electronic kinetic energy is increased above that of the free atom, and this increase is greater than its decrease between the nuclei, so the average, $\langle T_e \rangle$, is greater for electrons in the diatomic molecule than in the separated atoms. The total energy of a bound system with only Coulombic forces is related to the average kinetic energy and to the average potential energy: $E = -\langle T \rangle = \frac{1}{2}\langle V \rangle$ (the virial theorem as applied to a Coulombic system), and if the

---

[4] K. Ruedenberg, in O. Chalvet, ed., *Localization and Relocalization in Quantum Chemistry*, Vol. 1, (Reidel, Dordrecht, 1975), pp. 223–245; K. Ruedenberg, *Rev. Mod. Phys.* **34**, 326 (1962); M. J. Feinberg, K. Ruedenberg, and E. L. Mehler, *Adv. Quantum Chem.* **5**, 27 (1970); R. S. Mulliken and W. C. Ermler, *Diatomic Molecules, Results of ab initio Calculations* (Academic, New York, 1977), pp. 38–43.

diatomic molecule is at its equilibrium internuclear distance, $\langle T \rangle = \langle T_e \rangle$. Therefore, because $\langle T_e \rangle_{\text{molecule}} > \langle T_e \rangle_{\text{free atoms}}$, $E(R_e) < E(\infty)$. In short, the simple model based on the statics of electrostatic forces hides the delicate balance between potential and kinetic energy. Without the electrostatic forces of attraction between the nuclei and the electronic charge there would be no chemical bond, but the *effective* potential, the internal energy $E_{\text{elec}}(R)$, takes its form—slow drop with decreasing R to a minimum at R_e and a sharp rise for $R < R_e$—from the increasing kinetic energy of the electrons, as much as from their decreased potential.

Anticipating our discussion of vibrations in Chapter 7, we recognize that if q exceeds the value given in Eq. 6.15, then the net force between the nuclei is attractive, and if q is less than that value, the net force is repulsive. In effect, this is just how the electron distribution generates the chemical bond: when the distance R between two atoms that bond is moderately large, then the electron distribution has a small excess, above the value in Eq. 6.15, between those nuclei, generating an effective attractive force. When R is small, the electron distribution between the nuclei depletes (as a consequence of the exclusion principle), the nuclei become somewhat unshielded and repel each other more than the electrons attract them toward each other, giving a net repulsion.

We are now in a position to ask the following questions: First, under what conditions does negative charge distribute itself so as to create a stable bond? Second, is there a meaningful way to subdivide the electron density so that we can speak of bonding and antibonding electrons (or orbitals), rather than bonding and antibonding regions of space? And third, can one calculate the bonding forces with sufficient detail and accuracy to make reliable predictions of bond energies and molecular geometries?

The first of these three questions is really a rather large one. To know how the charge density in a molecule is distributed, we must presumably know the electronic wave function of the molecule. This means that we must solve the Schrödinger equation for the molecule at many internuclear distances, a task appreciably more formidable than that for an atom. Nevertheless, in strongly bound molecules the problem can be solved with reasonable accuracy by an approach similar to the one used for atoms, the method of one-electron orbitals. This method, in so far as it is valid, suffices to answer our second question in the affirmative. If each electron has its own wave function, we can indeed classify electrons and speak meaningfully of individual electrons as being responsible for particular bonds. The orbital method has its limitations, even with regard to electronic energies and charge distributions. In particular, it is not sufficiently accurate to describe weakly bound molecules in which the bonding forces are due in large part to correlation between the electrons. To the third question, on the possibility of predicting molecular properties, we can only answer, "Sometimes." Given a wave function, one can make such calculations; their accuracy depends on how good the wave function is. For very simple molecules, those with no more than three or four atoms, the theoretical calculations reach or exceed the accuracy of experimental measurements. For slightly larger molecules, calculations can be accurate enough to give useful assistance in the interpretation of complex experimental results. For quite large molecules, the prediction from theory of such things as bond angles and lengths still lies at the limits of our technical ability. However we rarely need to know these with the same accuracy that we demand for small molecules. Moreover we can rely on the very powerful generalization that bond lengths and bond angles among a given kind of atoms remain about the same in extremely large classes of compounds. This principle of *transferability* underlies a vast range of generalizations used throughout chemistry and molecular biology.

We have seen how small a charge is needed to produce a net bonding force; it should not be surprising that the delicately balanced attractive and repulsive forces must be known with considerable accuracy before one can say how the balance actually works out in any particular case. With this overview of the problem, we are now prepared to examine its solution in the simplest possible molecule.

6.2 The Simplest Molecule: The Hydrogen Molecule-Ion, H_2^+

Naturally, the simplest molecule is the one that contains the smallest number of particles. There must be two nuclei if we are to have a molecule at all, the simplest nuclei are protons, and there must be at least one electron to provide a bonding force. The species made from these constituents is called the *hydrogen molecule-ion,* H_2^+. It is a real species, known to exist in electric discharges; it is readily detected in a mass spectrometer. The spectrum of H_2^+ has been observed, and something is known of its chemistry. It is a fragile species, reacting almost any time it collides with a molecule of H_2 to produce the molecule-ion H_3^+ and a neutral H atom. Nevertheless, a great deal is known about H_2^+ because it is simple enough to be analyzed by reliable theoretical methods. We shall examine the H_2^+ molecule-ion in several ways, successively more precise and quantitative, each giving a slightly different viewpoint on how such a molecule can exist. In the present section we make a preliminary survey, before getting down to the business of finding the wave function.

In studying a molecule, a good starting point is to consider its properties when the atoms are very far apart, the limit at which we define the zero of energy. In the case of H_2^+, then, we begin with a hydrogen atom in its $1s$ ground state and a bare proton far away from it. Let us consider what happens when the distance R between the two protons is decreased.[5]

[5] This may sound as if we are talking about a dynamic (time-dependent) process. Actually, although we are considering a sequence of values of R, in each case we use the Born–Oppenheimer approximation and treat the nuclei as motionless. What we want is the solution to the time-independent Eq. 6.3 for each value of R.

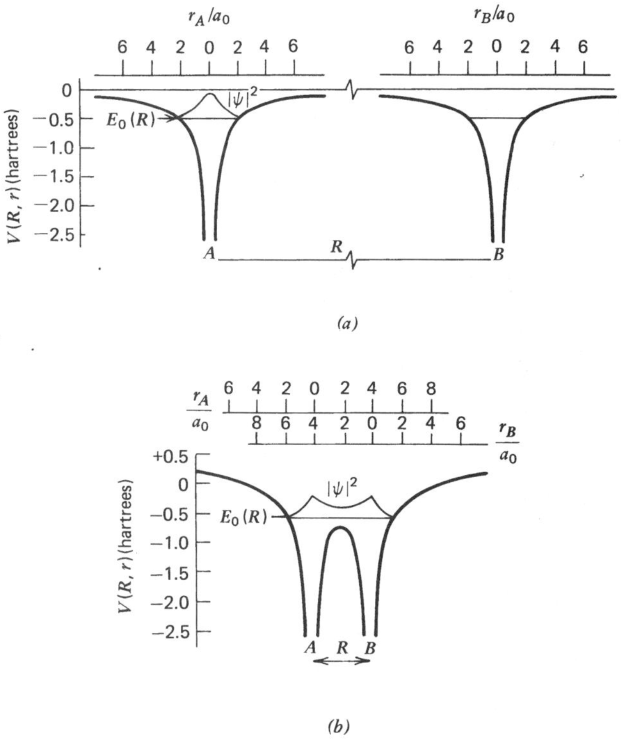

Figure 6.4 Potential energy of H_2^+,

$$V(R,r)=\frac{e^2}{4\pi\epsilon_0}\left(\frac{1}{R}-\frac{1}{r_A}-\frac{1}{r_B}\right),$$

along the internuclear axis. (*a*) H + H$^+$, nuclei far apart, electron localized around one nucleus. (*b*) H_2^+, $R = 4a_0$, electron equally likely to be around either nucleus. In each case the ground-state energy $E_0(R)$ and corresponding $|\psi|^2$ (arbitrary units) are shown; in part (*a*) these differ only slightly from those of a 1 *s* H atom. Note that $V(R,r)$ and $E(R)$ include the internuclear repulsion, which in 6.4(*b*) equals +0.25 hartree; this is the limit of $V(R,r)$ as r_A, $r_B \to \infty$.

When the two are very far apart, the energy $E(R)$ must be essentially independent of R. As the free proton approaches the atom, however, the repulsive force between the two protons and the attractive force between the electron and the free proton must begin to make themselves apparent. The electron will be more likely to be found on the side of the atom toward the free proton—in quantum mechanical terms, its wave function will be distorted from the symmetric 1*s* function to one with a greater amplitude in this direction. The centers of positive and negative charge no longer coincide, and we say that the atom is *polarized* by the free proton. The attractive force is greater than the repulsive force (since the electron is on the average closer to the free proton), and a net bonding force appears. According to Eq. 6.12, $dE(R)/dR$ is then positive, and $E(R)$ must be negative, since we have defined $E(\infty) = 0$. One can show (see Section 10.1) that $E(R) \propto -1/R^4$ in this fairly long-range region.

In the long-range limit of very large R, it is still legitimate to speak of the H atom and the H$^+$ ion as two distinct entities. Consider Fig. 6.4*a*, which shows the potential energy along a line through the two nuclei: There are two potential energy wells, separated by a high, very wide barrier. Although the probability of tunneling from one well to the other is not quite zero, at large R (say, over 20 Å) it is low enough to be neglected for practical purposes. We can thus treat a state with the electron localized near one nucleus as a stationary (time-independent) state of the system. The electron's potential energy and wave function differ only slightly from those in an isolated hydrogen atom.

Now suppose that the nuclei are fairly close together. As shown in Fig. 6.4*b*, there is no longer an effective barrier between the two potential wells. The electron is thus free to travel anywhere in the combined well, and is as likely to be found near one nucleus as the other in any given observation. In such a situation the two protons are completely equivalent with regard to their interaction with the electron. More than that, they are *indistinguishable.* This is a very important concept. Two macroscopic charged spheres might also exert identical forces, but we could distinguish them from one another in various ways—say, by painting them different colors. But there is no way to do this with protons. Two protons—or two electrons, or two of any fundamental particle—are absolutely indistinguishable by any kind of measurement. In our H_2^+ system, then, neither we nor the electron can distinguish proton A from proton B. Even if we prepare the H_2^+ molecule by bringing together atom A and free proton B, there is no experiment that can distinguish A from B once they are together. If we then draw the two apart again, there is no way to tell if the electron has ended up around the same nucleus as at the start.[6] Like the simultaneous position and momentum of a particle, these things are simply unknowable.

What does this indistinguishability imply about the wave function of H_2^+? We may "label" the nuclei A and B for convenience, but any physically measurable quantity must not be affected when the labels are interchanged. Since $|\psi|^2$, the probability density of finding an electron at a given point, is a measurable quantity, the value of $|\psi|^2$ at $r_A = a$, $r_B = b$ must be the same as at the corresponding point $r_A = b$, $r_B = a$. This can be summarized by writing

$$|\psi(A,B)|^2 = |\psi(B,A)|^2. \qquad (6.16)$$

For Eq. 6.16 to be valid, $|\psi|^2$ must be symmetric with respect to a plane equidistant from the nuclei, the plane normal to the bond at its midpoint.

[6] Unless the atoms passed each other so fast that their de Broglie wavelengths were much less than the shortest distance between them (though even then there is a small chance of what is called *exchange scattering*). Here we see the fundamental difference between classical and quantum mechanics. If classical mechanics were valid, one could distinguish the protons by following the trajectory of each particle with as much precision as required. But the uncertainty principle makes this impossible.

We can say a bit more about the form of ψ. Given the Born–Oppenheimer approximation, the electronic wave function must satisfy Eq. 6.3, which for H_2^+ becomes

$$\mathsf{H}\psi = \left[\frac{\mathbf{p}^2}{2m} + \frac{e^2}{4\pi\epsilon_0}\left(\frac{1}{R} - \frac{1}{r_A} - \frac{1}{r_B}\right)\right]\psi = E(R)\psi. \quad (6.17)$$

(From here on, H and ψ will stand for H_{elec} and ψ_{elec}, unless we indicate otherwise.) The potential energy goes to $-\infty$ at two points, the two nuclei. Near each nucleus it diverges in exactly the same way as does $V(r)$ in the hydrogen atom. We can thus carry over the reasoning of Section 4.3 to find the behavior of ψ in the vicinity of the nuclei. In particular, the ground-state wave function at each nucleus must decay exponentially from a cusp, like the corresponding hydrogen 1*s* function. The ground state $|\psi|^2$ thus has two equal peaks, one centered on each nucleus, as shown in Fig. 6.4*b*. However, since $\int|\psi|^2\,d\tau$ over the molecule must equal unity, each of these peaks must be only about half the size of the peak in a hydrogen atom. Note that most of the electron density in Fig. 6.4*b* appears to lie in the bonding region between the nuclei. We shall see that this is indeed the case, and that the net bonding force is sufficient for H_2^+ to be a stable molecule in its ground state.

Now let us look again at the long-range case. Since the arguments of the previous paragraph still apply, should we not expect to find two equal peaks even at large R? The answer is yes, as long as the two nuclei are still indistinguishable. If all we know about the system is that an electron is somewhere in the field of two protons, the probability density $|\psi|^2$ must be symmetric. But ordinarily we do know more than this. In Fig. 6.4*a* the presumption is that we have prepared the system in such a way that the electron is known to be around nucleus *A*—so that the system is H(*A*) + $H^+(B)$, rather than just H_2^+. This amounts to our imposing an additional condition that $r_A \ll r_B$. One can obtain an approximate solution of Eq. 6.17 with such a condition, giving a wave function ψ' like that in Fig. 6.4*a*. This solution is only as good as the approximation on which it is based, and obviously cannot be valid over the whole of space. Still, for large R it is at least as good as many of our other assumptions. One cannot expect the approximate ψ' to be valid at small R, however; it becomes a poor description when the potential barrier is so low that the electron is not likely to stay around one nucleus for the duration of a measurement.

Another way to look at the problem, then, is in time-dependent terms. Suppose that we have the Hamiltonian of Eq. 6.17, with fixed nuclei, but that we no longer try to assume a stationary state. We must solve the time-dependent Schrödinger equation, $\mathsf{H}\psi = -i\hbar(\partial\psi/\partial t)$. One possible solution is a wave function $\psi(r, t)$ that at $t = 0$ is almost entirely located around nucleus *A*. The function ψ' of the previous paragraph will be a good approximation to $\psi(r, t)$ in the region where its magnitude is significant. Both functions must have cusps at nucleus *B*, but the amplitudes of these cusps at *B* are far too small to be observed. Now suppose that we leave the system alone for a long time. Since tunneling from *A* to *B* has a nonzero (if very low) probability, the likelihood of the electron's being found at *B* will steadily increase. Eventually the system will approach a stationary state in which the electron is equally likely to be found on *A* or *B*, and the wave function will have two equal peaks. For large R, however, "eventually" may easily mean many times the age of the universe. On the other hand, when R is a few angstroms, it may take only 10^{-16} s (about a Bohr period) for the wave function to equalize itself, and the time-dependent process can be ignored. The distinction between our two cases is thus simply one of time scale: A stationary state is one that does not change measurably over the time in which we are interested.

All this raises another interesting question. Our three particles continue to interact, even if they are a meter or a mile apart. But what about *other* particles? Must the wave function of each electron somehow contain information on the positions of all the other particles in the universe? Stated in these grandiose terms, the question may be considered a philosophical one, but it symbolizes a real problem. As far as we know, all particles do interact with one another, no matter how far apart they are. If the principles of quantum mechanics are universally valid—and there is no reason to think otherwise—then strictly speaking the wave function of the universe (or as much of it as we care to consider) is not separable. It is only an approximation to speak of a wave function for any particular piece of matter; how good an approximation this is depends on how weak the interactions are. For a molecule in a gas the approximation is excellent; for an atom in a molecule it is rather poor, unless the atoms are very far apart; for an individual electron in an atom it is moderately good (the orbital approach, which we shall see is also useful in molecules). In general, one makes successive approximations of this sort until the interactions left out are too small to worry about.

6.3 H_2^+: Molecular Orbitals and the LCAO Approximation

We can now proceed to find the electronic wave function of the H_2^+ molecule. This wave function is what we call a *molecular orbital* (abbreviated MO): Just as an atomic orbital describes a single electron in an atom, a molecular orbital describes a single electron in a molecule. In H_2^+ the molecular orbital is identical with the true molecular wave function, but in many-electron molecules the orbital approach is again only an approximation.

As we have already seen, in describing molecules we have an important added complication that we did not face in atoms, the dependence of the wave function on the internuclear distance. We again start with the long-range case, where the problem is relatively simple; however, we wish to obtain a wave function that will join smoothly with the short-range solution. This means that we want the stationary-state solution to the time-independent Eq. 6.17 with indistinguishable nuclei. Even our long-range solution, then, must satisfy Eq. 6.16 and have equal amplitudes around the two nuclei: It must describe a stationary state of H_2^+, not of $H + H^+$. This is not a state that one would ordinarily observe, but its properties can be easily described.

Suppose, then, that we look at the system in the limit of infinitely large R. In the vicinity of nucleus A, the Schrödinger equation 6.17 then reduces to

$$\mathsf{H}\psi = \mathsf{H}_A\psi = \left(\frac{\mathbf{p}^2}{2m} - \frac{e^2}{4\pi\epsilon_0 r_A}\right)\psi = E(\infty)\psi. \qquad (6.18)$$

Since this is of the same form as the hydrogen atom wave equation, it will be satisfied by any function of the form $c_A\varphi_A$, where c_A is a constant and φ_A is a normalized hydrogen wave function centered at A. In other words, as $R \to \infty$, the wave function near A becomes identical to that in a hydrogen atom, except for a constant multiplier. By the same reasoning, the wave function near nucleus B must have the form $c_B\varphi_B$, with φ_B a normalized hydrogen wave function centered at B. If A and B are infinitely far apart, the regions "near A" and "near B" are the only places where ψ will differ significantly from zero; anywhere that φ_A has a reasonable amplitude, φ_B will be negligibly small, and vice versa, since each decreases exponentially with distance from its nucleus. We can thus describe the wave function everywhere in space by the equation

$$\psi = c_A\varphi_A + c_B\varphi_B. \qquad (6.19)$$

What this says is that the matter wave described by ψ is constructed by superposition (adding together the amplitudes) of two waves corresponding to states of atomic hydrogen. We say that the molecular orbital ψ is formed by *linear combination of atomic orbitals,* abbreviated LCAO. The atomic orbitals are what are called *basis functions.*

What can we say about the constants c_A and c_B? In accordance with Eq. 6.16, $|\psi|^2$ must be unaffected by interchange of the nuclei. This means that φ_A and φ_B must be the same function (e.g., both $1s$ functions), and that we must have $|c_A|^2 = |c_B|^2 \equiv |c|^2$. If we restrict the constants to real values, there are only two possible combinations, $c_A = c_B$ and $c_A = -c_B$, and ψ must have the form $c(\varphi_A \pm \varphi_B)$. The constant c is readily evaluated by normalizing the wave function, that is, setting the integral of $|\psi|^2$ over all space equal to unity:

$$\begin{aligned}\int|\psi|^2\,d\tau = 1 &= c^2\int|\varphi_A \pm \varphi_B|^2\,d\tau \\ &= c^2\left(\int|\varphi_A|^2\,d\tau \pm \int\varphi_A^*\varphi_B\,d\tau \right.\\ &\qquad \left.\pm\int\varphi_A\varphi_B^*\,d\tau + |\varphi_B|^2\,d\tau\right).\end{aligned} \qquad (6.20)$$

In the limit $R \to \infty$, either φ_A or φ_B is negligibly small at every point in space; we can thus write

$$\int\varphi_A^*\varphi_B\,d\tau = \int\varphi_A\varphi_B^*\,d\tau = 0, \qquad (6.21)$$

since the integrands are everywhere effectively zero. We defined φ_A and φ_B as normalized wave functions, so we have

$$\int|\varphi_A|^2\,d\tau = \int|\varphi_B|^2\,d\tau = 1. \qquad (6.22)$$

Substituting these results into Eq. 6.20, we find that

$$1 = c^2(1+0+0+1) = 2c^2 \quad \text{or} \quad |c| = \frac{1}{\sqrt{2}}. \qquad (6.23)$$

Thus $|\psi|^2$ around each nucleus is half as much as in a lone hydrogen atom, in agreement with our premise that the electron is equally likely to be around either nucleus.

What we have said thus far applies to any state of the H_2^+ molecule. The only restriction we have placed on φ_A and φ_B is that they must describe the same state of the hydrogen atom. The ground state of the molecule will of course be one in which φ_A and φ_B are $1s$ functions. The ground-state wave function at $R = \infty$ must thus be either

$$\psi_0(R=\infty) = \frac{\varphi_A(1s)+\varphi_B(1s)}{\sqrt{2}} \quad \text{or}$$
$$\psi_1(R=\infty) = \frac{\varphi_A(1s)-\varphi_B(1s)}{\sqrt{2}}. \qquad (6.24)$$

These two functions are in fact degenerate, with the energy

$$\begin{aligned}E_0(\infty) &= \int_{\text{all space}}\psi^*\mathsf{H}\psi\,d\tau \\ &= \int_{\text{near }A}\frac{\varphi_A^*(1s)}{\sqrt{2}}\mathsf{H}_A\frac{\varphi_A(1s)}{\sqrt{2}}\,d\tau \\ &\quad + \int_{\text{near }B}\frac{\varphi_B^*(1s)}{\sqrt{2}}\mathsf{H}_B\frac{\varphi_B(1s)}{\sqrt{2}}\,d\tau \\ &= \frac{1}{2}E_H(1s) + \frac{1}{2}E_H(1s) = E_H(1s),\end{aligned} \qquad (6.25)$$

where H_A, H_B are defined as in Eq. 6.18 and $E_H(1s)$ is the energy of a $1s$ hydrogen atom; just as in Eq. 6.20, the cross terms vanish. The ground-state energy of H_2^+ at $R = \infty$ is thus the same as that of $H(1s) + H^+$ in the same limit, the energy we defined to be zero for the molecule. Similarly, each higher-energy atomic wave function $\varphi(nl)$ gives rise to a degenerate pair of molecular wave functions with the energy $E_H(nl)$.

Thus, we have a complete solution—both wave functions and energies—for the H_2^+ molecule at infinite R. This is all very well, but what does it tell us about the real molecule at finite R? The answer is that it gives us a starting point for an approximate solution. For large but finite R, the wave function near each atom is very much like a hydrogen orbital, so we write the molecular wave function in the form of Eq. 6.19, which is exactly correct only in the limit $R \to \infty$. At finite R, however, it seems plausible that the wave function is still *something* like a sum of atomic orbitals. Suppose, then, that we retain the LCAO approach and assume the molecular wave function to be still given by Eq. 6.19, but with constants that vary with R. Since $|\psi|^2$ must be symmetric, we continue to have $|c_A|^2 = |c_B|^2$. The two functions formed from $1s$ orbitals are thus

$$\psi_0 = c_0[\varphi_A(1s)+\varphi_B(1s)] \tag{6.26}$$

and

$$\psi_1 = c_1[\varphi_A(1s)-\varphi_B(1s)], \tag{6.27}$$

corresponding to Eqs. 6.24. Unlike Eqs. 6.24, however, these are only *approximate* wave functions; later we shall see how good the approximation is.

The LCAO approximation is by far the most widely used method of constructing molecular orbitals. It has the advantage of combining maximum simplicity with reasonable accuracy. The use of atomic orbitals as a basis not only has a natural and intuitive appeal, but also greatly simplifies the calculations that must be made. This is because the atomic orbitals have already been computed with considerable accuracy, even for quite complex atoms; they merely have to be looked up and substituted into the appropriate equations. More fundamental is the fact that the formation of chemical bonds between atoms constitutes only a small perturbation of the electronic structures of those atoms, so the wave functions around atoms in molecules are very much like those around free atoms. After all, atoms generally retain their individual identities when they are constituents of molecules. The obvious exceptions are the atoms that become ions when they go into molecules, atoms such as Na and Cl when they make rocksalt, NaCl. We shall see that the LCAO method also yields an important insight into how chemical bonds are (or are not) formed, emphasizing the way in which constructive or destructive interference develops between the atomic wave functions. But remember that it is only an approximation. If one calculates molecular orbitals as accurately as possible, one finds that, even near nuclei, the functions are not really quite like atomic orbitals. In complex molecules, particularly in regions midway between nuclei, the molecular orbitals responsible for chemical bonds simply do not look like superpositions such as Eq. 6.19. Nevertheless, the LCAO method is nearly always adequate for an approximate understanding of chemical bonding. After all, the origin of the chemical bond lies in the fact that an electron can be in a region where it simultaneously attracts two nuclei closer together. Crudely, it is like an atomic electron held simultaneously in the field of each nucleus, and this is just what the LCAO approximation describes.

Analysis of the LCAO wave functions can give us a simple way of looking at how chemical bonds are formed. Consider the form of Eq. 6.26. Since s functions have no nodes, both $\varphi_A(1s)$ and $\varphi_B(1s)$ are everywhere positive. The sum of two positive numbers is a larger positive number. This means that the function ψ_0 corresponds to *constructive interference* of the two atomic wave functions. In any molecular orbital of this type the electron density is proportional to

$$\begin{aligned}|\varphi_A+\varphi_B|^2 &= |\varphi_A|^2+|\varphi_B|^2+\varphi_A^*\varphi_B+\varphi_B^*\varphi_A \\ &= |\varphi_A|^2+|\varphi_B|^2+2\,\mathrm{Re}(\varphi_A^*\varphi_B)\end{aligned} \tag{6.28}$$

("Re" ≡ real part of a complex quantity). For $1s$ functions, $\varphi_A^*\varphi_B$ is a positive real number, and Eq. 6.28 is everywhere greater than $|\varphi_A|^2 + |\varphi_B|^2$ (which would give the electron density if there were no interaction between the atoms). The increase is greatest where φ_A and φ_B both have appreciable magnitude; this occurs mainly in the region between the nuclei. Thus a higher fraction of the total electronic charge is found between the nuclei than would be there if the atoms did not interact. Since the molecular wave function is normalized ($\int|\psi|^2\,d\tau = 1$), this effect must be balanced by a relatively lower electron density in the region beyond the nuclei; in that region the increase in $|\varphi_A + \varphi_B|^2$ is outweighed by the decrease in the normalization constant c (which we shall evaluate later).

In contrast, let us now look at Eq. 6.27. Here the wave function ψ_1 is the difference of two positive numbers, and corresponds to *destructive interference* of the atomic orbitals. The electron density is proportional to

$$|\varphi_A+\varphi_B|^2 = |\varphi_A|^2+|\varphi_B|^2-2\,\mathrm{Re}(\varphi_A*\varphi_B), \tag{6.29}$$

which is everywhere less than $|\varphi_A|^2 + |\varphi_B|^2$; again the effect is greatest between the nuclei. Thus the electron density is decreased between the nuclei and increased beyond the nuclei, relative to the density in noninteracting atoms. In fact, since φ_A and φ_B are identical, they exactly cancel each other at all points where $r_A = r_B$; this means that the function ψ_1 has a nodal plane midway between the nuclei, a surface on which the value of ψ_1 is zero.

What is the significance of all this? Recall the bonding and antibonding regions of Fig. 6.3. We have shown that the wave function ψ_0 has a relatively high electron density in the bonding region between the nuclei, whereas ψ_1 has excess electron density in the antibonding region beyond the nuclei. It is thus clear that ψ_0 describes a "more bonding" state than ψ_1, that on the average the bonding force is greater in the state ψ_0. The Hellmann–Feynman theorem, Eq. 6.12, thus requires that $dE(R)/dR$ be greater for ψ_0 than for ψ_1. The two states have the same $E(\infty)$, Eq. 6.25, so ψ_0 must have a lower energy than ψ_1 for all finite values of R: $E_0(R) < E_1(R)$. We can reach the same conclusion by considering the potential energy curve of Fig. 6.4*b*. Since the electron in state ψ_1 is more likely to be found in the high-V regions beyond the nuclei, its total energy is higher in this state. We said earlier that one of the states ψ_0 and ψ_1 (which were degenerate at $R = \infty$) must be the ground state of the H_2^+ molecule. Now we have established that ψ_0 is the true ground state (or rather an approximation to it), whereas ψ_1 is an excited state of higher energy.

Note that we still have not proved that H_2^+ is a stable molecule. Although ψ_0 describes the state in which the bonding force is greatest, we do not yet know if that bonding force is sufficient to overcome the internuclear repulsion—if there is *enough* excess charge in the bonding region. To put it another way, we know that ψ_0 corresponds to the lowest $E(R)$ curve; but does that curve have the minimum necessary for a stable molecule (cf. Fig. 6.2)? To answer these questions, one must actually solve the Schrödinger equation for $E(R)$.

6.4 H_2^+: Obtaining the Energy Curve

First we must evaluate the normalization constants in our wave functions. It is convenient to define the integral

$$S_{AB} \equiv \int \varphi_A^* \varphi_B \, d\tau, \tag{6.30}$$

called the *overlap integral* of the orbitals φ_A and φ_B. We pointed out earlier that this integral vanishes when A and B are infinitely far apart. The principal contribution to the integral comes only from the regions where φ_A and φ_B are both appreciable, that is, the regions that dominate the bonding. Note that $S_{AB} = S_{BA}^*$ for any φ_A, φ_B.

We can normalize the ground-state wave function ψ_0, defined by Eq. 6.26. Integrating over all space, we have

$$\begin{aligned}\int |\psi_0|^2 \, d\tau = 1 = c_0^2\Big(\int |\varphi_A|^2 \, d\tau + \int |\varphi_B|^2 \, d\tau \\ + \int \varphi_A^* \varphi_B \, d\tau + \int \varphi_A \varphi_B^* \, d\tau\Big) \\ = c_0^2 (1 + 1 + S_{AB} + S_{AB}^*) \\ = 2c_0^2 (1 + S_{AB}).\end{aligned} \tag{6.31}$$

The first two terms equal unity because we choose φ_A, φ_B to be normalized. We have $S_{AB} = S_{AB}^*$ in this case because $1s$ orbitals are real functions, so that S_{AB} is also real. Solving for the constant c_0 gives

$$c_0 = [2(1 + S_{AB})]^{-1/2}, \tag{6.32}$$

and our ground-state molecular orbital is

$$\psi_0 = \frac{\varphi_A(1s) + \varphi_B(1s)}{[2(1 + S_{AB})]^{1/2}}. \tag{6.33}$$

If we carry out a similar calculation with Eq. 6.27, we obtain

$$\psi_1 = \frac{\varphi_A(1s) - \varphi_B(1s)}{[2(1 - S_{AB})]^{1/2}} \tag{6.34}$$

for the first-excited-state orbital.

Let us compare these results with our earlier discussion of bonding. In a hypothetical "nonbonded" molecule with no interaction between the atomic orbitals, the electron density would be simply $\frac{1}{2}(|\varphi_A|^2 + |\varphi_B|^2)$, as in the long-range case. By contrast, in the ground state, Eqs. 6.28 and 6.33 give

$$|\psi_0|^2 = \frac{|\varphi_A|^2 + |\varphi_B|^2 + 2\varphi_A\varphi_B}{2(1 + S_{AB})}, \tag{6.35}$$

since $\mathrm{Re}(\varphi_A^* \varphi_B) = \varphi_A\varphi_B$ for $1s$ orbitals. In this case S_{AB} is also necessarily positive, and the effect of division by $1 + S_{AB}$ is to lower the electron density everywhere. Counteracting this, however, is the positive term $2\varphi_A\varphi_B$ in the numerator, which tends to increase the electron density. The latter effect is dominant between the nuclei, where $|\psi_0|^2$ is higher than the nonbonded value. This reinforces our earlier conclusion that ψ_0 corresponds to excess electron density in the bonding region. We need not repeat the argument for ψ_1. These conclusions are illustrated by Fig. 6.5, which compares the actual electron densities.

Given the normalized LCAO wave functions, one can determine the corresponding energies $E(R)$. If ψ_0 were a true wave function, its energy (for a given R) would be given exactly by $\int \psi_0^* \mathsf{H} \psi_0 \, d\tau$, where H is the electronic Hamiltonian of Eq. 6.17. But in fact our ψ_0 and ψ_1 are only approximations to the true wave functions; we can still carry out the integration, but we must realize that the result will be an approximate energy. The variation principle, Eq. 5.11, tells us that this energy must be higher than the true value; thus if we obtain a minimum in the $E(R)$ curve, we can be sure that there is really an even deeper minimum. Bearing this in mind, let us proceed.

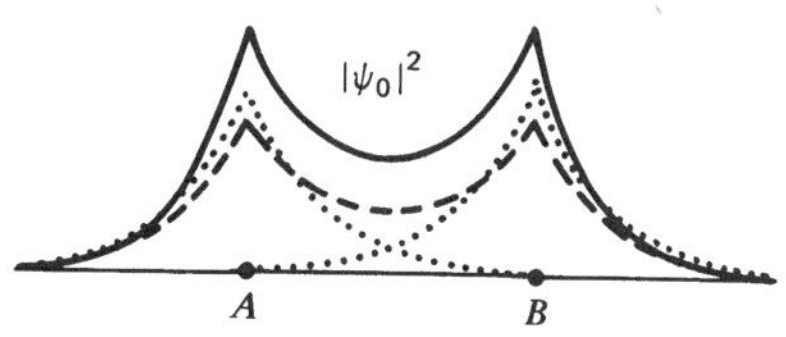

(a)

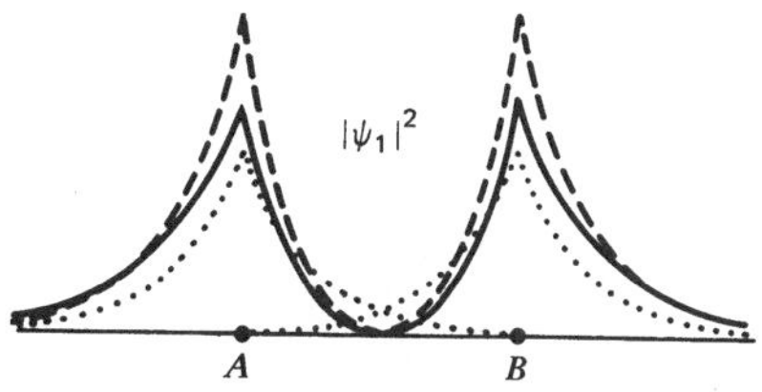

(b)

Figure 6.5 Electron density $|\psi|^2$ along the internuclear axis for the two lowest states of H_2^+. *(a)* The bonding ground state, ψ_0. *(b)* The antibonding excited state, ψ_1. The wave functions are calculated for the true equilibrium distance, $R_e = 2.00a_0$.
. . . . , noninteracting $1s$ atomic orbitals, $|\psi|^2 = (\frac{1}{2}|\varphi_A|^2 + |\varphi_B|^2)$;
----, LCAO approximation, Eqs. 6.33 and 6.34;
——, exact solution.

It is useful to define another kind of integral,

$$H_{ij} \equiv \int \varphi_i^* \mathsf{H} \varphi_j \, d\tau, \tag{6.36}$$

where H is the electronic Hamiltonian and the i, j stand for nuclei; when H is real, we have $H_{ji} = H_{ij}^*$. For a given internuclear distance, the energy corresponding to ψ_0 is

$$\begin{aligned} E_0(R) &= \int \psi_0^* \mathsf{H} \psi_0 \, d\tau \\ &= \frac{\int \varphi_A^* \mathsf{H} \varphi_A \, d\tau + \int \varphi_A^* \mathsf{H} \varphi_B \, d\tau + \int \varphi_B^* \mathsf{H} \varphi_A \, d\tau + \int \varphi_B^* \mathsf{H} \varphi_B \, d\tau}{2(1+S_{AB})} \\ &= \frac{H_{AA} + H_{AB} + H_{BA} + H_{BB}}{2(1+S_{AB})} \\ &= \frac{H_{AA} + H_{AB}}{1+S_{AB}}. \end{aligned} \tag{6.37}$$

Since H is symmetric in A and B, we must have $H_{AA} = H_{BB}$ and $H_{AB} = H_{BA}$. Evaluating the H_{ij}, with H from Eq. 6.17, we have[7]

$$\begin{aligned} H_{AA} &= \int \varphi_A^* \left(\frac{\mathbf{p}^2}{2m} - \frac{e^2}{4\pi\epsilon_0 r_A} \right) \varphi_A \, d\tau + \frac{e^2}{4\pi\epsilon_0 R} \int \varphi_A^* \varphi_A \, d\tau \\ &\quad - \int \frac{e^2}{4\pi\epsilon_0 r_B} \varphi_A^* \varphi_A \, d\tau \\ &= E_{\mathrm{H}}(1s) + \frac{e^2}{4\pi\epsilon_0 R} + J \end{aligned} \tag{6.38}$$

and

$$\begin{aligned} H_{AB} &= \int \varphi_A^* \left(\frac{\mathbf{p}^2}{2m} - \frac{e^2}{4\pi\epsilon_0 r_B} \right) \varphi_B \, d\tau + \frac{e^2}{4\pi\epsilon_0 R} \int \varphi_A^* \varphi_B \, d\tau \\ &\quad - \int \frac{e^2}{4\pi\epsilon_0 r_A} \varphi_A^* \varphi_B \, d\tau \\ &= \int \varphi_A^* E_{\mathrm{H}}(1s) \varphi_B \, d\tau + \frac{e^2}{4\pi\epsilon_0 R} S_{AB} + K \\ &= S_{AB} \left[E_{\mathrm{H}}(1s) + \frac{e^2}{4\pi\epsilon_0 R} \right] + K, \end{aligned} \tag{6.39}$$

where we define the two integrals

$$\begin{aligned} J &\equiv -\int \frac{e^2}{4\pi\epsilon_0 r_B} \varphi_A^* \varphi_A \, d\tau \quad \text{and} \\ K &\equiv -\int \frac{e^2}{4\pi\epsilon_0 r_A} \varphi_A^* \varphi_B \, d\tau; \end{aligned} \tag{6.40}$$

J is called the *Coulomb integral* (it gives the Coulomb interaction between the orbital around one atom and the nucleus of the other atom) and K the *exchange integral*. Substituting in Eq. 6.37, we obtain

$$\begin{aligned} E_0(R) &= \frac{E_{\mathrm{H}}(1s) + e^2/4\pi\epsilon_0 R + J + S_{AB}[E_{\mathrm{H}}(1s) + e^2/4\pi\epsilon_0 R] + K}{1+S_{AB}} \\ &= E_{\mathrm{H}}(1s) + \frac{e^2}{4\pi\epsilon_0 R} + \frac{J+K}{1+S_{AB}}. \end{aligned} \tag{6.41}$$

The first term is the ground-state energy of a hydrogen atom, which we have taken as the energy zero for the H_2^+ molecule, the second term is simply the internuclear repulsion at distance R, and the third term gives the electronic bonding energy. A similar calculation for the first excited state gives

$$\begin{aligned} E_1(R) &= \frac{H_{AA} - H_{AB}}{1 - S_{AB}} \\ &= E_{\mathrm{H}}(1s) + \frac{e^2}{4\pi\epsilon_0 R} + \frac{J-K}{1-S_{AB}}. \end{aligned} \tag{6.42}$$

To obtain the $E(R)$ curves, one must evaluate the integrals S_{AB}, J, K for each value of R and substitute the results in the above equations. This is a straightforward calculation, and we shall not bother with the details.[8] The results for $E_0(R)$ and $E_1(R)$ are plotted in Fig. 6.6, with the terms in Eqs. 6.41 and 6.42 shown separately. In the ground state the electron has a net bonding effect at all values of R, and at moderate distances this is enough to overcome the nuclear repulsion. Thus there is a shallow potential well, and H_2^+ can indeed exist as a stable molecule.[9] As R grows smaller, the electronic bonding energy increases, but only to a finite value; at the same time the nuclear repulsion increases without limit, so that the total energy curve must become repulsive. As for the excited state ψ_1, it is interesting that the electron still has a net bonding effect over most of the range, even though there is less charge in the bonding region. Here, however, the nuclear repulsion is greater for all values of R; thus the $E_1(R)$ curve is everywhere repulsive, and the excited state cannot be stable.

[7] The first terms in Eqs. 6.38 and 6.39 are, respectively, $\int \varphi_A^* \mathsf{H}_A \varphi_A \, d\tau$ and $\int \varphi_A^* \mathsf{H}_B \varphi_B \, d\tau$, where H_A and H_B are hydrogen-atom Hamiltonians. We can thus substitute $\mathsf{H}_A \varphi_A = E_{\mathrm{H}} \varphi_A$ and $\mathsf{H}_B \varphi_B = E_{\mathrm{H}} \varphi_B$ and take the constant E_{H} outside the integrals.

[8] If $\rho \equiv R/a_0$, one obtains

$$\begin{aligned} S_{AB} &= e^{-\rho}\left(1 + \rho - \frac{\rho^2}{3}\right), \\ J &= -\frac{e^2}{4\pi a_0}\left[\frac{1}{\rho} - e^{-2\rho}\left(1 + \frac{1}{\rho}\right)\right], \\ K &= -\frac{e^2}{4\pi a_0} e^{-\rho}(1+\rho). \end{aligned}$$

[9] Stability is relative, of course. The reason one cannot keep H_2^+ around is that it reacts very readily to form some *more* stable species. For example, if one brings H_2^+ into collision with H_2, the system immediately yields $H_3^+ + H$.

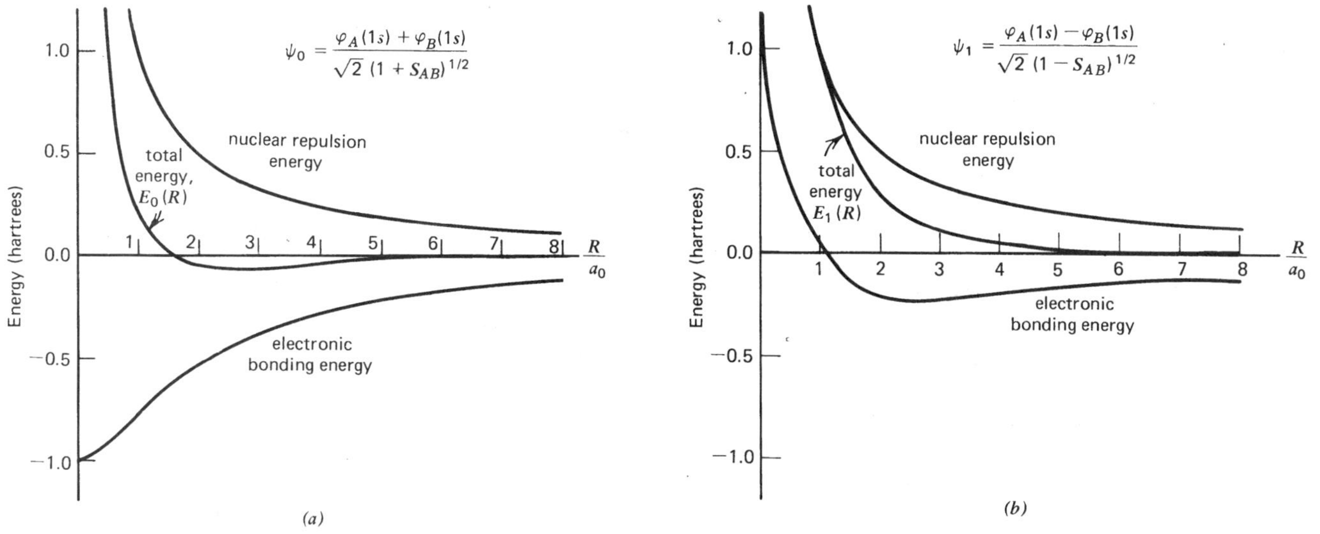

Figure 6.6 Energy of H_2^+ in the LCAO approximation. (*a*) The ground state. (*b*) The first excited state. In each graph we show the nuclear repulsion energy, $e^2/4\pi\epsilon_0 R$; the electronic bonding energy, the last term in Eqs. 6.41 and 6.42; and the total energy $E(R)$, the sum of the previous two. The energy zero is the energy of H(1*s*) + H^+ at infinite separation.

The eigenfunctions ψ_0 and ψ_1 are orthogonal, the functional equivalent to being "at right angles" or, more precisely, mutually exclusive. The mathematical condition for this is $\int \psi_1{}^* \psi_0 \, dV = 0$ analogous to the scalar product of vectors **a** and **b** to vanish; i.e., that $\mathbf{a} \cdot \mathbf{b} = 0$. This property of orthogonality is general; any two solutions of the Schrödinger equation (or, for that matter, of many differential equations) with different eigenvalues are orthogonal. Physically, this means that the states corresponding to those wave functions, which are necessarily states with different constants of motion and different quantum numbers, are mutually exclusive. If a system is in state *0*, then it is certainly not in state *1*. Appendix 6A discusses this property in more detail. Some methods of describing electronic states of molecules, and we shall discuss one of these, the Valence Bond method, use sets of wave functions that are not orthogonal and hence work conceptually to describe the physical system in terms of states that are not mutually exclusive. Such methods are sometimes computationally useful but at the expense of some loss of clarity of physical interpretation.

How good is the LCAO approximation? It can be improved by variational methods; for example, one can vary *Z* in the atomic orbitals, as in Eq. 5.13 for the helium atom. However, since there is only one electron to worry about, the H_2^+ molecule is one of the few quantum mechanical systems for which the Schrödinger equation can be solved exactly (in the Born–Oppenheimer approximation). In Fig. 6.7 we compare the LCAO and exact energies for the first two states; Fig. 6.5 gives the corresponding electron densities along the internuclear axis. In accordance with the variation principle, the exact energy is everywhere lower. The exact solution gives an equilibrium distance $R_e = 2.00a_0$ (1.06 Å) and a dissociation energy $D_e = -0.103$ hartree (–2.79 eV), compared to the LCAO values of $2.50a_0$ (1.32 Å) and –0.065 hartree (–1.78 eV). Since the nuclear repulsion is known exactly, the error of the LCAO method is entirely in the electronic binding energy (cf. Fig. 6.5): At $R = 2a_0$, the latter is off by only 8%, but the dissociation energy by 48%. This illustrates the basic problem in calculating dissociation energies, that they are small differences of large quantities. For other molecules the problem is even more difficult, because interactions between the electrons also come into play.

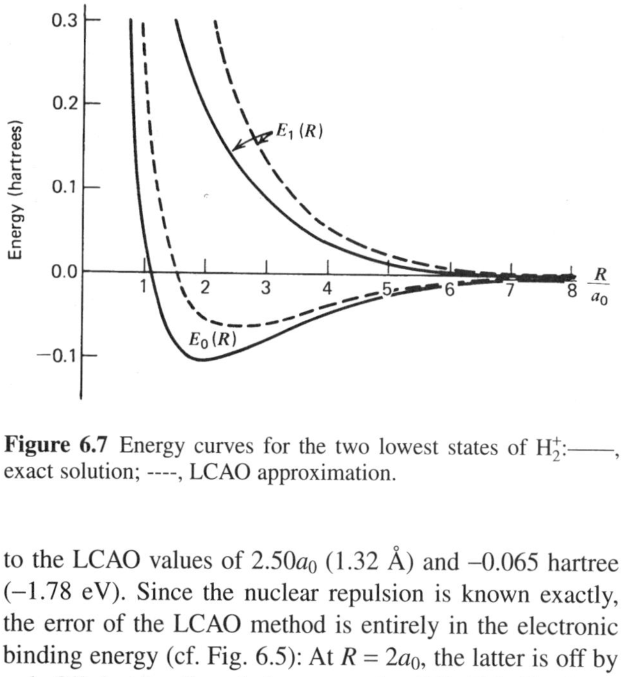

Figure 6.7 Energy curves for the two lowest states of H_2^+:——, exact solution; ----, LCAO approximation.

6.5 H_2^+: Correlation of Orbitals; Excited States

Thus far we have discussed only the lowest two energy levels of the H_2^+ molecule, but there are many higher levels. Our interest is mainly in how these states, and indeed those of any diatomic molecule, can be classified. We can best approach this problem by looking at what happens in the long-range and short-range limits.

We have already discussed the long-range or separated-atom limit. As $R \to \infty$ the molecular wave function goes smoothly into a sum or difference of atomic orbitals; for the two H_2^+ states we have studied, these are hydrogen 1*s* orbitals. At the other extreme, as $R \to 0$, the two nuclei must eventually collapse into one; this we call the *united-atom limit.* The united atom obtained from the H_2^+ molecule would have a nucleus with charge $+2e$ and a single electron, and would thus be a He^+ ion. (We overlook the fact that there is no bound nucleus composed of two protons; to make the picture truly realistic, we would have to use D_2^+ as our example.) What happens to the wave function in the united-atom limit? In the ground state, the two peaks in the function ψ_0 must coalesce as $R \to 0$. In the limit we have a nodeless wave function with a single cusp at the united nucleus; this clearly describes the 1*s* state of the hydrogenlike He^+ ion. Since there is a continuous transition of the molecular state from one limit to the other, we say that the ground state *correlates* smoothly with H(1*s*) in the separated-atom limit and $He^+(1s)$ in the united-atom limit.

What about ψ_1, the first excited state? This also goes into the H(1*s*) separated-atom limit, but to what state of He^+ does its united-atom limit correspond? The key is the fact that for all values of R the function ψ_1 has a nodal plane midway between the nuclei, perpendicular to the internuclear axis. This results from the symmetry condition (Eq. 6.16), which for ψ_1, becomes $\psi_1(A, B) = -\psi_1(B, A)$. The united-atom wave function must thus have a nodal plane through the nucleus, but no other nodes. It is clear from Fig. 4.8 that this must be a 2*p* orbital—more specifically, that 2*p* orbital whose long axis corresponds to the original $A - B$ axis. There is something puzzling here: A *p* orbital has $l = 1$ and angular momentum $\sqrt{2}\hbar$, yet this orbital was constructed from 1*s* orbitals with zero angular momentum. Since we have only pushed the nuclei together along a straight line, where did the angular momentum come from?

The answer to this question can give us a good deal of insight into the symmetry properties of a diatomic molecule. Recall the principle we stated in Section 3.10: If the energy is independent of some coordinate q_i, the conjugate momentum p_i is a constant of the motion. In an isolated atom the potential energy is spherically symmetric, and thus independent of all the angular coordinates, whose conjugate momenta are components of the angular momentum. In a classical atom, then, the angular momentum and all its components would be constants of the motion. As we showed in Section 3.11, the uncertainty principle makes it impossible for all these components to be known simultaneously, so they cannot all be constants of the motion in a real quantum mechanical system. However, the magnitude of the total angular momentum and any one of its components do remain constants of the motion in a system with spherical symmetry. Corresponding to these we have the two good quantum numbers l, m_l (for an electron) or L, M_L (for an atom).

In a diatomic system, which has cylindrical rather than spherical symmetry, the above argument no longer applies precisely. The energy of an electron depends very much on angular coordinates like θ_A and θ_B (Fig. 6.1). Thus the components of angular momentum corresponding to these angles cannot be constants of the motion, and neither can the magnitude of the orbital angular momentum: l (or L) is no longer a good quantum number. There is one exception to this breakdown. Since the *AB* or *z* axis is an axis of cylindrical symmetry, the energy is independent of the angle ϕ that describes rotation about this axis, and the component of angular momentum along this axis (p_ϕ or L_z) is thus a constant of the motion. The operator corresponding to L_z has the same form as for one component of angular momentum in an atom, $-i\hbar(\partial/\partial\phi)$, and the eigenvalues of L_z are given by

$$L_z = \pm\lambda\hbar \qquad (\lambda = 0,1,2,\ldots), \tag{6.43}$$

where λ is a quantum number corresponding to the absolute value of our earlier m_l. As in the atom, we use lowercase letters for a single electron and capital letters for a many-electron system. The axial angular momentum quantum number is designated by λ for an electron, Λ for a molecule, for which $L_z = \pm\Lambda\hbar$.

In spite of this formal similarity, there is a fundamental difference between L_z in an atom and L_z in a diatomic system. In an atom our choice of a *z* axis is entirely arbitrary, and L_z is quantized along whatever axis we choose. In the diatomic system this freedom of choice is gone. The *z* axis must be the system's unique symmetry axis, the axis joining the nuclei. Another difference involves the degeneracy of states. In a spherically symmetric potential the energy depends only on the quantum number l (or L), with $2l + 1$ degenerate states for each value of l. In the cylindrically symmetric potential we no longer have a quantum number l, and only λ (or Λ) is available to classify the energy levels. These levels are either nondegenerate, for $\lambda = 0$ ($\Lambda = 0$), or doubly degenerate, for any other value of λ (or Λ).

Now we can answer our original question, "Where did the angular momentum come from?" In the separated-atom H(1*s*) orbitals the angular momentum component is zero along any one axis, including the one that becomes the molecular *z* axis. Thus the ψ_1 state of H_2^+ has $L_z = 0$ in the separated-atom limit. Since L_z is a constant of the motion, it must also be zero for finite R, and even in the united-atom limit $He^+(2p)$. This is the case because a 2*p* orbital has zero angular momentum along its long axis, here taken as the *z* axis; that is, the orbital is $2p_z$, with $m_l = 0$, $L_z = 0$ (cf. Fig. 4.6). The $\sqrt{2}\hbar$ we mentioned above is the total angular momentum of the electron of $He^+(2p_z)$, which is no longer a constant when the spherical symmetry is removed.

It is customary to use the value of the axial angular momentum quantum number to classify the states of diatomic molecules. Just as atomic orbitals with $l = 0, 1, 2,$

3, . . . are respectively called *s, p, d, f*, . . . , so in diatomic molecules orbitals with $\lambda = 0, 1, 2, 3, \ldots$ are designated by analogy, with the corresponding Greek letters σ, π, δ, φ, . . . Capital letters are again used for the molecule as a whole: Molecular states with $\Lambda = 0, 1, 2, 3, \ldots$ are called Σ, Π, Δ, Φ, . . . states. The spin multiplicity $(2S + 1)$ of these states is indicated by a left superscript as in atoms. The two states of H_2^+ we have described both have a single electron in an orbital with $\lambda = 0$, a σ orbital; thus each of these states has $\Lambda = 0$, $S = \frac{1}{2}$, and can be called a $^2\Sigma$ state. When, as here, the molecule has a center of symmetry, a further distinction must be made. If the wave function is unchanged by inversion through the center of symmetry, that is, if $\psi(x, y, z) = \psi(-x, -y, -z)$, it is said to have *even parity;* if the reflection changes only the sign, $\psi(x, y, z) = -\psi(-x, -y, -z)$, the wave function has *odd parity.* States with even and odd parity are designated by the subscripts g and u (German *gerade* and *ungerade*), respectively. It is clear from Eqs. 6.26 and 6.27 that ψ_0 is even and describes a σ_g orbital, whereas ψ_1 is odd and describes a σ_u orbital; the corresponding molecular states are $^2\Sigma_g$ and $^2\Sigma_u$, respectively.

A molecular orbital may be labeled by either the united-atom orbital or the separated-atom orbital with which it correlates. The united-atom label precedes the species designation ($\psi_0 \rightarrow 1s\sigma_g$, $\psi_1 \rightarrow 2p\sigma_u$, etc.), whereas the separated-atom label follows it ($\psi_0 \rightarrow \sigma_g 1s$, $\psi_1 \rightarrow \sigma_u 1s$, etc.). This nomenclature is used mainly when one is emphasizing the two limits. For example, the "$\sigma_g 1s$" notation is appropriate for an LCAO wave function. When discussing the molecule itself, however, one can simply number the orbitals of each species (σ_g, σ_u, π_g, etc.) in order of increasing energy. Thus ψ_0 and ψ_1 correspond to the $1\sigma_g$ and $1\sigma_u$ orbitals, respectively. Figure 6.8 gives further examples of all these notations.

Now we are ready to examine the higher-energy orbitals of H_2^+. For variety, we shall start with the united-atom limit instead of the separated-atom limit. This makes good physical sense since the orbitals in question are very large. They have appreciable magnitudes in regions far beyond the two nuclei, regions in which the electron density must be quite similar to that in a He^+ ion. For this reason the LCAO approximation, based on the separated atoms, does not give an accurate representation of the excited states.[10] We start with He^+ in a given state, pull the nucleus apart into two singly charged nuclei, and see what kind of wave function we should obtain. We expect the higher states to be unstable,

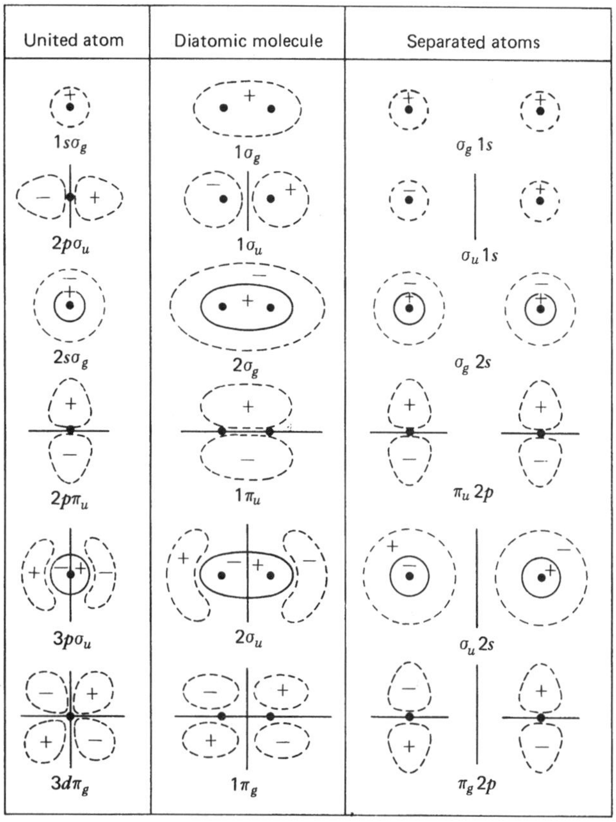

Figure 6.8 Correlation and shapes of orbitals in homonuclear diatomic molecules (for orbitals whose ϕ-dependent factors are real). The diagrams are purely schematic, with dashed lines bounding roughly the regions where the electron density is significant. The solid curves and lines are nodal surfaces; the sign of the wave function is given in each of the regions they separate. Each orbital is labeled by the three systems described in the text. Note that for H_2^+ the separated-atom limit of $2\sigma_g$ could be either $\sigma_g 2s$ or $\sigma_g 2p$, and that of $2\sigma_u$ either $\sigma_u 2s$ or $\sigma_u 2p$.

that is, to have repulsive $E(R)$ curves (curves with no minima), since they have even less electron density in the binding region of space than the state ψ_1. (Some states in fact have shallow $E(R)$ minima at large R, but for most practical purposes these are also unstable.)

We know that the $1s$ orbital of He^+ transforms into the $1\sigma_g$ molecular orbital (ψ_0), which at long range transforms into the sum of two $1s$ orbitals of atomic hydrogen. Into what states do the other united-atom orbitals transform? In speaking of such a "transformation" at all, we are invoking a very general principle of quantum mechanics, the *adiabatic principle.* Let us assume that the energy levels of a system vary smoothly with some parameter R, which, up to this point, we have taken to be the internuclear distance we treated as a parameter for the Born–Oppenheimer method. This gives us a series of $E(R)$ curves, each of which we can label as a particular state of the system. The adiabatic principle states that, if a system is initially in the ith energy level at a particular value of R, then for sufficiently slow changes

[10] For an illustration of how the LCAO approximation breaks down in the united-atom limit, consider the ground state. Since $\varphi(\mathrm{H}, 1s)$ varies as e^{-r/a_0}, the LCAO wave function (Eq. 6.33) also approaches e^{-r/a_0} when $R \rightarrow 0$ (and $r_1, r_2 \rightarrow r$). But in this limit the true wave function must approach that of He^+, which varies as e^{-2r/a_0} (since $Z = 2$). Similarly, at $R = 0$ the LCAO electronic bonding energy is 1.0 hartree (Fig. 6.5), whereas the true value must be

$$E_1(\mathrm{H}) - E_1(\mathrm{He}^+) = -0.5 \text{ hartree} - (-2.0 \text{ hartrees}) = 1.5 \text{ hartrees}.$$

in R it will remain in the same ith level (but as transformed by the change in value of R) unless otherwise disturbed.[11] For the particular case of the internuclear distance, this reduces to the Born–Oppenheimer approximation: The $E(R)$ curves calculated for fixed nuclei give the energy of the actual molecule in the limit of sufficiently slow nuclear motion. For rapid nuclear motion, as in a high-speed collision, both the adiabatic approximation and the very concept of an energy curve cease to apply, and transitions between electronic states become likely.

What we wish to know, then, is the behavior of the orbitals when the nuclei are pulled apart very slowly (adiabatically). There should be an $E(R)$ curve connecting each united-atom state with a corresponding separated-atom state; the two limiting states are said to *correlate* with each other. Correlated states, as well as the sequence of molecular states joining them, must be identical in those properties that are constants of the motion; these include the value of λ (or Λ) and the parity of the wave function. Thus σ_g state correlates only with a σ_g state, π_u, only with π_u, and so for each class of states. But each limit has many states of each symmetry class; how do we know which ones correlate with each other? The best answer is given by the *noncrossing rule:* Two $E(R)$ curves with the same invariant properties (λ, parity, spin, etc.) never cross each other. This is only another approximation, valid to the same extent as the Born–Oppenheimer approximation (from which it can be derived), but it is sufficiently accurate to be our main tool for analyzing the correlation of states.

Given these assumptions, the analysis of H_2^+ is easy. The lowest σ_g orbital derived from the united atom must correlate with the lowest σ_g orbital formed from the separated atoms, since it can cross no other σ_g curve; this is the ground state, for which we had already tacitly assumed the noncrossing rule. Similarly, the second lowest united-atom σ_g orbital correlates with the second lowest separated-atom σ_g orbital, the third lowest with the third lowest, and so forth. The same applies to each of the other symmetry classes (but note that there is nothing to prevent $E(R)$ curves of *different* symmetry classes—σ_g and π_g, say—from crossing each other).

But now we immediately have a problem: What classes of molecular states do we obtain from each united-atom or separated-atom state? Remember that the quantum number λ corresponds to $|m_l|$ in either limit. For an atomic state with a given value of l, the possible values of $|m_l|$ are 0, 1, . . . , l. Thus a united-atom s orbital ($l = 0$) can only transform into a σ orbital ($\lambda = 0$), a p orbital ($l = 1$) into either a σ or a π orbital ($\lambda = 0,1$), and so forth; in general the united-atom nl level gives rise to l distinct molecular states. The parity is easily assigned in the united-atom limit: All s orbitals are even (g), all p orbitals odd (u), all d orbitals even, and so on, as can easily be seen from Fig. 4.6. Combining these rules, we have the following correlations:

United-atom level:	s	p	d	f
Molecular orbitals:	σ_g,	σ_u, π_u,	$\sigma_g, \pi_g, \delta_g$,	$\sigma_u, \pi_u, \delta_u, \varphi_u, \ldots .$

The separated-atom limit is a little more complicated, since we are dealing with *pairs* of atomic orbitals. The relation between λ and $|m_l|$ remains, but now each pair of nl levels combines to form $2(2l + 1)$ molecular orbitals. The molecular orbitals also come in pairs, corresponding in the LCAO approximation to the sum and difference of two identical atomic orbitals. In each such pair of molecular orbitals, one is even (g) and the other odd (u); we have already discussed the example of the $\sigma_g 1s$ and $\sigma_u 1s$ orbitals. We thus have the correlations:

Separated-atom level:	s	p	d
Molecular orbitals:	σ_g, σ_u,	$\sigma_g, \sigma_u, \pi_g, \pi_u$,	$\sigma_g, \sigma_u, \pi_g, \pi_u, \delta_g, \delta_u, \ldots .$

Later we shall discuss the nature of the molecular orbitals in greater detail, but now we have enough information for a complete description of the orbital correlation.

The simplest way to display such a correlation is with a *correlation diagram;* Fig. 6.9*a* gives such a diagram for H_2^+. On the left and right sides of the diagram are the orbitals of the united atom and the separated atoms, respectively, each arranged in order of increasing energy. The energy scale is purely schematic; only the sequence of orbitals is significant. Lines are drawn connecting each united-atom orbital to the separated-atom orbital with which it correlates, following the correlation rules described above. Each of these lines can then be labeled with its appropriate molecular orbital designation; given the noncrossing rule, such designations as $1\sigma_g$, $2\sigma_g$, . . . for the lowest, second lowest, . . . σ_g states can be given unambiguously. Since the two limits correspond to $R = 0$ and $R = \infty$, the vertical sequence of orbitals across the diagram should approximately represent the sequence of orbital energies as a function of R. For H_2^+ it is possible to test this, since the exact $E(R)$ curves have been calculated for many of the molecular states; these are plotted in Fig. 6.9*b*. The picture is similar to that in the schematic diagram, but there are a number of differences in detail. In particular, the noncrossing rule is apparently violated by a number of the higher energy levels at large R, but this does not affect the correlation, since the states that thus cross are degenerate in one or the other limit.

Degeneracy leads to a special problem in the cases of H_2^+ and H_2. The $2\sigma_g$ orbital, derived from $He^+(2s)$, should correlate with the second lowest σ_g orbital of the separated atoms. But σ_g orbitals can be formed from both the $2s$ and

[11] "Adiabatic" comes from the Greek word for "uncrossable"; i.e., under adiabatic conditions the system will not "cross" from one $E(R)$ curve to another.

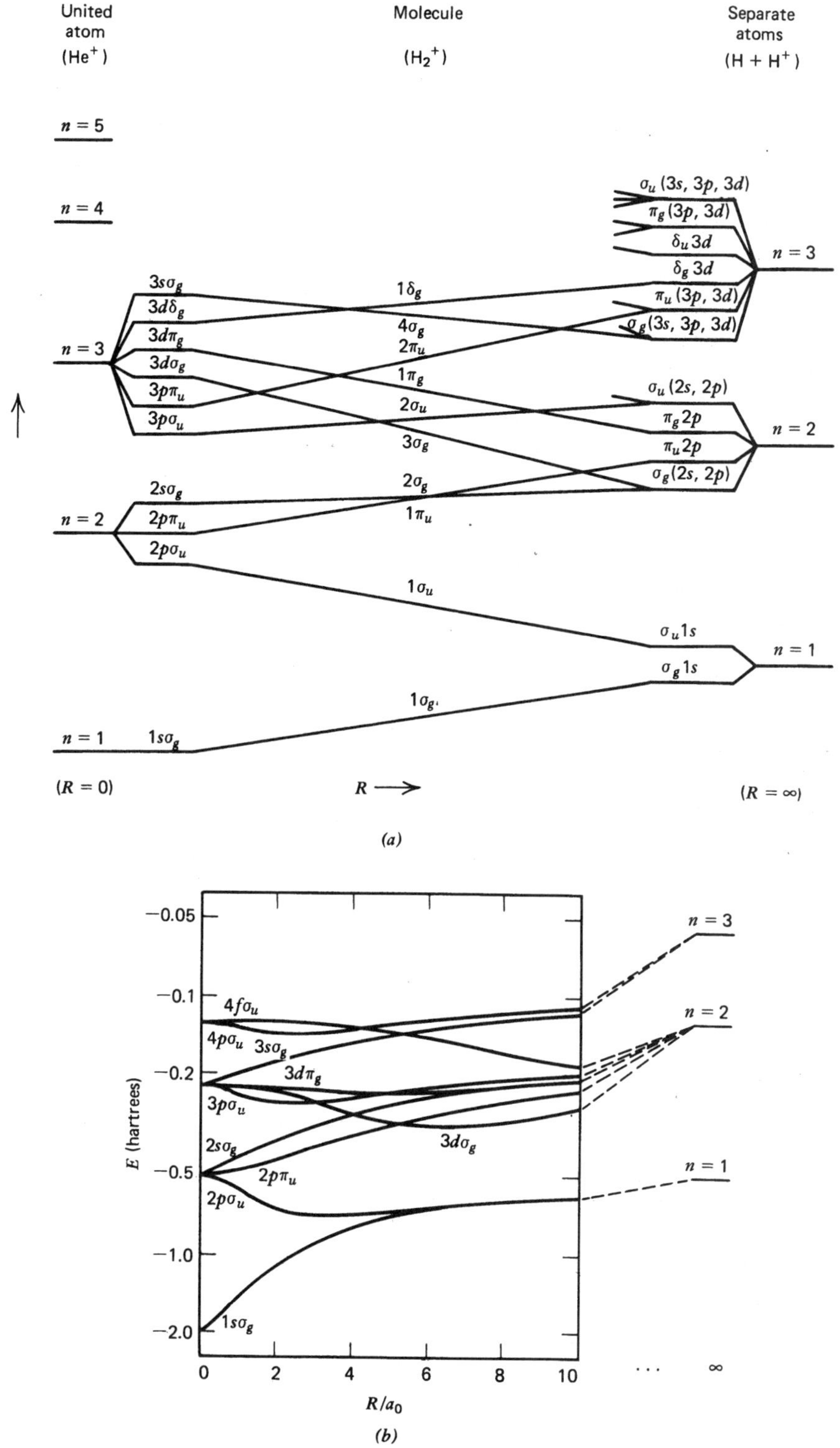

Figure 6.9 Correlation of energy levels for H_2^+. (*a*) Schematic correlation diagram, with degenerate energy levels spread apart for clarity. Correlations are shown for all united-atom levels through $n = 3$. See the text for a discussion of separated-atom levels such as "$\sigma_g(2s,2p)$." (*b*) Actual $E(R)$ curves for some of the energy levels, as calculated by Bates, Ledsham, and Stewart. The energy is measured relative to the free-electron limit, $H^+ + H^+ + e^-$. Since the calculations have not been carried beyond $R = 10a_0$, the curves are labeled in united-atom notation only.

the $2p$ levels of H + H⁺, and in the hydrogen atom these levels are degenerate.[12] We have thus used the notation "$\sigma_g(2s, 2p)$" to label the separated-atom limits of both the $2\sigma_g$ and the $3\sigma_g$ molecular orbitals; similar notation is used for other such degenerate orbitals of the same species. What this means is that, as far as satisfying the correlation rules goes, the separated-atom limit of the $2\sigma_g$ orbital could be any linear combination of the form $a(\sigma_g 2s) + b(\sigma_g 2p)$. The exact calculations have not been carried far enough to reveal the true nature of this limit. But the simplest approximation

[12] In heavier atoms $2s$ has lower energy than $2p$, so the σ_g orbital correlates with $\sigma_g 2s$; cf. Fig. 7.14.

to the $2\sigma_g$ orbital is an LCAO representation using only the 2*s* atomic orbitals:

$$\psi(\sigma_g 2s) = \frac{\varphi_A(2s) + \varphi_B(2s)}{[2 + 2S_{AB}(2s)]^{1/2}}, \tag{6.44}$$

analogous to Eq. 6.33. This approximation gives a fair description of the actual $2\sigma_g$ orbital at moderate *R;* if it is correct, then the $He^+(2s)$ radial node eventually splits into two H(2*s*) radial nodes, but only at very large *R* (cf. Fig. 6.8).

Let us now look at the nature of some other orbitals. There are two degenerate σ_g orbitals in the $n = 3$ level of He^+, but here there is no ambiguity. The calculations clearly show that the $3d\sigma_g$ has the lower energy and thus becomes the $3\sigma_g$ molecular orbital. It of course correlates with the other separated-atom orbital formed from the $\sigma_g(2s, 2p)$ degeneracy, which in the simplest LCAO approximation can be represented by $\psi(\sigma_g 2p)$. But the LCAO representations are hardly important for these higher-energy states, which near R_e are better described by the united-atom orbitals. As for the $3s\sigma_g$ level, it becomes the $4\sigma_g$ molecular orbital and correlates with the separated-atom $\sigma_g(3s, 3p, 3d)$ degenerate set of levels.

The first few σ_u orbitals derive from the *p* orbitals of He^+—specifically, those *p* orbitals with $m_l = L_z = 0$ (where the *z* axis becomes the internuclear axis). We have already described how the $1\sigma_u$ orbital (ψ_1) derives from the $He^+(2p, m_l = 0)$ orbital and correlates with the difference of two H(1*s*) orbitals. The similar $2\sigma_u$ orbital derives from the He^+ $(3p, m_l = 0)$ orbital and correlates with the $\sigma_u(2s, 2p)$ degeneracy. Its simplest LCAO representation is $\sigma_u 2s$, the difference of two H(2*s*) orbitals.

Thus far we have described only σ orbitals, with $L_z = 0$. In order to construct H_2^+ orbitals with nonzero angular momentum about the axis, we must begin with He^+ orbitals having such angular momentum themselves. For $\lambda = 1$ $(L_z = \pm\hbar)$, the simplest such orbitals are obviously the *p* orbitals with $m_l = \pm 1$ relative to the internuclear axis. Since the real and imaginary parts of each of these orbitals must have a nodal plane including the axis, symmetry requires that such nodal planes also exist in the molecule and the separated atoms; the separated atom limit must thus also consist of *p* orbitals with $m_l = \pm 1$. The simplest molecular orbitals of this kind are formed from the $He^+(2p, m_l = \pm 1)$ orbitals, and can be given the LCAO representation

$$\psi(\pi_u 2p, \lambda = 1) = \frac{\varphi_A(2p, m_l = \pm 1) + \varphi(2p, m_l = \pm 1)}{[2 + 2S_{AB}(2p, m_l = \pm 1)]^{1/2}}. \tag{6.45}$$

Since the value of λ is 1, this is a π orbital—$1\pi_u$ to be specific. It is an odd (*u*) orbital because it is formed from the sum of two *p* orbitals. If the orbitals are chosen real, or if we consider only their real or imaginary parts, the two atomic orbitals have their positive lobes on the same side of the axis, so reflection through the center of symmetry changes the sign of ψ. The difference of two $2p(m_l = \pm 1)$ atomic orbitals would give a π_g orbital, which turns out to correlate with a 3*d* orbital of the united atom. This may be made clearer by Fig. 6.8, which shows schematically the shapes of these and other orbitals.

Note that Eq. 6.45 defines two orbitals, one with $L_z = +\hbar$ and one with $L_z = -\hbar$. Such a pair of wave functions exists for any molecular state in which $\lambda \neq 0$. Since $\lambda\hbar$ is the magnitude of the angular momentum about the internuclear axis, the states with $L_z = \pm\lambda\hbar$ differ (in classical terms) only in the direction of the electron's rotation about the axis. If we changed our coordinate system from right-handed to left-handed, simply by reversing the direction of the *z*-axis, we would interchange the functions $\psi(L_z = +\hbar)$ and $\psi(L_z = -\hbar)$. But this would in no way change the physics of the situation, and the two states must thus be physically indistinguishable.[13] Being indistinguishable, they must have the same energy. In short, for every nonzero value of λ in a diatomic molecule there are two degenerate states, with identical energies and with wave functions differing only in orientation. This degeneracy, like that of the hydrogen atom's three 2*p* states, derives directly from the symmetry of space.

As usual when we have degenerate wave functions, we can take linear combinations of them to define new orbitals. For example, we can take the sum and difference of $\psi(L_z = +\hbar)$ and $\psi(L_z = -\hbar)$ to give functions analogous to Eqs. 3.145. These vary respectively as $\cos \lambda\phi$ and $\sin \lambda\phi$, where ϕ is the angle of rotation about the *z* axis, while the original functions varied as $e^{\pm i\lambda\phi}$. The new orbitals are no longer eigenfunctions of even one component of angular momentum, but they still have the same degenerate energy as the original orbitals. For axially symmetric molecules we are at liberty to use either description. As usual, the real functions are more convenient for graphic purposes, and have been shown in Fig. 6.8. Only in molecules that have no axis of cylindrical symmetry does the difference become physically significant. As for nomenclature, in common usage any linear combination of, say, π_u orbitals is still called a π_u orbital.

This completes our study of the H_2^+ molecule-ion. What have we learned that can be applied to other molecules? Like the electron in a hydrogen atom, that in H_2^+ has a series of distinct quantum states or orbitals, with wave functions and energies that vary with the internuclear distance. Only in the ground state is the electron density in the bonding region high enough to produce a very stable molecule. We have introduced the remarkably simple and useful LCAO approximation, which represents the molecular orbitals as

[13] This would not be true if we had a way of measuring the sign of L_z, that is, the direction of the electron's rotation. One can do this, for example, with a magnetic field, which can sense the direction of an electric current in a loop. Such a field would remove the degeneracy of the two states. However, we take it for granted in our discussion that external fields are absent. (There is a weak field from the various magnetic moments of the atoms, and this as usual produces a fine-structure splitting of energy levels.)

sums or differences of atomic orbitals, and shown how it can be used to obtain the $E(R)$ curve. But we have also seen that the LCAO method has its limitations for even this simplest of molecules. Finally, we have outlined how molecular orbitals can be classified according to their symmetry and angular momentum, and how they can be organized in a correlation diagram. Next we must find out what complications arise when a molecule has more than one electron.

6.6 The H_2 Molecule: Simple MO Description

The H_2^+ molecule-ion served as a simple example to help us grasp the physical concepts associated with chemical bonding. However, it is a rather exotic molecule compared with most of those we meet in everyday chemistry. (This is often the case in science: The most useful or familiar examples of a phenomenon are often terribly complex—water waves, for example—whereas those easiest to interpret tend to be rather abstruse.) The hydrogen molecule, H_2, is quite another matter. This is a perfectly stable species that exists as isolated gaseous molecules over a wide range of temperatures. Indeed, it is the most abundant molecule in the universe. Its physical and chemical properties are thoroughly known from experiment, and it is second only to H_2^+ in simplicity, having only two electrons. Thus it should be relatively easy both to apply a theory of molecular structure to H_2 and to test the results.

The H_2 molecule is similar in several ways to H_2^+. The symmetry is the same, the potential energy due to the nuclei is the same for a given R, and the molecule still resembles two H atoms in the limit $R \to \infty$. The fundamental difference is that there are two electrons rather than one. It is natural to make the same initial assumption as in a many-electron atom: that each electron in a molecule sees only the average field of the other electron(s), and that one can describe such an electron by a one-electron wave function (molecular orbital). This corresponds to the central-field approximation in an atom. Here, however, the field is no longer central, and we must do without the advantages of spherical symmetry. As before, the best approximate wave functions in this model are obtained by self-consistent-field methods; to improve on these results, one must apply corrections for the correlation between electrons.

Assume that we can get one-electron wave functions; how do we combine them to get the total molecular wave function? The simplest assumption is that one can assign to each electron a unique orbital, the aggregate of such assignments defining a unique electronic configuration of the molecule. For example, just as the boron atom has the configuration $1s^2 2s^2 2p$, we shall see (in Chapter 7) that the B_2 molecule has the ground-state configuration $(1\sigma_g)^2(1\sigma_u)^2(2\sigma_g)^2(2\sigma_u)^2(1\pi_u)^2$. Even in an atom this is only an approximation, though quite a good one, as we pointed out at the end of Section 5.7. Strictly, the interactions between electrons prevent the energy and angular momentum of a single electron from being conserved. One would not expect the single-configuration model to work any better for molecules. However, we can certainly use this model as a first approximation; later we shall introduce an approximation involving more than one configuration.

To the extent that electronic configurations are meaningful in molecules, the Pauli exclusion principle holds for individual electrons: If the electron spin is taken into account, only one electron can be assigned to a given quantum state. Each distinct spatial molecular orbital can then contain only two electrons, which must have opposite spins. One can thus apply the Aufbau principle, filling each orbital in turn in the order of increasing energy. Since H_2 has only two electrons, these rules would not seem to concern us here; yet the law behind them does entail a serious restriction on the possible wave functions of H_2, as we shall see in the next section.

The simplest kind of molecular wave function made up of one-electron orbitals is a Hartree product like Eq. 5.15. For a given state of H_2 such a wave function would have the form

$$\psi_{MO}(1,2) = \varphi_i(1)\varphi_j(2), \tag{6.46}$$

where i, j designate orbitals and 1, 2, electrons; each φ_i must include both spatial and spin coordinates. Spectroscopy tells us[14] that the ground state of H_2 is $^1\Sigma_g$. We thus say that the lowest-energy orbital is $1\sigma_g$ (as in H_2^+) and the ground-state configuration is $(1\sigma_g)^2$, with the two electrons assigned opposite spins. If we use the orbital designation to stand for the spatial part of the wave function, Eq. 6.46 for the ground state becomes

$$\psi_{MO}[(1\sigma_g)^2] = 1\sigma_g(1)\alpha(1)1\sigma_g(2)\beta(2), \tag{6.47}$$

where α and β are the two different Pauli spin wave functions introduced briefly in Chapter 5: α for $m_s = +\frac{1}{2}$ *and* β for $m_s = -\frac{1}{2}$. Because we are neglecting electron correlation, each of the spatial orbitals must be identical to the corresponding H orbital, and in the LCAO approximation will be given by Eq. 6.33. Combining these assumptions, we have

$$\begin{aligned}\psi_{MO}[(\sigma_g 1s)^2] &= \sigma_g 1s(1)\sigma_g 1s(2)\alpha(1)\beta(2) \\ &= \frac{[1s_A(1)+1s_B(1)][1s_A(2)+1s_B(2)]\alpha(1)\beta(2)}{2(1+S_{AB})}.\end{aligned} \tag{6.48}$$

[14] We know it is a $^1\Sigma$ state because it exhibits no Zeeman splitting at all in a magnetic field and thus must have both zero orbital angular momentum (so $\Lambda = 0$) and zero spin angular momentum (so $S = 0$, $2s + 1 = 1$). The g character is actually inferred from the intensities of spectral lines associated with specific rotational states, but we shall not pursue this point. For details, see G. Herzberg, *Molecular Spectra and Molecular Structure, Volume I, Spectra of Diatomic Molecules,* 2nd ed. (Van Nostrand, Reinhold, New York, 1950).

Here $1s_A$, $1s_B$ is a shorthand for our earlier $\varphi_A(1s)$, $\varphi_B(1s)$, normalized hydrogen atom eigenfunctions: $1s_A(1) = (\pi a_0^3)^{-1/2} e^{-r_{A1}/a_0}$, where r_{A1} is the distance from nucleus A to electron 1.

Equation 6.48 is the simplest molecular orbital wave function for the ground state of the H_2 molecule. Note that in its second form, this function can be interpreted as a product of two superpositions of atomic wave functions. The full wave function can be used to obtain an approximate $E(R)$ curve by the same method as in Section 6.4. In this case one evaluates $\iint \psi^* \mathsf{H} \psi \, d\tau_1 \, d\tau_2$ for each value of R, with the electronic Hamiltonian of Eq. 6.1 (including the $1/r_{12}$ term); the integration is lengthy but feasible. The calculated $E(R)$ is shown as curve 1 in Fig. 6.10. One obtains $R_e = 1.57a_0$ (0.84 Å) and $D_e = 0.0974$ hartree (2.65 eV), the latter relative to two $1s$ H atoms. The experimental values are $1.40a_0$ (0.74 Å) and 0.1744 hartree (4.75 eV). The electron density looks quite similar to that in the ground state of H_2^+ (Fig. 6.7), multiplied by 2. In the energy at least, this solution gives appreciably worse agreement than does the LCAO solution for H_2^+. This is not surprising, since here the flaws of the LCAO approach are combined with those of the one-electron-orbital and single-configuration assumptions.

One can improve upon this solution and still remain within the framework of the molecular orbital model. The greatest single improvement can be made by the technique we illustrated in Eq. 5.13, inserting a scale factor in the atomic orbitals. In this case one still uses Eq. 6.48 but sets $1s_A(1) = (\alpha^3/\pi)^{1/2} e^{-\alpha r_{A1}}$, and so on; α is then varied to minimize the energy for each value of R. This solution is shown as curve 2 in Fig. 6.10, and is obviously a considerable improvement over the initial, naive LCAO solution. One obtains $R_e = 1.38a_0$, $D_e = 0.1282$ hartree, with $\alpha(R_e) = 1.197/a_0$. The best possible wave function of the form of Eq. 6.46 is, of course, the self-consistent-field solution, which yields curve 3 of Fig. 6.10; but the SCF molecular orbitals can be represented only by complicated functions with no simple analytic relationship to atomic orbitals, and the $E(R)$ curve is only slightly improved over the scaled LCAO solution. We are still left with a correlation energy of 0.04 hartree (1.1 eV) at R_e, illustrating the limitations of the orbital assumption.

The hydrogen molecule illustrates a fundamental flaw that sometimes appears in the molecular orbital method, as we can see by looking at the separated-atom limit for H_2. What happens to the molecular orbital wave function as $R \to \infty$? The overlap integrals vanish, and Eq. 6.48 reduces to

$$\lim_{R\to\infty} \psi_{MO} = \frac{\alpha(1)\beta(2)}{2}[1s_A(1)1s_A(2) + 1s_A(1)1s_B(2) + 1s_B(1)1s_A(2) + 1s_B(1)1s_B(2)]. \quad (6.49)$$

In this limit, of course, there is negligible probability of finding an electron anywhere but near one of the two nuclei. If we square Eq. 6.49, each of the cross terms contains a factor like $1s_A(1)1s_B(1) \propto e^{-r_{A1}/a_0} e^{-r_{B1}/a_0}$, which vanishes since everywhere either r_{A1} or r_{B1} must be very large. Thus, all the cross terms drop out, and we have

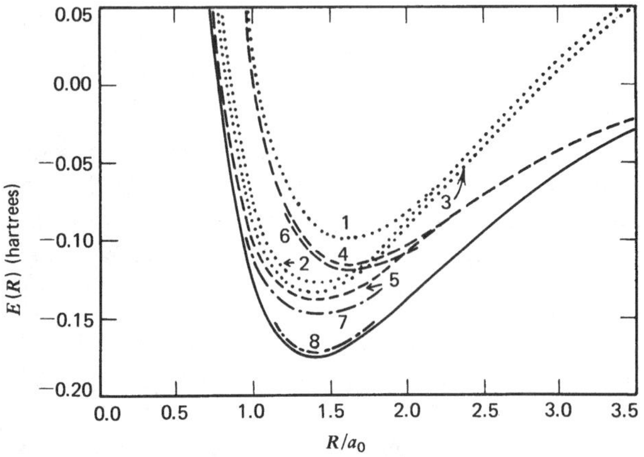

Figure 6.10 Potential energy curves for the ground state of H_2, calculated by various approximations; the energy zero corresponds to H(1s) + H(1s). The heavy solid line is the experimental $E(R)$ curve. Molecular orbital calculations (······):
1. Simple LCAO, Eq. 6.48, with $1s = (\pi a_0^3)^{-1/2} e^{-r/a_0}$,
2. Same with scaled atomic orbitals: $1s = (\alpha^3/\pi)^{1/2} e^{-\alpha r}$, with α varied to minimize $E(R)$,
3. Self-consistent-field solution.

Valence bond calculations (- - - -):
4. Heitler–London, Eq. 6.72,
5. Same with scaled atomic orbitals.

Other calculations (-·-·-·-):
6. Mixed MO and VB, Eq. 6.76,
7. Same with scaled atomic orbitals,
8. Eleven-term variation function, including terms in r_{12}.

$$\lim_{R\to\infty} (\psi_{MO})^2 \propto [1s_A(1)]^2[1s_A(2)]^2 + [1s_A(1)]^2[1s_B(2)]^2 + [1s_B(1)]^2[1s_A(2)]^2 + [1s_B(1)]^2[1s_B(2)]^2. \quad (6.50)$$

We have assumed that the location of each electron is independent of that of the other. The probability that two independent events occur simultaneously is the product of the probabilities of the two events. Thus, in the integral of Eq. 6.50, $\iint \psi^2 \, d\tau_1 \, d\tau_2$, the first term gives the probability that both electrons are in atom A; the next two terms give the probability that one electron is in each atom; and the fourth term gives the probability that both electrons are in atom B. Since all the $1s$ functions have the same form, each of the four terms contributes the same amount to the integral, and each of the four electron arrangements has the same probability, 25%. This means that the wave function describes a state in which one is equally likely to observe two neutral atoms, H(1s) + H(1s), or a negative ion and a bare proton, $H^-(1s^2) + H^+$.

But for two hydrogen atoms infinitely far apart, the state of lowest energy is clearly one in which each atom is in *its* ground state, H(1*s*) + H(1*s*). It takes 13.60 eV or 0.5 hartree, the ionization potential, to remove an electron from a hydrogen atom, and only 0.75 eV, the electron affinity, is gained by adding that electron to another neutral hydrogen atom. Thus the state $H^-(1s^2) + H^+$ has an energy 12.85 eV higher than H(1*s*) + H(1*s*), and the mixed state described by Eq. 6.49 should have an energy halfway between the two, +6.42 eV. If one extends the $E(R)$ curve obtained from the LCAO function (Eq. 6.48) to $R = \infty$, one in fact obtains an energy 8.50 eV (0.312 hartree) higher than the ground state; the 2.08-eV discrepancy is half the correlation energy of H^-, for which the Hartree function $1s(1)1s(2)$ is only an approximation. As one can guess from Fig. 6.10, the SCF solution is not much better; it also approaches an unphysical limit at large *R*. The molecular orbital approach is thus clearly inadequate because of its behavior at the separated-atom limit. A wave function of this type may be reasonably accurate when *R* is near its equilibrium value,[15] but it certainly does not describe the ground state—or any stationary state—of two distant H atoms.

What, then, is the wave function really like in the adiabatic, separated-atom limit? Since the electron densities on the two atoms are independent there, we might expect to find a simple product of two atomic orbitals like

$$\lim_{R\to\infty} \psi[(\sigma_g 1s)^2] \propto 1s_A(1)1s_B(2). \tag{6.51}$$

This wave function will obviously give the correct energy for two separated 1*s* H atoms. It describes a state in which there is a very high degree of correlation between the two electrons: The probability density $|\psi|^2$ has a significant magnitude only when electron 1 is near nucleus *A* and electron 2 is simultaneously near nucleus *B*. This is in contrast to a molecular orbital wave function like Eq. 6.48, in which there is by definition no correlation between the electrons.

Here we have two mathematical descriptions representing sharply contrasting physical situations, yet each is a reasonably good description of the H_2 molecule in one range of *R*. The MO description (Eq. 6.48) is never exactly correct, but for small *R* it does give a qualitatively correct $E(R)$ curve; as $R \to \infty$ it becomes completely wrong. Equation 6.51 is an exact solution at $R = \infty$ (apart from the considerations discussed in the next section), but is inadequate to describe bonding at finite *R*. The real ground-state wave function must somehow make a transition between these two extremes, and we wish to find a function that behaves in the same way. Before we begin this search, however, we must study the consequences of an apparently trivial fact—that all electrons are identical.

[15] The wave function near R_e does have a significant amount of ionic character (contribution of terms corresponding to H^-H^+), because Coulomb attraction drastically lowers the energy of the ionic state. Cf. Section 6.9.

6.7 Symmetry Properties of Identical Particles

We have several times discussed the importance of symmetry in quantum mechanics. The constants of motion of a system are directly determined by the symmetry of the Hamiltonian; thus in diatomic molecules the equivalence of all orientations about the *z* axis requires that L_z be conserved. In H_2, as in H_2^+, the indistinguishability of the two nuclei means that $|\psi|^2$ must be symmetric about a plane midway between the nuclei. It should not be surprising that the indistinguishability of the electrons also has important consequences. We suggested some of these consequences in our discussion of atoms (which also contain identical electrons); now we shall look at the fundamental principles involved.

Here is the situation that confronts us. All electrons are absolutely indistinguishable. The same is true for all protons, and for all particles of any particular kind. The value of any real (measurable) physical quantity must thus be independent of the way we name or number such identical particles. Now, the physical description of a system depends only on the absolute square of the wave function. This means that, if $\psi(1, 2)$ is a wave function involving the identical particles 1 and 2, we must have

$$|\psi(1, 2)|^2 = |\psi(2, 1)|^2, \tag{6.52}$$

where $\psi(2, 1)$ is obtained from $\psi(1, 2)$ by interchanging particles 1 and 2—that is, rewriting the wave function with the names of particles 1 and 2 exchanged. Equation 6.16 says the same thing for the special case of identical nuclei.

What limitations does Eq. 6.52 place on the wave function? It tells us that $\psi(2, 1)$ must differ from $\psi(1, 2)$ by no more than a phase factor of the form $e^{i\delta}$, where δ is a constant. This is the only kind of factor that has no effect on $\psi^*\psi$, in that $(e^{i\delta})^* e^{i\delta} = e^{-i\delta}e^{i\delta} = e^0 = 1$. We can be more specific about the value of $e^{i\delta}$. If we define an operator P_{12} to mean "permute (exchange) particles 1 and 2," then by what we have just said we must have

$$\psi(2,1) \equiv \mathsf{P}_{12}\psi(1,2) = e^{i\delta}\psi(1,2). \tag{6.53}$$

If we carry out the permutation of 1 and 2 a second time, we must be back where we started. We can thus write

$$\begin{aligned}\psi(1,2) &= \mathsf{P}_{12}\psi(2,1) = \mathsf{P}_{12}[e^{i\delta}\psi(1,2)] \\ &= e^{i\delta}\mathsf{P}_{12}\psi(1,2) \\ &= e^{2i\delta}\psi(1,2).\end{aligned} \tag{6.54}$$

This means that

$$e^{2i\delta} = 1 \quad \text{or} \quad e^{i\delta} = \pm 1, \tag{6.55}$$

or, in terms of the wave functions,

$$\psi(2,1) = \pm\psi(1,2). \tag{6.56}$$

This argument is completely general: It holds for systems with not just two, but any number of identical particles, because we can always interchange them two at a time.

According to Eq. 6.56, the wave function for any system containing identical particles must either remain unchanged or change sign if we interchange two of the identical particles. There are two physically distinct cases here. If the wave function is unchanged, $\psi(2, 1) = \psi(1, 2)$, we say that it is *symmetric* with respect to particles of the type in question; if the wave function changes sign, $\psi(2, 1) = -\psi(1, 2)$, we say it is *antisymmetric* with respect to those particles. The particles found in nature are sharply divided into two classes, those with symmetric and those with antisymmetric wave functions. A given particle never changes from one class to the other (unless the phenomenon called "supersymmetry" allows a bit of crossover under **very** high-energy conditions, as in the big bang). The names of these classes are based on the statistical behavior of the particles, which we shall discuss in Part II. Those with symmetric wave functions obey what is called Bose–Einstein statistics, and are thus named *bosons.* These include photons, deuterons, α particles, and all other particles with integral spin (photons have spin 1). The particles with antisymmetric wave functions obey Fermi–Dirac statistics, and are called *fermions.* These include electrons, protons, neutrons, and all other particles with half-integral spin.

We are concerned principally with the wave functions describing electrons. Since the electron is a fermion, any wave function $\psi(1, 2, \ldots)$ involving two or more electrons must change sign if we interchange the numbers on any two of the electrons. This is clearly not true of our molecular orbital wave functions for the H_2 molecule. Equations 6.47 through 6.49 all contain the factor $\alpha(1)\beta(2)$, specifically describing a state in which electron 1 has the spin wave function α and electron 2 has the spin wave function β. If we interchange the electrons we get a new wave function, containing $\alpha(2)\beta(1)$, rather than the same wave function with reversed sign. Since α and β here correspond to different spins, $\alpha(1)\beta(2)$ and $\alpha(2)\beta(1)$ describe what appear to be physically distinct states. Thus these wave functions are not really physically admissible descriptions of a two-electron system.

Very well, then, let us write a wave function that is antisymmetric with respect to interchange of electrons. We assume that the spatial and spin coordinates are separable, so that $\psi = \psi_{\text{spatial}}\psi_{\text{spin}}$. (For a two-electron system, this can always be done.) For the product to be antisymmetric, if one of the two factors is symmetric the other must be antisymmetric. We already have symmetric spatial functions, such as $1\sigma_g(1)1\sigma_g(2)$ in Eq. 6.47; thus we must look for an antisymmetric spin function. The trouble with our original $\alpha(1)\beta(2)$ was that interchange of electrons transformed it to the quite different function $\alpha(2)\beta(1)$. But suppose that we try a linear combination of these two functions, with opposite signs to give the antisymmetry:

$$\psi_{\text{spin,anti}} = \alpha(1)\beta(2) - \alpha(2)\beta(1). \tag{6.57}$$

This function does indeed change sign when we exchange 1 and 2. We can thus combine it with any symmetric spatial function to give an antisymmetric total wave function. The molecular orbital wave function for the ground state of H_2 then becomes

$$\psi_{\text{MO,anti}}[(1\sigma_g)^2] = \frac{1\sigma_g(1)1\sigma_g(2)}{\sqrt{2}}[\alpha(1)\beta(2) - \alpha(2)\beta(1)], \tag{6.58}$$

in which the $\sqrt{2}$ maintains the normalization. This function is antisymmetric with respect to interchange of the electrons and is thus physically admissable. Unfortunately, since the spatial part is the same as before, it is just as poor an approximation as Eq. 6.47. One can also obtain an antisymmetric wave function by combining an antisymmetric ψ_{spatial} with a symmetric ψ_{spin}; we shall see that such a function describes the first excited state of H_2.

What physical interpretation can we give to the wave function of Eq. 6.58? Both $\alpha(1)\beta(2)$ and $\alpha(2)\beta(1)$ are legitimate solutions of the wave equation, each describing a conceivable eigenstate of the system. The principle of superposition assures that a linear combination of such solutions is also a solution, with the squared coefficients of the individual terms giving the relative probabilities of observing the corresponding eigenstates.[16] Here the coefficients of the two terms are equal in magnitude, so the two states should be equally likely. In other words, if we make measurements on an H_2 molecule in its ground state, half the time we should find electron 1 with $m_s = +\frac{1}{2}$ and electron 2 with $m_s = -\frac{1}{2}$; the other half of the time we should find electron 1 with $m_s = -\frac{1}{2}$ and electron 2 with $m_s = +\frac{1}{2}$. Since we cannot tell electrons 1 and 2 apart anyway, this result is just what we should expect. In any case, the function of Eq. 6.58 specifies that we will definitely observe the two electrons to have opposite spins; whichever spin electron 1 may have, electron 2 has the opposite. So long as this function describes the spins, however far apart the electrons may be, they have opposite spins. This is sometimes taken to be a paradox, that observation of $m_s = +\frac{1}{2}$ for electron 1

---

[16] That is, if for some state of a system $\psi = \sum_i c_i\varphi_i$, where the φ_i are eigenfunctions of an operator Q, then in that state the probability of observing the ith eigenvalue Q_i is

$$P(Q_i) = |c_i|^2 = c_i{}^*c_i.$$

in Boston forces electron 2 in Los Angeles to have $m_s = -\frac{1}{2}$, by what is sometimes called "action at a distance." The resolution is of course that (a) if the two electrons are very far apart, it is unrealistic to expect that the function of Eq. 6.58 will describe them properly, but (b) so long as that function does describe the electrons, it specifies that the system of the two of them is in a state in which they always have opposite spins, so, whatever we observe for the spin of one *implies* what the spin of the other must be—not that the physics of measuring one spin in Boston causes a physical effect in Los Angeles. There is no mystery, only a condition set when the electrons are prepared in the state described by the function of Eq. 6.58.

Given the antisymmetry of electrons, one can derive the Pauli exclusion principle, which states that no two electrons can occupy the same quantum state. To see why this is so, let us try writing a wave function for two electrons in exactly the same state. This means that we assign to both electrons the same orbital φ_i, including both spatial and spin coordinates; for example, $\varphi_i(1)$ might be $1\sigma_g(1)\alpha(1)$. The antisymmetrized orbital-product wave function corresponding to Eq. 6.58 is then

$$\psi_{\text{anti}}(1,2) = \frac{1}{\sqrt{2}}[\varphi_i(1)\varphi_i(2) - \varphi_i(2)\varphi_i(1)] = 0. \qquad (6.59)$$

A wave function of zero magnitude describes a state of zero probability; in other words, though the state (Eq. 6.59) satisfies the antisymmetry requirement, it can never be observed. This proves the exclusion principle, which applies not just to electrons, but to all fermions. (It is possible to infer antisymmetry from the exclusion principle as well, but the argument is beyond the scope of this book.) By contrast, there is no such restriction on the quantum states of bosons. Since the wave function for two bosoms must be symmetric, suitable spin functions to accompany a symmetric spatial function include

$$\psi_{\text{spin,symm}} = \begin{cases} \alpha(1)\alpha(2), \\ \beta(1)\beta(2), \\ \alpha(1)\beta(2) + \beta(1)\alpha(2). \end{cases} \qquad (6.60)$$

The symmetric wave function for two bosons in exactly the same state can thus be written, for example, as

$$\psi_{\text{symm}}(1,2) = \varphi_i(1)\varphi_i(2)\alpha(1)\alpha(2). \qquad (6.61)$$

Since this function is in general nonzero, there is nothing to prevent two bosons, or indeed any number of bosons, from occupying the same quantum state.

Note that the exclusion principle as we have stated it is meaningful only when electrons can be assigned to individual atomic or molecular orbitals, with the wave function written as a linear combination of orbital products. This is logical, since only in the orbital approximation does each electron have its own set of quantum numbers. But this form of the exclusion principle is only a special case of the general rule, which we can add to the postulates of Section 3.8:

POSTULATE VII. Any eigenfunction of a many-particle system must be antisymmetric with respect to interchange of any two identical fermions, and symmetric with respect to interchange of any two identical bosons.

Which particles are fermions and which bosons is intimately connected to their spins: Particles with integral values of spin are bosons and particles with half-integral spin are fermions. The connection is well established, and within the context of chemistry appears to be rigorous. However, in the realm of physics of elementary particles, there are conjectures that under extreme conditions, it may be possible to observe mixed fermion-boson behavior; such mixing is called "supersymmetry."

How does one write a wave function for more than two electrons? We again use the orbital approximation, with φ_i including both spatial and spin coordinates. The antisymmetric wave function for two electrons in the orbitals φ_i and φ_j, respectively, must be of the form

$$\psi_{\text{anti}}(1,2) = \frac{1}{\sqrt{2}}[\varphi_i(1)\varphi_j(2) - \varphi_i(2)\varphi_j(1)] = \frac{1}{\sqrt{2}}\begin{vmatrix} \varphi_i(1) & \varphi_i(2) \\ \varphi_j(1) & \varphi_j(2) \end{vmatrix}, \qquad (6.62)$$

equivalent to a determinant of the orbital functions, in which the orbitals fix the row indices and the electrons the column indices. This can be generalized to any number of electrons: The appropriate wave function for N electrons in N orbitals is

$$\psi_{\text{anti}}(1,2,\ldots,N) = (N!)^{-1/2}\begin{vmatrix} \varphi_1(1) & \varphi_1(2) & \cdots & \varphi_1(N) \\ \varphi_2(1) & \varphi_2(2) & \cdots & \varphi_2(N) \\ \cdots & \cdots & \cdots & \cdots \\ \varphi_N(1) & \varphi_N(2) & \cdots & \varphi_N(N) \end{vmatrix}, \qquad (6.63)$$

a determinant in which each row corresponds to a given orbital and each column to a given electron. The $(N!)^{-1/2}$ is needed as a normalization constant, since the expansion of an $N \times N$ determinant has $N!$ terms. This function clearly satisfies the antisymmetry requirement, since any determinant changes sign when two columns are interchanged.

Now we can understand certain cryptic statements in Chapter 5. The trouble with a Hartree wave function like 5.15 is that it is not antisymmetric; instead one writes the determinantal function of expression 6.63. If the orbitals are chosen to give the lowest possible energy, within the approximation that each electron moves in the *average* potential of all the others, the result is called the

Hartree–Fock wave function for N electrons.[17] (Further discussion of the Hartree–Fock method is postponed until Chapter 8.) We can also see why the phenomena associated with the exclusion principle are often called "exchange" effects; the apparent repulsion ("exchange force") between electrons of the same spin merely expresses the fact that they must have different spatial orbitals, or else the total wave function would vanish by Eq. 6.59. We could have explained all this in Chapter 5, but we preferred to use the two-electron case of H_2 to illustrate the concepts, which are equally applicable to atoms and molecules.

6.8 H_2: The Valence Bond Representation

Now we can return to the problem of the H_2 molecule. Making the molecular orbital wave function antisymmetric does not keep it from going wrong as $R \to \infty$, so we still need another approach. Suppose we try to find an approximate wave function that joins smoothly with the separated-atom limit. What is the wave function in this limit? We suggested earlier that it should resemble Eq. 6.51 (which does give the correct energy), but that function is obviously not antisymmetric. Nevertheless, it does illustrate correctly that the wave function at $R = \infty$ must be made up of atomic orbitals. The simplest method of doing this was devised by W. Heitler and F. London in 1927, in what was historically the first quantum mechanical treatment of the chemical bond.

We begin by considering two completely independent hydrogen atoms, with one electron in each atom. Since the electrons are indistinguishable, we may have either electron 1 around nucleus A and electron 2 around nucleus B, or vice versa. Neglecting for the moment the symmetry problem, we have two possible spatial wave functions, $1s_A(1)1s_B(2)$ and $1s_A(2)1s_B(1)$. As for the spin, when the atoms are completely separated, we may assign either spin ($m_s = \pm\frac{1}{2}$) to each electron, independent of the other. This gives us four possible two-electron spin wave functions: $\alpha(1)\alpha(2)$, $\beta(1)\beta(2)$, $\alpha(1)\beta(2)$, and $\beta(1)\alpha(2)$. But in fact none of these spatial or spin functions is antisymmetric. How can we obtain an antisymmetric total wave function?

Let us write the total wave function as a product of spatial and spin functions. As in our molecular orbital functions, this form is chosen for mathematical and physical simplicity;[18] it certainly is not a correct form for the true wave function. For the product of spatial and spin functions to be antisymmetric, one must be symmetric and the other antisymmetric. As before, let us make symmetric and antisymmetric linear combinations of our simple functions. For the spatial functions we have

$$\psi_{\text{spat,symm}} = 1s_A(1)1s_B(2) + 1s_A(2)1s_B(1) \tag{6.64}$$

and

$$\psi_{\text{spat,anti}} = 1s_A(1)1s_B(2) - 1s_A(2)1s_B(1). \tag{6.65}$$

There are three simple symmetric spin functions,

$$\psi_{\text{spin,symm}} = \begin{cases} \alpha(1)\alpha(2), & (6.66a) \\ \beta(1)\beta(2), & (6.66b) \\ \alpha(1)\beta(2) + \alpha(2)\beta(1), & (6.66c) \end{cases}$$

and one antisymmetric spin function,

$$\psi_{\text{spin,anti}} = \alpha(1)\beta(2) - \alpha(2)\beta(1). \tag{6.67}$$

Combining these to give an antisymmetric product, we have, in this approximation, four possible wave functions generated from hydrogen $1s$ orbitals:

$$\psi_1 = C_1[1s_A(1)1s_B(2) + 1s_A(2)1s_B(1)] \quad [\alpha(1)\beta(2) - \alpha(2)\beta(1)], \tag{6.68}$$

$$\psi_2 = C_2[1s_A(1)1s_B(2) - 1s_A(2)1s_B(1)] \quad \alpha(1)\alpha(2), \tag{6.69}$$

$$\psi_3 = C_3[1s_A(1)1s_B(2) - 1s_B(2)1s_B(1)] \quad \beta(1)\beta(2), \tag{6.70}$$

$$\psi_4 = C_4[1s_A(1)1s_B(2) - 1s_B(2)1s_B(1)] \quad [\alpha(1)\beta(2) + \alpha(2)\beta(1)]. \tag{6.71}$$

The C_i's are normalization constants; for $R = \infty$ one readily evaluates them as $C_1 = C_4 = 1/2$, $C_2 = C_3 = 1 \neq \sqrt{2}$.

What is the significance of the four functions 6.68–6.71? They are all satisfactory wave functions for two $1s$ hydrogen atoms infinitely far apart—satisfactory in that they both give the correct energy and are antisymmetric. There are other such functions, but these are the only simple ones that are constructed from H($1s$) orbitals and can be factored into spatial and spin parts. Now we make the *approximation* that these exact solutions for $R = \infty$ can be applied to the H_2 molecule at finite R, with appropriate normalization constants. This is exactly analogous to the reasoning by which we wrote the MO wave functions of Eqs. 6.26 and 6.27 for H_2^+. Obviously the present wave functions cannot be factored into one-electron molecular orbitals. They are basically two-electron functions, and were specifically designed by Heitler and London to represent the electron-pair bond. Their purpose was to find a quantum mechanical basis for G. N. Lewis' concept that chemical bonds consist of electron pairs

[17] In an open-shell atom or molecule, with N electrons in *more than N* orbitals, one must replace Eq. 6.63 by a linear combination of determinants, one for each possible distribution of the electrons among the orbitals.

[18] In fact, this is possible only for the two-electron system. For three or more electrons there is no way to write an antisymmetric total wave function that can be factored into spin and spatial parts. There are systematic methods for dealing with such cases, but we need not consider them here.

shared between two atoms. Similar approximations can be made for more complicated molecules if one assumes that the molecular wave function is a product of two-electron functions, each representing a conventional chemical bond. Wave functions of this type are called *valence bond* (VB) functions, since only the valence electrons and their bonds are taken into account.

The functions of Eqs. 6.68 through 6.71 all correspond to the same energy at $R = \infty$, but this can no longer be true at finite R. The spatial functions of Eqs. 6.64 and 6.65 obviously correspond to different electron densities at any given point in space, and thus to different energy levels. We can neglect the effect of the electron spins on the energy. We conclude that, of our four wave functions, ψ_1 corresponds to one state of the H_2 molecule, whereas ψ_2, ψ_3, and ψ_4 all correspond to a single other state, or rather to three degenerate states. Two 1s hydrogen atoms thus give rise to both a singlet and a triplet term of the H_2 molecule.

We can analyze these terms in much the same way that we treated atomic terms in Section 5.6. Let α correspond to $m_s = +\frac{1}{2}$, so that in the state corresponding to ψ_2 the molecule has $M_s = \sum m_s = 1$. Similarly, ψ_3 describes a state in which each electron has $m_s = -\frac{1}{2}$ and the molecule $M_S = -1$. These two wave functions differ only in the direction of the spin vector, which is not physically significant in the absence of an external field. Thus they are degenerate, corresponding to the same energy at any value of R. Since a two-electron system can never have $|M_S| > 1$, the states described by ψ_2 and ψ_3 must be components of a term with $S = 1$. This is, of course, a triplet term, with $2S + 1 = 3$, and must thus have a third component.

The three components of a triplet have $M_S = 1, 0, -1$ ($S_z = \hbar, 0, -\hbar$). The third component must therefore be a state with $M_s = 0$, corresponding to one electron with $m_s = +\frac{1}{2}$ and one with $m_s = -\frac{1}{2}$. Since both the electrons and the atoms are indistinguishable, the wave function for this state should logically contain an equal mixture of $\alpha(1)\beta(2)$ and $\alpha(2)\beta(1)$. This condition is satisfied by the function ψ_4, which is also degenerate with ψ_2 and ψ_3 since it has the same spatial wave function. According to our approximate wave functions, then, the components of this triplet are exactly degenerate if one neglects spin–orbit interactions. To complete our term analysis, we are left with the function ψ_1, which also has $M_S = 0$; this can obviously describe the single component of a singlet term ($S = 0$).

Which of these two terms, the triplet or the singlet, has the lower energy? One could, of course, find out by solving for the $E(R)$ curves, but we can deduce the answer without doing this. Note that the spatial part of ψ_1 is the sum of two positive terms, which add constructively at every point in space. In contrast, the spatial part of the triplet wave functions has a nodal plane midway between the nuclei. Thus the singlet state has excess electron density in the bonding region, and can be assumed to have a minimum in the $E(R)$ curve, whereas the triplet state is certainly antibonding. This conclusion is confirmed by both calculation and experiment. The ground state is indeed a singlet state, with a fairly deep energy minimum. The first excited state is a triplet state in which the two atoms repel each other very strongly. Because both these states correlate with 1s atomic states, they are both Σ states ($\Lambda = 0$); the singlet is $^1\Sigma_g$ and the triplet is $^3\Sigma_u$. The exact $E(R)$ curves for these states are shown in Fig. 6.11; we shall discuss the excited states of H_2 in Section 6.10.

As for the ground state, one can obtain the valence bond $E(R)$ curve in the usual way. Normalization of Eq. 6.68 gives for the wave function

$$\psi_{\text{VB}} = \frac{[1s_A(1)1s_B(2)+1s_A(2)1s_B(1)][\alpha(1)\beta(2)-\alpha(2)\beta(1)]}{2(1+S_{AB}^2)}, \tag{6.72}$$

where S_{AB} is the overlap integral defined in Eq. 6.30. Evaluating $\iint \psi^* \mathsf{H} \psi \; d\tau_1 \, d\tau_2$ for this function, with normal atomic

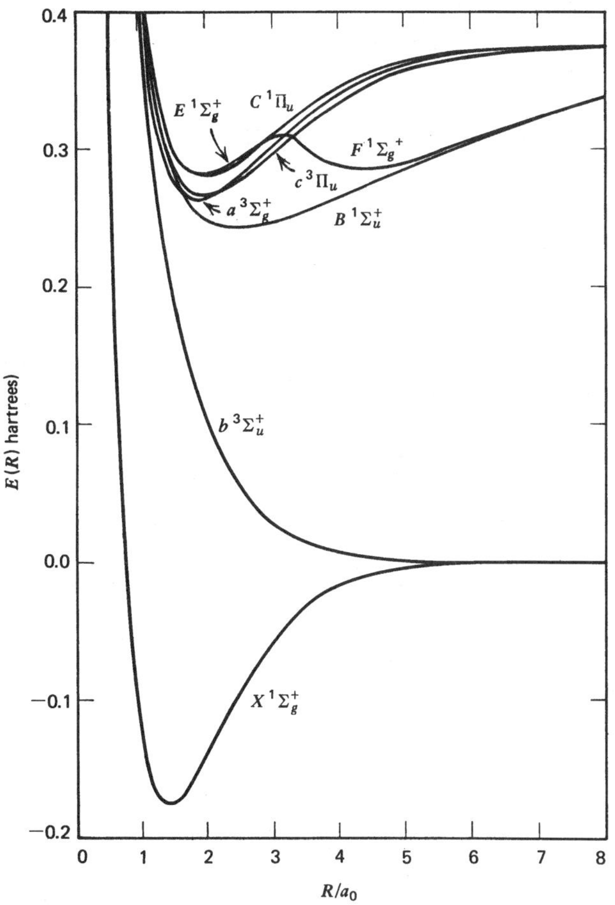

Figure 6.11 Potential energy curves for the ground state and the first few excited states of H_2; the energy zero corresponds to two separated H(1s) atoms. Each state is labeled with its spectroscopic term symbol, the initial letter of which is an arbitrary identifier (X always designates the ground state, excited singlets are usually assigned A, B, . . . , in order of increasing energy, and excited triplets a, b, . . . , similarly).

orbitals, gives curve 4 of Fig. 6.10. This is appreciably better than the simple MO solution (curve 1), and yields $R_e = 1.64a_0$ (0.87Å) and $D_e = 0.1154$ hartree (3.14 eV). As before, one can obtain considerable improvement by using scaled atomic orbitals. This gives curve 5 of the figure, with $R_e = 1.40a_0$, $D_e = 0.1383$ hartree, $\alpha(R_e) = 1.166$. Since this is better than even the self-consistent-field result, the valence bond method definitely seems preferable to the molecular orbital method in this case. However, the MO method is more easily extended to systems with more than two electrons. But the state of the art has long since gone beyond either approximation in its simple form, as we shall now see.

6.9 H_2: Beyond the Simple MO and VB Approximations

Thus far we have what seem to be two quite distinct kinds of wave function for the H_2 molecule. In the MO method we assume that the molecular wave function is a product of one-electron orbitals, then approximate each of these orbitals by a sum of atomic orbitals. In the VB method we form a two-electron function directly as an antisymmetrized combination of atomic orbital products. Although we have given plausible arguments for both methods, neither gives any more than a crude approximation to the true molecular wave function. In this section we shall look at some of the ways to obtain better approximations.

To begin with, let us examine the relationship between the MO and VB wave functions. For the ground state of H_2 these functions are given by Eqs. 6.58 and 6.72, respectively. Let us disregard the spin functions, which are the same in both cases, and concentrate on the atomic orbitals from which the spatial functions are composed. Expanding the molecular orbitals according to Eq. 6.33, we have

$$\psi_{\text{MO,spat}} = \frac{[1s_A(1)+1s_B(1)][1s_A(2)+1s_B(2)]}{2(1+S_{AB})} = \frac{1s_A(1)1s_A(2)+1s_B(1)1s_B(2) + [1s_A(1)1s_B(2)+1s_B(1)1s_A(2)]}{2(1+S_{AB})}, \tag{6.73}$$

whereas the corresponding VB function is

$$\psi_{\text{VB,spat}} = \frac{1s_A(1)1s_B(2)+1s_B(1)1s_A(2)}{\sqrt{2}(1+S_{AB}^2)}. \tag{6.74}$$

To understand the physical and essential mathematical content of these equations, we can ignore their denominators, which are merely normalization constants. If we compare the two functions, we see at once that the numerator of Eq. 6.74 is identical to the two terms in brackets in the numerator of Eq. 6.73. Thus the MO function contains the VB function. But the MO function contains two other terms that do not appear in the VB function.

How can we interpret this relationship? Each of the first two terms in Eq. 6.73 (and not in Eq. 6.74) describes a state in which both electrons are on the same atom, $H^- + H^+$, whereas the two terms in brackets describe states with one electron on each atom. It is customary to call these "ionic" and "covalent" terms, respectively. Since all the terms in each function have equal coefficients, one says that the MO function is an equal mixture of covalent and ionic terms, whereas the VB function contains only the covalent terms. We have gone through this argument before, in discussing Eq. 6.50: At $R = \infty$ the MO function literally describes a state that is 50% covalent and 50% ionic. This is *not* true at finite R, since the ionic–covalent cross terms in $(\psi_{\text{MO}})^2$ do not vanish. At finite distances, especially at the atomic scale, the atomic wave functions on A and B are not *orthogonal:* $\int \varphi_A^*(1)\varphi_B(1)\,d\tau_1 \neq 0$, which is a mathematical expression of the physical condition that the one-electron states $\varphi_A(1)$ and $\varphi_B(1)$ are not mutually exclusive states. (See Appendix 6A.) This in turn means that states represented by such product functions as $\varphi_A(1)\varphi_B(2)$ and $\varphi_A(1)\varphi_A(2)$ are also not mutually exclusive. Thus in this approximation one cannot divide the bond (i.e., the electron density) in any unique way into "covalent" and "ionic" parts. If we say that the wave function is 50% ionic, this is merely a convenient shorthand to describe the mathematical form of our approximate function.

To compare the two approximations further, let us consider how they treat the correlation between the positions of the two electrons. The valence bond function (Eq. 6.74) is highly correlated. For $R >> R_e$, ψ_{VB} is negligibly small unless the two electrons are near different nuclei; as we pointed out in connection with Eq. 6.51, this indeed describes the correct long-range behavior. But even at values of R near R_e, ψ_{VB} is quite small whenever the two electrons are close to each other.[19] In contrast to this, the ionic terms in Eq. 6.73 ensure that ψ_{MO} has appreciable magnitude even when both electrons are near the same nucleus. In fact, writing ψ as an orbital product explicitly assumes that there is *no* Coulombic correlation between the electrons. (On the other hand, both wave functions do contain what we have called exchange correlations between electrons of the same spin, as a direct result of the wave functions' being antisymmetric.) So one approximation to the singlet state contains no Coulombic correlation, whereas the other assumes strong correlation that becomes absolute at long range. A naive but sensible guess would say that the true extent of correlation is somewhere between these extremes.

[19] Each term in the VB function has the form $e^{-\alpha r_{1A}}e^{-\alpha r_{2B}}$, or its equivalent with 1 and 2 exchanged. If the nuclei are not close together, then at any point space either r_A or r_B (or both) must be large, and the corresponding $e^{-\alpha r}$ small. So if both electrons are near the same point, at least one of the factors in $e^{-\alpha r_{1A}}\,e^{-\alpha r_{2B}}$ is small.

Assuming that this is so, can we devise a wave function intermediate between the two extreme representations? This is quite easy to do. If we define

$$\psi_{\text{covalent}} \equiv 1s_A(1)1s_B(2)+1s_B(1)1s_A(2),$$
$$\psi_{\text{ionic}} \equiv 1s_A(1)1s_A(2)+1s_B(1)1s_B(2), \tag{6.75}$$

then it is immediately clear that both ψ_{MO} and ψ_{VB} are of the form

$$\psi_{\text{mixed}} = a(R)\psi_{\text{covalent}} + b(R)\psi_{\text{ionic}}, \tag{6.76}$$

where $a(R)$ and $b(R)$ are constants for any given value of R. In the MO approximation we impose the condition $a(R) = b(R)$; in the VB approximation we set $b(R) = 0$. But in accord with the variation principle, we can obviously obtain a better wave function by varying the ratio of $a(R)$ to $b(R)$ until $E(R)$ is minimized. As usual, the more variable parameters one introduces into the wave function, the more closely one can approximate the true solution of the Schrödinger equation. The results of this calculation are shown in Fig. 6.10 as curve 6 (using normal atomic orbitals) and curve 7 (using scaled atomic orbitals). The latter gives $R_e = 1.415a_0$ (0.749 Å), $D(e) = 0.1470$ hartree (4.00 eV), our closest approach yet to the true values. The ratio $b(R)/a(R)$ for curve 7 is 0.256 at R_e, and of course approaches zero as $R \to \infty$.

We are still a long way from the true wave function, but already we have gone beyond the simple MO and VB approaches. However, one can extend the language of either method to cover mixed functions like Eq. 6.76. In VB language, one describes the wave function as a linear combination of functions corresponding to hypothetical "structures," a structure being a particular arrangement of electrons and bonds. Thus ψ_{covalent} describes the covalent structure written as H:H or H—H, ψ_{ionic} is a combination of the ionic structures $H_A^-H_B^+$ and $H_A^+H_B^-$ (which must be equally likely), and ψ_{mixed} combines all three structures.

We can also reach Eq. 6.76 by an extension of the MO method. In the LCAO approximation, the ground state is one in which both electrons are in the orbital $1\sigma_g$, defined by Eq. 6.33 as a sum of 1*s* orbitals. But the difference of the same atomic orbitals gives rise to the excited (and antibonding) orbital $1\sigma_u$, defined by Eq. 6.34. Consider a state in which both electrons are in the orbital $1\sigma_u$. The LCAO molecular wave function, corresponding to Eq. 6.48, is then

$$\psi_{\text{MO}}[(\sigma_u 1s)^2] = \frac{[1s_A(1)-1s_B(1)][1s_A(2)-1s_B(2)]}{2(1+S_{AB})} \times \frac{[\alpha(1)\beta(2)-\alpha(2)\beta(1)]}{\sqrt{2}}. \tag{6.77}$$

As in the ground state, the spatial function is symmetric in the electrons, so the spin function must be antisymmetric. Expanding the spatial function as in Eq. 6.73, we have

$$\psi_{\text{MO,spat}}[(\sigma_u 1s)^2] = \frac{\begin{array}{l}[1s_A(1)1s_A(2)+1s_B(1)1s_B(2)\\ \quad -1s_A(1)1s_B(2)-1s_B(1)1s_A(2)]\end{array}}{2(1+S_{AB})} = \frac{\psi_{\text{ionic}} - \psi_{\text{covalent}}}{2(1+S_{AB})}, \tag{6.78}$$

which is again of the form of Eq. 6.76. Like the ground state, the state described by Eq. 6.77 is $^1\Sigma_g$, the product of two odd (u) functions giving an even (g) function. Since functions 6.58 and 6.77 have the same electronic symmetry, either *could* describe the ground state—and so, of course, could any linear combination of the two,

$$\psi_{\text{MO,mixed}} = x(R)\psi_{\text{MO}}[(\sigma_g 1s)^2] + y(R)\psi_{\text{MO}}[(\sigma_u 1s)^2]. \tag{6.79}$$

We can again improve the wave function by varying the ratio $x(R)/y(R)$ to minimize $E(R)$. But substitution of Eqs. 6.73 and 6.78 in Eq. 6.79 gives

$$\begin{aligned}\psi_{\text{MO,mixed}} &= \frac{\alpha(1)\beta(2)-\alpha(2)\beta(1)}{2\sqrt{2}(1+S_{AB})} [x(R)(\psi_{\text{ionic}}+\psi_{\text{covalent}}) + y(R)(\psi_{\text{ionic}}-\psi_{\text{covalent}})]\\ &= \frac{\alpha(1)\beta(2)-\alpha(2)\beta(1)}{2\sqrt{2}(1+S_{AB})} \{[x(R)+y(R)]\psi_{\text{ionic}} + [x(R)-y(R)]\psi_{\text{covalent}}\}\\ &= a'(R)\psi_{\text{covalent}} + b'(R)\psi_{\text{ionic}},\end{aligned} \tag{6.80}$$

an equation of the same form as Eq. 6.76. So since ψ_{covalent} and ψ_{ionic} are exactly the same functions in both cases, the variation to minimize $E(R)$ must give identical results in both cases: $a(R) = a'(R)$, $b(R) = b'(R)$. Thus the mixed wave function obtained is the same whether it is derived according to Eq. 6.76 or Eq. 6.79. It is purely a matter of taste and convenience which form one chooses for computation.

In MO language, a wave function like Eq. 6.79 or Eq. 6.80 is said to involve *configuration interaction* (CI). A "configuration" here has the same meaning as in our theory of atomic structure,[20] a particular distribution of electrons among orbitals. In this case our ground-state wave function is a mixture of the two configurations $(\sigma_g 1s)^2$ and $(\sigma_u 1s)^2$. One can improve the wave function still further by adding

[20] One can also carry out CI calculations on atoms, writing the total wave function as a sum of determinantal functions. This is the usual way of improving on the ordinary SCF calculations. As we saw in Chapter 5, however, the single-configuration model works rather well for atoms—at least for those with closed shells or a single electron outside closed shells.

additional configurations, as long as they have the correct symmetry properties. Given enough configurations, one may approach arbitrarily close to the true wave function, although after the first few important configurations are included, this convergence is usually rather slow. One is not restricted to LCAO orbitals, of course, and it is possible to define other orbitals that give more rapid convergence.

With orbitals chosen to give the most rapid convergence of a CI series, the so-called *natural* orbitals, the wave function for H_2 accounts for over 90% of the energy associated with electron correlation.[21] By using wave functions containing the interelectronic distance explicitly, it is possible to predict energies of dissociation and ionization for H_2 that lie within about 10^{-4} eV or less of the experimental values.[22]

There were, for a period, extended controversies over whether MO or VB wave functions were better for any particular molecule. In retrospect these arguments lose their force, because the two methods converge with the next approximation. This was recognized very early, of course, but for molecules larger than H_2 people thought, until the mid-1950s, that the next level would be too difficult to carry out. Since about 1960, however, computers have made quite extensive CI calculations feasible. Yet there are still many instances—in large molecules, or when only an approximate result is needed—in which the simple MO and VB approximations are used. Each has its advantages, and we shall use both approaches in the next few chapters. The simple VB method gives more accurate results in cases where correlation is very important; the MO calculations are much easier, and offer convenient ways to interpret spectra and bonding pictorially. Excited states are quite awkward to handle by VB methods. The "structures" of the VB method are easily visualizable chemical species, so that one can often use "chemical intuition" to guess the properties of the ground states of molecules that involve several such structures. Used naively, the VB method can cause misunderstandings because one can, in general, draw many more "structures" than the number of independent wave functions. Moreover the functions that represent different "structures" in the sense of representing each bond by a pair of electrons with opposite spins, are almost never orthogonal to one another. This means that different "structures" do not represent different, mutually-exclusive states. On the side of the MO method, electronic spectra are more easily interpreted in terms of single-electron transitions between molecular orbitals. The choice of method thus depends on what one wants to know, how accurate a result one wants, and how much work one is prepared to do.

Before leaving the ground state of H_2, we should say something about the most accurate calculations that have been performed. The best available CI calculations account for only about 97% of the correlation energy. One can do better by writing a function that explicitly contains electron correlation in the form of r_{12} terms, like Eq. 5.14 for the He atom. Given enough adjustable parameters to vary, one can get quite close to the true $E(R)$ curve, and thus presumably to the true wave function. Curve 8 of Fig. 6.10 was obtained by varying an 11-parameter function. Calculations have been made with as many as 50 parameters, giving an $E(R)$ curve that is correct within experimental error. Just as in atoms, unfortunately, it is impractical to extend calculations of this type to more than a few electrons.

6.10 H_2: Excited Electronic States

Thus far we have examined in detail only the ground state of the H_2 molecule. We have mentioned two excited states: the ${}^3\Sigma_u$ state whose VB approximation is given by Eqs. 6.69–6.71, and the ${}^1\Sigma_g$ state whose MO approximation is given by Eq. 6.77. In this section we shall consider the excited states of H_2 more systematically; $E(R)$ curves for the ground state ($X\ {}^1\Sigma_g^+$) and several excited states are given in Fig. 6.11.[23]

To begin with, let us find the MO representation of the first excited state. This is obviously a state in which one electron is in the lowest-energy orbital, $\sigma_g 1s$, whereas the other electron is in the lowest excited orbital, $\sigma_u 1s$. Previously we have considered the configurations $(\sigma_g 1s)^2$ (the ground state) and $(\sigma_u 1s)^2$, both of which can give only singlet states. The exclusion principle tells us that if both electrons are in the same orbital they must have opposite spins, giving $S = 0$. However, for the configuration $(\sigma_g 1s)(\sigma_u 1s)$ both singlet and triplet states are possible. The symmetries of these states follow the same rules as we developed in Section 6.8: The singlet state has a symmetric spatial function and antisymmetric spin function, the triplet states the reverse. The singlet state is thus represented by

$$\begin{aligned}\psi_1[(\sigma_g 1s)(\sigma_u 1s)] \\ = C_1[\sigma_g 1s(1)\sigma_u 1s(2) + \sigma_g 1s(2)\sigma_u 1s(1)] \\ [\alpha(1)\beta(2) - \alpha(2)\beta(1)],\end{aligned} \tag{6.81}$$

[21] S. Hagstrom and H. Shull, *Rev. Mod. Phys.* **35,** 624 (1963); E. R. Davidson and L. L. Jones, *J. Chem. Phys.* **37,** 2966 (1962); W. D. Lyons and J. O. Hirschfelder, *J. Chem. Phys.* **46,** 1788 (1967).

[22] W. Kolos and L. Wolniewicz, *J. Chem. Phys.* **41,** 3663, (1964); **43,** 2429 (1965); **49,** 404 (1968); **51,** 1417 (1969); **45,** 509 (1966); **48,** 3672 (1968); *Chem. Phys. Lett.* **24,** 457 (1974).

[23] The superscript $^+$ in the Σ term symbols means that the wave function is unchanged by reflection in a plane containing the internuclear axis. We shall say more about this symmetry property in the next chapter.

and the triplet state by

$$\begin{aligned}
&\psi_2[(\sigma_g 1s)(\sigma_u 1s)] \\
&\quad = C_2[\sigma_g 1s(1)\sigma_u 1s(2) - \sigma_g 1s(2)\sigma_u 1s(1)]\alpha(1)\alpha(2), \\
&\psi_3[(\sigma_g 1s)(\sigma_u 1s)] \\
&\quad = C_3[\sigma_g 1s(1)\sigma_u 1s(2) - \sigma_g 1s(2)\sigma_u 1s(1)]\beta(1)\beta(2), \\
&\psi_4[(\sigma_g 1s)(\sigma_u 1s)] \\
&\quad = C_4[\sigma_g 1s(1)\sigma_u 1s(2) - \sigma_g 1s(2)\sigma_u 1s(1)] \\
&\qquad [\alpha(1)\beta(2) + \alpha(2)\beta(1)].
\end{aligned} \tag{6.82}$$

These functions are analogous to Eqs. 6.68–6.71, but with the atomic orbitals of the latter replaced by molecular orbitals. In fact, if we expand the spatial part of Eqs. 6.82 in atomic orbitals, we obtain

$$\begin{aligned}
&\sigma_g 1s(1)\sigma_u 1s(2) - \sigma_g 1s(2)\sigma_u 1s(1) \\
&= \frac{[1s_A(1)+1s_B(1)][1s_A(2)-1s_B(2)] - [1s_A(2)+1s_B(2)][1s_A(1)-1s_B(1)]}{2(1+S_{AB})} \\
&= \frac{1s_A(2)1s_B(1) - 1s_A(1)1s_B(2)}{1+S_{AB}},
\end{aligned} \tag{6.83}$$

which differs only by a constant multiplier from the spatial part of Eqs. 6.69–6.71; all the "ionic" terms like $1s_A(1)1s_A(2)$ cancel out. In short, for the first excited state of H_2 the simplest MO and VB wave functions turn out to be identical.

As we pointed out earlier, this first excited state has the symmetry $^3\Sigma_u$. Its $E(R)$ curve, labeled $b\,^3\Sigma_u^+$ in Fig. 6.11, is everywhere repulsive. Like the ground state it gives H(1s) + H(1s) in the limit $R \to \infty$. The repulsive nature of this state is vividly exhibited by a hydrogen discharge lamp, the ultraviolet light from which consists largely of continuous radiation. Any state with a minimum in the potential energy curve is a bound state, and thus has quantized energy levels (in this case vibrational–rotational energy levels, which we shall discuss in the next chapter); but a state with no minimum is free and unquantized. Electrons dropping from higher orbitals into the $\sigma_u 1s$ orbital to give this triplet state thus produce continuous emission.

What about the singlet state represented by $(\sigma_g 1s)(\sigma_u 1s)$? If we expand the spatial part of Eq. 6.81 in atomic orbitals, we obtain

$$\begin{aligned}
&\sigma_g 1s(1)\sigma_u 1s(2) + \sigma_g 1s(2)\sigma_u 1s(1) \\
&= \frac{1s_A(1)1s_A(2) - 1s_B(1)1s_B(2)}{1+S_{AB}};
\end{aligned} \tag{6.84}$$

here the "covalent" terms cancel out, and we have the odd-parity equivalent of the ψ_{ionic} of Eq. 6.75. This state has the symmetry $^1\Sigma_u$, and is labeled $B^1\Sigma_u^+$ in Fig. 6.11. Its energy is relatively high: Eq. 6.84 predicts that it should give $H^-(1s^2) + H^+$ in the long-range limit. Actually, it gives H(1s) + H(2s), the energy of which is lower than that of two ions, but still quite high. Nevertheless, there is a minimum in the $E(R)$ curve, so that this configuration can exist as a stable excited state.

As for the $^1\Sigma_g$ state whose simplest MO representation is the doubly excited $(\sigma_u 1s)^2$, its energy is obviously still higher. The simple MO expansion (Eq. 6.78) again gives the wrong energy as $R \to \infty$, in fact the same wrong energy as we obtained for the ground state. The lowest excited $^1\Sigma_g$ state actually has two minima in the $E(R)$ curve, which appear spectroscopically to be distinct states (since each has its own set of vibrational–rotational levels); these are labeled $E^1\Sigma_g^+$ and $F^1\Sigma_g^+$ in Fig. 6.11. A single MO configuration cannot give a complicated curve like this; a much better representation is obtained by taking a mixture of the configurations $(\sigma_g 1s)(\sigma_g 2s)$ and $(\sigma_u 1s)^2$. If we wrote such a wave function in a form like Eq. 6.79, we would find that the coefficients $x(R)$ and $y(R)$ are rather rapidly varying functions of R. States of this type are fairly rare and can be considered as pathological cases. In most such cases, including this one, the outer minimum is described by a wave function with a large amount of ionic character—in this case the configuration $(\sigma_u 1s)^2$ which like $(\sigma_g 1s)^2$ is half "ionic" at long range.

We have now described the ground state and three excited states (one of them a triplet), and these are only the states that can be constructed from 1s atomic orbitals. This by no means exhausts the excited states of the H_2 molecule; there are, in fact, an infinite number of such states. Obviously we can also construct molecular states from H atoms one or both of which are themselves in excited states. Each combination of atomic orbitals can give two or more distinct molecular wave functions, corresponding to all the possible symmetric and antisymmetric combinations of electrons and nuclei.[24] Thus there are many more possible electronic states

[24] For example, consider the states formed from H(1s) + H(2s). There are four possible spatial wave functions, which in the valence bond approximation can be written:

$$1s_A(1)2s_B(2) + 1s_A(2)2s_B(1) + 1s_B(1)2s_A(2) + 1s_B(2)2s_A(1)$$
(symmetric in both electrons and nuclei, $^1\Sigma_g$),

$$1s_A(1)2s_B(2) - 1s_A(2)2s_B(1) + 1s_B(1)2s_A(2) - 1s_B(2)2s_A(1)$$
(symmetric in nuclei, antisymmetric in electrons, $^3\Sigma_g$),

$$1s_A(1)2s_B(2) + 1s_A(2)2s_B(1) - 1s_B(1)2s_A(2) - 1s_B(2)2s_A(1)$$
(symmetric in electrons, antisymmetric in nuclei, $^1\Sigma_u$),

$$1s_A(1)2s_B(2) - 1s_A(2)2s_B(1) - 1s_B(1)2s_A(2) + 1s_B(2)2s_A(2)$$
(antisymmetric in both electrons and nuclei, $^3\Sigma_u$).

Each of these describes a physically distinct state of the molecule (three degenerate states for the triplets).

of the molecule than of the H atom. Of course, only a limited number of the excited states have actually been observed, and not all of those have been fully analyzed.

To the extent that the molecular orbital approximation is valid, we can speak of singly and doubly excited states of H_2, the singly excited states being those in which one electron remains in the lowest ($1\sigma_g$) orbital. We have seen that the first excitation requires a large amount of energy, even for the lowest stable excited states. Just as in the helium atom (Section 5.3), a doubly excited state would have an energy well above the ionization limit, which in this case is the ground-state energy of $H_2^+ + e^-$. On the scale of Fig. 6.11, the minimum in the H_2^+ $E(R)$ curve lies at +0.397 hartree. Since all the stable singly excited states have energies not much below this,[25] one would expect them to bear some resemblance to the H_2^+ molecule-ion. In fact the $1\sigma_g$ orbital is very like the H_2^+ wave function, extending over a quite small region of space around the nuclei, whereas the excited electron occupies a very large orbital, which must resemble an excited orbital of the helium atom.[26] The higher the excitation, the less the orbitals overlap and the better the orbital approximation works. In effect, one has an H_2^+ ion with an electron orbiting it at long range. A highly excited electron must see the H_2^+ core as something close to a point charge. The orbital energies are then given reasonably well by a "Rydberg formula" like Eq. 2.68, and we speak of these highly excited states as molecular *Rydberg states.* There must be an infinite number of Rydberg states for any molecule having a stable molecule-ion like H_2^+. The excited electron is far from the nuclei, so it has very little effect on the bonding force between them. Thus the $E(R)$ curves for excited states of H_2 are often quite similar to that for H_2^+, as can be seen by comparing Figs. 6.6 and 6.11.

What about the doubly excited states, in which both electrons are energized? Not just for H_2 but for most molecules, doubly excited states generally have energies higher than the energy needed to remove a single electron. In the case of H_2, this means that the excited state has an energy higher than at least some states of H_2^+. Such an excited state can thus lose energy by getting rid of (ionizing) one of its electrons, the excess energy becoming kinetic energy of the ionized electron. This process is called *autoionization* or *preionization,*[27] and normally takes place with no emission of radiation. Autoionization thus competes with the various kinds of radiative transitions (Section 4.5) as a way for some excited states to decay. Although the relative efficiency of these decay processes varies widely with the states involved, it is clear that the possibility of autoionization shortens the lifetime of an excited state. In practice, it is mostly doubly excited states that can undergo autoionization, and many such states have very short lifetimes ($\approx 10^{-11} - 10^{-12}$s) compared with typical singly excited states ($10^{-7} - 10^{-8}$s). When it can occur, ionization is usually the mode selected by an excited molecule or atom to relieve itself of excess energy.

This completes our discussion of the electronic states of the H_2 molecule. But each of these states encompasses many vibrational and rotational energy levels. In the next chapter we investigate these levels and the spectra to which they give rise, and we extend our analysis to diatomic molecules more complicated than H_2.

[25] Those shown in Fig. 6.11 have $E(\infty) = +0.375$ hartree, corresponding to H(1s) + H (2s or 2p), whereas H_2^+ has $E(\infty) = +0.500$ hartree, corresponding to $H(1s) + H^+ + e^-$.

[26] As we pointed out for H_2^+, highly excited states are best described in terms of the united-atom limit. The united-atom, descriptions of the states in Fig. 6.11 are as follows (cf. Fig. 6.8):

$X^1\Sigma_g^+$	$E^1\Sigma_g^+$	$a^3\Sigma_g^+$	$B^1\Sigma_u^+$,	$b^3\Sigma_u^+$	$c^1\Pi_u$,	$c^3\Pi_u$
$(1s\sigma_g)^2$	$(1s\sigma_g)$	$(2s\sigma_g)$	$(1s\sigma_g)$	$(2p\pi_u)$	$(1s\sigma_g)$	$(2p\pi_u)$

[27] The corresponding process in excited atoms is called the *Auger effect.*

APPENDIX 6A

Orthogonality

The concept of orthogonality is a generalization of the geometric concept of perpendicularity. Two vectors **a** and **b** with real components are perpendicular, or orthogonal, if their scalar product is zero:

$$\begin{aligned}\mathbf{a}\cdot\mathbf{b} &= a_x b_x + a_y b_y + a_z b_z \\ &= |\mathbf{a}||\mathbf{b}| \cos\theta_{ab} \\ &= 0 \quad \text{if and only if (iff) } \mathbf{a} \text{ and } \mathbf{b} \text{ are orthogonal.}\end{aligned} \qquad (6A.1)$$

The first of these equations suggests (but does not prove) that the two vectors have no basis vectors in common. The value of a scalar product is independent of the choice of orientation of the coordinate system. If **a** and **b** are orthogonal, one can always choose an orientation for the coordinate system so that **a** and **b** have no common basis vectors; for example, **a** and **b** can be chosen to lie along the x and y axes.

The generalization of orthogonality from ordinary vectors in three-dimensional space to vectors in an n-dimensional abstract space is straightforward: If $\mathbf{a} = \{a_1, a_2, \ldots, a_n\}$ and $\mathbf{b} = \{b_1, b_2, \ldots, b_n\}$, then $\mathbf{a}\cdot\mathbf{b} = a_1b_1 + a_2b_2 + \cdots a_nb_n$; **a** and **b** are orthogonal iff $\mathbf{a}\cdot\mathbf{b} = 0$. If the components of the vectors are complex numbers instead of real numbers, then one writes $\mathbf{a}\cdot\mathbf{b} = a_1^*b_1 + \cdots a_n^*b_n$ for the scalar product. Again, if **a** and **b** are orthogonal, they are built from different, mutually exclusive sets of basis vectors of the space.

To generalize the idea of orthogonality from vectors to functions we replace the summation $a_1b_1 + \cdots + a_nb_n$ with an integration over the continuous variable or variables of the functions. Thus, if $\phi_A(x)$ and $\phi_B(x)$ are real and exist for $-\infty < x < \infty$, then the equivalent of the scalar product of ϕ_A and ϕ_B is the integral of their product over the range of their argument. If x ranges from $-\infty$ to ∞,

$$S_{AB} = \int_{-\infty}^{\infty} \phi_A(x)\phi_B(x)\,dx. \qquad (6A.2)$$

If $\phi_A(x)$ and $\phi_B(x)$ are complex and $-\infty < x < \infty$, then we write, for the equivalent of their scalar product, Eq. 6.30:

$$S_{AB} = \int_{-\infty}^{\infty} \phi_A^*(x)\phi_B(x)\,dx \qquad (6.30)$$

The functions $\phi_A(x)$ and $\phi_B(x)$ are orthogonal if and only if $S_{AB} = 0$.

Sometimes the range of the argument is finite. For example, consider $\phi_A = \sin mx$ and $\phi_B = \sin nx$, for $0 \le x \le 2\pi$:

$$\begin{aligned}S_{AB} = \int_0^{2\pi} \sin mx \sin nx\,dx &= 0 \quad \text{if } m \neq n, \\ &= \tfrac{1}{2} \quad \text{if } m = n,\end{aligned} \qquad (6A.3)$$

so $\sin mx$ and $\sin nx$ are orthogonal on the interval $0 \le x \le 2\pi$.

The second equality of Eq. 6A.1 shows that $\mathbf{a}\cdot\mathbf{b}$ is the projection of the vector **a** on the vector **b**, which, as the first equality shows, is the same as the projection of **b** on **a**. We can likewise think of Eq. 6A.1 or Eq. 6.30 as the projection of function ϕ_A on ϕ_B. The projection of one vector or function on another is the same, geometrically, as the overlap of one vector or function with the other. But the overlap of functions has a physical interpretation that goes beyond the simple geometric model. In physical terms, if two functions representing states of a system are orthogonal, then the states are mutually exclusive: A 1s state and a 2s state of the hydrogen atom represent two mutually exclusive states of that atom. The wave functions ψ_0 and ψ_1 of Eqs. 6.26 and 6.27 or 6.33 and 6.34 are orthogonal, and correspond to mutually exclusive states of an electron in the H_2^+ molecule.

APPENDIX 6B

Hermitian Operators

In Section 3.8, Postulate III supposes that every variable of classical mechanics can be represented by a linear mathematical operator. The statement of that postulate now needs to be made a bit stricter: To every variable of classical mechanics there corresponds a linear *Hermitian* operator (after the French mathematician of the nineteenth century, Charles Hermite). The reason: Hermitian operators have two properties that we would like observables to exhibit. For Hermitian operators, the eigenvalues, which correspond to the values the property can assume (Postulate IV), are *real,* and the eigenfunctions corresponding to different eigenvalues are *orthogonal.* Thus the observables are real rather than complex numbers, as we want them to be, and, from the interpretation of orthogonality in Appendix 6A, the eigenstates with different eigenvalues must be mutually exclusive states of the system.

An operator R is Hermitian iff (if and only if)

$$\int \phi_A^* \mathrm{R} \phi_B \, d\tau = \int (\mathrm{R}\phi_A)^* \phi_B \, d\tau. \tag{6B.1}$$

To give this a physical interpretation, note that $\mathrm{R}\phi_B$ is some new function ϕ_C, corresponding to the state of the system initially described by ϕ_B but then acted upon by the process represented by the operator R. Hence $\int \phi_A^* \mathrm{R} \phi_B \, d\tau$ is the projection of the state of the system in state *C,* represented by ϕ_C, onto the state *A,* represented by ϕ_A. If *C* and *A* are mutually exclusive, $\int \phi_A^* \phi_C \, d\tau \equiv \int \phi_A^* \mathrm{R} \phi_B \, d\tau = 0$; otherwise *A* and *C* share some aspects and have nonzero overlap.

The state represented by $\mathrm{R}\phi_A$ corresponds to a system initially in *A* and then acted upon by the operation corresponding to the operator R. Call this state *D,* and represent it with ϕ_D. Then $\int (\mathrm{R}\phi_A)^* \phi_B \, d\tau$ is the same as $\int \phi_D^* \phi_B \, d\tau$, the projection or overlap of state *B* onto state *D.* The operator R is Hermitian if the projection of *C* on *A,* $\int \phi_A^* \phi_C \, d\tau$, is the same as the projection of *B* on *D,* $\int \phi_D^* \phi_B \, d\tau$. In other words, an operator is Hermitian if and only if the same result occurs (but with proper allowance for complex conjugation, where $\sqrt{-1}$ appears) whether the operation R is performed on *B* and the result is projected onto *A,* or *B* is projected onto the result of performing R on *A.*

Let us show that the eigenvalues of a Hermitian operator are real. If Eq. 6B.1 holds and if $\mathrm{R}\phi_A = R_A \phi_A$, so ϕ_A is an eigenfunction of R, then

$$\int \phi_A^* \mathrm{R} \phi_A \, d\tau = R_A \int \phi_A^* \phi_A \, d\tau$$
$$= \int (\mathrm{R}\phi_A)^* \phi_A \, d\tau = R_A^* \int \phi_A^* \phi_A \, d\tau. \tag{6B.2}$$

so $R_A = R_A^*$ and R_A must be real. Similarly, if $\mathrm{R}\phi_B = R_B \phi_B$, so that both ϕ_A and ϕ_B are eigenfunctions of R, and if $R_A \neq R_B$, then $\int \phi_A^* \phi_B \, d\tau = 0$, since

$$\int \phi_A^* \mathrm{R} \phi_B \, d\tau = R_B \int \phi_A^* \phi_B \, d\tau$$
$$= \int (\mathrm{R}\phi_A)^* \phi_B \, d\tau = R_A \int \phi_A^* \phi_B \, d\tau. \tag{6B.3}$$

The equalities of 6B.3 can hold only if the overlap integral of ϕ_A and ϕ_B vanishes. Therefore the Hermitian operators do indeed have eigenvalues that are real and can correspond to observable properties, and the physical states corresponding to different observable values, that is, to different eigenvalues, are mutually exclusive.

• FURTHER READING

Herzberg, G., *Molecular Spectra and Molecular Structure, Volume I, Spectra of Diatomic Molecules,* 2nd Ed. (Van Nostrand Reinhold, New York, 1950).

Hurley, A. C., *Introduction to the Electron Theory of Small Molecules* (Academic, London, 1976).

Karplus, M., and Porter, R. N., *Atoms and Molecules* (Benjamin, Menlo Park, Calif., 1970), esp. Chapters 5, 6 and 7.

Kauzmann, W., *Quantum Chemistry* (Academic, New York, 1957), Chapter 11A, B.

Kondratyev, V., *The Structure of Atoms and Molecules* (Noordhoof, Groningen, The Netherlands, 1964), Chapter 7.

Mulliken, R. S., and Ermler, W. C., *Diatomic Molecules, Results of ab initio Calculations* (Academic, New York, 1977), esp. Chapters I–III.

Slater, J. C., *Quantum Theory of Molecules and Solids,* Volume I (McGraw-Hill, New York, 1963), esp. Chapters 1–4.

• PROBLEMS

1. In making the Born–Oppenheimer approximation, one supposes that the molecular wave function is separable

(Eq. 6.2), and thus that the electronic wave function $\psi_{elec}(r, R)$ is an eigenfunction of an electronic Hamiltonian H_{elec} for each fixed R. If one supposes that R is a quantum mechanical variable, then H_{nuc} can act on $\psi_{elec}(r, R)$ as well as on $\psi_{nuc}(R)$. When $\psi(r, R)$ is separable but H_{nuc} is allowed to act on ψ_{elec}, what terms occur in $\int \psi^*(r, R)\mathsf{H}\psi(r, R)\, d\tau$ that do not appear if H_{nuc} is only allowed to act on $\psi_{nuc}(R)$?

2. The binding energy of the deuteron, ^2H (D), is 2.2 MeV. The binding energy of the α particle is approximately 28.5 MeV. The first ionization potential of He is 24.5 eV; the second is 54.14 eV. Using a logarithmic scale for the energy, draw a curve of the total energy, including the nuclear binding energy, for the D_2 molecule going from the equilibrium internuclear distance down to $R = 0$. Follow the nuclear repulsion up to $R \sim 10^{-6} R_e$ and sketch the rest of the curve schematically. You may wish to put R on a nonlinear scale.

3. Show that if, in the diatomic molecule AB, the nuclear charges Z_A and Z_B are unequal, with $Z_A > Z_B$, then the region far enough from the molecule and out beyond B is a bonding region. At what distance beyond B is any electronic charge on the internuclear axis exactly nonbonding?

4. Construct the surfaces on which the bonding force is zero for the diatomic molecule LiF, that is, for $Z_A = 3$, $Z_B = 9$.

5. Figure 6.3 shows contours on which the bonding force of Eq. 6.8 is zero. For the system of two protons, calculate enough points to sketch contours for curves on which the bonding force is not zero; develop the curves for two negative (antibonding) values of $F_{bonding}$ and two positive values of $F_{bonding}$. Express your values in terms of $F_{bonding} \times (4\pi\epsilon_0/eq)$; give the value of $(4\pi\epsilon_0/eq)$, so a reader could translate your plot into SI units of force. (Eliminate one distance, r_A or r_B and one cosine, $\cos\theta_A$ or $\cos\theta_B$, from Eq. 6.8. You may want to use a programmable calculator or computer to map $F_{bonding}$, and then construct contours from the map.)

6. Compute the charge density in electrons per cubic angstrom and the expectation value of the electron's kinetic energy in the neighborhood of a nucleus and at the midpoint of the H—H axis in H_2^+ for the two normalized molecular orbitals Eqs. 6.33 and 6.34. Estimate the local kinetic energy from the slopes of the functions at their cusps and at $R/2$.

7. A proton B approaches a hydrogen atom A. At what internuclear distance R is the most probable electron-nuclear distance r_A of an undisturbed hydrogen atom equal to R? Where is the distance R equal to the mean value of r_A? (You may wish to look back to Chapter 4.)

8. The graphs of Fig. 6.7 show the charge density of the orbitals of Eqs. 6.33 and 6.34 along the internuclear axis. Construct similar curves for a line parallel to the internuclear axis but a distance $0.1a_0$ away from it.

9. The overlap integral of Eq. 6.30 can be evaluated easily by making the coordinate transformation $\lambda = (r_A + r_a)/R$, $\mu = (r_A - r_a)/R$.
(a) Show that surfaces of constant λ are ellipses and surfaces of constant μ are hyperbolas.
(b) Show that Eq. 6.33 and 6.34 take the forms

$$\psi_0 = \text{constant} \times e^{-\lambda R/2}(e^{\mu R/2} + e^{-\mu R/2})$$

and

$$\psi_1 = \text{constant} \times e^{-\lambda R/2}(e^{\mu R/2} - e^{-\mu R/2}),$$

or

$$\psi_0 = \text{constant} \times e^{-\lambda R/2} \cosh\left(\frac{\mu R}{2}\right)$$

and

$$\psi_1 = \text{constant} \times e^{-\lambda R/2} \sinh\left(\frac{\mu R}{2}\right).$$

(c) Show that the variable λ ranges from 1 to ∞, and that μ ranges from -1 to $+1$.

10. Show that the overlap integral

$$S_{AB} = \frac{1}{\pi}\int_{\substack{\text{all}\\ \text{space}}} e^{-\alpha r_A} e^{-\alpha r_B}\, d\tau$$
$$= e^{-\alpha R}\left(1 + \alpha R + \tfrac{1}{3}\alpha^2 R^2\right).$$

Use the fact that

$$\int_0^\infty x^n e^{-ax}\, dx = \frac{n!}{a^{n+1}};$$

to evaluate the term involving $e^{-\alpha(r_A + r_B)}$, use the transformation of Problem 9 and the fact that the volume element $dx\, dy\, dz$, when transformed to the coordinates λ, μ, ϕ, is $(R^3/8)(\lambda^2 - \mu^2)\, d\lambda\, d\mu\, d\phi$.

11. Show that atomic orbitals follow the rule that they are even (g) with respect to their nuclei if l is zero or even, and odd (u) if l is odd. In an H_2^+ molecule we can construct both g and u molecular orbitals from sums and differences of $p\sigma$ orbitals on the separated atoms. Resolve this apparent paradox.

12. Draw a correlation diagram connecting the orbital energies of the separated-atom limit with those of the united-atom limit for the one-electron molecule HeH^{2+}.

13. Write the analogs of Eqs. 6.58, 6.62, 6.68, and 6.69 if particles 1 and 2 are bosons.

14. What are the charge densities of the molecular orbital wave function, Eq. 6.73, and the valence bond wave function, Eq. 6.74, at the midpoint between A and B and at the nucleus A, when R has the value $1.415a_0$? Which function would you say delocalizes the charge more?

15. The molecule-ion HeH^+ has been studied both experimentally and theoretically. Its equilibrium internuclear distance is $1.4632a_0$. Construct a potential curve for the electrons along the He—H axis analogous to Fig. 6.4*a* and indicate the energies of the ground-state (1*s*) levels of the separated atoms. Sketch the free-atom 1*s* orbitals of H and of He^+, each centered at its appropriate nucleus. The lowest molecular orbital of HeH^+ is *not* well expressed by the form constant $\times$ $[\varphi_H(1s) + \varphi_{He^+}(1s)]$. Explain in physical terms why the lowest state of this species contains much more of one 1*s* orbital than of the other. Which predominates? What are the energies of the orbitals of He^+ with $n = 2$? What atomic orbitals of hydrogen and of the helium ion are likely to be important in the first excited state of HeH^+?

16. Draw a curve of the interference term of Eq. 6.28, $2\,\mathrm{Re}(\varphi_A^*\varphi_B)$, for the midpoint between the two nuclei in H_2^+, as a function of the internuclear distance R.

17. From Eqs. 6.41 and 6.42, and footnote 8 on page 223, compute the internuclear distance at which $E_0(R) = 0$, and the distances at which $E_0(R) = +1$ eV and $E_1(R) = +1$ eV. The latter correspond to the classical distances of closest approach for a proton colliding with a hydrogen atom along the $1\sigma_g$ and $1\sigma_u$ potential curves, respectively.

18. One of the dissociation limits of H_2 is the ion pair $H^+ + H^-$. Locate this dissociation limit on an energy scale, relative to the neutral states of the form $H(1s) + H(nl)$ and $H(2s) + H(nl)$. Where does the ionic curve of $-e^2/4\pi\epsilon_0 R$ fall on the diagram of Fig. 6.11? Can you associate any of the potential curves or portions of the curves of Fig. 6.11 with an electrostatic attraction $E(R)$ in Eq. 6.11?

19. If two nuclei move at a relative velocity of order 1% or more of the average speed of the electrons attached to them, one can be concerned about the validity of the adiabatic or Born–Oppenheimer approximation. What energy must a proton have, in its motion relative to a hydrogen atom, to achieve roughly 1% of the average speed of the electron in the transient H_2^+ molecule? Use the energy of the H_2^+ molecule at its equilibrium distance to estimate the electron's speed. (For a particle bound by Coulomb forces, $\langle \mathrm{K.E.}\rangle = -\frac{1}{2}\langle \mathrm{P.E.}\rangle$.)

20. Recall from Section 5.5 that the hydrogen atom forms a stable negative ion, H^-, and that the electron affinity of the hydrogen atom is approximately 0.75 eV. The H_2 molecule in its ground electronic and vibrational state does not form a stable negative ion. Give a physical interpretation of why no stable H_2^- can be made by attaching a slow electron to H_2, and explain what occurs when a hydride ion (H^-) and a hydrogen atom undergo a close, slow collision. Sketch a potential curve for $H^- + H$ together with the curve for the ground state of $H^- + H + e^-$ as given in Fig. 6.11.

21. By using the curves of Fig. 6.11, assuming that electronic excitation of a molecule of H_2 in its ground state occurs from the lowest point of the ground-state potential and that neither the position nor the momentum of the nuclei can change during the excitation process, estimate the wavelengths at which one might expect to observe transitions to the $b^3\Sigma_u^+$ and the $B^1\Sigma_u^+$ states.

22. From your knowledge of the spatial dimensions of the $1\sigma_g$ orbital of H_2 and of the atomic orbitals of the hydrogen atom, estimate at approximately what principal quantum number n the excited-state orbitals of H_2 could be well represented by atomic orbitals. How will the accuracy of this representation depend on the angular momentum $l\hbar$?

23. Perform a HF–SCF calculation for the ground electronic state of H_2 to investigate the effect of the basis set representation on computed values of D_e and R_e. Use the (a) STO–3G, (b) 3–21G, and (c) 6–31G basis sets. (See Problem 5.26 for basis set descriptions and software packages.)

24. In the HF–SCF method, the wave function is represented by a single electronic configuration. In Configuration Interaction (CI) methods, the wave function is represented by a sum of different configurations. A singly-excited configuration (single excitation) is formed by promoting a single electron from an occupied orbital to a previously unoccupied orbital. A doubly excited configuration (double excitation) is formed by promoting two electrons to previously unoccupied orbitals. The CISD method expresses the wave function as a sum of the Hartree–Fock configuration and all single and double excitations. Use the CISD method along with the (a) STO–3G, (b) 3–21G, and (c) 6–31G basis sets to calculate the values of D_e and R_e for the ground electronic state of the H_2 molecule.

25. An alternative approach (to CI) for estimating the correlation energy is perturbation theory. While perturbation theory calculations are size consistent, they are not variational (as are CI methods). Use the MP2 (second-order Møller–Plesset perturbation theory) method along with the (a) STO–3G, (b) 3–21G, and (c) 6–31G basis sets to calculate the values of D_e and R_e for the ground electronic state of H_2.

26. The potential energy surface (PES) of the excited electronic state $B\ {}^1\Sigma_u^+$ of the hydrogen molecule possesses a minimum, which can support bound states (see Fig. 6.11). Use the CISD method along with the (a) STO–3G, (b) 3–21G, and (c) 6–31G basis sets to calculate the values of D_e and R_e for this excited electronic state of H_2.

CHAPTER 7

More about Diatomic Molecules

Our study of H_2^+ and H_2 has given us some idea of the basic principles of chemical bonding and the electronic properties associated with the simple chemical bond. There is much more to learn about molecular structure, even for the relatively simple case of diatomic molecules. In this chapter we consider the general properties of diatomic molecules.

In Chapter 6 we were concerned only with the distribution and energy of the electrons; we only mentioned the motion of the atomic nuclei. Nuclei are never fixed in space. In a molecule, they oscillate around their equilibrium positions as if the bond were a spring; they can revolve about the molecule's center of mass; and the molecule as a whole can move in space. There are energy levels associated with all these motions, and we must find out what they are. Fortunately, we have already done most of the work to accomplish this, in terms of simple models. As we develop a description of these energy levels, we shall be able to analyze the full richness of molecular spectroscopy.

We are by no means through with electrons either. Although H_2 does illustrate the nature of the chemical bond, there are many phenomena that appear only in more complicated molecules. We shall discuss the general properties of diatomic molecules, mainly in terms of the molecular orbital approach, and go on to study several interesting classes of molecules in some detail. These include homonuclear molecules such as N_2, which have the same symmetry properties as H_2; first-row hydrides such as OH, the simplest heteronuclear molecules; and several series of related molecules in which trends can be discerned. We shall also look in a little more detail at how one may carry out calculations of electronic structure.

Let us examine more carefully a molecule of H_2 in its ground electronic state. We know the $E(R)$ curve, but just what does this curve mean? According to our interpretation in Chapter 6, $E(R)$ is the electronic energy of the molecule at internuclear distance R, calculated *with the nuclei fixed.* But the uncertainty principle ensures that no real molecule can have fixed nuclei. The more precisely we know the positions of the nuclei, the more uncertain become their momenta. Thus the best we can say (in classical language) is that the nuclei oscillate or vibrate within a certain range about their equilibrium positions. This language still assumes the molecule as a whole (i.e., its center of mass) to be fixed in space, but this cannot be generally true either: The molecule almost certainly has some translational momentum. Finally, the orientation of the molecule cannot be fixed in space, or the uncertainty in its angular momentum would be infinite. Thus we must also consider the molecule's rotation about its center of mass. We wish to describe all these types of motion.

7.1 Vibrations of Diatomic Molecules

Our starting point is again the Born–Oppenheimer approximation, which says that the nuclear and electronic motions are separable. We rationalized this on the basis that electrons move much faster than nuclei, and indeed Born and Oppenheimer derived the approximation from the full Schrödinger equation, cf. footnote 2 of Chapter 6. Just as we could consider the nuclei at rest when treating the electronic motion, we can consider the electrons as a smoothly averaged distribution that adjusts at once to the nuclear motion. The Born–Oppenheimer approximation is expressed by Eqs. 6.2 and 6.3; for the nuclear motion it yields Eq. 6.4, which we can rewrite as

$$\left(\frac{\mathbf{p}_A^2}{2m_A}+\frac{\mathbf{p}_B^2}{2m_B}\right)\psi_{\text{nucl}}(R_n)=[E-E(R)]\psi_{\text{nucl}}(R_n). \qquad (7.1)$$

Here R_n is shorthand for all the nuclear coordinates, E is the molecule's total energy, and $E(R)$ is its electronic energy—which is also the *effective* potential energy field in which the nuclei move.

Now we make another assumption, that the translational, vibrational, and rotational motions of the nuclei are also separable from one another. (This is not quite accurate, and later we shall mention some corrections to this approximation.) We thus have

$$\mathsf{H}_{\text{nucl}} \equiv \frac{\mathbf{p}_A^2}{2m_A} + \frac{\mathbf{p}_B^2}{2m_B} = \mathsf{H}_{\text{trans}} + \mathsf{H}_{\text{vib}} + \mathsf{H}_{\text{rot}}, \quad (7.2)$$

$$E - E(R) = E_{\text{trans}} + E_{\text{vib}} + E_{\text{rot}}, \quad (7.3)$$

and

$$\psi_{\text{nucl}}(R_n) = \psi_{\text{trans}}\psi_{\text{vib}}\psi_{\text{rot}}; \quad (7.4)$$

$\mathsf{H}_{\text{trans}}$ and ψ_{trans} depend only on the center-of-mass coordinates, H_{vib} and ψ_{vib} only on the internuclear distance R, and H_{rot} and ψ_{rot} only on the angles giving the molecule's orientation in space. For each type of motion we obtain a wave equation in the form $\mathsf{H}_j\psi_j = E_j\psi_j$. In a gas, the translational wave equation is simply that for a free particle in three dimensions, the solution of which we already know. The rotational motion will be considered in the next section; let us now analyze the vibrational motion.

Classically, the total energy of the nuclei can be written as

$$E - E(R) = \left[\frac{p_X^2 + p_Y^2 + p_Z^2}{2M}\right] + \left[\frac{p_R^2}{2\mu} + V(R)\right] + \left[\frac{1}{2\mu R^2}\left(p_\theta^2 + \frac{p_\phi^2}{R^2 \sin^2\theta}\right)\right], \quad (7.5)$$

where X, Y, Z are the coordinates of the center of mass, M is the total molecular mass, μ is the reduced mass $m_A m_B/(m_A + m_B)$, and p_R, p_θ, p_ϕ have the same meaning as in Eq. 3.147. When the p's are converted to the corresponding operators, the three expressions in square brackets become $\mathsf{H}_{\text{trans}}$, H_{vib}, and H_{rot}, respectively. According to the Hellmann–Feynman theorem, the potential energy $V(R)$ for vibration (motion along the internuclear axis) must be the same as our $E(R)$. Separating the variables in Eq. 7.1 in the usual way, we find that the wave equation for vibrational motion is simply

$$\mathsf{H}_{\text{vib}}\psi_{\text{vib}}(R) = \left[\frac{\mathsf{p}_R^2}{2\mu} + E(R)\right]\psi_{\text{vib}}(R) = E_{\text{vib}}\psi_{\text{vib}}(R), \quad (7.6)$$

where $p_R \equiv -i\hbar(\partial/\partial R)$, the operator for the relative momentum of the nuclei. Given the $E(R)$ curve for a molecule, one can solve Eq. 7.6 to obtain the vibrational eigenfunctions $\psi_{\text{vib}}(R)$ and eigenvalues E_{vib}. But the function $E(R)$ is in general quite complicated and known only in numerical form. Although Eq. 7.6 can be solved by numerical methods, it is useful to obtain a simpler function that will give a good approximate solution.

As we have indicated before, one of the scientist's most powerful tools for understanding a system is to find a parallel to it in a system already understood. By "parallel" we mean "similar in mathematical structure." If the parallelism is exact, then one has solved the new problem. Even if the parallelism is only approximate, one can generally learn a great deal about the new system by asking just how it deviates from the old one. The point of all this, of course, is that such a parallel model exists for molecular vibrations; not surprisingly, the model is our old friend the harmonic oscillator. Let us see why this is a good model.

The most significant characteristic of a molecule like H_2 is the existence of a stable minimum in the potential energy curve. Remember that the bonding force is $dE(R)/dR$; thus at the minimum R_e, where the slope of the $E(R)$ curve is zero, there is no net bonding force on the nuclei. If we squeeze or stretch the molecule, there will be a force tending to restore the internuclear distance to R_e. This restoring force is what gives rise to molecular vibrations, with the distance oscillating back and forth around the value R_e. This motion is mathematically equivalent to that of a single particle of mass μ in a one-dimensional potential well defined by $E(R)$. We have described the properties of such potential wells in Chapter 4. Bound states (those with $E < 0$, in terms of Fig. 6.2) have quantized energies; their wave functions are oscillatory where $E > V$, but decay rapidly toward zero in the classically forbidden region where $E < V$. The actual number of bound states and the energies at which they fall depend very specifically on the form of the potential energy curve. A molecule in a given electronic state may exist in any one of the vibrational states associated with the potential curve for that state. For the H_2 molecule in its ground electronic state, for example, spectroscopy reveals the existence of 14 bound vibrational states,[1] shown in Fig. 7.1.

We still have not introduced our model. We know that there is a restoring force tending to bring R to its equilibrium value. The simplest assumption one can make about the restoring force is that it is proportional to the displacement from equilibrium, $F(R) \propto |R - R_e|$. This is just what we have defined as a harmonic oscillator, for which Eqs. 4.4 and 4.5 give us

$$F(R) = -k(R - R_e) \quad \text{and} \quad V(R) = \tfrac{1}{2}k(R - R_e)^2, \quad (7.7)$$

with $R - R_e$, replacing the displacement x; here $F(R)$ is the magnitude of the force tending to increase R, and $V(R)$ is zero at the bottom of the potential well. To the extent that the harmonic oscillator model is valid, we can apply all the results of Section 4.2 to molecular vibrations. Obviously it is not really valid for any molecule, because $E(R)$ always levels off at large R, where the atoms dissociate and no longer interact. However, it is approximately valid for all stable molecular states near equilibrium. In other words, any $E(R)$ curve with a stable minimum can be approximated by a parabola in the vicinity of that minimum; this is illustrated for the H_2 molecule in Fig. 7.1.

---

[1] In the case of H_2, the complete (non-Born–Oppenheimer) wave equation can be solved with sufficient accuracy to confirm this experimental result.

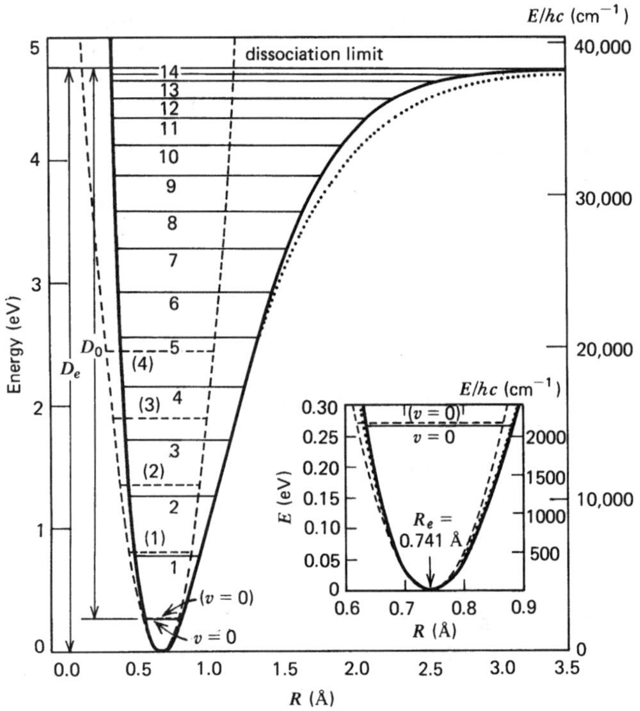

Figure 7.1 Vibrational energy levels of the H_2 molecule. Solid lines (——) represent the experimental $E(R)$ curve and energy levels; dashed lines (- - - -) give the parabolic approximation described in the text and the first few of the corresponding harmonic oscillator levels; and the dotted line (· · · · · ·) is a Morse potential fit. The inset shows an enlargement of the region near the minimum. The two "dissociation energies" are indicated: $D_e = 4.748$ eV, $D_0 = 4.477$ eV.

How good this approximation is depends on the spacing of the vibrational energy levels. For bound states at energies for which the true and parabolic curves nearly coincide, the harmonic oscillator model should give an excellent description. But how many states fall in this range of energies, and how likely is the molecule to be in these states? To find out, let us make an order-of-magnitude calculation.

For most stable diatomic molecules in their ground states, the potential well has a depth of about 5 eV, or 8×10^{-19} J. To keep the numbers simple, let us say that the best parabola fitted to the minimum is 1 Å wide ($R - R_e = 0.5$ Å) when $V(R) = 8 \times 10^{-19}$ J; this is approximately correct for H_2. Solving Eq. 7.7 for the force constant k, we obtain

$$k = \frac{2V(R)}{(R-R_e)^2} = \frac{2 \times 8 \times 10^{-19}\ \mathrm{J}}{(0.5 \times 10^{-10}\ m)^2} = 640\ \mathrm{N/m}.$$

The force constant of a harmonic oscillator is related to the frequency by Eq. 2.38; here, for the mass m we must use the molecule's reduced mass μ. For the H_2 molecule μ is just half the mass of a single hydrogen atom, or about 8×10^{-28} kg. The oscillator frequency in our approximate model should thus be

$$\nu(H_2) = \frac{1}{2\pi}\left(\frac{k}{\mu}\right)^{1/2} \approx \frac{1}{2\pi}\left(\frac{640\ \mathrm{N/m}}{8 \times 10^{-28}\ \mathrm{kg}}\right)^{1/2}$$
$$= \frac{(80 \times 10^{28}\ \mathrm{s}^{-2})^{1/2}}{2\pi} \approx 1.5 \times 10^{14}\ \mathrm{s}^{-1},$$

equivalent to a wavenumber $\nu / c \equiv \tilde{\nu} \approx 5000\ \mathrm{cm}^{-1}$. Since the spacing of harmonic oscillator energy levels is $h\nu$, this corresponds to

$$\Delta E(H_2) = h\nu(H_2) \approx (6.6 \times 10^{-34}\ \mathrm{J\ s})(1.5 \times 10^{14}\ \mathrm{s}^{-1})$$
$$\approx 10^{-19}\ \mathrm{J},$$

which is one-eighth of our assumed well depth. The lowest energy level is $\frac{1}{2}h\nu$ above the bottom of the well (the zero-point energy), the next level is $h\nu$ higher, and so on. Looking at Fig. 7.1, we can see that the harmonic oscillator model should be reasonably good for at least the first level or two. For many processes at room temperature it is only these levels that will be significantly occupied, as we shall see in Part II.[2] All other molecules have greater reduced masses than H_2, and none have significantly larger force constants. Thus the harmonic oscillator model gives a fairly good description of the energy-level spacing in nearly all stable diatomic molecules. (There are a few exceptions—molecules with very shallow or highly nonparabolic potential wells, such as HgH or Ar_2.)

To check the accuracy of our calculations, let us now look at some data. The vibrational energy levels of a diatomic molecule can be fitted (see below) to a power series in the quantum number v, beginning with a harmonic oscillator term $(v + \frac{1}{2})h\nu_e$. For $^1H_2$ the frequency ν_e is found to be $1.3192 \times 10^{14}\ \mathrm{s}^{-1}$, which is quite close to our crude estimate.[3] The various isotopic forms of the H_2 molecule presumably have the same electronic structure and thus the same potential curve.[4] Then the best parabolic fits to $E(R)$

[2] The relative population of two energy levels in thermal equilibrium is given by

$$N_2 / N_1 = e^{-(E_2 - E_1)/k_B T},$$

where $k_B = 1.38 \times 10^{-23}$ J/K and T is the absolute temperature. At room temperature (300 K) we have, for two levels 10^{-19} J apart,

$$\frac{N_2}{N_1} = e^{-10^{-19}\ \mathrm{J}/(1.38\times10^{-23}\ \mathrm{J/K})(300\ \mathrm{K})} \approx e^{-24} \approx 4 \times 10^{-11};$$

in other words, each level will contain only 4×10^{-11} as many molecules as the one below it. Since there are about 3×10^{22} gas molecules/liter under ordinary conditions, only about 50 molecules/liter will be in the third vibrational level.

[3] Working backward from this frequency gives a force constant

$$k = 4\pi^2 \mu \nu_e^2 = 4\pi^2 (8.3676 \times 10^{-28}\ \mathrm{kg})(1.3192 \times 10^{14}\ \mathrm{s}^{-1})^2$$
$$= 574.9\ \mathrm{N/m},$$

the value used to draw the parabolic curve in Fig. 7.1.

[4] This is not quite true. As in atoms (see footnote 24 on page 208), there is a very small isotopic shift in the electronic energy, but this effect can be ignored here.

Table 7.1 Test of Harmonic Oscillator Model with Vibrational Constants of H_2 Isotopes

(Atomic masses: ^{1}H, 1.00782 amu; D, 2.01410 amu)

Molecule	Reduced Mass, μ (amu)	Vibrational Frequency, $\nu_e \times 10^{-14}$(s^{-1})	$\mu^{1/2}\nu_e$ 10^{14}amu$^{1/2}$s^{-1}
$^1H_2$	0.50391	1.3192	0.9365
^{1}HD	0.67171	1.1429	0.9367
D_2	1.00705	0.9345	0.9378

should all have the same k, and Eq. 2.38 shows the harmonic oscillator frequency should be proportional to $\mu^{-1/2}$. We can thus test the validity of the harmonic oscillator model with the measured values of the vibrational constant ν_e. If the model were valid, ν_e would be the oscillator frequency and all the isotopes would have the same value of $\mu^{-1/2}\nu_e$. The data are given in Table 7.1. The deviations from exact correspondence to $\nu_e \propto \mu^{-1/2}$ are largely due to the deviation of the real curve from its parabolic approximation.

Since the harmonic oscillator model does work rather well for most diatomic molecules, we can apply the results of Section 4.2. The vibrational energy levels should thus be given by

$$E_{\text{vib}}(\upsilon)+\left(\upsilon+\tfrac{1}{2}\right)h\nu_0 \quad (\upsilon=0,1,2,\ldots), \tag{7.8}$$

where ν_0 is the oscillator frequency and υ is the conventional symbol for the vibrational quantum number in a diatomic molecule. The vibrational wave functions in this model are $\psi_\upsilon(z)$, as listed in Table 4.1, in terms of the dimensionless displacement variable $z \equiv (4\pi^2\mu\nu_0/h)^{1/2}(R - R_e)$. For example, the ground vibrational state is described by a wave function with a single maximum at R_e; the first excited state, by a wave function with two maxima and a node at R_e, and so on. The vibration is not strictly harmonic, so these are only approximations to the true wave functions (for which the ground state's maximum and the next state's node will not lie exactly at R_e); but the qualitative behavior is given correctly. As υ increases, the harmonic model becomes less accurate. But as υ increases, the molecule behaves more and more like a classical oscillator, in that the difference between adjacent energy levels becomes a small fraction of the total vibrational energy.

Note in Fig. 7.1 the distinction between the two "dissociation energies," D_e and D_0. The well depth D_e is fundamentally more interesting to the theoretician. However, a measurement of the minimum energy required to dissociate a molecule in its ground state yields D_0, since no molecule can have less than the zero-point vibrational energy. For the same reason, when one calculates the energy change of a chemical reaction in which a given bond is formed or broken, it is D_0 that must be taken into account. For example, the energy required to bring about the reaction

$$H_2(g)+Cl_2(g)\rightarrow 2HCl(g),$$

with reactants and product in their ground states, is

$$\Delta E_{\text{reac}} = D_0(H_2)+D_0(Cl_2)-2D_0(HCl).$$

Thus, only D_0 is properly called the dissociation energy, and, to be strict, we should but will not always refer to it by that name. When a distinction must be made, D_e can be called the well depth. In the harmonic oscillator model, the two quantities are related by the equation $D_e = D_0 + \frac{1}{2}h\nu_0$.

Suppose that we want to describe molecular vibrations with more accuracy than the harmonic oscillator model can yield, but without going all the way to a numerical solution. How can we do this? The harmonic model approximates the potential curve by a parabola. The actual curve, however, rises more steeply at small R and falls off more slowly at large R. We can thus improve the fit by adding to $V(R)$ a correction that is positive for $R < R_e$ and negative for $R > R_e$. An obvious choice is a term proportional to $(R - R_e)^3$, which changes sign at $R = R_e$. We can thus replace Eqs. 7.7 by

$$F(R)=-kx+k'x^2 \quad \text{and} \quad V(R)=\tfrac{1}{2}kx^2-\tfrac{1}{3}k'x^3, \tag{7.9}$$

with $x \equiv R - R_e$. This is an *anharmonic oscillator* model, with the added term called the *anharmonicity*. Equations 7.9, of course, fit the experimental data appreciably better near R_e than does the harmonic model. For most vibrational levels observed at room temperature the anharmonicity effect simply shifts E_{vib} by a small amount, generally less than 1%. However, this model is clearly unsuitable for describing the vibrational levels high in the potential well (which might be occupied in a very hot gas), since the predicted $V(R)$ diverges at large R instead of leveling off. One can obtain a still better fit by adding a term in x^4, a term in x^5, and as many terms as the data can justify; given enough terms, one can approximate the potential curve to any desired accuracy. In general, the energy eigenvalues can be written in the form

$$E_{\text{vib}}(\upsilon) = h\nu_e\left[\left(\upsilon+\tfrac{1}{2}\right)-x_e\left(\upsilon+\tfrac{1}{2}\right)^2+y_e\left(\upsilon+\tfrac{1}{2}\right)^3+\cdots\right] \quad (\upsilon=0,1,2,\ldots), \tag{7.10}$$

where ν_e, x_e, y_e, . . . are constants that can be fitted to the spectroscopic data;[5] this equation defines the ν_e we introduced in Table 7.1.

[5] Since most spectroscopists express their data in wavenumbers, one usually finds tabulated the values of $\tilde{\nu}_e \equiv \nu_e/c$, $\tilde{\nu}_e x_e$, $\tilde{\nu}_e y_e$, ... The wavenumber corresponding to $E_{\text{vib}}(\upsilon)$ is called $G(\upsilon) \equiv E_{\text{vib}}(\upsilon)/hc$. A general expansion of the energy of nuclear motion in powers of the quantum numbers is called a Dunham expansion.

The experimental $E(R)$ curve can be obtained by numerical calculation from the observed $E(v)$ values, but this is difficult. However, there are a number of empirical functions in closed form that can be used to give potential energy curves that offer adequate approximations for many purposes. The best known of these is the *Morse potential,*[6]

$$V(R) = D_e[1 - e^{-a(R-R_e)}]^2, \tag{7.11}$$

where D_e is the well depth and a is a constant; to a good approximation one has $a = \pi\nu_e(2\mu/D_e)^{1/2}$. Note that this function correctly rises steeply at small R and levels off at large R, giving $V(0) = \infty$ and $V(\infty) = D_e$. In fact the Schrödinger equation can be solved with $V(R)$ given by Eq. 7.11. The energy eigenvalues of the Morse potential are given by the first two terms of Eq. 7.10 with $x_e = h\nu_e/4D_e$, and can be used to approximate the true energy levels all the way to the dissociation limit. A Morse potential curve for H_2 is included in Fig. 7. 1; note that the agreement is rather poor at large R.

One can learn much about the vibrational energy levels of molecules by observing transitions between these levels. Such transitions can be induced by radiation, giving rise to discrete spectral lines in the usual manner (Section 4.5). Molecular vibrations are most commonly studied by observing the vibrational absorption spectra, which for diatomic molecules are found primarily in the near-infrared region. Most observed values of ν_e lie in the range $(0.6–12) \times 10^{13}$ s^{-1}, corresponding to wavenumbers of 200–4000 cm^{-1} or wavelengths of 2.5–50 μm (cf. Table 7.2). The molecules observed can be in gas, pure liquid, solution, or solid form; the effects of the surrounding medium on ν_e are usually small but observable and not negligible. But not all molecules display such spectra. In fact, isolated homonuclear molecules such as H_2 or O_2 have no pure vibrational spectra at all. To see the reasons for this and other features of vibrational spectra, let us examine the excitation process in greater detail.

We consider excitation in terms of a classical model, essentially the same as that we applied to the hydrogen atom in Section 4.5. Any heteronuclear diatomic molecule has a dipole moment, since no two elements have exactly the same electronegativity. Thus such a molecule can be considered as an oscillating dipole, as illustrated in Fig. 7.2*a* (where $+q$ and $-q$ indicate the positive and negative ends of the dipole, respectively). Suppose that the molecule is bathed in infrared radiation, the wavelength of which is far greater than any dimension of the molecule. We can thus assume the electric field of the radiation to be uniform in space near the molecule, but oscillatory in time. At a given time, the component of this field parallel to the bond axis[7] exerts an instantaneous force on the dipole that tends to either stretch or compress the bond (Fig. 7.2*a*).

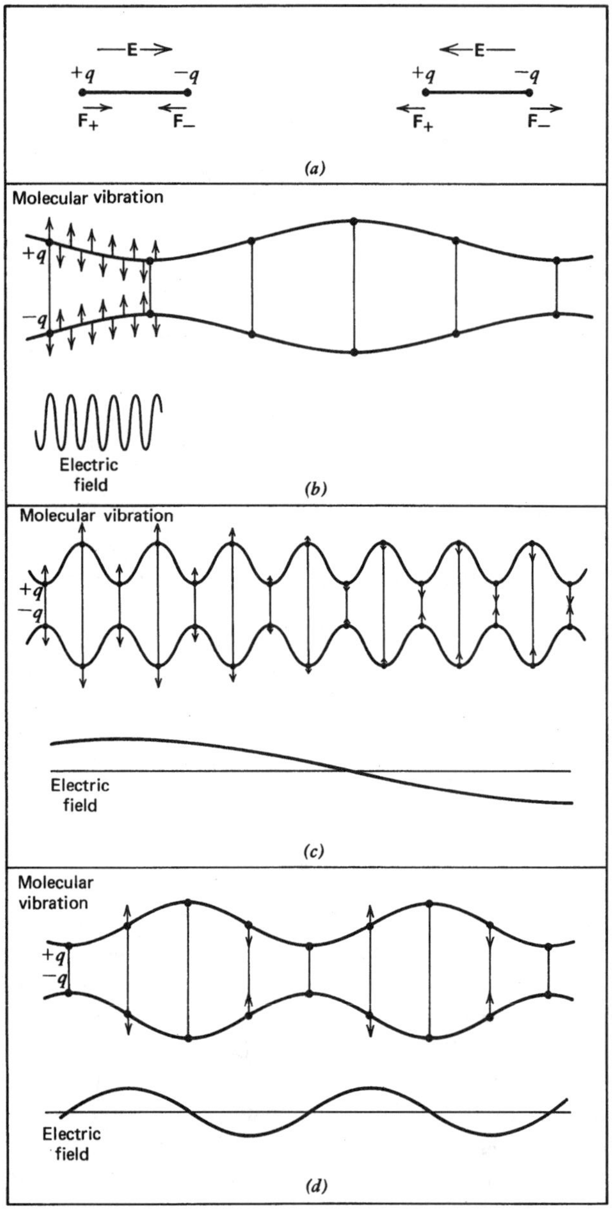

Figure 7.2 Interaction between an electric field and an oscillating (molecular) dipole. (*a*) The forces exerted on the dipole by the field, which alternately tend to compress or stretch the dipole as the field direction changes. (*b*) Field varies much faster than the molecular vibration ($\nu >> \nu_e$); here and in the subsequent diagrams, the instantaneous forces on the dipole are indicated by arrows. (*c*) Field varies much slower than the molecular vibration ($\nu << \nu_e$). (*d*) Field and dipole oscillate at the same frequency ($\nu = \nu_e$), but 90° out of phase. As shown here, the dipole absorbs energy from the field; for a field phase 180° different, the forces are reversed and the dipole gives up energy to the field.

The effect of this oscillating force on the molecular vibration is the same as in our earlier analysis. Let the field frequency be ν and the molecule's natural vibration frequency be ν_e. If $\nu >> \nu_e$ (Fig. 7.2*b*), the field reverses itself

[6] P. M. Morse, *Phys. Rev.* **34,** 57 (1929).

[7] Since the force exerted by an electric field acts in the same direction as the field itself (**F** = q**E**), the field components perpendicular to the bond axis do not affect the bond length and can be neglected here.

many times during a single vibration period, and can only impose a slight quivering onto the normal vibration. If $\nu << \nu_e$ (Fig. 7.2*c*), the field varies slowly compared to the vibration rate, and can only slightly increase or decrease the average value of R for some number of cycles. In either of these extremes, the field cannot easily transfer energy to the molecule. But when $\nu = \nu_e$, and the phase relationship is right (Fig. 7.2*d*), the field stretches the bond when it is expanding and compresses it when it is contracting, or vice versa. These are just the conditions that maximize energy exchange between the field and the dipole, which absorbs energy in the first case and gives it up in the second. Such an exchange of energy is an *electric dipole* transition.

Thus far we have been considering a heteronuclear molecule, which is necessarily an oscillating dipole. But what if the two ends of the molecule are identical, as in H_2 or any other homonuclear diatomic? Can such a molecule, with no dipole moment, absorb electromagnetic radiation? Our model says no, provided that the electric field is uniform over the length of the molecule; for then the forces on the two ends of the molecule are equal at all times, and the oscillations of the field can have no effect on the bond length. In our simple model, this is the explanation for the absence of vibrational spectra of homonuclear molecules.

This classical model is inadequate for a full explanation of vibrational spectra. For one thing, a classical dipole can absorb energy in any amount from an oscillating electric field, with its own vibration amplitude varying continuously; we know that this is not true of a quantized oscillator. To go beyond the simple model, we refer to Eq. 4.33, which gives the probability of a transition between any two quantum states. Here we are interested only in the vibrational part of the wave function, for which Eq. 4.33 reduces to

$$\mu_{v'v} = \int_0^\infty \psi_{v'}{}^*(R)\mu(R)\psi_v(R)\, dR, \tag{7.12}$$

where $\mu(R)$ is the molecule's instantaneous dipole moment and ψ_v and ψ_v' are the vibrational wave functions in states v and v'. The probability of a transition between these two states is proportional to $|\mu_{v'v}|^2$. It is immediately clear what happens in a homonuclear molecule: The dipole moment is zero for all values of R, the integral $\mu_{v'v}$ vanishes for all v, v', and there is thus zero probability of any vibrational electric dipole transition.

We have thus shown (on two levels) why homonuclear diatomic molecules have no pure vibrational spectra. But if that is the case, then how does one obtain vibrational energy levels like those in Fig. 7.1? One way is to look at the vibrational "structure" of *electronic* spectra. Transitions between different bound electronic states can be associated with specific initial and final vibrational states. In particular, transitions from a single vibrational level in an upper electronic state to a set of vibrational levels in a lower electronic state involve a series of quanta (a *band* of spectral lines) whose energies differ by the vibrational energy interval in the lower state. We shall consider spectra of this type in Section 7.3. One can also observe vibrational transitions in homonuclear diatomics directly by bombarding the molecules with electrons of precisely known energy and measuring the characteristic energy losses of the scattered electrons; this is essentially a variation of the Franck–Hertz experiment. In such a collision, of course, the electric field due to the electron does vary over molecular dimensions, and our previous arguments prohibiting vibrational transitions no longer apply.

Now let us return to Eq. 7.12 and see what transitions are allowed in heteronuclear molecules. For the integral to be nonzero it is not enough that $\mu \neq 0$; the dipole moment of the molecule must vary[8] with R. But this is no problem, because the dipole moments of all polar molecules do vary with the internuclear distance R. An especially simple result is obtained with the harmonic oscillator model, the vibrational wave functions of which are defined by Eq. 4.8. Setting $|\mu| = qR$ (with q constant), one can show without much difficulty that the integral of Eq. 7.12 is nonzero only when v and v' differ by 1. In other words, for the harmonic oscillator we have the *selection rule* for electric dipole transitions

$$\Delta v = \pm 1; \tag{7.13}$$

all other transitions are forbidden. This means that the quantum mechanical harmonic oscillator can absorb or give up only one quantum at a time.

The mathematical basis for the selection rule 7.13 can be seen easily if we consider the situation of small-amplitude oscillations, which are the only ones for which the harmonic approximation and the $\Delta v = \pm 1$ rule apply. If we so restrict our model, then we can write our wave functions $\psi_v(R)$ as functions of the displacement of R from the equilibrium distance R_e (μ is the reduced mass, not to be confused with the dipole moment μ):

$$z = (\mu\omega / \hbar)^{1/2}(R - R_e), \tag{7.14}$$

and the vibrational wave function is given by Eq. 4.8. Moreover, $\mu(R)$ can also be expanded about R_e, in a Taylor series—i.e., in powers of the displacement and successively higher derivatives of μ:

$$\mu(R) = \mu(R_e)\frac{d\mu(R)}{dR}\bigg|_{R_e} z + \frac{1}{2}\frac{d^2\mu(R)}{dR^2}\bigg|_{R_e} z^2 + \cdots. \tag{7.15}$$

[8] If μ is constant, it can be taken out of the integral, which then becomes $\int_0^\infty \psi_{v'}{}^* \psi_v \, dR = 0$, vanishing because the vibrational eigenfunctions are orthogonal (see Appendix 6A).

Note that $\mu(R_e)$ and all the derivatives of $\mu(R)$ evaluated at $R - R_e$ are constants. Hence the *transition dipole,* Eq. 7.12, becomes

$$\mu_{v'v} = \mu(R_e)\int \psi_{v'}(z)\psi_v(z)\,dz + \left.\frac{d\mu(R)}{dr}\right|_{R_e}\int \psi_{v'}(z)z\psi_v(z)\,dz + \frac{1}{2}\left.\frac{d^2\mu(R)}{dR^2}\right|_{R_e}\int \psi_{v'}(z)z^2\psi_v\,dz + \cdots. \quad (7.16)$$

The limits of integration can be taken as $\pm\infty$ so long as we make the harmonic approximation. Because $\psi_{v'}(z)$ and $\psi_v(z)$ are eigenfunctions of the same Hermitian operator and have different eigenvalues (if we are considering a transition), they are orthogonal. Hence $\int \psi_{v'}(z)\psi_v(z)\,dz$ vanishes. The next term does not vanish. First, note that neither $\int \psi_v{}^* \psi_v dz$ nor $\int e^{-z^2} H_v^2(z)\,dz$ vanishes. Next, from the definition of the Hermite polynomials, we find by direct substitution that

$$2zH_v(z) = 2vH_{v-1}(z) + H_{v+1}(z). \quad (7.17)$$

In words, multiplication of $H_v(z)$ by z transforms $H_v(z)$ into a sum of one higher and one lower $H_{v'}(z)$, on the scale of eigenvalues. This in turn means that the integral multiplying $d\mu/dR$ in Eq. 7.16 does not vanish provided that either $v' = v + 1$ or $v' = v - 1$. Therefore the harmonic oscillator satisfies Eq. 7.13. We shall not show here why the harmonic oscillator does not exhibit transitions with $|\Delta v| > 1$. Real molecules, which are never strictly harmonic, do exhibit transitions with $|\Delta v| > 1$, but these are generally much less probable than those with $\Delta v = \pm 1$ provided that the molecule has a nonvanishing dipole moment. The higher terms of Eq. 7.16 for general anharmonic oscillators are not zero but are small.

The form of the vibrational spectrum is governed by the spacing of the energy levels. Since the energy levels of the harmonic oscillator are equally spaced, the selection rule 7.13 limits its vibrational spectrum to a single line, with $\Delta E = \pm h\nu_0$; this is illustrated in Fig. 7.3*a*. The spectrum of a real molecule is of course more complicated. First, as mentioned above, the selection rule (Eq. 7.13) does not hold rigorously for an anharmonic oscillator, although transitions with $\Delta v = \pm 1$ do remain far more likely than those with $\Delta v = \pm 2, \pm 3, \ldots$. Also, the spacing of the energy levels is no longer constant. According to our reasoning in Section 4.2, since the real potential is "flatter" (i.e., widens more rapidly) than that of the harmonic oscillator, the spacing should decrease with increasing energy; this is indeed what one observes (cf. Fig. 7.1) for almost all stretching vibrations. Thus the transition $(v = 0) \to (v = 1)$ requires more energy than the transition $(v = 1) \to (v = 2)$, and so on upward. The vibrational spectrum therefore consists of many lines rather than one, as shown in Fig. 7.3*b*. If one takes an absorption spectrum of a gas sample, molecules in the ground state absorb light at one frequency (close to ν_e), molecules in the first excited state absorb at a slightly lower frequency, and so forth. The intensity of an absorption line is of course proportional to the number of molecules in the initial state. For light molecules at moderate temperatures virtually all the molecules are in the ground vibrational state (cf. Chapter 21 and footnote 2 on page 184), and only the line corresponding to $(v = 0) \to (v = 1)$ is normally observed; at higher temperatures one can use the intensity ratios to determine the distribution of molecules among the vibrational states. (This, too, is discussed in Chapter 21.)

The above discussion and Fig. 7.3 refer to a pure vibrational spectrum. Real molecular spectra are complicated by the fact that a molecule's rotational energy may also change when the vibrational energy changes. As a result, each of the

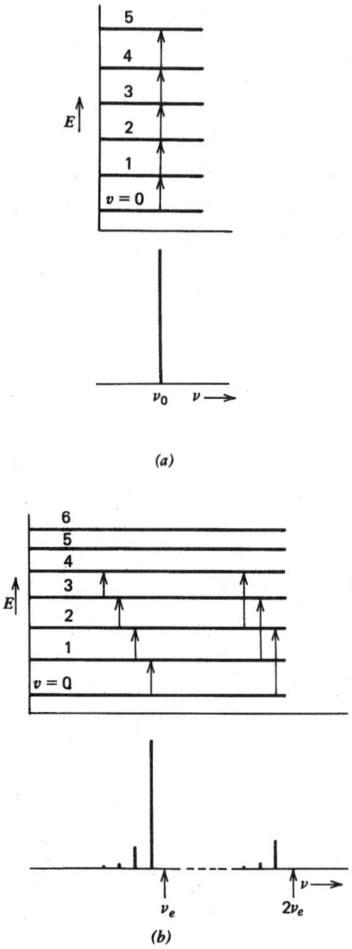

Figure 7.3 Schematic representation of molecular vibrational absorption spectra (neglecting rotational effects). (*a*) Harmonic oscillator model: $\Delta v = \pm 1$ only, all transitions at the same frequency ν_0. (*b*) Real diatomic molecule: The energy levels are not evenly spaced, so one can observe a band of lines beginning near ν_e; since the selection rule is not rigorous, there is a weaker band ($\Delta\nu = \pm 2$) near $2\nu_e$, and still weaker bands at $3\nu_e$, $4\nu_e$, ... Within each band, the intensity of a given line is proportional to the population of the initial state; the intensities as drawn here correspond to a gas with $h\nu_e/k_BT \approx 1.8$ (cf. Chapter 21).

Table 7.2 Vibrational and Rotational Constants of Some Diatomic Molecules

Molecule	μ (amu)	R_e (Å)	$\tilde{\nu}_e$ (cm^{-1}) ($\tilde{\nu}_e \equiv \nu_e/c$)	$\tilde{\nu}_e x_e$ (cm^{-1})	$\tilde{\nu}_e Y_e$ (cm^{-1})	D_0 (eV)	B_e (cm^{-1})	α_e (cm^{-1})
$^{1}H_2$	0.50391	1.7412	4400.39	120.815	0.7242	4.4773	60.864	3.0764
HD ($^{1}H^{2}H$)	0.67171	0.7412	3812.29	90.908	0.504	4.5128	45.663	2.0034
D_2 ($^{2}H_2$)	1.00705	0.7412	311.70	61.82	0.562	4.5553	30.457	1.0786
First-row homonuclear molecules								
$^{7}Li_2$	3.50800	2.6725	351.44	2.592	−0.0058	1.12	0.6727	0.00704
$^{11}B_2$	5.50465	1.590	1051.3	9.4		2.9	1.212	0.014
$^{12}C_2$	6.00000	1.2425	1854.71	13.340	−1.172	6.24	1.8198	0.01765
$^{14}N_2$	7.00154	1.094	2358.07	14.188	−0.0124	9.7598	1.9987	0.01781
$^{16}O_2$	7.99745	1.2075	580.19	11.98	0.0475	5.1156	1.4456	0.01593
$^{19}F_2$	9.49910	1.409	919.0	13.6		1.604	0.8901	0.0146
Other homonuclear molecules								
$^{23}Na_2$	11.4949	3.0786	159.23	0.726	−0.0027	0.75	0.1547	0.00079
$^{39}K_2$	19.48185	3.923	92.64	0.354		0.51	0.0562	0.00022
$^{85}Rb_2$	42.4558	4.20	57.28	0.96	−0.0008	0.47	0.0127	0.0000264
$^{133}Cs_2$	66.9525	4.58	41.99	0.080	−0.0002	0.45		
$^{35}Cl_2$	17.48222	1.9878	559.71	2.70		2.484	0.2441	0.00153
$^{79}Br^{81}Br$	39.9524	2.2809	323.33	1.081		1.9708	0.0811	0.00032
$^{127}I_2$	63.4502	2.6666	214.52	0.607	−0.0013	1.5437	0.0374	0.00012
Hydrides								
$^{7}Li^{1}H$	0.88123	1.5954	1405.65	23.200	0.1633	2.429	7.5131	0.2132
$^{12}C^{1}H$	0.92974	1.124	2859.1	63.3		3.47	14.448	0.530
$^{16}O^{1}H$	0.94808	0.9706	3735.21	82.81		4.392	18.871	0.714
$^{1}H^{19}F$	0.95705	0.9168	4139.04	90.05	0.932	5.86	20.9560	0.7958
$^{1}H^{35}CI$	0.97959	1.2746	2991.09	52.82	0.2244	4.4361	10.5936	0.3072
$^{1}H^{81}Br$	0.99511	1.4145	2649.21	45.22	−0.0029	3.755	8.4651	0.2333
$^{1}H^{127}I$	0.99988	1.6090	2308.09	38.981	−0.1980	3.053	6.5108	0.1686
Other heteronuclear molecules								
$^{7}Li^{19}F$	5.12381	1.5638	910.34	7.929		5.94	1.3454	0.02030
$^{9}Be^{16}O$	5.76432	1.3310	1487.32	11.830	0.0224	4.60	1.6510	0.0190
$^{11}B^{14}N$	6.16351	1.281	1514.6	12.3		3.99	1.666	0.025
$^{12}C^{14}N$	6.46219	1.1720	2068.70	13.144		7.567	1.8991	0.01735
$^{12}C^{16}O$	6.85621	1.1283	2169.82	13.294	0.0115	11.09	1.9313	0.01751
$^{14}N^{16}O$	7.46676	1.1508	1904.03	13.97	−0.0012	6.50	1.7046	0.0178
$^{23}Na^{35}Cl$	13.8707	2.3606	366	2.05		4.25	0.2181	0.00161
$^{39}K^{79}Br$	26.0850	2.8207	213	0.80	0.0011	3.925	0.0812	0.00040

lines in the vibrational spectrum resolves into a band of closely spaced rotational–vibrational lines. We shall consider this and other complications in Section 7.3.

Let us now say a little about how the vibrational constants vary among diatomic molecules. In the harmonic oscillator model, as we have already noted, the oscillator frequency is proportional to $\mu^{-1/2}$, where μ is the molecule's reduced mass. This is a fairly good approximation for real molecules. For homonuclear molecules μ is one-fourth the molecular weight (half the atomic weight), so heavy molecules have relatively low vibrational frequencies. However, the reduced mass of a heteronuclear molecule cannot exceed the mass of the lighter atom, which does most of the actual vibrating relative to the center of mass; all diatomic hydrides thus have $\mu < 1$ amu, and high vibrational frequencies. The variation of force constants (k) among molecules is not nearly so great as that of μ. In spite of these variations, the actual magnitudes of molecular vibrational frequencies (or energies) fall within a rather short and well-defined range. This can be seen in Table 7.2, in which $\tilde{\nu}_e$ ($\equiv \nu_e/c$), x_e, y_e are the constants in Eq. 7.10: The wavenumber $\tilde{\nu}_e$ ranges from over 4000 cm^{-1} for $^{1}H_2$ to below 50 cm^{-1} in some heavy molecules, but this covers only two decades of the whole electromagnetic spectrum.

Finally, as can also be seen in Table 7.2, there is at least a qualitative correlation between the length of a chemical bond (R_e) and its fundamental vibration frequency ν_e. This is not difficult to rationalize. Chemical bonds fall within a relatively narrow range of energies and lengths, so most ground-state $E(R)$ curves are similar in shape. This is why they can almost all be approximated by a given function like a Morse potential. The bond length is the distance at which the short-range repulsive force balances the more slowly varying attractive force. The smaller the value of R_e, the more steeply both forces vary with R, and the narrower is the potential well near R_e. But a narrow well is one with a high force constant and thus a high vibration frequency. There we have our qualitative correlation.[9] Potential wells at small R tend to be not only steeper, but also deeper, because almost all the bonding forces fall off with distance. Thus, we also have a correlation between bond energy and vibration frequency. To sum up: Long bonds are weak and correspond to slow, low-frequency vibrations with large amplitudes; short bonds are strong, with deep potential wells, and correspond to high vibration frequencies with small amplitudes of motion.

7.2 Rotations of Diatomic Molecules

Next we consider the rotational behavior of diatomic molecules. We continue to assume that the translational, vibrational, and rotational motions of the nuclei are separable, Eqs. 7.2–7.4. Thus we begin by considering rotation in the absence of vibration, that is, with fixed internuclear distance. Just as vibration is well described by the harmonic oscillator model, the rotational motion of the molecule is well described by another model familiar to us, that of the rigid rotator (see Section 3.12).

[9] There are a number of empirical formulas correlating vibrational constants with bond lengths. These are accurate enough to be used in estimating bond lengths in new compounds from their infrared spectra. One of the most widely used is *Badger's rule,*

$$k = a(R_e - d_{ij})^{-3},$$

where k is the force constant, R_e is the usual equilibrium internuclear distance, $a \approx 186$ (N/m)Å^3 (a universal "constant"), and d_{ij}, which is not universal, depends on the rows of the periodic table in which the atoms i and j fall. The following table gives d_{ij} in angstroms for various rows of the periodic table:

	j			
i	H	Row 1	Row 2	Row 3
H	0.025	0.335	0.585	0.650
Row 1	0.335	0.680	0.900	
Row 2	0.585	0.900	1.180	

More specifically, to a good approximation we can treat any diatomic molecule as a rigid symmetric top. This, it will be recalled, is a rigid body with two equal moments of inertia, $I_0 \equiv I_x = I_y \neq I_z$; the z axis is as usual the bond axis. We have already analyzed this case in detail. The rotational Hamiltonian of the rigid symmetric top is

$$\left| \mathsf{H}_{\text{rot}} = \frac{1}{2I_0}\mathbf{L}^2 + \frac{1}{2}\left(\frac{1}{I_z} - \frac{1}{I_0}\right)\mathsf{L}_z^2 \right|, \tag{7.18}$$

where $\mathbf{L}^2$ and L_z^2 are angular momentum operators; the eigenvalues of H_{rot} are given by Eq. 3.177. Now, in a diatomic molecule the only contributions to $I_z[\equiv \Sigma_i m_e(x_i^2 + y_i^2)]$ are those due to the electrons and the nonzero radii of the nuclei, both very small; I_0, on the other hand, is essentially μR^2, where μ is the molecule's reduced mass and R is the internuclear distance. Thus we have $I_z << I_0$, and the coefficient of L_z^2 in Eq. 7.18 is much larger than that of $\mathbf{L}^2$. The energy term associated with L_z^2 is either very large (when $L_z \neq 0$ for the electrons, i.e., in all except $^1\Sigma$ states) or negligibly small (in $^1\Sigma$ states). In either case it is a constant in any given electronic state, and can conveniently be included in the electronic energy.

Thus, we can disregard the second term in Eq. 3.177 and write the rotational energy of the molecule as simply

$$E_{\text{rot}} - \frac{J(J+1)\hbar^2}{2I_0} = \frac{J(J+1)\hbar^2}{2\mu R_e^2} \quad (J = 0,\ 1,\ 2,\ \ldots), \tag{7.19}$$

where J (replacing the earlier l) is the conventional symbol for the rotational quantum number of a molecule; we have evaluated I_0 at the equilibrium internuclear distance, R_e. This equation is identical in form to Eq. 3.170, which gives the energy levels of the pure rigid rotator. Each energy level is thus $(2J + 1)$-fold degenerate: The angular momentum about a given axis through the center of gravity is $M_J\hbar$, where for each J the quantum number M_J can assume the values $J, J - 1, \ldots, -J + 1, -J$. And the rotational wave functions are simply the spherical harmonics $Y_{J,M_J}(\theta, \phi)$.

This is all quite straightforward, but how valid is the model? The assumption of fixed internuclear distance seems questionable, since the molecule is certainly vibrating (at least with the zero-point energy) at the same time as it rotates. But let us compare the rates of the two motions. If the molecule is rotating with angular velocity ω, its angular momentum for fixed R is $\mu R_e^2\,\omega$; combining this with the eigenvalue equation $L^2 = J(J + 1)\hbar^2$, we obtain

$$\omega = \frac{[J(J+1)]^{1/2}\,\hbar}{\mu R_e^2} = \left(\frac{2E_{\text{rot}}}{\mu R_e^2}\right)^{1/2} \tag{7.20}$$

for the angular velocity. We shall see in Chapter 21 that the average rotational energy of gaseous diatomic molecules is k_BT, where $k_B = 1.381 \times 10^{-23}$ J/K and T is the absolute temperature. For $^1H_2$ molecules at room temperature, the average period of a rotation is thus

$$\tau_{rot} = \frac{2\pi}{\omega} \approx 2\pi\left(\frac{\mu R_e^2}{2k_B T}\right)^{1/2}$$
$$= 2\pi\left[\frac{(8.368\times10^{-28}\ \text{kg})(0.7412\times10^{-10}\ \text{m})^2}{2(1.381\times10^{-23}\ \text{J/K})(300\ \text{K})}\right]^{1/2}$$
$$= 2\pi(5.55\times10^{-28}\ \text{s}^2)^{1/2} = 1.48\times10^{-13}\ \text{s}.$$

On the other hand, the period of a harmonic oscillator is simply ν_{-1}, so for $^1H_2$ we have (cf. Table 7.1)

$$\tau_{vib} \approx \nu_e^{-1} = (1.3192\times10^{14}\ \text{s}^{-1})^{-1} = 7.58\times10^{-15}\ \text{s}.$$

Thus, the $^1H_2$ molecule at room temperature goes through about 20 vibrational periods in the course of a single rotation; for most molecules the ratio τ_{rot}/τ_{vib} is even greater.[10] As a first approximation, then, it seems reasonable to assume that the vibrational effects average out, justifying our calculation of rotational energy with fixed R. This is exactly the same reasoning as that leading to the Born–Oppenheimer assumption.[11]

Although the simple rigid-rotator model is adequate for qualitative purposes, more accurate expressions are needed to meet the demands of quantitative spectroscopy. One must allow for the effect of vibration on the rotational motion. Strictly, the R_e^{-2} in Eq. 7.19 should be replaced by the average value of R^{-2} over a vibration; because of the asymmetry of the $E(R)$ curve, the average value of R must increase with υ, so that we have $\langle R\rangle_\upsilon > R_e$, $\langle R^{-2}\rangle_\upsilon < R_e^{-2}$. The actual rotational energies are thus somewhat less than those given by Eq. 7.19, and can be approximated by[12]

$$E_{rot} = J(J+1)hcB_\upsilon$$
$$= J(J+1)hc\left[B_e - \alpha_e\left(\upsilon+\frac{1}{2}\right)+\cdots\right],$$
$$\left(B_e \equiv \frac{\hbar}{4c\mu R_e^2}\right), \tag{7.21}$$

where α_e is a constant, B_υ is the rotational constant for vibrational state υ, and B_e is the rigid-rotator value. The factor hc is inserted to give B_e and α_e the units of wavenumbers. For $^1H_2$ we readily obtain

$$B_e = \frac{1.0546\times10^{-34}\ \text{J s}}{4\pi(2.9979\times10^{-8}\ \text{m/s}) \times (8.3676\times10^{-28}\ \text{kg})(0.7412\times10^{-10}\ \text{m})^2}$$
$$= 60.9\ \text{cm}^{-1},$$

in agreement with experiment, whereas α_e is found empirically to be about 3.0 cm^{-1}. (Actually, as we shall see, the value of R_e is obtained from the measured B_e.) Other molecules have larger moments of inertia ($I = \mu R_e^2$) and thus smaller B_e's; values of B_e and α_e for a number of molecules are listed in Table 7.2.

In writing Eq. 7.19 for a diatomic molecule, we have effectively assumed the atoms to be point masses, so that we can neglect rotation about the internuclear axis. This is valid enough for a molecule in a $^1\Sigma$ state, where the electrons have zero angular momentum relative to the nuclei, but suppose that this is not the case. Let us go back to the symmetric-top model. We can rewrite the eigenvalue equation (Eq. 3.177) as

$$E_{rot} = J(J+1)hcB_\upsilon + M_J^2 hc(A - B_\upsilon),$$
$$(J = 0,1,2,\ldots;\ M_J = J, J-1,\ldots,-J+1,-J), \tag{7.22}$$

where $A \equiv \hbar/4\pi cI_z$ and B_υ is the same as in Eq. 7.21. The quantum numbers J and M_J are defined by the equations

$$|\mathbf{J}|^2 = J(J+1)\hbar^2 \quad \text{and} \quad J_z = M_J\hbar, \tag{7.23}$$

which respectively give the square and the z component of the molecule's total angular momentum $\mathbf{J}$. Although both electronic and nuclear motions contribute to $\mathbf{J}$, its z component consists almost entirely of the angular momentum of the electrons about the bond axis. There are a number of ways in which the orbital and spin electronic angular momenta (which we call $\mathbf{L}$ and $\mathbf{S}$, respectively) can couple with each other and with the nuclear rotation; we shall look at only the simplest case. In most electronic states the orbital electronic angular momentum is strongly coupled to the bond axis, and only its z component is quantized; thus we have $|L_z| = \Lambda\hbar$, as we assumed in the last chapter. In singlet states ($\mathbf{S} = 0$), L_z is the only contribution to J_z, so we also have $|J_z| = \Lambda\hbar$, $|M_J| = \Lambda$. Thus for most singlet states Eq. 7.22 can be written in the form

$$E_{rot} = J(J+1)hcB_\upsilon + \Lambda^2 hc(A - B_\upsilon),$$
$$(\Lambda = 0,1,2,\ldots;\ J = \Lambda, \Lambda+1, \Lambda+2,\ldots). \tag{7.24}$$

Similar equations can be derived for other cases. Since the second term can be included in the electronic energy, what is the point of obtaining these equations? The answer is that

[10] For example,

	$^1H^{35}Cl$	$^{35}Cl_2$	$^{127}I_2$
τ_{rot}(300 K)/s	4.05×10^{-13}	2.34×10^{-12}	5.98×10^{-12}
τ_{vib}/s	1.12×10^{-14}	5.90×10^{-14}	1.55×10^{-13}

[11] Note that both τ_{rot} and τ_{vib} are appreciably longer than the Bohr period of about 10^{-16} s (Table 2.2), which is typical of the time scale of electronic motions.

[12] There is also a smaller correction for centrifugal distortion, i.e., the stretching of the bond by centrifugal force. This appears in E_{rot} as a term $D_\upsilon hcJ^2(J+1)^2$, where, typically, $D_\upsilon < 10^{-4} B_\upsilon$.

they reveal restrictions on the allowable energy levels. In an electronic state described by Eq. 7.24, for example, we can only have $J \geq \Lambda$, since the total angular momentum, $[J(J+1)]^{1/2}\,\hbar$, cannot be less than its z component $\Lambda\hbar$. Since $A >> B_v$, electronic states with different Λ's can have very different total rotational energies, often differing by several electron volts. Within a given electronic (and vibrational) state, however, adjacent rotational levels may differ by as little as 10^{-5} eV.

This brings us to rotational spectra. In the next section we shall consider the complications that arise when a molecule simultaneously undergoes rotational and vibrational (or electronic) transitions; here we are concerned only with pure rotational spectra, which are found in the far infrared and microwave regions. As in the case of vibration (pages 186–188), rotational electric dipole spectra appear only for heteronuclear diatomic molecules, since a homonuclear molecule has a dipole moment $\mu = 0$ in all orientations. For heteronuclear molecules the moment μ varies in direction, though not magnitude, in the course of a rotation. The selection rule for electric dipole transitions of a heteronuclear rigid rotator is

$$\Delta J = \pm 1, \tag{7.25}$$

which also usually holds for a real molecule in a given electronic state. (The logic leading to this rule is very much like that following Eq. 7.13, with rotational wave functions replacing the Hermite polynomials and the permanent dipole operator, rather than its derivative, playing the role of the operator inducing the transition. Another way of interpreting Eq. 7.25 is as a consequence of the fact that the photon has an intrinsic angular momentum of one unit of $\hbar$.) Substituting in Eq. 7.21, we find that the only allowed pure rotational transitions are those with

$$\Delta E = 2J'hcB_v \quad \text{or} \quad \tilde{\nu} = 2J'B_v, \tag{7.26}$$

where J' is the quantum number of the *upper* state. We thus obtain a series of equally spaced spectral lines, beginning at $\tilde{\nu} = 2B_v$ and proceeding to higher wavenumbers at intervals of $2B_v$. This is illustrated in Fig. 7.4. Since $2B_e$ ranges from 120 cm^{-1} ($^1H_2$) down to less than 0.1 cm^{-1}, corresponding to 0.08×10^{-10} cm in wavelength, at least the beginning of this series may be in the microwave region. However, in experiments at room temperature, one frequently observes transitions in which the quantum number J' is quite large, giving rise to higher-frequency lines well into the infrared.[13]

The spacing of rotational lines is one of the best methods for determining internuclear distances (bond lengths). It is worth describing here how one carries out such a determination. The pure rotational spectrum, as we have said, lies in the far infrared or microwave region. To obtain such a spectrum, one can pass radiation through a rather rarefied gaseous sample of the molecule in question, vary the frequency of the applied radiation, and measure the fraction of energy transmitted from the power source to a detector.[14] The source may be electrons oscillating in a vacuum (a klystron tube), electrons accelerated in a nonmetallic solid by a sufficiently high voltage, or a variable-frequency laser operating in the microwave region of the spectrum; the detector is ordinarily a crystal-diode rectifier. One can carry out such measurements with exceedingly high accuracy—so high, in fact, as to outstrip the ability of existing theories to interpret all the details, which include interactions with vibration, electronic motion, and nuclear spins. In first approximation, however, the lines in a given band are equally spaced, and B_v is simply half the wavenumber spacing. Given such experimental values of B_v for several vibrational states, one simply uses Eq. 7.21 to calculate B_e and thus R_e.

We mention in passing the other principal method of determining bond lengths of gaseous molecules, which utilizes electron diffraction (see Sections 3.1 and 11.6). This method is less precise and often less accurate than

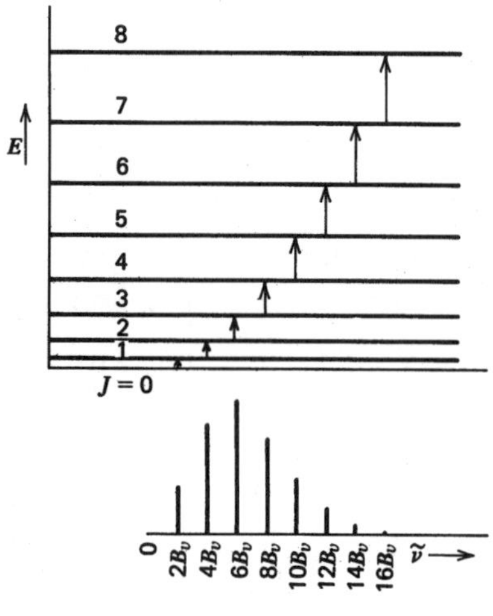

Figure 7.4 Pure rotational absorption spectrum of a diatomic molecule. Given the selection rule $\Delta J = \pm 1$, there is a series of lines with the constant wavenumber spacing $2B_v$. The intensities of the lines are proportional to the populations of the initial states, and as drawn here correspond to a gas with $B_v hc/k_BT \approx 0.1$ (cf. Chapter 21).

[13] As we mentioned earlier, in a gas the average value of E_{rot} is k_BT, which at 300 K is 4.14×10^{-21} J (0.0258 eV), corresponding to a wavenumber of 208 cm^{-1}. Thus at room temperature the average value of $J(J+1)$ for $^1H_2$ is about 3.5 ($J \approx 1-2$), for I_2 about 5500 ($J \approx 75$).

[14] In practice, rotational spectroscopy usually involves a technique somewhat more complicated than direct absorption. In a laboratory system, the amount of energy absorbed directly is always an extremely small fraction of the total incident radiation, so that the direct spectral signal would be very weak. Instead, one alternately applies and turns off an external electric or magnetic field, which splits and shifts the spectral lines (by the Stark or Zeeman effect), moving them on and off the frequency of the applied radiation. The observed signal is then the difference between the amounts of microwave power detected with the external field on and off. The effect of taking such a difference signal is a strong enhancement of the signal strength relative to the background "noise" reaching the detector.

microwave spectroscopy, but it does give a direct measure of R_e. A beam of electrons is allowed to impinge on a gas of diatomic molecules; because the electrons are deliberately given quite high energies, they are scattered primarily by the massive atomic nuclei. Since the molecules move randomly, one obtains a quite complex pattern of scattering intensity averaged over all possible molecular orientations. Nevertheless, this pattern contains information on the internuclear spacing. Each nucleus in a molecule produces a set of scattered wavelets, and the wavelets from each pair of nuclei interfere with each other. A single diatomic molecule thus scatters electrons much as two pinholes in a screen produce a diffraction pattern with light from a point source. Despite the random orientation, the net effect of all the molecules is to produce a diffraction pattern which depends on the scattering angle in a very specific way. One can compare this pattern with that corresponding to an assumed R_e, and adjust the latter until the two match. Similar calculations can be made for even very complex molecules, but for diatomic molecules the computation is simple and limited only by the accuracy with which one can measure the shapes of the diffraction peaks.

7.3 Spectra of Diatomic Molecules

Most of our knowledge of molecular structures derives from spectroscopic measurements of one kind or another. Suppose that one subjects a gas of diatomic molecules to electromagnetic radiation, varying the frequency ν across the spectrum. Whenever $h\nu$ equals the difference between two molecular energy levels, a transition may be induced. What kinds of transitions are these? In the microwave and far infrared regions, as just described, one observes pure rotational transitions, in which only the quantum number J changes. In the near and middle infrared one sees vibrational transitions. In Section 7.1 we described the pure vibrational spectrum, in which only v changes; unfortunately, things are not really that simple, since rotational transitions of various kinds usually occur at the same time. Radiation in the visible and ultraviolet regions is associated with transitions between different electronic states, again complicated by simultaneous vibrational and rotational transitions. At still higher energies, x-rays produce transitions among the energy levels of the atomic cores (see Chapter 2). Figure 7.5, which illustrates some of the energy levels in a diatomic molecule, should give an idea of the energy differences involved in various kinds of transitions.

These transitions are observed by the standard techniques of experimental spectroscopy, which we shall review here briefly. In *emission spectroscopy,* the sample is subjected to thermal, electric, or other excitation, and the excited atoms or molecules emit energy as they drop to lower energy levels. The emitted radiation is collected at a detector and its intensity is measured as a function of wavelength. An emission spectrum of atomic iron was shown in Fig. 2.9. In *absorption spectroscopy,* a sample of the material under study is subjected to radiation in an appropriate part of the spectrum, and the molecules absorb energy, undergoing

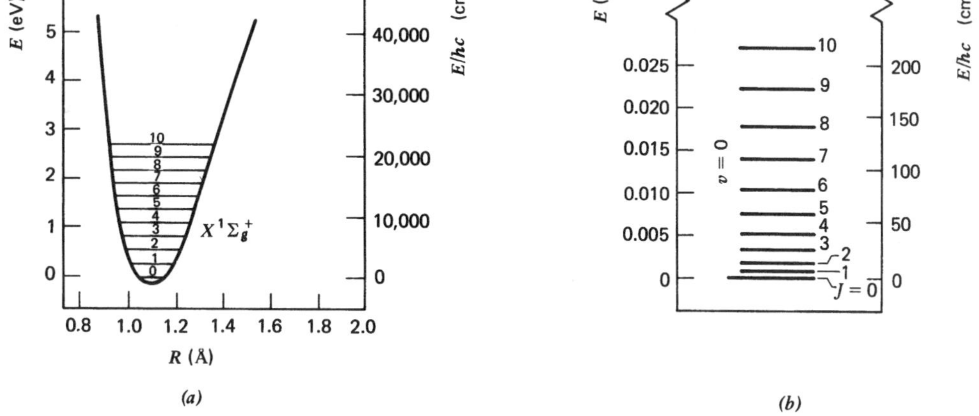

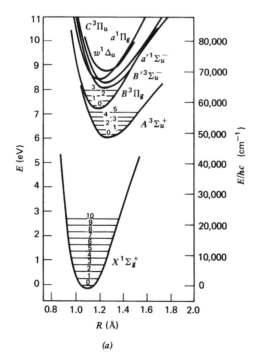

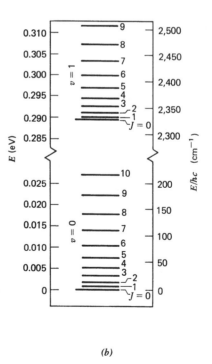

Figure 7.5 Energy levels of the N_2 molecule. (*a*) Electronic states, with vibrational levels ($v = 0, 1, 2, \ldots$) shown in the lowest three. [Not shown is the $W^3\Delta_u$ state; this state has about the same minimum energy as the $B^3\,\Pi_g$ (see Table 7.6), but its R_e is not known.] The energy zero is the ground electronic-vibrational state ($X^1\Sigma_g^+$, $v = 0$). After W. Benesch, J. T. Vanderslice, S. G. Tilford, and P. G. Wilkinson, *Astrophys. J.* **142,** 1227 (1965). (*b*) Rotational structure of the two lowest vibrational levels of the ground state; the energy zero is the same as in (*a*).

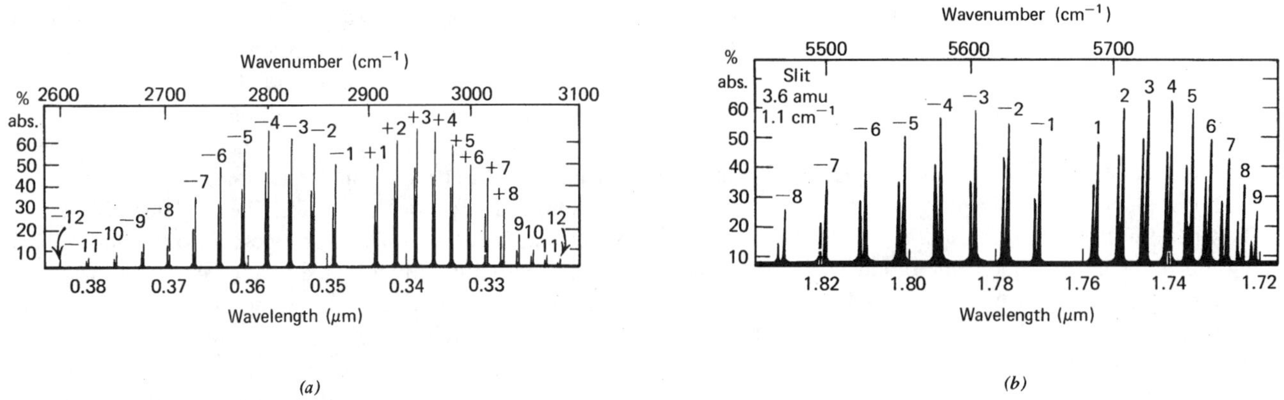

Figure 7.6 The infrared spectrum of HCl vapor associated with transitions (*a*) $v=0 \rightarrow v=1$ and (*b*) $v=0 \rightarrow v=2$. The individual rotation–vibration lines are all doublets because of the presence of the isotopic species $H^{35}Cl$ and $H^{37}Cl$; the former is the more intense. Numbers above spectral lines specify ΔJ. From C. F. Meyer and A. A. Levin, *Phys* Rev. **34,** 44 (1929).

transitions to higher energy levels. This energy transfer depletes the incident radiation at specific wavelengths (one for each transition), giving rise to a set of spectral lines. The radiation passing through the sample is collected at a detector, where its intensity is measured. The signal of interest is the amount of radiation absorbed, that is, the difference between incident and transmitted intensities. What we usually call a "spectrum" is a representation of the amount of absorption as a function of wavelength or frequency; two regions of the infrared spectrum of HCl are shown in Fig. 7.6. *Fluorescence spectroscopy* is a process of exciting emission spectra, usually by excitation with radiation; in this process, one may measure the distribution of wavelengths of emitted radiation, as in any other kind of emission spectroscopy, or the dependence of the intensity of emission (at a selected wavelength or at all wavelengths simultaneously) on the frequency of the exciting radiation, to obtain an *excitation* or *action spectrum.*

All experimental spectrometers have the same basic components, but their details can vary greatly, especially with the region of the spectrum under study. These components include the following.

1. *Sample.* When the sample is gaseous or liquid, it must be placed in a holder of some kind; such a sample holder must be reasonably transparent to the radiation one is using. In the ultraviolet below 2000 Å even air becomes opaque, and the whole apparatus must be placed in (or be) a vacuum chamber.
2. *Radiation source.* For absorption spectra in the visible and ultraviolet regions older sources were the standard types of broad-band, continuous light sources: tungsten filament lamps, electric discharges and arcs, and so on. Usual infrared sources were heated rods of refractory material such as SiC. Klystron tubes are still used in the microwave region, and standard radiofrequency generators at longer wavelengths. In all regions of the spectrum, tunable (variable-wavelength) lasers are now the most commonly-used radiation sources. In emission spectroscopy the radiation source is the sample itself; it may be excited by simple heating, by electric discharge, by chemical reaction (as in a flame), or by laser-induced excitation.
3. *Optical system.* The optical system consists of whatever devices are used to select or disperse radiation of different wavelengths and collect it at detectors. In most spectral regions of interest, the separation of radiation of many wavelengths can be performed by *dispersing elements,* either prisms or diffraction gratings; the separated radiation can be focused with mirrors and lenses. Prisms and lenses can be used, of course, only where they are transparent. With variable-wavelength lasers, one can dispense with dispersing elements. In the long-wavelength microwave and radio regions, different wavelengths are also obtained by "tuning" the radiation source.
4. *Detector.* The simplest type of radiation detector used to be a photographic emulsion; the intensity is determined by measuring the extent of darkening. Now one typically uses photosensitive semiconductors, photoelectric cells (visible and ultraviolet), thermocouples and bolometers (infrared), and crystal diodes (microwave). The apparatus is usually so designed to record and display a record of the relation between intensity and wavelength or frequency.

We saw that any transition between bound states of a molecule has a definite energy and frequency ($\Delta E = h\nu$), and should thus give rise to a sharply defined spectral line. Yet in practice one always observes "lines" of nonzero width (in wavelength or energy), at best resembling narrow spike-like curves. Sometimes this is an apparatus effect, due to the limited resolution of the optical system, but with sufficient resolving power one obtains line widths characteristic of the sample itself. This broadening of spectral lines has several causes, principally the following. (1) The *natural line width* is the consequence of the uncertainty principle, as stated in Eq. 3.87. Any measurement carried out in a finite time Δt has an energy uncertainty of the order of $\hbar/\Delta t$. But the natural line width is usually extremely small, about 10^{-4} Å for spontaneous emission of visible light. (2) *Doppler broadening* results from the fact that the molecules of the sample are in motion in various directions, so that the frequency of the radiation absorbed or emitted is shifted up or down by the well-known Doppler effect. (3) *Pressure broadening* results from the perturbation of energy levels by intermolecular forces, especially during collisions. In a gas, pressure broadening increases rapidly with density, and is the dominant contribution to line widths at high pressures. Normally, visible and ultraviolet lines from gaseous samples have linewidths dominated by Doppler broadening. Infrared and microwave lines have widths due primarily to pressure broadening at all but the lowest pressures.

In the remainder of this section we shall discuss some of the major aspects of spectroscopy that we have not yet described: (1) simultaneous vibration–rotation spectra; (2) the Raman effect, in which light is scattered inelastically rather than absorbed; (3) electronic spectra.

Since a real molecule may vibrate and rotate simultaneously, it can undergo transitions in which both quantum numbers change. In fact, in the ground electronic states of most diatomic molecules a pure vibrational transition is forbidden. In any such simultaneous transition ΔE_{vib} is likely to be far greater than ΔE_{rot} (because of the energy-level spacings), so vibration–rotation spectra are found in the same part of the spectrum as pure vibrational spectra. For Σ electronic states, which include most stable ground states, the selection rules are those we have already introduced:

$$\Delta v = \pm 1 \quad \text{and} \quad \Delta J = \pm 1, \qquad (7.27)$$

with $\Delta v = \pm 1$ only approximately true for anharmonic oscillators.

Figures 7.6 and 7.7 illustrate the kind of spectrum obtained when molecules undergo such transitions. Each line of the pure vibrational spectrum is replaced by a whole band of vibration–rotation lines. These lines are nearly equally spaced, as in the pure rotational spectrum; however, there is a gap in the center of the band. This gap corresponds to the pure vibrational transition with $\Delta J = 0$, which is forbidden for Σ states by the selection rule of Eq. 7.27. The position of the missing line is referred to as the *null line* or *band origin,* with wavenumber $\tilde{\nu}_0$. The lines of the band on each side of the origin are called the *P branch* and the *R branch,* with $\tilde{\nu} < \tilde{\nu}_0$ and $\tilde{\nu} > \tilde{\nu}_0$, respectively. It is conventional to designate the upper state by a prime (v', J', etc.) and the lower state by a double prime (v'', J'', etc.). As is shown in Fig. 7.7, the *P branch* corresponds to transitions with $J' = J'' - 1$ ($\Delta E < hc\tilde{\nu}_0$), and the *R branch* to transitions with $J' = J'' + 1$ ($\Delta E > hc\tilde{\nu}_0$). With the rotational energy in each state given by Eq. 7.21, the wavenumbers of the lines in the band are given by

$$\tilde{\nu} = \frac{E' - E''}{hc} = \tilde{\nu}_0 + B'_v J'(J'+1) - B''_v J''(J''+1), \qquad (7.28)$$

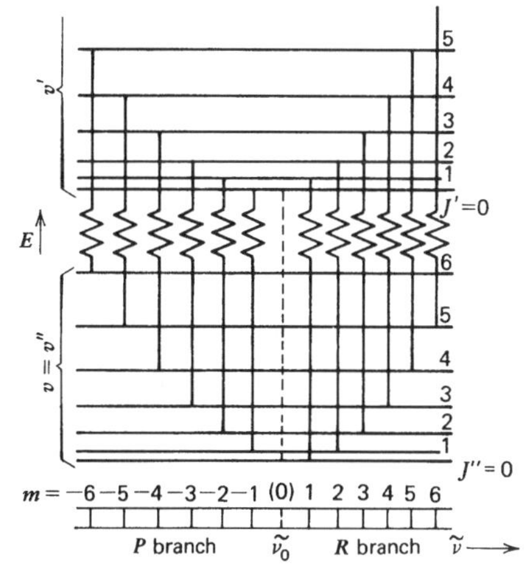

Figure 7.7 Schematic representation of a vibration–rotation band in the infrared spectrum of a diatomic molecule in a Σ electronic state. Each line in the band corresponds to a transition between a lower level v'', J'' and an upper level v', J', with $J' = J'' - 1$ in the *P* branch and $J' = J'' + 1$ in the *R* branch. The lines are labeled with the running number *m,* defined in Eq. 7.29. The band origin $\tilde{\nu}_0$ corresponds to the forbidden transition v'; $J' = 0 \leftrightarrow v''$; $J'' = 0$. The spectrum is drawn for $B_{v'} < B_{v''}$ so that the band converges to a head (not shown) in the *R* branch.

where $\tilde{\nu}_0 \equiv (E'_{\text{vib}} - E''_{\text{vib}})/hc$. The relative intensities of the lines are as usual proportional to the populations of the initial rotational states; the intensity distribution in each branch (Fig. 7.6*b*) resembles that in a pure rotational band (Fig. 7.4).

Equation 7.28 can be rewritten in the form

$$\tilde{\nu} = \tilde{\nu}_0 + (B'_v + B''_v)m + (B'_v - B''_v)m^2, \qquad (7.29)$$

where the *running number m* equals $-J''$ in the *P* branch and $J'' + 1$ in the *R* branch. For a rigid rotator the spectral lines are equally spaced, but because of the quadratic term in Eq. 7.29 this cannot be true in a vibrating rotator. From Eq. 7.21 we know that B_v decreases with increasing v within a given electronic state, so B'_v must be a little smaller than B''_v in any infrared transition.[15] Thus the quadratic term is negative, and the spacing between successive lines must decrease with increasing *m;* this effect is illustrated in Fig. 7.7 in slightly exaggerated form (since B'_v and B''_v are nearly equal). Eventually a point should be reached at which the spacing decreases to zero ($d\tilde{\nu}/dm = 0$), after which $\tilde{\nu}$ decreases with increasing *m.* There should thus be a sharp upper limit to the spectrum at some wavenumber $\tilde{\nu}_b$, which we call the *band head.* Band heads in vibration–rotation spectra always occur in the *R* branch. However, B'_v and B''_v are usually so close together that these heads appear at very high values of *m* (or *J*), and thus are usually too weak to

[15] Another way to say this is that the upper state has a larger moment of inertia, since $\langle I \rangle = \mu \langle R^2 \rangle$ increases with v.

observe in spectra taken at ordinary temperatures. We shall see that the situation is different for electronic spectra.

Our description thus far applies only to molecules in Σ electronic states, that is, molecules with $\Lambda = 0$. When $\Lambda \neq 0$ there is a change in the selection rule, with $\Delta J = \pm 1$ replaced by $\Delta J = 0, \pm 1$. (This is a consequence of the *vector* addition of the angular momenta of the photon and the molecule.) One thus observes an additional series of transitions with $\Delta J = 0$, known as the Q branch. The best-known example is in the ground state of NO, which is $^2\Pi$. For the Q branch, Eq. 7.29 is replaced by

$$\tilde{\nu}_Q = \tilde{\nu}_0 + (B'_v - B''_v)J + (B'_v - B''_v)J^2 . \qquad (7.30)$$

Since the spacing increases steadily in both directions, there is no band head. The coefficient $B'_v - B''_v$ is very small, so all the observed lines of the infrared Q branch are very close to $\tilde{\nu}_0$, and under low or moderate resolution appear as a single intense line.

Up to this point we have been talking only about absorption and emission spectra, but one can also observe the spectrum of radiation *scattered* by molecules. By scattered radiation (cf. Section 2.3) we simply mean radiation that leaves the sample in a direction different from its incident direction. This involves a two-step process on the molecular level: An incident photon strikes a molecule and excites it to a highly unstable condition (in classical terms, sets up a forced, nonresonant oscillation); the molecule then attains a new stationary state by emitting a second photon, which may depart in any direction. But the two steps are virtually simultaneous, and no stationary excited state exists in the interval between them. Scattering should thus not be confused with fluorescence or phosphorescence, in which the molecule absorbs one photon, forms an excited state that lasts long enough to be characterized as a stationary state (anything from picoseconds to hours), then decays by emitting a second photon. (Very rapid fluorescence merges into scattering; it is not important to make a sharp distinction in such cases.) In either kind of process the two photons need not have the same energy; if they do not, of course, the final state of the molecule is different from the initial state, and a spectrum is observed. If the incident and scattered photons do have the same energy, one speaks of *Rayleigh scattering,* corresponding to the *elastic* scattering of particles. The effect is the same as if a single photon bounced off the molecule. If the energies are different, the process is called *Raman*[16] *scattering,* corresponding to inelastic scattering of particles. It is the *Raman* effect that gives rise to a spectrum.

A schematic Raman spectrum is illustrated in Fig. 7.8. To keep the spectrum simple, one ordinarily uses a monochromatic beam of incident light, that is, light of a single wavelength, usually in the visible or ultraviolet. The scattered

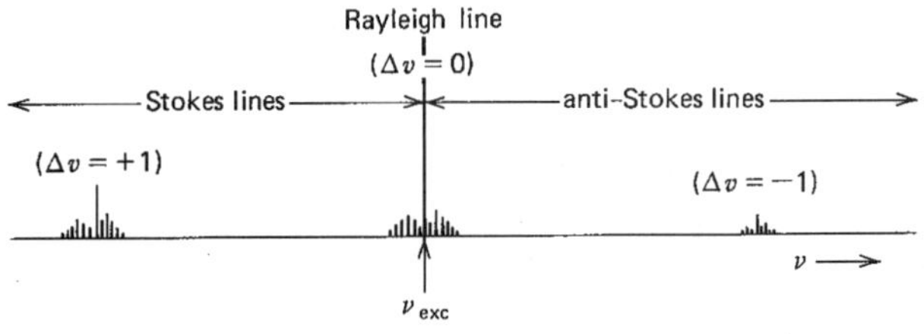

Figure 7.8 Schematic Raman spectrum of a diatomic gas. Three vibrational bands are shown, each with its rotational fine structure. The intense line at the exciting frequency ν_{exc} is due mainly to Rayleigh scattering; in the other bands the center line is the unresolved Q branch.

light is customarily observed at right angles to the incident beam. Since Rayleigh scattering is by far the more likely effect,[17] the scattered light is dominated by a single intense line, the *Rayleigh line,* with the same wavelength as the incident light. All the other lines in the spectrum are the result of Raman scattering. If ν_{exc} is the exciting frequency, a line of frequency ν corresponds to a molecular transition of energy

$$\Delta E = h(\nu_{exc} - \nu). \qquad (7.31)$$

Note that the frequency shift $\nu - \nu_{exc}$ for a given transition is independent of ν_{exc} itself, depending only on ΔE. The shift may be to either lower or higher frequency, corresponding respectively to net absorption and net emission of energy by the molecule. The lines with $\nu < \nu_{exc}$ are called *Stokes lines,* and those with $\nu > \nu_{exc}$ are called *anti-Stokes lines.* As in the infrared spectrum, each vibrational transition gives rise to a closely spaced band of lines corresponding to different rotational transitions. Since at ordinary temperatures most molecules are in the vibrational ground state, the Stokes band for a given $|\Delta v|$ is much more intense than the corresponding anti-Stokes band.

What kinds of transitions do we see in the Raman spectrum? Let us look again at Eq. 4.33, from which we find the transition probability. This equation involves the molecular dipole moment $\boldsymbol{\mu}$, which we have thus far taken to be the moment in the absence of external fields; let us call the latter $\boldsymbol{\mu}_0$. But the incident light beam has an electric field oscillating at frequency ν_{exc}, and this must in general induce an additional dipole moment in the molecule. (The electrons follow the field more easily than the heavy nuclei, so the centers of positive and negative charge oscillate relative to one another.) The total molecular dipole moment is thus $\boldsymbol{\mu}_0 + \boldsymbol{\mu}_{ind}$, with the induced moment given by

$$\boldsymbol{\mu}_{ind} = \alpha \mathbf{E} = \alpha \mathbf{E}_0 \cos 2\pi \nu_{exc} t, \qquad (7.32)$$

[16] The phenomenon was predicted in 1923 by A. Smekal and first demonstrated by C. V. Raman and K. S. Krishnan in 1928.

[17] Incidentally, the probability of Rayleigh scattering is proportional to λ^{-4}, so that from a beam of white light far more blue than red will be scattered. This is why the sky is blue.

where **E** is the oscillating field and α is the molecular *polarizability* (cf. Section 10.1). We can thus rewrite Eq. 4.33 as

$$\mu_{n'n} = \int \ldots \int \psi_{n'}{}^* \mu_0 \psi_n \, d\tau + \mathbf{E} \int \ldots \int \psi_{n'}{}^* \alpha \psi_n \, d\tau. \quad (7.33)$$

(Remember that the transition probability between states n and n' is proportional to $|\mu_{n'n}|^2$.) Consider a vibrational–rotational transition for which we define $\nu_{nn'} \equiv |E_{n'} - E_n|/h$. This transition can occur in either of two ways. If the first term in Eq. 7.33 is nonzero, there can be ordinary absorption or emission of radiation with frequency $\nu_{nn'}$ in the infrared or microwave regions; but this can occur only when there is a permanent dipole moment μ_0, and thus is not possible for homonuclear molecules. Here we are interested in the second possibility. If the second term in Eq. 7.33 is nonzero, a Raman scattering process can occur in which the scattered light has the frequency $\nu_{exc} \pm \nu_{nn'}$. When the incident light is in the visible or ultraviolet, we have $\nu_{exc} >> \nu_{nn'}$, and the scattered light is in the same region. Just as the first integral vanishes unless μ_0 changes in the course of the vibration or rotation, the second integral vanishes unless α changes. But all diatomic molecules have a nonzero polarizability, which varies with both bond length and orientation to the field: Thus both homonuclear and heteronuclear molecules display Raman spectra.

Since the mechanisms are so different, we need not be surprised if different selection rules apply for Raman and ordinary infrared transitions. The selection rule for vibration is in fact the same: $\Delta v = \pm 1$ for the harmonic oscillator, with weaker bands corresponding to $\Delta v = \pm 2, \pm 3, \ldots$ for anharmonic molecules; we also have a band with $\Delta v = 0$, corresponding to the pure rotational spectrum. But the rotational selection rule for diatomic molecules in Σ states is quite different from the rule for simple absorption or emission; for Raman transitions,

$$\Delta J = 0, \ \pm 2. \quad (7.34)$$

Thus, each band contains an *S branch* ($\Delta J = +2$), an *O branch* ($\Delta J = -2$), and a *Q branch* ($\Delta J = 0$); the *Q*-branch lines are again given by Eq. 7.30 and are hard to resolve. For molecules with $\Lambda \neq 0$ the selection rule becomes $\Delta J = 0, \pm 1, \pm 2$, and *P* and *R* branches are also observed.

Raman spectroscopy offers a useful complement to infrared spectroscopy for studying molecular vibrations and rotations. Raman spectra are observed for transitions that do not appear at all in the infrared, including the entire vibration–rotation spectrum of homonuclear diatomic molecules. This technique has one significant drawback, in that Raman transitions are relatively weak; their intensities are low compared with typical infrared transitions. This is to be expected, since a Raman transition involves two fairly unlikely processes rather than one. Note also that the measurement of Raman spectra is not limited to the infrared region. Any convenient exciting line can be used, since the frequency shift is independent of the exciting frequency. In fact, ν_{exc} is usually chosen to be in the visible or ultraviolet regions, where $\nu_{exc} >> \nu_{nn'}$ and the whole Raman spectrum can be found close to the exciting line. Normally, Raman spectra are far less intense than infrared spectra, because they result from a scattering process, which is second-order in the dipole moment; i.e., the intensity depends on the fourth power of matrix elements of that operator, rather than first-order, dependent on the square of such matrix elements. Intensities of Raman spectra can be enhanced either by using exciting radiation of a frequency near a resonant absorption frequency of the scatterer (*resonance* Raman spectroscopy) or by exciting the scatterers when they are on the surface of a metal, where the polarizability of the conduction electrons of the metallic substrate enhances the effect (*surface-enhanced* Raman spectroscopy).

Finally we come to electronic spectra. Since the electronic states of a molecule typically differ in energy by several electron volts, transitions between them are generally observed in the visible and ultraviolet regions. To a given electronic transition there corresponds a series of bands, one for each accessible vibrational transition; within a band there is a line for each possible rotational transition. A typical electronic spectrum is shown in Fig. 7.9.

The most prominent features of an electronic band spectrum are the band heads, the sharply defined edges on one side of the bands. We have already explained the origin of band heads in the context of vibration–rotation spectra. The most important selection rule for electronic transitions is that for the rotational quantum number *J*, namely

$$\Delta J = 0, \ \pm 1, \quad (7.35)$$

with the following exceptions: $\Delta J = 0$ (the *Q* branch) does not occur if $\Lambda = 0$ in both initial and final states, that is, if both are Σ states; and the transition $(J' = 0) \leftrightarrow (J'' = 0)$ is always forbidden. Thus there are always *P* and *R* branches, and in most cases also a *Q* branch. The wavenumbers in these branches are again given by Eqs. 7.29 and 7.30, but the

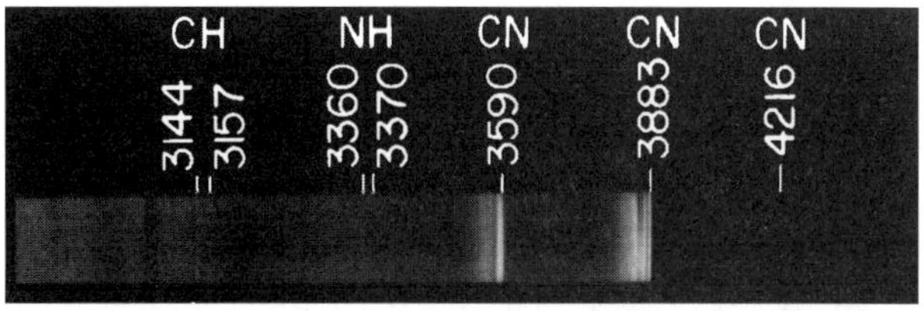

Figure 7.9 A typical electronic absorption spectrum of a gas undergoing a rapid reaction. The spectrum shows the presence of three diatomic molecules, CH, NH, and CN, none of which is stable under ordinary conditions. The numbers are wavelengths, in angstroms, of the band heads. This spectrum was taken with what is called a "medium dispersion" prism spectrograph, exhibiting the spectral range from about 200 to 700 nm (2000 to 7000 Å) on a 10-in. photographic plate. The photograph is printed in negative so absorption lines appear white. From D. W. Cornell, R. S. Berry, and W. Lwowski, *J. Am. Chem.* Soc. **88,** 544 (1966).

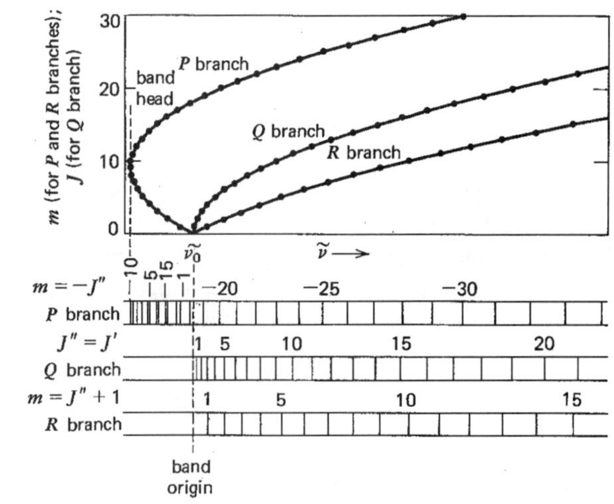

Figure 7.10 Structure of an electronic band spectrum. The upper part of the figure is what is known as a *Fortrat diagram,* showing the parabolic curves describing the three branches according to Eqs. 7.29 and 7.30. Below are shown the corresponding spectral lines of the three branches (which would be superimposed in the actual spectrum). The spectrum illustrated has a head in the P branch, at about $m = -9$.

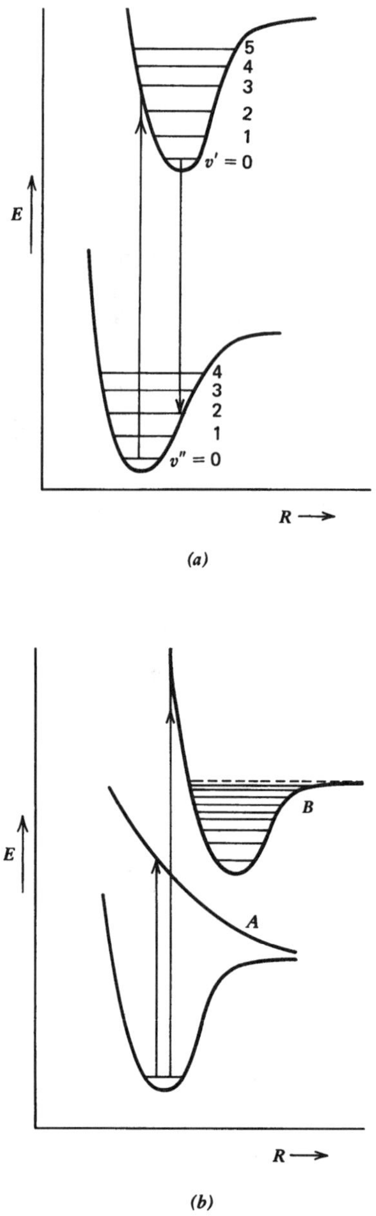

Figure 7.11 Illustration of the Franck–Condon principle. (a) Stable excited state: Any $v' - v''$ transition is possible, but the most likely are between states having maximum values of $|\psi_{\text{vib}}|^2$ (cf. Fig. 4.4) at nearly the same R. Here, for example, an excitation from $v'' = 0$ is most likely to yield $v' = 3$, and the state $v' = 0$ is most likely to decay to $v'' = 2$, as indicated by the vertical arrows. (b) Two ways to obtain a continuous spectrum: excitation to an unstable state (A), or to a stable state (B) at a level above its dissociation limit.

physical situation is quite different. In infrared transitions B_v' and B_v'' are nearly equal, so heads appear only at high m and the Q branch is very narrow. In electronic spectra this is no longer the case. Since two electronic states are involved, with completely different $E(R)$ curves and values of R_e there is no reason why B_v' and B_v'' should be close to each other. Thus sharp band heads can appear at low values of m, and the Q branch can be as broad as the P and R branches; an example is analyzed in Fig. 7.10.

Another difference from the infrared spectrum is that either B_v' or B_v'' may be the larger, depending on which state has the higher moment of inertia, so a head may appear in either the P or the R branch. Because the band heads are so well defined, they are commonly used for identification and analysis of spectra.

There is no vibrational selection rule in electronic transitions; its place is taken by the *Franck–Condon principle.* This assumes that, during an electronic transition, the nuclei tend to retain their initial positions and momenta. The concept is based on the rapidity with which the very light electrons make their transition, relative to the time and impulse required for the nuclei to change their positions and momenta. Retention of position means that the transition in a single molecule can be represented schematically by a vertical line between the potential curves of the upper and lower electronic states, $E'(R)$ and $E''(R)$, respectively. Retention of momentum means that the lower terminus of this "transition line" should be as far above $E''(R)$ as the upper terminus is above $E'(R)$. Such lines are normally drawn in the simplest way, connecting one potential curve with the other. The vertical lines in Fig. 7.11a illustrate such transitions. Strictly, position and momentum cannot both be preserved in a transition between bound states; the most probable transitions should be those for which both conditions are most nearly satisfied. In addition, different vibrational states have their maximum values of $|\psi_{\text{vib}}|^2$ at different values of R. The more nearly these values coincide for a given $v' - v''$ transition, the more likely that transition is to be observed. (Remember that the maximum of $|\psi_{\text{vib}}|^2$ lies at R_e for $v = 0$, but near the extremes of the vibration for higher states; cf. Fig. 4.4.) Each electronic transition gives rise to a set of vibrational bands, the relative intensities of which vary in accord with these principles. If the final state lies above the dissociation limit, as in Fig. 7.11*b*, a continuous spectrum will be observed. This is always the case when the electronic state in question is unstable, like the H_2 $b^3\Sigma_u^+$ state we discussed in Section 6.10.

To make the concept of the Franck–Condon principle more precise, we write out the electronic counterpart of Eq. 7.12 for a diatomic molecule. The wave function for the initial state is $\psi_{\text{nucl}}(R)\psi_{\text{elec}}(r;\, R)$, and for the final state, $\psi'_{\text{nucl}}(R)\psi'_{\text{elec}}(r;\, R)$. We neglect the rotational wave functions, which are not relevant at this point, and use r to represent all electronic coordinates, as we did in Eqs. 6.2 and 6.3. The transition is electronic; for convenience, suppose that it is induced by a uniform electromagnetic field, so that the transition operator is the electronic dipole moment operator $\boldsymbol{\mu}(r)$. Then the transition amplitude is

$$A = \int dR \int \psi^*_{\text{nucl}}(R)\psi^*_{\text{elec}}(r,R)\,\boldsymbol{\mu}(r)\psi'_{\text{nucl}}(R)\psi'_{\text{elec}}(r,R)\; dr. \tag{7.36}$$

We frequently approximate the value of A by assuming that the electronic factor—the electronic transition dipole—is given by its value at $R = R_e$, so that the electronic factor of the transition dipole takes the form

$$\langle \mu(R) \rangle_{R_e} = \int \psi^*_{\text{elec}}(r,R_e)\,\boldsymbol{\mu}(r)\psi'_{\text{elec}}(r,R_e)\; dr. \tag{7.37a}$$

This is called the Condon approximation. With this approximation, the full transition dipole is

$$\langle \boldsymbol{\mu} \rangle_{\text{Condon}} = \langle \boldsymbol{\mu}(R) \rangle_{R_e} \int \psi^*_{\text{nucl}}(R)\psi'_{\text{nucl}}(R)\; dR. \tag{7.37b}$$

Alternatively, we can call upon the mean value theorem and our knowledge that the expectation of $\boldsymbol{\mu}(r)$ is bounded to tell us that there is some mean value $\langle\boldsymbol{\mu}(r)\rangle_0$ of $\boldsymbol{\mu}(r;\, R)$ for which the accurate dipole moment, $\langle\boldsymbol{\mu}\rangle_{\text{acc}}$, is

$$\langle \boldsymbol{\mu} \rangle_{\text{acc}} = \langle \boldsymbol{\mu}(r) \rangle_0 \int \psi^*_{\text{nucl}}(R)\psi'_{\text{nucl}}(R)\; dR. \tag{7.38}$$

In words, the transition amplitude can be written to be proportional to the overlap of the vibrational wave functions of the initial and final states (see Appendix 6A). If $\psi_{\text{nucl}}(r)$ and $\psi'_{\text{nucl}}(R)$ have amplitudes distributed in the same range of R *and* if they have similar wavelengths, that overlap is large. The first of these conditions corresponds to the preservation of location; the second, to the preservation of momentum. If neither is well met, the integral $\langle\boldsymbol{\mu}\rangle_{\text{acc}}$ of Eq. 7.37 is small and the transition probability, proportional to $|\langle\boldsymbol{\mu}\rangle_{\text{acc}}|^2$, is correspondingly low.

Not all electronic transitions are allowed; here again we find selection rules. The total angular momentum **J** has the magnitude $[J(J+1)]^{1/2}\hbar$, where the rotational quantum number J obeys the selection rule of Eq. 7.35. For a diatomic molecule, **J** is the sum of orbital and spin electronic components and the angular momentum of nuclear rotation. As in the atom (Sections 5.6 and 5.7), all these components interact with one another, but some kinds of coupling are more important than others. We shall assume that one can separately define the total electronic orbital angular momentum **L**, with z component $\Lambda\hbar$, and the total electronic spin angular momentum **S**, of magnitude $[S(S+1)]^{1/2}\,\hbar$. When spin-orbit coupling can be neglected,[18] one finds the selection rules

$$\Delta\Lambda = 0,\ \pm 1 \tag{7.39}$$

($\Sigma \leftrightarrow \Sigma$ or $\Sigma \leftrightarrow \Pi$ allowed, but $\Sigma \leftrightarrow \Delta$ forbidden) and

$$\Delta S = 0 \tag{7.40}$$

($^1\Sigma \leftrightarrow {}^1\Sigma$ allowed, $^1\Sigma \leftrightarrow {}^3\Sigma$ forbidden). There are additional selection rules for various coupling conditions, but we shall ignore these. For homonuclear molecules one must also consider the symmetry of the wave function, since the parity of the wave function always changes in an electronic transition; we write this rule as

$$g \leftrightarrow u, \quad g \not\leftrightarrow g, \quad u \not\leftrightarrow u, \tag{7.41}$$

where "$\not\leftrightarrow$" designates a forbidden transition. (In the orbital approximation, the same rule holds for the orbitals if only one electron takes part in the transition.) On the other hand, the wave function's symmetry under interchange of electrons or nuclei is fundamentally associated with the indistinguishability of identical particles, and must always be conserved:

$$s \leftrightarrow s, \quad a \leftrightarrow a, \quad s \not\leftrightarrow a \tag{7.42}$$

(s = symmetric, a = antisymmetric). This consequence of particle identity is probably the strictest of all molecular selection rules.

We have been able to give only the barest outline of molecular spectroscopy. To finish our treatment of the subject, let us mention one additional complication: the *isotope effect*. The spacing of both vibrational and rotational energy levels depends on the molecule's reduced mass μ, which of course varies from one isotopic species to another. The effect is quite large for hydrogen, as we illustrated in Table 7.1. It is still present in even the heaviest molecules. For example, a given vibration–rotation band of BrCl appears in the spectrum at four different (but overlapping) positions, corresponding to $^{79}Br^{35}Cl$, $^{81}Br^{35}Cl$, $^{79}Br^{37}Cl$, and

[18] Spin–orbit coupling does occur, of course, and as in atoms gives the spectrum a fine structure. In addition, when $\Lambda > 0$ electronic-rotational interaction splits the otherwise degenerate energy levels with $M_J = \pm|M_J|$ (Λ-*type doubling*). We shall say more about these effects in Section 7.9.

$^{81}Br^{37}Cl$; in the pure rotational spectrum there are four closely spaced lines for each transition. This is one more complexity that the spectroscopist must take into account when analyzing a spectrum. But this very complexity makes it possible to excite selectively a spectral line or band of a single isotopic species. Such selective excitation can be used as a first step for starting a photo-induced chemical reaction or ionization that allows one to separate one isotope from another, a process that is otherwise difficult and costly.

7.4 The Ionic Bond

We can now return to the problem of chemical bonding. In Chapter 6 we introduced the basic principles of covalent bonding, for the simple cases of one- and two-electron molecules; later in this chapter we shall extend these principles to the general problem of the electronic structure of diatomic molecules. First, however, we must examine a class of molecules constituting such an extreme case that they are best described in terms of a different (and simpler) model. These are the molecules, best typified by the diatomic alkali halides, in which we can say that there is an *ionic bond.*

The characteristic feature of such a molecule is that its properties—especially electron distribution and bonding energy—closely resemble those of two oppositely charged ions in close proximity. The NaCl molecule, for example, is quite like the ion pair Na^+Cl^-. Of course, some redistribution of electronic charge occurs in the formation of any diatomic molecule from the atoms. In a covalent bond this redistribution is rather subtle: A quite small amount of charge is shifted from antibonding to bonding regions, increasing the net bonding force enough to hold the molecule together. But in a homonuclear molecule like H_2 the total amount of charge in each half of the molecule must be the same as in the separated atoms. In the alkali halides we have a charge redistribution of quite a different order. Charge roughly equivalent to one electron shifts from one atom to the other, i.e., from the region surrounding one nucleus to that surrounding the other. This is clearly illustrated by Fig. 7. 12, which shows that the electron distribution in the LiF molecule is much more like the ion pair $Li^+ + F^-$ than the neutral atoms Li + F.

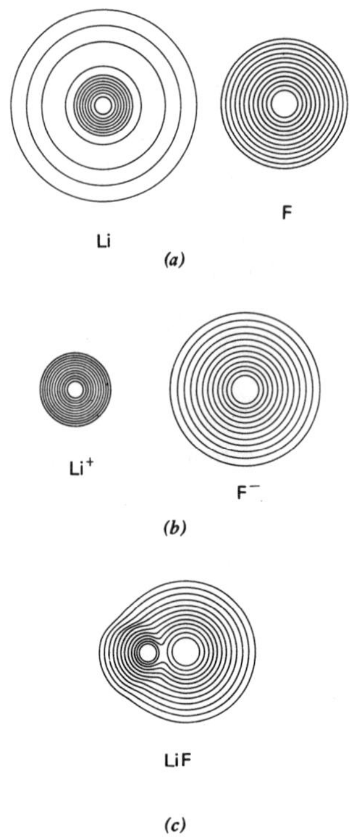

Figure 7.12 Electron distribution in the lithium-fluorine system. (*a*) Separated atoms, Li + F. (*b*) Separated ions, $Li^+ + F^-$. (*c*) The LiF molecule at the equilibrium internuclear separation, $R_e = 1.564$ Å. Diagrams taken from A. C. Wahl, *Sci. Am.* **322,** No. 4, 54 (April 1970).

Why does a charge shift of this magnitude occur? As always in bonding problems, because the resulting charge distribution gives the lowest energy for the molecule as a whole. Thus ionic bonding is most likely to occur when an electron can be easily removed from one atom and easily added to the other—"easily" in both cases refers to the energy involved. In the terminology we introduced in Section 5.5, one atom must have a low ionization potential and the other a high electron affinity. It is obvious from Figs. 5.6 and 5.7 that these conditions are best satisfied for the alkali halides. This reasoning is confirmed by detailed calculations of the molecular wave functions, such as those leading to Fig. 7.12. But calculations of this type are essentially the same as those used in covalently bonded molecules. We brought up ionic bonding at this point mainly because there is a *simpler* way to get useful results. Let us see now what it is.

The model we use could not be simpler. We assume that the molecule *is* an ion pair, and calculate the energy of interaction between two ions at close range. Why is this easier than calculating the interaction between two neutral atoms? The answer is clear when we consider the two cases at long range. There is virtually no interaction between two neutral atoms at long range;[19] the dominant terms in the bonding energy result from the reorganization of charge that occurs when the atoms come close together, and must thus be calculated in terms of the forces between individual electrons and nuclei. But two oppositely charged ions do interact strongly at quite long range because of our old friend, the attractive Coulomb force. In first approximation the ions are spherically symmetric charge distributions, which interact as if all their charge were concentrated at the nuclei. Thus

[19] The van der Waals attraction that exists between all atoms or molecules at long range (see Chapter 10, especially Section 10.1) is negligible in comparison with the energies involved in bonding.

the long-range force between the ions is the same as that between two point charges: The force is given by Eq. 1.5 and the corresponding potential energy by Eq. 2.55. Since this interaction involves the whole charge of the ions, even at R_e it is much larger than the corrections resulting from charge reorganization. One can readily confirm this assertion, since the Coulomb energy ($e^2/4\pi\epsilon_0 R$) is remarkably easy to calculate. It will be seen in Table 7.3 that the Coulomb energy at R_e by itself gives a fairly good approximation to the observed bond energy (last column). (Note that to carry out this confirmation, we must know R_e from some other source.)

The existence of an R_e implies that another force, in addition to the attractive Coulomb force, contributes to the bond. As in all molecules, the equilibrium bond length R_e is the distance at which there is a balance between attraction and repulsion. The repulsion here is essentially the same as that which occurs when any two atoms or ions "come in contact," that is, when there is significant overlap between the atomic (ionic) wave functions. There is no simple expression that would enable us to calculate this repulsive force as easily as the attractive Coulomb force. A detailed calculation can of course be made in terms of individual electrons and nuclei, but that is just what we are trying to avoid. Fortunately, there is an empirical expression that gives a good description of the short-range repulsion: The potential energy is assumed to fall off exponentially with distance,

$$V_{\text{rep}}(R) = Be^{-R/\rho}, \tag{7.43}$$

where B and ρ are constants. This is the repulsive part of the *Born–Mayer potential,* originally introduced to describe the similar interaction in ionic crystals. One can at least get a qualitative idea of why the repulsive potential has this steep form. For very small R the potential is dominated by the direct internuclear repulsion, which has the relatively slowly decreasing Coulombic form ($V \propto R^{-1}$); yet for $R \geq R_e$ the repulsive part of the potential becomes extremely small. The repulsive potential must thus decrease appreciably faster than R^{-1} in the intermediate region, and the exponential is one of the simplest functions that has this behavior.[20] The reason for this steep decrease is the increased screening of the nuclei from one another as the amount of electronic charge between them increases with increasing R. This process is essentially complete at a distance slightly greater than R_e, as the overlap of the atomic wave functions becomes negligible.

In the ionic bond model, then, the total potential energy of a diatomic molecule must include a Coulombic ion-attraction term and a short-range repulsion term like

Table 7.3 Test of the Ionic Bond Model for Some Alkali Halide Molecules*

Molecule	R_e (Å)	$e^2/4\pi\epsilon_0 R_e$ (eV)	Ionic Dissociation Energy, $D_e(MX \to \text{ions})$(eV) Calculated	Observed
LiF	1.564	9.21	7.9996	7.983
LiCl	2.021	7.14	6.513	6.648
NaCl	2.361	6.10	5.616	5.750
KF	2.171	6.63	5.993	6.036
KI	3.048	4.72	4.458	4.601
RbCl	2.7869	5.17	4.835	4.917
CsCl	2.906	4.96	4.692	4.870

* Data taken from P. Brumer and M. Karplus, *J. Chem. Phys.* **58,** 3903 (1973) and references therein.

Eq. 7.43. Additional corrections can be added; the most important of these is an attractive term due to the *polarization* of each ion by the charge on the other. That is, the field of each ion tends to push the other ion's nucleus in one direction and its electrons in the opposite direction, creating an induced dipole on which the first ion exerts an additional attractive force. This process is analyzed in Section 10.1. The magnitude of the polarization energy for each ion is given by Eq. 10.14. Combining this with the other terms, we find a total ionic bond interaction energy of

$$E(R) = -\frac{q_1 q_2}{4\pi\epsilon_0 R} + Be^{-R/\rho} - \frac{\alpha_1 q_2^2 + \alpha_2 q_1^2}{32\pi^2\epsilon_0^2 R^4}, \tag{7.44}$$

where q_i is the charge and α_i the polarizability of the ith ion; for the alkali halides we of course have $q(M^+) = -q(X^-) = e$.

The validity of the ionic bond model can be tested by evaluating Eq. 7.44 at $R = R_e$ and comparing the result with the experimental dissociation energy. But here we must be careful. Equation 7.44 gives the energy of the ionic molecule MX relative to the separated ions $M^+ + X^-$; this corresponds to the "ionic dissociation energy"

$$D_e(\text{MX} \to \text{M}^+ + X^-) = E(\text{M}^+ + X^-, R = \infty) - E(MX, R = R_e). \tag{7.45}$$

However, the ground state of the system at large R corresponds to the neutral atoms in their ground states, not to the ions. This is true for any pair of atoms: Since the lowest ionization potentials (I) are greater than the highest electron affinities (A), it always requires energy to make a pair of separated ions from the ground-state neutral atoms. This energy of ion-pair formation is

$$E(\text{M}^+ + \text{X}^-, R = \infty) - E(\text{M} + \text{X}, R = \infty) = I(\text{M}) - A(\text{X}). \tag{7.46}$$

[20] An even more suitable (but less tractable) repulsive potential would be $V_{\text{rep}}(r) = BR^{-1}e^{-R/\rho}$, which gives the correct Coulomb behavior as $R \to 0$. This "screened Coulomb potential" is also known as the *Debye potential* in the context of ionic solutions, as we shall see in Chapter 26.

The true equilibrium dissociation energy[21] of the MX molecule is thus that into the separated neutral atoms,

$$D_e(\mathrm{MX} \to \mathrm{M} + \mathrm{X}) = D_e(\mathrm{MX} \to \mathrm{M}^+ + \mathrm{X}^-) - I(\mathrm{M}) + A(\mathrm{X}). \quad (7.47)$$

The predictions of the ionic bond model for a number of alkali halides are compared with experiment in Table 7.3. The "ionic dissociation energies" are those defined by Eq. 7.45. The calculated values are essentially the values of $-E(R_e)$ predicted by Eq. 7.44, with some minor corrections (van der Waals interaction, dipole–dipole interaction, etc.). The values of B and ρ are fitted to the repulsive part of the $E(R)$ curve, and the values of the α's to the observed dipole moment. It turns out that the repulsive and polarization terms are both so small near R_e that the ionic Coulomb potential alone ($e^2/4\pi\epsilon_0 R_e$) gives a fairly good approximation to D_e, as we noted earlier. With the other terms included, the agreement with experiment is within a few percent. We conclude that the ionic bond model gives a satisfactory description of the alkali halide molecules. This is not surprising, since Eq. 7.44 is basically the same as the equation used to calculate the lattice energy of an MX crystal, which is well known to be a lattice of M^+ and X^- ions (see Chapter 11).

For an additional test of the ionic bond model, let us consider the dipole moments of ionic molecules. If such a molecule did consist simply of two spherical nonoverlapping ions with charges $\pm q$ a distance R_e apart, the magnitude of the dipole moment would be given by Eq. 4.29 as $\mu = qR_e$, or eR_e for the alkali halides. As can be seen from Table 7.4, the actual dipole moments[22] are appreciably less. This discrepancy is due to the polarization effect already mentioned. By attracting the electrons of a negative ion away from the nucleus and toward itself, a positive ion has the effect of reducing the dipole moment; the negative ion attracts the positive ion's nucleus and repels its electrons. The more polarizable the ions are, the more the molecular dipole moment is reduced. In fact, this effect can be used to evaluate the α_i's of Eq. 7.44. The polarizability of an ion is roughly proportional to its size. In most of the alkali halides the X^- ions are significantly larger and thus more polarizable than the M^+ ions, so the principal effect is the distortion of the halide ion; the alkali ion is effectively very stiff.

Table 7.4 Dipole Moments of Some Ionic Molecules

Molecule	Ionic Charge, q	R_e (Å)	qR_e (D)	Measured Dipole Moment, μ (D)
LiH	± e	1.595	7.66	5.882
LiF	± e	1.564	7.51	6.325
LiCl	± e	2.018	9.69	7.126
LiBr	± e	2.170	10.42	7.265
LiI	± e	2.392	11.49	7.428
NaCl	± e	2.361	11.34	9.001
KCl	± e	2.667	12.81	10.269
RbCl	± e	2.787	13.39	10.510
CsCl	± e	2.906	13.96	10.387
HCl	± e	1.275	6.12	1.109
TlCl	± e	2.485	11.94	4.543
SrO	± 2e	1.920	18.44	8.900
BaO	± 2e	1.940	18.64	7.954

The extent of polarization is of course also affected by the charge doing the polarizing. The larger the value of q, the more the dipole moment is reduced. This is made strikingly clear by the example of SrO and BaO in Table 7.4. Naive chemical notions (based on the stability of closed-shell structures) suggest that these molecules might have the structure $M^{2+}O^{2-}$, but examination of the second ionization potentials (Fig. 5.6) shows the unlikelihood of this idea. In fact, the alkaline earth oxides have dipole moments close to what one would expect for pairs of singly charged ions. One can still describe these compounds with an ionic bond model, but it takes a more elaborate version than we have used for the alkali halides. The reason is that the ions involved, Ba^+ and O^-, for example, have open shells and are thus far more easily distorted by polarization than closed-shell ions (cf. Pacchioni et al., Phys. Rev. B 48, 11573 [1993]).

It might seem that the ionic bond model is of limited importance, since it is directly applicable to only a small class of diatomic molecules. But the utility of the concept actually extends over a far wider range. For one thing, it can be used with little change to describe the bonding in a vast number of ionic crystals. (In fact, it works better for crystals than for single molecules, since the presence of neighbors on all sides of each ion reduces the polarization effects.) The model displays very clearly the distinction between the long-range attractive and short-range repulsive forces, whose balance is crucial for the existence of stable chemical bonds. Moreover, the ionic bond provides us with the extreme case of a chemical bond having the maximum asymmetry of charge distribution. The homonuclear covalent bond, with its perfectly symmetric charge distribution, lies at the other end of this scale, and the bonds in other heteronuclear diatomic molecules fall between these extremes. One can usefully interpret most bonds as some kind of mixture or superposition of ionic and symmetric covalent contributions.

[21] That is, the well depth, measured from the $E(R)$ minimum, not from the ground vibrational state.

[22] Dipole moments can be measured in several ways. If an electric field is applied to a gas of polar molecules, the molecules will tend to align themselves with the field. The extent of this alignment can be related to the gas's dielectric constant, and thus to the (measured) capacitance of the system. The electric field also splits the degenerate rotational energy levels with the same J and different M_J (Stark effect); this splitting can be observed either directly in the microwave spectrum or by measuring the deflection of molecular beams by the field.

Finally, we can use the ionic model to gain an insight into the nature of bonding that applies even to covalent molecules. Consider any two atoms that form a bond. In the long-range limit, each atom consists of one or more valence electrons outside a closed-shell core—"outside" in the sense that the core electrons have very low charge density at the radii where the valence electrons have their maxima. As the atoms move closer together, bonding effects occur as the valence electrons begin to overlap. But even near R_e there is no significant overlap between one atom's valence electrons and the other atom's core. An atomic core is essentially the same as a spherically symmetric ion. If an ionic bond model is valid for two ions in "contact" with each other, it should also apply reasonably well for the core part of the interaction in an ordinary bond. Given the distance between the cores, only the Coulombic part of the interaction should be important; in other words, the cores act much like point charges. This means that it is a fairly good approximation to neglect the atomic cores and consider only the valence electrons[23] in describing chemical bonding. This is of course precisely the approximation made in elementary treatments of chemistry, and now we can see why it works as well as it does.

7.5 Homonuclear Diatomic Molecules: Molecular Orbitals and Orbital Correlation

To study the electronic structure of covalent diatomic molecules, we must further develop the theory introduced in Chapter 6. We shall base the next part of our discussion on the molecular orbital approach, the simplest and clearest for our present purposes. Thus we assume that each electron sees only the average field of the others and can be described by a one-electron wave function. Since each orbital can "contain" only two electrons (with opposite spin), one must in general use several orbitals to describe a molecule.

Suppose that one finds a suitable set of molecular orbitals; what does one do with them? The most important thing to know is how the energies and shapes of the orbitals vary as some characteristic parameter of the molecule is changed. For a diatomic molecule the key parameter is the internuclear distance R. Given the sequence of orbital energies, one can develop an Aufbau principle like that used for atoms and derive the equilibrium electronic configuration of the molecule. The orbital energies as functions of R are most simply displayed in a correlation diagram, like the one we drew for H_2^+ in Section 6.5. We shall see that these and other properties of orbitals vary systematically in series of molecules.

[23] In a transition metal atom the "valence electrons" must include the partly filled inner d or f subshells, which are quite close to the atom's "surface."

The first tool we need for the description of molecular orbitals is a systematic way of classifying them. Just as atomic orbitals are given the names of the analogous states of the one-electron H atom, molecular orbitals are named after the states of the one-electron H_2^+ molecule. We have already introduced most of the terminology needed, but we shall review it here. Both the orbitals and the states of the molecule as a whole are classified in terms of the wave function's symmetry properties.

We have introduced two kinds of symmetry thus far, the *permutational* symmetry associated with the indistinguishability of elementary particles, and the *spatial* symmetry associated with the indistinguishability of coordinate systems. It is spatial symmetry that concerns us here; we classify molecules according to the invariance of their physical description under specific kinds of coordinate changes—rotation, reflection, and inversion. We must keep in mind that a change in a molecule's coordinate frame is equivalent to an equal but opposite change in the molecule itself; translating the origin of coordinates a distance $+x_0$ along the x axis with the molecule fixed is equivalent to translating the molecule a distance $-x_0$ in a fixed coordinate system. For convenience we describe spatial transformations in terms of moving the molecule rather than the coordinates, but the two are equivalent. Each of us can think of symmetry and invariance in whichever way seems clearer. Now let us list the kinds of symmetry properties that apply to diatomic molecules.

1. The most important symmetry property, at least for the molecular energy, describes the behavior of the electronic wave function under rotation about the internuclear axis. This is given by the quantum number λ (Λ for the many-electron wave function), which specifies the angular momentum about the internuclear axis: $L_z = \pm\lambda\hbar$. The corresponding part of the wave function is $e^{\pm i\lambda\phi}$, where ϕ is the angle of rotation about the z axis (Fig. 6.1). Any rotation through an angle ω simply changes the wave function to $e^{\pm i\lambda(\phi+\omega)}$, introducing a constant phase factor $e^{\pm i\lambda\omega}$ but leaving the angular momentum unchanged. This corresponds to the fact that all values of ϕ are physically indistinguishable. States with $L_z = +\Lambda\hbar$ and $L_z = -\Lambda\hbar$ (for $\Lambda > 0$) are doubly degenerate. We identify states with $L_z = +\Lambda\hbar$; λ, $\Lambda = 0, 1, 2, 3, \ldots$ by the already familiar designations $\sigma, \pi, \delta, \varphi, \ldots$ (orbitals), $\Sigma, \Pi, \Delta, \Phi, \ldots$ (molecules).
2. The molecule is also physically unchanged by reflection in a plane containing the z axis, the axis of the bond. The electronic wave function must either remain unchanged or reverse its sign; the two cases are designated by the superscripts + and –, respectively. For $\Lambda > 0$, one of the degenerate molecular states with $L_z = \pm\Lambda\hbar$ is +, the other –, and the superscript is usually omitted; Σ states must be either Σ^+ or Σ^-, but all those we discussed in Chapter 6 are Σ^+.

Symmetry properties (1) and (2) apply to all diatomic (or other linear) molecules. In a homonuclear molecule the two ends are indistinguishable, imposing an additional symmetry condition:

3. The electronic wave function must either remain unchanged (even parity) or reverse its sign (odd parity) when the molecule is inverted through its center of symmetry, $\psi(-x, -y, -z) = \pm\psi(x, y, z)$. As described in Section 6.5, we designate even and odd states by the subscripts g and u, respectively.

We have spoken here only of the electronic wave function (assuming the Born–Oppenheimer approximation), but there are also symmetry conditions on the total molecular wave function. The most important of these is its symmetry or antisymmetry (denoted by s or a) under interchange of identical nuclei, which divides the rotational levels into two groups. We shall postpone analysis of this subject to Chapter 21, which treats the relation between symmetry and the distribution of molecules over energy levels.

Now we ask how the orbitals vary as a function of internuclear distance. The limiting cases are simple enough: as $R \to \infty$ the molecule behaves like two separated atoms; as $R \to 0$ it collapses into a simple united atom. We generally know the states, orbitals, and orbital energies of both the separated-atom and united-atom limits. Given the Born–Oppenheimer approximation, we know that the total electronic energy of a molecule varies smoothly with R. If the orbital approximation holds for all R, we can say the same about the orbital energies, which we designate as $\epsilon_i(R)$. Thus we can join the united-atom and separated-atom orbital energies by a set of smooth $\epsilon(R)$ curves, making up what we have already defined as a correlation diagram. In the one-electron diagram for H_2^+ (Fig. 6.9) many of the energy levels are degenerate. A more typical example is the N_2 molecule (Fig. 7.13), whose united-atom limit is the silicon atom. It is often unnecessary in such a diagram to give the energies quantitatively for intermediate values of R. Frequently the ordering of connecting lines as functions of R, with the limiting energies, is sufficient for analysis of physical phenomena.

How, then, does one know where to draw the connecting lines between the separated-atom and united-atom limits? The rules are quite straightforward, all taking the form of conservation laws:

1. Angular momentum about the internuclear axis (defined by the quantum number λ) is conserved, so that a π orbital, for example, remains a π orbital for all values of R. Since $\lambda\hbar$ gives the absolute value of L_z, in the two limits λ goes over into $|m_l|$, the absolute value of the atomic quantum number m_l. In the united-atom limit, for each subshell with quantum numbers n, l there are $2l + 1$ degenerate atomic orbitals, whose angular momentum components along a given axis have the values $L_z = m_l\hbar$ ($m_l = -l, -l + 1, \ldots, l - 1, l$); we of course take the z axis to be the same as the internuclear axis for $R > 0$. As soon as R does have a value greater than zero, these degenerate orbitals become physically distinguishable and have distinct energies, except that each pair with $m_l = \pm|m_l|$ remains degenerate. Thus the $2l + 1$ molecular orbitals have only l distinct energy levels, with values of λ ($= |m_l|$) ranging from 0 to l. For example, the $3d$ subshell of the united atom has five degenerate orbitals ($l = 2$, $m_l = -2, \ldots, 2$) which in the molecule become a single $3d\sigma$ orbital ($\lambda = 0$), a degenerate pair of $3d\pi$ orbitals ($\lambda = 1$), and another degenerate pair of $3d\delta$ orbitals ($\lambda = 2$).

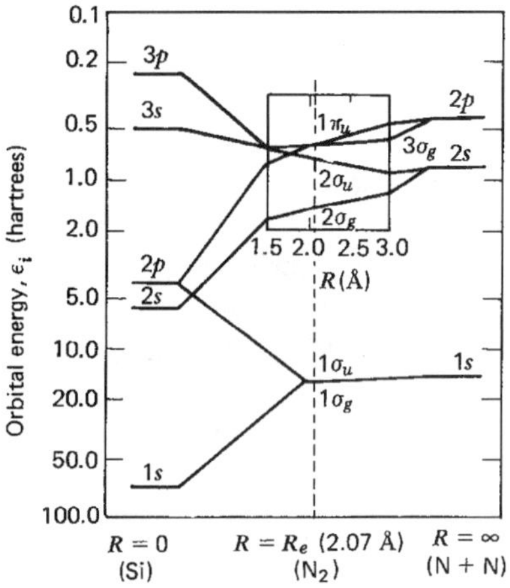

Figure 7.13 Correlation diagram for orbitals of N_2. The orbital energy curves in the inset rectangle are those calculated by P. E. Cade, K. D. Sales, and A. C. Wahl, *J. Chem. Phys.* **44,** 1973 (1966); elsewhere the correlation curves are straight lines connecting the atomic and molecular orbital energies.

2. In homonuclear molecules, the parity of each orbital is also conserved. Consider the separated-atom limit. Each atomic subshell yields the same orbitals as in the united atom, and each of these orbitals has its exact counterpart on the other atom. By exactly the same reasoning as in our analysis of H_2^+ (Section 6.3), the symmetry of a homonuclear molecule requires that as $R \to \infty$ all its orbitals take on the form of a sum or difference of two atomic orbitals. Thus for each atomic orbital φ_i there are two molecular orbitals, $(\varphi_i)_A \pm (\varphi_i)_B$, which as $R \to \infty$ have the same energy; the sum is obviously of even parity (g) and the difference of odd parity (u). As R decreases the LCAO approximation no longer holds exactly and the two orbitals become nondegenerate, but the parity remains the same. Corresponding to our example above, the $3d$ subshells of identical separated atoms yield 10 molecular orbitals: $\sigma_g\,3d$, $\sigma_u\,3d$, and a degenerate pair each of $\pi_g\,3d$, $\pi_u\,3d$, $\delta_g\,3d$, $\delta_u\,3d$. In the united-atom limit the parity of the orbitals is easily derived from the form of the spherical harmonics (Table 3.1): All orbitals with even l are g, since they are polynomials of even degree in $\sin\theta$ and $\cos\theta$; similarly, all orbitals with

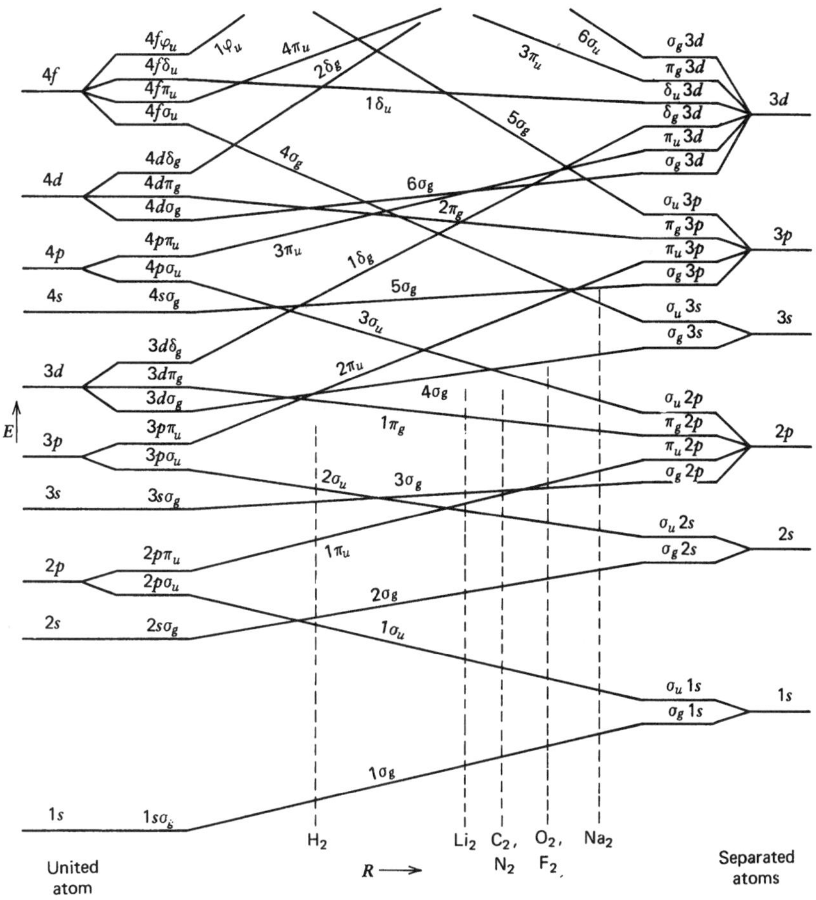

Figure 7.14 Orbital correlation diagram for homonuclear diatomic molecules. Each orbital is labeled by the three systems described in Section 6.5. The energy scale is schematic, but adjusted so that the sequence of orbital energies in several molecules can be shown by vertical dashed lines (corresponding roughly to $R = R_e$). The orbitals whose lines slant upward from left to right are bonding; those whose lines slant downward are antibonding.

odd l are u. Thus our five united-atom $3d$ orbitals, for example, are all g (having $l = 2$).

Now we can define all the molecular orbitals near the united-atom and separated-atom limits, with notations of the form $1s\sigma_g$ and $\sigma_g 1s$, respectively, as illustrated at the left and right sides of Fig. 7.14. We still need to know how to correlate these limits. Given that λ and parity are conserved, we must correlate only orbitals of the same symmetry type σ_g with σ_g, π_u with π_u, and so on. But there are many orbitals (an infinite number, in fact) of each type. To decide which correlates with which, we need a third and final rule:

3. Orbitals are connected in order of increasing energy. In the two limits, the order of orbital energies is that of the isolated atoms (cf. Fig. 5.2). Two connecting lines may cross if they refer to orbitals that differ in λ or parity, but must *not* cross if they refer to orbitals of the same type. This is the *noncrossing rule,* which we introduced in Section 6.5. It has its origin in the fundamental but unproven assumption that there are no *accidental* degeneracies in nature: There is always some perturbation that keeps two states of a system from having exactly the same energy, unless some fundamental symmetry underlies the degeneracy. The spherical symmetry of an atom is responsible for the $(2l + 1)$-fold degeneracy of states with quantum number l, and the axial symmetry of a diatomic molecule for the twofold degeneracy of states with $\lambda > 0$; in the fine structure, even these degeneracies are removed. Among the commonest perturbations that can cause crossings to be avoided is the kinetic energy of the nuclei affecting the electronic energy. Expressed in quantum mechanical terms, this is precisely the breakdown of the Born–Oppenheimer approximation. Its magnitude and effect appears when we compute the matrix elements of the *nuclear* kinetic energy operator $\mathbf{H}_{\text{nucl}}$ acting on the electronic wave functions $\psi_{\text{elec}}(r, R)$, precisely the contributions explicitly neglected in the discussion of footnote 2 of Chapter 6. The effects of perturbations of this kind are typically important only when two potential curves are very close. Appendix 7A discusses the way this can be analyzed and presents an introduction to perturbation theory.

Given the noncrossing rule, the construction of the correlation diagram is simple. We connect the lowest σ_g orbital on one side with the lowest σ_g orbital on the other, giving the $1\sigma_g$ molecular orbital, which is always the lowest-energy orbital. Similarly, we connect the second-lowest σ_g orbitals to give the $2\sigma_g$ molecular orbital, and so forth. The same

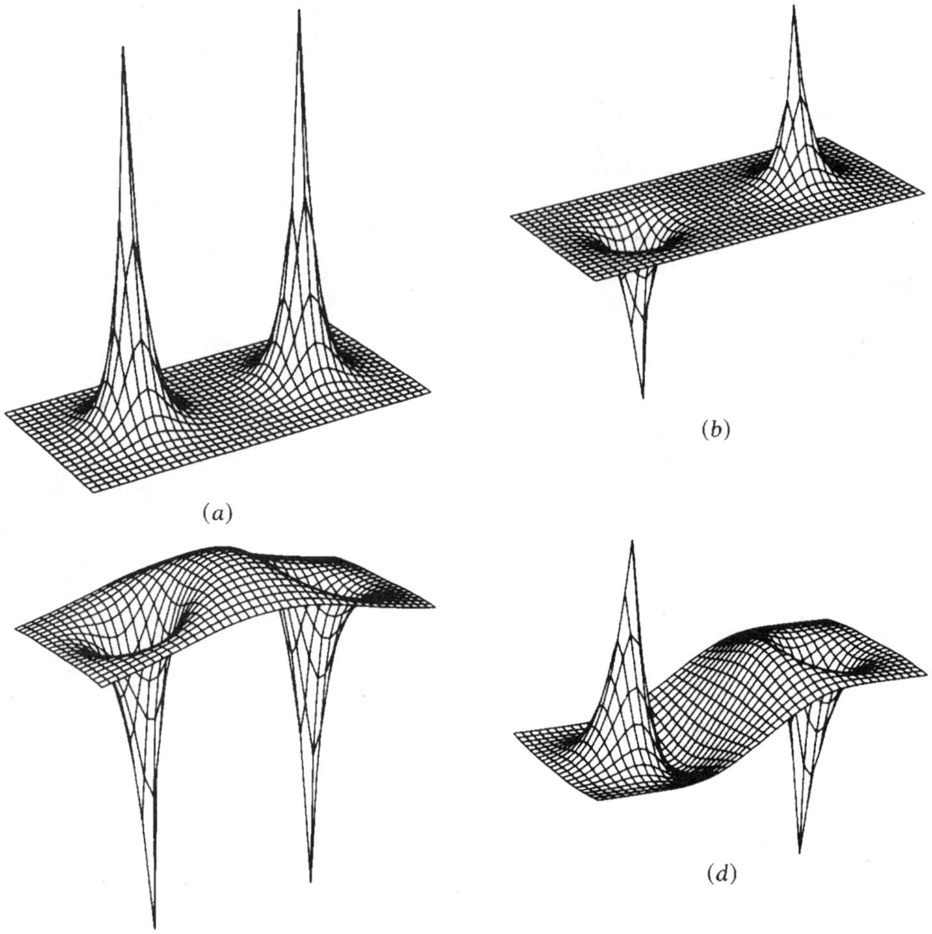

Figure 7.15 Amplitudes of the lowest orbitals of the Li_2 molecule. The graphs are all drawn for an Li–Li distance of 2.67 Å or 5.047 bohrs, the equilibrium distance. The representation gives a projection of a three-dimensional plot with the vertical scale linear in (electron density)$^{1/2}$. The first three orbitals, shown as (*a*), (*b*), and (*c*), are the $1\sigma_g$ ($1s_A + 1s_B$, nearly), the $1\sigma_u$ ($1s_A - 1s_B$, nearly), and the $2\sigma_g$ (roughly $2s_A + 2s_B$) and are normally occupied. The $2\sigma_u$ orbital, (*d*), is normally empty. Calculations were carried out by Mary Dolan and plotted by Mary Dolan and David Campbell.

process is carried out for each of the other symmetry types. We thus obtain the diagram of Fig. 7.14. The energy scale of Fig. 7.14 is schematic only, and not intended to represent the true spacing of energy levels.

The quantum number λ, equivalent to $|m_l|$, is a good quantum number within the orbital approximation for all values of R, and corresponds to the one conserved component of angular momentum. By contrast, the quantum numbers n and l for a given orbital are often different in the united-atom and separated-atom limits. The angular momentum quantum number l simply loses its meaning in the molecule because the total angular momentum of an individual electron is not conserved when the electron moves in a nonspherical potential. We could say that the quantum number n retains some validity, as an index to the number of nodes ($n - 1$) in the orbital wave function at a given R; but this number varies with R, and is not ordinarily used to identify orbitals. An orbital for which n is higher in the united atom than in the separated atoms is said to be *promoted*. The energies of promoted orbitals are generally also higher in the united-atom limit, because their promotion more than compensates for the larger nuclear charge of the united atom. Correspondingly, the energies of nonpromoted orbitals are always lower in the united-atom limit. The reason for calling attention to the distinction is this. Electrons in nonpromoted orbitals tend to stabilize, and those in promoted orbitals to destabilize the formation of a bond. In other words, promoted orbitals are usually antibonding, and nonpromoted orbitals are generally bonding.

Finally, we should say something about the shapes of the molecular orbitals. These are basically the same as the orbitals in H_2^+, several of which were shown in Fig. 6.8. Figures 7.15 and 7.16 show the amplitudes of several orbitals for Li_2 and for N_2 and CO. The nodal surfaces of orbitals are of particular interest because they are closely related to the symmetry of the orbital. For example, an orbital with $\lambda \neq 0$ has λ nodal planes containing the internuclear axis, which remain nodal planes for all values of R. All σ_u, π_g, δ_u, φ_g, . . . orbitals have an additional nodal plane through the center of symmetry, perpendicular to the bond axis. The shapes of other nodal surfaces, such as those radial nodes that are spheres in free atoms, vary with R, usually in a complicated manner. Two nodal surfaces of the separated atoms are likely to merge into one at small R, as in the $2\sigma_g$ orbital (Fig. 6.8).

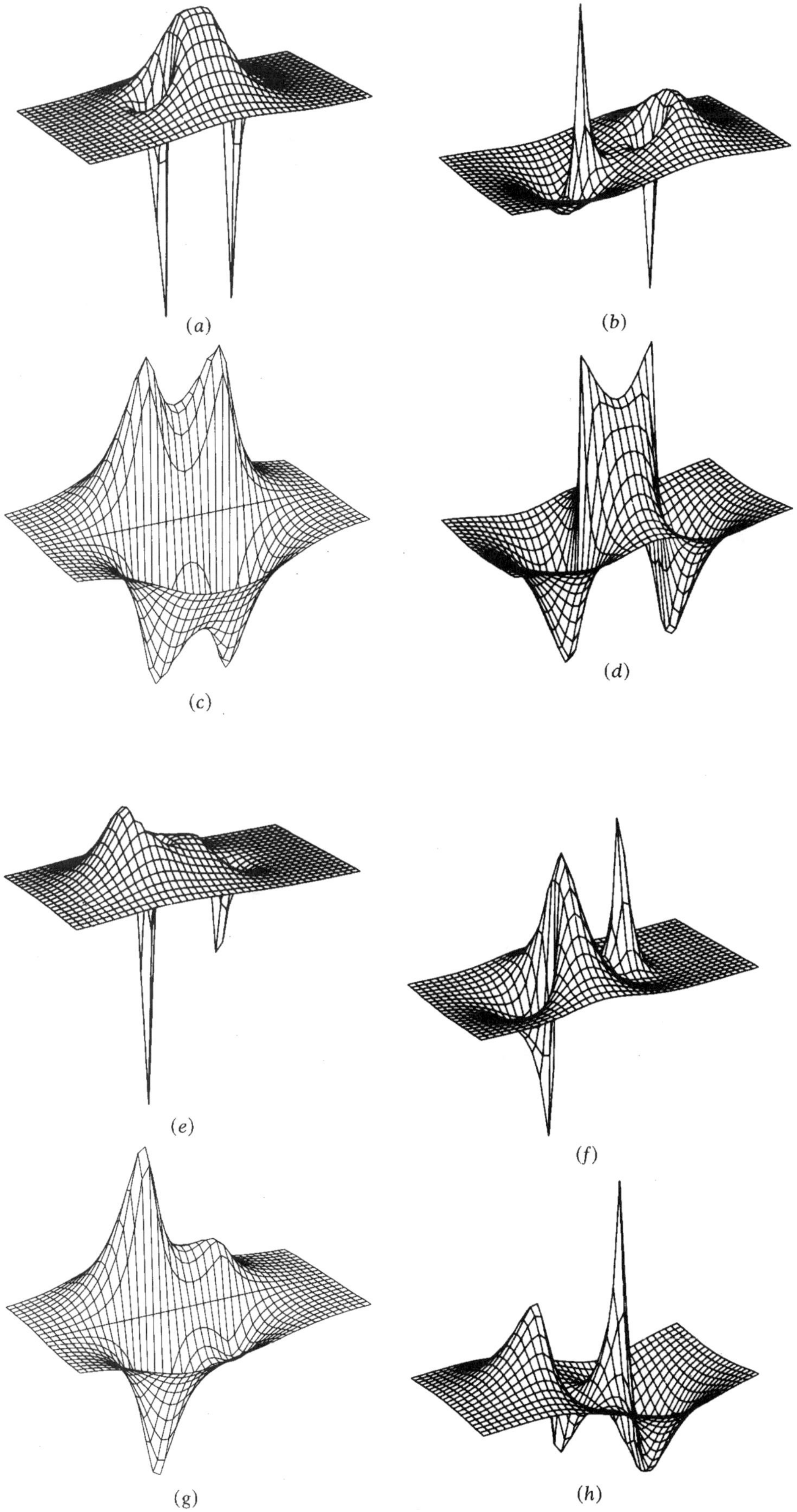

Figure 7.16 Amplitudes of the normally occupied orbitals of N_2 and CO that correlate with 2*s* and 2*p* orbitals of the separated atoms. The N—N bond distance is 1.10 Å or 2.08 bohrs; the C—O bond distance is 1.13 Å or 2.14 bohrs. All amplitudes are plotted with a vertical scale linear in ψ itself. (*a*) N_2, $2\sigma_g$ (largely $2s_A + 2s_B$). (*b*) N_2, $2\sigma_u$ (largely $2s_A - 2s_B$). (*c*) N_2, $1\pi_u$ (largely $2p\pi_A + 2p\pi_B$). (*d*) N_2, $3\sigma_g$ (roughly $2p\sigma_A - 2p\sigma_B$ if both atoms are drawn with right-handed coordinate systems). (*e*), (*f*), (*g*), (*h*) are the corresponding orbitals for CO. Calculations were carried out by Mary Dolan and plotted by Mary Dolan and David Campbell.

7.6 Homonuclear Diatomic Molecules: Aufbau Principle and Structure of First-Row Molecules

To the extent that the molecular orbital model is valid, we can describe the electronic structure of a diatomic molecule in terms of an Aufbau principle. Just as in atoms, we assign the electrons to orbitals in order of increasing energy until all the electrons are accounted for. But it is clear from Fig. 7.14 that the order of orbital energies varies with R. The sequence for a given molecule must be determined empirically, using spectroscopic and chemical evidence. We have drawn Fig. 7.14 so that the orbital sequences for a number of molecules (at R_e) can be indicated by vertical lines: One simply reads up each dashed line to find the orbitals in order of increasing energy. The representation is still only schematic and tells us nothing about the relative spacing of the energy levels, but note the agreement with the more quantitative Fig. 7.13.

The Pauli exclusion principle still applies: Each orbital can hold only two electrons, which must be assigned opposite spins. Remember that the lines in Fig. 7.14 represent not individual orbitals but orbital energy levels. Each σ level represents a single orbital, holding two electrons, but each π, δ, . . . level represents two degenerate orbitals (with $L_z = \pm\lambda$) and can thus hold four electrons. Applying this rule, we obtain the ground-state configurations listed in Table 7.5 for the homonuclear diatomic molecules through the first row of the periodic table.

The fact that we can write a configuration for a molecule does not mean that such a molecule really exists as a stable, observable entity. For example, the Aufbau principle implies that the ground state of He_2 should clearly be $(1\sigma_g)^2\,(1\sigma_u)^2$; yet there is no stable He_2 molecule corresponding to this configuration.[24] We must thus look into the conditions governing the stability of a molecule.

The key point here is that some orbitals are bonding, others antibonding. The principles developed in the last chapter indicate that if $R > R_e$, the bonding orbitals normally have a higher electron density between the nuclei than in the separated atoms, whereas the antibonding orbitals have a lower electron density between the nuclei than in the separated atoms. It is clear that all the orbitals with nodal planes midway between the nuclei (cf. Fig. 6.8) must be antibonding. Thus all σ_u, π_g, δ_u, . . . orbitals are antibonding. These are molecular orbitals in which there is destructive interference between the atomic orbitals, and may be referred to as "minus" orbitals. The "plus" orbitals with constructive interference (σ_g, π_u, δ_g, . . .) are ordinarily bonding. Note that the "plus" and "minus" orbitals can be classified into pairs, with each orbital of the separated atoms producing one of each type, $\varphi_A \pm \varphi_B$ in the simple LCAO approximation. This pairing is precise and unambiguous in homonuclear molecules, but loses some of its clarity in molecules composed of two very different atoms. The electronic energies of bonding and antibonding orbitals generally increase and decrease, respectively, with increasing R; the energy levels in Fig. 7.14 have been drawn to reflect this. (In fact, He_2 does exist, bound by exceedingly weak forces, by far the weakest among all known molecules. These are van der Waals forces, not covalent forces.)

Let us apply this information to the ground states of the simplest few molecules. The H_2^+ molecule-ion has one electron in the bonding $1\sigma_g$ orbital, and a dissociation energy (D_e) of 2.79 eV. The H_2 molecule has the configuration $(1\sigma_g)^2$ with both electrons in the bonding orbital; the dissociation energy, $D_e = 4.75$ eV, is nearly twice that for H_2^+. But the excited state $(1\sigma_g)(1\sigma_u)$ (i.e., $b^3\Sigma_u^+$, cf. Fig. 6.11), with one bonding and one antibonding electron, is unstable. There exists a He_2^+ molecule-ion, which is stable in the sense that its $E(R)$ curve has a true minimum, at $R_e = 1.08$ Å. The dissociation energy of He_2^+, the difference between $E(R_e)$ and the energy of $He(1s^2) + He^+(1s)$, is about 2.5 eV, close to that of H_2^+. The configuration should be $(1\sigma_g)^2(1\sigma_u)$, which, like H_2^+, has one more bonding than antibonding electron. And He_2, $(1\sigma_g)^2(1\sigma_u)^2$, with two bonding and two antibonding electrons, only barely exists as a stable molecule.

We can draw a fairly obvious generalization from these data: The bonding power of an electron in a bonding ("plus") orbital is roughly equal to the antibonding power of an electron in the complementary antibonding ("minus") orbital. In fact, the antibonding effect is usually slightly greater. There are a number of plausible ways to account for this imbalance. Perhaps the most physically explicit explanation is that electron–electron repulsion tends to limit the amount of charge in the strongly bonding region. Thus any "molecule" in which equal numbers of bonding and antibonding orbitals are filled has no net bonding force, and does not exhibit a chemical bond. We predict that Be_2 and Ne_2 as well as He_2 should not exist as stable molecules; the same conclusion holds for any species in which the separated atoms have only filled orbitals (such as Ar_2, Kr_2, Xe_2). This is the molecular orbital explanation for the "inertness" of closed-shell atoms.

We conclude that the strength of the bonding in a homonuclear diatomic molecule should be roughly proportional to the *bond order,* defined as the number of bonding pairs of electrons minus the number of antibonding pairs. Bond orders of 1, 2, and 3 correspond to what are conventionally called single, double, and triple bonds. Table 7.5 includes the bond orders and dissociation energies of the first-row molecules, which can be seen to support our hypothesis.

In general, when a molecule is best described in the separated-atom scheme, the molecular orbitals deriving from the same subshell of the separated atoms tend to be very close in energy. This is the case for $R \geq R_e$ in first-row molecules, and is especially noticeable when one compares

[24] But there are many stable excited states, the lowest of which is $(1\sigma_g)^2(1\sigma_u)(2\sigma_g)$.

Table 7.5 Ground States of the Homonuclear Diatomic Molecules from H_2 to Ne_2*

Molecule	Electron Configuration[a]	Term Symbol	Bond Order	D_e (eV)	R_e (Å)
H_2	$(1\sigma_g)^2$	$^1\Sigma_g^+$	1	4.7478	0.7412
He_2	$(1\sigma_g)^2(1\sigma_u)^2$	$^1\Sigma_g^+$	0	$<10^{-5}$	>50
Li_2	$KK(2\sigma_g)^2$	$^1\Sigma_g^+$	1	1.14 ± 0.3	2.6725
Be_2	$KK(2\sigma_g)^2(2\sigma_u)^2$	$^1\Sigma_g^+$	0	—	—
B_2	$KK(2\sigma_g)^2(2\sigma_u)^2(1\pi_u)^2$	$^3\Sigma_g^-$	1	3.0 ± 0.2	1.590
C_2	$KK(2\sigma_g)^2(2\sigma_u)^2(1\pi_u)^4$	$^1\Sigma_g^+$	2	6.24 ± 0.22	1.2425
N_2	$KK(2\sigma_g)^2(2\sigma_u)^2(3\sigma_g)^2(1\pi_u)^4(1\pi_g)^2$	$^1\Sigma_g^+$	3	9.7559 ± 0.0017	1.094
O_2	$KK(2\sigma_g)^2(2\sigma_u)^2(3\sigma_g)^2(1\pi_u)^4(1\pi_g)^2$	$^3\Sigma_g^-$	2	5.116 ± 0.004	1.2075
F_2	$KK(2\sigma_g)^2(2\sigma_u)^2(3\sigma_g)^2(1\pi_u)^4(1\pi_g)^4$	$^1\Sigma_g^+$	1	1.604 ± 0.1	1.409
Ne_2	$KK(2\sigma_g)^2(2\sigma_u)^2(3\sigma_g)^2(1\pi_u)^4(1\pi_g)^2(3\sigma_u)^2$	$^1\Sigma_g^+$	0	—	—

*Taken from B. de B. Darwent, *Bond Dissociation Energies in Simple Molecules,* NSRDS-NBS 31, U.S. Dept. of Commerce (U.S. Govt. Printing Office, Washington, D.C., 1970).

[a] The designation *KK* refers to the filled inner-shell configuration $(1\sigma_g)^2(1\sigma_u)^2$, corresponding to filled K ($n = 1$) shells in the separated atoms.

the $1\pi_u(\pi_u 2p)$ and $3\sigma_g$ ($\sigma_g 2p$) orbitals. In the ground states of B_2 and C_2 only the $1\pi_u$ orbitals are occupied, which implies that in these species $1\pi_u$ has lower energy than $3\sigma_g$. The experimental evidence[25] indicates that this is also true for N_2 where both levels are first occupied in the ground state, but that the $3\sigma_g$ electrons are more tightly bound in O_2 and F_2. This behavior is represented by the crossover of the $1\pi_u$ and $3\sigma_g$ lines between N_2 and O_2 in Fig. 7.14. The evidence from which these conclusions are drawn is primarily that of photoelectron spectroscopy.[26] It follows from Koopman's Theorem (Section 5.4) that the energy of an orbital should equal the energy required to ionize an electron from it.

Let us survey how the physical and chemical properties of the molecules in Table 7.5 reflect their configurations. Hydrogen has been sufficiently covered in the last chapter. As already noted, He_2, Be_2, and Ne_2 do not exist as stable molecules in their ground states; at ordinary temperatures helium and neon are monatomic gases, whereas beryllium is a metal that vaporizes (b.p. 2970°C) to a monatomic gas.[27] But we shall now consider each of the other stable species individually.

Lithium is also a metal at ordinary temperatures. Its vapor is primarily monatomic, containing only a few percent of Li_2 molecules. This can be attributed to the relative weakness of the Li–Li bond. Similar behavior is found for the other alkali metals; the bonds become longer and weaker in successively heavier molecules in the series. The bonds are weak for the same reason that the alkali atoms are large (Section 5.5). The valence (*ns*) electrons are loosely bound and diffuse in the atoms, and even in the molecules the electron density between the nuclei is low. In contrast to H_2^+ and H_2^+, the one-electron bonds in the Li_2^+, Na_2^+, . . . ions are stronger than the bonds in the corresponding neutral molecules. Apparently the electron–electron repulsion (including that between valence and core electrons) outweighs the bonding effect of the second valence electron.

Not as much is known about the vapor of boron, which remains a black semiconducting solid to well over 2000°C. However, spectroscopy has shown that the B_2 molecule is a reasonably stable species at high temperatures. The bond is of a "normal" length, and correspondingly much stronger than the bond in Li_2.

As for carbon, C_2 molecules are well known in flames, shock waves, arcs, or indeed any hot system containing an excess of carbon. But the study of carbon vapor is complicated by the presence of larger polymers. Up to around 5000°C the vapor is mainly C_3, with significant amounts of C_4, C_5, . . . , after C_2, particularly the odd-numbered members; it is not easy to get a large concentration of C_2. This situation reflects the ease with which carbon forms chains or networks of bonds, as in the solid forms of diamond and graphite and the enormous variety of organic compounds. One reason for this behavior is revealed by the energy spectrum of C_2 (Table 7.6): The energies of the $1\pi_u$ and $3\sigma_g$ orbitals are so close together that C_2 always contains a significant number of molecules in the lowest excited configuration, $KK(2\sigma_g)^2(2\sigma_u)^2(1\pi_u)^3(3\sigma_g)$.[28] The

[25] The calculation shown in Fig. 7.13 predicts that $3\sigma_g$ should be lower than $1\pi_u$ in N_2, but by a very small amount; the experimental energies are in the reverse order.

[26] In this technique a sample is ionized by irradiation with photons of known energy, usually in the vacuum ultraviolet or x-ray regions, and the energies of the emitted electrons are measured; subtraction gives the ionization energy for each electron. See Section 7.10.

[27] However, there are stable Mg_2, Ca_2, . . . molecules, indicating that the simple MO theory is not adequate to describe these molecules.

[28] It is much easier to observe transitions to or from this excited state than transitions involving the closed-shell ground state. As a result, for many years the . . . $(1\pi_u)^3(3\sigma_g)$ configuration was thought to be the true ground state. The correct analysis of the spectrum and the ordering of states was only made in 1963 by Ballik and Ramsay.

Table 7.6 Some Low-Lying Excited States of First-Row Homonuclear Diatomic Molecules

Molecule	State	Configuration	Energy above Ground State (eV)
Li_2	$A^1\Sigma_u^+$	$KK(2\sigma_g)(2\sigma_u)$	1.744
	$B^1\Pi_u$	$KK(2\sigma_g)(1\pi_u)$	2.534
	$C^1\Pi_u$	$KK(2\sigma_g)(2\pi_u)$ or $KK(1\pi_u)^2$	3.788
	$D^1\Pi_u$	?	≤4.233
B_2	$A^3\Sigma_u^-$	$KK(2\sigma_g)^2(2\sigma_u)^2(3\sigma_g)(3\sigma_u)$	3.791
C_2	$a^3\Pi_u$	$KK(2\sigma_g)^2(2\sigma_u)^2(1\pi_u)^3(3\sigma_g)$	0.089
	$B^3\Sigma_g^-$	$KK(2\sigma_g)^2(2\sigma_u)^2(1\pi_u)^2(3\sigma_g)^2$	0.789
	$A^1\Pi_u$	$KK(2\sigma_g)^2(2\sigma_u)^2(1\pi_u)^3(3\sigma_g)$	1.040
	$c^3\Sigma_u^+$	$KK(2\sigma_g)^2(2\sigma_u)(1\pi_u)^4(3\sigma_g)$	1.651
	$d^3\Pi_g$	$KK(2\sigma_g)^2(2\sigma_u)(1\pi_u)^3(3\sigma_g)^2$	2.482
	$C^1\Pi_g$	$KK(2\sigma_g)^2(2\sigma_u)(1\pi_u)^3(3\sigma_g)^2$	4.248
	$C'^1\Pi_g$	$KK(2\sigma_g)^2(2\sigma_u)^2(1\pi_u)^2(3\sigma_g)(1\pi_g)$	4.643
	$e^3\Pi_g$	$KK(2\sigma_g)^2(2\sigma_u)^2(1\pi_u)^2(3\sigma_g)(1\pi_g)$	5.058
N_2	$A^3\Sigma_u^+$	$KK(2\sigma_g)^2(2\sigma_u)^2(1\pi_u)^3(3\sigma_g)^2(1\pi_g)$	6.169
(cf. Fig. 7.5)	$B^3\Pi_g$	$KK(2\sigma_g)^2(2\sigma_u)^2(1\pi_u)^4(3\sigma_g)(1\pi_g)$	7.353
	$W^3\Delta_u$	$KK(2\sigma_g)^2(2\sigma_u)^2(1\pi_u)^3(3\sigma_g)^2(1\pi_g)$	7.356
	$B'^3\Sigma_u^-$	$KK(2\sigma_g)^2(2\sigma_u)^2(1\pi_u)^3(3\sigma_g)^2(1\pi_g)$	8.165
O_2	$a^1\Delta_g$	$KK(2\sigma_g)^2(2\sigma_u)^2(3\sigma_g)^2(1\pi_u)^4(1\pi_g)^2$	0.977
	$b^1\Sigma_g^+$	$KK(2\sigma_g)^2(2\sigma_u)^2(3\sigma_g)^2(1\pi_u)^4(1\pi_g)^2$	1.627
	$c^1\Sigma_u^-$	$KK(2\sigma_g)^2(2\sigma_u)^2(3\sigma_g)^2(1\pi_u)^4(1\pi_g)^2$	4.050
	$C^3\Delta_u$	$KK(2\sigma_g)^2(2\sigma_u)^2(3\sigma_g)^2(1\pi_u)^4(1\pi_g)^2$	4.255
	$A^3\Sigma_u^+$	$KK(2\sigma_g)^2(2\sigma_u)^2(3\sigma_g)^2(1\pi_u)^3(1\pi_g)^3$	4.340
	$B^3\Sigma_u^-$	$KK(2\sigma_g)^2(2\sigma_u)^2(3\sigma_g)^2(1\pi_u)^3(1\pi_g)^3$	6.120
F_2	$^3\Pi_u$	$KK(2\sigma_g)^2(2\sigma_u)^2(3\sigma_g)^2(1\pi_u)^4(1\pi_g)^3(3\sigma_u)$	~ 1–1.5 (not observed)
	$A^1\Pi_u$	$KK(2\sigma_g)^2(2\sigma_u)^2(3\sigma_g)^2(1\pi_u)^4(1\pi_g)^3(3\sigma_u)$	repulsive
	$B^1\Pi_g$	$KK(2\sigma_g)^2(2\sigma_u)^2(3\sigma_g)^2(1\pi_u)^3(1\pi_g)^4(3\sigma_u)$	?

unpaired electrons can readily form bonds with other C atoms, gaining more than enough energy to compensate for the excitation. (Carbon poses a very special situation: Under some conditions, e.g., in an electric arc and ca. 100 torr of He, carbon forms the soccer-ball-like molecule C_{60}, a shell of 60 carbon atoms, called "buckminsterfullerene" and larger members of the series of "cage compounds" which are generically called "fullerenes." The C_{60} molecule is a remarkably stable species found in 1984 and made in sufficient quantity for a definitive determination of its structure in 1991. Its structure is precisely that of a standard soccer ball, with carbon atoms at the vertices of all the polyhedral faces. Besides their conventional chemistry, the fullerenes can hold atoms of other substances—e.g., He, Ar, or some metal atoms, within their cages.)

This illustrates the general principle that low-lying excited states facilitate bond formation. We introduced this idea (in a negative sense) in Section 5.5 to account for the "inertness" of the rare gas atoms, but it also applies to molecules. The molecules Li_2, B_2, and C_2 all have fairly low excited states (Table 7.6), and can thus exist only in the vapor phase; the solid forms of these elements (and of most other elements) have structures in which each atom is bonded to many nearest neighbors. In contrast, the first excited state of N_2 is very high (cf. Fig. 7.5), and nitrogen exists as discrete diatomic molecules even in the solid phase, which is thus called a *molecular crystal.* The ground state of N_2 is effectively a closed-shell structure, with all the occupied orbitals filled, and the physical properties of nitrogen are in many ways similar to those of an inert gas. In particular, N_2 exists as a gas to well below room temperature.

The properties of oxygen are unusual in several respects. The ground state of O_2 has the configuration . . . $(1\pi_g)^2$. Since there are two degenerate $1\pi_g$ orbitals, the $1\pi_g$ electrons can go in either the same or different orbitals, leading to singlet and triplet states, respectively. In analogy to Hund's rule for atoms, and for the same reasons, the triplet state is of lower energy. Thus the ground state of O_2 has $S = 1$ and is paramagnetic; the explanation of this phenomenon was one of the early triumphs of molecular orbital theory.[29] Additional confirmation is provided by the ions derived from O_2, in which the bond length (and thus strength) varies in accordance with the MO bond order:

Species	Ground-State Configuration	Bond Order	R_e (Å)	D_0 (eV)
O_2^+	. . . $(3\sigma_g)^2(1\pi_u)^4(1\pi_g)$	$2\frac{1}{2}$	1.117	6.662
O_2	. . . $(3\sigma_g)^2(1\pi_u)^4(1\pi_g)^2$	2	1.208	5.116
O_2^-	. . . $(3\sigma_g)^2(1\pi_u)^4(1\pi_g)^3$	$1\frac{1}{2}$	1.33	4.07
O_2^{2-}	. . . $(3\sigma_g)^2(1\pi_u)^4(1\pi_g)^4$	1	1.49 (in solid Na_2O_2, etc.)	

But in O_2 the ground state itself has unpaired electrons. How is it that additional bonds do not form[30] as in carbon? The answer may be that the electrons in question are in the antibonding $1\pi_g$ orbitals; the next available bonding orbital is the much higher $4\sigma_g$. Yet S_2, with a similar orbital structure, readily polymerizes to ring molecules such as S_8 and S_6. Whatever the reason, the O_2 molecule is the stable form of oxygen at all temperatures below those at which the molecules dissociate. Like nitrogen, oxygen is a room-temperature gas and a low-temperature molecular crystal.

Finally, the F_2 molecule is similar to N_2 in that it has a closed-shell ground state and relatively inaccessible excited states. Fluorine shows the same "inert" behavior as nitrogen and oxygen—inert with regard to physical properties only, since F_2, like O_2, is highly reactive chemically. The energy of the first excited state is probably quite low, but transitions from the ground state are strongly forbidden.

Thus far we have discussed only electron configurations and the qualitative inferences one can draw from them. Of course the configurations are only the starting point for molecular orbital calculations. We need not go into the details of these calculations here; the methods are those we outlined in Chapter 6. One sets up a trial wave function in some way and computes the energy as a function of R by Eq. 5.11. The trial function ordinarily contains adjustable parameters which are varied to minimize the energy. Within the orbital approximation, the molecular wave function has the form of the antisymmetrized Hartree-Fock product (Eq. 6.63), in which the φ_i's are molecular orbitals. The simplest trial MOs are LCAO functions of the type already discussed, for example,

$$\varphi(2\sigma_g) = \varphi(\sigma_g\, 2s) = C[\varphi_A(2s) + \varphi_B(2s)], \quad (7.48)$$

but such simplistic forms give unsuitably inaccurate results for anything more complicated than H_2. The next step is to go beyond the correlation diagram and use mixtures of atomic orbitals of appropriate symmetry to generate "better" atomic orbitals, as in

$$\varphi(2\sigma_g) = C_1\varphi(\sigma_g\, 2s) + C_2\varphi(\sigma_g\, 1s) + C_3\varphi(\sigma_g\, 2p) + \cdots . \quad (7.49)$$

This process is called *hybridization* and will be discussed in greater detail in the next chapter. However the trial MOs are set up, the energy is then minimized by the self-consistent-field method. As in atomic calculations, the speed of convergence depends largely on what functions are chosen to represent the atomic orbitals. The best SCF MO calculations now give essentially accurate Hartree–Fock results for first-row molecules, and are being extended to larger molecules as larger computers are introduced. The Hartree–Fock solution, we recall, is the best possible solution *within the one-electron–orbital approximation;* it is adequate for many purposes. The correlation energy, the difference between the true energy and the Hartree–Fock energy, can be calculated by configuration interaction or similar techniques (Section 6.9). For simple molecules, the best calculations lead to predictions of spectral line positions within about 10–20 cm^{-1} of the observed lines. This is accurate enough to be very useful for identification, but considerably less accurate than the measurement of spectral lines.

The energy is not the only molecular property that can be computed. We can gain insights from examining the electron distributions in homonuclear diatomic molecules. In Figs. 6.7, 7.15, and 7.16, we looked at orbital amplitudes. Now we turn to total densities. Figure 7.17 shows typical "three-dimensional" pictures of total densities for three molecules. Figure 7.18 shows electron-density contour maps for a larger set. All were obtained from SCF calculations. Electron densities are given in Fig. 7.18 for both individual orbitals and the molecule as a whole. In the H_2 molecule all but the innermost few contours are nearly circular, that is, the outer portions of the electron distribution resemble that in the spherical He atom; even between the nuclei the electron density is only slightly below that at the two nuclear peaks. In contrast, the weakly bound Li_2 molecule is quite like two separate atoms, with a very deep valley between the two peaks. For the other first-row molecules, the stronger the bond, the more nearly the total electron density resembles that in a single atom. The "saddle" between the nuclei is still quite low in B_2, reaches about half the peak height in N_2, and has fallen again by the time we reach F_2. However, the outer reaches of the electron distribution shrink steadily with increasing Z, the same trend one finds for atomic sizes across a period. As for the orbitals, note that all the inner-shell ($1\sigma_g$ and $1\sigma_u$) orbitals of the first-row molecules do not differ significantly from the core densities in two separate atoms. Although the calculations leading to Fig. 7.18 neglect electron correlation, one could hardly tell the difference on the scale of these figures.

[29] The simple valence bond theory would predict one of the structures $:\ddot{O}=\ddot{O}:$, which is not paramagnetic, or $:\dot{\ddot{O}}-\dot{\ddot{O}}:$, which is too weakly bonded. The MO interpretation straightforwardly accounts for both the paramagnetism and the bond order, $:\dot{\ddot{O}}-\dot{\ddot{O}}:$.

[30] Oxygen does form the O_3 (ozone) molecule, but no further polymerization has been observed. Indeed, O_3 is quite unstable relative to O_2; the reaction $2O_3 \rightarrow 3O_2$ occurs readily if a suitable catalyst is present.

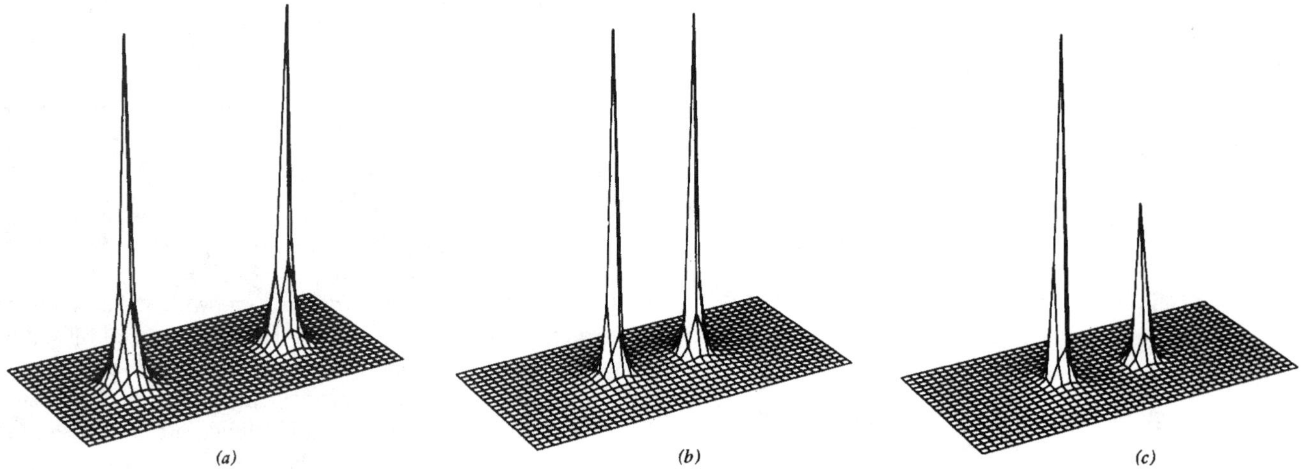

Figure 7.17 Total electron densities for (*a*) Li_2; (*b*) N_2; (*c*) CO. Each molecule is taken at its equilibrium internuclear distance. Calculations were carried out by Mary Dolan and plotted by Mary Dolan and David Campbell.

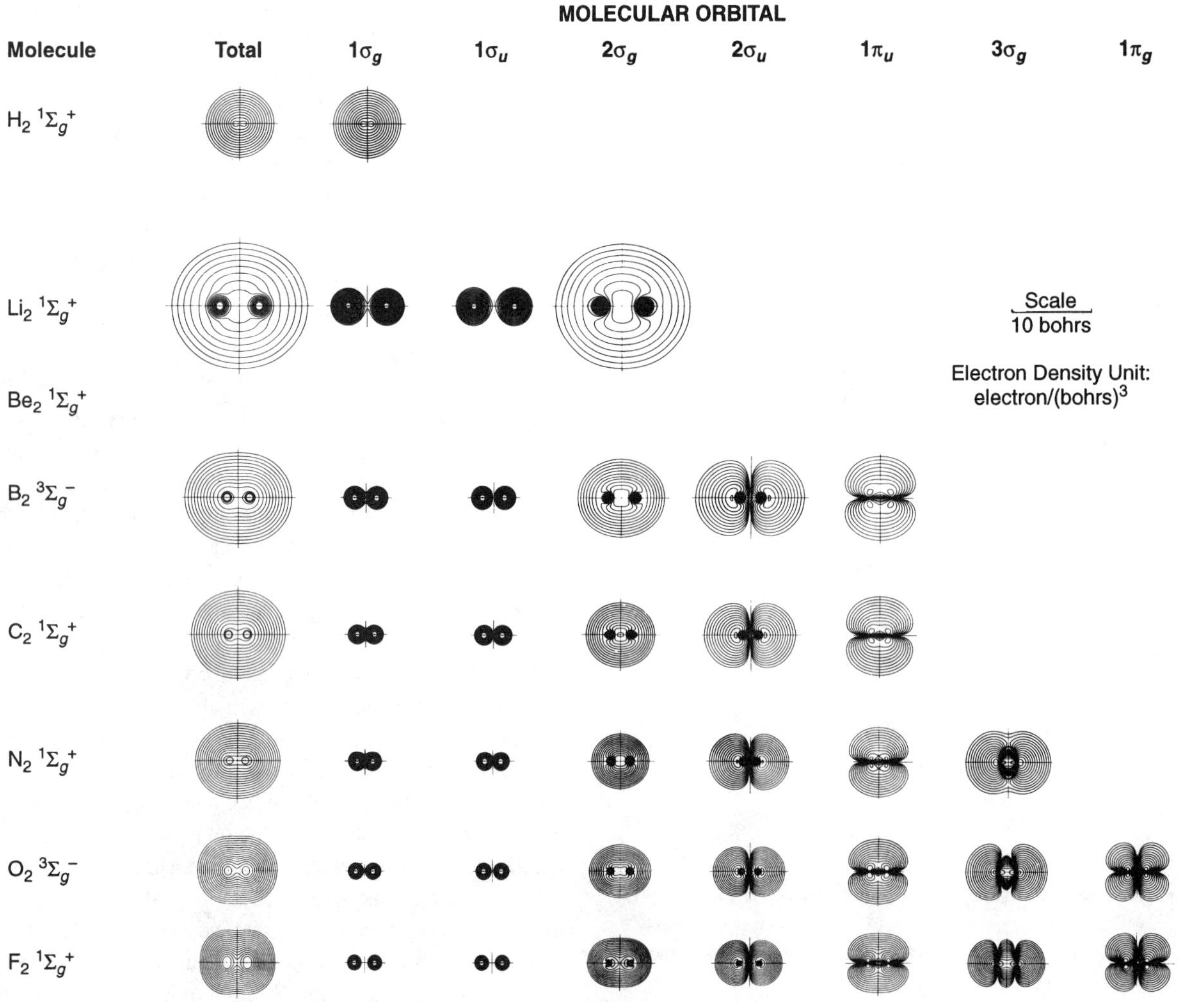

Figure 7.18 Contour maps of electron densities of orbitals and of total electron densities for the seven lightest stable homonuclear molecules. From A. C. Wahl, *Sci. Am.* **322,** No. 4, 54 (April 1970).

7.7 Introduction to Heteronuclear Diatomic Molecules: Electronegativity

A heteronuclear molecule has no end-to-end symmetry. Thus symmetry condition 3 of Section 7.5 is not applicable: Neither the orbitals nor the molecular wave function can be described as g or u. Otherwise our previous nomenclature can be retained. In the orbital approximation, the molecular orbitals tend to localize predominantly around one nucleus or the other; the ionic molecules of Section 7.4 are an extreme case of this. Only when the molecule is very close to the united-atom limit do the normally occupied orbitals have comparable amplitudes around both nuclei. This asymmetric charge distribution carries over to the molecule as a whole, and heteronuclear molecules are therefore polar, that is, have nonzero dipole moments.

As in homonuclear molecules, we can use a correlation diagram to represent the sequence of orbital energies. Figure 7.19a is a general correlation diagram for heteronuclear diatomic molecules. Since the two atoms are different, they

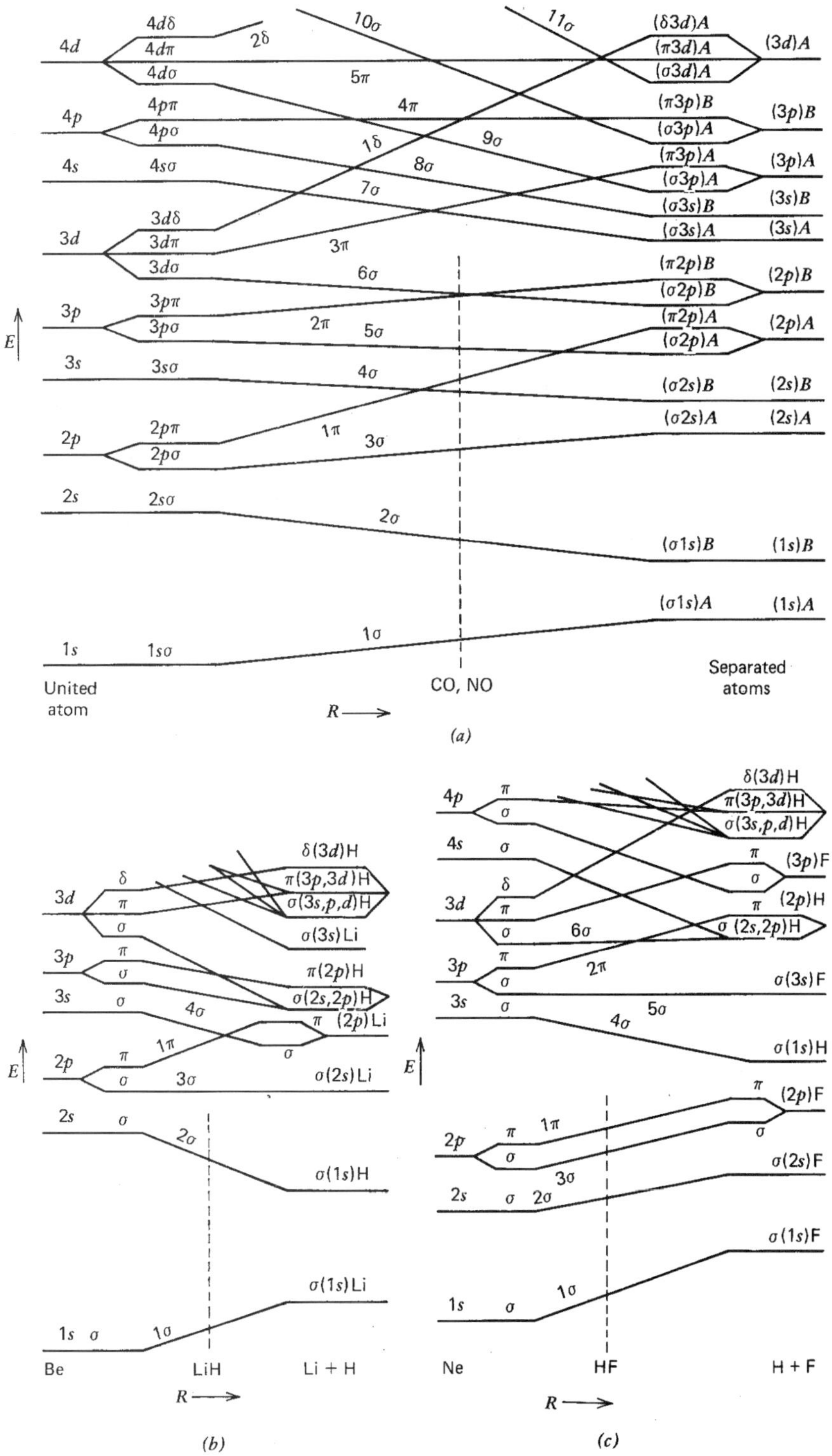

Figure 7.19 Correlation diagrams for heteronuclear diatomic molecules: (a) for atoms of nearly equal size; (b) for LiH; (c) for HF. (In the latter two diagrams the degenerate H levels are grouped as in Fig. 6.9.)

have different energy levels in the separated-atom limit. Our assumption is that in this limit each molecular orbital correlates only with an atomic orbital of one or the other atom (the atom around which its amplitude is greatest). As we shall see, this is an oversimplification. The resulting correlation diagram is somewhat simpler than that for homonuclear molecules, primarily because there are no orbital parities that must be matched. However, it contains an additional element of arbitrariness, in that the relationship between A and B separated-atom levels varies with the nature of these atoms: The higher an atom's nuclear charge Z, the lower are that atom's energy levels of a given n and l. The effect is especially pronounced when one of the atoms is hydrogen. In Figs. 7.19*b* and 7.19*c* we illustrate this with correlation diagrams for LiH and HF; the other first-row diatomic hydrides are intermediate between these two.

What is the nature of the orbitals in heteronuclear molecules? In the LCAO approximation they consist of sums or differences of atomic orbitals in unequal amounts. Thus one might write a given molecular orbital φ_n as

$$\varphi_n = \alpha(R)(\varphi_i)_A + \beta(R)(\varphi_i)_B, \tag{7.50}$$

where $(\varphi_i)_A$ and $(\varphi_i)_B$ are atomic orbitals of atoms A and B. The coefficients $\alpha(R)$ and $\beta(R)$, like those in Eqs. 6.76 and 6.79, are constants for any given value of R. The correlation diagrams imply that either $\alpha(R)$ or $\beta(R)$ vanishes in the separated-atom limit; at R_e one coefficient is usually much greater than the other.[31] Results such as this are obtained by the usual SCF techniques, and can be improved by using hybridized atomic orbitals, as in Eq. 7.49. One can again obtain orbital and whole-molecule electron densities: Compare the "charge densities"—strictly, total probability densities—shown in Fig. 7.20 for the first-row diatomic hydrides with Fig. 7.18 for homonuclear molecules. Although one cannot yet determine the electron distribution experimentally with comparable precision, these results are generally in agreement with chemical intuition and the measured properties of molecules.

In Section 5.5 we said that "electronegativity" describes an atom's power to attract electrons. In Fig. 7.20 we see a clear illustration of what this means. In a heteronuclear molecule the electrons that form the chemical bond—that is, those in the highest occupied bonding orbital—are found primarily near the more electronegative atom.[32] Among the first-row hydrides, calculations like those leading to Fig. 7.20 show clearly that hydrogen is intermediate in electronegativity between metals and nonmetals. In HF the 3σ bonding orbital is mainly centered around the F atom, whereas in LiH the 2σ orbital is almost entirely around the H atom. In fact, the alkali metals are so much more electropositive than hydrogen that all the alkali hydrides can be regarded as ionic molecules (M^+H^-), just like the alkali halides; in the solid state these compounds consist of separate M^+ and $H^-(1s^2)$ ions. Similar behavior is found for CaH_2 and the heavier alkaline earth hydrides. We shall see that the empirical electronegativity scales agree with these conclusions.

The simplest electronegativity scale is that proposed by R. S. Mulliken. It is based on the idea that an atom's ability to hold its outermost electrons is proportional to its ionization potential (I), whereas its ability to attract additional electrons is proportional to its electron affinity (A). The Mulliken electronegativity is thus defined by the average of the two quantities,

$$x^M = K^M\left(\frac{I+A}{2}\right), \tag{7.51}$$

where $K^M = (3.15 \text{ eV})^{-1}$. The numerical factor is chosen to give a convenient scale, one in which the highest electronegativity (that of F) is about 4.0. The chief drawback of the Mulliken scale is that it uses isolated-atom properties, whereas "electronegativity" is meant to describe the behavior of an atom *in a molecule.* Other scales have thus been devised using molecular properties; we shall describe only one of these.

Linus Pauling suggested that the bond in a heteronuclear molecule could be regarded as the sum of two contributions: a covalent part (which is the only contribution in a homonuclear molecule) and an ionic part. In Pauling's valence bond interpretation, the bond in the molecule AB is a hybrid of the structures A—B and A^+B^-, if A is the more electropositive atom. On purely empirical grounds, the covalent contribution to the bond energy is taken to be the geometric mean of the covalent A—A and B—B bond energies. By subtraction the ionic contribution is

$$\Delta = D_e(\text{A—B}) - [D_e(\text{A—A})D_e(\text{B—B})]^{1/2}, \tag{7.52}$$

where the D_e to be used here is the measured or inferred single-bond energy.[33] The apparent electronegativity difference between atoms A and B is found to correlate well with the square root of Δ, so the Pauling electronegativity is defined by the equation

$$(x^P)_B - (x^P)_A = K^P \Delta^{1/2}, \tag{7.53}$$

[31] In particular, inner-shell MOs are almost pure atomic orbitals: For example, in all the first-row hydrides (HA) the 1σ orbital is found to be over 99% $(1s)_A$. This further confirms our conclusion in Section 7.4 that atomic cores take little part in bonding.

[32] The lowest normally empty orbital—such as 3σ in LiH or 4σ in HF—is commonly centered around the electropositive atom. Such an orbital is normally antibonding: Since it is usually a promoted orbital, the separated-atom electrons must be forced "uphill" to occupy it. If overlap is neglected and the bonding orbital is approximated by $\alpha\varphi_A + \beta\varphi_B$, orthogonality requires that the corresponding antibonding orbital be $\beta\varphi_A - \alpha\varphi_B$.

[33] This is not necessarily the dissociation energy of the diatomic molecule. For example, N_2 has $D_e = 9.8$ eV but contains a triple bond; the N—N single-bond energy of 1.65 eV is obtained from measurements on molecules such as $H_2N—NH_2$. See Chapter 14 for a discussion of bond-energy calculations.

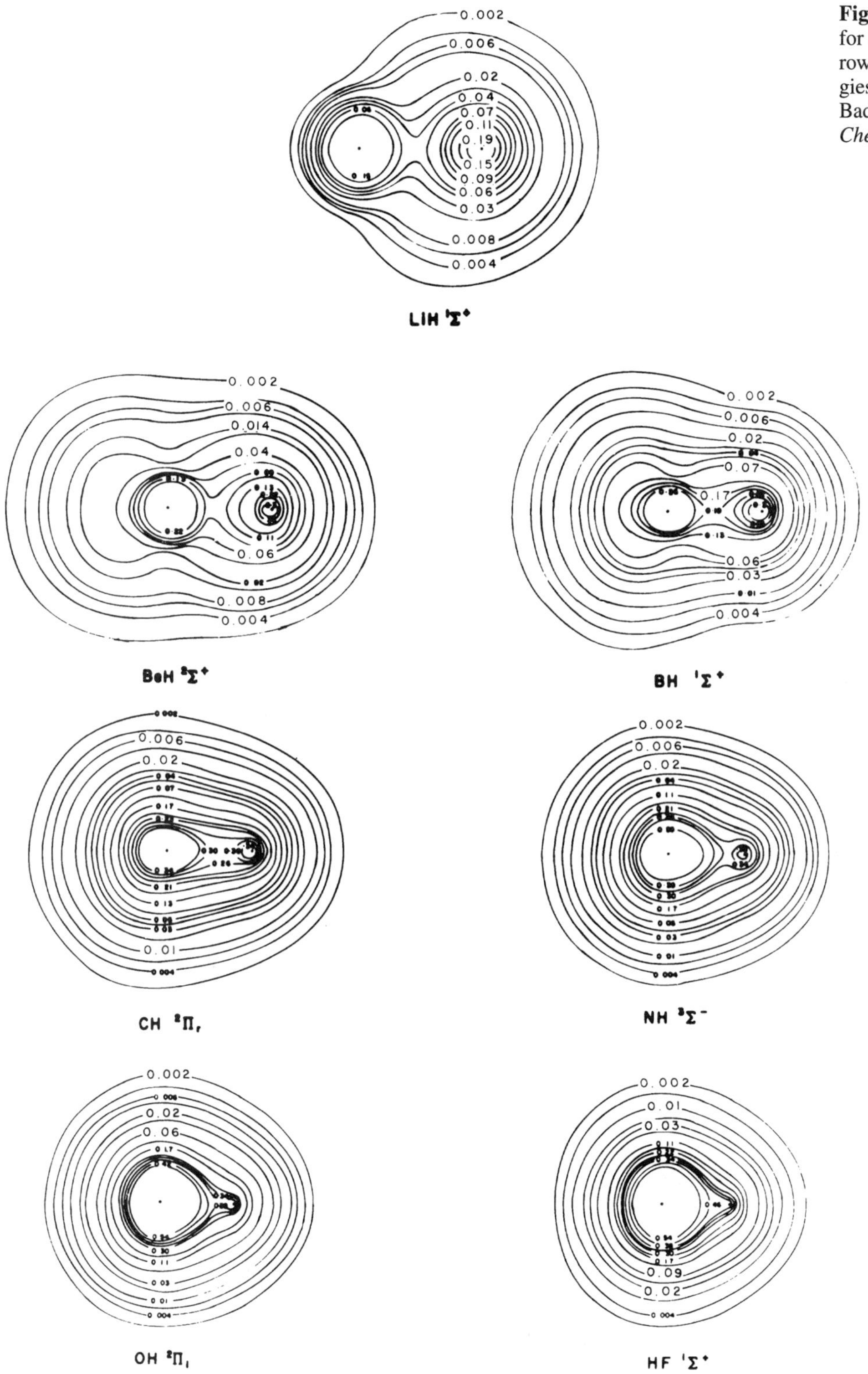

Figure 7.20 Electron-density contours for total probability densities of the first-row diatomic hydride molecules. Energies are given in hartrees. From R. F. W. Bader, I. Keaveny, and P. E. Cade, *J. Chem. Phys.* **47,** 3381 (1967).

with $K^P = (1\ \text{eV})^{-1/2} = (96.5\ \text{kJ/mol})^{-1/2}$; the scale is anchored by again setting $(x^P)_F \approx 4.0$.

The numerical factors of the Pauling and Mulliken scales were deliberately chosen to bring the two scales into conformity with each other. The two do indeed agree quite closely, as can be seen from the values in Table 7.7 for first-row elements:

Table 7.7 Mulliken and Pauling Electronegativities for First-Row Atoms

	Li	Be	B	C	N	O	F
x^M	0.94	1.46	2.01	2.63	2.33	3.17	3.91
x^P	0.98	1.57	2.04	2.55	3.04	3.44	3.98

The Pauling electronegativities listed in Table 7.7 are actually based on the Coulomb force felt by an electron at the atom's covalent radius, and are probably the best available comprehensive set. It can be seen that the values confirm our qualitative conclusions above.

7.8 Bonding in LiH; Crossing and Noncrossing Potential Curves

Let us take a closer look at lithium hydride, the simplest stable heteronuclear molecule. At ordinary temperatures this species is a white crystalline solid (m.p. 680°C), but individual LiH molecules exist in the gas phase. The separated-atom limit is Li($1s^22s$) + H($1s$); the united-atom limit is Be($1s^22s^2$). Examining Fig. 7.19*b*, we see that the lowest orbital, the 1σ, correlates naturally with the $1s$ orbitals of both Li and Be; this orbital is thus closely centered around the Li nucleus for all values of *R*. The next orbital, the 2σ, is much higher on the energy scale (about 60 eV at R_e). In the separated-atom limit, the H($1s$) orbital has a binding energy (ionization potential) of 13.6 eV, whereas the Li($2s$) orbital has a binding energy of only 5.36 eV (cf. Fig. 7.21). Because of this large energy difference, the 2σ orbital even near R_e is primarily like H($1s$); that is, by far the largest coefficient[34] in Eq. 7.50 is that of $(\varphi_{1s})_H$. The 3σ orbital, on the other hand, is mainly like Li($2s$). The LiH molecule has four electrons; applying the Aufbau principle, we conclude that the ground-state configuration near R_e should be $1\sigma^22\sigma^2$, corresponding to a $^1\Sigma$ state. Since the valence (2σ) electrons are centered near the H nucleus, the molecule should be largely ionic, with the effective structure Li^+H^-. Spectroscopic and dipole-moment measurements confirm these conclusions.

As usual, however, the simple MO method breaks down at large *R*. Suppose that we take the molecule at R_e and slowly move the nuclei away from each other. If the electrons remained in the same orbitals (as represented by the correlation diagram), the separated-atom limit would be the ionic $Li^+(1s^2) + H^-(1s^2)$. But we know that any two neutral atoms A and B have a lower total energy than the corresponding ions A^+ and B^-. Eq. 7.46 gives the energy of ion-pair formation as

$$Q(A^+B^-) \equiv E(A^+ + B^-) - E(A+B) = I(A) - A(B), \quad (7.54)$$

which is always positive. For LiH we have $I(\text{Li}) = 5.36$ eV, $A(\text{H}) = 0.75$ eV, $Q(\text{Li}^+\text{H}^-) = 4.61$ eV. The problem here is rather akin to that we found for H_2 in Section 6.6. There (and for any other homonuclear molecule) a single-configuration MO method predicts a 50%-ionic limit; here (and for any other heteronuclear molecule with an even number of electrons) it predicts a 100%-ionic limit. Within the MO framework, the paradox can again be resolved by the use of configuration interaction. Such a description of the LiH molecule would be given by the wave function

$$X(R)\psi(1\sigma^2 2\sigma^2) + Y(R)\psi(1\sigma^2 2\sigma 3\sigma),$$

in which we must have $X(R) >> Y(R)$ for $R \lesssim R_e$, $Y(R) >> X(R)$ as $R \to \infty$, so that at long range we have the configuration $1\sigma^22\sigma3\sigma \to \text{Li}(1s^22s) + \text{H}(1s)$.

Accurate multiconfiguration calculations have been carried out for LiH and other diatomic hydrides, cf. Meyer and Rosmus, *J. Chem. Phys.* **63,** 2356 (1975) and for first-row species such as N_2, c.f. Andersson et al., *J. Chem. Phys.* **96,** 1218 (1992), O_2, cf. Vahtras et al., *J. Chem. Phys.* **96,** 2118 (1992), and CO, cf. Docken and Liu, *J. Chem. Phys.* **66,** 4309 (1977). However results of the same accuracy have not yet been reported for many other heteronuclear molecules. It is thus useful to know the limits of the single-configuration model. Fortunately, we can get a good idea of these limits from a very simple and crude calculation of the potential energy. Consider the two long-range states, Li + H and $Li^+ + H^-$; what happens in each case if we decrease *R* slowly, neglecting configuration interaction? In the ionic case the model of Section 7.4 should be applicable, with the interaction energy given by Eq. 7.44. This energy is dominated by the Coulomb term, $-e^2/4\pi\varepsilon_0 R$. Even at R_e this term gives a fairly good approximation to D_e (cf. Table 7.3), and at a somewhat larger distance, large enough to neglect overlap of the electron clouds, it should be the only significant term. Let us then assume that $E_{\text{ionic}}(R) = -e^2/4\pi\varepsilon_0 R$ for large *R*. To the same order of approximation, the two neutral atoms

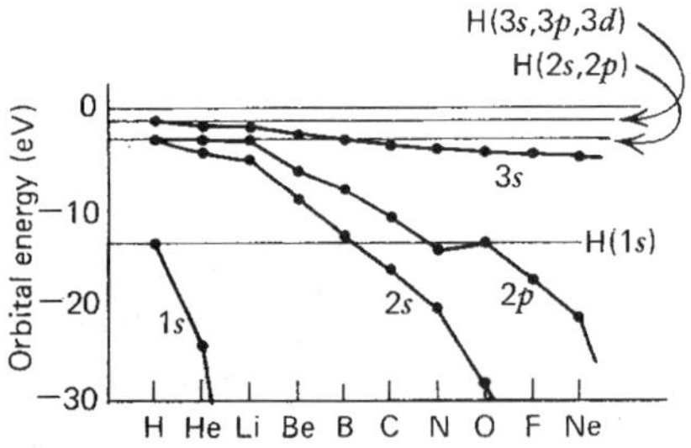

Figure 7.21 Orbital energies of first-row atoms. The energy plotted is that required to remove an electron from the orbital in question, as obtained from spectroscopic data.

[34] Here are the results of a SCF-MO calculation on LiH, using hybridized atomic orbitals on the Li atom. The numbers tabulated are the coefficients C_{ni} in the expression $\varphi_n = \sum_i c_{ni}\varphi_i$, where φ_n is a molecular orbital and the φ_i are atomic orbitals.

	φ_i			
φ_n	Li(1*s*)	Li(2*s*)	Li(2*p*σ)	H(1*s*)
1σ	0.997	0.016	−0.005	0.006
2σ	0.131	−0.323	−0.231	−0.685
3σ	0.134	−0.805	0.599	0.148

Better calculations have been performed, using many more atomic orbitals, but the basic pattern remains the same.

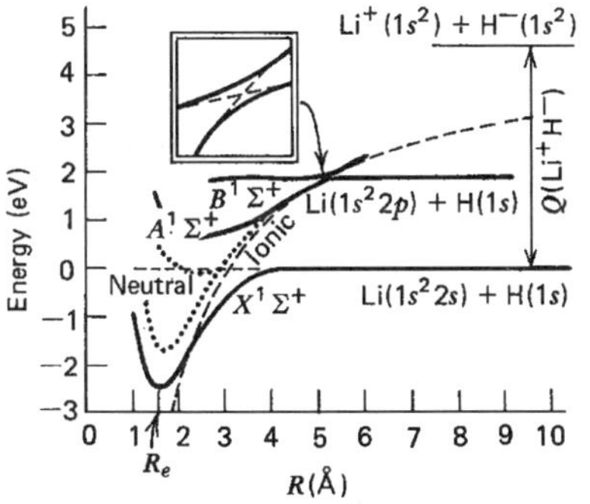

Figure 7.22 Potential energy curves for LiH: - - - - simple model described in text,

$$E_{\text{ionic}}(R) = E(Li^+ + H^-) - e^2/4\pi\epsilon_0 R,$$
$$E_{\text{neutral}}(R) = E(Li + H);$$

single-configuration MO calculation for the configurations $1\sigma^2 2\sigma^2$ (ionic) and $1\sigma^2 2\sigma 3\sigma$(neutral); —— experimental curves for the two lowest states ($X^1\Sigma^+$ and $A^1\Sigma^+$), and the approximate location of the next state ($B^1\Sigma^+$). The inset illustrates how one can join segments of single-configuration curves (- - - -) to obtain curves that do not violate the noncrossing rule (——).

should have no significant interaction energy at large R. In Fig. 7.22 we have plotted these approximate potential-energy curves, which can be seen to cross at about $2R_e$. A single-configuration MO calculation gives essentially the same result, demonstrating the validity of this extremely crude model.

What is the significance of these results? Both single-configuration states must have the symmetry $^1\Sigma^+$. (There is also a $^3\Sigma^+$ state derived from the neutral atoms, but as in H_2 it should be entirely repulsive.) But the noncrossing rule is applicable to molecular states as well as individual orbitals: Two nondegenerate states of the same symmetry cannot have exactly the same energy, so their potential-energy curves cannot cross. Thus a single-configuration calculation must break down at the point where it predicts such a crossing, though it may be reasonably accurate elsewhere. One can obtain reasonably good results by joining the curve segments away from the crossing, as shown in the inset to Fig. 7.22. Curves obtained in such a way would be qualitatively similar to the actual curves for the two lowest states of LiH, shown by the solid lines in Fig. 7.22.[35] Similar effects occur in any of the "ionic molecules" of Section 7.4: Although the ground state is predominantly ionic near R_e, it goes over to a pair of neutral atoms at long range.

When we introduced the noncrossing rule, we mentioned that it is only approximately true; let us see why. To determine a potential-energy curve we must assume the nuclei to be at rest for each value of R (the Born–Oppenheimer approximation). The energy of a real system is thus given exactly by the $E(R)$ curve only in the limit of infinitely slow nuclear motion, that is, in an *adiabatic* process (to use the term we introduced in Section 6.5). In an adiabatic process the noncrossing rule can be shown to hold rigorously. But no real molecule is an adiabatic system. Since vibration is always present, R is always changing at a nonzero rate. How does this affect the noncrossing rule? In most cases where the rule applies, the two potential curves involved approach very close to each other (cf. the $A^1\Sigma^+ - B^1\Sigma^+$ "intersection" in Fig. 7.22). Let us say that they are within ΔE of each other over a range ΔR. Now the uncertainty principle enters the picture: If the uncertainty in a molecule's energy is greater than the separation between two such curves, there is a good chance of the molecule's crossing from one curve to the other, that is, entering the transition region ΔR in one electronic state and leaving it in another. According to Eq. 3.87, the noncrossing rule should apply only when $\Delta t \gtrsim \hbar/\Delta E$, where Δt is the time the molecule spends within the transition region; this is called the *Massey adiabatic criterion.* If the relative speed of the two atoms is $v = |dR/dt|$, we have $\Delta t = \Delta R/v$, and the criterion becomes $v \gtrsim \Delta E\ \Delta R/\hbar$. The speed v, of course, increases with increasing vibrational excitation. For a given crossing, the adiabatic criterion is usually satisfied for low vibrational states but not for higher states. In terms of the Born–Oppenheimer approximation, the separability of nuclear (vibrational) and electronic motions breaks down when the vibrational energy is sufficiently great.

Transitions of the type just described can also occur between states of different symmetry classes, whose potential curves *can* cross one another. The usual conservation laws and selection rules of course apply in such transitions. Processes of either kind give rise to a number of interesting phenomena, some of which are illustrated in Fig. 7.23. In ordinary dissociation (Fig. 7.23*a*) a molecule is excited into a repulsive state and the atoms immediately fly apart; the process takes about as long as a single vibrational period. *Predissociation* (Fig. 7.23*b*) is a slower process, in which the molecule is excited to a bound state A, remains there at least long enough to execute a few vibrations, then undergoes a transition to a repulsive state B which crosses[36] the bound state, and dissociates. *Autoionization* or *preionization* (Fig. 7.23*c*), which we discussed at the end of Section 6.10, is a similar transition from a bound neutral state to a state of molecule-ion plus free electron. Finally, in *phosphorescence* (Fig. 7.23*d*) the transition is to a lower-energy stable excited state from which decay to the ground state is forbidden (usually because it is a triplet-singlet transition, violating the rule $\Delta S = 0$). Since "forbidden" is not an absolute term, the decay does occur with emission of light, but over a long period after the initial excitation.[37]

---

[35] Similar crossings (only one of which is shown) occur every time the "ionic" curve intersects the energy of an excited $^1\Sigma^+$ state of Li + H.

[36] The figure has been drawn for two states that actually cross. If they have the same symmetry, then the lower state instead has a maximum at the "intersection," and dissociation occurs by ordinary tunneling through this potential barrier.

[37] Phosphorescence should not be confused with *fluorescence,* which is the emission of light by nonforbidden transitions, most commonly singlet–singlet, and thus occurs mainly within nanoseconds (or at most microseconds) after excitation.

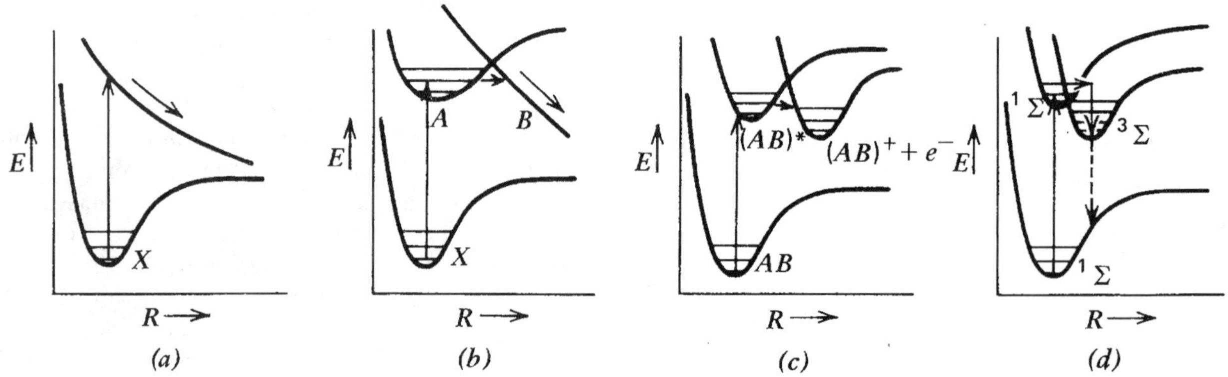

Figure 7.23 Excited-state transitions. (*a*) shows ordinary dissociation, whereas the other figures show processes involving crossing potential curves: (*b*) predissociation; (*c*) autoionization; (*d*) phosphorescence. See discussion in the text.

7.9 Other First-Row Diatomic Hydrides

Except for LiH and HF, all the first-row diatomic hydrides are highly reactive species observed mainly in high-temperature systems. They are common reaction intermediates: CH can be detected in nearly all hydrocarbon flames, and OH (hydroxyl) in all flames containing oxygen and hydrogen in any form. They can also be detected by radio astronomy in interstellar space, where molecules are usually too far apart to react. The first molecule of any kind so detected was OH, in 1963. Since most of space is very cold, virtually all the molecules there must be in their ground electronic states. The states thus observed are indeed the ground states predicted by *a priori* calculations. The member of this series most commonly encountered in the laboratory is of course HF, the only one that exists as stable diatomic molecules at room temperature: HF is a colorless liquid boiling at 19.4°C.

By combining the information in Figs. 7.19 and 7.21, it is not difficult to deduce the ground-state MO configurations of the other first-row hydrides as we have done for LiH. These configurations are given in Table 7.8, along with other data on the molecules. Over this series the energy sequence of the occupied orbitals does not change, but the atomic orbitals with which they correlate do. For example, in LiH and BeH the 2σ orbital correlates with and thus largely resembles the H(1*s*) orbital. Since H(1*s*) and B(2*s*) have nearly the same energy, the 2σ orbital in BH is a roughly equal mixture of the two, and in CH and beyond the 2σ orbital is predominantly the heavy-atom 2*s*, which lies below the H(1*s*). Similar analyses can be made for the other orbitals. Although the 3σ and 1π orbitals (from CH on) are degenerate in both limits, the 3σ in each case is found to have a lower energy at R_e and thus is occupied first.

What is the bonding nature of these orbitals? Remember that a bonding orbital is expected to have a higher electron density in the bonding region than one finds in the separated atoms. The 1σ orbital in all these molecules is virtually identical to the heavy-atom 1*s*, and like most inner-core orbitals is essentially nonbonding. The 1π orbital is also nonbonding, since it consists of only the heavy-atom 2*p* with no significant contribution from hydrogen. (The lowest *p* orbital of hydrogen is at a much higher energy.) The 2σ orbital is clearly bonding in LiH, but takes less and less part in the bonding as it becomes more like the heavy-atom 2*s;* in HF it can also be considered nonbonding. But the slack is taken up by the 3σ orbital, which, from CH onward is the main constituent of the bond. In simple chemical terms, one can think of the combination $2\sigma^2 3\sigma^2$ as adding to a single bond and a nonbonding lone pair, as in the Lewis formulas for :B:H or H :F̤̈:, but the bond is really made up of contributions from both orbitals.

The strength of the bonding, as indicated by the value of D_e, increases fairly steadily from LiH to HF. Even more clear-cut is the upward trend in the vibrational frequency $\tilde{\nu}_e$. This cannot be due to the relation $\tilde{\nu}_e \propto \mu^{-1/2}$, since the reduced mass μ varies very little over this series ($\mu_{LiH} = 0.88\ m_H$; $\mu_{HF} = 0.96\ m_H$). But $\tilde{\nu}_e$ does correlate well with the decrease in the bond length R_e, which is of course due to the decreasing size of the heavy atoms (Section 5.5). An increase in $\tilde{\nu}_e$ corresponds to a more sharply curved potential minimum, and we conclude that both the depth and the curvature of the potential wells tend to increase as R_e decreases.[38] The long-range attractive forces have much the same form between any two atoms, depending mainly on the electronegativity difference, but the short-range repulsive forces depend strongly on the atomic "sizes." Thus the position of the potential minimum in the hydrides must be sensitive mainly to the repulsive forces. This is in fact generally true, and is what makes the concept of a "covalent radius" meaningful.

Now let us look at the dipole moments. It can be seen that the magnitude of μ decreases sharply from LiH, then increases slightly toward HF. Presumably this is due to a

[38] This agrees with the empirical Badger's rule (see footnote 9 on page 190).

Table 7.8 Ground States of First-Row Diatomic Hydrides

Molecule	Electron Configuration	Term Symbol	D_e (eV)	R_e (Å)	$\tilde{\nu}_e$ (cm^{-1})	$\lvert\mu\rvert$ (D)
LiH	$1\sigma^2 2\sigma^2$	$^1\Sigma^+$	2.515	1.5954	1406	5.88
BeH	$1\sigma^2 2\sigma^2 3\sigma$	$^2\Sigma^+$	2.4 ± 0.3	1.297	2058	(0.3 calc.)
BH	$1\sigma^2 2\sigma^2 3\sigma^2$	$^1\Sigma^+$	3.54	1.236	2367	1.27
CH	$1\sigma^2 2\sigma^2 3\sigma^2 1\pi$	$^2\Pi$	3.65	1.124	2859	1.46
NH	$1\sigma^2 2\sigma^2 3\sigma^2 1\pi^2$	$^3\Sigma^-$	3.40	1.045	3126	(1.0–1.9 calc.)
OH	$1\sigma^2 2\sigma^2 3\sigma^2 1\pi^3$	$^1\Pi$	4.621	0.9706	3735	1.66
HF	$1\sigma^2 2\sigma^2 3\sigma^2 1\pi^4$	$^1\Sigma^+$	6.11	0.9168	4139	1.82

change in polarity, as the heavy atom changes from being more electropositive to more electronegative. However, there are no direct measurements of the direction of μ in these molecules. Apart from LiH, how can we tell whether a given molecule is primarily H^+X^- or X^+H^-? A good idea of this can be obtained from the long-range limit. If we set the energy of two separated atoms A + B as zero, then the energy of the ion pair $A^+ + B^-$ is simply the $Q(A^+B^-)$ of Eq. 7.54, that is, $I(A) - A(B)$. If we calculate $Q(H^+X^-)$ and $Q(X^+H^-)$, the smaller of these quantities must correspond to the lowest ionic state at $R = \infty$, which is probably the principal ionic contribution to the ground state at $R = R_e$. For example, in CH we have $Q\ (C^+H^-) = I(C) - A(H) = 11.26\text{ eV} - 0.75\text{ eV} = 10.51\text{ eV}$, $Q(H^+C^-) = I(H) - A(C) = 13.60\text{ eV} - 1.25\text{ eV} = 12.35\text{ eV}$; thus $C^+ + H^-$ has an energy 1.84 eV lower than $H^+ + C^-$, and the CH molecule should be primarily C^+H^-. For the series of first-row hydrides such calculations give:

X	Li	Be	B	C	N	O	F
$Q(X^+H^-)$(eV)	4.61	8.57	7.55	10.51	13.78	12.86	16.67
$Q(H^+X^-)$(eV)	12.98	13.22	13.42	12.35	13.8	12.13	10.15

Thus, the heavy atom should be the positive end of the molecule in LiH, BeH, BH, and CH, the negative end in OH and HF, whereas the NH molecule should be very nearly nonpolar. These results are plausible, and are supported by the electron-distribution calculations of Fig. 7.20.

So far we have spoken only of the ground states of the diatomic hydrides. Although we shall make no detailed study of the excited states, it is worthwhile to survey the relationship between molecular states and those of the separated atoms. This can be done systematically by extending the vector model of Section 5.6 from atoms to molecules. The method is applicable to both heteronuclear and homonuclear molecules, though complicated in the latter case by degeneracies.

As an example, let us consider the manifold of states of the CH molecule derived from the ground states of the separated atoms. As we showed in Section 5.6, these ground states are $C(^3P)$ $(L = 1, S = 1)$ and $H(^2S)$ $(L = 0, S = \frac{1}{2})$. The spin and orbital angular momenta of the atoms add as vectors to give those of the molecule. The reasoning is the same as we applied to obtain the values of J in connection with Eq. 5.30. Since S is quantized in integral steps, its possible values for a diatomic molecule are

$$S = S_A + S_B, S_A + S_B - 1, \ldots, |S_A - S_B|, \qquad (7.55)$$

where S_A and S_B are the atomic values. In this case we can have $S = \frac{3}{2}$ or $S = \frac{1}{2}$, that is, quartet or doublet states. As for the orbital angular momentum, we are concerned only with its component along the internuclear axis. The quantum number Λ is given by

$$\Lambda = |(M_L)_A + (M_L)_B|, \qquad (7.56)$$

where each M_L can have any of its possible values L, $L - 1, \ldots, -L$. In CH we have $(M_L)_C = 1, 0, -1$, and $(M_L)_H = 0$, giving the possible values $\Lambda = 0$ (Σ states) or $\Lambda = 1$ (degenerate Π states). Since the spin and orbital angular momenta add independently, we should have all told from $C(^3P) + H(^2S)$ the states $^2\Pi$, $^2\Sigma$, $^4\Pi$, and $^4\Sigma$, in order of increasing energy. The order is that predicted by Hund's rules, with Λ replacing L. The state $^2\Pi$ is in fact the ground state of CH, and $^2\Sigma$ is the lowest observed excited state; the other two states are as yet unobserved. Similar analyses can be made for any other pair of separated-atom states.

It is generally true, as it is for CH, that some of the states generated from a given separated-atom limit remain unobserved. The reason for this is that many of the states have no stable minima in their $E(R)$ curves. Such states have no characteristic band spectra by which they can be identified, but only continuum radiation. Indeed, their potential curves are sometimes so strongly repulsive that molecules with R near the ground-state R_e are not encountered at all. To see how such repulsive states come about, let us consider the $^4\Pi$ state of CH. This state has $S = \frac{3}{2}$, and thus must have at least three singly occupied orbitals; to have $\Lambda = 1$, there must be an odd number of electrons in π orbitals. The lowest-energy configuration meeting these specifications is $1\sigma^2 2\sigma^2 3\sigma 1\pi 4\sigma$, where one electron is promoted from the 3σ to the 4σ orbital, which is so strongly antibonding that it overcomes the remaining bonding forces. (In the $^2\Sigma$ state, with configuration $1\sigma^2 2\sigma^2 3\sigma 1\pi^2$, electron to the *non*bonding 1π

orbital is enough to reduce the binding energy to only 0.4 eV.) Despite the repulsive nature of the $^4\Pi$ state, it must become the lowest-energy state at some small value of R, since it correlates with the united-atom ground state, N(^4S), which also has three singly occupied orbitals. As $R \to 0$ we have $1\sigma \to 1s$, $2\sigma \to 2s$, $3\sigma \to 2p\sigma$, $1\pi \to 2p\pi$.

We mentioned previously that the diatomic hydrides have been observed by radio astronomy. But just what kind of transitions can these molecules have in the radiofrequency region of the spectrum? For OH, for example, even the pure rotational spectrum is well into the infrared region, with B_e = 18.9 cm^{-1}. First of all, the energy levels of molecules show fine-structure splitting, which, as in atoms (Section 5.7), is due primarily to spin–orbit interaction. When the spin and orbital angular momenta are strongly coupled, analogous to *LS* coupling in atoms, the component of total angular momentum along the internuclear axis is $\Omega\hbar$, where

$$\Omega = |\Lambda + \Sigma| \qquad (\Sigma = S, S-1, \ldots, -S). \tag{7.57}$$

The quantum numbers Λ, Σ, Ω correspond to the atomic M_L, M_S, M_J, respectively. By "strongly coupled" we mean that the orbiting electrons produce a magnetic field that tends to align the spin magnetic moment with the axis. There are other types of coupling, but we need consider only this one. The ground-state term of the OH molecule is $^2\Pi(S = \frac{1}{2}, \Lambda = 1)$, with the possible values $\Sigma = \frac{1}{2}, -\frac{1}{2}$ and $\Omega = \frac{3}{2}, \frac{1}{2}$. Thus there are two states, designated as $^2\Pi_{3/2}$ (the ground state) and $^2\Pi_{1/2}$. But the energy difference between these states is 0.017 eV (corresponding to 140 cm^{-1}), which is even greater than the rotational spacing.[39] One thus observes two distinct though overlapping rotational bands in laboratory spectra, so we must seek further for a radiofrequency transition. Besides the coupling between spin and orbital electronic angular momenta, there is a much weaker interaction between electronic and *rotational* angular momenta. This interaction destroys the degeneracy between the two states differing only in the orientation of **L** (with $L_z = \pm\Lambda\hbar$; cf. Section 6.5), and is thus known as *Λ-type doubling.*[40] The splitting is proportional to $J(J+1)$, where J is the rotational quantum number. For the $J = 1$ state of OH ($^2\Pi_{3/2}$) it equals 7.8×10^{-6} eV or 0.063 cm^{-1}, which is indeed in the radiofrequency region. Thus we see how successively weaker interactions can give us finer and finer probes of molecular structure.

The final topic we shall discuss in this section takes us out of the realm of pure intramolecular forces; this is the *hydrogen bond,* a phenomenon of which HF provides the simplest illustration. The subject is introduced here, rather than in Chapter 10, because it is so characteristic of HF. A number of compounds in which hydrogen is bonded to very electronegative elements (mainly F, O, and N) show strong attractive forces between molecules. For example, although one expects boiling points in a family of compounds to increase with molecular weight as in the inert gases, HF, H_2O, and NH_3 go counter to this trend:

NH_3	−33.4	H_2O	0.0	HF	19.5	Ne	−245.9
PH_3	−133	H_2S	−60.7	HCl	−84.9	Ar	−185.7
AsH_3	−55	H_2Se	−41.5	HBr	−67.0	Kr	−152.3
SbH_3	−17.1	H_2Te	−2.2	HI	−35.4	Xe	−107.1

(boiling points in °C). The heats of vaporization vary in the same way; those for HF, H_2O, and NH_3 are higher than expected by 20–40 kJ/mol (0.2–0.4 eV/molecule). This is about one order of magnitude weaker than a normal covalent bond, but still much stronger than ordinary intermolecular attractions (see Chapter 10); for example, the potential well for two Ar atoms is about 0.01 eV deep. Data from spectroscopy, neutron diffraction, and other sources clearly show that in these substances the hydrogen atoms are normally located between two electronegative atoms, but closer to one than the other. This is usually indicated by a formula such as H—F · · · H—F, with the "long bond" shown by dots.

How does a hydrogen atom between two electronegative atoms have a bonding effect? We know that in an HF molecule the bond is strongly polar, so that there is an excess of negative charge on the F atom and an excess of positive charge on the H atom. As in our model for ionic bonds, one can to a good approximation treat each atom as a point charge but with a charge less than e. Just as negative charge in the region between two positive nuclei tends to draw them together (see Section 6.1), so a positively charged H atom between two negatively charged F atoms exerts a bonding force—but a much weaker one, because the charges are only partial and the distances greater. In an MO treatment, one might consider the ordinarily nonbonding 2σ orbital of one HF molecule to include an H($1s$) component from the other molecule. The hydrogen bond tends to pull the H atom away from its nearest-neighbor fluorine, with the net effect of weakening the restoring force in the H—F oscillator (in which the H does nearly all the moving). This effect lowers $\tilde{\nu}_e$ from 4138 cm^{-1} in free HF to about 3400 cm^{-1} in $(HF)_2$ showing that the environment of the H atom is rather drastically changed by hydrogen bonding.

The simplest example of a "molecule" showing hydrogen bonding is the dimer $(HF)_2$ which exists in the gas at relatively low temperatures. Two possible structures come to mind, neither of which corresponds to the true structure. One might expect either the linear H—F · · · H—F, or the ring

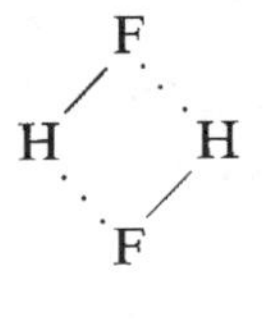

[39] The average temperature of interstellar space is probably about 3 K, corresponding to an average molecular energy of only 4×10^{-4} eV (3 cm^{-1}); thus, virtually all OH molecules in space should be in the $^2\Pi_{3/2}$ state if they are in thermal equilibrium.

[40] When $\Lambda = 0$, that is, in Σ states, there can still be an interaction between spin and rotational angular moments. But $^1\Sigma$ states (like atomic ^1S states) have no fine structure except that due to nuclear spin.

with two hydrogen bonds. In reality, the HF dimer appears to have the bent structure

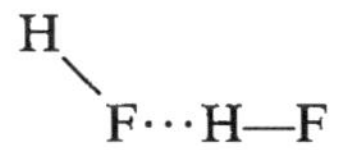

The location of the central H is very close to the F—F axis. More highly polymerized forms also exist, and liquid and solid HF are basically made up of long zigzag chains:

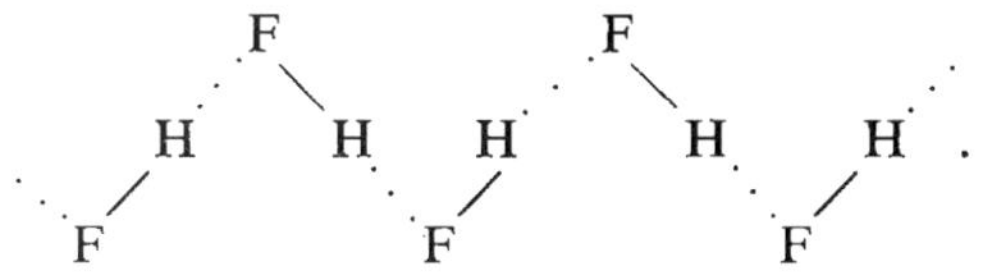

In the next chapter we shall see why the chains are bent rather than linear. If equimolar amounts of HF and KF are crystallized together, one obtains a well-defined crystalline species KHF_2, made of K^+ and FHF^- ions; in the FHF^- ion the hydrogen atom is found to be exactly midway between the two F atoms. This carries the hydrogen bond to its ultimate form, with the H atom associated with no particular molecule.[41]

7.10 Isoelectronic and Other Series

To interpret the differences among molecules, we must find or invent concepts that characterize the important changes from one molecule to another. Which concepts we choose will depend on the molecules under consideration. Thus far the key factor in our analysis of diatomic molecules has been the number of electrons, added one by one to a relatively stable set of orbitals. In heteronuclear molecules we needed the additional concept of orbital polarity, leading to the distinction between ionic and covalent bonding. To study the polarity effect in relative isolation, let us now consider some sets of molecules with the same total number of electrons—what we call *isoelectronic* series of molecules.

Among the simplest and most informative of such series is that isoelectronic with C_2, including BN, BeO, and LiF, in order of increasing polarity. All are known as diatomic molecules in the vapor phase, with properties varying from C_2, which is homonuclear and thus covalently bound, to LiF, a very ionic molecule which we described in Section 7.4. It is worth noting that a similar transition from covalent to ionic bonding is found in the solid forms of these species, from the covalent structures of carbon (diamond and graphite) to the ionic lattice of Li^+F^-.

[41] The valence bond theory treats FHF^- as a hybrid of the structures

$$(F—H\cdots F)^- \quad \text{and} \quad (F\cdots H—F)^-;$$

in MO theory one must use "three-center orbitals," with contributions from all three atoms.

One would expect all these molecules to have ground-state electron configurations equivalent to that of C_2, that is, $1\sigma^2 2\sigma^2 3\sigma^2 4\sigma^2 1\pi^4$. The ground states of C_2, BeO, and LiF are all $^1\Sigma$, corresponding to this configuration. Although the lowest known state of BN is $^3\Pi$, corresponding to what was long thought to be the ground state of C_2 (see footnote 28 on page 209), the low-lying $^1\Sigma$ state has not yet been observed, and the ground state may yet prove to be this $^1\Sigma$ state. Within the series, the orbitals change character regularly as the difference between the nuclear charges grows. The 1σ and 2σ orbitals, which we approximate by $1s_A \pm 1s_B$ in C_2, become the $1s$ orbitals of the high-Z and low-Z atom, respectively: They do not mix significantly with the $n = 2$ orbitals, since even in LiF the Li($1s$) orbital has appreciably lower energy than the F($2s$). The 3σ orbital changes from $2s_A + 2s_B$ in C_2 to the high-Z (N, O, or F) $2s$, and the 1π from $2p_A + 2p_B$ to the $2\pi p$ orbital on the high-Z atom. But the 4σ orbital, which correlates with the atomic $2s$ in C_2, becomes more and more like the $\sigma 2p$ of the high-Z atom as one goes to LiF. (Cf. the atomic orbital energies in Fig. 7.21.)

Some of the properties of molecules in this and other series are listed in Table 7.9. From C_2 to LiF the bond lengths increase and the vibrational frequencies decrease, but the dissociation energy varies in a more complicated

Table 7.9 Properties of Some Series of Diatomic Molecules

	Ground State	D_0 (eV)	R_e (Å)	$\tilde{\nu}_e$ (cm_{-1})	$\lvert\mu\rvert$ (D)
Isoelectronic Series					
C_2	$^1\Sigma_g^+$	6.24	1.2425	1855	0
BN	$^3\Pi$ (?)	3.99	1.281	1515	(1.4 calc.)
BeO	$^1\Sigma^+$	4.60	1.331	1487	(7.3 calc.)
LiF	$^1\Sigma^+$	5.94	1.564	910	6.33
N_2	$^1\Sigma_g^+$	9.760	1.094	2358	0
CO	$^1\Sigma^+$	11.09	1.128	2170	0.112
BF	$^1\Sigma^+$	7.85	1.262	1401	0.5 ± 0.2
Families of the Periodic Table					
HF	$^1\Sigma^+$	5.86	0.917	4139	1.826
HCl	$^1\Sigma^+$	4.446	1.275	2991	1.109
HBr	$^1\Sigma^+$	3.755	1.414	2649	0.828
HI	$^1\Sigma^+$	3.053	1.609	2308	0.448
Li_2	$^1\Sigma_g^+$	1.12	2.672	351.4	0
Na_2	$^1\Sigma_g^+$	0.75	3.079	159.2	0
K_2	$^1\Sigma_g^+$	0.51	3.923	92.6	0
Rb_2	$^1\Sigma_g^+$	0.47	4.20	57.3	0
Cs_2	$^1\Sigma_g^+$	0.45	4.58	42.0	0
F_2	$^1\Sigma_g^+$	1.604	1.409	919.0	0
Cl_2	$^1\Sigma_g^+$	2.484	1.988	559.7	0
Br_2	$^1\Sigma_g^+$	1.971	2.281	323.3	0
I_2	$^1\Sigma_g^+$	1.544	2.667	214.5	0

manner. This is the result of a balance between two opposing trends: The covalent bonding power weakens as the bond grows longer and the orbitals become more nonbonding, but the ionic contribution to the bond increases as the charge distribution becomes more polarized. The dipole moment of course increases with the electronegativity difference between the two atoms.

Another interesting set of isoelectronic molecules consists of N_2, CO, and BF. Both N_2 and CO exist as stable diatomic gases at room temperature, and even consist of diatomic molecules in the solid state; BF, however, is unstable.[42] The properties listed in Table 7.9 show the same conflicting trends as in the previous series. All three compounds have $^1\Sigma$ ground states, corresponding to the configuration $1\sigma^2 2\sigma^2 3\sigma^2 4\sigma^2 1\pi^4 5\sigma^2$. In CO, calculations show that the 1σ and 2σ orbitals are essentially atomic $1s$ orbitals, the 3σ and 5σ orbitals are largely concentrated on O and C, respectively, whereas the 4σ and doubly degenerate $1s$ orbitals furnish the bulk of the bonding. This corresponds approximately to the Lewis formula :C≡O:. That the dissociation energy of CO is higher than that of N_2 can be attributed to a small amount of ionic character added to what is still essentially a triple bond. In BF, ionic bonding is presumably important, but not enough to make ionic B^+F^- crystals stable.

There are a number of striking similarities in the physical properties of N_2 and CO. (Their chemical properties are of course rather different, but even there the two exhibit comparable inertness toward some reagents—notably *not* toward hemoglobin!) The gas densities, boiling points, viscosities, and thermal conductivities of the two species are almost the same. This similarity is due in part to their near-identical molecular weights, but the intermolecular forces also reflect the similar internal structure of the two molecules. The latter is particularly apparent if one looks at the orbital binding energies, which have been measured by photoelectron spectroscopy (values in eV):

	1σ	2σ	3σ	4σ	1π	5σ
N_2	409.9	409.9	37.3	18.6	16.8	15.5
CO	542.1	295.9	38.3	20.1	17.2	14.5

Except for the atomic-core 1σ and 2σ orbitals, the two sets of energies are almost identical; the energies obtained by molecular orbital calculations are somewhat different, but show the same pattern. Results of this sort clearly justify our treating the members of an isoelectronic species as closely related.

We must say something more here about photoelectron spectroscopy, a conceptually simple and powerful method for determining molecular energy levels. The nature of the photoelectric effect has been described in Section 2.2; here we apply it to free molecules. Specifically, suppose that radiation of frequency ν (energy $h\nu$) strikes a molecule and releases an electron with binding energy ϵ. If we neglect the small recoil effects, the binding energy should be given by

$$\epsilon = h\nu - T, \tag{7.58}$$

where T is the kinetic energy of the released electron. By measuring or selecting ν and then measuring T as in the Franck–Hertz experiment, for example, one can determine ϵ, which by Koopmans' theorem should equal the electron's orbital energy in the molecule. The energy $h\nu$ must of course be greater than ϵ. One most commonly uses radiation in the vacuum ultraviolet to release the valence electrons, and x-rays for the core electrons. Typically one irradiates a sample, gaseous or solid, with monochromatic radiation such as that of the He $2p \rightarrow 1s$ transition at 584 Å, and measures the kinetic energies of the photoelectrons. Ultraviolet radiation is associated with states whose natural lifetimes are of order 10^{-9} s and so, according to the uncertainty principle, can provide energy resolution at best with $\Delta E = \hbar/\Delta t$ or about 4×10^{-6} eV (0.03 cm^{-1}), somewhat better than the energy of the electrons can be determined. With x-rays, the energy resolution is typically 1–5 eV, due in large part to the short lifetimes of the excited states from which they are emitted. Hence x-ray photoelectron spectroscopy can locate the approximate energy of shells, but one must use ultraviolet radiation to probe the separations of valence orbitals. With ultraviolet photoelectron spectroscopy it is quite straightforward to distinguish different vibrational levels, especially of the final ion, and even rotational levels of very light molecules have been resolved.

Thus far we have not carried our analysis beyond the first row of the periodic table. According to the principles outlined in Section 5.5 we expect the elements below the first row to exhibit bonding behavior similar to that of the first element in each family (alkali metals, halogens, etc.). And in fact, each family does bond in generally similar ways. But what systematic changes may we expect to find as we go down a family of diatomic molecules (varying one or both atoms)? The properties of several such series are given in Table 7.9. One can readily name others (the alkali halides, interhalogen compounds such as ClF, interalkali compounds such as NaK, the analogs of N_2 and O_2, etc.), but essentially the same trends are found in all cases.

These trends largely reflect the effects of atomic size, which of course increases slowly as one goes down each column of the periodic table. In a similarly bonded series of molecules, the bonds must become longer and thus weaken with increasing atomic size. An additional effect is found in heteronuclear molecules: Since the nuclei become better shielded as more electrons are added, the ionization potentials, electron affinities, and thus electronegativities also decrease with atomic size. Depending on the nature of the series, these changes will either increase or decrease the ionic contribution to the bonding. There is one other effect of atomic size: The core electrons are by no means com-

[42] Even CO has a tendency to disproportionate, $2CO \rightarrow C + CO_2$, but the reaction is *very* slow at room temperature.

pletely shielded or nonbonding, and do take some part in bond formation; as we noted earlier, in many transition metals the inner-shell *d* electrons are almost as important as the valence electrons in bonding.

The trends in Table 7.9 are for the most part clear and consistent, with the exception of the anomalously low dissociation energy of F_2. This recalls the fact that the electron affinity of F is less than that of Cl. The low dissociation energy of F_2 has puzzled scientists for many years. The explanation, as with the electron affinity, seems to be that the inner core electrons (1*s* and, to some degree, 2*s*) form a relatively more important fraction of the total electron cloud in fluorine than in larger halogen atoms, so that the repulsive contribution to the F—F bond is more important than in other halogen–halogen bonds.

The macroscopic properties of molecular families often also show clear-cut trends—for example, the boiling points tabulated in the last section. However, these depend on intermolecular forces, which have more to do with the overall size of a molecule than with the nature of its bonding except where hydrogen bonding or other electrostatic effects are significant. We shall consider intermolecular forces in Chapter 10, and macroscopic behavior in Part II. For now, though, let us proceed to molecules with more than two atoms.

APPENDIX 7A

Perturbation Theory

The effects of small perturbations lend themselves to a set of related mathematical tools that, together, comprise *perturbation theory.* Here we outline the basic ideas of perturbation theory, just in its simplest forms, which we shall see are equivalent to its lowest orders. In the context of quantum mechanics, we suppose that there is a parameter κ that measures the strength of the perturbation of interest, so that we can write the Hamiltonian H as the sum of a part H_0 we know and understand well, and a complicated but small part $\kappa\mathsf{H}_1$, which constitutes the perturbation. We define H_1 to make $\kappa < 1$ so that the effects of the perturbation can be expressed by expansions in powers of κ, which means that both the energy and the wave functions can be so expanded:

$$E = E_0 + \kappa E_1 + \kappa^2 E_2 + \ldots \tag{7A.1}$$

and

$$\psi = \psi_0 + \kappa\psi_1 + \kappa^2\psi_2 + \ldots. \tag{7A.2}$$

The problem is one of finding E_1, E_2, . . . , ψ_1, ψ_2, . . . and so forth, from a Schrödinger equation whose Hamiltonian is $\mathsf{H} = \mathsf{H}_0 + \kappa\mathsf{H}_1$ and whose eigenvalues and corresponding eigenfunctions are those of Eqs. 7A.1 and 7A.2. The simple equation $\mathsf{H}\psi = \mathsf{E}\psi$ becomes

$$(\mathsf{H}_0 + \kappa\mathsf{H}_1)(\psi_0 + \kappa\psi_1 + \kappa^2\psi_2 + \cdots) = (E_0 + \kappa E_1 + \kappa^2 E_2 + \cdots)(\psi_0 + \kappa\psi_1 + \kappa^2\psi_2 + \cdots); \tag{7A.3}$$

a cumbersome equation indeed. However, we use a trick of mathematical reasoning here to simplify: If the parameter is truly arbitrary and may take any value in a range, say between 0 and 1, then we must be able to carry out the multiplications indicated in Eq. 7A.3, collect the terms on each side corresponding to specific powers of κ, and then write *separate equalities for each power of* κ. If we do this, we obtain a hierarchy of equations that begins with these:

$$\mathsf{H}_0\psi_0 = E_0\psi_0, \tag{7A.4}$$

$$\mathsf{H}_1\psi_0 + \mathsf{H}_0\psi_1 = E_1\psi_0 + E_0\psi_1, \tag{7A.5}$$

$$\mathsf{H}_2\psi_0 + \mathsf{H}_1\psi_1 + \mathsf{H}_0\psi_1 = E_2\psi_0 + E_1\psi_1 + E_0\psi_2, \tag{7A.6}$$

$$\vdots$$

and we shall not need to go further here. (As we have defined the perturbation here, H_2 vanishes.) The first equation here is just the Schrödinger equation for the system in the absence of the perturbation, in effect what we assume we know to be true. The next step is evaluating the *expectation value* of the first-order perturbation, $\langle E_1\rangle$ (the change linear in κ) because the change in energy of the system due to the perturbation is, to a first approximation, just $\kappa\langle E_1\rangle$. To find $\langle E_1\rangle$, we need only multiply on the left by and integrate over all the coordinates in the wave functions:

$$\langle E_1\rangle = \int \psi_0^* E_1\psi_0 \, d\tau = \int \psi_0^* (\mathsf{H}_1\psi_0 + \mathsf{H}_0\psi_1 - E_0\psi_1)\, d\tau, \tag{7A.7}$$

which we can simplify now by invoking the requirements that H_0 is Hermitian and therefore acts on its eigenfunction ψ_0^* simply to multiply it by the real number E_0 and that the functions ψ_0, ψ_1, ψ_2, . . . are all orthogonal and normalized. This means that the second and third terms on the right side of Eq. 7A.7 vanish, and we have simply that the first-order perturbation of the energy is the expectation value of the perturbing part of the Hamiltonian,

$$\langle E_1\rangle = \int \psi_0^*\mathsf{H}_1\psi_0 \, d\tau. \tag{7A.8}$$

This expression tells us that the perturbation causes a first-order change in the energy if and only if the integral of Eq. 7A.8 is nonvanishing. If H_1 represents the action of a uniform electric field on an atom then $\langle E_1\rangle$ vanishes because the energy of interaction contained in has the form of the scalar product of the electric field $\mathbf{E}$ and the atomic dipole moment $\boldsymbol{\mu}$: $\mathsf{H}_1 \propto \mathbf{E}\cdot\boldsymbol{\mu}$, and only $\boldsymbol{\mu}$ contains the coordinates appearing in the wave function. But no atom has a dipole moment, so there is no first-order perturbation of an atom's energy by a uniform electric field; i.e., $\langle E_1\rangle = \mathbf{E}\cdot\int \psi_0^*\boldsymbol{\mu}\psi_0\, d\tau$ vanishes because the integral vanishes. There are many important perturbations of the energy that do not vanish in first order, but, like this example, many do. Note that to find the first-order perturbation of the energy, we only need to know the wave function ψ_0 of the original, unperturbed problem and not any of the corrections to the wave function.

While the first-order contribution vanishes in the above example, its second-order contribution, $\langle E_2\rangle$, does not. Next we evaluate $\langle E_2\rangle$ and along the way find the first-order correction to the wave function. We can start this by returning to the equation for the first-order corrections, Eq. 7A.6, and rearranging it just a bit:

$$\begin{aligned} \mathsf{H}_1\psi_0 - E_1\psi_0 &= -(\mathsf{H}_0\psi_1 - E_0\psi_1), \quad \text{or} \\ (\mathsf{H}_1 - E_1)\psi_0 &= -(\mathsf{H}_0 - E_0)\psi_1, \end{aligned} \tag{7A.9}$$

the first step to finding ψ_1. Next, we suppose that we would like to know and express ψ_1 in terms of the eigenfunctions of H_0, which we presumably know or can find. That means we will write ψ_1 as a series expansion of the form

$$\psi_1 = \sum_{k=1} c_k \varphi_k, \tag{7A.10}$$

in which each of the functions φ_k satisfies a Schrödinger equation for the zero-th order Hamiltonian,

$$\mathsf{H}_0\varphi_k = E_0^{(k)}\varphi_k. \tag{7A.11}$$

Note that this means that $\psi_0 = \varphi_0$. Our task is then to find expressions for the unknown coefficients c_k. If we substitute the expansion of Eq.7A.10 into Eq. 7A.9, multiply that expression by any specific φ_j^*, and integrate over the variables of the wave function and then use both Eq. 7A.11 and the orthogonality condition on the φ_k's, we obtain an explicit, calculable expression for the contribution of that particular φ_j to the first-order perturbation function,

$$c_j = -\frac{\int \varphi_j^* \mathsf{H}_1 \varphi_0 \, d\tau}{E_0^{(j)} - E_0}. \tag{7A.12}$$

This expression tells us two things: first, the numerator shows that the more H_1 transforms the unperturbed function φ_0 into φ_j, the more important is the contribution of φ_j to the perturbed wave function; second, the denominator shows that the more the eigenvalues corresponding to φ_0 and φ_j differ, the less φ_j contributes to the perturbed wave function.

Next we turn to the second-order equation and rearrange it, recognizing that H_2 vanishes:

$$-E_2\psi_0 + (\mathsf{H}_1 - E_1)\psi_1 + (\mathsf{H}_0 - E_0)\psi_2 = 0, \tag{7A.13}$$

which we then multiply by ψ_0^* and integrate to find $\langle E_2 \rangle$ or just E_2, since they are the same here. The first term gives us just E_2, and the third term vanishes. This leaves us with

$$E_2 = -\int \psi_0^* \mathsf{H}_1 \psi_1, \tag{7A.14}$$

which becomes useful when we substitute Eqs. 7A.10 and 7A.12. With just a bit of rearranging, we find the second-order correction to the energy,

$$E_2 = \sum_k \frac{\left| \int \varphi_k^* \mathsf{H}_1 \varphi_0 \, d\tau \right|^2}{E_0^{(k)} - E_0}. \tag{7A.15}$$

This expression has content similar to Eq. 7A.12, in that the more the perturbation transforms the unperturbed state into a particular excited state, the greater is that state's contribution, in this instance, to the perturbed energy, and the more distant the unperturbed and excited states are in energy, the less the excited state affects the unperturbed state.

We close with a remark about the one situation in which Eqs. 7A.12 and 7A.15 cannot be used. This is the bothersome case in which the unperturbed state is degenerate with another state with which it mixes under the action of the perturbation. If this happens, the two (or more) "offending" states, call them ψ_0^A and ψ_0^B, must be given special treatment; we consider only the example of two degenerate functions with a common eigenvalue of H_0, namely E_0. We can write the first-order wave functions as linear combinations of ψ_0^A and ψ_0^B,

$$\begin{aligned} \psi_I &= a\psi_0^A + b\psi_0^B, \\ \psi_{II} &= a'\psi_0^A + b'\psi_0^A \end{aligned} \tag{7A.16}$$

and require that $a^2 + b^2 = a'^2 + b'^2 = 1$ (and that $\int \psi_I^* \psi_{II} \, d\tau = 0$, if we wish to include it), that is, that the two functions we seek must be normalized and orthogonal. This gives us two simultaneous linear equations for the two independent parameters of (only one, if we include all the conditions) which have a solution if and only if the determinant of the coefficients of the unknowns is zero. The equation stating this is

$$\begin{vmatrix} \int \psi_0^{A*} \mathsf{H} \psi_0^A \, d\tau - E & \int \psi_0^{A*} \mathsf{H} \psi_0^B \, d\tau \\ \int \psi_0^{B*} \mathsf{H} \psi_0^A \, d\tau & \int \psi_0^{B*} \mathsf{H} \psi_0^B \, d\tau - E \end{vmatrix} = 0, \tag{7A.17}$$

in which we can compute all the integrals and need only to find the eigenvalues E, which are the solutions of the quadratic equation which Eq. 7A.17 represents. We can make this equation look simpler by writing the four integrals in a shorthand notation, $\int \psi_0^{A*} \mathsf{H} \psi_0^A \, d\tau \equiv \mathsf{H}_{AA}$, $\int \psi_0^{B*} \mathsf{H} \psi_0^A \, d\tau \equiv \mathsf{H}_{BA}$, etc.,

$$\begin{vmatrix} \mathsf{H}_{AA} - E & \mathsf{H}_{AB} \\ \mathsf{H}_{BA} & \mathsf{H}_{BB} - E \end{vmatrix} = 0 \tag{7A.18}$$

or

$$(\mathsf{H}_{AA} - E)(\mathsf{H}_{BB} - E) - \mathsf{H}_{BA}\mathsf{H}_{AB} = 0 \tag{7A.19}$$

which has the solution

$$\begin{aligned} E = {} & \tfrac{1}{2}(\mathsf{H}_{AA} - \mathsf{H}_{BB}) \\ & \pm \tfrac{1}{2}\sqrt{(\mathsf{H}_{AA} + \mathsf{H}_{BB})^2 - (4\mathsf{H}_{BA}\mathsf{H}_{AB})}, \end{aligned} \tag{7A.20}$$

an equation giving the energies of the two states that are degenerate in zero-th order but not in second order. One of the most important examples of this situation is the way the energies of two electronic states (with the same symmetry) of a diatomic molecule "split" when their Born–Oppenheimer potential curves cross. The dominant perturbation in this case is most probably the action of the nuclear kinetic energy on the electronic wave functions, as discussed in Section 7.5. The full Hamiltonian is meant to be included in Eq. 7A.17, but the part that contributes to the off-diagonal elements, H_{AB} and H_{BA}, is the nuclear kinetic energy operator, with ψ_0^A and ψ_0^B as two electronic states determined within the Born–Oppenheimer approximation.

● FURTHER READING

Bernath, P. F., *Spectra of Atoms and Molecules* (Oxford, New York, 1995).

Gaydon, A. G., *Dissociation Energies and Spectra of Diatomic Molecules,* 3rd Ed. (Chapman and Hall, London, 1968).

Herzberg, G., *Molecular Spectra and Molecular Structure, Volume I. Spectra of Diatomic Molecules,* 2nd Ed. (Van Nostrand-Reinhold, Princeton, N.J., 1950).

Hurley, A. C., *Introduction to the Electron Theory of Small Molecules* (Academic Press, London, 1976).

Karplus, M., and Porter, R. N., *Atoms and Molecules* (W. A. Benjamin, Inc., Menlo Park, Calif., 1970), Chapters 5, 6, and 7.

Kauzmann, W., *Quantum Chemistry* (Academic Press, Inc., New York, 1957), Chapters 11C–G and 12.

Kondratyev, V., *The Structure of Atoms and Molecules* (P. Noordhoff N. V., Groningen, The Netherlands, 1964), Chapters 7–10.

Morrison, M. A., Estle, T. L., and Lane, N. F., *Quantum States of Atoms, Molecules and Solids* (Prentice-Hall, Inc., Englewood Cliffs, N.J., 1976), Chapters 12, 14–17.

Mulliken, R. S., and Ermier, W. C., *Diatomic Molecules, Results of* ab initio *Calculations* (Academic Press, New York, 1977), esp. Chapters IV, V, and VI.

Slater, J. C., *Quantum Theory of Molecules and Solids,* Volume I (McGraw-Hill Book Co., Inc., New York, 1963), esp. Chapters 5, 6, and 7.

Streitweiser, A., and Owens, P. H., *Orbital and Electron Density Diagrams. An Application of Computer Graphics* (The Macmillan Company, New York, 1973).

Yarkony, D. R., Ed., *Modern Electronic Structure Theory, Parts I and II* (World Scientific, Singapore, 1995).

● PROBLEMS

1. Because the vibrational spacings of diatomic molecules generally diminish with increasing energy, it is possible to extrapolate the vibrational spacing, as a function of vibrational energy, to zero, and thereby obtain a moderately accurate estimate of the dissociation energy of the molecule. Such graphs are known as Birge–Sponer plots, after R. Birge and H. Sponer. Using values of $\tilde{\nu}_e$, $\tilde{\nu}_e x_e$, and $\tilde{\nu}_e y_e$ from Table 7.2, construct such plots for Na_2, CH, and HCl, then evaluate D_e and D_0, the dissociation energies from the bottom of the potential and

from the ground vibrational state. Compare these values of D_0 with those given in Table 7.2.

2. Find the outer classical turning points and thus the classical zero-point amplitudes of vibration for H_2, LiH, and HCl, from the data in Table 7.2. (Refer to Problems 21 and 22 in Chapter 4 if you need help finding the classical turning point.)

3. Using Table 7.2, determine the vibrational quantum numbers at which the actual vibration energy level spacings of F_2 and Cl_2 deviate 1% and 10% from the harmonic spacings based on $\tilde{\nu}_e$ alone.

4. The dipole moment of HCl is 1.109 D (Table 7.4), its equilibrium bond length is 1.27 Å, and its vibration frequency is approximately 2991 cm^{-1} (Table 7.2). Suppose that a spatially uniform laser beam of 1 W, with precisely this frequency and a cross section of 0.01 cm^2, is incident on a sample of HCl vapor. What is the maximum instantaneous stretching force exerted by this field on an HCl molecule with $R = R_e$? Based on the force constant of Table 7.2 and the assumption that the dipole moment μ increases directly with R very near R_e, compute the effective restoring force of the chemical bond on the nuclei at their classical outer turning point in the ground vibrational state. Compare the restoring force of the bond with the stretching force of the electric field.

5. Compute the average moment of inertia for the ground vibrational-electronic state of O_2, assuming that this molecule is a harmonic oscillator, so that the lowest vibrational function of Table 4.2 can be used for the probability amplitude. Compare this moment of inertia with the value implied by the B_e of 1.4456 cm^{-1} in Table 7.2.

6. Most common diatomic molecules have lowest vibrational spacings that are much larger than their lowest rotational spacings, yet vibrational spacings diminish and rotational spacings increase with increasing quantum number. At what vibrational and rotational quantum numbers do these spacings become roughly equal for H_2, N_2, and HBr? What are the total energies of the molecules at these levels? Compare these total energies with the corresponding dissociation energies.

7. Calculate the frequencies of the vibration–rotation transitions of OH from data in Table 7.2, for the transitions $\upsilon = 0 \rightarrow 1$ and $\upsilon = 1 \rightarrow 2$, for $\Delta J = -1$, 0, and +1 for J from 0 to 10. Plot the transitions on an energy scale that displays the entire set of lines and still allows one to resolve them. What spectral resolution would be required to distinguish the lines, if one requires resolution of at least half the spacing of the most closely spaced lines?

8. As the rotational energy of a molecule increases, the centrifugal force on the nuclei adds a repulsive term to the effective potential as given in Eq. 7.19. Find the value of J for which the centrifugal potential brings the lowest point of the rotationless curve up to D_0, that is, find the first rotational state for which $E_{rot} + V(R) > 0$ everywhere, for H_2, Na_2, and I_2.

9. Compute a rotational analog of Problem 4. That is, assume that an assembly of HCl molecules is in a laser beam whose power density is 1 W/cm^2 and whose frequency is precisely resonant with the first rotational transition of HCl. Given that its dipole moment is 1.109D, compute the total force on the nuclei and the torque (force times lever arm from the center of mass) when the force is a maximum.

10. From the rotational line frequencies given below, compute the bond length of the diatomic molecule NaCl in its ground and first excited vibrational state. [Data are from A. Honig, M. Mandel, M. L. Stitch, and C. H. Townes, *Phys. Rev.* **96,** 629 (1954).] All transitions are $J = 1$ to $J = 2$.

	$v = 0$ (MHz)	$v = 1$ (MHz)
$Na^{35}Cl$	26051.1 ± 0.75	25857.6 ± 0.75
$Na^{37}Cl$	25493.9 ± 0.75	25307.5 ± 0.75

How much effect do the uncertainties in the spectral line frequencies have on the inferred bond lengths?

11. The rotational spectrum of a polar diatomic molecule is observed in the microwave region of the electromagnetic spectrum. In the case of RbBr, in the vibrational state $v = 0$, the $J = 8 \to 9$ transition is observed at:

Molecule	(MHz)
$^{85}Rb^{79}Br$	25 596.03
$^{87}Rb^{79}Br$	25 312.99
$^{85}Rb^{81}Br$	25 268.84

Assume that RbBr behaves as a rigid rotator and calculate the internuclear separations in the various isotopic molecules. The atomic masses are $^{79}Br = 78.94365$ amu, $^{81}Br = 80.93232$ amu, $^{85}Rb = 84.93920$ amu, $^{87}Rb = 86.93709$ amu. Are you surprised by your results? How do you interpret them?

12. The bond lengths of Na_2 are 3.0782 Å and 3.63 Å in the ground and first excited electronic states, respectively. Based on this difference, compute the separation in cm^{-1} between the band origin ($J' = 0$ to $J'' = 0$) and the band head (where the rotational lines turn around and start returning on themselves) for a transition between these states.

13. One speaks of rotational bands in electronic spectra as being "degraded toward the red" or "degraded toward the blue" depending on whether the rotational line spacings eventually increase toward longer or shorter wavelengths. Such "degradation" is usually immediately apparent to the eye. Show that one can tell immediately whether R_e is greater in the ground or excited state depending on the direction of the degradation or shading.

14. Doppler broadening is very important in many regions of the spectrum and in a variety of situations, including spectroscopy of stars and the interstellar medium. The Doppler shift is the result of wave crests and peaks reaching the observer faster or slower than they would if the source of radiation and the observer were at rest relative to each other. Show that the shift in wavelength $\Delta\lambda$ of radiation sent by a source moving with velocity v relative to the observer is given approximately by

$$\Delta\lambda = \frac{\lambda v}{c}.$$

Calculate the Doppler shift for radiation of 1000 MHz and 1000 cm^{-1} if $v = 10^{-5}$ cm/s.

15. Compare the bonding in Cl_2 and Cl_2^+. How does it differ? Would you expect Cl_2^- to exist as a stable species? Why?

16. Show the forms of the nodal surfaces for the σ, π, and δ orbitals constructed as sums and differences of $3d$ orbitals in the homonuclear diatomic molecule Si_2.

17. The dissociation energy of F_2 is 1.60 eV, whereas those of Cl_2 and Br_2 are 2.48 eV and 1.97 eV, respectively. Give at least one interpretation of this apparent anomaly.

18. The molecules N_2 and C_2H_2 are isoelectronic. From considerations based on the electronic structure of N_2, predict the geometry (i.e., bent or linear) of C_2H_2, and discuss its electronic structure in terms of the types of orbitals occupied, nature of the orbitals, and so on.

19. Write the electron configurations for the ground states of Mg_2, Fe_2, MgH, and HCl. Which of these would you expect to be stable on the basis of the electron configuration alone?

20. Predict the dissociation energy, equilibrium internuclear distance, and vibration frequency $\tilde{\nu}_e$ by extrapolation from the data in Table 7.9, for each of the following:
(a) At_2
(b) DAt
(c) Fr_2

21. Rationalize the empirical facts that the ionization potential of H_2 is greater than that of atomic H, whereas that of O_2 is less than that of atomic O.

22. Rationalize the exponential form of the repulsive contribution to the energy of an ionic molecule, Eq. 7.43. Recall that the united-atom limit has a finite energy, achieved when the two nuclei coalesce. Hint: Consider the effect of the Pauli principle. Explain why a form $Be^{-R/\rho}/R$ might be even more plausible.

23. Consider the reaction, in a gas-phase collision,

$$K + HI = KI + H.$$

Suppose that the initial velocity of the K atom is 5×10^4 cm/s, and that of the HI molecule is negligibly small. Suppose that the rotational energy of HI is negligibly small. Further, suppose that the K would have approached to within 5 Å of the center of mass of the HI if no reaction occurred. Finally, assume that the velocity of the H atom leaving after the reaction is negligibly small. The energy of the KI bond is 3.34 eV and that of the HI bond is 3.06 eV. The internuclear separation in KI is 3.048 Å, and the vibrational frequency is 173 cm^{-1}. What do the conditions for conservation of energy, linear momentum, and angular momentum imply about the vibrational and rotational state of the product molecule KI? Estimate the rotational quantum number J and the vibrational quantum number v that characterize the product KI. (*Hint:* Look back at Rutherford scattering to see how the angular momentum of a particle moving toward another molecule is related to its velocity and the impact parameter; cf. Appendix 2A.)

24. The hydrogen halides have continuous electronic absorption spectra, the onsets being about 2500 Å for HCl, 2650 Å for HBr, and 3270 Å for HI.
(a) How do you interpret these observations? Draw likely potential energy curves for the states in question in HI.
(b) If the difference in energy between the $^2P_{1/2}$ and $^2P_{3/2}$ states of the I atom is 1 eV, must the onsets of production of these two dissociated iodine atoms from HI also be separated by 1 eV?
(c) What is the likely consequence of irradiating HI with light of wavelength 2537 Å? 1849 Å?

25. The occurrence of predissociation is often inferred from spectra by the sudden appearance of wide, diffuse lines where at lower energies the lines are narrow and sharp. Explain why this is an indication of predissociation or autoionization. How would you distinguish predissociation from autoionization experimentally?

26. Compare the dissociation energies, bond lengths, and vibrational frequencies of a number of diatomic molecules, for example, from Table 7.2. Examine these for correlations among bond length, dissociation energy, and vibration frequency. What correlations do you find? (Other sources of data should also be consulted.)

27. Use (a) the HF–SCF, (b) the CISD, and (c) the MP2 methods along with a 6–31G(d) basis set (which adds a set of d-type polarization functions to nonhydrogen atoms) to compute values of D_e and b R_e for the ground electronic states of the homonuclear diatomic molecules N_2 and F_2. Compare the calculated values with the experimental values found in Table 7.5.

28. Using the computed value of R_e (from Problem 7.27) and Eq. 7.21, calculate the rotational constant, B_e, for the N_2 and F_2 molecules. Compare the calculated values with the experimental values found in Table 7.2.

29. Use the same electronic structure methods and basis set utilized in Problem 7.27 to compute the fundamental harmonic vibrational frequency of N_2 and F_2 in their ground electronic states. Using the fundamental harmonic vibrational frequency and the value of D_e calculated in Problem 7.27, compute a theoretical estimate of the values of D_0 for the two homonuclear diatomics. Compare with the experimental values found in Table 7.2.

30. Use (a) the HF–SCF, (b) the CISD, and (c) the MP2 (second-order Møller–Plesset perturbation theory) methods along with a 6–31G(d,p) basis set (which adds a set of d-type polarization functions to nonhydrogen atoms and a set of p-type polarization functions to hydrogen) for the heteronuclear diatomic molecules HF and LiF in their ground electronic states, computing D_e, R_e, the fundamental harmonic vibrational frequency, and D_0. Additionally, compute the dipole moments of HF and LiF. Compare the calculated results with values given in Tables 7.2, 7.4, and 7.8.

31. The third excited electronic state of F_2 has a symmetry $^1\Pi_g$. Use the CISD method along with a 6–31G(d) basis set to compute values of D_e and R_e for F_2 in this excited electronic state. Estimate the energy difference between the ground electronic state and the third excited electronic state of F_2 by using the results from Problem 7.27. Compare with the experimental difference.

CHAPTER 8

Triatomic Molecules

Starting with the fundamental quantum laws, we have examined the properties of first atoms, then diatomic molecules. Now it is time to take another step in building up our picture of the microstructure of matter. In this chapter we extend our analysis to triatomic molecules. These molecules are interesting in themselves, including as they do some of the species most important in chemistry and biology, especially H_2O and CO_2. But in addition they display most of the complexity involved in considering larger molecules.

The principal feature of this added complexity is, of course, *molecular geometry.* Diatomic molecules have only one internal coordinate, the internuclear distance; in triatomic and larger molecules we must consider the distances and directions between many pairs of atoms—in simple language, we wish to know the bond lengths and bond angles. In the triatomic case there are three independent internal coordinates: We can expect the values of all these variables to depend on the molecule's electronic structure, and in the first part of the chapter we consider this relationship. The problem of the relation between molecular geometry and electronic structure is a partially solved puzzle that continues to tantalize theorists, but is well enough understood to permit making many powerful generalizations.

In the second part of the chapter we shall be concerned with the vibrations and rotations of triatomic molecules. In the triatomic molecule we have three or four vibrational *degrees of freedom*—that is, three or four independent types of vibration going on at the same time. Such molecules also have rotational degrees of freedom, two if the molecule has four kinds of vibration and three if the molecule has only three vibrational modes. This results in a complicated energy spectrum, which plays a crucial role in all fields concerned with the capacity of molecules to absorb, emit, or store energy. Our investigation of vibrations and rotations begins with Section 8.6.

We shall sometimes depart from fundamental concepts in this chapter. Our discussion has reached a level where the complexity of the problems forces us to develop semiempirical guidelines, particularly for the interpretation of structure. This is even more necessary in larger molecules, but the triatomic molecules are still simple enough for us to see the connections between these semiempirical rules and the underlying physical laws responsible for their validity.

8.1 Electronic Structure and Geometry in the Simplest Cases: H_3 and H_3^+

Our first examples of triatomic molecules are the simplest, which, as in diatomic molecules, are those made up of hydrogen atoms. The one-electron H_3^{2+} ion is quite unstable and need not be considered. However, the two-electron H_3^+ ion is a stable species, produced whenever an H_2 molecule collides with an H_2^+ molecule-ion, as occurs in electric discharges in gaseous hydrogen. The H_3^+ ion appears in interstellar clouds, as a consequence of just such collisions; the spectrum of this species is so well understood that its identification is completely unambiguous. The neutral molecule H_3 is unstable with respect to decomposition into H_2 + H, but since the exchange reaction

$$H + H_2 \rightarrow H_2 + H$$

probably involves transient configurations similar to H_3, its nature is of considerable interest to students of chemical reaction mechanisms. Let us therefore consider the properties of the H_3^+ and H_3 molecules, in particular their potential-energy surfaces and the resulting equilibrium shapes.

How many variables do we need to describe the shape of a triatomic molecule? We begin with three atoms, each of which has three independent coordinates. (We neglect the electrons here; the "structure" of a molecule refers to the relative positions of the nuclei.) Forming a molecule might effectively freeze one or more of these coordinates, but it cannot increase or decrease the number of independent coordinates—or *degrees of freedom.* Hence, however we describe our triatomic molecule, we have precisely nine coordinates with which to make our description. Three of these coordinates are essentially always used to specify the location of the center of mass of the molecule. Three more specify the molecule's orientation, unless the molecule is linear, in which case only two coordinates suffice to fix the orientation in space. This leaves three remaining coordinates for a nonlinear triatomic molecule and four for a linear triatomic molecule, with which we can describe the geometry and the motions of the nuclei. While some coordinate systems and variables are easier to use than others, at this point there is considerable arbitrariness about how we choose to describe the structure and vibrations of our molecule.

By the term "the structure" of a molecule, we mean its equilibrium geometry. The geometry of the H_3^+ and H_3 molecules—and all other triatomic molecules—may be either linear or triangular. To specify the molecule's "structure," we must know the equilibrium bond lengths and the equilibrium angles between the bonds. Chemical intuition leads us to describe the geometry of the H_3 molecule in terms of two H—H bonds and a single angle between them. Our interest here is primarily in the equilibrium value of this angle; the forces determining the bond lengths are much like those described in Chapter 7 for diatomic molecules. (There are, however, subtle and sometimes important consequences of the interactions between bonds and between nonbonded atoms; we shall address these later.) Thus, we need to examine how the molecule's energy varies as a function of the bond angle.

For a straightforward approximate interpretation we again employ the molecular orbital approach, based on the use of one-electron wave functions and energies. Furthermore, we again employ the LCAO approximation, which treats the molecular orbitals as sums and differences of atomic orbitals. Even though much more elaborate calculations are necessary to describe molecular properties quantitatively, the crude LCAO model is easy to use and gives results that are vivid and, for the most part, qualitatively correct. In particular, we shall see that it gives the correct answer for the shape of H_3^+.

We are still using the Born–Oppenheimer approximation, so we wish to know the electronic energy for each possible position of the nuclei. Let us begin with the most symmetric linear structure, that with the nuclei equally spaced, so that we have $R_{AB} = R_{BC}$, $\angle ABC = 180°$, where A, B, C designate the three nuclei (cf. Fig. 8.1). In the LCAO approximation the lowest-energy molecular orbital should be formed entirely from hydrogen 1s orbitals, and with as few nodes as possible. The simplest orbital that satisfies these requirements is

$$\psi_1 = a_1(1s_A) + b_1(1s_B) + c_1(1s_C), \qquad (8.1a)$$

where a_1, b_1, c_1 are constants. We can immediately simplify this further: Given the equal nuclear spacing, the nuclei A and C are exactly equivalent physically, so their atomic orbitals must contribute equally to any molecular orbital. Thus we have $a_1 = c_1$ and

$$\psi_1 = a_1(1s_A + 1s_C) + b_1(1s_B). \qquad (8.1b)$$

As in the diatomic molecule where we could form two molecular orbitals from a given pair of atomic orbitals, here

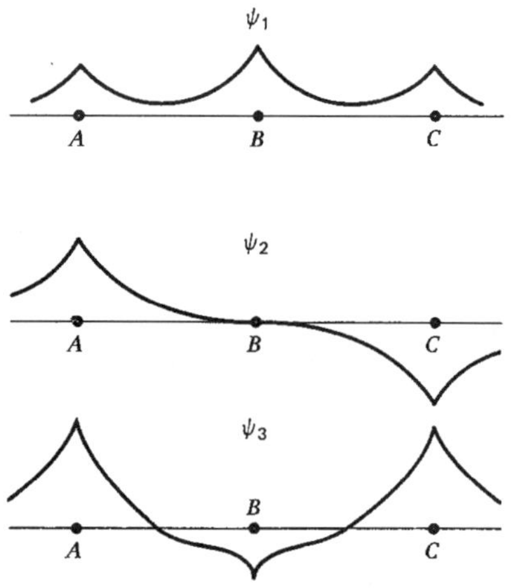

Figure 8.1 Orbitals in linear H_3^+ or H_3 (equally spaced, $R_{AB} = R_{BC}$). The graphs show schematically the variation of each orbital along the internuclear axis.

we can form three MOs from a given set of three AOs—in this case $1s_A$, $1s_B$, $1s_C$. Since the lowest orbital, ψ_1, has no nodes, we expect the other two to have respectively one and two nodes. Since in our model nuclei A and C are equivalent, the one-node orbital must have its node exactly midway between them, that is, on a plane bisecting the A–C axis. But this nodal plane passes directly through the B nucleus, so the coefficient of $1s_B$ in this orbital must be zero. Thus we can write the one-node orbital as

$$\psi_2 = a_2(1s_A - 1s_C). \qquad (8.2)$$

Note that this orbital, ψ_2, is orthogonal to ψ_1—no matter what coefficients we choose, so long as they are consistent with Eqs. 8.1 and 8.2, $\int \psi_1^* \psi_2 \, d\tau = 0$. The reason is that these two orbitals have *different symmetries.* The nodeless ψ_1 does not change if we redefine the coordinates by making x, y, and z into $-x$, $-y$, and $-z$, while the function ψ_2 does change sign if we so redefine the coordinates. This property, in which functions of different symmetries are orthogonal, has its counterpart physical interpretation: Two states of different symmetry, described by the two orthogonal wave functions, are *mutually exclusive states.* We discuss spatial symmetries in more detail later in this section.

Now we turn to the third of the three molecular orbitals. This one has the same symmetry as ψ_1. A suitable and general way to write the two-node orbital built from 1s atomic orbitals is

$$\psi_3 = a_3(1s_A + 1s_C) - b_3(1s_B), \qquad (8.3)$$

with the A and C nuclei again equivalent. The values of a_1, b_1, a_3, and b_3 are not independent. We normally require ψ_1 and ψ_3 to be orthogonal, which gain is equivalent to requiring that states 1 and 3 be mutually exclusive—a condition we are free to impose, but which is not required, in this case, by symmetry. (Of course ψ_2 is orthogonal to ψ_3 because they have different symmetries.) The orthogonality condition gives us one independent equation that allows us to reduce the number of coefficients to three; normalization of ψ_1 and ψ_3 allows us to remove two more. Hence ψ_1 and ψ_3 have only one independent coefficient once the atomic orbitals are chosen. Although we have made no systematic attempt to select the best LCAO orbitals, Eqs. 8.1–8.3 are convenient functions with the correct symmetry and nodal properties. They can be optimized by varying the constants a_i, b_i to minimize the energy. The orbitals ψ_1, ψ_2, ψ_3 are shown schematically in Fig. 8.1.

As in all our previous discussions of orbital models, we obtain the ground state of the molecule as a whole by an Aufbau-principle approach. That is, the electrons are assigned in succession to the one-electron states (orbitals) in accordance with the exclusion principle, up to two electrons, of opposite spin, in each orbital, starting with the lowest energy level and proceeding upward in energy until all the electrons are assigned. Thus the ground-state configuration of the linear H_3^+ ion is $(\psi_1)^2$ with both electrons in the lowest orbital; and the ground state of the linear neutral H_3 molecule is $(\psi_1)^2\psi_2$. The actual molecular wave functions must of course be properly antisymmetrized, with electron spin included, as we explained in Section 6.7. The energy of each of these species can then be calculated in the usual way, by evaluation of the integral $\int \psi^* \mathsf{H} \psi \, d\tau$.

Now we can ask how the molecular energy varies with the bond angle ($\theta \equiv \angle ABC$; cf. Fig. 8.2*a*). Let us assume for simplicity that we continue to have $R_{AB} = R_{BC}$, so that the molecule passes from a linear structure ($\theta = 180°$) into an isosceles triangle and eventually an equilateral triangle ($\theta = 60°$, $R_{AB} = R_{AC}$). We can continue to use the orbitals Eqs. 8.1–8.3: Only the coefficients a_i, b_i and the orbital energies vary as θ changes. What predictions can we make about the energy changes?

Consider first the ground state of H_3^+ in which only the orbital ψ_1 is occupied. Recall our naive concept of bonding force as arising from the constructive interference of electron waves in regions between nuclei. The orbital ψ_1, with all its atomic orbital coefficients positive, exhibits such constructive interference between all three pairs of nuclei. If we bend the molecule, the overlap between the electrons on atoms *A* and *C* increases, creating an incipient bond between these two atoms. There is thus a net bonding force that lowers the orbital (and molecular) energy as R_{AC} decreases. For sufficiently small R_{AC}, as in any other bond, the repulsive forces between the nuclei become dominant. Hence an equilibrium structure occurs for the value of θ and R_{AC} at which the attractive and repulsive forces just balance and the energy has its minimum value. Given the symmetry of the problem, it should not surprise us if the equilibrium geometry of H_3^+ is an equilateral triangle, and detailed calculations show that this is indeed the case. Figure 8.2*a* shows how the potential energy of H_3^+ varies with the bond angle when the bond lengths are fixed. Diagrams such as those in Fig. 8.2, displaying orbital or molecular energies as functions of bond angles, are often called *Walsh diagrams,* after A. D. Walsh, an early exploiter of such diagrams for systematically interpreting molecular structures.

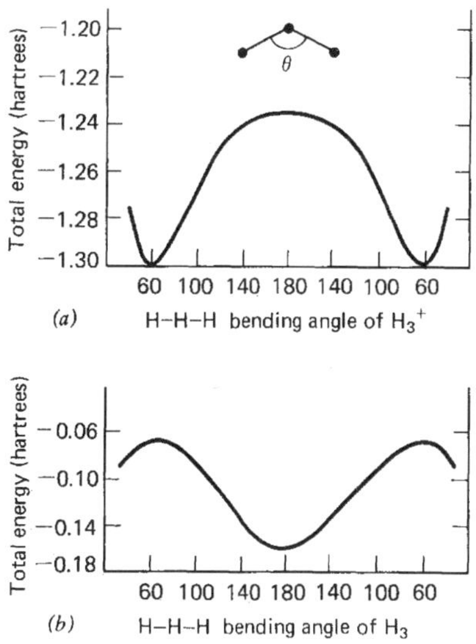

Figure 8.2 Potential energy of bending in H_3^+ and H_3. (*a*) Definition of the bond angle θ and $V(\theta)$ in H_3^+. Energy values from R. E. Christofferson, *J. Chem. Phys.* **41,** 960 (1964). H—H bond distance is 1.625 bohrs. (*b*) $V(\theta)$ in H_3. Figures from graphs given by C. W. Eaker and C. A. Parr, *J. Chem. Phys.* **65,** 5155 (1977). H—H bond distance is 1.72 bohrs. Both potential energy curves calculated for fixed bond lengths, with $R_{AB} = R_{BC}$.

Although the equilateral triangle is the equilibrium structure of H_3^+, the linear arrangement is also one of special symmetry. The reason is that any displacement made by varying θ from linearity has a physically equivalent counterpart on the "other side." That is, bending H_3^+ (or H_3) to reduce θ from 180° to 170° is exactly the same physical operation as increasing θ from 180° to 190°; similarly, any angle $180° - x$ is physically equivalent to $180° + x$. Because of this symmetry, any such molecule's orbital and molecular energies as functions of θ for fixed internuclear distance must be symmetric around 180°, which must thus be an *extremal* point—a maximum or minimum with respect to the bending angle. In the case of H_3^+ we have argued that the extremum should be a maximum, and the calculations used to generate Fig. 8.2*a* confirm this.

The linear arrangement of H_3^+ is thus what would be called a position of *unstable equilibrium* in classical mechanics. At $\theta = 180°$ there is no force tending to bend the molecule, but once θ is even infinitesimally different from 180°, the "force"[1] $-\partial V/\partial\theta$ tends to increase the bending.

[1] Although $-\partial V/\partial\theta$ does not have the dimensions of a true force, it can be considered the generalized force associated with θ. If a mechanical system is completely described by a set of variables q_i (distances, angles, etc.), then the work performed in an infinitesimal displacement can always be written in the form $\Sigma_i Q_i \, dq_i$, where the Q_i are generalized forces. For systems whose energy is conserved (those for which a potential energy can be defined) this means that $Q_i = -\partial V/\partial q_i$, and there is an effective "force" tending to lower the value of V.

The equilateral triangle, in contrast, is a position of *stable equilibrium,* in that the potential energy has a minimum, any displacement from which creates a restoring force. That the equilibrium position is an *equilateral* triangle (rather than some other isosceles triangle) is not something that we can infer from considerations of symmetry alone. The equilateral arrangement does have a higher symmetry, but symmetry is not the only factor governing molecular geometries. The equilibrium shape is the outcome of the detailed balance among the various interactions (electron–electron, electron–nucleus, and nucleus–nucleus). There are other triatomic molecules in which the three atoms are identical but the equilibrium geometry is only an isosceles triangle; an example is ozone (O_3), for which the equilibrium value of θ is 117°.

Although we were able to deduce easily that H_3^+ should be bent, the neutral H_3 molecule is more problematical. The same reasoning we just used implies that the linear arrangement must correspond to an extremum in the energy—but an extremum of which kind? The ground-state electronic configuration is $(\psi_1)^2\psi_2$, with a molecular energy crudely given by $2\epsilon_1 + \epsilon_2$. (Equation 8.26 will give a much more accurate expression.) As in H_3^+, the energy of ψ_1 decreases with bending, because ψ_1 is bonding between atoms *A* and *C*. But ψ_2 is antibonding between *A* and *C*, having a node midway between these atoms; hence the energy of ψ_2 goes *up* when the *A–C* distance is reduced. Whether the energy of the whole molecule goes up or down with bending depends on whether the increase in ϵ_2 is greater or less than twice the decrease in ϵ_1 ("twice," because of the two electrons in ψ_1). But we cannot deduce this by simple arguments; one must actually carry out the energy calculations. Detailed calculations in fact show that the total energy increases as θ deviates from 180°, so that the linear structure is the equilibrium geometry for H_3. The potential energy of H_3 as a function of θ is shown in Fig. 8.2*b*.

However, the important point here is not the result itself, but what it shows us about the limitations of various approaches. In the case of H_3^+ we could infer the structure correctly from qualitative considerations. In H_3 such a simple approach does not suffice, and either measurements or elaborate calculations are necessary to learn the true structure. Indeed, the molecular orbital model itself does not suffice. One must take electron correlation into account to obtain a conclusive result. One who studies molecular structures must become skilled in judging which of these kinds of situations applies in a given case. We shall see that simple MO considerations, combined with some generalizations based on experiment, often do provide a powerful base for inferring molecular structures—but that one must be careful to avoid applying the method beyond its range of validity.

Thus far we have considered only the case $R_{AB} = R_{BC}$, but a complete determination of the molecular geometry must include the dependence of the energy on bond lengths as well as on bond angle. Such calculations have also been made. The results are similar to those we found for diatomic

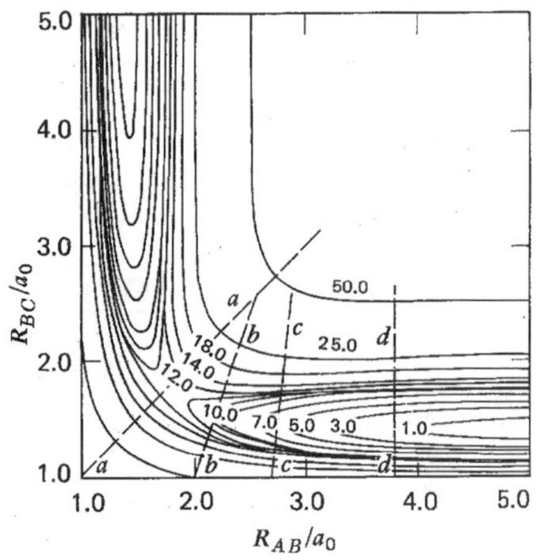

Figure 8.3 Contour map for the potential energy surface of linear H_3. The energy contours are labeled in kilocalories per mole (1 eV = 23.06 kcal/mol), relative to H_2 + H at infinite separation. The calculated saddle point is at $R_{AB} = R_{BC} = 1.765a_0$ (0.934 Å), $V(R_{AB}, R_{BC}) = 11.35$ kcal/mol.

molecules. For each bond length the energy becomes infinite as $R \to 0$ (if we neglect the collapse of the protons into a single nucleus), levels off as $R \to \infty$, and may have a minimum somewhere between. For a given bond angle, the energy as a function of R_{AB} and R_{BC} can best be represented by a "contour map"; Fig. 8.3 is such a map for linear neutral H_3. (In Part III we shall see how such diagrams are used in the study of chemical kinetics.) Note that the line $R_{AB} = R_{BC}$ does not include any kind of equilibrium position, but rather a *saddle point* (named after the shape of the potential energy surface): $V(R_{AB}, R_{BC})$ decreases as one goes down from the lowest point for which $R_{AB} = R_{BC}$ into either "valley," that is, as the H_3 molecule splits into H_2 + H. This is consistent with the observation that H_3 does not exist as a stable species.

We have used the term "symmetry" a number of times in this section. Symmetry is extremely useful in analyzing the properties of polyatomic molecules, and we must say something more about the subject. The symmetry properties of a system are constraints, additional knowledge about a system beyond what we know about the "ordinary" unconstrained system. Consequences of symmetry—the theorems of *group theory*—often provide us with the tools to make complicated problems much easier, making the difference between a problem that can be solved with pencil and paper and a problem requiring a large computer. We shall not develop any of these theorems here; we shall only provide enough illustrations of spatial symmetries to allow us to draw some simple, qualitative (but mathematically correct) inferences from the structures of some symmetrical molecules.

Of the two kinds of symmetry that concern us now, permutational and spatial, that were discussed in Section 7.5, it is clearly spatial symmetry that is relevant here. By the symmetry properties of a given molecular structure we mean the spatial transformations one can perform on the molecule (or, equivalently, on the coordinate system to which it is

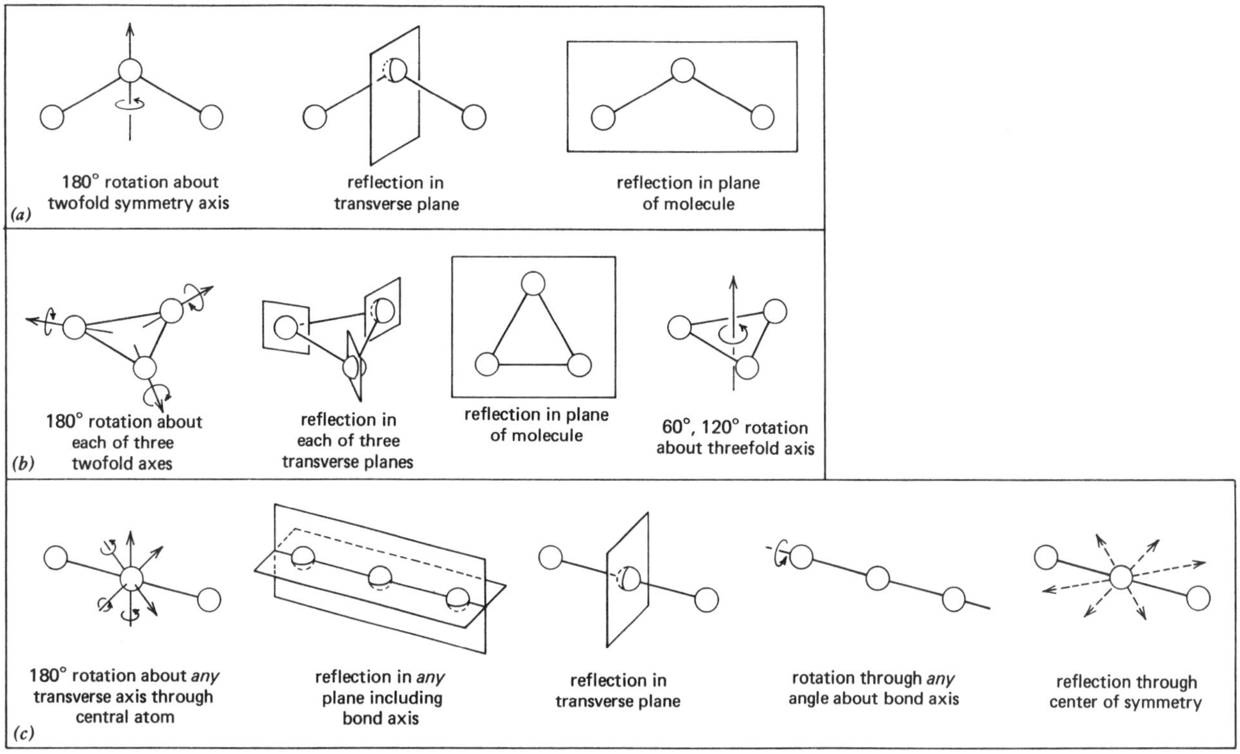

Figure 8.4 Symmetry operations of the H_3 molecule for various structures. (*a*) isosceles triangle; (*b*) equilateral triangle; (*c*) linear.

referred) that leave the molecule in a condition physically indistinguishable from its initial state. Let us see what these properties are for an H_3 molecule.

Suppose that we restrict ourselves to the case that $R_{AB} = R_{BC}$. There are three possible H_3 structures with at least two equivalent hydrogen atoms: isosceles triangle, equilateral triangle, and linear. We have listed these in what we say is the order of increasing symmetry, on the following basis. The isosceles triangle, as shown in Fig. 8.4*a,* could be transformed in three ways without our being able to detect that the transformation had been made: Rotation by 180° about an axis through the central H atom lying in the molecular plane, reflection in a plane passing through the same atom, equidistant between the other two and thus perpendicular to the molecular plane, and reflection in the plane of the molecule. In the equilateral triangle, Fig. 8.4*b,* all three atoms are equivalent, so we have a 180° rotation axis and a transverse reflection plane through *each* of them. The plane of the molecule remains a symmetry plane, as it must be in all triatomic molecules; in addition we have a symmetry axis for rotation by $\pm 2\pi/3$—a "threefold rotation axis"—perpendicular to this plane, passing through the center of the triangle. Thus the equilateral triangle has a higher symmetry (more symmetry operations) than the isosceles triangle. But the linear structure, Fig. 8.4*c,* has even greater symmetry, indeed, an infinite number of symmetry operations: *Any* plane including the bond axis is a symmetry plane; *any* axis through the central atom perpendicular to the bond axis is a twofold symmetry axis; and the bond axis itself is an ∞-fold symmetry axis for rotation, in the sense that rotation through *any* angle about it leaves the molecule unchanged. There is also a single symmetry plane perpendicular to the bond axis, and the central atom is a center of symmetry through which the coordinates of all the particles in the molecule can be reflected. In fact, the linear H_3 molecule has all the symmetry properties of a homonuclear diatomic molecule.

How does symmetry allow us to apply constraints to the solution of the quantum mechanical problem? These constraints often sharply reduce the amount of calculation required, and in some cases—those of continuous symmetry, such as rotation about the axis of a linear molecule—allow us to infer constants of the motion, quantum numbers, and even energy-level schemes without having to solve complicated equations explicitly. As we have mentioned several times, the conservation of linear and angular momentum can be interpreted as consequences of the translational and rotational symmetry of physical systems. Because the physically measurable properties of an isolated system are unchanged by any translation or rotation of external coordinates, the momenta conjugate to those coordinates must be conserved. We have seen a similar effect of continuous symmetry in the internal coordinates, with cylindrical symmetry (in diatomic and other linear molecules) leading to conservation of one component of angular momentum. *Point* symmetry, such as that of the equilateral triangle, is not powerful enough to force the constancy of any dynamical quantity, but it does place a limitation on the possible wave functions, which must have forms that keep the physically equivalent sites indistinguishable. For example, we were able to make the inference that the linear configuration of a triatomic molecule A—B—C must correspond to either a maximum or a minimum energy with respect to changes in θ. This deduction is based on the observation that all directions in which the molecular A—B—C angle can bend are equivalent, which is a way of saying that the linear triatomic molecule has symmetry higher than that of the bent triatomic. If our triatomic molecule is homonuclear, the equilateral triangle is also a geometry of high symmetry and therefore its energy must be a maximum or minimum or *saddle point,* a point that is a minimum along one direction and a maximum along a perpendicular direction, like the center of a saddle or the top of a mountain pass. More general inferences concerning the forms of orbitals, of vibrational motions, or of magnetic interactions among atomic nuclei, for example, can be drawn from knowledge (or supposition) of the symmetries of molecules or solids.

Another sort of inference concerns what sorts of wave functions can be coupled by a physical process represented by an operator. If ψ_A and ψ_B are the wave functions of states A and B and R is the operator, then the process corresponding to R can couple states A and B if the corresponding integral $I = \int \psi_A^* \mathsf{R} \psi_B \, d\tau$ does not vanish. Physically, this means that the operation R transforms the state B into something new. But this, in turn, puts a condition on the *integrand,* $\psi_A^* \mathsf{R} \psi_B$: For I to differ from zero, it must not have equivalent positive and negative parts that cancel one another. But in systems of relatively high symmetry, such as an atom or a small molecule, for which atomic or molecular orbitals are natural functions, the integrands $\psi_A^* \mathsf{R} \psi_B$ themselves often do have cancelling parts, such as the equivalent positive and negative regions of the $1\sigma_u$, $1\pi_u$, $2\sigma_u$, and $1\pi_g$ functions of Fig. 6.8. These integrands give vanishing integrals I. When an integral between ψ_A and ψ_B is zero for general reasons of symmetry, one says that a *selection rule* prohibits the coupling of states A and B. By contrast, integrands with forms such as that of the $1\sigma_g$ or $2\sigma_g$ orbitals of Fig. 6.8 give rise to nonzero integrals, in general, because they do not have equivalent cancelling parts. Such integrands are said to be totally symmetric or invariant, meaning that they are unchanged, even in sign, if the coordinate system or the orientation of the molecule is transformed by any of the symmetry operations (rotation, reflection, etc.) that leave the molecule itself indistinguishable from what it was prior to the transformation. If ψ_A and ψ_B give rise to a nonzero integral with the operator R, then we say that the R-type coupling of states A and B is allowed. Specifically, this means that the physical process associated with the operation R transforms state B into some state that has something in common with state A. For example, the electric dipole coupling of vibrational states of a heteronuclear diatomic molecule is allowed for states separated by one vibrational quantum, but is forbidden for the corresponding states of a homonuclear diatomic molecule. The electric dipole operator acts on any state ψ_n of a harmonic oscillator by transforming it into a mixed state composed of ψ_{n+1}, the state above the original, and ψ_{n-1}, the state below; hence one-quantum transitions are allowed for harmonic oscillators provided they have dipole moments—which homonuclear diatomics do not. The coupling of vibrational states separated by two quanta in a heteronuclear diatomic molecule is forbidden in the approximation that the molecule is harmonic, but this rule does not apply for real anharmonic molecules, as Fig. 7.6*b* shows. In Section 8.8 we shall make use of the symmetry properties of the H_2O molecule to determine the form of its molecular orbitals.

8.2 Dihydrides: Introduction to the Water Molecule

Now we consider the next major step in molecular complexity beyond H_3, the triatomic molecules of the form AH_2. These dihydrides are all symmetrical, with the A atom in the center; some are linear and some are bent. Many of the dihydrides are known only as transient, unstable, or at least very reactive species. Among these are BH_2, CH_2, and NH_2, which have been detected only by spectroscopy, mass spectrometry, or the appearance of specific products testifying to their presence during a chemical reaction. The alkaline earth metals form ionic hydrides (CaH_2, etc.) much like those of the alkalis, and some other metals form dihydrides that are stable only in their solid states, but we are not concerned with these here. However, the elements of one family of the periodic table do form stable, covalently bound dihydrides. These are, of course, the well-known series H_2O, H_2S, H_2Se, H_2Te (and H_2Po, which has been prepared in only minute quantities). Our discussion here will concentrate on the water molecule, H_2O, which is both the simplest member of the series and by far the most important to chemistry.

Water, as we all know, is a liquid between 0 and 100°C at atmospheric pressure. The other dihydrides in the series are all evil-smelling, toxic gases under the same conditions. At least one important reason for the differences in the volatility of these species is the tendency of H_2O molecules to form hydrogen bonds with one another. This property is also found in HF (cf. Section 7.9) and in NH_3, but not to the extent observed in water. In solid water (ice) the strength of the hydrogen bonds is sufficient to create a completely ordered array of oxygen atoms: Each oxygen atom is surrounded by a tetrahedron of four other oxygen atoms; one hydrogen atom lies between each pair of oxygens, joined by a normal covalent bond to one and a hydrogen bond to the other (schematically, O—H · · · O). Some but not all forms of ice also have ordered arrays of hydrogen atoms; the others have disorder in the sense that there is no repeating pattern of normal covalent and hydrogen bonds. Much of the order of the ice structures is retained in liquid water. We shall discuss the structure of water in more detail in Chapters 23 and 26.

How is it that H_2S, H_2Se, and H_2Te are so much less likely to form hydrogen bonds? As usual in trends within a family of the periodic table, the effect is basically one of atomic size, both directly and as reflected in electronegativity: As the size of the central atom (A) increases, its nucleus becomes better shielded by the core electrons and the atom becomes less electronegative. Thus from H_2O to H_2Te the H—A bond becomes less ionic, the electron density around the H atom greater, and the H nucleus itself better shielded. Being less like a bare proton than in H_2O, the H atom is less likely to bond to another atom. In addition, with increasing size of A both H—A and H · · · A bonds necessarily become longer and thus weaker.

Let us now turn to the electronic structure of the water molecule. The molecule has 10 electrons, one from each of the hydrogen atoms and eight from the oxygen atom, which has the ground-state electronic configuration $1s^22s^22p^4$. It is thus isoelectronic with the neon atom ($1s^22s^22p^6$) and the hydrogen fluoride molecule. The HF molecule (Section 7.9), although chemically very reactive, is much like neon in its electronic structure. Its configuration is $1\sigma^22\sigma^23\sigma^21\pi^4$, but most of the orbitals are essentially the same as in the fluorine atom. Only the 3σ bonding orbital is primarily a mixture of hydrogen and fluorine orbitals, so that the configuration of HF can be approximated as[2]

$$(1s_F)^2(2s_F)^2(2p\pi_F)^2(2p\sigma_F+1s_H)^2.$$

One might expect the H_2O molecule to have a closely related structure. Specifically, it is reasonable to suppose that two equivalent O—H σ bonds are formed when different $2p$ orbitals of oxygen, say, the $2p_x$ and $2p_y$, combine with the $1s$ orbitals of the two hydrogen atoms (Fig. 8.5). The configuration of H_2O in terms of localized bonding orbitals would then be

$$(1s_O)^2(2s_O)^2(2p_{z,O})^2(2p_{x,O}+1s_H)^2(2p_{y,O}+1s_{H'})^2$$

(by "localized" we mean that each bond is described by an orbital of its own).

The assumption we make in writing this description of H_2O is that the $1s$, $2s$, and $2p_z$ orbitals of oxygen are nonbonding. How reasonable is this? The $1s$ electrons are certainly deep in the core and thus out of the picture. The same

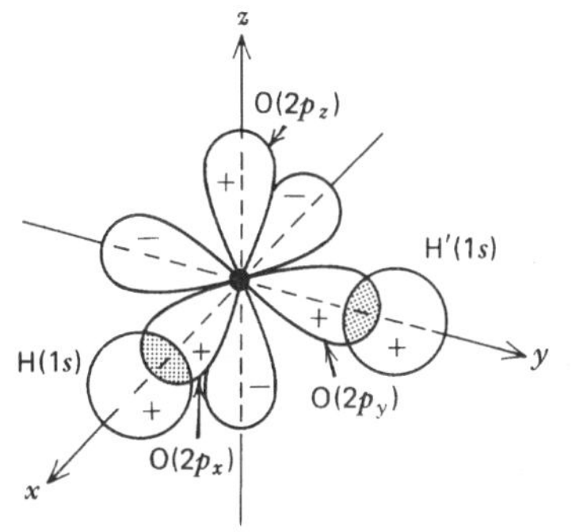

Figure 8.5 Localized-bond picture of H_2O. The two bonds are formed by overlap (indicated by shading) of the oxygen $2p_x$ and $2p_y$ orbitals with one hydrogen $1s$ orbital each. The oxygen $2p_z$ orbital is nonbonding, as are the oxygen $1s$ and $2s$ orbitals (not shown).

is largely true of the $2s$ orbital. The binding energy of an O($2s$) electron is 28.5 eV, compared with 13.6 eV for O($2p$), whereas the O—H bond energy in H_2O is only about 5 eV; thus the $2s$ orbital is too deep to take much part in the bonding. The $2p_z$ orbital, on the other hand, is essentially nonbonding since it has a nodal plane in the plane of the molecule (Fig. 8.5), and thus provides very little charge density in the O—H bonding regions.

The angle between the O—H bonds in this localized-bond model would be expected to be 90°, corresponding to the right angle between any two of the oxygen p orbitals. But spectroscopic measurements that determine the moments of inertia show that the bond angle in H_2O is actually 104.52°. Although this is not far from 90°, it is surely enough to make us ask about the cause of the deviation. One can rationalize the larger angle by considering the interaction between the two bonds. The electrons in one bond repel those in the other, and the two hydrogen nuclei also repel each other; this repulsion should cause the bond angle to be somewhat larger than in the pure localized-bond model. The attraction between the electrons in one bond and the proton of the other bond acts in the opposite direction, but the proximity of the electron clouds makes the repulsion dominate, and the net effect should tend to increase the angle. This model is intuitively pleasing, and can even be quantified with sufficiently elaborate calculations. It also lends itself to the kind of argument that rationalizes experimental observations such as the value of the dipole moment, for example. But the method requires skill and judgment for effective application, and it is not very useful for obtaining more information than one puts in. We shall see that there are better ways to go beyond the localized-bond model.

Although the assumption of simple localized σ bonds is thus not quite adequate for the water molecule, it does a better job of accounting for the structures of the other molecules in the series. As the size of the central atom increases, the bond angle becomes closer to the predicted 90° value.

[2] The notation $2p\sigma_F + 1s_H$ implies only some linear combination of the two orbitals, in general not an equal mixture. A more exact notation for the orbital would be

$$\varphi = c_1(2p\sigma_F) + c_2(1s_H),$$

with the constants c_1, c_2 varied as usual to minimize the energy.

This can again be interpreted as a size effect. The bonds are longer and thus farther apart, so that the interbond repulsion just described becomes less important. The following table illustrates these trends:

Molecule (AH_2)	H_2O	H_2S	H_2Se	H_2Te
A—H bond length (Å)	0.9572	1.33	1.46	1.69
Bond angle (deg)	104.52	92.3	91.0	89.5

8.3 Hybrid Orbitals

To improve the simple localized-bond model, one commonly uses what are called *hybrid orbitals*—atomic orbitals that are mixtures (linear combinations) of the orbitals used in describing isolated atoms. The hybrid atomic orbitals can then be combined in the usual way to construct molecular orbitals, usually still of the localized variety corresponding to individual bonds. We mentioned this method previously in connection with Eq. 7.49; now we shall see how it works.

In a systematic, rigorous theory, it is difficult to find a place for hybridized bond orbitals. The more straightforward course is to construct delocalized molecular orbitals extending over the entire molecule; this is what we did for H_3, and in the next section we shall outline a similar method for H_2O. But in such calculations one loses sight of the intuitively appealing concept of the isolated chemical bond. Like an individual orbital, an individual bond in a polyatomic molecule has no "real" existence in any rigorous sense; yet the concept of the individual bond, with moderately well-defined characteristics of its own, is too useful to chemistry to be abandoned. In particular, experience shows that many molecular properties can be well represented by adding contributions associated with individual bonds, bond energies, for example; cf. Section 14.8. Thus there is still a useful role for a theory involving localized bond orbitals. More advanced methods of this type can be developed as extensions of either the MO or the VB method, with the localized orbitals chosen subject to the constraint that individual orbitals have as little exchange interaction with one another as possible. The construction of hybrid orbitals is in effect a way of approximating and short-cutting these approaches, by making educated guesses as to the atomic orbital contributions to the localized molecular orbitals. In practice this is rather effective, provided that the molecular geometry is known.

For an example of the simplest type of hybrid orbitals, let us consider a linear molecule of the form AH_2, where A is a first-row atom. (There may be no stable example of this species: It used to be thought that the ground state of CH_2 was linear, but CH_2 is now known to be bent. However, we shall see that the orbitals introduced here are useful in more complicated molecules.) We wish to generate two collinear A—H bonding orbitals, each made up of a H(1*s*) orbital and some orbital of the A atom. If we use one of the three A(2*p*) orbitals, the other two have their maximum densities at right

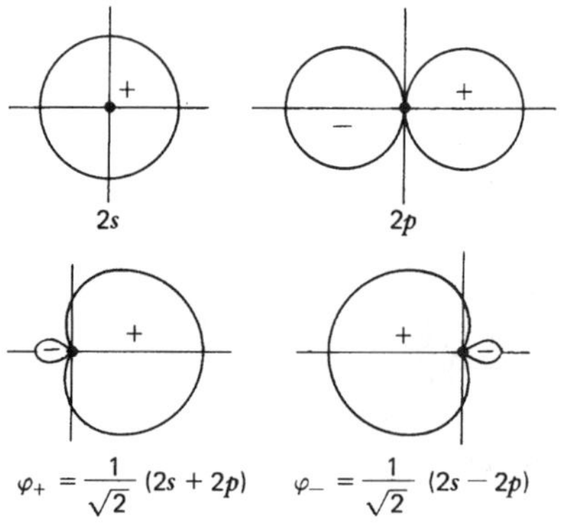

Figure 8.6 Construction of *sp* hybrid orbitals. The diagrams represent the angular parts of the wave functions, shown as in Fig. 4.6.

angles to the internuclear axis and cannot take part in the bonding; thus to make a pair of bonds we must also call upon the A(2*s*) orbital. We could say that the 2*s* orbital forms one bond and the 2*p* orbital the other, but this would be unrealistic. Such bonds would have different properties, and experiment shows that the bonds in any AH_2 molecule are identical. Thus we take the next step and say that each bond contains contributions from both A(2*s*) and A(2*p*), that is, that the A orbitals taking part in the bonds are hybridized. This is not unreasonable if the 2*s* and 2*p* orbitals have fairly similar binding energies.

How do we construct such hybrid orbitals? Although any linear combination of 2*s* and 2*p* might do, let us take for simplicity the case of an equal mixture. The simplest such orbitals that satisfy the requirements of equivalence and collinearity are

$$\varphi_+ = \frac{1}{\sqrt{2}}(2s+2p) \quad \text{and} \quad \varphi_- = \frac{1}{\sqrt{2}}(2s-2p), \qquad (8.4)$$

where "2*s*" and "2*p*" stand for the ordinary atomic orbitals. The two hybridized orbitals are identical except in orientation, as can be seen in Fig. 8.6. They are strongly localized on opposite sides of the atom, with their greatest extensions lying exactly 180° apart. Thus each can combine with a H(1*s*) orbital, to produce two collinear A—H bonds.[3] Since the orbitals φ_+ and φ_- contain equal contributions from *s* and *p* orbitals, they are called *sp hybrids*.

We see here the characteristic directionality that motivates us to use hybrid orbitals. This property makes them especially useful in interpreting stereochemistry. Frequently a particular atom in a molecule forms several bonds symmetrically distributed in space. In such cases one can always define a set of hybrid orbitals projecting in the bond directions; however accurate they may be otherwise, they at least give the correct geometry. To construct such orbitals, one

[3] Or with suitable orbitals of other atoms; $MgCl_2$ is one example of a stable linear AX_2 molecule for which *sp* hybridization should give a good description.

must in general start with as many conventional atomic orbitals as the number of hybrids desired.

Consider next the case of *three* equivalent orbitals. The BH_3 molecule is known as a reactive transient species (detected by mass spectrometry); although its structure has never been determined experimentally, theoretical calculations point to its having a planar, equilateral triangle geometry. We thus wish to construct three equivalent boron orbitals whose axes point to the vertices of an equilateral triangle. We can use the $2s$ orbital and two different $2p$ orbitals, say, $2p_x$ and $2p_y$ (defining the xy plane as the plane of the molecule). The resulting hybrid orbitals are called *trigonal* or sp^2.

It is instructive to go through the construction of trigonal orbitals. To begin with, if the orbitals are equivalent, all three must have the same proportions of s- and p-orbital characteristics. Thus each should have an s–p ratio of 1:2 in the electron density, so that the ratio of s to p amplitudes must be $1:\sqrt{2}$ in the wave function itself. Similarly, since each orbital contains one-third of the total s-orbital contribution, the 2s coefficient in each must have an absolute value of $1/\sqrt{3}$ for normalization. Combining this information, we can immediately write one of the hybrid orbitals as

$$\varphi_1 = \sqrt{\frac{1}{3}}(2s) + \sqrt{\frac{2}{3}}(2p_x). \tag{8.5}$$

We are left with the $2p_y$ orbital and the remaining third of the $2p_x$ orbital to make up the p contribution in the still-undefined hybrids φ_2 and φ_3. If each is divided evenly between the two hybrids, the coefficients of $2p_x$ and $2p_y$ should have absolute values of $\sqrt{1/6}$ and $\sqrt{1/2}$, respectively. It remains to determine the signs of these coefficients. If φ_1 points along the positive x axis, then by symmetry φ_2 and φ_3 must both extend toward negative x, one above and the other below the x axis. Thus both $2p_x$ coefficients should be negative, whereas the $2p_y$ coefficients should have opposite signs. We can thus write the remaining hybrids as

$$\varphi_2 = \frac{1}{\sqrt{3}}(2s) - \frac{1}{\sqrt{6}}(2p_x) + \frac{1}{\sqrt{2}}(2p_y) \tag{8.6}$$

and

$$\varphi_3 = \frac{1}{\sqrt{3}}(2s) - \frac{1}{\sqrt{6}}(2p_x) - \frac{1}{\sqrt{2}}(2p_y). \tag{8.7}$$

The trigonal orbitals of Eqs. 8.5–8.7 are illustrated in Fig. 8.7. It can be seen that they are indeed equivalent and symmetrical, with their long axes 120° apart.

The localized-bond representation of the BH_3 molecule is obtained by combining each of the trigonal boron orbitals with a H(1s) orbital, to obtain a bond orbital of the form

$$\psi_i = \alpha(\varphi_{i,\mathrm{B}}) + \beta(1s_{\mathrm{H}}). \tag{8.8}$$

The constants α and β are unequal, simply because the electronegativities of boron and hydrogen are unequal. Just as in heteronuclear diatomic molecules, these localized bonds are somewhat polar. Although BH_3 itself is unstable with respect to B_2H_6 or the elements, the bonding is similar in the boron trihalides and their analogs (BF_3, $AlCl_3$, etc.) and some other compounds: In each case the observed plane-triangular structure can be described in terms of sp^2-hybridized orbitals on the central atom. On the other hand, one must not use plane trigonal orbitals for the ammonia (NH_3) molecule, which is known to be pyramidal rather than planar. In NH_3 we have a total of eight valence electrons, two more than in BH_3, so that four roughly equivalent orbitals are needed to hold them; the system is best described in terms of tetrahedral or sp^3 hybridization, which we shall discuss in the next chapter.

Let us now return to the H_2O molecule and see how hybridization applies in this case. As before, the O(1s) electrons can be ignored, whereas the O(2p_z) electrons remain out of the molecular plane and nonbonding. This leaves us with six electrons, four from the oxygen atom and one from each of the hydrogens, which we wish to assign to three orbitals. Suppose that we assume trigonal hybridization on the oxygen atom, combining the $2s$, $2p_x$, and $2p_y$ orbitals to obtain three equivalent sp^2 orbitals. Then two of these orbitals can combine with H(1s) orbitals to form the O—H bonds, whereas the third contains a nonbonding "lone pair" of oxygen electrons. There is one major difficulty with this interpretation: it again predicts the wrong bond angle. Bonds formed with sp^2 hybrids would be 120° apart, whereas the pure p bonds discussed in the previous section would be 90°

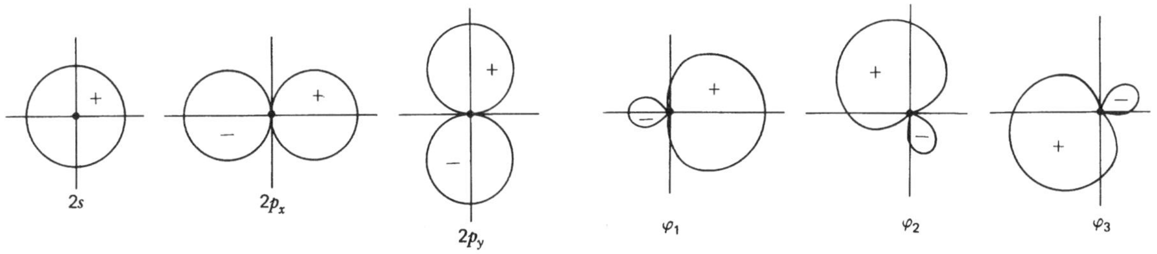

Figure 8.7 Construction of the trigonal (sp^2) hybrid orbitals φ_1, φ_2, φ_3, defined in Eqs. 8.5–8.7. As in Fig. 8.6, only the angular parts of the wave functions are shown.

apart. Since the bond angle is actually 104.5°, the bonding is apparently best described by something midway between the sp^2 and pure-p models.

There is a way to handle this problem within the hybrid orbital model, since we are not restricted to equivalent hybrids. The bond angle obviously increases with the amount of s-orbital character (pure p 90°, sp^2 120°, sp 180°). Suppose, then, that the two bond-forming hybrids are intermediate between pure p and sp^2, say, about 20% s and 80% p. The nonbonding hybrid would then have to be 60% s and 40% p to make the sums come out right; this is reasonable, since we expect the more strongly bound 2s electrons to take a smaller part in the bonding than the 2p electrons. By varying the s–p ratio, one can adjust the bond angle to exactly match the observed value; the percentages just stated happen to be about right. But we have apparently extracted more information from the model only by feeding more information in; the next section will show us a better way to analyze the H_2O molecule.

We have not yet exhausted the varieties of hybridization. Besides the hybrids of s and p orbitals that we have discussed, one can also construct hybrids with contributions from d orbitals. These are useful in analyzing the bonding of the larger atoms, especially the transition metals. With suitable orbital combinations one can reproduce a wide range of molecular geometries. We shall have more to say on this subject in the next chapter.

8.4 Delocalized Orbitals in H_2O: The General MO Method

An alternative method of analyzing the electron distribution in polyatomic molecules involves the use of *delocalized* molecular orbitals extending over the entire molecule (or some large section thereof) rather than orbitals localized in individual bonds. In this model it is assumed that each electron responds to the field of the entire set of nuclei and other electrons; the one-electron orbitals are the solutions of the one-electron Schrödinger equation corresponding to this field. This approach has become increasingly common with the development of large electronic computers. In particular, it has been used in almost all extensive *ab initio* calculations on the electronic structure of H_2O. The first such calculations that included all the occupied orbitals were those of F. O. Ellison and H. Shull, completed in 1955; for simplicity, we shall refer to their results in the following illustrative discussion. Many more recent calculations, using increasing numbers of nominally unoccupied orbitals in the set included in the configuration mixing, have reached an impressive level of agreement with experiment: The O—H bond length found in experiments is approximately 0.9575 Å and the highest-level computations yield 0.9594 Å; the computed vibration frequencies differ from the observed frequencies by ca. 1 cm^{-1}, about one part in 3000; the observed bond angle is 104.5° and the computed value is 104.2°; even the dipole moment, a very sensitive quantity to derive from theory, is given well by theory, 1.847 D from experiment and 1.92 D from the computations; cf. D. R. Yarkony, *Modern Electronic Structure Theory, Part 1* (World Scientific, Singapore, 1995), Chapter 1.

As usual we construct the molecular orbitals by taking linear combinations of a *basis set* of orbitals. For a basis set in the H_2O molecule we could use the fundamental s and p atomic orbitals, the hybrid orbitals described in the last section, or some other combination of localized bond orbitals. The net result would be the same by any of these ways, provided that one carried out full calculations to obtain the best delocalized orbitals possible. Accepting this, we may just as well use the basic atomic orbitals themselves: the 1s, 2s, and three 2p oxygen orbitals directly, and two orbitals (χ_4 and χ_7) representing the sum and difference of the two H(1s) orbitals. These basis orbitals are shown in Fig. 8.8.

We want both the basis set and the eventual molecular orbitals to be *symmetry orbitals,* orbitals that share some of the symmetry properties of the molecule. The basis orbitals in Fig. 8.8 are thus classified into three symmetry classes, called a_1, b_1, and b_2. The labels indicate properties thus: An a orbital remains unchanged (i.e., looks exactly the same) when the molecule is rotated 180° about the y axis, whereas a b orbital changes sign under this rotation. Similarly, an orbital with subscript 1 remains unchanged when the molecule is reflected in the xy plane, whereas one with subscript 2

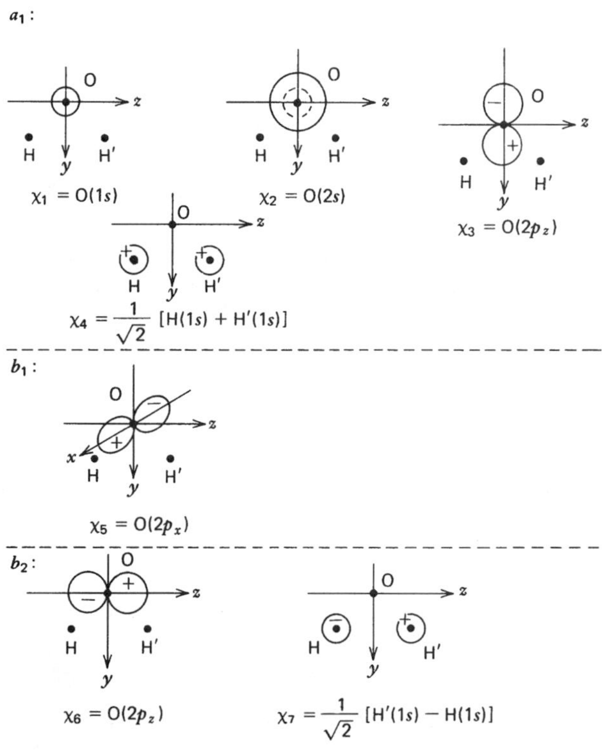

Figure 8.8 Basis set of orbitals for the H_2O molecule, classified by symmetry types (a_1, b_1, b_2). All orbitals except the O(2p_x) are shown in cross section in the plane of the molecule; the O(2p_x) orbital extends perpendicular to the molecular plane.

changes sign in this operation. Since the molecule (and thus $|\psi|$) remains unchanged under these operations, the wave functions must either remain unchanged or reverse their signs. As can be seen in the figure, the *yz* plane is defined as the plane of the molecule, and the *xy* plane is the symmetry plane midway between the two hydrogen atoms; the origin is at the oxygen nucleus, with the hydrogen atoms at $y > 0$.

The significance of this classification is that symmetry orbitals of a given class can mix only with others of the same class to generate *molecular symmetry orbitals* of that class. In Section 8.1, we outlined the geometric basis of this statement, and we shall not attempt to prove it in any more detail. We earlier assumed the validity of this theorem for diatomic molecules, where the symmetry classes are σ_g, π_u, etc. The noncrossing rule is one of its consequences. We shall now see how, in polyatomic molecules, the use of symmetry orbitals makes possible a considerable simplification in our calculations.

The seven independent basis orbitals shown in Fig. 8.8 can be combined to give seven independent molecular orbitals, which we shall designate as φ_i. According to the mixing rule just stated, there must be four orbitals of the symmetry class a_1, each of which can be written as a linear combination of the a_1 basis orbitals,

$$\varphi_i = c_{i1}\chi_1 + c_{i2}\chi_2 + c_{i3}\chi_3 + c_{i4}\chi_4 \quad (i = 1,2,3,4); \tag{8.9}$$

one is of the class b_1, which must be identical to the oxygen $2p_x$ orbital,

$$\varphi_5 = c_{55}\chi_5 \tag{8.10}$$

(although we include c_{55} for consistency, it must equal 1); and two are of the class b_2,

$$\varphi_i = c_{i6}\chi_6 + c_{i7}\chi_7 \quad (i = 6,7). \tag{8.11}$$

In these equations the c_{ij} are constants and the χ_j are the orbitals defined in Fig. 8.8. The computational task is now to determine the best values of the c_{ij} and the corresponding energy eigenvalues. The techniques are the standard ones we have described in our earlier discussions of molecular wave functions, but it is worthwhile to review the process.

First, we must define the molecular (many-electron) wave function. We have seven molecular orbitals and only 10 electrons to assign to them, so we must specify a configuration. From spectroscopy one can deduce that in the ground state of H_2O three a_1 orbitals, one b_1, and one b_2 are occupied by two electrons each. Let us define these occupied orbitals to be the φ_1, φ_2, φ_3, φ_5, and φ_6. To each of these orbitals we assign one electron with spin function $\alpha(m_s = +\frac{1}{2})$ and one with spin function $\beta(m_s = -\frac{1}{2})$. We designate a spin orbital (product of spatial and spin functions) with α spin by φ_i alone, the corresponding spin orbital with β spin by $\overline{\varphi}_j$ (with a superscript bar). The total wave function corresponding to a given assignment of electrons to orbitals could then be a Hartree product like Eq. 5.15, for example,

$$\psi = \varphi_1(1)\overline{\varphi}_1(2)\varphi_2(3)\overline{\varphi}_2(4)\varphi_3(5)\,\overline{\varphi}_3(6)\varphi_5(7)\overline{\varphi}_5(8)\varphi_6(9)\overline{\varphi}_6(10) \tag{8.12}$$

where the numbers in parentheses designate electrons: for example, $\overline{\varphi}_2(4)$ is the wave function of electron 4 in spatial orbital 2 with β spin.

But Eq. 8.12 is not suitable for a molecular wave function. As we showed in Section 6.7, the indistinguishability of electrons requires that the total wave function be antisymmetric under interchange of any pair of electrons. We must thus include in the wave function all the possible products like Eq. 8.12. There are 10! (=3,628,000) such products,[4] each of which can be constructed from Eq. 8.12 by a series of permutations (exchanges) of pairs of electrons. To make the total wave function antisymmetric, we must put a *minus* sign before each product obtained by an *odd* number of permutations, a *plus* sign before each product obtained by an *even* number of permutations, and add all the results together. Fortunately for the simplicity of our notation, this unwieldy sum of products is identical to the expansion of the 10×10 determinant $\|\varphi_i(k)\|$, in which the rows (i) are indexed according to the spin orbital and the columns (k) according to the label on the electron. The total molecular wave function thus becomes

$$\psi = (10!)^{-1/2}\,\mathsf{A}\varphi_1(1)\cdots\overline{\varphi}_6(10) = (10!)^{-1/2}\det\varphi_1(1)\cdots\overline{\varphi}_6(10)$$

$$= (10!)^{-1/2}\begin{vmatrix} \varphi_1(1)\varphi_1(2)\cdots\varphi_1(10) \\ \overline{\varphi}_1(1)\overline{\varphi}_1(2)\cdots\overline{\varphi}_1(10) \\ \cdots\cdots\cdots\cdots\cdots \\ \cdots\cdots\cdots\cdots\cdots \\ \overline{\varphi}_6(1)\overline{\varphi}_6(2)\cdots\overline{\varphi}_6(10) \end{vmatrix}, \tag{8.13}$$

where A is the antisymmetrization operator and the factor $(10!)^{-1/2}$ normalizes the wave function.[5] This is of the same form as the general many-electron wave function we wrote as Eq. 6.63, and is called a *Slater determinant.*

Although we thus have a formal expression for the total wave function in terms of the $\varphi_i(k)$, we still do not know the orbitals themselves. As usual, this problem is solved by applying the variation principle. In accord with Eq. 5.11, we wish to find the orbitals that minimize the energy $\int \psi^* \mathsf{H} \psi \, d\tau$, where ψ is given by Eq. 8.13. The integration

[4] Any of the 10 electrons can be assigned to φ_1; for each of these choices any of the remaining nine can be assigned to $\overline{\varphi}_1$; and so forth, giving a total number of assignments $10! \equiv 10 \cdot 9 \cdot 8 \cdot \cdots \cdot 2 \cdot 1$.

[5] If all the $\varphi_i(j)$ are orthonormal (as we assume), each product like Eq. 8.12 would contribute unity to $\int |\psi|^2 \, d\tau$ since there are 10! such products in our total wave function, we must divide $|\psi|^2$ by 10!, ψ by $(10!)^{1/2}$, to obtain $\int |\psi|^2 \, d\tau = 1$.

is carried out over all 10 sets of electron coordinates, with the electronic Hamiltonian

$$\mathsf{H} = \sum_{k=1}^{10} \frac{\mathsf{p}_k^2}{2m_e} - \sum_{K=1}^{3}\sum_{k=1}^{10} \frac{Z_K e^2}{4\pi\epsilon_0 r_{Kk}} + \frac{1}{2}\sum_{k=1}^{10}\sum_{l\neq k} \frac{e^2}{4\pi\epsilon_0 r_{kl}}, \qquad (8.14)$$

where the index K is used for nuclei and the indices k, l for electrons. (We assume the Born–Oppenheimer approximation, with the nuclei fixed in some chosen geometry.)

When the wave function is of the form of Eq. 8.13, one can show that the orbitals that minimize the total energy are the *Hartree–Fock orbitals.* These are defined as the solutions of the equations

$$\mathsf{H}_{\mathrm{HF}}(1)\varphi_i(1) = \epsilon_i \varphi_i(1) \quad (i = 1,2,\ldots,10), \qquad (8.15)$$

where H_{HF}, is the Hartree–Fock effective Hamiltonian defined below, φ_i is a one-electron orbital, ϵ_i is the corresponding energy eigenvalue ("orbital energy of φ_i"), and the argument "1" of H_{HF} and φ_i indicates the coordinates of a single electron. The operator $\mathsf{H}_{\mathrm{HF}}(1)$ is defined by

$$\mathsf{H}_{\mathrm{HF}}(1)\varphi_i(1) = \left\{ \frac{\mathsf{p}_1^2}{2m_e} + \sum_{K=1}^{3} \frac{Z_K e^2}{4\pi\epsilon_0 r_{K1}} + \frac{1}{2} \right.$$
$$\left. \sum_j \left[\int \varphi_j^*(2) \frac{e^2}{4\pi\epsilon_0 r_{12}} \varphi_j(2)\, d\tau_2 \right] \right\} \varphi_i(1)$$
$$-\frac{1}{2}\sum_j \left[\int \varphi_j^*(2) \frac{e^2}{4\pi\epsilon_0 r_{12}} \varphi_i(2)\, d\tau_2 \right] \varphi_j(1) \qquad (8.16)$$

the sums over j are taken over all the occupied orbitals, unbarred (spin α) and barred (spin β). The argument "2" designates the coordinates of a second electron over which the two kinds of integrals in square brackets are taken. The first two terms on the right-hand side of Eq. 8.16 are simply the kinetic energy of electron 1 and its Coulomb interaction with the nuclei. The remaining two sums give, respectively, the Coulomb and exchange interactions of electron 1 with all the other electrons. Each of these sums contains a term, with $j = i$, representing the electron's interaction with itself, but these two terms cancel each other; they are retained to assure that the operator H_{HF} is identical for all the orbitals, which are thus all part of the same set of solutions to Eq. 8.15. Note that the jth exchange term differs from the jth Coulomb term precisely in the exchange of places and arguments between φ_j and φ_i. Also note that if φ_i and φ_j correspond to different spin functions, their *exchange* interaction vanishes[6] because $\alpha(1)\beta(1) = 0$. The Coulomb term is just what would appear in a classical treatment of the energy of interaction of two interpenetrating clouds of electric charge whose shapes are given by $|\varphi_i(1)|^2$ and $|\varphi_j(2)|^2$. The exchange term has no classical counterpart; in the formalism we use here, it is a necessary consequence of the condition that the electrons be indistinguishable and *antisymmetric* with respect to exchange.

As we explained in Section 5.4, the Hartree–Fock equations treat the motion of each electron as if it were governed by the average field of all the other electrons and the nuclei, with no correlation between the motions of different electrons. This neglect of correlation results from the replacement of the exact wave function by a product, albeit antisymmetrized, of one-electron orbitals. The correlation effects can be allowed for by configuration interaction, writing the total wave function to include electron configurations (i.e., Slater determinants) in addition to the initial Hartree–Fock configuration: cf. Section 6.9. As mentioned previously, there are many computations of the H_2O molecule that include elaborate configuration interaction. However in the present section, we shall restrict ourselves to the Hartree–Fock model.

In principle, then, all one has to do is solve Eq. 8.15 for the orbitals. This is manageable with high-speed computers, but is still a substantial task. Since the equation for each φ_i contains all the other orbitals (in the Hartree–Fock operator), it can only be done by successive approximation; an exact solution is impossible for systems with more than one electron. What one always does is to expand the orbitals in a set of basis functions; when the basis functions are atomic orbitals, this is the LCAO approximation. The more basis functions one uses, the better is the approximation. For a given set of basis functions, the best molecular orbitals are obtained by the self-consistent-field method we outlined in Section 5.4.

In general, we assume that our molecular orbitals are of the form

$$\varphi_i(1) = \sum_j c_{ij}\chi_j(1), \qquad (8.17)$$

where the sum is taken over all the basis functions χ_j. We have already specified these orbitals for H_2O, Eqs. 8.9–8.11, but let us develop the general case. In the first stage of the calculation, one must guess the values of the c_{ij}, and introduce a trial set of orbitals which we call $\varphi_i^{(0)}$. Suppose now that we define the quantity

$$\epsilon_i^0 = \frac{\int \varphi_i^{0*}(1)\mathsf{H}_{\mathrm{HF}}(1)\varphi_i^0(1)\, d\tau_1}{\int \varphi_i^{0*}(1)\varphi_i^0(1)\, d\tau_1}. \qquad (8.18)$$

If the φ_i were the solutions of the Hartree–Fock equations (Eq. 8.15), the ϵ_i^0 would be the eigenvalue ϵ_i; but when the φ_i are only trial functions, the variation principle tells us that $\epsilon_i^0 \geq \epsilon_i$. If one uses the trial orbitals $\varphi_i^{(0)}$ to define $\mathsf{H}_{\mathrm{HF}}(1)$

[6] This means that the exchange interaction vanishes for the ground states of two-electron systems such as He, H_2, and H_3^+. This, in turn, implies that the total energies of such states can be written as the sum of two orbital energies, something that does not hold for systems of three or more electrons.

by Eq. 8.16, the next approximation $\varphi_i^{(1)}$ can be obtained by varying the φ_i to minimize the ϵ_i^0. Then a new $\mathsf{H}_{\mathrm{HF}}(1)$ is defined in terms of the $\varphi_i^{(1)}$ and the ϵ_i^0 are minimized again to obtain the $\varphi_i^{(2)}$. The process is repeated—iterated—until no further change in the φ_i is observed.

The key step in the above process is the variation of the φ_i—that is, of the c_{ij}—to minimize the energy. How is this done? Let us expand the orbitals in Eq. 8.18 in terms of the basis functions, giving

$$\epsilon_i^0 = \frac{\int \left[\sum_j c_{ij}\chi_j^*(1)\right] \mathsf{H}_{\mathrm{HF}}(1) \left[\sum_k c_{ik}\chi_k(1)\right] d\tau_1}{\int \left[\sum_j c_{ij}\chi_j^*(1)\right] \left[\sum_k c_{ik}\chi_k(1)\right] d\tau_1} = \frac{\sum_j \sum_k c_{ij}c_{ik}H_{jk}}{\sum_j \sum_k c_{ij}c_{ik}S_{jk}}, \tag{8.19}$$

where

$$H_{jk} \equiv \int \chi_j^*(1)\mathsf{H}_{\mathrm{HF}}(1)\chi_k(1)\, d\tau_1 \quad \text{and}$$
$$S_{jk} \equiv \int \chi_j^*(1)\chi_k(1)\, d\tau_1 . \tag{8.20}$$

We assume that all the c_{ij} are real, $c_{ij}^* = c_{ij}$, and that all the χ_j are expressed as functions of electron 1; the indices j and k are used only to distinguish the double summations over a common set of functions. We can rearrange Eq. 8.19 as

$$\sum_j \sum_k c_{ij}c_{ik}(H_{jk} - \epsilon_i^0 S_{jk}) = 0. \tag{8.21}$$

We now vary the c_{ij} to minimize ϵ_i^0. Let us differentiate Eq. 8.21 with respect to *one* of the c_{ij}, say c_{il}, holding all the others constant. We obtain

$$\sum_j c_{ij}H_{jl} + \sum_j c_{ij}H_{lj} - \epsilon_i^0\left(\sum_j c_{ij}S_{jl} + \sum_j c_{ij}S_{lj}\right) - \frac{\partial \epsilon_i^0}{\partial c_{il}}\sum_j\sum_k c_{ij}c_{ik}S_{jk} = 2\sum_j c_{ij}H_{jl} - 2\epsilon_i^0 \sum_j c_{ij}S_{jl} - \frac{\partial \epsilon_i^0}{\partial c_{il}}\sum_j\sum_k c_{ij}c_{ik}S_{jk} = 0, \tag{8.22}$$

where we have used the relationships $H_{jl} = H_{lj}$ (because $\mathsf{H}_{\mathrm{HF}}(1)$ is a Hermitian operator; cf. Appendix 6B) and $S_{jl} = S_{lj}$. If ϵ_i^0 is a minimum with respect to variation of c_{il}, it must satisfy $\partial\epsilon_i^0/\partial c_{il} = 0$. Eq. 8.22 can then be valid only if

$$\sum_j c_{ij}(H_{jl} - \epsilon_i^0 S_{jl}) = 0. \tag{8.23}$$

We obtain an equation like 8.23 for each value of l, that is, for each of the basis functions that make up the orbital φ_i. If there are n basis functions, we have n equations in n unknowns, the c_{ij} (j = 1, 2, . . . , n). It can be shown (Cramer's rule) that such a set of linear homogeneous equations has a nontrivial solution if and only if the determinant of the coefficients vanishes; thus, we must have

$$\left\| H_{jl} - \epsilon^0 S_{jl} \right\| = 0. \tag{8.24}$$

A determinant of this form is called a *secular determinant.* Note that we have dropped the subscript on ϵ^0: Since all the orbitals are of the same form (8.17), varying the c_{ij} to minimize the energy must lead to the same determinant in every case. Expansion of Eq. 8.24 gives an nth-degree equation in ϵ_i^0, which of course has n roots. These roots are the orbital energies corresponding to our n molecular orbitals φ_i—or rather the approximations to the orbital energies at this stage in the SCF process. Once these ϵ_i^0 are obtained, one substitutes each of them in Eq. 8.23 to get n sets of c_{ij}, then substitutes the c_{ij} in Eq. 8.17 to get the new set of φ_i; given the new φ_i, one recalculates $\mathsf{H}_{\mathrm{HF}}(1)$ and the H_{jl}, solves Eq. 8.24 over again, and so forth.

In the preceding paragraphs our language has been completely general, applying to any molecular orbitals of the form of Eq. 8.17. Now let us make a number of simplifications. First, it is convenient to choose the basis functions to be normalized and orthogonal (cf. Appendix 6A). This means that $S_{jk} = \delta_{jk}$, where δ_{jk} is a Kronecker delta. The diagonal elements in the secular determinant become $H_{jl} - \epsilon^0$, and the off-diagonal elements are simply H_{jl}. Next we assume that the basis functions (and the molecular orbitals constructed from them) are symmetry orbitals. It can be shown by direct integration or group theory that the integrals H_{jk}, defined in Eq. 8.20, must vanish whenever χ_j and χ_k are of different symmetry types. As a result, the only nonzero terms in the secular determinant appear in blocks, one for each symmetry type. For the H_2O molecule, with the basis functions defined in Fig. 8.8, these simplifications reduce Eq. 8.24 to[7]

[7] There may be some confusion as to how we got from the 10×10 determinant (Eq. 8.13) to the 7×7 determinant (Eq. 8.25). The orders of the two determinants are actually quite independent. If we choose to form our molecular orbitals from n basis functions, we can have n independent linear combinations of the form of Eq. 8.17, leading to an $n \times n$ secular determinant (Eq. 8.24); n can be as large as one wishes (and can handle), as long as it is greater than the number of actually occupied spatial orbitals. Here we have $n = 7$, giving seven MOs, of which only five are occupied in the ground state of H_2O. The molecular wave function for a given configuration, Eq. 8.13, is formed from only the occupied orbitals—in this case the 10 electrons give us five doubly occupied spatial orbitals and thus 10 spin orbitals, leading to a 10×10 determinant.

$$\begin{Vmatrix} H_{11}-\epsilon^0 & H_{12} & H_{13} & H_{14} & 0 & 0 & 0 \\ H_{21} & H_{22}-\epsilon^0 & H_{23} & H_{24} & 0 & 0 & 0 \\ H_{31} & H_{32} & H_{33}-\epsilon^0 & H_{34} & 0 & 0 & 0 \\ H_{41} & H_{42} & H_{43} & H_{44}-\epsilon^0 & 0 & 0 & 0 \\ 0 & 0 & 0 & 0 & H_{55}-\epsilon^0 & 0 & 0 \\ 0 & 0 & 0 & 0 & 0 & H_{66}-\epsilon^0 & H_{67} \\ 0 & 0 & 0 & 0 & 0 & H_{76} & H_{77}-\epsilon^0 \end{Vmatrix}$$

$$= \begin{Vmatrix} H_{11}-\epsilon^0 & H_{12} & H_{13} & H_{14} \\ H_{21} & H_{22}-\epsilon^0 & H_{23} & H_{24} \\ H_{31} & H_{32} & H_{33}-\epsilon^0 & H_{34} \\ H_{41} & H_{42} & H_{43} & H_{44}-\epsilon^0 \end{Vmatrix} (H_{55}-\epsilon^0) \begin{Vmatrix} H_{66}-\epsilon^0 & H_{67} \\ H_{76} & H_{77}-\epsilon^0 \end{Vmatrix} = 0. \tag{8.25}$$

Expansion of the original determinant in minors thus enables us to factor it into three smaller determinants, each of which can separately be set equal to zero. Thus, instead of having to solve a seventh-degree equation in ϵ^0, one need solve only one linear equation, one quadratic, and one quartic. This greatly reduces the work of calculation and illustrates the advantage of using symmetry orbitals. (If we do *not* assume the χ_j orthogonal, then the nonzero terms in Eq. 8.25 are all of the form $H_{jl} - \epsilon S_{jl}$, but the determinant still factors.)

What we have outlined here is the molecular orbital method used in the great majority of molecular calculations. One sets up the MOs, evaluates the integrals, constructs the secular determinant, solves the latter for the orbital energies, and repeats the process as many times as needed to yield self-consistency. In practice, not only the c_{jk} but constants within the χ_j themselves may be varied. (It should now be obvious why large computers are needed for calculations on many-electron systems!) Once the SCF orbitals and energies are obtained, one applies the Aufbau principle, putting the available electrons into the orbitals in order of increasing energy; in the H_2O molecule the 10 electrons fill the lowest five of our seven spatial orbitals. The molecular wave function is then given by Eq. 8.13 or its equivalent, using only the occupied spin orbitals. The final step is to determine the total molecular energy. This cannot be taken as simply the sum of the orbital energies, since we would then be counting each of the electron–electron interactions twice. When the orbitals are orthogonal, the total energy is actually

$$E = 2\sum_i \epsilon_i + \sum_i \sum_{j>i} (2J_{ij} - K_{ij}), \tag{8.26}$$

where ϵ_i is the orbital energy,

$$J_{ij} \equiv \iint \varphi_i^*(1)\varphi_j^*(2)\frac{e^2}{4\pi\epsilon_0 r_{12}}\varphi_i(1)\varphi_j(2)\, d\tau_1\, d\tau_2, \tag{8.27}$$

and

$$K_{ij} \equiv \iint \varphi_i^*(1)\varphi_j^*(2)\frac{e^2}{4\pi\epsilon_0 r_{12}}\varphi_i(2)\varphi_j(1)\, d\tau_1\, d\tau_2; \tag{8.28}$$

J_{ij} and K_{ij} are referred to as the two-electron Coulomb and exchange integrals, respectively. The sums in Eq. 8.26 are taken over all the occupied *spatial* orbitals, with the factors of 2 appearing because of the double occupancy of these orbitals; K_{ij} remains undoubted, because the exchange interaction takes place only between electrons of the same spin.

The results of Ellison and Shull's SCF LCAO calculation on the H_2O molecule are shown in Table 8.1, for a bond angle of 105° (less than 1° from the experimental value). The basis orbitals are labeled as in Fig. 8.8; the molecular orbitals are given their spectroscopic designations, with the notation of Eqs. 8.9–8.11 in parentheses. The orbital energies are in the order $1a_1 < 2a_1 < 1b_2 < 3a_1 < 1b_1 < 4a_1 < 2b_2$, with the first five occupied by two electrons each; this agrees with the spectroscopic results we mentioned earlier. Of the occupied orbitals, the $1a_1$ is essentially the O(1s); the $2a_1$ is predominantly O(2s) with a bit of O(2p_y) and H(1s) + H′(1s); the $3a_1$, which is strongly bonding, is mainly O(2p_y) with large admixtures of O(2s) and H(1s) + H′(1s); the $1b_2$, also strongly bonding, is a mixture of O(2p_z) and H(1s) − H′(1s) with the latter predominating; and the $1b_1$, is the nonbonding O(2p_x) orbital. Ellison and Shull's calculations were not accurate enough to give the equilibrium bond angle in H_2O, since the total energy varies very slightly with angle; they made calculations at several angles from 90° to 180°, and found the energy minimum to be at somewhat

Table 8.1 Molecular Orbitals for the Ground State of the H_2O Molecule with R(O—H) = 0.9581 Å, ∢ H—O—H = 105°

	COEFFICIENTS IN $\varphi_i = \Sigma_i C_{ij}\chi_i$							
Molecular Orbital	$\varphi_1(1s)$	$\varphi_2(2s)$	$\varphi_3(2p_y)$	$\varphi_4(\mathrm{H+H'})$	$\varphi_5(2p_x)$	$\varphi_6(2p_z)$	$\varphi_7(\mathrm{H'-H})$	**Orbital Energy ϵ_i (eV)**
$1a_1(\varphi_1)$	1.0002	0.0163	0.0024	−0.0033	0	0	0	−557.27
$2a_1(\varphi_2)$	−0.0286	0.8450	0.1328	0.1781	0	0	0	−36.19
$3a_1(\varphi_3)$	−0.0258	−0.4601	0.8277	0.3441	0	0	0	−13.20
$4a_1(\varphi_4)$	−0.086	−0.833	−0.642	1.061	0	0	0	13.7
$1b_1(\varphi_5)$	0	0	0	0	1	0	0	−11.79
$1b_2(\varphi_6)$	0	0	0	0	0	−0.5428	0.7759	−18.55
$2b_2(\varphi_7)$	0	0	0	0	0	−1.013	1.230	15.9

[a] From F. O. Ellison and H. Shull, *J. Chem. Phys.* **23,** 2348 (1955).

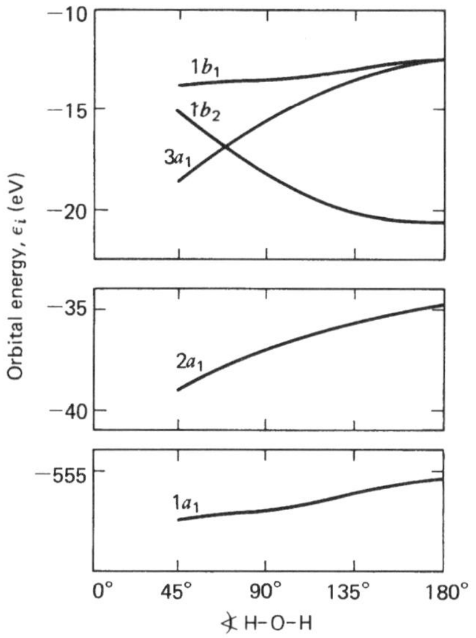

Figure 8.9 Energies of the occupied orbitals of H_2O as functions of bond angle.

over 120°. As we pointed out, more recent calculations have improved the results considerably. Figure 8.9 gives a Walsh diagram based on one such calculation, which used a total of 22 basis functions; this calculation gave 108.4° for the bond angle, not as close to experiment as the SCF result or as those from later, more elaborate computations. The reason is that values of observables are determined, in such calculations, by the balances of many countervailing effects, and different levels of elaboration inevitably distribute emphases differently among these effects. Hence only when the computation has come quite close to the exact solution can we hope that it will yield values of observables that are uniformly close to those we find from experiments.

The variation of orbital energies with bond angle can be predicted approximately by a simple qualitative argument due to A. D. Walsh. The energies of the $1a_1$, $2a_1$, and $1b_1$ orbitals (like the orbitals themselves) are rather insensitive to the bond angle. The $3a_1$ and $1b_2$ energies, on the other hand, are strongly angle-dependent and chiefly govern the variation of the total energy. If the molecule were linear, the H + H′ basis orbital would overlap as much with the negative as with the positive lobe of the $O(2p_z)$. Thus these orbitals could not mix at all in forming the $3a_1$ MO, the energy of which would therefore be quite high at 180°. As the bond angle decreases from 180°, we expect the $3a_1$ energy to decrease monotonically as the mixing of H + H′ and $O(2p_z)$ increases. Conversely, we expect H′—H and $O(2p_y)$ to exhibit a maximum of mixing at 180°, so that the $1b_2$ energy should increase with decreasing bond angle. It can be seen that Fig. 8.9 bears out these predictions approximately, although the $2a_1$ orbital energy in particular seems to vary with angle much more than Walsh expected. However, Walsh's method seems to work better for the total energy than for individual orbitals.

Similar reasoning can be applied to other dihydrides. In any AH_2 molecule in which the bonding involves only *s* and *p* orbitals of the A atom, the valence-shell MOs should resemble the $2a_1$ and higher orbitals of H_2O. Thus a dihydride with only four valence electrons, such as BeH_2, has the $3a_1$ orbital empty; the $1b_2$ orbital dominates the angular dependence of the energy, and the total energy has its minimum at 180°—in other words, BeH_2 should be linear. On the other hand, in all dihydrides with five or more valence electrons (BH_2, CH_2, NH_2, H_2S, etc.), the $3a_1$ orbital has a sufficiently great effect to shift the minimum energy to some lesser angle—these molecules should all be bent. These conclusions, known as Walsh's rules, are at best semiempirical, but they seem to work in virtually all cases; they can also be extended to predict the shapes of molecules in excited states.

One can also consider the variation of orbital energies with bond length, the only parameter we had to worry about in diatomic molecules. For H_2O or any other AX_2 molecule, one natural way to do this is to assume a fixed bond angle and hold the two bond lengths equal while varying their magnitude; keeping the lengths equal ensures that the orbitals retain their symmetry properties. One can readily obtain correlation diagrams similar to those we drew for diatomics. In dihydrides the heavy-atom core orbitals—the $1a_1$, and $2a_1$ in H_2O, corresponding to the oxygen $1s$ and $2s$ in the separated-atom limit—of course remain at the bottom of the energy scale for all values of *R;* as in diatomics, their energies decrease as they approach the united-atom limit (because of the higher *Z*). The $3a_1$, $1b_1$, and $1b_2$ orbitals have energies fairly close together at all *R*, eventually converging to become the triply degenerate $2p$ level of the united atom, but the energies of the bonding $3a_1$ and $1b_2$ orbitals drop more rapidly with decreasing *R* than does that of the nonbonding $1b_1$, showing the effect of increased bond formation on the energy. The first excited orbital, the $4a_1$, is promoted to the relatively high-energy united-atom $3s$ level and is thus antibonding. All this is illustrated in the correlation diagram of Fig. 8.10, which is generally applicable to dihydrides.

To close this section, let us see how the orbital energies calculated for H_2O compare with those determined experimentally. Photoelectron spectroscopy (Section 7.10), with either ultraviolet or x-ray radiation, is the most straightforward technique for making such measurements. In Table 8.2

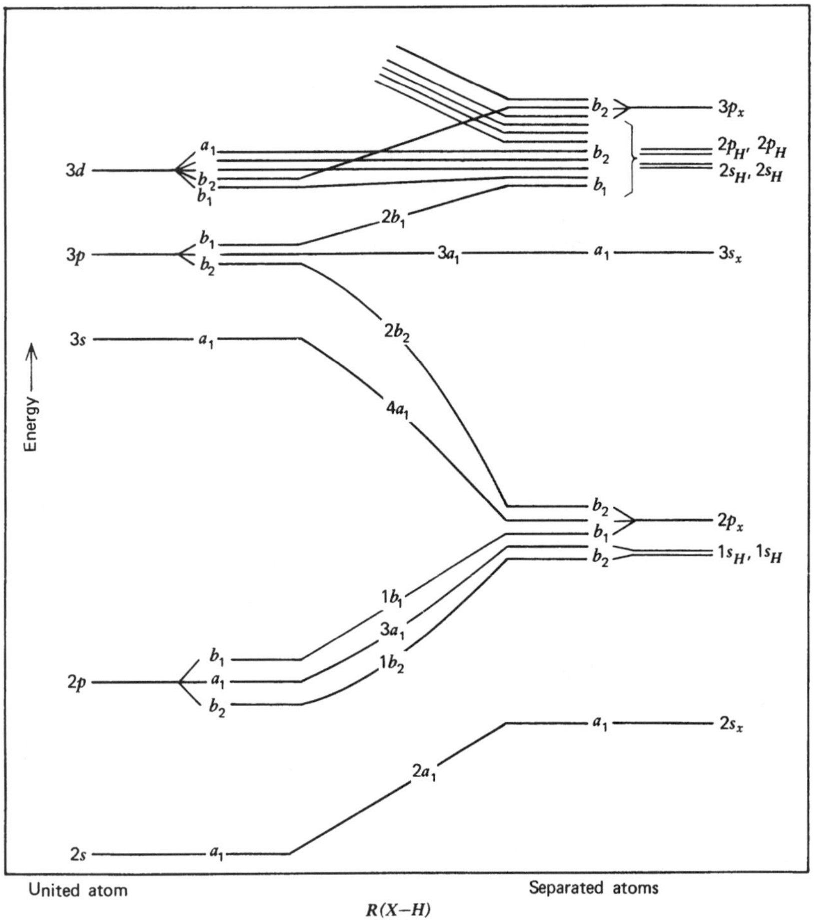

Figure 8.10 Correlation diagram for bent XH_2 molecules with equal X—H bond lengths. The $1a_1$ orbital (united-atom $1s$, separated-atom $1s_x$) has a very low energy and is omitted. It is assumed that the X($2p$) electron has a smaller ionization energy than the H($1s$) electron, and so on. From G. Herzberg, *Electronic Spectra and Electronic Structure of Polyatomic Molecules* (Van Nostrand Reinhold, New York, 1966).

Table 8.2 Observed and Calculated Binding Energies of H_2O Orbitals

	MEASURED ENERGY, ϵ_i (eV)		CALCULATED ENERGY, ϵ_i (eV)	
Orbital	**X-ray**	**Ultraviolet**	**Ellison and Shull**	**Neumann and Moskowitz**
$1a_1$	–539.7	Insufficient energy	–557.27	–559.40
$2a_1$	–32.2	Insufficient energy	–36.19	–36.79
$1b_2$	–18.4	–18.55	–18.55	–19.56
$3a_1$	–14.7	–14.73	–13.20	–15.84
$1b_1$	–12.6	–12.61	–11.79	–13.80
Total molecular energy	–2080.6		–2064.3	–2069.54
				("Best," 1995: 2077.7)

the measured binding energies of the occupied orbitals of H_2O are compared with two sets of theoretical values, those of Ellison and Shull and a more recent calculation by Neumann and Moskowitz,[8] both for approximately the experimental geometry.

8.5 Bonding in More Complex Triatomic Molecules

The electronic structures of more complex triatomic molecules can be analyzed in ways analogous to those we have described for H_2O. The calculations are more demanding and extensive, but the principles remain the same. In this section we shall merely give a rapid survey of such molecules, with comments on the nature of the bonding. The most interesting question to be considered here is why some triatomic molecules are linear and others bent.

Next in simplicity after the dihydrides are the triatomic monohydrides. Those that are known, like most dihydrides, are largely unstable or highly reactive transient species such as HNO or HCO (which are found in flames).[9] One very important exception to this generalization is the molecule HCN, hydrogen cyanide. This substance is a stable, volatile liquid with an almondlike odor; its moderately high toxicity is well known. As in H_2O, HF, and so on, the anomalously high boiling point (25.6°C) is due to hydrogen bonding. In aqueous solution HCN behaves as a weak acid, again like HF. In fact, the tightly bound CN (cyanide) group acts in many ways like a halogen atom; its electron affinity of 3.8 eV is higher than those of the halogens (Fig. 5.7), so CN^- ions are readily formed. Cyanide salts such as NaCN are quite similar to the corresponding halides, and cyanogen halides such as ClCN resemble interhalogen compounds.

The isolated HCN molecule bears a close relationship to the isoelectronic N_2 molecule. The molecule is linear, corresponding to *sp* hybridization on the carbon atom (Section 8.3). That is, one can interpret the C—H bond as a localized σ orbital composed of the H(1*s*) orbital and a mixture of the carbon 2*s* and $2p\sigma$ orbitals, with the latter orbitals also combining with N($2p\sigma$) to form the C—O σ bond; the remaining C and N 2*p* orbitals form two π bonds as in N_2. Delocalized MO calculations give an electron distribution not far from this description. The excited states are also similar to those of N_2; the absorption spectra for excitation in both molecules lie in the vacuum ultraviolet, in similar patterns. The known excited states of HCN all have bent structures.

There are many stable triatomic species containing no hydrogen. The most common (and most important to life) is surely carbon dioxide, a relatively unreactive substance; CO_2 molecules are linear, with the shape O=C=O. Other well-known stable triatomic molecules include those of ozone (O_3), sulfur dioxide (SO_2), carbonyl sulfide (OCS), carbon disulfide (CS_2), nitrous oxide (N_2O), and nitrogen dioxide (NO_2). All of these are gases under ordinary conditions, as is usual for small molecules without hydrogen bonding. Their structures follow patterns consistent with the similarity of elements in the same family of the periodic table.[10] Thus O_3, SO_2, and S_2O all have bent structures, with bond angles of 116.8°, 119.5°, and 118.0°, respectively; and CO_2, OCS, CS_2, OCSe, and so on, are all linear. The N_2O molecule, isoelectronic with CO_2, is also linear, but NO_2 has a bond angle of 134.1°.

To interpret the relationship between geometry and electronic structure in these molecules, let us consider only the three examples CO_2, NO_2, and O_3. We can neglect the 1*s* and 2*s* atomic electrons, which in each case give rise to six molecular orbitals with relatively little net bonding power (four σ_g and two σ_u, in the linear geometry, four a_1 and two b_2 in the bent geometry). The orbitals formed from the atomic 2*p* orbitals offer a richer and structurally more important picture. In Fig. 8.11 we show one set of symmetry orbitals that can be formed from atomic *s* and *p* orbitals in bent triatomic molecules; the best MOs, as indicated, would be mixtures of these (but for full accuracy within the Hartree–Fock approach, augmented with additional, normally empty orbitals).

Consider first the 2*p*-derived orbitals in the linear geometry. Those formed from $2p_z$ ($2p\sigma$) orbitals become σ MOs, including: A bonding σ_g orbital with large overlap along both bond axes; a σ_g orbital that is normally nonbonding because its maxima occur near the end atoms; and an antibonding σ_u orbital with nodes crossing each bond. The atomic $2p_x$ and $2p_y$ orbitals similarly each give rise to a strongly bonding π_u orbital, a nonbonding π_g orbital, and an antibonding π_u orbital; in the linear case the $2p_x$ and $2p_y$ sets are degenerate with each other. Now consider the molecules CO_2, NO_2, and O_3 in turn in linear form. In CO_2 we have four 2*p* electrons from each of the two oxygens and two from the carbon, a total of 10. Let us apply the Aufbau principle. Six of the 10 electrons occupy the bonding σ_g and two π_u orbitals, leaving four to go into the nonbonding σ_g and two π_g orbitals. In the ground state the last four electrons in fact occupy the π_g orbitals, consistent with the idea that these are predominantly atomic electrons localized on the oxygen atoms (Lewis formula) :Ö=C=Ö: . In a linear NO_2 molecule we would presumably have the same configuration with one more electron, which would go into the nonbonding σ_g orbital; and in linear O_3 this σ_g orbital would be filled by two electrons.

Let us compare these descriptions with the corresponding electronic structures for molecules with bent geometries. As Fig. 8.11 indicates, the orbitals that form degenerate π pairs

[8] D. Neumann and J. W. Moskowitz, *J. Chem. Phys.* **49,** 2056 (1968).

[9] The hypohalous acids (HOCl, HOBr, HOI) are well known in aqueous solution, but are so unstable that they cannot be prepared in the pure state.

[10] The geometry (and presumably the bonding) is also little changed when halogen atoms replace hydrogen. One finds the bond angles OF_2 103.2°, OCl_2 110.8°, SCl_2 101°, all similar to H_2O; whereas ClCN, etc., are linear like HCN.

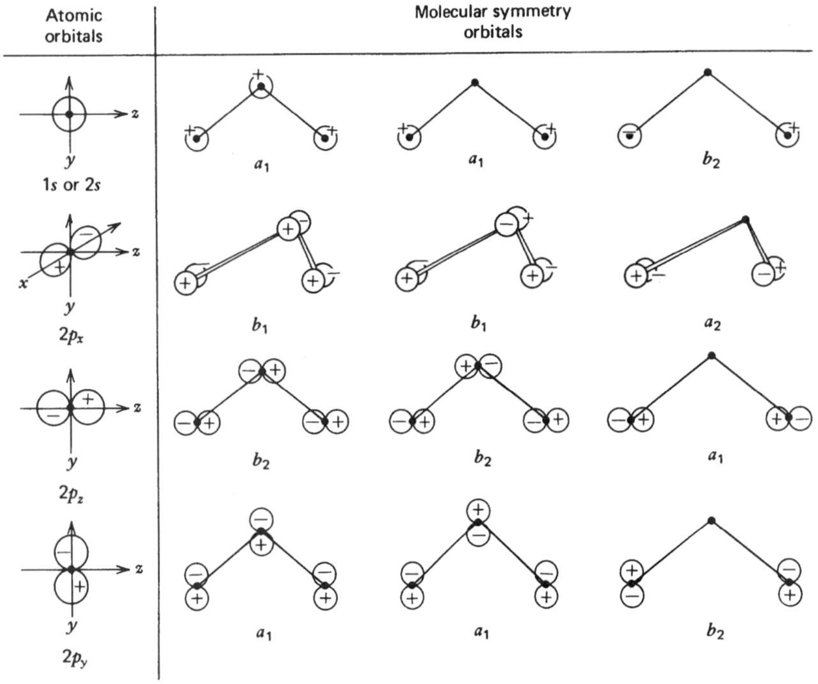

Figure 8.11 Schematic representation of a possible set of molecular symmetry orbitals for first-row bent triatomic molecules. Those shown here would be suitable as basis orbitals; the most realistic MOs of each symmetry type would be a mixture of all those shown here of that type. The coordinate system is the same as that used in Fig. 8.8. In the linear geometry (with all three atoms along the z axis), the symmetries of the orbitals corresponding to those shown are as follows: $1s$, $2s$ – σ_g, σ_g, σ_u; $2p_x$ – π_u, π_u, π_g; $2p_y$ – π_u, π_u, π_g; $2p_z$ – σ_u, σ_g, σ_g.

in the linear molecule split upon bending. For example, the first orbitals shown as formed from $2p_x$ and $2p_y$, atomic orbitals, are physically indistinguishable, and thus degenerate π_u orbitals in the linear molecule. In the bent geometry they are both physically inequivalent and of different symmetry classes, and thus have different energies. The $2p_x - b_1$ orbital continues to show π-like bonding (with a node on the bond axis, and its maximum amplitude perpendicular to the bond axis), and its energy is only weakly affected by bending until the outer atoms begin to overlap (lowering the orbital energy). In contrast, the $2p_y - a_1$ orbital starts to lose bonding character as soon as bending begins, so that its energy increases sharply as the bond angle decreases from 180°. Much the same thing happens with the $2p_z - b_2$ orbital which is the bonding σ_g orbital in the linear geometry. What of the orbitals that are nonbonding in the linear molecule? The $2p_z - a_1$ orbital becomes more bonding and decreases in energy as the outer atoms begin to overlap; so do the originally antibonding $2p_x - b_1$ and $2p_y - a_1$ orbitals; the $2p_x - a_2$ and $2p_y - b_2$ orbitals become antibonding between the outer pair of atoms and thus increase in energy.

This analysis is enough to let us account for the structures of CO_2, NO_2, and O_3. In the linear CO_2 molecule the highest occupied levels (and the only ones very sensitive to angle) are the bonding π_u and nonbonding π_g orbitals. Upon bending these become four distinct orbitals, three of which increase in energy. Thus it is not surprising that the total energy is lowest in the linear geometry. In bent NO_2 the "extra" electron occupies a mixture of the originally nonbonding $2p_z - a_1$ and originally antibonding $2p_y - a_1$ orbitals, both of whose energies decrease sharply with bending. This lowering apparently affects the total energy more than the simultaneous energy increase in the π-like orbitals, since the evidence is clear that NO_2 is bent. In bent O_3 there are *two* electrons in the newly bonding a_1 orbital, so the energy lowering—and the bending at equilibrium—is even more pronounced. (The structure of this bonding a_1 orbital tends to be dominated by the central atom's $2p_y$ contribution, becoming more localized on this atom as its atomic number increases.)

Reasoning such as that we have described here was introduced by R. S. Mulliken and fully developed by A. D. Walsh.[11] It is summarized by a set of "Walsh's rules" like those we discussed for the dihydrides: Molecules of the types AB_2 and BAC should be linear if they have up to 16 valence electrons (CO_2 and N_2O have 16); those with 17 to 20 valence electrons should be bent, with progressively smaller bond angles (17: NO_2, 134°; 18: O_3, 117 °; 19: ClO_2, 117°; 20: OF_2, 103°); and those with 22 valence electrons should again be linear (I_3^-). Similarly, for HAB molecules Walsh obtained the rules: 10 valence electrons, linear (HCN); 11–14, bent (HCO, HNO, HO_2, HOCl); 16, linear (FHF^-).

An alternative way of rationalizing molecular geometries is based on the localized-orbital model. Let us look again at CO_2, NO_2, and O_3. In CO_2 we can say that the $2p$ electrons of the central carbon atom, sp-hybridized with the $2s$ electrons, are used in σ bonding orbitals, with the oxygen atoms supplying electrons for the π orbitals. In NO_2

[11] Walsh's rules for triatomic and larger molecules were derived in a series of articles in *J. Chem. Soc.* (1953), p. 2260ff.

and O_3 the a_1 orbital holding the extra electron(s) is interpreted as an atomic $2p$ orbital on the central atom; we have just noted that this is not far from the truth. Now we apply an empirical rule of thumb: A nonbonding atomic p electron tends to remain localized and repel electrons in adjacent chemical bonds. The effect is even more pronounced with two such electrons, which occupy a localized hybrid orbital and repel adjacent bonds even more strongly than the bonding pairs repel one another. Thus in NO_2 one electron is repulsive enough to reduce the bond angle to 134°; in O_3 the "lone-pair" reduces the angle to less than 120°, the angle one would expect for equivalent sp^2 orbitals. Like the ordinary hybridization model, this concept of lone-pair repulsion is useful for qualitative interpretation but not for exact calculations.

In Table 8.3 we summarize the geometric properties of a number of triatomic molecules, including examples of all the classes we have discussed. Also tabulated are the fundamental vibration frequencies, which we shall take up in the remainder of this chapter.

Table 8.3 Constants of Some Triatomic Molecules[a]

				FUNDAMENTAL VIBRATION FREQUENCIES		
Molecule, A—B—C	R_{AB}(Å)	R_{BC}(Å)	**Bond Angle**	$\tilde{\nu}_1$(cm^{-1})	$\tilde{\nu}_2$(cm^{-1})	$\tilde{\nu}_3$(cm^{-1})
Dihydrides						
H—O—H	0.9572	0.9572	104°31′	3657	1595	3776
H—O—D	0.9572	0.9572	104°31′	2724	1403	3708
D—O—D	0.9572	0.9572	104°31′	2666	1179	2787
H—S—H	1.334	1.334	91°16′	2611	1183	2626
H—Se—H	1.46	1.46	91°0′	2260	1074	2350
Monohydrides						
H—C—O	1.08	1.20	119°30′	2700	1083	1820
H—N—O	1.063	1.212	108°36′	2596	1562	1110
H—O—Cl	0.975	1.689	102°30′	3626	1242	739
H—C—N	1.064	1.156	180°	2096	712	3312
Other Linear Molecules						
O—C—O	1.62	1.162	180°	1388	667	2349
O—C—S	1.164	1.558	180°	859	522	2050
S—C—S	1.554	1.554	180°	655	397	1510
Cl—C—N	1.629	1.163	180°	714	396	2213
N—N—O	1.126	1.191	180°	1285	589	2224
Other Bent Molecules						
O—N—O	1.197	1.197	134°15′	1306	755	1621
O—O—O	1.278	1.278	116°49′	1110	705	1043
O—S—O	1.433	1.433	119°33′	1151	518	1362

[a] The geometric constants are for the ground vibrational state rather than the bottom of the potential well (R_0, not R_e).

8.6 Normal Coordinates and Modes of Vibration

In the remainder of this chapter we shall consider the dynamical behavior of triatomic molecules—that is, the phenomena associated with the motions of the atomic nuclei, especially the molecular vibrations. This behavior is more complex than the simple vibrations of diatomic molecules, but can be thought of in a similar way. The concepts we shall introduce here are generally applicable to larger molecules as well as to triatomics.

We begin as usual by splitting up the overall problem of molecular structure into tractable pieces. Thus we assume that the electronic energy of the molecule can be separated from the energy of nuclear motion (the Born–Oppenheimer approximation), and that the electronic structure and energy can be calculated to any desired degree of accuracy for any assignment of nuclear coordinates along lines such as those we described in the preceding sections. We further assume that the translational, vibrational, and rotational motions of the nuclei are separable. Without these assumptions, the problem can be solved approximately but quite accurately. However the calculations are messy.

In general, an N-atom molecule has $3N$ nuclear coordinates, but we know that not all of these affect the molecular energy. In the diatomic molecule we needed to consider only one coordinate, the internuclear distance R. The reason is that, in the absence of external fields, the internal energy of a free molecule is independent of the location of the molecule's center of mass and of the molecule's orientation in space. It takes three coordinates to locate the center of mass; the orientation of a linear object is defined by two coordinates, and of a nonlinear object, by three coordinates. This is illustrated in Fig. 8.12 with possible sets of such coordinates. Thus the internal energy of the N-atom molecule is a function of $3N - 5$ (if linear) or $3N - 6$ (if nonlinear) coordinates. It is these coordinates, which can be chosen in many ways, that we call the vibrational coordinates of the molecule; we say that the molecule has $3N - 5$ or $3N - 6$ vibrational *degrees of freedom.*

If we were dealing with a very large molecule like a protein (or with a crystal), then N would be so large that the difference between $3N - 6$ and $3N$ could be neglected. On the other hand, in a diatomic molecule (which is of course linear) we have $N = 2$, so $3N - 5 = 1$, the single vibrational coordinate R. Our present concern is with the case $N = 3$, and we find that we have three vibrational coordinates for a nonlinear molecule, four vibrational coordinates for a linear molecule. This is clearly not as simple as the diatomic case. However, we shall see that the vibrations of a triatomic molecule present a quite tractable problem, containing all the conceptual essentials for understanding more complex species.

Let us digress to dispose of the nonvibrational degrees of freedom. The three coordinates of the center of mass of course describe the translational motion of the molecule as

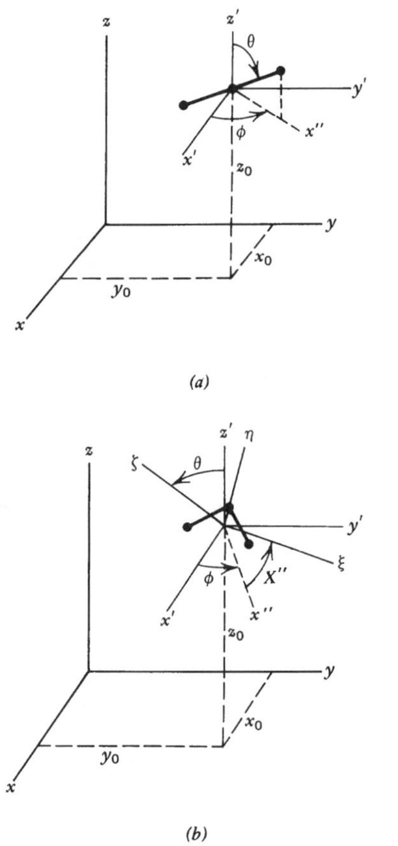

Figure 8.12 Possible sets of coordinates for triatomic molecules. The Cartesian coordinates x_0, y_0, z_0 give the location of the center of mass, and the angles θ, ϕ, χ give the molecule's orientation relative to the Cartesian axes for (*a*) linear molecules, (*b*) nonlinear molecules. In (*a*) the angles θ, ϕ are the standard spherical coordinates giving the direction of the molecule's symmetry axis. In (*b*) the *Eulerian angles* θ, ϕ, χ are obtained as follows: Rotate the coordinate system through an angle ϕ about the z' axis, with the x' axis becoming the x'' axis; then through an angle θ about the x'' axis (the *line of nodes*), with the z' axis becoming the ζ axis; and finally through an angle θ about the χ axis. The new center-of-mass-fixed coordinates ξ, η, ζ thus defined are chosen to fit the symmetry of the molecule, in this case assumed to lie in the $\xi\eta$ plane.

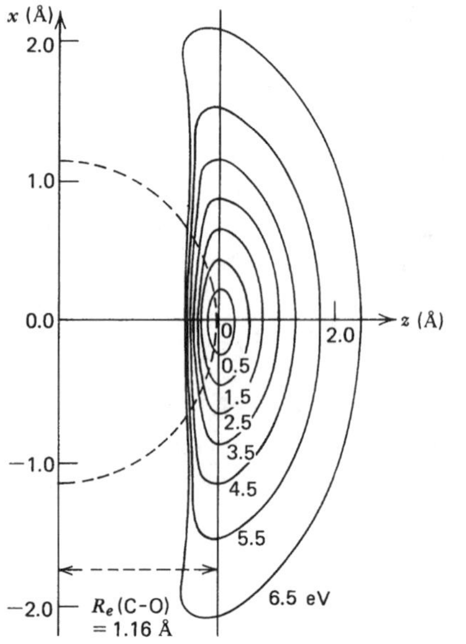

Figure 8.13 Potential energy contours for CO_2, drawn for the special case in which one C—O distance is fixed at its equilibrium value of 1.16 Å. The oxygen on the right is allowed to move relative to the clamped carbon atom at the origin and left-hand oxygen atom. Motion of the oxygen atom along the dashed semicircle would correspond to pure bending of the O—C—O angle, with no change in bond length. Note that the natural motion of the oxygen along the bottom of the potential trough would not follow the dashed semicircle; it would involve a small but significant amount of bond stretching as the O—C—O angle decreases from 180°. From G. Herzberg, *Electronic Spectra and Electronic Structure of Polyatomic Molecules* (Van Nostrand Reinhold, New York, 1966).

a whole, which is again that of a free particle. A rigid linear triatomic molecule is a highly prolate symmetric top with the same rotational properties as a diatomic molecule, and the results of Section 7.2 can be applied with little change; only the definition of I_0 ($\equiv \sum_i m_i z_i^2$, where the z axis is the symmetry axis) is different. We again have the selection rule $\Delta J = \pm 1$, and observe a pure rotational spectrum only when the molecule has a permanent dipole moment, as do HCN **and** OCS, but not CO_2. As for nonlinear triatomic molecules, they are all asymmetric tops (Section 3.12); although many are close to being symmetric tops, the others have very complex sets of rotational energy levels, about which we shall say no more.

In a triatomic molecule, then, the internal energy is a function of three (or four) independent coordinates R_j. By "internal energy" we mean the equivalent of the $E(R)$ of the diatomic molecule, the total electronic energy plus the potential energy of the nuclei. Recall that this energy can be interpreted as the effective potential energy of nuclear motion. Instead of a potential energy curve we now have a *surface* in a space of four (or five) dimensions, with peaks and valleys and saddles displaying the behavior of E as a function of the R_j. We cannot show such a surface in full, but we can give cross sections showing how E varies with one or two coordinates when the others are held constant. Examples of these for CO_2 are shown in Figs. 8.13 and 8.14.

But we are getting ahead of ourselves. The first question we must ask is, what is the best way to select the vibrational coordinates R_j? Out of the infinite number of coordinates we could define, is there one set preferable to all the others? The answer is yes. We want a set of coordinates that are linearly independent so none are redundant, and in terms of which the vibrational motion can be described in the simplest possible way. Such a set exists, and can be found in a relatively straightforward way, provided that we make one assumption about the molecular structure: We must assume that the molecule has a well-defined equilibrium geometry and that the nuclei perform only small oscillations about their equilibrium positions, in a coordinate system fixed with respect to that equilibrium geometry, a so-called molecular frame.

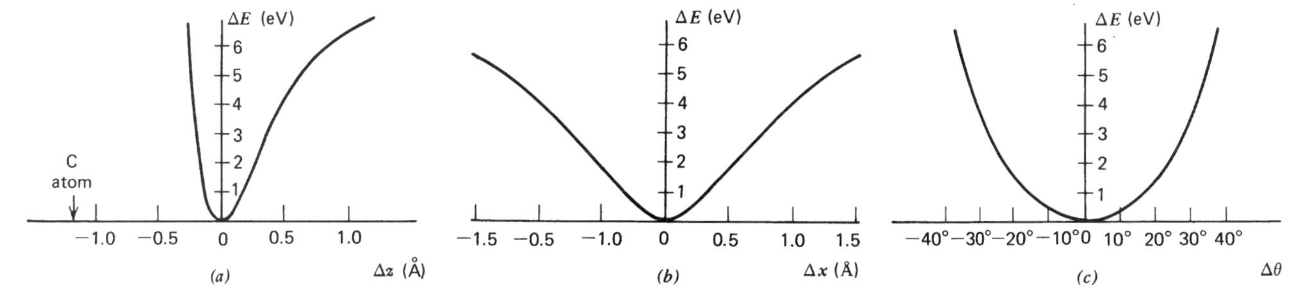

Figure 8.14 Potential energy curves showing how the energy of the CO_2 molecule varies with (*a*) stretching of one C—O bond (variation along *z* in Fig. 8.13); (*b*) displacement of one O atom perpendicular to the symmetry axis (variation of *x* in Fig. 8.13); (*c*) change in the O—C—O bond angle for fixed bond lengths (minimum at 180°).

In the space defined by the vibrational coordinates, the complete potential energy surface for a polyatomic molecule has a many-dimensional well; the bottom of this well is the equilibrium point corresponding to the geometry in which the energy is a minimum. If the potential were one-dimensional (as in a diatomic molecule), we could approximate the energy levels and wave functions by assuming the well to be parabolic. This is the harmonic oscillator model we described in Section 4.2. We extend this model by assuming that an n-dimensional well can be approximated by a set of n noninteracting harmonic oscillators—in other words, that the cross section along each of the coordinates in our "best set" can be approximated by a parabola. The coordinates we call the *normal coordinates* of the molecule are the set for which this approximation is most nearly true. The directions of the normal coordinates in the n-dimensional space are thus the directions along which the bottom of the well is best represented by parabolic curves. The basic problem of describing molecular vibrations is to find that combination of stretches, bends, torsions, and vibrations that defines these directions.

We can state the properties of normal coordinates more precisely in mathematical language. The potential energy V can be expanded as a power series in any complete set of linearly independent coordinates. Generally, for an N-atom molecule with $3N - 6$ vibrational coordinates χ_i, the potential energy can be written as the Taylor series

$$V(X_1, X_2, \ldots, X_{3N-6}) = V_0 + \sum_i \left(\frac{\partial V}{\partial X_i}\right)_{X_{i0}} (X_i - X_{i0}) + \frac{1}{2}\sum_i \sum_j \left(\frac{\partial^2 V}{\partial X_i \partial X_j}\right)_{X_{i0}, X_{j0}} (X_i + -X_{i0})(X_j - X_{j0}) \cdots, \tag{8.29}$$

where the X_{i0} are the coordinates for the equilibrium geometry, and $V_0 = V(X_{i0})$. The linear terms drop out at once, since the equilibrium point—the bottom of the well—is by definition the point where all first derivatives vanish, $(\partial V/\partial X_i)_{X_{i0}} = 0$. The expression becomes even simpler if we choose our coordinates to have their origins at the equilibrium values, $X_{i0} \equiv 0$ (the corresponding procedure for a diatomic molecule would be to define a coordinate $X \equiv R - R_0$), and take V_0 to be the energy zero; Eq. 8.29 then becomes

$$V(X_1, X_2, \ldots, X_{3N-6}) = \frac{1}{2}\sum_i \sum_j \left(\frac{\partial^2 V}{\partial X_i \partial X_j}\right)_0 X_i X_j + \cdots. \tag{8.30}$$

Now we can introduce the harmonic oscillator approximation, which consists of simply dropping all terms in the Taylor series beyond those written explicitly above, that is, beyond the quadratic terms. As in diatomic molecules, the harmonic approximation is useful only near the bottom of the potential well, and anharmonicity corrections must be added for an exact solution. For convenience we define the second derivatives of the potential energy at the equilibrium point as *force constants,*

$$k'_{ij} \equiv \left(\frac{\partial^2 V}{\partial X_i \partial X_j}\right)_{X_i = X_j = 0}; \tag{8.31}$$

the potential energy in the harmonic approximation then has the form

$$V(X_1, X_2, \ldots, X_{3N-6}) = \frac{1}{2}\sum_i \sum_j k'_{ij} X_i X_j. \tag{8.32}$$

What we have said thus far is applicable to any set of nuclear coordinates, but to define the kinetic energy we must be more specific. We may always express the coordinates X_j as linear combinations of the Cartesian coordinates of the nuclei in the molecule-fixed system. This is rigorously valid for coordinates that are internuclear distances, and also valid in the limit of small displacements for any other choice of coordinates, such as bond angles. It is not difficult to show that the vibrational kinetic energy must then have the form

$$T = \frac{1}{2}\sum_i \sum_j m'_{ij} \left(\frac{dX_i}{dt}\right)\left(\frac{dX_j}{dt}\right), \tag{8.33}$$

where the m'_{ij} are constants with dimensions of mass, analogous to the reduced mass we met in treating the diatomic molecule.[12] The Hamiltonian is simply the sum of the kinetic energy operator, Eq. 8.33, and the potential, Eq. 8.32: $\mathsf{H} = T + V$.

One could obtain the Hamiltonian operator by combining Eqs. 8.32 and 8.33 and converting all the X_i to the corresponding operators. But the resulting Schrödinger equation would be very awkward to solve, because of all the cross terms involving two variables. As always, we prefer a situation in which the variables are separable. If there is a set of variables Q_α in terms of which the Hamiltonian can be written as a sum of noninteracting parts,

$$\mathsf{H}(Q_1, Q_2, \ldots, Q_{3N-6}) = \sum_\alpha \mathsf{H}_\alpha(Q_\alpha) \tag{8.34}$$

then the wave function can be written as a product,

$$\psi(Q_1, Q_2, \ldots, Q_{3N-6}) = \prod_\alpha \psi_\alpha(Q_\alpha), \tag{8.35}$$

and the energy as a sum,

$$E \equiv \sum_\alpha \epsilon_\alpha,$$

$$\mathsf{H}_\alpha(Q_\alpha)\psi_\alpha(Q_\alpha) = \epsilon_\alpha \psi_\alpha(Q_\alpha). \tag{8.36}$$

All this is very well, but can it be done? The answer is yes. Given any coordinate system in which Eqs. 8.32 and 8.33 hold, one can always transform to a new coordinate system in which the cross terms vanish and the Hamiltonian is separable. (We shall not prove this general theorem here,[13] but in the next section we show how the transformation is performed in a specific case.)

The new coordinates Q_α thus defined are the normal coordinates of the molecule. When the kinetic and potential energies of vibration are expressed in terms of them, they take the diagonal forms

$$T = \frac{1}{2}\sum_\alpha m_\alpha \left(\frac{dQ_\alpha}{dt}\right)^2 = \sum_\alpha \frac{p_\alpha^2}{2m_\alpha}, \qquad \left(p_\alpha \equiv m_\alpha \frac{dQ_\alpha}{dt}\right) \tag{8.37}$$

and

$$V = \frac{1}{2}\sum_\alpha k_\alpha Q_\alpha^2, \tag{8.38}$$

where the m_α and k_α are new reduced masses and force constants, respectively.[14] The Hamiltonian can then clearly be separated according to Eq. 8.34, with

$$\mathsf{H}_\alpha = \frac{\mathsf{p}_\alpha^2}{2m_\alpha} + \frac{1}{2}k_\alpha Q_\alpha^2 \qquad \left(\mathsf{p}_\alpha \equiv i\hbar \frac{\partial}{\partial Q_\alpha}\right). \tag{8.39}$$

But this is identical in form to the Hamiltonian of the one-dimensional harmonic oscillator, Eq. 4.6. Thus the wave functions $\psi_\alpha(Q_\alpha)$ are all simple harmonic oscillator functions of the sort with which we are familiar, and the energy eigenvalues are given by

$$\epsilon_\alpha = \left(v_\alpha + \tfrac{1}{2}\right)h\nu_\alpha = \left(v_\alpha + \tfrac{1}{2}\right)\hbar\omega_\alpha, \quad (v_\alpha = 0, 1, 2, \ldots), \tag{8.40}$$

where

$$\omega_\alpha = 2\pi\nu_\alpha = \left(\frac{k_\alpha}{m_\alpha}\right)^{1/2}. \tag{8.41}$$

For sufficiently small displacements, then, the vibrations of a polyatomic molecule or any other system of coupled oscillators reduce to those of a set of independent harmonic oscillators, one for each vibrational degree of freedom. Each of these independent oscillations is referred to as a *normal mode of vibration.*

How do normal coordinates look, in terms of atomic displacements? In general, each of the Q_α is a linear combination of the positions of all the nuclei in terms of, say, their Cartesian coordinates. They are thus most effectively represented by diagrams showing the relative motions of the nuclei. In Fig. 8.15 we illustrate the three normal modes of vibration of the SO_2 molecule. In each normal mode the nuclei oscillate in the directions indicated by the arrows, with relative displacements from their equilibrium positions indicated by the lengths of the arrows, and all with the same frequency ν_α. Any actual vibration of the molecule is a superposition (linear combination) of the normal modes. If the molecule's structure has elements of symmetry, the normal coordinates reflect that symmetry. The normal coordinates of an "ABA" molecule such as CO_2, SO_2 or NO_2 include a symmetric stretching mode like the top sketch in Fig. 8.15, in which the two outer atoms move synchronously out and in, and an antisymmetric stretching mode like the

---

[12] In the diatomic molecule we have only one vibrational coordinate, X_1; the obvious choice for this coordinate is

$$X_1 \equiv R - R_0 = |z_A - z_B| - R_0,$$

where the A–B axis is the z axis of the molecule-fixed system. The sums in Eqs. 8.32 and 8.33 then reduce to one term each ($j = k = 1$), with $k_{ij}' = k$, $m_{ij}' = \mu$:

$$V = \tfrac{1}{2}kX_1^2, \qquad T = \tfrac{1}{2}\mu\left(\frac{dX_1}{dt}\right)^2.$$

Since X_1 is already a normal coordinate, no further transformation is necessary.

[13] The transformation involved here is identical to the principal-axis transformation of rigid-body mechanics. Texts on mechanics may be consulted for details.

[14] Note that Eq. 8.38 corresponds to our earlier definition of the normal coordinates as those in which the potential energy well is most nearly parabolic. (Remember that this result holds exactly only in the limit of the small displacement.)

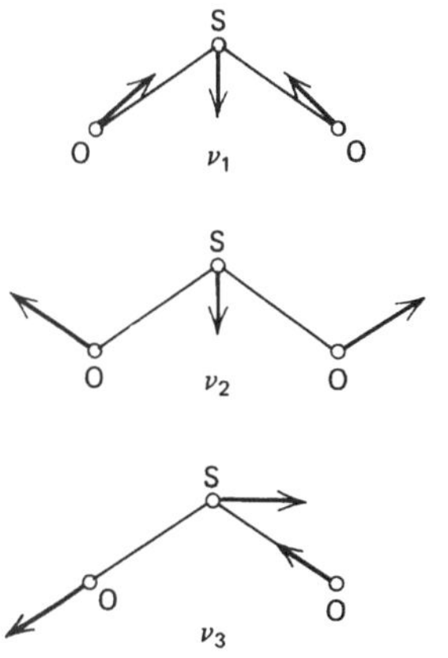

Figure 8.15 Normal modes of vibration of the SO_2 molecule.

bottom sketch in that figure, in which one of the outer atoms moves toward the central atom while the other outer atom moves away. The third mode is primarily a bending motion of the bond angle. In a linear molecule, with four vibrational degrees of freedom, there are two identical bending modes, since there are two independent ways in which the molecule may bend. We can picture these as bending toward either of two perpendicular axes, both perpendicular also to the molecular axis. If the molecule does not have such symmetry—e.g., N_2O—then the normal coordinates do not have such simple schematic interpretations but are completely and unambiguously determined, nonetheless, by the conditions of Eq. 8.35.

Each normal mode motion must conserve not only energy but also linear and angular momentum. Hence, displacements of some atoms in one direction must always be compensated by motion of other atoms in other directions so as to satisfy these conservation laws. These are the constraints that determine the quantitative relationships among the displacements of the atoms.

In general, it is a routine calculation to determine the normal modes of vibration if one knows the potential energy surface. But potential energy surfaces cannot be observed directly. What one measures spectroscopically are the frequencies corresponding to energy-level spacings, which can be analyzed to yield the normal vibration frequencies ν_α. The usual problem is thus to obtain as much of the surface as possible from the normal modes, rather than vice versa. Furthermore, the normal coordinates themselves are not always the most interesting physically; we are frequently concerned more with how the energy varies with particular bond lengths and angles. This means that in first approximation we want to know the "valence" force constants k'_{ij} of Eq. 8.32, which give the curvature of the potential surface for such variations.

But here a major difficulty arises. For a nonlinear molecule there are $3N - 6$ independent vibrational frequencies, but, at first glance, apparently $(3N - 6)^2$ parameters k'_{ij} (in a given set of coordinates). However there is no force constant between an atom and itself, so there are no terms k'_{ii}, and Eq. 8.31 requires that $k'_{ij} = k'_{ji}$. We are left with a total of $\frac{1}{2}(3N-6)(3N-5)$ independent force-constant parameters.[15] Thus in a nonlinear triatomic molecule one must try to determine six parameters from three fundamental vibration frequencies, in a nonlinear four-atom molecule, 21 parameters from six frequencies, and so forth. How, then, can one obtain enough information about the force constants to construct the potential energy surface from experimental data?

The answer to this seemingly unsolvable problem lies in the experimenter's control over the *masses* of the nuclei, through the use of isotopes. One assumes that each potential energy surface depends only on the nuclear charges, and not on the masses. We know that this is a good assumption in the H_2 molecule (Table 7.1), and it works even better for molecules containing heavier atoms. Replacing one isotope with another should not change any of the force constants. But such isotopic substitution does change the reduced mass m_α for any vibrational mode involving motion of the substituted nucleus, and Eq. 8.41 implies that the vibration frequency ν_α must also be changed. For example, replacing one ^{16}O atom in CO_2 by ^{18}O produces a molecule whose three fundamental vibration frequencies are different from those of $^{12}C^{16}O_2$. Each such substitution provides a new set of frequencies,[16] and it is not difficult to obtain enough data to determine all the unknown force constants. Indeed, for small molecules one can often overdetermine the force constants, and thus check the validity of assuming them mass-independent.

It is useful to recognize also that usually not all the $\frac{1}{2}(3N-6)(3N-5)$ independent force constants are of equal importance. We are usually much more concerned with knowing the force constants associated with bonds and bond angles than with the interactions between distant atoms. In SO_2, the force constants for stretching and shrinking the S—O bonds and bending the O—S—O angle usually concern us more than the force constant associated with varying the distance between the oxygen atoms.

[15] One k'_{ii} for each coordinate and one k'_{ij} for each pair of different coordinates; for $3N - 6$ coordinates, this adds up to

$$3N-6+\frac{(3N-6)[(3N-6)-1]}{2}=\frac{(3N-6)(3N-5)}{2}.$$

[16] To illustrate isotopic effects, here are the wave numbers corresponding to the fundamental vibration frequencies for some of the isotopic forms of CO_2:

	$\tilde{\nu}_1$ (cm^{-1})	$\tilde{\nu}_2$ (cm^{-1})	$\tilde{\nu}_3$ (cm^{-1})
$^{12}C^{16}O_2$	1388	667	2349
$^{13}C^{16}O_2$	1370	648	2283
$^{16}O^{12}C^{18}O$	1259	663	2332
$^{16}O^{13}C^{18}O$	1342	643	2266
$^{12}C^{18}O_2$	1230	657	2314

(See also the isotopic forms of H_2O in Table 8.3.)

8.7 A Solvable Example: The Vibrational Modes of CO_2

The CO_2 molecule is one case in which the forms of the normal modes of vibration can be easily deduced from the molecular structure. The treatment here is simplified, but it can be done rigorously. We assume (and know from experimental evidence) that at equilibrium the molecule is linear and symmetric, with both C—O distances equal to R_0 (1.162A). We define the coordinate system shown in Fig. 8.16, with the equilibrium molecular axis taken as the x axis. The *local displacement coordinates* x_i, y_i, z_i for each nucleus are then its Cartesian coordinates measured from its equilibrium position.

We have a total of nine local displacement coordinates, but only four (= $3N - 5$) of these can vary independently. Thus there must exist five relationships among the local displacement coordinates, which we can derive from the momentum conservation laws. With our coordinate system fixed in the molecule, the total linear momentum of the nuclei is not only conserved but equal to zero. Thus we have for the three Cartesian components

$$\sum_{i=1}^{3} p_{xi} = \sum_{i=1}^{3} p_{yi} = \sum_{i=1}^{3} p_{zi} = 0, \tag{8.42}$$

with the sums taken over the three atoms. Similarly, since the coordinate system rotates with the molecule, conservation of angular momentum about the center of mass gives us

$$\sum_{i=1}^{3} L_{xi} = \sum_{i=1}^{3} L_{yi} = \sum_{i=1}^{3} L_{zi} = 0. \tag{8.43}$$

The first of Eq. 8.43 is of no use to us: Since the nuclei remain near the x axis at all times, the individual L_{xi} are negligibly small. But the other five of the above equations can give us useful interrelationships.

Note that the momentum components are defined by $p_{xi} \equiv m_i(dx_i / dt)$, and so on, and that for a harmonic oscillation the velocity dx_i / dt is proportional to the displacement x_i a quarter-cycle later. Thus we can replace Eq. 8.42 by $\sum_i m_i x_i = 0$, and so on, which for the CO_2 molecule becomes

$$\begin{aligned} m_O(x_1 + x_3) + m_C x_2 &= 0, \\ m_O(y_1 + y_3) + m_C y_2 &= 0, \\ m_O(z_1 + z_3) + m_C z_2 &= 0. \end{aligned} \tag{8.44}$$

Similarly, the only significant angular momentum components are those that can be approximated for small displacements as $L_{yi} = \pm m_i R_0(dz_i / dt)$ and $L_{zi} = \pm m_i R_0(dy_i / dt)$ (for the end atoms only). Making the same substitution as above, we convert the second and third of Eq. 8.43 to

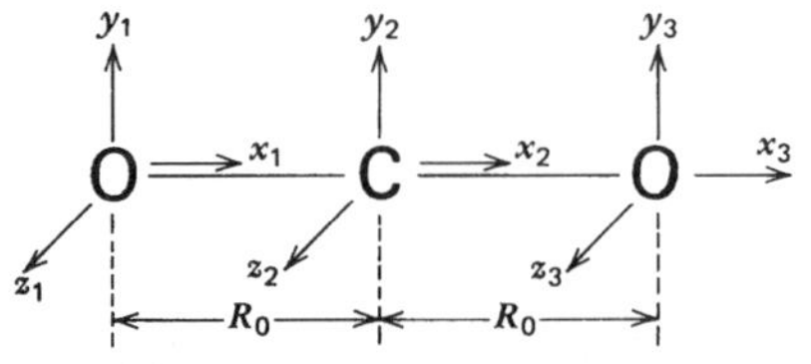

Figure 8.16 Coordinate system for the CO_2 molecule. The local coordinates x_i, y_i, z_i for each nucleus are measured relative to its equilibrium position. The molecular center of mass is at the equilibrium position of the C nucleus.

$$\begin{aligned} m_O R_0 (z_1 - z_3) &= 0, \\ m_O R_0 (y_3 - y_1) &= 0. \end{aligned} \tag{8.45}$$

Equations 8.44 and 8.45 are the five relationships among the local displacement coordinates that we sought. Combining them to solve for the displacements, we obtain unique solutions for the y and z coordinates,

$$y_1 = y_3, \qquad y_2 = -\frac{2m_O}{m_C} y_1, \tag{8.46}$$

and

$$z_1 = z_3, \qquad z_2 = -\frac{2m_O}{m_C} z_1. \tag{8.47}$$

For the x coordinates we have $x_2 = -(m_O/m_C)(x_1 - x_3)$, which yields two solutions consistent with the symmetry of the molecule (since atoms 1 and 3 are indistinguishable):

$$x_1 = -x_3, \qquad x_2 = 0, \tag{8.48}$$

or

$$x_1 = x_3, \qquad x_2 = -\frac{2m_O}{m_C} x_1. \tag{8.49}$$

Note that each of Eqs. 8.46–8.49 involves only a single set of displacement coordinates (x_i, y_i, or z_i). This implies that the x's, y's, and z's can vary independently of one another, and that each of these equations alone can describe a possible vibration of the molecule for small displacements from equilibrium. We can thus identify Eqs. 8.46–8.49 with the four normal modes of vibration of the CO_2 molecule; a detailed calculation would show that they indeed have all the properties of normal vibrations outlined in the previous section. The normal modes of CO_2 are illustrated in Fig. 8.17. Two of these modes, those in which only the x coordinates vary, are called *stretching* modes, one symmetric, described by Eq. 8.48, and the other antisymmetric, described by Eq. 8.49. The other two modes are *bending* modes in the y and z directions; since the two bending modes differ from

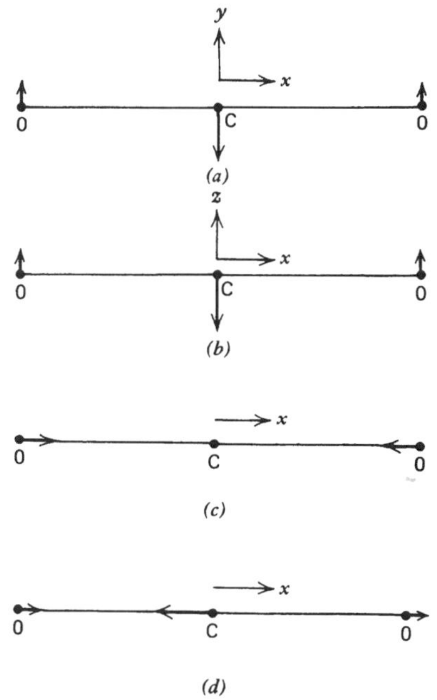

Figure 8.17 Normal modes of vibration of the CO_2 molecule: (*a*), (*b*) bending, Eqs. 8.46 and 8.47, respectively; (*c*) symmetric stretching, Eq. 8.48; (*d*) asymmetric stretching, Eq. 8.49.

each other only in their direction, they are physically indistinguishable and thus degenerate.

In each mode the relative motions of the nuclei are given by the appropriate one of Eqs. 8.46–8.49. Thus, in the y-bending mode, if the oxygen atom in position 1 vibrates away from equilibrium by an amount described by the vector

$$(x_1, y_1, z_1) = (0, y_1, 0),$$

then Eq. 8.46 tells us that the other oxygen atom must simultaneously be at

$$(x_3, y_3, z_3) = (0, y_1, 0)$$

and the carbon atom at

$$(x_2, y_2, z_2) = \left(0, -\frac{2m_{\mathrm{O}}}{m_{\mathrm{C}}} y_1, 0\right) \approx \left(0, -\frac{8}{3} y_3, 0\right).$$

The normal coordinate for this mode is thus a combination of y_1, y_2, and y_3 in the ratio $1{:}(-\frac{8}{3}){:}1$. Similar analyses are easily made for the other modes. The symmetric stretching mode involves motion of only the oxygen atoms, but in all three of the other modes the light carbon atom moves nearly three times as far as the heavier oxygens. In each mode the three nuclei remain in phase with one another, oscillating with the same frequency ν_α. Remember, however, that all this is true only for the normal modes themselves: an ordinary molecular vibration is in general a linear combination of all the normal modes, with the nuclei not in phase.

A qualitative picture of the classical motions corresponding to these modes helps understand the physical phenomenon. In the symmetric stretching mode of CO_2, only the outer atoms move. In the antisymmetric stretching mode, the inner and outer atoms move; Eq. 8.49 makes clear that the light, central carbon atom moves much more than the heavier oxygen atoms. Likewise, in the bending modes, the light carbon atom moves much further from its equilibrium position than the oxygen atoms do. In CH_2, the hydrogen atoms do most of the moving in all the modes: symmetric stretch, antisymmetric stretch, and bend.

To calculate the normal frequencies ν_α themselves, one must know the potential energy surface, or at least the force constants, which give the curvature of the surface along the normal coordinates. However, our previous knowledge of the behavior of oscillators at least permits us to order the modes by frequency. The two bending modes must have identical frequencies, since they are physically indistinguishable. For a given amount of atomic displacement, the stretching modes involve greater changes in nearest-neighbor distances, and thus in bond energies, than do the bending modes. Hence the stretching modes must have higher frequencies than the bending modes. Finally, since lighter particles can move more rapidly, the asymmetric stretching mode, mainly a motion of the C atom, has a higher frequency than the symmetric stretching mode, motion of O atoms only. Thus we expect the order of the frequencies to be

$$\nu(\text{asymm. stretch}) > \nu(\text{symm. stretch}) > \nu(y\text{-bending}) = \nu(z\text{-bending}).$$

Experiment bears this out. The characteristic frequencies of $^{12}C^{16}O_2$, obtained from spectroscopy, are assigned as follows[17] ($\tilde{\nu} = \nu/c$):

$$\begin{aligned} &\tilde{\nu}_3\,(\text{asymmetric stretch}) = 2349\ \text{cm}^{-1}, \\ &\tilde{\nu}_1\,(\text{symmetric stretch}) = 1388\ \text{cm}^{-1}, \\ &\tilde{\nu}_2\,(\text{bending}) = 667\ \text{cm}^{-1}. \end{aligned}$$

This kind of ordering is generally unambiguous. Occasionally one cannot predict whether the stretching motion of a heavy atom will have a higher or lower frequency than the

[17] The numbering of the fundamental modes (ν_1, ν_2, ν_3) is conventional, corresponding to that shown for SO_2 in Fig. 8.15, and tabulated for other molecules in Table 8.3: ν_1 and ν_3 are primarily bond-stretching modes, with the end atoms moving in opposite directions in ν_1, the same direction in ν_3; whereas ν_2 is primarily a bending mode. Symmetric vibrations are always listed first, in order of decreasing frequency.

bending motion of one or more hydrogen atoms. But the rules of thumb that

$$\nu(\text{stretch}) > \nu(\text{bend}) \quad \text{and} \quad \nu(\text{light}) > \nu(\text{heavy}) \qquad (8.50)$$

are reliable generalizations for most molecules.

In the harmonic oscillator approximation, the individual normal modes of oscillation act independently. Each mode may contain any number of vibrational quanta from zero upward. Thus the vibrational energy level of the CO_2 molecule is defined by specifying all four quantum numbers, the v_α of Eq. 8.40. There are clearly many ways to distribute a given amount of energy among several modes, so the polyatomic molecule has a much higher energy-level density or number of levels per unit energy than does the diatomic molecule. This property is illustrated in Problem 5 at the end of the chapter; its significance will become clear in Chapter 21, where we discuss heat capacities in terms of the distribution of energy over molecular energy levels.[18] On simple probability grounds, in most of the possible states the quanta are approximately evenly distributed among the normal modes, just as if we were to throw a large number of dice, the great majority of possible results would have the dice evenly distributed among their six "modes." This is the so-called equipartition of energy. In the next section we shall look into the rules governing transitions among these numerous vibrational energy levels.

8.8 Transitions and Spectra of Polyatomic Molecules: Rotations and Vibrations

We have already had a good deal to say about the mechanisms governing transitions from one energy level to another. The general principles were outlined in Section 4.5, and in Section 7.1 we applied these principles to the vibrations of diatomic molecules. Let us now see how they can be extended to molecules with three or more atoms.

Recall that a pure vibrational transition can occur in a diatomic molecule only when the dipole moment varies in the course of a vibration. A similar rule holds for polyatomic molecules, except that one must speak of individual modes of vibration. Pure vibrational transitions can occur only in those modes for which the molecular dipole moment varies with the normal coordinate. If there is no such change, then an oscillating electromagnetic field exerts a net force of zero on the molecule at all times, just as in a homonuclear diatomic molecule, and there can be no interaction between the vibrational mode and the field. But if the dipole moment does vary with a given normal coordinate, then the molecule can exchange energy with a field whose frequency is the same as that of the normal mode in question, by a mechanism like that illustrated in Fig. 7.2. Modes in which the dipole moment varies are called *infrared-active;* those that leave the dipole moment unchanged are *infrared-inactive.*

To illustrate the foregoing argument, let us again consider the CO_2 molecule. Of the normal modes shown in Fig. 8.17, the only one in which the dipole moment does not change with the normal coordinate is the symmetric stretching mode, for which μ is at all times identically zero. This mode thus cannot absorb electromagnetic radiation, at least not within the approximations we have been using. The asymmetric stretching mode, on the other hand, clearly has a time-varying molecular dipole moment. The average value of the moment is, of course, zero, corresponding to the equilibrium configuration about which the oscillation takes place. If the molecule is placed in an electromagnetic field oscillating at $\tilde{\nu} = 2349$ cm^{-1} ($\nu = 7.04 \times 10^{13}$ s^{-1}), then the field reverses direction at the same frequency as the transient dipole moment, and energy can be transferred between the molecule and the field. This energy transfer must, of course, involve a gain or loss of vibrational quanta in the mode in question. The two bending modes also involve variation of the instantaneous molecular dipole moment, and can thus exchange energy with a field oscillating at $\tilde{\nu} = 667$ cm^{-1}.

The selection rules for such vibrational transitions are again derived from Eq. 7.12, and are similar to those for diatomic molecules. That is, to the extent that the harmonic oscillator approximation applies, we have for each normal mode the rule that only one quantum may be absorbed or emitted in a single transition:

$$\Delta v_\alpha = \pm 1, \qquad (8.51)$$

corresponding to Eq. 7.13 for the diatomic case. Also as in the diatomic case, each allowed vibrational transition gives rise to a band of vibration rotation lines. Since real vibrations are not truly harmonic, one also observes *overtone bands* with $\Delta v_\alpha = \pm 2, \pm 3$, and so on, as well as something not possible for diatomic molecules, *combination bands* in which two or more normal modes undergo transitions simultaneously;[19] however, overtones and combinations are in general much weaker than the fundamental bands.

In Section 7.3 we described the nature of Raman scattering. Analogous to the rule for ordinary vibrational transitions, a given normal mode of a polyatomic molecule is Raman-active only if the polarizability varies with the normal coordinate. For any molecule with a center of symme-

[18] Another application will be in our treatment of the kinetics of unimolecular reactions (Sections 30.14 and 31.1).

[19] A mode that is inactive by itself can give a band in combination with another mode. For example, CO_2 has a band at 2076 cm^{-1} corresponding to transitions with $\Delta v = \pm 1$, in both the bending mode and the (normally inactive) symmetric stretching mode.

try, infrared-active modes are Raman-inactive, whereas Raman-active modes are infrared-inactive (the *rule of mutual exclusion*). Thus in CO_2 and other linear AB_2 molecules the symmetric stretching mode (ν_1) is Raman-active only, and the other modes (ν_2 and ν_3) are infrared-active only. This rule does not apply if the molecule has no center of symmetry. The selection rule of Eq. 8.51 also applies to Raman vibrational transitions.

As for rotational transitions, the same rule holds as for diatomic molecules: A pure rotational spectrum is observed only for molecules with a permanent dipole moment. We have already noted (Section 8.6) that linear polyatomic molecules have the same rotational properties as diatomic molecules; the selection rules are again $\Delta J = \pm 1$ for the infrared spectrum, $\Delta J = 0, \pm 2$ for the Raman spectrum. Other types of molecules (Section 3.12) have more complicated rotational spectra; spherical tops (such as CH_4) and nonlinear symmetric tops (such as NH_3) are of course not found among triatomic molecules. Since a molecule's rotational constants are inversely proportional to its moments of inertia, one can determine bond lengths by comparing the rotational spectra of isotopically substituted molecules, just as one uses the vibrational spectra of such molecules to determine the force constants.

8.9 Transitions and Spectra of Polyatomic Molecules: Magnetic Transitions

Magnetic transitions, described in Section 5.1, may occur whenever a system has an intrinsic magnetic moment μ_m, which may be due to electronic orbital angular momentum and its moment μ_L, to electron spin and its moment μ_S, or to a nuclear spin and its magnetic moment μ_I. Equation 5.8 gives the energy levels of the electrons in a magnetic field **B**; an equivalent equation, with μ_I replacing $\mu_L + \mu_S$, so that its energy is

$$E - E_0 = -\boldsymbol{\mu}_I \cdot \mathbf{B} = -(\mu_I)_z B = \gamma \hbar (m_I)_z B, \qquad (8.52)$$

where we have introduced the *gyromagnetic ratio,* a commonly used quantity that is simply the ratio between the magnetic moment and the angular momentum associated with that moment, so that for the z-component of a magnetic moment $(\mu_m)_z$, associated with the angular momentum component $\hbar m_z$, $\gamma = (\mu_m)_z/\hbar m_z$. Since $\hbar m_z$, the z-component of the total angular momentum $\hbar[L(L+1)]^{1/2}$ has values $-L \le m_z \le +L$ with only integral intervals, this system has $2L + 1$ equally spaced energy levels whose spacing is directly proportional to the applied magnetic field **B**. (Here we use L to indicate any kind of angular momentum, not just orbital angular momentum.) The only transitions normally allowed between these levels as a result of absorption of energy from an applied oscillatory magnetic field are those that change m_z by one unit—i.e, $\Delta m_z = \pm 1$. Atoms, molecules, and nuclei exhibit a wide range of angular momenta and magnetic moments. However, if the angular momentum we are observing is that due to the spin of an electron or a proton, then $m_z = \pm \frac{1}{2}$. In these two cases, which are very important in practice, we have

$$\boldsymbol{\mu}_e = g_e \left(\frac{-e}{2m_e} \right) \mathbf{S} = \gamma_e \mathbf{S} \qquad (5.7')$$

and

$$\boldsymbol{\mu}_p = g_p \left(\frac{e}{2m_p} \right) \mathbf{S} = \gamma_p \mathbf{S}. \qquad (8.53)$$

The spins of these elementary particles are both $\mathbf{S} = \hbar \cdot [\frac{1}{2}(\frac{1}{2}+1)]^{1/2}$, but their magnetic moments differ by a factor of approximately 1800 because of the dependence of the gyromagnetic ratios on the mass of the source of the spin. The two additional complications, which make nuclear magnetic resonance spectroscopy a powerful probe in chemistry, arise from the nature of the *local* magnetic field at each nucleus of a molecule. The field to which any particular nucleus is subject consists of whatever external field we may apply, plus local modifications to that field that are specific to the molecular environment. The two modifications are the fields due to other magnetic species—electrons or nuclei—in the vicinity and the shielding due to all the paired (and hence, nonmagnetic or, more strictly, diamagnetic; cf. Section 5.1) electrons, which give almost a unique signature to each local environment in which a proton may be found as the nucleus of a hydrogen atom. This uniqueness makes it possible to determine structures of very complex molecules, even proteins, from their nuclear magnetic resonance spectra. In practice, one uses magnetic resonance spectra not only of protons but of virtually all magnetic nuclei; ^{13}C is one commonly studied isotope, partly because it is rare enough that it is very unlikely that any small molecule would contain more than one such atom. Even ^3He is now used, particularly in situations where one wants a probe unaffected by a chemical bond; e.g., the interior of a porous medium or a "hollow molecule," such as the soccer-ball-like C_{60} molecule called "buckminsterfullerene."

A classic example of a nuclear magnetic resonance spectrum is the proton spectrum of ethanol, CH_3CH_2OH, in water. This spectrum, taken with a constant-frequency oscillatory field and a slowly varying uniform field **B**, is shown in Fig. 8.18. The three parts of the spectrum correspond, from left to right or from low to high magnetic field, to the proton attached to the oxygen or hydroxyl proton, the two methylene or CH_2 protons, and to the three methyl or CH_3 protons. The different values of the external field at which the responses appear are due to the differences of local magnetic shielding experienced by the three kinds of protons. The hydroxyl proton is the least shielded, and the methyl

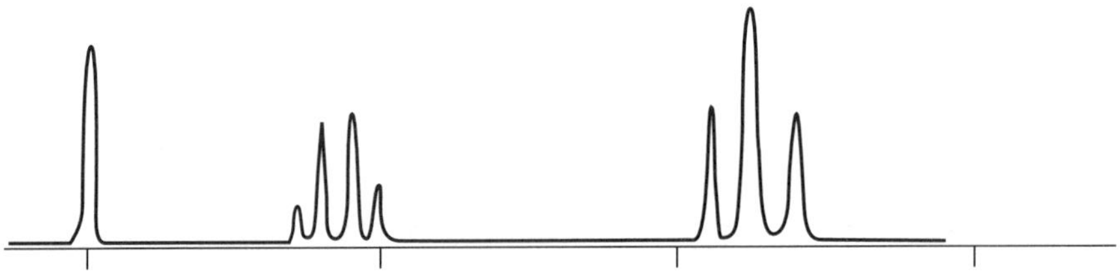

Figure 8.18 Nuclear magnetic resonance spectrum of the protons of the hydrogen atoms of ethanol, CH_3CH_2OH, showing absorption as a function of the applied magnetic field **B**. The single line at lowest field is due to the single proton of the hydroxyl or OH group; its integrated intensity may be taken to have the value 1. It appears at lowest field because it is the least shielded and as a single line because these protons exchange rapidly with the protons of the surrounding water solvent. As a result, they never remain fixed long enough to respond to the quantized fields of other nearby protons. The middle group, arising from the two equivalent methylene or CH_2 protons, has a relative intensity of 2 and appears at a field strength higher than the hydroxyl proton because these protons are more shielded than the hydroxyl. The group at highest field, from the three protons of the methyl or CH_3 group, is the most shielded and has a total relative intensity of 3. The origin of the patterns is described in the text.

protons, the most. The amount of shielding is almost a unique constant for each of many kinds of sites; the most natural, convenient way to express these constant amounts is in terms of the parts per million (ppm) of the applied field **B** by which a resonance is shifted from a chosen standard, usually taken as the protons of the methyl groups of tetramethyl silane, "TMS." The methyl group of protons in ethanol is shifted 1.0 ppm to lower field than the methyl protons of TMS; the methylene protons are shifted in the same direction approximately 3 ppm, and the hydroxyl proton, approximately 4 ppm. The different patterns of lines are due to the different but quantized fields produced by other protons in a molecule at the site of each kind of proton. The net effect of other protons on the hydroxyl proton is null, because this proton exchanges with the protons of water too rapidly to recognize the quantized magnetic fields produced by its neighbors in an ethanol molecule. The methylene and methyl protons do not exchange, and do see these local fields, which combine with the external field. The equivalent methyl protons experience the external field added vectorially to the magnetic field of the two methylene protons, which, together, may produce a field of +1, 0, or −1 units of field from the magnetic moment μ_p, in the direction of the external field. This is because the two methylene proton spins, each with $s = 1/2$, take on a combined state with $S = 1$, so M_s may be +1, 0, or −1. The two combinations yielding 0 together have twice the statistical weight of the nonzero spin states, so the central line of the methyl triplet has twice the intensity of the two outer lines. Likewise, the methylene protons experience **B** added vectorially to the field of the three methyl protons. These couple to form a tight system with a spin state of $S = 3/2$, so M_s may be +3/2, +1/2, −1/2, or −3/2, with relative statistical weights 1:2:2:1. Hence, any of four net local field strengths may affect the methylene protons, so the methylene proton spectrum shows four lines with intensities reflecting the probabilities of these four field strengths.

8.10 Transitions and Spectra of Polyatomic Molecules: Electronic Transitions

Finally, of course, we come to transitions between electronic states of a molecule. We know that each such state can be represented by a potential energy surface in multidimensional space. Let us say more about the properties of such surfaces.

The classical vibration of a diatomic molecule corresponds to the motion of a particle of mass μ back and forth along the molecule's $V(R)$ curve. A horizontal line represents the molecule's fixed total energy; the height of this line above the $V(R)$ curve gives the instantaneous kinetic energy as a function of R. All this can be represented in two dimensions because the potential energy of a diatomic molecule is a function of a single variable, the distance R.

Polyatomic molecules require us to be more sophisticated in visualizing the vibrational energy levels. If a molecule has n degrees of freedom, its potential energy "surface" requires a space of $n + 1$ dimensions: n for the normal coordinates or equivalent variables, one for the energy. For an example that we can visualize, consider a hypothetical system with only two degrees of freedom, a CO_2 molecule somehow constrained to prevent bending. Such a molecule would have only the two stretching modes, and its energy could be represented in three dimensions. If we choose the x axis to represent displacement in the symmetric stretching mode and the y axis to represent the asymmetric stretch displacement, leaving the z axis for the energy, we obtain a surface like that shown in Fig. 8.19. Within the harmonic approximation the surface is an elliptic paraboloid, extending upward from a single minimum at the molecule's equilibrium position. Each horizontal cross section of the paraboloid is an ellipse bounding the classically allowed region for a given total energy. The principal axes of all these

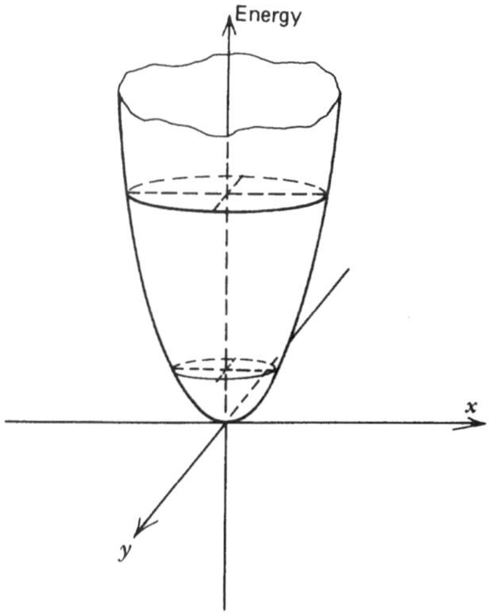

Figure 8.19 Potential energy surface for a system with two vibrational degrees of freedom.

ellipses correspond to the directions of maximum and minimum vertical curvature, and thus to the normal coordinates of the molecule. Quantum mechanically, only certain energy levels are allowed, and in each level the vibrational wave function is a product of harmonic oscillator functions for each coordinate. All these results are easily extended mathematically (though not pictorially) to the case of three or more degrees of freedom.

What we have been describing, of course, is the potential surface for only the ground state of CO_2, and indeed only an approximation to that. Each electronic state of a molecule has its own potential surface, just as each state of a diatomic molecule has its own potential curve. As in the diatomic case, we know that the vibrations must become anharmonic for large displacements; the true potential surfaces are thus not parabolic, but either "level off" or go toward infinity at extreme values of the normal coordinates. The surfaces for different electronic states need not have even approximately the same shapes. For example, some of the excited states of CO_2 are bent rather than linear in their equilibrium geometries.

All this should make it clear that the spectra associated with electronic transitions in polyatomic molecules can be extremely complicated. Whether a given transition is allowed depends, at the lowest level of interpretation, on the relationship between the symmetry classes of the two electronic states. This is a subject we cannot go into here, except to say that the rules are generalizations of those for diatomic molecules. As in diatomic molecules, the spectrum of a given electronic transition consists of a band with a complex vibrational and rotational fine structure. Electronic transitions can often be interpreted well in molecular orbital terms, on the assumption that one electron at a time changes orbitals, with the others left unchanged.

In our discussion of diatomic molecules we mentioned a number of phenomena that can occur when two potential curves cross (Section 7.8). The same can occur in polyatomic molecules when two surfaces intersect: The molecule can move from one surface to the other, that is, make a radiationless transition from one electronic state to the other. Such crossings are in fact far more important in polyatomic molecules than in diatomics. One reason for this, which we have already mentioned, is the high density of vibrational energy levels in a polyatomic molecule. For a transition between two bound states, the smaller the interval between energy levels, the more likely it is that levels in two electronic states will interact. In addition, if the new electronic state has a lower potential surface than the initial state, it also has many more vibrational levels into which the molecule can drop (by emission or collision) after the transition. Such subsidiary processes tend to make the initial transition irreversible. Processes such as predissociation and autoionization are also more likely for polyatomic molecules, since a given molecule can usually split into many different sets of fragments.

The mechanism just described is one of the most important ways in which energy is degraded, that is, becomes less available. Suppose that an excited state is initially produced by a single quantum of high-energy radiation; a radiationless transition to a lower bound state (*internal conversion*) then transforms much of the electronic energy into vibrational energy, and collisions further degrade the vibrational energy into rotational and finally translational energy until that energy is completely equipartitioned among all the available degrees of freedom, i.e., dissipated into heat. The original energy increment is split into smaller and smaller quanta, the reassembly of which becomes progressively less likely. Furthermore, the conversions become easier as the spacing between the energy levels involved becomes smaller. Vibrational energy can be converted into thermal motion much more readily than could the original electronic excitation.

We have touched here upon collisions, which provide one of the major mechanisms for inducing transitions between energy levels. Another of course is absorption or emission of electromagnetic radiation. Indeed, collisions are the dominant mechanism for vibrational and rotational energy exchange in solids, liquids, and even gases at atmospheric pressure. A third is the radiationless conversion process just mentioned. Radiative processes, which we have stressed because of their importance in studying molecular energy levels, become dominant only in extremely rarefied gases (as in the upper atmosphere and outer space). Radiationless processes become increasingly important with the complexity of the system. Their probabilities (or rates) increase with the densities of energy levels, and hence, with the number of particles in each molecule and with the extent the molecule deviates from being harmonic. The separation of modes expressed by Eqs. 8.32 and 8.34 is never perfect, so there is always some mixing of the modes—i.e., exchange of energy among different degrees of freedom, so long as it is consistent with conservation of energy. If the system is quite nonrigid, so the amplitudes of atomic motion may be large, we can expect the atoms to travel into very anharmonic, nonparabolic regions of their potential surfaces. In such regions, energy flow among modes can occur readily.

As for the collisional processes, we shall discuss the mechanisms of such energy transfer at some length in Part III of this book; here let us note only some general principles. Any kind of molecular transition can be induced by collision; the greater the energy difference involved, the more impulsive a collision is required, that is, the more energy, either kinetic or internal, the molecules must have and the closer they must approach each other. Transfer of energy can also occur between one type of motion (translational, rotational, vibrational, electronic) in one molecule and the same or another type in the other molecule. As we have already indicated, the probability of such energy transfer is greatest when the energy levels involved are closest together; thus in general the probabilities fall in the order (T = translation, R = rotation, V = vibration)

$$T-T > T-R > R-R > T-V \approx R-V > V-V.$$

One noteworthy exception to this ranking occurs when two transition processes are *resonant,* that is, have ΔE's that match very closely. Thus vibrational energy transfer between identical molecules is often very fast. Similarly, there may be "accidental" resonances, such as that between the vibrational frequency of $N_2(\tilde{\nu} = 2360\ \text{cm}^{-1})$ and the asymmetric stretching mode of CO_2 ($\tilde{\nu} = 2349\ \text{cm}^{-1}$), and energy transfer between such modes is easy.

• FURTHER READING

Avery, J., *Quantum Theory of Atoms, Molecules and Photons* (McGraw-Hill Book Co., Inc., New York, 1964), Chapter 7.

Herzberg, G., *Electronic Spectra and Electronic Structure of Polyatomic Molecules* (D. van Nostrand and Co., Princeton, N.J., 1966).

Herzberg, G., *Infrared and Raman Spectra* (D. van Nostrand and Co., Princeton, N.J., 1945).

Kondratyev, V., *The Structure of Atoms and Molecules* (P. Noordhoff N. V., Groningen, The Netherlands, 1964), Chapter 8.

Molecular Spectroscopy, A Specialist Periodical Report, Vols. 1 ff. (The Chemical Society, London, 1973 et seq.).

Perić, M., Engels, B., and Peyerimhoff, S. D., "Theoretical Spectroscopy on Small Molecules: *Ab initio* Investigations of Vibronic Structure, Spin–Orbit Splittings and Magnetic Hyperfine Effects in the Electronic Spectra of Triatomic Molecules" in *Quantum Mechanical Electronic Structure Calculations with Chemical Accuracy,* Langhoff, S. R., ed. (Kluwer Academic Publishers, Dordrecht, 1995), pp. 261–356.

Slater, J. C., *Quantum Theory of Molecules and Solids, Volume 1. Electronic Structure of Molecules* (McGraw-Hill Book Co., Inc., New York, 1963), Chapters 5–7.

Wilson, E. B., Decius, J. C., and Cross, P. C., *Molecular Vibrations* (McGraw-Hill Book Co., Inc., New York, 1955), Chapters 2, 4, 6, and 8.

Yarkony, D. R., ed., *Modern Electronic Structure Theory, Parts I and II* (World Scientific, Singapore, 1995).

• PROBLEMS

1. What are the symmetry operations of an equilateral H_3 molecule? Which correspond to the symmetry operations of an isosceles H_3 molecule? In what sense is the symmetry of the equilateral triangle higher than that of the isosceles?

2. Show how the orbitals φ_{2s}, φ_{2px}, φ_{2py}, φ_{2pz}, and $\varphi_{2px} \pm i\varphi_{2py}$ transform if we take
   (a) $x \to -x,\ y \to -y,\ z \to -z$ (inversion);
   (*b*) $x \to x,\ y \to y,\ z \to -z$ (reflection in the x, y plane);
   (*c*) $x \to -x,\ y \to -y,\ z \to z$ (reflection in the y, z plane).

3. Construct three normalized hybrid orbitals of the form $a\varphi_{2s} + b\varphi_{2px} + c\varphi_{2p\pi}$, with two of the orbitals equivalent and the third different. Give *a, b,* and *c* explicitly in terms of the angle between the two equivalent hybrids.

4. Figure 8.8 shows orbitals of types a_1, b_1, and b_2, but none of symmetry type a_2. Construct or sketch an orbital that would be of symmetry type a_2 for the H_2O molecule.

5. Construct molecular orbitals with symmetry appropriate to a linear geometry for the molecule MgF_2, using the atomic orbitals with $n = 3$ for Mg and with $n = 2$ for F as the basis set. First suppose that only the $3s$ and $3p$ orbitals of Mg need to be included; then suppose that the $3d$ orbitals should also be included. Which orbitals have the same symmetry properties? If the orbitals are required to be normalized and orthogonal, how many mixing parameters still remain to be computed?

6. Construct molecular symmetry orbitals for a nonlinear PF_2 radical (FPF). Use only the atomic $2p$ orbitals for fluorine and the orbitals with $n = 3$ for phosphorus. First assume only the $3p$ orbitals are important; then include the $3d$ orbitals as well.

7. Write out the full meaning—that is, the explicit sum—represented by the abbreviated notation for the determinantal representation of the electronic ground state of the H_3 molecule

$$\Psi(\mathbf{r}_1, \mathbf{r}_2, \mathbf{r}) = \text{const.} \cdot \left\| \varphi_A(1)\bar{\varphi}_B(2)\varphi_C(3) \right\|,$$

   where φ_A, φ_B, and φ_C are $1s$ atomic orbitals on hydrogen atoms *A, B,* and *C,* respectively.

8. What is the geometric meaning of the statement

$$\frac{\partial^2 V(X_1, X_2)}{\partial X_1 \partial X_2} = \frac{\partial^2 V(X_1, X_2)}{\partial X_2 \partial X_1}$$

for all X_1, X_2 in some region? How is this connected to the smoothness of V?

9. Show from the properties of determinants or by explicit calculation that the determinantal form of an n-electron wave function automatically satisfies the Pauli exclusion principle, in that no determinantal wave function exists in which two or more electrons occupy the same orbital and have the same spin.

10. Write out the explicit Hartree–Fock equations for the normally occupied molecular orbitals of a triangular H_3 molecule. How many independent equations are there? Be sure to include all the terms of the potential, with each orbital, spin, and variable of integration specified. (You may use vector notation for the coordinate variables.)

11. Obtain the exact solutions to the equation for E:

$$\begin{Vmatrix} H_{11}-E & H_{12} \\ H_{21} & H_{22}-E \end{Vmatrix} = 0.$$

Now suppose $H_{12} = H_{21}$. What is the form of E if $H_{11} = H_{22}$? Suppose H_{12} can be controlled by the observer; how does the difference between the two values of E depend on H_{12}? Now suppose $|H_{12}| << H_{11} - H_{22}$. How does the difference between the two values of E depend on $|H_{12}|$ in this case? How do the two kinds of behavior connect? (Use your general solution.)

12. A subject of some controversy at one time was the pair of related questions regarding CH_2: Is the ground state a singlet state or a triplet state, and is the ground state linear or bent? Explain, by using an argument based on the molecular orbitals of this molecule, why both pairs of alternatives are plausible and why one might expect the answer to the first question to imply the answer to the second.

13. Consider the 4×4 factor of Eq. 8.25. Make the approximating assumption that one root can be found by supposing that only the diagonal elements and the first row and column of the array are important. By expanding the determinant so obtained, find the approximate value of E. Explain the physical assumption inherent in this approximation by discussing the meaning of the off-diagonal quantities H_{ij}, with $i \neq j$. The formula you obtain is called the second-order perturbation expression for E.

14. Using Walsh's rules and the molecular orbitals of NH_2, predict the spin and structure—linear or bent—of this molecule. Would you expect any change in the bond angle in the series NH_2, PH_2, AsH_2, SbH_2?

15. When it was discovered that the water molecule has a dipole moment, electrostatic models were proposed to account for a low enough symmetry in this molecule to permit a nonzero moment. Show that a "suitable" model is one in which the protons are simple positive point charges at fixed equal distances from the oxygen, which is a doubly negative polarizable ion, and that the interaction between the two induced dipoles together with the Coulomb attractions and repulsion give rise to a finite H—O—H angle α, as shown.

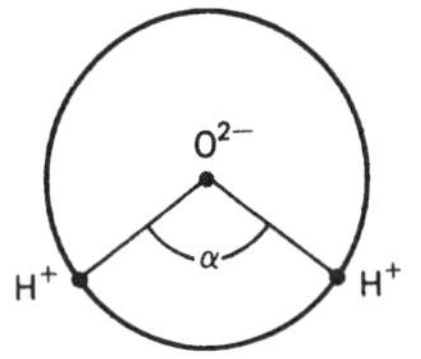

16. Construct the occupied molecular orbitals of the bent ozone molecule and show qualitatively how the energy of each orbital changes as the molecule is (a) bent toward a liner configuration and (b) bent further away from a linear configuration.

17. Make a diagram showing all the vibrational levels of the CO_2 molecule for which the total vibrational energy is less than 5000 cm^{-1}. What is the total number of levels in the neighborhood of 5000 cm^{-1} of excitation? (Each combination, such as 1 quantum in the y-bend, 2 quanta in the z-bend or 1 quantum in the symmetric stretch, counts as a *single energy* level.) Plot the mean density of levels, choosing a wide enough interval of energy for averaging to make a reasonably smooth curve. Compare this with the mean density of levels of a single one-dimensional harmonic oscillator with frequency 667 cm^{-1}.

18. Describe the potential surface for a CO_2 molecule constrained to bend in the xy and xz planes. Show that, with two quanta appropriately assigned, the motion can be construed as a sort of rotation. What degeneracy is associated with this rotation? What is the classical picture of the way this rotation occurs? How would the potential surface change if the molecule were permitted to undergo symmetric stretching also?

19. Describe the potential surface for a molecule such as NO_2 whose equilibrium geometry is that of an isosceles triangle and which is constrained to undergo only bending (nondegenerate for a nonlinear molecule) and symmetric stretching.

20. Show, by using a classical argument, that, within the approximation of a spatially uniform field, a diatomic molecule whose dipole moment is independent of internuclear distances does not absorb electromagnetic radiation to undergo a vibrational transition.

21. Which modes of vibrational motion of the NO_2 molecule can absorb energy from an oscillating, spatially uniform electromagnetic field?

22. Show that CO_2 may undergo rotational transitions and exchange energy with an oscillatory electromagnetic field if the field has significant variation in strength over a range comparable with the size of the molecule.

23. The pure rotational absorption spectrum of the linear molecule OCS shows the following transitions:

		ν(MHz)	B_0 (MHz)
$^{16}O^{12}C^{32}S$	$J = 1 \rightarrow 2$	24325.92	6081.480
	$3 \rightarrow 4$	48651.7	
	$4 \rightarrow 5$	60814.1	
$^{16}O^{12}C^{34}S$	$J = 1 \rightarrow 2$	23731.33	5932.840
	$3 \rightarrow 4$	47462.3	

Use these data to deduce the O—C and C—S bond lengths in OCS. In the case of this linear triatomic molecule,

$$I = \frac{1}{m_O + m_C + m_S} [m_O m_C d_{CO}^2 + m_C m_S d_{CS}^2 + m_O m_S (d_{CO} + d_{CS})^2],$$

where d_{CO} and d_{CS} are the CO and CS bond lengths.

24. Sketch the potential for the stretching modes of a linear triatomic molecule ABA, whose equilibrium geometry has two different A—B distances.

25. Use (a) the HF–SCF, (b) the CISD, and (c) the MP2 methods along with a 6–31G(d,p) basis set to determine the following parameters for the ground electronic state of the H_2O molecule. (See Problems 5.26 and 7.27 for a description of the methods and basis sets.)

(a) The equilibrium geometry, comparing the bond lengths and bond angle to the experimental values in Table 8.3;

(b) The fundamental harmonic vibrational frequencies, comparing them to the values in Table 8.3;

(c) The dipole moment, comparing it to the experimental value of 1.854 Debye.

26. Use (a) the HF–SCF, and (b) the MP2 methods along with a 6–31G(d,p) basis set to determine the following parameters for the ground electronic state of the CO_2 molecule. (*Note:* This molecule is nonpolar.) (See Problems 5.26 and 7.27 for basis set descriptions and software packages.)

(a) The equilibrium geometry, comparing the bond lengths and bond angle to the experimental values in Table 8.3;

(b) The fundamental harmonic vibrational frequencies, comparing them to the values in Table 8.3.

CHAPTER 9

Larger Polyatomic Molecules

In this chapter we shall survey the properties of molecules with four or more atoms. We shall emphasize the structural characteristics of these polyatomic molecules, especially the ways in which bonding affects the spatial arrangement of atoms or groups. One of the key concepts in the study of polyatomic molecules is *separability.* Such molecules can often be treated as being composed of subunits (radicals, functional groups, ligands) bonded together so that each retains a chemical identity of its own, almost independent of its surroundings. We shall discuss how this separability comes about, and why it sometimes does not.

We begin by considering small polyatomic molecules, such as the tetrahedral molecule CH_4 (methane). Species like this can be understood easily if we extend the ideas we developed in Chapter 8. Larger molecules have extended structures that can be classified into two main groups, although many species have characteristics of both. These groups are the *catenated* and the *polyhedral* molecules.

Catenated ("chainlike") molecules are those in which a chain of atoms linked together gives the molecule its essential structure. The chain may be straight, branched, or bent into a ring, and usually other atoms or groups are attached to it. Typical examples are the alkanes, C_nH_{2n+2}, the simple alcohols, $C_nH_{2n+1}OH$, and a large fraction of the other compounds of carbon, the subject matter of organic chemistry. Among inorganic substances, catenated species include the extended structures formed by boron, and by silicon or aluminum with oxygen. Some of the discussion of very extended structures will be reserved for Chapter 11, on solids, because some kinds of solids are nothing more than indefinitely large catenated molecules. A polyhedral compound is a species in which a central atom (or ion) is bound to several *ligands,* which may be atoms, ions, molecules, or molecular fragments. Usually the central atom can be thought of as being at the center of a polyhedron, the vertices of which are defined by the atoms to which the central atom is bound. Many simple molecules and ions like CH_4 or SO_4^{2-} are of this polyhedral type; the term "coordination compound" is sometimes used for weakly bound polyhedral species—in fact, those in which the central atom and the ligands are separable in solution reactions near room temperature. The best known of these are the complexes composed of a metal cation surrounded by a nearest-neighbor shell of neutral or charged ligands, for example, the ion $Pt(NH_3)_4Cl_2^+$. We shall have much to say about the bonding and other properties of these complexes. Proceeding to still weaker bonds, we shall look at such loosely defined species as solvated ions, and finally (in the next chapter) at the interactions between distinct molecules.

9.1 Small Molecules

The first group of polyatomic molecules we shall examine are analogous to those discussed in Chapter 8, where, in fact, we gave some consideration to the bonding in species like BH_3. These are what we have just been calling polyhedral molecules, with a central atom bound to other atoms at the vertices of a polyhedron. Some examples are the plane triangle BH_3, the tetrahedron CH_4, the triangular pyramid NH_3, and the trigonal bipyramid PF_5, each with a characteristic set of bond angles. As in Chapter 8, we wish to be able to interpret and predict molecular shapes in order to see what factors determine the bond angles.

The most familiar and in many ways the simplest interpretation of bond angles is that based on hybridization. If we start with this approach, we assume the bonding electrons around the central atom occupy hybrid orbitals, linear combinations of the atomic orbitals so defined as to have a strong directional character. In Section 8.3 we outlined this method and described the construction of *sp* (linear) and sp^2 (trigonal) orbitals; the latter gives a good description of the bonds and structure of the BH_3 molecule. Extending the method to a combination of the 2*s* and

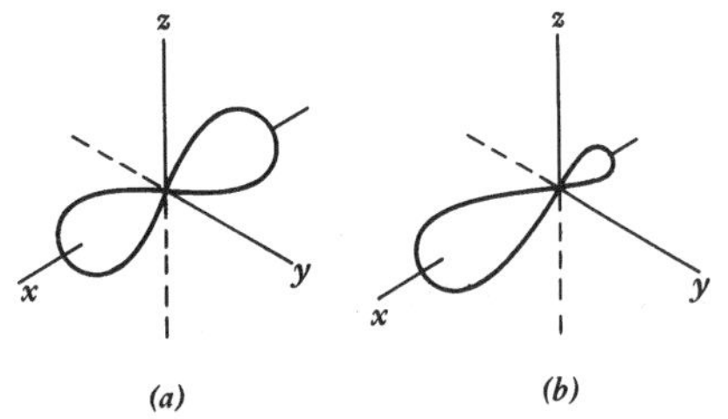

Figure 9.1 (*a*) The spatial coordinates for the orbitals of Eq. 9.1 and the spatial distribution for the $2p_x$ orbital, to exemplify the basic *p* orbital. (*b*) The spatial distribution of the hybrid φ_2 of Eq. 9.1.

all three 2*p* orbitals (sp^3 hybridization), one can show that the four orbitals

$$\begin{aligned}\varphi_1 &= \tfrac{1}{2}(2s+2p_x+2p_y+2p_z),\\ \varphi_2 &= \tfrac{1}{2}(2s+2p_x-2p_y-2p_z),\\ \varphi_3 &= \tfrac{1}{2}(2s-2p_x+2p_y-2p_z),\\ \varphi_4 &= \tfrac{1}{2}(2s-2p_x-2p_y+2p_z),\end{aligned} \tag{9.1}$$

are completely equivalent and symmetric and directed toward the corners of a regular tetrahedron. Figure 9.1 indicates their distribution. (The choice of linear combinations of Eq. 9.1 is not unique; each choice corresponds to a specific orientation of the hybrid orbitals with respect to the *x*, *y*, and *z* axes.) Thus, sp^3 orbitals are appropriate to describe the bonding in the tetrahedral CH_4 molecule. In the localized-bond approximation, each of these orbitals can be combined with a H(1*s*) orbital to form two σ-type pair orbitals, one bonding and normally occupied, the other antibonding and normally empty. Experimental determination of the structure confirms that the four C—H bonds are equivalent and arranged tetrahedrally, with a bond angle of 109°28′. Their equivalence insures that they all have the same energy as well. Nearly all other compounds[1] in which a carbon atom forms four single bonds have bond angles within a few degrees of this value, and can be described in terms of sp^3 hybridization. If the four substituents on such a carbon atom are not all the same, then of course the bonding orbitals need not, and in general will not have, the same energies.

One can similarly devise hybridization schemes to correspond to all the other bond geometries commonly observed. For larger central atoms this often requires the use of *d* orbitals; this is clearly the case when the atom forms more than four bonds because more than four atomic orbitals must participate. For the hybridization model to be applicable, all the atomic orbitals contributing to the hybrid must be fairly close in energy. Figure 9.2 illustrates some of the most common molecular geometries and the corresponding types of hybridization.

Combinations of hybrid orbitals are ordinarily used to portray localized bond orbitals. This is basically a valence bond approach, and it was in the early development of the VB theory that hybrid orbitals were first introduced. In a molecular orbital treatment at the Hartree–Fock level, as we outlined it in Section 8.4, one obtains delocalized orbitals extending over the whole molecule. In sharp contrast to the hybrid orbitals we have just constructed, the occupied molecular orbitals of the valence shell of tetrahedral CH_4 consist of one *s*-like orbital ψ_1, of the form $\alpha\varphi_1 + \beta(h_1 + h_2 + h_3 + h_4)$ where h_j is the 1*s* orbital on hydrogen *j*, and, at another energy, three degenerate *p*-like orbitals ψ_2, ψ_3, ψ_4, each comprised of the orbitals ϕ_2, ϕ_3, and ϕ_4, with the hydrogen 1*s* orbitals.[2] What has happened to the tetrahedral symmetry? We can regenerate it by taking four linear combinations of the MOs, of exactly the same form as the combinations of atomic orbitals in Eq. 9.1, but with ψ_1 replacing 2*s* and ψ_2, ψ_3, and ψ_4 in place of the atomic 2*p* orbitals. The resulting *equivalent orbitals* are functions that have their contours of maximum electron density oriented along the directions of the four bonds, and are thus more appropriate for calculations of bond properties. The individual localized equivalent orbitals do not have the symmetry of the entire molecule that characterizes the natural stationary-state solutions of simple Hartree–Fock one-electron Schrödinger equations. They are solutions of slightly more complicated equations, which are still consistent with the full Schrödinger equation for all the electrons.

Hybridization need not embrace all the bonds formed by a given atom. Consider the ethylene (C_2H_4) molecule, which has the planar skeletal structure

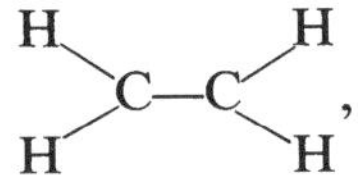

with three 120° bond angles around each carbon. This structure is interpreted as the result of sp^2 hybridization like that in BH_3 (Section 8.3), using the 2*s* and two of the 2*p* orbitals on each carbon to form three trigonal σ bonds. The remaining C(2*p*) orbitals, which have their axes perpendicular to the plane of the molecule, combine to form a bonding π orbital like that in the isoelectronic O_2 molecule. Thus we

[1] The exceptions are discussed in Section 9.4.

[2] The remaining two electrons are in a nonbonding "inner-shell" orbital virtually identical to the C(1*s*) orbital. The reason the molecular orbital method gives *s*- and *p*-like orbitals lies in the symmetry of the Hamiltonian of the one-electron Schrödinger equation. For the CH_4 problem, the equilibrium geometry of the regular tetrahedron determines that symmetry. With only *s* and *p* orbitals of the central atom and the four hydrogenic 1*s* orbitals, the stationary-state functions must have the same angular dependence as the free atom.

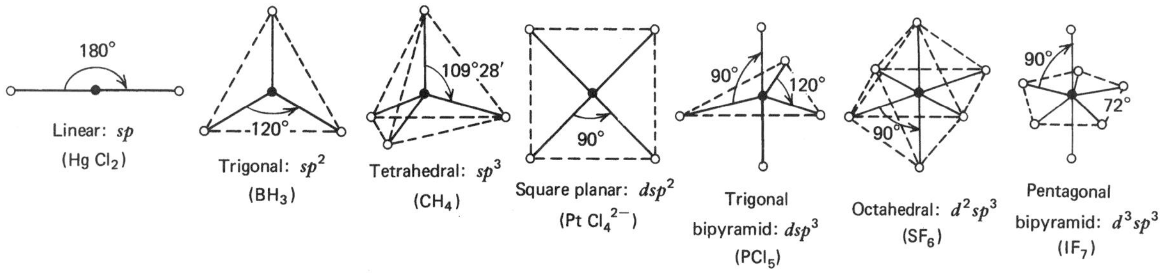

Figure 9.2 Some common types of small symmetric molecules. In each diagram the solid lines represent the bonds formed by the central atom (●), whereas the dashed lines indicate the polyhedron (or plane figure) formed by the atoms (○) to which the central atom is bound. Bond angles are indicated; in each case except the trigonal and pentagonal bipyramids, all the angles are the same and the bonds equivalent. Each symmetry type is labeled with its name, a hybridization scheme that gives the appropriate bond directions, and an example of a molecule with that symmetry.

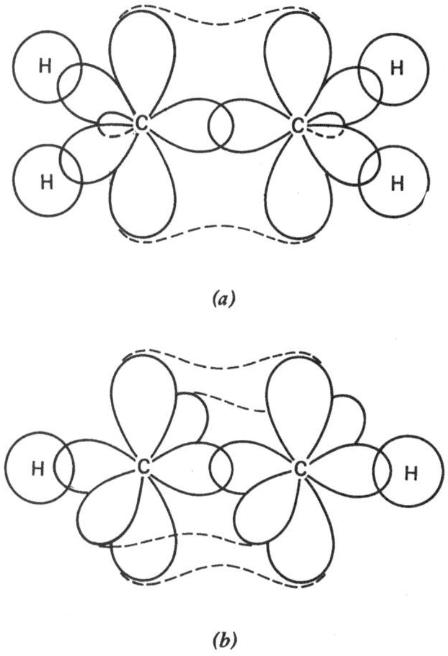

Figure 9.3 Schematic representation of orbitals in (*a*) ethylene and (*b*) acetylene, with the C—H and C—C σ bonds interpreted in terms of hybrid orbitals and the π bonding orbitals described by molecular Hartree–Fock-like orbitals.

have one σ bond and one π bond between the two carbon atoms, corresponding to the double bond of the conventional formula $H_2C{=}CH_2$. Similarly, the linear acetylene molecule (H—C≡C—H) is interpreted as having *sp* (linear) hybridization on each carbon; the leftover 2*p* orbitals form two bonding π orbitals at right angles to each other, as in the isoelectronic N_2, and the single C—C σ bond and two π bonds constitute the conventional triple bond. The orbitals in ethylene and acetylene are shown in Fig. 9.3.

Thus far we have spoken only of equivalent hybrids, but these are applicable only in the most highly symmetric molecules. Consider the ammonia (NH_3) molecule. The molecule is pyramidal, with the N atom at the apex and H—N—H angles of 106.7°. Since NH_3 is isoelectronic with CH_4, there must again be four occupied valence-level orbitals, but only three of these correspond to bonds. If the bonding orbitals involved the N(2*p*) electrons only, without the participation of the N(2*s*) orbital, the bond angle would be only 90°. On the other hand, an equivalent mixture as in CH_4 (each orbital $\frac{1}{4}s$ and $\frac{3}{4}p$) would give the tetrahedral angle[3] 109.5°. Thus the hybrid orbitals should be intermediate between these two extremes. A mixture of about 22% *s* and 78% *p* for the bonding orbitals gives the observed angle. In Section 8.3 we used a similar argument to account for the bond angle of the H_2O molecule. In a quantitative treatment, one should properly vary the amount of mixing to optimize the energy, then see how well this predicts the bond angle. In practice, one must frequently go to extensive MO calculations—at or often well beyond the Hartree–Fock mean field level, incorporating correlation in the form of configuration interaction—to make quantitatively accurate predictions of bond angles (cf. Fig. 8.9 for H_2O).

The hybridization model is generally useful in describing simple first-row compounds. When boron is the central atom, one usually picks the same sorts of hybrids as for carbon. In compounds of nitrogen, oxygen, and fluorine, the contribution of the 2*s* orbital becomes successively less important. The reason is that, as the nuclear charge increases, the 2*s* electrons become more tightly bound and farther from the 2*p* in energy (cf. Fig. 7.21). Recall how we could infer from perturbation theory that the further apart are the energies of two approximate states, the less they mix in an accurate calculation. In HF, isoelectronic with CH_4, NH_3, and H_2O, one can to a good approximation ignore hybridization and speak of a simple $\sigma 2p$ bond.

Hybridization without optimization does not really explain anything. It gives us a convenient way to interpret molecular shapes, but we must know the geometry in order to determine the hybridization. One of the simplest

[3] In fact, the NH_4^+ ion, with the same number of electrons but one more nucleus, is tetrahedral.

techniques for actually predicting shapes (but not necessarily values of angles) was developed by Mulliken and Walsh: One estimates how the energies of the molecular orbitals vary with bond angle, then judges on semiempirical grounds how these variations balance out to minimize the total energy. In Sections 8.4 and 8.5 we outlined the reasoning leading to "Walsh's rules" in triatomic molecules; here we shall merely state some of the results[4] (in terms of the number of valence electrons):

AH_3: ≤ 6 valence electrons, planar (BH_3); 7–8 valence electrons, pyramidal (CH_3, NH_3, H_3O^+); 10 valence electrons, planar (none known).

HAAH: 10, linear (C_2H_2); 12, bent planar (N_2H_2); 14, bent nonplanar (H_2O_2).

AB_3: ≤ 24, planar (BF_3, CO_3^{2-}, NO_3^-, SO_3); 25–26, pyramidal (ClO_3, IO_3^-, PF_3); 28, planar (ClF_3).

H_2AB: ≤ 12, planar (H_2CO); 14, nonplanar (H_2NF).

The concepts involved here are at best somewhat diffuse,[5] but Walsh's rules are a useful guide, and do lead us to the correct answer far more often than not.

An alternative approach, based on the localized-electron-pair viewpoint, is also useful in predicting molecular shapes qualitatively, particularly for larger molecules and molecules with very low symmetry. This method (which we also mentioned at the end of Section 8.5), was developed primarily by Gillespie and Nyholm,[6] and is sometimes called the valence-shell electron-pair repulsion (VSEPR) theory. Its basic premises are (1) that electrons in an atom's valence shell can be considered as localized pairs, whether bonding or nonbonding (lone pairs); (2) that all these electron pairs repel one another, and are distributed in space so as to minimize the repulsion; (3) that lone pairs repel more strongly than bonding pairs, so the repulsive interactions decrease in the order lone pair–lone pair > lone pair–bonding pair > bonding pair–bonding pair.

Premise (2) predicts that an AX_n molecule in which all the electron pairs around A are bonding will have its bonds as far apart as possible. One can easily show that the molecule must then have one of the following symmetric structures:

Number of Electron Pairs	Structure	Example
2	Linear	$HgCl_2$
3	Triangular	BCl_3
4	Tetrahedral	CH_4
5	Trigonal bipyramid	PCl_5
6	Octahedral	SF_6
7	Pentagonal bipyramid	IF_7

Most of these are illustrated in Fig. 9.2. Molecules with lone pairs are assumed to have essentially the same structures, but with the lone pairs occupying some of the vertices; the shapes of most such types of simple molecules are shown in Fig. 9.4. The structures are of course distorted by the greater repulsion of the lone pairs; this effect is in accord with the decreasing bond angles in the sequences

$$CH_4\ 109.5°,\ NH_3\ 106.7°,\ H_2O\ 104.5°;$$
$$SiH_4\ 109.5°,\ PH_3\ 93.3°,\ H_2S\ 92.2°;$$
$$GeH_4\ 109.5°,\ AsH_3\ 91.8°,\ H_2Se\ 91.0°;$$
$$SnH_4\ 109.5°,\ SbH_3\ 91.3°,\ H_2Te\ 89.5°.$$

The decrease in angle as one goes from NH_3 to SbH_3, or from H_2O to H_2Te, is associated with the increasing size (or decreasing electronegativity) of the central atom: The farther out the bonding pairs are, the smaller the angle at which their mutual repulsion reaches equilibrium.

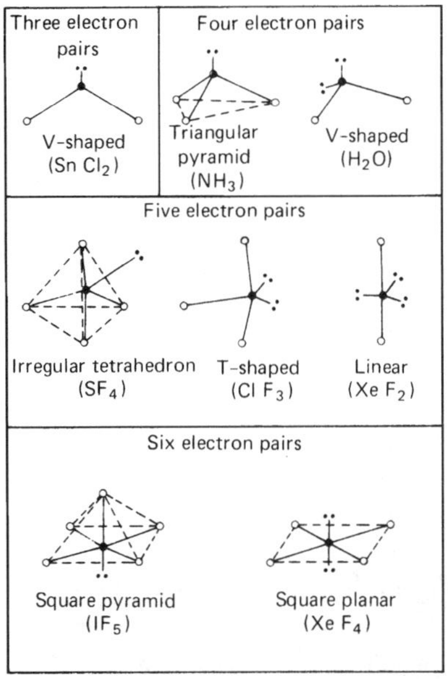

Figure 9.4 Shapes of small molecules with lone pairs of electrons. The diagrams are drawn as in Fig. 9.2, with the lone pairs indicated by —: symbols. The cases are classified by the number of electron pairs around the central atom, and for each shape a molecular example is given. The shapes shown here are those actually observed, and do not include all the conceivable configurations; see the text for discussion of this point.

[4] See A. D. Walsh, *J. Chem. Soc.* (1953), p. 2260ff.

[5] The problem is, as we pointed out in connection with Fig. 8.9, that the "energy" of Walsh's arguments often does not correspond closely to the actual orbital energy. Moreover, the total energy is not simply a sum of orbital energies: cf. Eq. 8.25. Coulson and Deb have argued that the quantity most appropriate to be represented by the vertical axis in a Walsh diagram is the work done when, say, a bond angle is varied. This approach has been used to obtain a set of molecular-shape rules including but more extensive than Walsh's; see B. M. Deb et al., *J. Am. Chem. Soc.* **96,** 2030, 2044 (1974). The highest occupied MO seems to have a dominant effect on the molecular geometry.

[6] See R. J. Gillespie and R. S. Nyholm, *Quart. Rev.* **11,** 339 (1957); R. J. Gillespie, *J. Chem. Educ.* **40,** 295 (1963).

Useful as the VSEPR theory is, it cannot predict molecular structures in all cases without additional information. The model naturally tends to fail when two or more alternative structures lie rather close in energy. A very good example of this is the molecule ClF_3. The valence shell of the central Cl atom, according to a localized-pair picture, consists of three Cl—F bonding pairs and two lone (nonbonding) pairs. Applying premise (2) above, we would predict that the five pairs are arranged approximately along the axes of a trigonal bipyramid, as in the PCl_5 molecule (Fig. 9.2). But not all the positions in a trigonal bipyramid are equivalent: In PCl_5, for example, the three *equatorial* P—Cl bonds are 2.04 Å long, whereas the two *axial* bonds are 2.19 Å long. Taking this difference into account, we can formulate three possible structures for ClF_3, as shown in Fig. 9.5; these structures should differ in energy and other properties. The angles between lone pairs in structures I, II, and III are about 180°, 120°, and 90°, respectively. If the lone pair–lone pair repulsion were the only significant effect, we would expect I to be the true structure. But the lone pair–bonding pair and bonding pair–bonding pair repulsions are less in structure II, and there are more of these; there is no way to tell *a priori* from a qualitative model which of these effects is dominant. In fact, experiment shows that the true structure is II, with all the atoms in a plane but one F atom distinctly different from the other two. Measurements on other five-electron-pair molecules show that we can interpret the structure of ClF_3 as though the lone pairs always occupy the equatorial positions in the trigonal bipyramid, as we have indicated in Fig. 9.4. Apparently the lone pair–lone pair repulsion does not vary significantly at angles beyond 120°. In six-electron-pair molecules, by contrast, the six positions are all equivalent, and two lone pairs always occupy *trans* positions as shown for XeF_4.

The localized-pair model has yet to be quantified in any comprehensive way. We do know that both Coulomb and exchange forces contribute to the repulsion between electron pairs. Calculations for specific cases show that the exchange contribution is often very small when a molecule is near its equilibrium geometry. This suggests that a pure Coulomb repulsion model might be adequate for predicting molecular structures. The simplicity of such a model makes it attractive

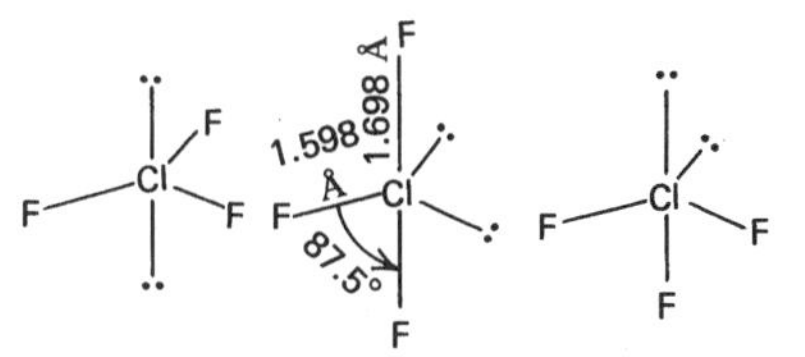

Figure 9.5 Possible structures of the ClF_3 molecule. The experimentally observed structure is closest to II, with the dimensions shown on the figure.

for computations. One simply treats the ligands as point charges sliding over the surface of a sphere, and determines the geometry of lowest energy. This method usually does give a correct picture, in agreement with experiment or more elaborate calculations.

Either Walsh's rules or the VSEPR method is adequate if one merely wishes to know the approximate equilibrium structure of a molecule. But more accurate calculations are required for many purposes, especially in studying the dynamics of rearrangements, as in molecular vibrations or reaction mechanisms. In such cases one most commonly uses delocalized molecular orbitals, which are convenient for computations. The variations of the MO method should be familiar by now. In order of increasing accuracy, one can use a semiempirical approach with variable parameters, *ab initio* self-consistent-field calculations, or a model including configuration interaction to account for electron correlation. Yet in some ways, the more one refines the calculation for quantitative accuracy, the more one loses the interpretive simplicity of elementary models. For instance, the complete electron density (that is, $|\psi|^2$) of a molecule cannot be partitioned in any obvious way into bonds and lone pairs, nor even into one-electron orbitals. Much research in recent years has been devoted to finding methods of extracting simple interpretive concepts and generalizations from the exact wave function.

In the future one can expect to see the structures of larger and larger molecules interpreted in terms of the microscopic forces between electrons and nuclei. The calculation of accurate bond lengths requires primarily a knowledge of the very strong interactions between nearest-neighbor atoms. Bond angles are governed largely by the interactions between next-nearest neighbors, and structural factors like the "staggered" ends of the H_2O_2 and C_2H_6 molecules require consideration of still more distant interactions. As the distance increases the strength of the interaction decreases, and the computations needed to account for it become more lengthy as small effects add up. Meanwhile, chemists and spectroscopists can do much to interpret these structural factors with simple models like the stretching and bending motions described in the last chapter. But we must remember that bonds are the result of electron distributions around nuclei in space, not the springs holding rigid balls together that molecular-vibration theory implies. The simplifications of such models are powerful tools, but they merely extract the most important factors; they do not provide an accurate representation of reality.

9.2 Catenated Carbon Compounds; Transferability

Most of the molecules we have examined thus far are simple polyhedra. However, there are several classes of compounds whose molecules have large, extended structures, made possible by the capacity of atoms to link to one another in long chains of consecutive bonds. The compounds to which we refer have definite structures and compositions; they consist of well-defined molecules, and are not to be confused with solids, some of which are infinitely extended molecules. We begin with the compounds of carbon, for which the problems of bonding and electronic structure have been studied extensively. In the next section we shall look at a variety of inorganic extended structures, the study of which is less well developed.

It is a commonplace that more compounds containing carbon are known than of any other element except hydrogen, itself contained in most carbon compounds. This myriad of carbon compounds exists primarily because carbon atoms can bond to one another to produce chains of any length, and because a variety of other atoms or groups can be attached at many positions along such a chain. Such compounds are referred to as *catenated* or chainlike. Only sulfur exhibits an ability for "self-catenation"—bonding to like atoms to form chains—comparable to carbon's. Liquid sulfur consists mainly of a mixture of S_8 rings and very long chains of ($\sim 10^5$) S atoms which also comprise "plastic sulfur." But since such sulfur chains have no free valences to attach other groups, their chemistry is comparatively impoverished. One might expect silicon to form structures like those of carbon, but silicon so readily binds oxygen that Si—O—Si rather than Si—Si is the basis for its extended structures.[7] Carbon is unique in its chemical versatility—and fortunately so, for we literally live on its chemistry.

The bonding in carbon compounds is most commonly described in terms of localized bonding orbitals. For useful approximate descriptions of the contribution of each carbon atom to these bond orbitals, we turn to the various hybrid atomic orbitals already discussed: sp^3 for 4-coordinated carbon, with four bonds arranged tetrahedrally as in CH_4; sp^2 for three of the bonds around a 3-coordinated (planar) carbon, with the remaining electrons forming a $2p\pi$ bond as in C_2H_4; and sp for two bonds on a 2-coordinated (linear) carbon, with two $2p\pi$ bonds as in C_2H_2. The hybrid orbitals on adjacent carbon atoms overlap constructively to produce the localized bond orbitals. The structures of most organic compounds can be understood without going beyond this level of approximation.

We know that electrons are not really localized in bonds, and that a correct model of a molecule's electronic structure should allow the electrons to wander anywhere in the molecule. If we wanted to use a delocalized MO description, a different set of orbitals would be required for every compound. Yet the localized description is extremely powerful. For example, the C—C and C—H bonds in ethane (C_2H_6) are quite similar to those in pentane (C_5H_{12}), decane ($C_{10}H_{22}$), or any other hydrocarbon with no double or triple bonds. As long as one compares groups of atoms that have

Table 9.1 Characteristic Parameters of Common Structural Units

Group	Molecule	Bond Length (Å) or Angle (°)	Vibration Frequency (cm^{-1})
C—C	C_2H_6	1.54	993
	C_2H_5Cl	1.54	972
	$CH_2Cl—CH_2Cl$	1.55	1052
C—H	C_2H_6	1.10	2975
	C_2H_5Cl	1.1	2890–2983
	$CH_2Cl—CH_2Cl$	(assumed 1.09)	2950–3005
	CH_3OH	1.096	2834–2980
	$H_2C{=}O$	1.09	2843
C—O	CH_3OH	1.43	1030
C=O	$H_2C{=}O$	1.22	1746
O—H	CH_3OH	0.96	3328
C—C—Cl	C_2H_5Cl	110°	336 (bend)
	$CH_2Cl—CH_2Cl$	112°	300 (bend)

the same kinds of bonds, the properties of a group in one compound can be used to predict the properties of the same group in another compound, almost without regard to the rest of either molecule. Properties of this sort are said to exhibit *transferability.*

Some examples of transferability are given in Table 9.1. The bond lengths and angles are established by the usual methods, based on either rotational spectra or x-ray and electron diffraction. The vibration frequencies come directly from the infrared spectrum, with their assignments to particular groups based partly on normal coordinate analyses, but mainly on correlations among the spectra of related compounds. Indeed, the latter is one of the clearest and most important applications of transferability. The vibration frequencies associated with many functional groups are so characteristic, and often so easy to measure, that one can readily use them for purposes of identification. For example, carbonyl (> C=O) groups always have a vibration-rotation band at about 1800 cm^{-1}, as indicated in Table 9.1; other clear-cut examples include the hydroxyl stretching frequency at about 3600 cm^{-1}, the nitrile (—C≡N) stretching band in the region 2100–2250 cm^{-1}, and the acetylenic (—C≡C—) band near 2100 cm^{-1} but varying from about 2020 to 2300 cm^{-1}. In fact, all the groups common in organic chemistry have their own characteristic patterns of infrared bands. Infrared spectroscopy has thus become an indispensable physical tool for the rapid characterization of materials. Besides the properties in Table 9.1, others that show transferability include bond energies (see Section 14.8) and even chemical reactivities. Many macroscopic properties such as heat capacity, critical constants and surface tension have also been analyzed empirically as sums of group contributions.[8]

[7] Some long-chain silicon compounds (up to at least $Si_{14}F_{30}$) are known, but these react with oxygen so readily that they are unstable in the presence of air or water.

[8] Many such empirical correlations can be found in Robert C. Reid and Thomas K. Sherwood, *The Properties of Gases and Liquids,* 2nd ed., (McGraw-Hill, New York, 1966).

How does transferability come about? How can it be that electrons behave as though they were localized when we know that ultimately they are not? The answer lies largely in the ease with which delocalization occurs—more specifically, in the smallness of the energy differences between the delocalized orbitals. For example, the valence-electron MOs in CH_4 have an energy range not much greater than that between the carbon $2s$ and $2p$ orbitals. If certain of the occupied delocalized MOs are not only built from the same types of atomic orbitals but are very similar in energy as well, then to a reasonably good approximation one can consider the delocalized orbitals as degenerate. The significance of this lies in an important property of any degenerate set of eigenfunctions of an operator: One can always construct from those functions linear combinations that are also eigenfunctions—but that may have quite different spatial properties from the original functions. This is just the method used to construct the localized "equivalent orbitals" of CH_4, which we mentioned in the last section. If the valence MOs were actually degenerate, the localized and delocalized descriptions would be equally valid. As it is, the energies are close enough for the localized description to be a useful approximation. In all saturated hydrocarbons and many other organic compounds, the normally occupied MOs based primarily on carbon $2s$ and $2p$ orbitals have energies similarly close together, and the properties of (for example) the C—C and C—H bonds are largely transferable.

We can express this reasoning in more formal terms, in a way that will be useful when we study the structure of solids. Suppose that we begin our description of a molecule with a set of atomic orbitals (probably hybridized), from which we construct a set of localized bonding orbitals φ_j. These localized orbitals are clearly different from the delocalized molecular orbitals ψ_k, which are the eigenfunctions of a one-electron Hartree–Fock operator $\mathsf{H}(1)$ like that defined in Eq. 8.16: $\mathsf{H}(1)\psi_k = \epsilon_k\psi_k$. The localized φ_j's, on the other hand, are not eigenfunctions of $\mathsf{H}(1)$; thus, the operation of $\mathsf{H}(1)$ on φ_j does not multiply φ_j by an eigenvalue, but creates a new function $\mathsf{H}(1)\varphi_j$. The degree to which $\mathsf{H}(1)$ "spoils" the functional form of φ_j gives the "amount" by which φ_j is not a true eigenfunction. This "amount" is measured by the values of integrals of the form $\int\varphi_i^* \,\mathsf{H}(1)\varphi_j\, d\tau$, which would vanish[9] (for $i \neq j$) if φ_i and φ_j were eigenfunctions of $\mathsf{H}(1)$. One can then determine the true (stationary-state) eigenfunctions ψ_k by finding linear combinations of the φ_j's for which the integrals $\int\varphi_k^* \,\mathsf{H}(1)\varphi_j\, d\tau$ vanish. This is just the standard MO calculation with a different kind of basis function. The process described in the previous paragraph is simply the same calculation carried out in reverse, the construction of localized φ_j's as linear combinations of the delocalized ψ_k's.

The interesting thing about the localized bond orbitals φ_j is that those between two atoms of any given elements typically have nearly the same expectation value of the energy. This is particularly so if the hybridization or coordination number is specified. For example, the integral $\int\varphi_i^* \,\mathsf{H}(1)\varphi_j\, d\tau$ has nearly the same value for all single-bond orbitals φ_j between pairs of carbon atoms, especially when the carbons have the same coordination number. The corresponding integrals for φ_j's representing C—H bonds have a different value, but are again all similar to one another. In short, the localized orbitals fall into groups corresponding to different kinds of bonds, with the functions in each such group nearly degenerate with one another. This corresponds to the empirical observation of characteristic bond energies.

Thus, we can form localized orbitals φ_j that fall into nearly degenerate sets. If the delocalized MOs are written as linear superpositions of these φ_j's,

$$\psi_k = \sum_j c_{jk}\varphi_j, \tag{9.2}$$

then the expectation values of the MO energies are given by

$$\epsilon_k = \frac{\int\psi_k{}^*\mathsf{H}(1)\psi_k\, d\tau}{\int\psi_k{}^*\psi_k\, d\tau} = \frac{\sum_i\sum_j c_{ik}c_{jk}\int\varphi_i{}^*\mathsf{H}(1)\varphi_j\, d\tau}{\sum_i\sum_j c_{ik}c_{jk}\int\varphi_i{}^*\varphi_j\, d\tau}$$

$$= \frac{\sum_i\sum_j c_{ik}c_{jk}H_{ij}}{\sum_i\sum_j c_{ik}c_{jk}S_{ij}}, \tag{9.3}$$

with H_{ij} and S_{ij} defined as in Eq. 8.20. The magnitude of the "off-diagonal" H_{ij}'s ($i \neq j$) corresponds roughly to the rate at which electrons exchange between bonds. The smaller these integrals are, on the other hand, the more nearly the localized orbitals approach stationary states. But if the φ_j's for a given set of bonds are nearly degenerate and the off-diagonal H_{ij}'s can be neglected, then Eq. 9.3 implies that the ψ_k's built mainly from those bonds must also be nearly degenerate. In other words, the better the approximation that the electrons behave like localized pairs, the smaller are the energy separations between the delocalized MO states.

What happens as the size of the molecule increases? The more atoms there are in a molecule, the more φ_j's and ψ_k's we have. Now the exchange integrals $H_{ij}(i \neq j)$ are usually of appreciable magnitude only for adjacent (nearest-neighbor) bonds; and the number of adjacent-bond pairs increases much more slowly than the total number of possible pairings of bonds. For example, in a straight-chain hydrocarbon C_nH_{2n+2}, there are $n - 1$ C—C bonds, $n - 2$ pairs of adjacent C—C bonds, but a total of $(n - 1)(n - 2)$ pairs of C—C bonds, adjacent or not. Since the energy spacings between the delocalized MOs (of a given type) depend mainly on the H_{ij}, the average spacing between the ϵ_k's decreases as the size of the molecule increases. When the energy levels in a given set become very closely spaced (compared, say, with

[9] This is the property of orthogonality of eigenfunctions, for which see Appendices 6A and 9A.

the mean thermal energy), they can be treated as a continuum—or rather as a band of states, a continuum with upper and lower energy limits. This is just what happens in a solid, and we shall see in Chapter 11 that this is the key to the interpretation of the electronic structure of solids.

Localized bond orbitals, then, are rather good substitutes for the delocalized stationary-state molecular orbitals ψ_k. If one can construct localized orbitals representing not individual bonds but entire fragments of molecules, these "group orbitals" should be still better approximations to the ψ_k. In fact, it is possible to obtain good wave functions for the polyatomic building blocks of large molecules. These functions not only describe the group properties well, but as a set are quite good for representing all but the most global properties of the entire molecule. One is not even restricted to one-electron functions; it is quite feasible to use many-electron functions that include electron correlation. This type of calculation is called "molecules in molecules," and is an extension of the similar "atoms in molecules" method in which accurate wave functions for free atoms are used as basis functions. Both methods lend themselves well to semi-empirical parameterization when accurate wave functions for the building blocks are not available. The formulation of wave functions of large molecules in terms of the wave functions of well-defined constituents is the mathematical expression of the assumption of transferability of group properties.

For a further illustration of the transferability of properties of chemical subunits, let us consider the electronic binding energies, as measured by photoelectron spectroscopy. Table 9.2 gives some examples. As with vibration frequencies, the assignments to particular groups are based on correlation of measurements for many compounds, supplemented by observing the effects of specific chemical changes, and reinforced by some theoretical interpretation of why the spectra appear where they do. According to Koopman's theorem the binding energies should approximately equal the MO energies, and thus should be quite close to the energies of group orbitals. In fact, the energies for particular groups remain clustered about a common average energy throughout a series, shifting only by small, systematic amounts.

In certain classes of molecules the concept of localized properties breaks down badly. Some of these breakdowns are due to what are called steric strain effects, consequences of the physical interference or blockage of one group by another, which will be discussed in Section 9.4. In other instances a significant part of the electron density belongs to one-electron states that spread over many atoms and offer high mobility for electrons to move from one region of the molecule to another. This phenomenon is generally a property of molecules that have both multiple and single bonds in a formal valence bond picture; one might say that some of the electron density "leaks" out of the multiple bonds.

Delocalization is most striking in the so-called *aromatic* carbon compounds, and occurs to some degree throughout

Table 9.2 Binding Energies of Electrons in Particular Groups as Determined by Photoelectron Spectroscopy

Molecule	Group	Binding Energy (eV)
Ethane		
C_2H_6	C—C bond	11.5[a] ($p\sigma$ bond)
		20.42, 23.9[b] ($s\sigma_u$, $s\sigma_g$ bonds)
	C—H bond	14.7[a]
n-Butane		
C_4H_{10}	C—C bond	10.6[a] (probably several C—C bonds)
	C—H bond	14.4[a]
n-Hexane		
C_6H_{14}	C—C bond	10.3[a]
Ethylene		
C_2H_4	C=C π electron	10.51[b]
	C—C σ electron	12.38[b] ($p\sigma$ bond)
		19.1, 23.7[b] ($s\sigma_u$, $s\sigma_g$ bonds)
	C—H	14.3[a]
1-Butene		
C_4H_8	C=C π electron	9.6[a]
	C—C (probably both true single bonds and σ of double bond)	11.3[a]
	C—H	14.5[a]
cis-2-Butene		
C_4H_8	C=C π electron	9.1[a]
	C—C	11.2, 12.4[a]
Acetylene (Ethyne)	C≡C π electron	11.40[b]
	C—C σ electron	16.7[b]
Benzene		
C_6H_6	C=C (aromatic π electrons)	9.25, 12.2[b]
	C—C electrons (5 of the 6 orbitals)	11.5, 15.45, 19.02, 22.97, 25.9[b]
Methanol		
CH_3OH	O—H	10.8[a]
Ethanol		
C_2H_5OH	O—H	10.5
Phenol		
C_6H_5OH	C=C (aromatic π electrons)	8.5, 9.4, 11.3
	O—H	
Acetone		
$(CH_3)_2C{=}O$	C=O	9.7
2-Butanone		
$CH_3COCH_2CH_3$	C=O	9.5

[a] From M. J. S. Dewar and S. D. Worley, *J. Chem. Phys.* **50,** 694 (1969).

[b] From E. Heilbronner and J. P. Maier, in *Electron Spectroscopy: Theory, Techniques and Applications,* C. R. Brundle and A. D. Baker, eds. (Academic Press, London, 1977), 1.

the broader class of unsaturated compounds of carbon. The classic example is, of course, the simplest aromatic hydrocarbon, benzene (C_6H_6). The cyclic benzene molecule contains six equivalent carbons, one hydrogen atom attached to each carbon, and *six equivalent carbon–carbon bonds.* The simplest valence picture of benzene is the classic pair of Kekulé structures, written with alternating double and single bonds:

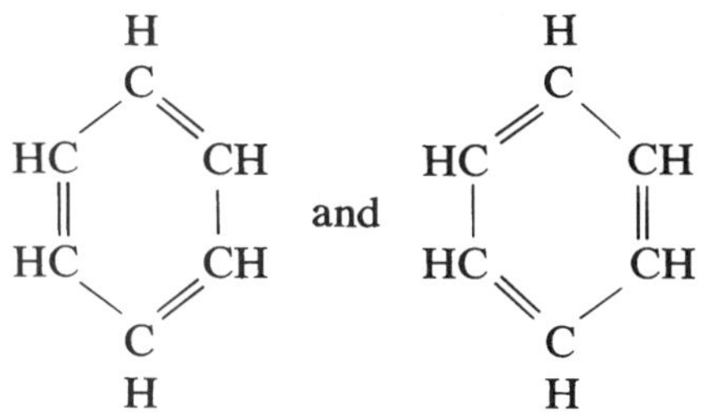

Yet experiment shows clearly that all the carbon–carbon bonds have identical properties; for example, all are 1.397 Å long, intermediate between the "normal" lengths of C—C (1.537 Å) and C═C (1.335 Å) bonds. The simplest way to represent this equivalence theoretically is to make the molecular wave function an equal mixture (superposition) of the wave functions corresponding to the two Kekulé structures. The equivalence of the bonds is taken to mean that the π electrons constituting the formal double bonds in these structures are so mobile that, on any conceivable experimental time scale for an experiment, the π electrons are best represented as a superposition of wave functions corresponding to the two Kekulé structures.

Our most significant clue to the properties of benzene comes from the molecule's energy. The total energy of all the bonds (the energy required to separate the molecule into its constituent atoms) is measured to be about 1.6 eV greater than the total energy of the bonds in a single Kekulé structure, calculated from "standard" C—C and C═C bond energies derived from molecules such as propylene (CH_3—CH═CH_2). In Section 14.8 we describe how these standard bond energies are obtained. The difference between the real and hypothetical energies of all the bonds is called the *resonance energy.* The historical basis is that, in a valence bond representation, the most important contribution to the difference of real and hypothetical energies is the delocalization achieved by superposing two physically equivalent—and therefore resonant—Kekulé structures. It is generally accepted that delocalization is the most important factor giving rise to resonance energies. As in the H_2 molecule, however, the superposition of only two valence bond structures, although a good first approximation, is quite inadequate for a quantitative description of benzene.

A second clue to the special character of benzene and other aromatic hydrocarbons is their relative chemical inertness compared with hydrocarbons containing "normal" double bonds, such as ethylene (CH_2═CH_2) or butadiene (CH_2═CH_2—CH═CH_2). Molecules such as ethylene, the unsaturated or olefinic hydrocarbons, are quite reactive toward the addition of halogen molecules,

$$CH_2{=}CH_2 + Br_2 \rightarrow CH_2Br{-}CH_2Br,$$

or of hydrogen halides,

$$CH_2{=}CH{-}CH_3 + HBr \rightarrow CH_3{-}CHBr{-}CH_3.$$

Although olefinic molecules undergo such reactions readily, aromatic molecules such as benzene, naphthalene, and anthracene add halogens, hydrogen halides, and other species only under rather severe conditions. In other words, the π electrons of the aromatic systems appear on chemical grounds to be more strongly held than those of olefins.

A third property of aromatic systems that suggests a special character for their π electrons comes from the behavior of these molecules in magnetic fields. As we mentioned in Section 5.1, one can use the protons of hydrogen atoms in molecules as probes of the local magnetic field in their vicinity (proton magnetic resonance or nuclear magnetic resonance spectroscopy). Such measurements show that the protons of benzene experience a local magnetic field significantly smaller than that observed at protons in typical nonaromatic hydrocarbons. The interpretation that best describes the origin of this lower field is that the external magnetic field induces a ring current, in which the delocalized π electrons circulate rapidly around the ring. This ring current in turn induces a local magnetic field that acts to *oppose* the external field outside the current loop, where the hydrogen atoms of benzene lie. Aromatic compounds are virtually unique in exhibiting this property, which is so characteristic of delocalized electrons.

Before we go on to interpret the electronic structure of aromatic molecules, we should point out two quantities that can be used only with great caution as indications of aromatic behavior. One is the ionization potential; the other is bond length. Ionization potentials of aromatic and olefinic molecules are in fact rather similar. By careful choice one can arrive at suitable comparisons. For example, the first ionization potential of a hypothetical cyclohexatriene, corresponding to a single Kekulé structure, would surely be less than the first ionization potentials of cyclohexene, cyclopentene, or cyclopentadiene (8.9, 9.1, and 8.5 eV, respectively), yet the first ionization potential of benzene is 9.25 eV. This further illustrates the higher binding energy (of the π electrons) as a result of delocalization. Yet care must be used in such comparisons; in particular, one must compare compounds whose sizes are comparable. The reason is that the ionization potentials of small molecules tend to be higher than those of their larger homologs. For example, ethylene has a first ionization potential of 10.5 eV, propylene 9.7 eV.

As for bond lengths, we have noted that the 1.397 Å carbon–carbon bond of benzene is shorter than the 1.537 Å C—C bond of ethane but longer than the 1.338 Å C—C bond of ethylene. It is thus frequently said that the amount of double-bond character (or π-bond order) in benzene is intermediate between that in ethane (single bond, or π-bond order of 0) and that in ethylene (double bond, corresponding to a π-bond order of 1). The π-bond order is the bond order p minus the σ-bond order, normally 1. There is a very good correlation between bond length R_e and MO bond order p (cf. Section 7.6) for carbon–carbon bonds, proposed by C. A. Coulson,

$$p = \frac{1.02 - 0.53R_e}{0.235R_e - 0.16}, \tag{9.4}$$

where R_e is expressed in angstroms. In benzene the MO calculation gives a bond order of $1\frac{2}{3}$, in agreement with Eq. 9.4. The difficulty in the present context is in the level of subtlety required to isolate a connection between bond length or bond order and aromatic character, itself a concept more qualitative than quantitative. To some extent the gross behavior of carbon–carbon bond lengths can be interpreted in terms of the hybridization of the two carbon atoms, as Table 9.3 indicates. Several "single" C—C bonds, such as those of vinyl cyanide, CH_2═CH—C≡N, and cyanoacetylene, HC═C—C≡N, are comparable to the "aromatic" bond lengths of benzene and graphite. Presumably this is because the π electrons nominally in the adjacent multiple bonds are actually somewhat delocalized. Delocalization is especially likely to occur when "single" and "multiple" bonds alternate (*conjugated* multiple bonds), as in butadiene, vinyl cyanide, and cyanocetylene—and in the nominal structures of benzene and other aromatic ring compounds. A generalization better than saying, "All single bonds have about the same lengths; all double bonds likewise" is the more detailed statement, "All single bonds between carbons whose other bonds have the same nominal order have about the same length; likewise for double and triple bonds." The single C—C bonds in butadiene, hexatriene, and higher homologs are much more alike than the single bonds of ethane, propene, and butadiene.

Table 9.3 Carbon–Carbon Bond Lengths for Various Compounds[a]

Molecule	Bond Type	C—C Bond Length (Å)
Ethane, CH_3—CH_3	sp^3–sp^3	1.537
Propene, CH_2═CH—CH_3	sp^3–sp^2	1.51
Methyl acetylene, CH≡C—CH_3	sp^3–sp	1.459
Butadiene, CH_2═CH—CH═CH_2	sp^2–sp^2	1.476
Vinyl cyanide, CH_2═CH—C≡N	sp^2–sp	1.426
Cyanoacetylene, CH≡C—C≡N	sp–sp	1.376
Benzene, C_6H_6 (any bond)	sp^2–sp^2	1.397
Graphite (any bond)	sp^2–sp^2	1.421
Ethylene, CH_2═CH_2	sp^2–sp^2	1.338
Allene, CH_2═C═CH_2	sp^2–sp	1.309
Butatriene, CH═C═C═CH	sp–sp	1.285
Acetylene, CH≡CH	sp–sp	1.205

[a] From H. J. Bernstein, *Trans. Faraday Soc.* **57**, 1649 (1961).

The simplest theory for aromatic hydrocarbons, the *Hückel molecular orbital* (HMO) *theory*, is virtually an archetype for simple, phenomenological models. One treats explicitly only the π electrons, which are assumed to be responsible for such chemical and physical properties as the optical spectrum, the ease of attack of particular sites by other reagents, the local charge densities on atoms and in bonds, and the bond orders. The rest of the electrons and the nuclei, which define the geometry of the usually planar molecular skeleton but not necessarily the precise values of bond lengths, appear only as part of an effective potential field for the π electrons. The π electrons are described by molecular orbitals ψ_j which are represented as linear combinations of the carbon $2p\pi$ orbitals φ_k, the $2p$ orbitals extending perpendicular to the molecular plane. For benzene we have

$$\psi_j = \sum_{k=1}^{6} c_{jk}\varphi_k; \tag{9.5}$$

note that each of these π MOs extends over the entire ring. The coefficients c_{jk} may always be determined by the variation method, as outlined in Section 8.4. To simplify the process, the HMO theory makes further assumptions. The only interactions large enough to require inclusion in the Hamiltonian are those (1) between an electron on a given carbon atom and the other electrons and nucleus of that atom, and (2) between an electron on a given carbon atom and the electrons and nuclei of adjacent atoms to which that atom is bonded. In benzene each of these classes of interactions is the same for all the carbon atoms, so we can define

$$\alpha \equiv \int \varphi_j{}^* \mathsf{H}_{\text{eff}} \varphi_j \, d\tau \qquad (\text{for all } j), \tag{9.6}$$

$$\beta \equiv \int \varphi_j{}^* \mathsf{H}_{\text{eff}} \varphi_k \, d\tau \qquad (\text{atoms } j \text{ and } k \text{ bonded}), \tag{9.7}$$

where H_{eff} is the effective one-electron Hamiltonian for a π electron. In terms of Eq. 8.20, we set $H_{jj} = \alpha$ for all j, $H_{jk} = \beta$ for bonded atoms, $H_{jk} = 0$ otherwise. The values of α and β are empirical parameters to be determined from experiment. The overlap integrals S_{jk} of Eq. 8.20 are taken as unity on a given atom, zero between different atoms, or $S_{jk} = \delta_{jk}$. Substituting these approximations into the secular Eq. 8.24, we have for benzene

$$\begin{Vmatrix} \alpha-\epsilon & \beta & 0 & 0 & 0 & \beta \\ \beta & \alpha-\epsilon & \beta & 0 & 0 & 0 \\ 0 & \beta & \alpha-\epsilon & \beta & 0 & 0 \\ 0 & 0 & \beta & \alpha-\epsilon & \beta & 0 \\ 0 & 0 & 0 & \beta & \alpha-\epsilon & \beta \\ \beta & 0 & 0 & 0 & \beta & \alpha-\epsilon \end{Vmatrix} = 0, \tag{9.8}$$

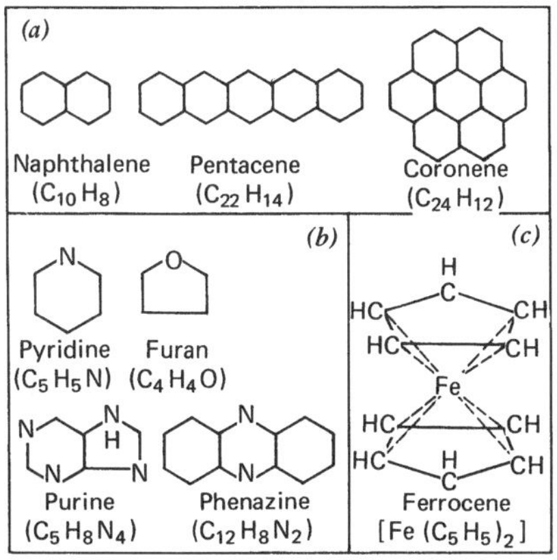

Figure 9.6 Some aromatic compounds. (*a*) Fused benzene rings. (*b*) Heterocycles. (*c*) Ferrocene. Only the skeletons are shown; π bonds are omitted.

the lowest three roots of which are $\epsilon_1 = \alpha + 2\beta$, $\epsilon_2 = \epsilon_3 = \alpha + \beta$, because $\beta < 0$. Putting six electrons in these three orbitals gives a total π-electron energy of $6\alpha + 8\beta$; in the same approximation a localized π bond has the energy[10] $\alpha + \beta$, so one of the Kekulé structures has the π-electron energy $6\alpha + 6\beta$. The *delocalization energy* of benzene is thus -2β, which should equal the empirical resonance energy; thus, we estimate $\beta \approx -0.8$ eV. Similar calculations for a number of aromatic compounds give values of β that cluster closely around –0.69 eV (–67 kJ/mol). This indicates clearly that in these compounds delocalization is too important to be neglected.

Benzene is of course not the only type of aromatic compound. Many such compounds consist of fused benzene rings, with the π electrons effectively delocalized over the entire ring system; some examples are shown in Fig. 9.6*a*. Note that the fraction of hydrogen atoms decreases as the size of the system increases. In the limit we have pure carbon in the form of graphite, which consists of parallel infinite sheets of fused six-carbon rings (see Chapter 11). The C—C distance in graphite is 1.421 Å, closer to the 1.395 Å typical of aromatic compounds than to the 1.54 Å of a single bond or the 1.335 Å of an ethylenic double bond. The parallel sheets, on the other hand, are 3.35 Å apart and bound together very weakly by overlap of the π-electron clouds. Hence the layers slide over or off one another quite easily, the property that makes graphite so suitable for pencil leads. There are also aromatic systems with rings containing one or more atoms other than carbon (heterocyclic compounds); Fig. 9.6*b* shows several, including pyridine (isoelectronic with benzene), furan (also with six π electrons), and some fused systems. Also isoelectronic with benzene is the $C_5H_5^-$ cyclopentadienyl) ion, which forms the remarkable series of "sandwich" compounds typified by *ferrocene* (dicyclopentadienyliron), Fig. 9.6*c;* the bonding in the rings is aromatic, and the π electrons bond to iron by overlapping with empty 3*d*-orbitals on the iron atom. Benzene forms similar "sandwich" compounds, such as $Cr(C_6H_6)_2$.

We have outlined the principles that underlie the great number and variety of carbon compounds. A particularly rich assortment of these compounds are found in living systems, each evolved to fit a specific function. In addition to the variations in the carbon skeleton, the many kinds of active functional groups that can be attached to that skeleton result in a wide range of chemical behavior. Another significant factor is the stability of so many carbon compounds under what we call "ordinary conditions." Chemistry would surely be very different in an atmosphere containing much hydrogen rather than oxygen, or at an ambient temperature of 500°C. But as things are, only carbon appears to form such a myriad of compounds with such a varied chemistry.

9.3 Other Extended Structures

Carbon may be unique in the variety of its extended structures, but it is not the only element that forms them. Let us now look at some examples from among the inorganic substances.

We have already mentioned the Si—O—Si linkage so important to silicon chemistry. Many silicon–oxygen compounds have structures built from interlinked SiO_4 tetrahedra. Such compounds typically have a framework of polymeric silicate anions, with cations simply inserted in gaps in the structure. These anions may form infinite chains, flat or pleated sheets, or three-dimensional networks. Typical structures for the first two cases are shown in parts *a* and *b* of Fig. 9.7. These structures are reflected in the physical properties of the materials. Asbestos, a chain silicate, readily splits into thin fibers that are easily blown about or carried into the lungs, which is apparently part of the reason it is such a health hazard, whereas mica, a sheet silicate at the molecular level, cleaves into thin sheets. A three-dimensional silicon–oxygen network in which every oxygen atom is bound to two silicon atoms is simply silica, SiO_2, the substance of common sand.

If some of the silicon atoms are replaced by aluminum atoms, the resulting structures are polymeric anions. The charge-balancing cations in these structures are rather loosely attached to the oxide sites, whereas the anionic framework is both rigid and rather open. As a consequence, these aluminosilicates make good cation exchangers. Silicate and aluminosilicate minerals make up most of the earth's crust.

Other extended structures are formed by organosilicon compounds, in which carbon atoms are directly bonded to silicon. These are called *siloxanes* when the framework is still built up from the basic Si—O—Si linkage, with organic groups replacing the singly bound oxygen atoms of the

[10] The localized-bond secular equation is

$$\begin{Vmatrix} \alpha-\epsilon & \beta \\ \beta & \alpha-\epsilon \end{Vmatrix} = 0,$$

giving $\epsilon = \alpha \pm \beta$, with $\epsilon = \alpha + \beta$ lower in energy.

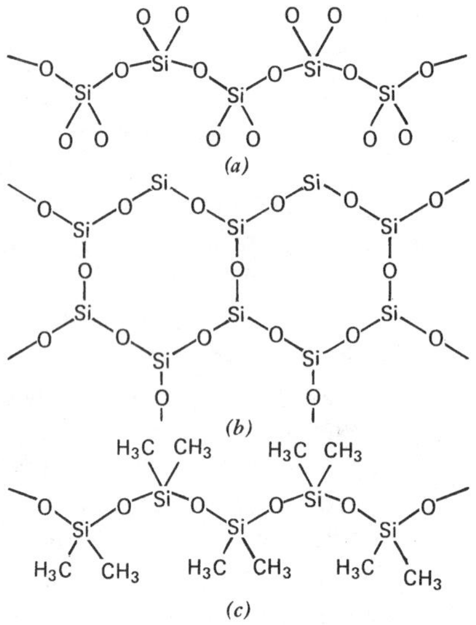

Figure 9.7 Structures of some silicon compounds. (*a*) A chain silicate $(SiO_3^{2-})_n$. (*b*) A sheet silicate $(Si_2O_5^{2-})_n$. Each Si atom is bound to four O atoms, the fourth being alternatively above or below the sheet. (*c*) A linear polysiloxane $[(CH_3)_2SiO]_n$.

polysilicates. In Fig. 9.7*c* we show such a linear polysiloxane, the structure of which is quite analogous to the linear silicate of Fig. 9.7*a;* similarly, there are sheet polysiloxanes with structures like Fig. 9.7*b*. These polymers are the materials known commercially as "silicones."

Let us mention briefly some extended structures formed by other nonmetallic elements, starting with sulfur. We have already referred to the polymeric forms of the element. Sulfur vapor contains all the species from S_2 to S_{10}, most of which are probably rings. Liquid sulfur at temperatures near the normal melting point consists primarily of S_8 rings which, with increasing temperature, polymerize into very long chains (—S—S—S—S—). In aqueous solution sulfur forms polysulfide, ions, S_n^{2-}, and polythionate ions, $(O_3S—S_n—O_3)^{2-}$, both of which have extensive chemistry. "Silicon disulfide," $(SiS_2)_n$, has a structure rather like the linear polysilicates, consisting of SiS_4 tetrahedra joined by pairs of sulfur atoms:

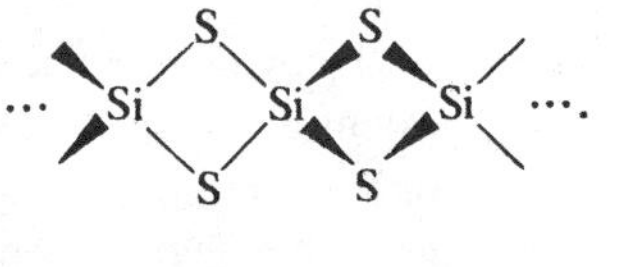

Although neither ozone nor SO_2 polymerizes, the analogous SeO_2 forms infinite chains,

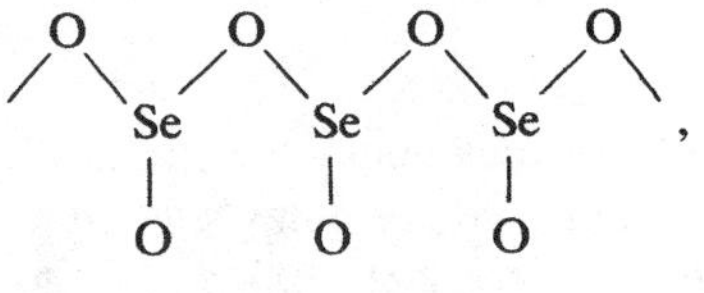

and solid SO_3 has a number of polymeric forms including the silicate-like

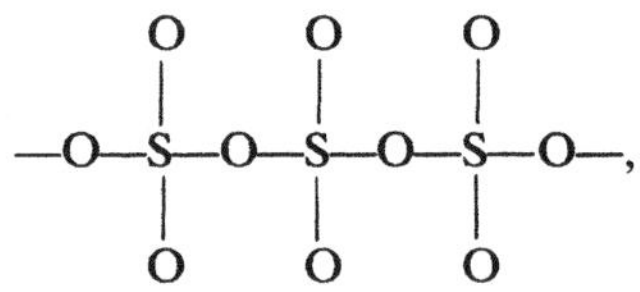

which, as one might expect, resembles asbestos in its properties.

Phosphorus forms a number of interesting structures. The vapor, the liquid, and the "white" form of the solid all consist of tetrahedral P_4 molecules, Fig. 9.8*a;* the other solid forms are polymers of various kinds. Since the P—P—P angle in P_4 is only 60°, the structure is highly strained (see Section 9.4); this is associated with the high reactivity of phosphorus. The tetrahedral arrangement of P atoms is retained in the common oxides, shown in parts *b* and *c* of Fig. 9.8, and in a similar but more complex series of phosphorus sulfides. The orthophosphate anion, PO_4^{3-}, is also tetrahedral; as with the silicates, these tetrahedra can join to form many types of condensed phosphates with P—O—P linkages, some forming infinite chains. There are a number of phosphonitrilic compounds of the general formula $(PNX_2)_n$, where X is a halogen atom, an organic group, and so on; for large *n* these are linear polymers with alternating P and N atoms, having a structure like the polysiloxanes. The trimers such as $(PNCl_2)_3$, are cyclic and nearly planar, with the structure

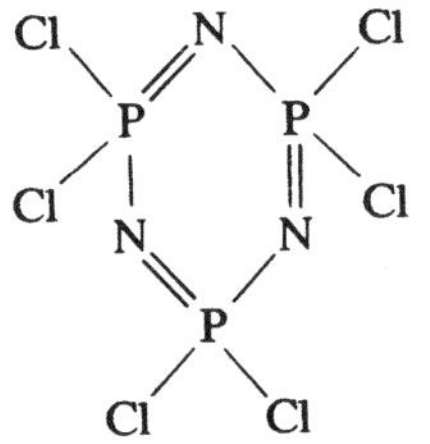

in which the ring is quasi-aromatic, with delocalized π bonding rather like that in benzene.

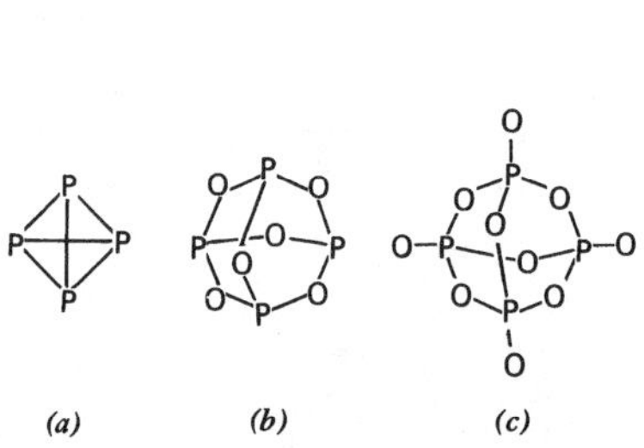

Figure 9.8 Phosphorus and its oxides. (*a*) Tetrahedral P_4 molecule. (*b*) "Phosphorus trioxide," P_4O_6. (*c*) "Phosphorus pentoxide," P_4O_{10}. The "trioxide" and "pentoxide" names derive from empirical formulas, P_2O_3 and P_2O_5, respectively.

Another quasi-aromatic compound is borazine, $B_3N_3H_6$, with the structure

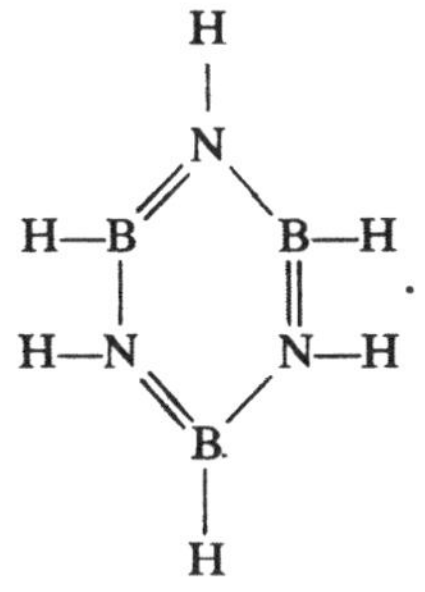

The compound is isoelectronic with benzene and has somewhat similar properties, but is more reactive because of its polar B—N bonds. There are many other compounds in which carbon is replaced by the boron–nitrogen combination. The most notable of these is boron nitride, BN, which in the solid state has two structures exactly paralleling those of diamond and graphite; the diamondlike material ("borazon") is even harder than diamond itself.

Although all the compounds discussed thus far are entirely nonmetallic, many metallic elements can also form extended structures. For example, tungsten forms a series of elaborate anions with oxygen; some examples are $W_6O_{21}^{6-}$ and $W_{12}O_{41}^{10-}$. These are called *isopolytungstates,* and compounds of this type are generally referred to as anions of *isopoly acids.* (The "iso" means that only a single element is present besides oxygen and hydrogen.) Similar *heteropoly* anions also exist in which two or more metals (or nonmetals such as Si, As, and P) are present, for example, $Co_2W_{12}O_{42}^{8-}$ or $PMo_{12}O_{40}^{3-}$; however, a single metal always predominates, most commonly tungsten or molybdenum. These anions are typically built up from 6-coordinated metal atoms at the centers of octahedra, with the WO_6 or MoO_6 octahedra joined in much the same way as the SiO_4 and PO_4 tetrahedra described earlier: Shared vertices correspond to metal atoms sharing single oxygen atoms (W—O—W), whereas shared edges correspond to the sharing of pairs of oxygen ligands

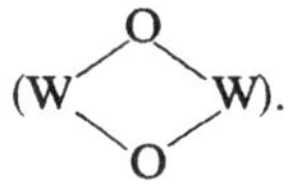

Remarkably, a simple electrostatic model largely accounts for the relative sizes and stabilities of these aggregates of octahedra, some of which are shown in Fig. 9.9, and even for the nonexistence of certain species. The electrostatic calculations similar to those we shall describe for ionic crystals in Sections 11.8 and 11.9 indicate that structures in which octahedra share edges increase in stability as more octahedra join on, themselves arranged into polyhedra. The increasing amount of attractive energy from oppositely charged nearest-neighbor atoms then outweighs the repulsive contributions from next-nearest neighbors—up to the point at which a closed "superpolyhedron" is formed, with maximum symmetry and the largest possible number of shared edges, but with much less stability than its less complex congeners. These predictions are in excellent accord with what is observed in nature. "Open" clusters of polyhedra are common, whereas closed, simple polyhedra are very rare.

Although the species we have discussed in this section have a variety of strange structures, the bonding in all of them is relatively conventional. That is, in each case one can construct bonding orbitals corresponding to the bonds in ordinary chemical formulas, and there are enough valence electrons available to fill all these bonding orbitals. Sometimes the best orbitals are delocalized, but this does not affect their bonding capacity. For example, in benzene the three delocalized π orbitals hold the same six electrons as the three π bonds of the Kekulé formula. Now we shall look at some species for which this is no longer true, the so-called electron-deficient molecules, best typified by the boron hydrides.

The boron hydrides (or *boranes*) range from the unstable BH_3 through B_2H_6 (diborane), B_4H_{10}, B_5H_9, B_5H_{11}, B_6H_{10}, B_6H_{12}, B_8H_{12}, B_8H_{18}, B_9H_{15}, $B_{10}H_{14}$ (decaborane, the most stable member of the series), $B_{10}H_{16}$, and $B_{18}H_{22}$ (two isomers) to $B_{20}H_{16}$. At room temperature B_2H_6 and B_4H_{10} are gases, the other species below B_{10} are liquids, and the rest are solids. Although diborane is thermally stable at room temperature, most others below B_{10} are not, and all the species below B_8 react readily—sometimes violently—with air or water, largely because of boron's strong affinity for oxygen. Boron and hydrogen also form such anions as BH_4^- (the tetrahedral borohydride ion), $B_3H_8^-$, $B_{10}H_{10}^{2-}$, $B_{12}H_{12}^{2-}$, and $B_{20}H_{18}^{2-}$.

The structures of some of the boron hydrides are illustrated in Fig. 9.10. These structures have been interpreted as based on an icosahedral frame of 12 boron atoms, most clearly seen in the regular icosahedron of the $B_{12}H_{12}^{2-}$ ion. The smaller species are largely incomplete icosahedrons, whereas the larger ones consist of these fragments fused in various ways. We may also note here the *carboranes,* compounds in which the framework is made of both boron and carbon atoms. The best known is *ortho*-carborane, $B_{10}C_2H_{12}$, which is isoelectronic with $B_{12}H_{12}^{2-}$ and has essentially the same structure.

Now what do we mean by calling the boron hydrides "electron-deficient molecules"? Consider the B_2H_6 molecule: Since there are eight atoms, at least seven conventional two-electron bonds—14 electrons—would be needed to join the molecule together; but there are only 12 valence electrons available. Thus some of the bonding orbitals must either be unfilled or extend over more than two atoms; we shall see that the latter is the case in B_2H_6. Such species naturally tend to be less tightly bound than molecules with conventional bonds, that is, molecules in which all the bonding orbitals are normally occupied. To interpret how bonding is

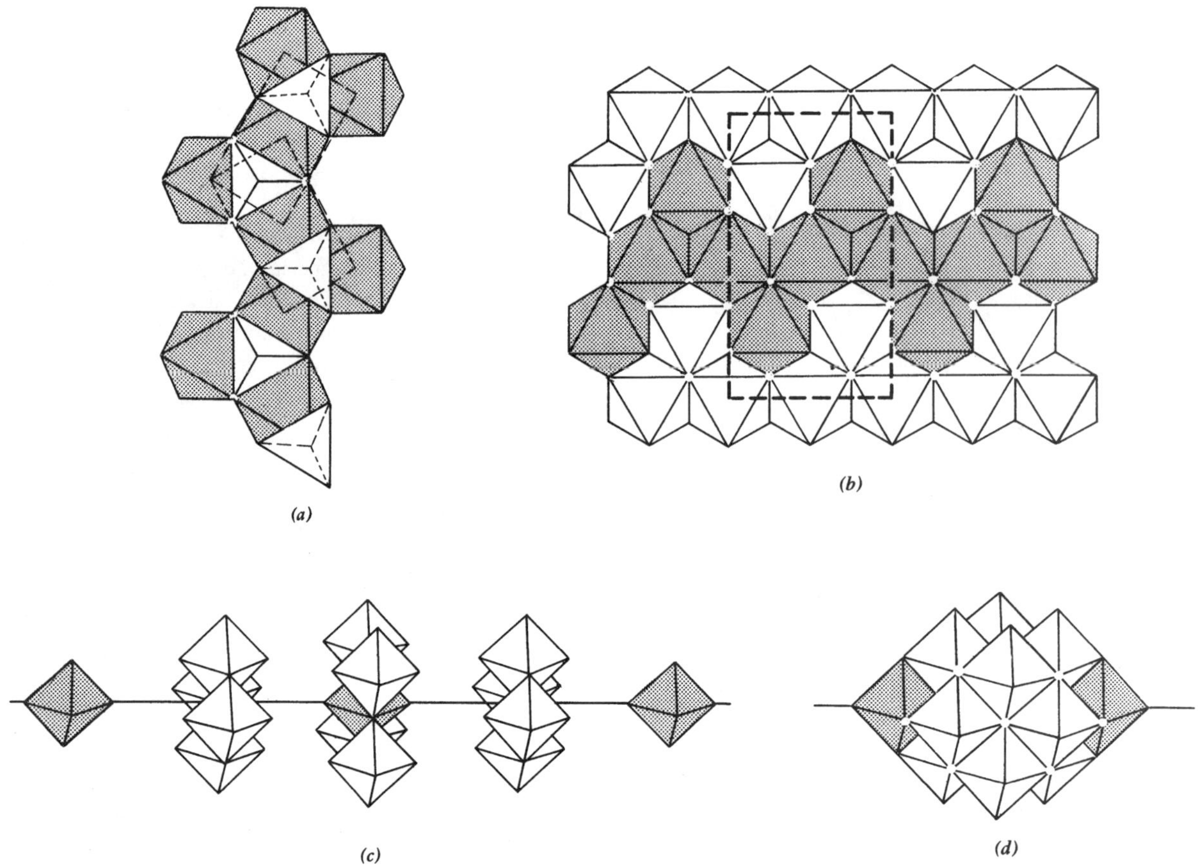

Figure 9.9 Typical structures that are well represented by octahedra with shared edges; (*a*) and (*b*) represent olivine, the basaltic material comprising approximately 65% of the earth's upper mantle, whose composition is (Mg, Fe)$_2$SiO$_4$—that is, a variable ratio of Mg/Fe, but always two metal ions per silicate; (*c*) and (*d*) represent an exploded view and a condensed picture of the synthetic mineral $Mg_3V_{12}O_{40}$, in which the shaded octahedra represent the magnesium ions with their oxygen ligands and the unshaded octahedra represent the vanadium ions. In (*a*), the repeat units of olivine are shown; in (*b*), one sees the extended structure. Figure provided by P. B. Moore, with permission.

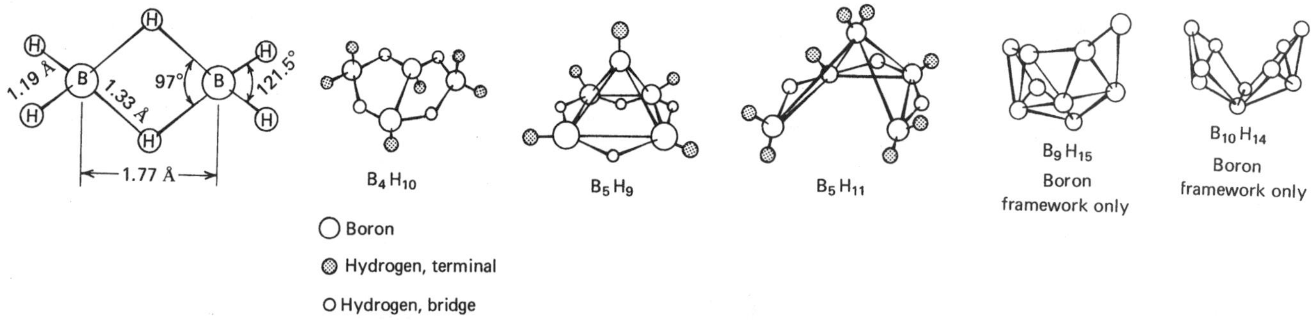

Figure 9.10 Structures of representative boron hydrides. [Cf. F. A. Cotton and G. Wilkinson, *Advanced Inorganic Chemistry,* 2nd ed. (Wiley-Interscience, New York, 1966), p. 276; and W. N. Lipscomb, in H. J. Emeléus and A. G. Sharpe (Eds.), *Advances in Inorganic Chemistry and Radiochemistry,* Vol. 2 (Academic, New York, 1960), p. 279.]

accomplished in the boron hydrides, W. N. Lipscomb found it useful to distinguish three kinds of bonding orbitals. One of these is the conventional two-electron bond between adjacent atoms, thus called a *two-center bond.* These bonds are formed in the same way as any of the other two-center bonds we have discussed. The terminal B—H bonds in B_2H_6 are of this type.

The second kind of bond is the three-center, two-electron bond involving two borons and a bridging hydrogen atom. Bonding of this type is necessary to account for the B—H—B bridges in B_2H_6. Normally, two electrons occupy the lowest-energy molecular orbital that can be constructed from the participating atomic orbitals of the three atoms. We have already looked at one such three-center orbital in

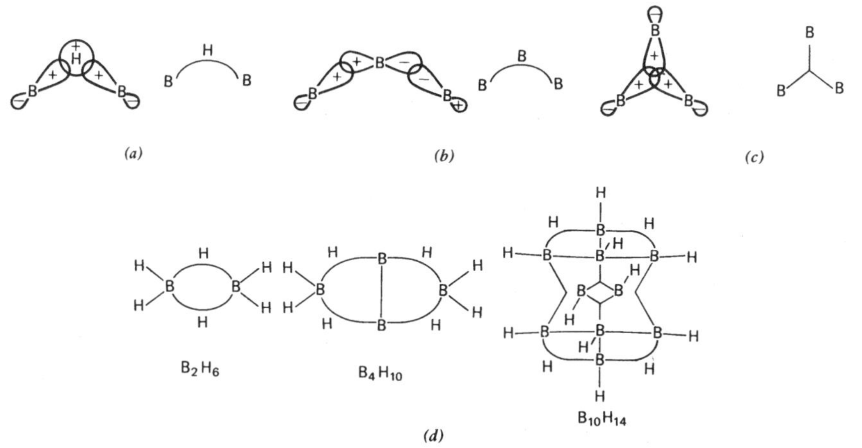

Figure 9.11 Bridge bonding in boron hydrides. (*a*) B—H—B bonding orbital, formed from B sp^3 orbitals and an H 1*s* orbital, as in B_2H_6. (*b*) B—B—B bonding orbital with properties of the central atom different from those of the outer two, as in B_4H_{10}. (*c*) B—B—B bonding orbital between three nearly equivalent atoms, as in $B_{10}H_{14}$. (*d*) Bonding diagrams of some boron hydrides, using the notation indicated in parts (*a*)–(*c*).

detail, the lowest orbital of the H_3 molecule (Section 8.1), which is simply the sum of 1*s* orbitals on the three H atoms. The lowest-energy three-center MO has only as many nodes as the constituent atomic orbitals (in H_3, no nodes at all); that is, the atomic orbitals combine to give only *constructive* interference.[11] The B—H—B bonding orbital is best considered as the sum of two tetrahedral sp^3 orbitals, one from each boron atom, and the 1*s* orbital of the bridging hydrogen, as shown in Fig. 9.11*a;* the B_2H_6 molecule has two such orbitals, each occupied by two electrons. The same atomic orbitals can combine to give two higher-energy orbitals with interatomic nodes, but these are normally empty. The average energy of the B—H—B bond is only about 0.75 eV (72 kJ/mol), so that, doubly occupied, it contributes the 1.5 eV needed to hold two BH_3's as a B_2H_6. Note how weak such bonds are relative to the typical two-center bond (~3–5 eV).

The third type of bond found in boron hydrides is the three-center, two-electron homonuclear (B—B—B) bond. Two forms of this bond are distinguished. In one form, Fig. 9.11*b,* the three atoms form an isosceles triangle, and the central atom has properties quite different from the other two. In the other form, Fig. 9.11*c,* the three atoms are nearly equivalent. In Fig. 9.11*d* we illustrate how the bonding in some of the boron hydrides is accounted for in terms of the various types of bonds we have described.

Three-center bridge bonds are by no means restricted to the boron hydrides. The ordinary hydrogen bond (Section 7.9) can be interpreted in terms of a three-center orbital to which one of the end atoms makes a much greater contribution than the other—except in the symmetric FHF^- ion, which has a bonding energy (~1.2 eV) comparable to that of the B—H—B bond. In the free H_2F_2 molecule, apparently the two hydrogen atoms act alternately as bridges, exchanging roles rapidly; this contrasts with the boron hydrides, where the hydrogen atoms are rather rigidly fixed. Three-center bonds also occur in a variety of dimeric and polymeric species in which halogen atoms act as bridges. The dimers Al_2Cl_6 and Be_2Cl_4 are examples of this, with two Cl atoms in each forming bridges and the others forming normal metal-halogen bonds; and solid beryllium chloride has the polymeric structure

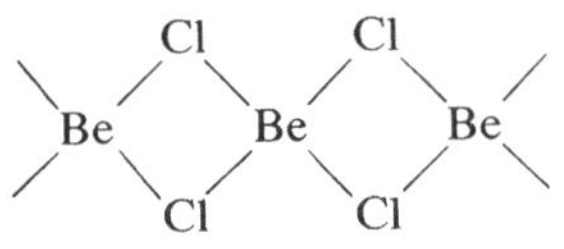

made up of joined tetrahedra like the (conventionally bound) SiS_2.

9.4 Some Steric Effects

In this section we consider a number of effects that influence molecular structure on a scale larger than that of localized bonding. Small molecules such as H_2O or CH_4 have nearly rigid structures, in that the distances between all pairs of atoms, not just those directly bonded to each other, remain constant, except for small variations due to vibration. In larger molecules this is frequently not the case, and non-bonded atoms can assume various positions relative to one another. Such molecules can assume different geometric *conformations* of the atoms in space that can be reached from one another without breaking bonds. Sometimes the molecules are quite flexible, and can assume many different conformations with nearly equal ease. Other large molecules prefer quite specific conformations as the result of many individually weak interactions, such as hydrogen bonds or the simple fact that atoms get in each other's way. When one conformation is appreciably more stable than others, it is useful to think of the molecule as having a *secondary structure,* as opposed to the *primary structure* defined by the chemical bonds. Let us see what factors govern the capacity of molecules to take on different conformations.

Certain general pattems are immediately obvious: In general, closed structures—rings, three-dimensional polyhedra, etc.—are particularly rigid, whereas open chainlike structures are flexible. What is the nature of the internal motions

[11] Cf. Fig. 8.11 for examples of possible MOs in a homonuclear three-center system.

associated with this flexibility? It certainly is not a matter of bond stretching, since the force constants for stretching are the largest among all types of molecular vibrations; because of this "stiffness," the energy levels for stretching modes are typically quite widely spread, of the order of 0.1–0.3 eV apart. The bending of bond angles is easier, but not enough so to account for the loose nature of open-chain molecules. Indeed, bending modes must be moderately stiff to account for the rigidity of closed structures. The answer to our question lies in a third kind of motion, which we have not considered up to this point, *intemal rotation,* the rotation of two groups of atoms relative to each other about the bond joining them. If the electron density in the bond has cylindrical symmetry we would expect this rotation to occur freely; this is generally true of single (σ-type) bonds,[12] the electron density in which is concentrated symmetrically along the bond axis.

Examined more closely, internal rotation is never completely free except in linear molecules. Even molecules with linear skeletons, such as dimethylacetylene, $H_3C—C{\equiv}C—CH_3$, have small barriers to rotation of the two CH_3 groups, relative to one another. Even if the bond itself is cylindrically symmetric, the interactions between more distant atoms cannot be. These interactions are always small, but not always negligible. Consider the ethane ($H_3C—CH_3$) molecule, in which the relevant interactions are the repulsions between the hydrogen atoms of the two methyl groups. These repulsions create a potential barrier to rotation about the C—C bond, as shown in Fig. 9.12*a*. The equilibrium conformation is one in which one methyl group is "staggered" with respect to the other, so that we could see all six hydrogen atoms if we looked down the C—C axis. The "eclipsed" configuration, in which the hydrogen atoms are lined up, has an energy about 0.127 eV higher (1024 cm^{-1} in E/hc). Thus there is a potential energy barrier to rotation every 120°. This does not mean that rotation cannot occur. As with any barrier of finite height, the system can cross over or even tunnel through these barriers (Section 4.1). In each of the potential wells there are harmonic oscillator-like energy levels corresponding to the energy of torsional oscillation about the minimum. The true stationary states of the

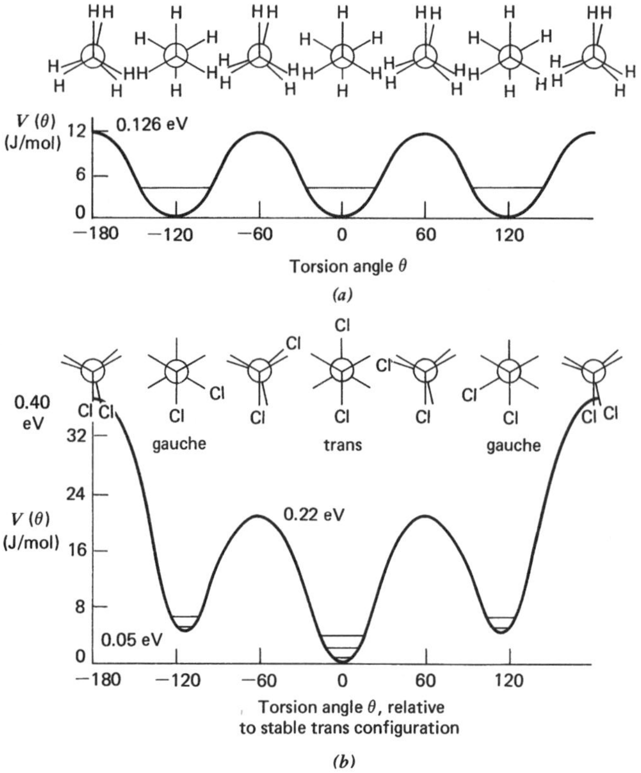

Figure 9.12 (*a*) Hindered rotation in ethane. The graph shows the internal effective potential energy $V(\theta)$ as a function of the torsion angle θ. The diagrams above the graph show the conformation at intervals of 60°, as viewed down the C—C axis. Only the "near" carbon is shown; the other is hidden behind it. The horizontal lines indicate the lowest torsional vibrational levels, but do not show the splittings associated with the multiple wells. The torsional vibration frequency $\tilde{\nu}$ is 289 cm^{-1} or 0.0358 eV. (*b*) Hindered rotation in 1,2-dichloroethane, $CH_2Cl—CH_2Cl$. Two of the potential wells in this molecule are equivalent; the third is 0.05 eV deeper, enough to support three vibrational levels lying below the lowest point of the other two wells.

system are combinations of these localized states. In ethane, the barrier is so low that we can think of this molecule as having rather free rotation about the C—C bond—except at very low temperatures, at which only the ground state for torsional oscillation is populated (cf. Chapter 21).

What we have said about the ethane molecule is generally applicable to internal rotation about single bonds. However if there are at least two kinds of substituents in each end of the bond, as in 1,2-dichloroethane, the three potential minima analogous to those of Fig. 9.12*a* are not equivalent; the internal energy of 1,2-dichloroethane is shown in Fig. 9.12*b*. The three *rotamers,* or rotational isomers, are different, even though they are interconvertible at ordinary temperatures. The first has an energy 0.5 eV lower than the second and third, which differ only in being mirror images of one another. (The mirror image aspect is discussed further in Section 9.6.) The substituents may interact strongly enough to make one rotamer truly stable, in the sense that its potential is deep enough to support a vibrational quantum

---

[12] But not of multiple bonds: Rotation cannot occur about a π bond without destroying the overlap of p orbitals that creates the bond, i.e., without breaking the π bond. This is why

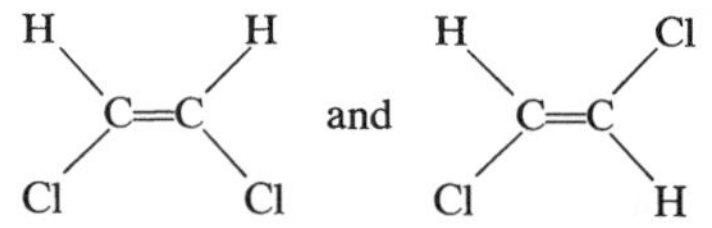

are chemically distinct species with different properties (geometrical isomers), whereas at ordinary temperatures we distinguish only one 1,2-dichloroethane,

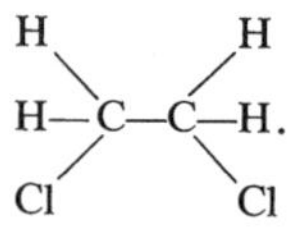

state below the minima of the less stable rotamers. This situation occurs with 1,2-dichloroethane, as the vibrational levels in Fig. 9.12*b* indicate. A more extreme example is the ring compound cyclooctene, which has a stable *cis* form and an unstable *trans* form:

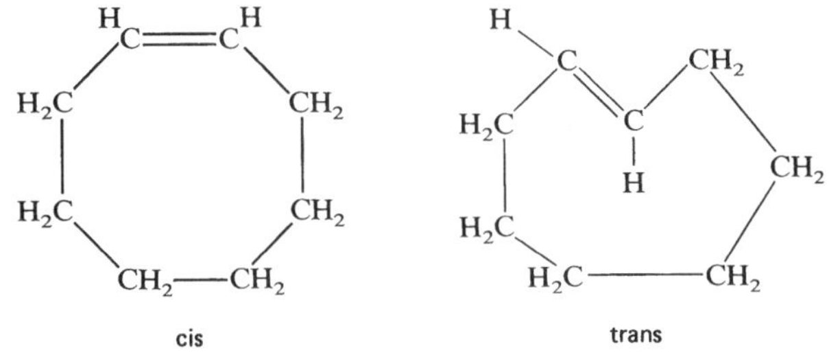

Rotation about the double bond is highly hindered in this system, as it is in ethylene, but here we also see the additional feature of an asymmetric situation in which the ground state of the stable form has an energy well below that of the unstable form.

Even if the rotameric forms of a molecule have the same energy, it may be that we cannot observe the conversion of one form into another. If the groups are very large, the quantum levels for torsional oscillation (libration) must be closely spaced, so that many levels lie well down in the potential well. Recall from our discussion in Chapter 4 that the quantum mechanical wave function penetrates the classically forbidden region where the total energy is less than the potential energy, but that the wave amplitude must diminish as the wave penetrates. If the potential is relatively steep, the wave amplitude diminishes to a small fraction, say e^{-1} of its amplitude at entry within a range of about half a wavelength or less. Therefore, for a quantum state with a low quantum number, and thus a long effective wavelength and a large spatial extent, the wave function can penetrate relatively far into the classically forbidden region in a potential such as that of Fig. 9.12*b*. However the wave function of a quantum state with a high quantum number but the same total energy, corresponding to a larger mass and momentum and shorter wavelength than in the first case, must penetrate only a short distance before its amplitude becomes very small. In fact it is their effective wavelengths that determine the extent to which different waves penetrate the same barrier; penetration lengths are typically much less than a quarter-wavelength for all but the smallest barriers. This in turn implies that massive substituents are much less effective than very light substituents at penetrating barriers far enough to tunnel through them. The only way to observe conversion of one rotamer into another in such a case is to excite the species to an energy higher than the potential barrier, where it can rotate essentially as a free rotor. Hence, if the barrier is high, rotational motion about one bond may well be a rare event.

A long-chain molecule may nevertheless change its shape if it can incur small rotations about many bonds. (Molecular ball-and-stick models give persuasive demonstrations of this behavior.) In general, unless there are specific stiffening forces between formally nonbonded atoms, such as intramolecular hydrogen bonds, long-chain molecules tend to be quite flexible in the liquid or gaseous state.

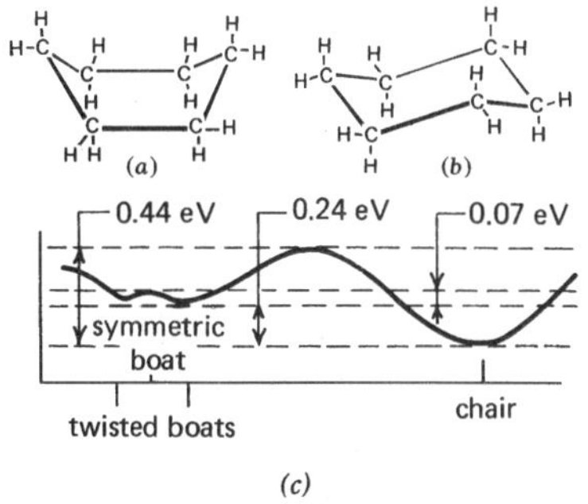

Figure 9.13 The two conformations of cyclohexane. (*a*) Boat. (*b*) Chair. (*c*) The effective potential for boat–chair interconversion, showing the slight relative minimum for twisting in the boat conformation and the large absolute minimum for the chair, with a high barrier between the minima. The coordinate of the abscissa is a complicated mixture of bond bending and bond stretching, corresponding to the lowest-energy path from one minimum to the next.

As the examples of cyclohexane and cyclooctene suggest, even ring compounds need not have rigid structures. In such molecules as cyclopentane or cyclohexane, rotation about the C—C bonds is sufficiently free to allow the molecules to take on a range of conformations. Bond angles and bond lengths are well preserved throughout these motions. But although open-chain compounds often prefer no definite conformation until they are crystallized, ring compounds usually have certain well-defined shapes that are more stable than others. The best-known examples are the "boat" and "chair" forms of cyclohexane, shown in Fig. 9.13. These two forms have almost equal energies, with the "chair" form about 0.24 eV (23 kJ/mol) more stable. The "chair" and "boat" forms can also interconvert quite readily; the barrier between them is only 0.44 eV (42 kJ/mol) high. This barrier results largely from the fact that interconversion cannot occur without *some* change in bond angles. The difference between the energies of "boat" and "chair" forms is thought to be due partly to the greater repulsion between nonbonded hydrogen atoms in the former. Repulsive interactions of this type (one speaks of "steric forces") play an increasingly important role in determining the preferred conformations as molecules become larger and less rigid.

The "natural" bond angle for the carbon atom is the tetrahedral 109°28′ (or 120° for trigonal carbon). If rotation about single bonds can occur easily, then rings of almost any size can be constructed with all the C—C—C angles close to the natural value. However, there are a few exceptions. For example, the carbon skeleton of the C_3H_6 (cyclopropane) ring is necessarily planar; in a ball-and-stick picture cyclopropane thus has internal bond angles of only 60°. As a result, the molecule is severely strained, that is, it has considerably more internal energy per CH_2 group than does a larger ring such as cyclohexane. This strain energy

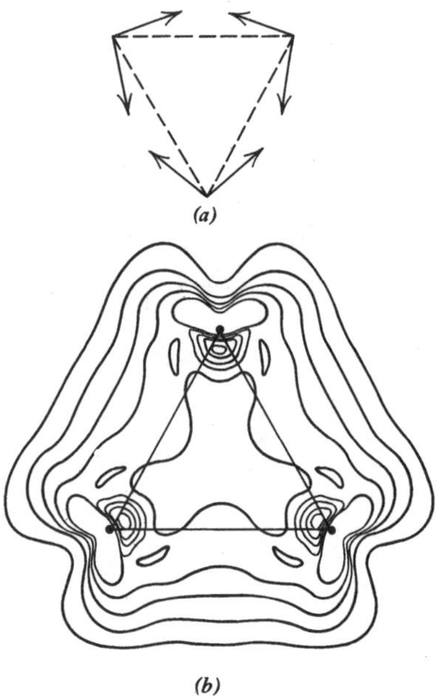

Figure 9.14 The bonds of cyclopropane, showing how the regions of maximum electron density fall outside the ring of C—C internuclear axes. (*a*) The directions of the optimized hybrid C—C bonding orbitals. (*b*) Contour maps of the corresponding electron density, with the C—C internuclear axes drawn in to emphasize the "bent" nature of the bonds, especially near the carbon atoms.

equivalent to the work required to bend the bond angle from its natural value of 109°28′ to 60°. One may speculate that the bending energy is stored in the system of bonding electrons in two ways: Partly as an increase in electron–electron repulsion due to the proximity of the skeletal C—C bonds, and partly as a spatial redistribution of electron density away from the bond axis. In unstrained molecules we assume that the electron density in a bond is largely in an ellipsoid whose axis is the internuclear axis of the bonded atoms. But apparently in the C_3H_6 ring the "ridge" of the C—C bonding electron density contours follows a curve bent outward from the ring, as shown in Fig. 9.14. This displacement reduces both electron–electron repulsions and electron–nuclear attractions. Strained bonds of this type are sometimes called "banana bonds" or "bent bonds" because of the shape of their electron density contours.[13]

The cyclobutane (C_4H_8) ring is also quite strained. In rings with more than four atoms, however, the strain is unimportant, because they can attain strain-free conformations by bending out of the plane. Planar cyclobutane would be square, with 90° bond angles; the equilibrium is actually highly puckered, with about the same carbon strain—bond angles of 88°—but the hydrogen atoms are farther apart in the bent structure than in the planar. The benzene ring is planar but quite strain-free, because the C—C—C angle is the natural bond angle of 120° for trigonal carbons. The planarity of the ring is apparently necessary for its aromatic behavior, because this creates the maximum overlap of the $2p\pi$ orbitals and therefore the maximum delocalization of the electrons in these orbitals. In contrast, the cyclooctatetrane ring,

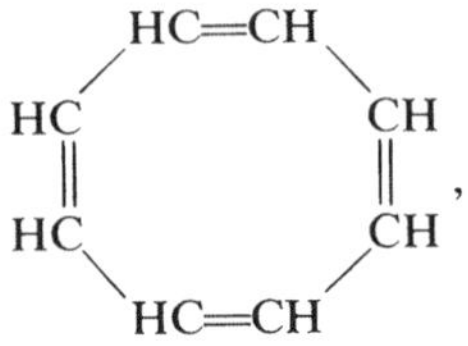

,

which one might expect to resemble benzene, is nonplanar (since a planar ring would have 135° bond angles and much strain) and is not like aromatic molecules in its properties. This is both because of its nonplanarity and because it has $4n$ electrons, rather than $4n + 2$, which affects the orbital energy-level structure. The cyclopentane molecule is a ring with the structure of four coplanar carbons and one carbon that lies out of that plane. This gives the molecule a strain-free form with almost normal "tetrahedral" bond angles, in contrast to what it would be if the carbons were coplanar, making the angles in the ring all 72°. Yet this molecule is nonrigid enough that each of the carbons are, at one time or another, the out-of-plane atom. When one atom surrenders this role to its neighbor, it appears to an observer, since all the carbon atoms are equivalent but their sites are not, that the molecule has rotated through $2\pi/5$ when in fact one atom has moved into the plane of three others, and an adjacent atom has moved out of that plane. Such a motion is strictly a rotation combined with a permutation and is called a "pseudorotation."

Molecules such as cyclopentane and cyclohexane, which can move easily from one structure to another, are often called "floppy," or "fluxional." They may pseudorotate readily, as with C_5H_{10}, or isomerize, as with C_6H_{12}. There are other kinds of pseudorotation, such as that of trigonal bipyramidal molecules like PF_5. This molecule has three fluorine atoms in one plane with the phosphorus atom, and two others on an axis through the phosphorus and perpendicular to the plane of the other three, as shown in Fig. 9.15*a*. The geometry implies that there are two physically different sites for the fluorine atoms, so we might suppose that the NMR spectrum of the fluorine nuclei of this substance would show two lines at different frequencies or field strengths, corresponding to the two kinds of fluorines. In fact, at room temperature, the fluorine NMR spectrum consists of a single sharp line! The reason is that at 300 K, this molecule can readily undergo a pseudorotation that rearranges the fluorine atoms by bending the F—P—F axial bond angle from 180° to 120° and one of the equatorial F—P—F bond angles from 120° to 180°. Halfway through

[13] The use of "banana bonds" is not limited to strained molecules. For example, the triple bond in N_2 can be interpreted as three equivalent "banana bonds" rather than the conventional one σ and two π bonds. For a symmetric case such as N_2, in the orbital approximation the two descriptions are mathematically equivalent: It is simply a matter of how one finds it convenient to partition the total electron density. There is, however, a real physical meaning to the "bent bonds" of cyclopropane.

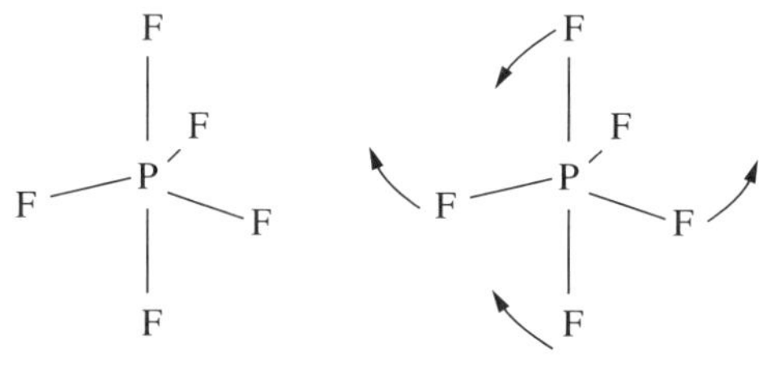

Figure 9.15 (*a*) The structure of PF_5, a trigonal-bipyramidal molecule, (*b*) and its mode of rearrangement by pseudorotation through a square pyramid to a new trigonal bipyramid structure in which the atoms that were axial in the initial structure become equatorial, and (*c*) two of the original equatorial atoms become axial.

the process, the molecule has the structure of a square-based pyramid. (At very low temperatures, the spectrum does show two lines because the molecule does not have enough energy to penetrate the barrier to pseudorotation.) This process is shown in Figs. 9.15*a, b,* and *c.* Pseudorotation is more than an arcane curiosity: This is the molecular process by which phosphate esters, such as adenosine di- and triphosphate interconvert in living systems, in the process called "phosphate ester hydrolysis." Each phosphorus in one of these molecules normally lies at the center of a tetrahedron, with only four bonds. However, when a hydrolysis reaction starts, a new substituent, such as an —OH group, attaches itself to a phosphorus atom, giving that atom five bonds, at least temporarily. The incoming group goes into an axial position; then the molecule undergoes a pseudorotation, so the incoming group becomes equatorial, and two of the original substituents become axial. When one of these leaves, that phosphorus again has only four bonds and has substituted an —OH for a larger group; the original ester of phosphoric acid has hydrolyzed.

We close this section with a brief discussion of the structures that arise when molecules are so large that there are important interactions between segments far from each other along any sequence of bonds. The simplest effect to discuss—though one of the most difficult to describe quantitatively—is that of *excluded volume.* Suppose that we grow a long-chain polymer in solution. As the molecule grows, so long as it has the flexibility associated with internal rotation about single bonds, it will form what is called a *random coil.* That is, it will have no particular preferred shape, and can be compared to a piece of soft spaghetti in water. The only constraint on its shape and growth is then the prohibition against two pieces of the chain occupying the same space—the excluded-volume effect. This is the simple result of the short-range repulsion always present between nonbonded atoms. With random-coil molecules, one is forced to use statistical calculations to determine the volume occupied per unit chain length and other properties related to the physical form of the molecules. However, there are other large molecules with quite rigid shapes, and these we consider next.

Three kinds of interactions are particularly important in giving a large molecule a well-defined rigid shape. One is the presence of a network of strong bonds, as in condensed aromatic ring compounds; we have already discussed interactions of this type in some detail. If the network is essentially unlimited in extent, then the "molecule" is in fact a covalently bound solid; examples we have mentioned include graphite and the polysilicates. A second type is weak bonding between one part of a large molecule and another that is near in space but far away along the sequence of (strong) bonds; the most common interactions of this type involve hydrogen bonds. And third is the effect of interactions between the large molecule and the solvent medium that contains it, especially when different parts of the solute molecule interact in different ways.

The possibilities of forming hydrogen bonds between otherwise unbonded parts of large molecules are rich indeed, whenever the molecules contain both acidic (proton-donating) and basic (proton-accepting) functional groups. Of particular interest are biological macromolecules, in which the proton acceptors are most commonly carbonyl [(> C=O)] groups, whereas the hydrogen atoms are most commonly furnished by hydroxyl (—OH), amino ($—NH_2$), or imino [(> NH)] groups. Perhaps the most important of such interactions for us are those in deoxyribonucleic acid (DNA), the material making up the genes that determine our heredity. The DNA molecule consists of two long chains, each made of alternating ribose (a five-carbon sugar) and phosphate groups, with one of the nitrogen-containing bases, purines or pyrimidines, adenine, thymine, guanine, or cytosine, attached to each ribose ring. The two chains are held together only by hydrogen bonds joining the bases—adenine to thymine, and guanine to cytosine—in such a way that the entire molecule forms the famous double helix (Fig. 9.16). The hydrogen bonding is weak enough to allow a living cell's reproduction system to separate the strands gently and thus replicate each chain one link at a time, yet strong enough to keep the DNA in the form of a rigid rod of coiled strands when it is not being replicated. (At high temperatures the hydrogen bonds are broken, and the rigid helix degenerates into a random coil.) Hydrogen bonds perform similar functions in many other biological systems, notably the helical proteins that form hydrogen bonds to their water solvent. Their function, it seems, is to stabilize well-defined structures of very large molecules, both for storage purposes and to keep specific active sites in suitable positions for performing specific tasks such as catalyzing reactions.

As for interactions between solvent molecules and large solute molecules, these are often similar in their effects to intramolecular hydrogen bonding, but far less specific. A familiar old saw in chemistry is *similia similibus solvuntur*—"like dissolves like," where "like" usually refers to the classes of polar and nonpolar molecules. This is the result of the attractive interactions between polar groups (often involving hydrogen bonds) and of the much weaker attractions or even repulsions between polar and nonpolar

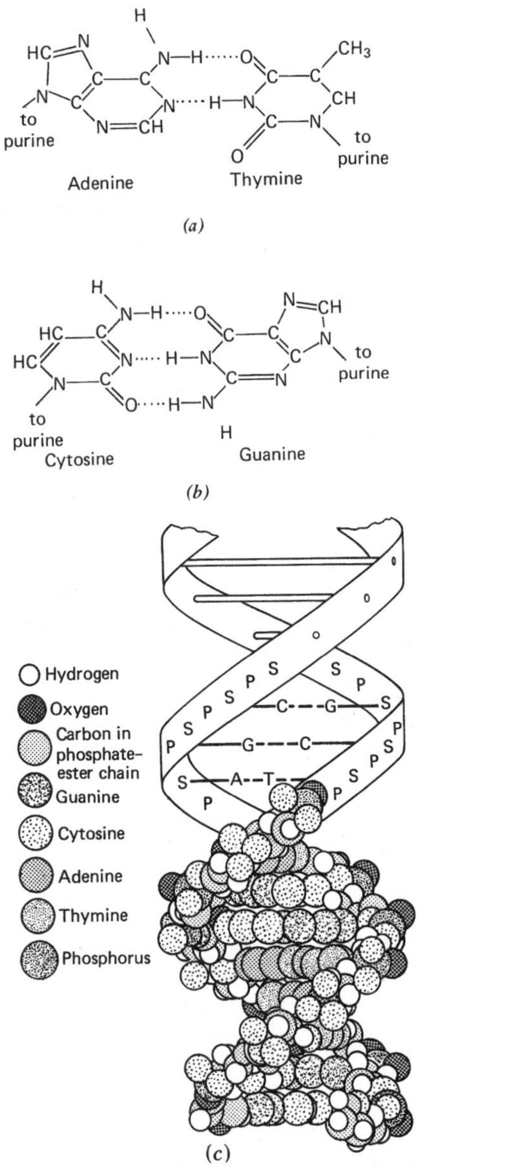

Figure 9.16 Hydrogen bonding between *(a)* adenine and thymine, *(b)* cytosine and guanine, resulting in *(c)* the double-helix structure of deoxyribonucleic acid.

groups. Consider a molecule containing both polar and nonpolar regions—for example, a long hydrocarbon chain with a polar functional group at one end, such as $CH_3(CH_2)_{16}$ COOH—in a polar solvent such as water. There will be a strong tendency for the molecule to arrange itself so as to maximize the contact between its polar segments and the solvent, and simultaneously to minimize the contact between nonpolar segments and the solvent. This may result in the solute molecules' aggregating into droplets or films with their polar segments forming the interface with the solvent. Such interactions can have strong effects on the conformation of the solute molecules. A hydrogen-bonded solvent such as water has a structure of its own (Chapter 10 and Chapter 26), and this will also be affected, in general, by the *hydrophilic* (water-attracting) or *hydrophobic* (water-repelling) nature of the solute molecules. These systems are complex enough that molecular mechanics is insufficient to describe their behavior; to go further one needs the statistical methods that we shall consider in Part II of this book.

9.5 Complex Ions and Other Coordination Compounds: Simple Polyhedra

Now we shall begin our consideration of the second major class of polyatomic molecules, the so-called *coordination compounds* or *complexes*. Structurally, these species are characterized by a central atom or ion and a set of surrounding *ligands,* which may be atoms, ions, or neutral molecules. Most complexes have well-defined geometric structures, and even those that can rearrange internally have strongly preferred geometries. These structures are ordinarily polyhedral, with each of the ligand atoms occupying a vertex and bonded to the central atom. In coordination compounds the bonds between ligands and the central atom are generally weaker than the corresponding bonds in such polyhedral species as CH_4 or SO_4^{2-}; the ligands thus more readily undergo substitution reactions. In these respects the coordination compounds are intermediate between ordinary covalent molecules and the loose ion-molecule clusters we shall discuss in Section 10.3.

Virtually all positive ions exist as coordination compounds of some kind in aqueous solution (and in a number of other solvents); many of these species also exist as complex ions in the solid phase. At one extreme we have the weakly bound, labile structures of the hydrated alkali metal ions. The larger ones, of Rb^+ and Cs^+, are particularly labile. At the other extreme are the ordinary covalent molecules. Many complexes differ from covalent molecules in that the central ion and the ligands are capable of independent stable existence, usually in solution. The most studied complexes are those formed by the transition metal cations, which combine with a host of ligands, including all the common polar solvents. Transition metal complexes offer especially rich challenges, in that they have several properties—most notably magnetism and electronic absorption spectra in the visible and near-ultraviolet regions—that can be used to probe the electronic levels involved in their bonding. In subsequent sections we shall use these properties to analyze the electronic structures of the complex ions, and then see how this information can help us to interpret their geometric structures.

First, however, we must survey the types of structures that these compounds exhibit. The most important factor in classifying structures is the *coordination number,* the number of ligands bonded directly to the central atom (or the number of nearest neighbors, if they are all equivalent); these ligands constitute the *coordination shell.* The most common coordination numbers in the solid or liquid phase are 4 and 6, although 5 is reasonably common and values ranging from 2 to 9 are known. We shall for the most part limit ourselves to the species with values of 4, 5, and 6.

The geometric structures of complexes were first deduced by counting the number of *isomers*—species with the same formula, but different chemical or physical properties—of a given complex. This analysis was carried out primarily by Alfred Werner, who, in a series of studies from 1893 until 1911, showed that (1) certain complexes formed series from which the number of ligands could be inferred, and (2) the number of isomers of each complex could be used to determine the geometry, if one assumed that all the possible isomers were known. (Werner received the 1913 Nobel Prize in Chemistry for this work.)

Werner's reasoning is best illustrated by the classic case of the complexes of Pt(IV)[14] with Cl^- and NH_3. These were formed by adding ammonia to aqueous solutions of $PtCl_4$. Five such complexes were known to Werner; they were isolated and studied by direct chemical analysis, by conductance measurements to find the number of ions present in the solution, and by titration with Ag^+ to find the number of free chloride ions present. The five complexes crystallized into salts with the compositions $PtCl_4 \cdot 2NH_3$, $PtCl_4 \cdot 3NH_3$, $PtCl_4 \cdot 4NH_3$, $PtCl_4 \cdot 5NH_3$, and $PtCl_4 \cdot 6NH_3$. Their solutions respectively had conductivities corresponding to 0, 2, 3, 4, and 5 ions per platinum atom, whereas the amounts of AgCl precipitated immediately corresponded to 0, 1, 2, 3, and 4 "free" Cl^- ions per platinum atom. All the Cl^- will be precipitated as AgCl if one waits long enough, but Werner was wise enough to recognize that chlorine was present in two states of binding, one that reacted slowly with silver ions and one that reacted immediately. The conclusions he drew were that there are six ligands coordinated with each platinum and that the five complexes can be represented as

$$[Pt(NH_3)_2Cl_4],$$
$$[Pt(NH_3)_3Cl_3]^+ \cdot Cl^-,$$
$$[Pt(NH_3)_4Cl_2]^{2+} \cdot 2Cl^-,$$
$$[Pt(NH_3)_5Cl]^{3+} \cdot 3Cl^-,$$
$$[Pt(NH_3)_6]^{4+} \cdot 4Cl^-.$$

The species in square brackets presumably maintain their identities in solution, with the Cl^- made available for precipitation only by slow exchange reactions. These are referred to as complex ions. The complex ions move about freely as entities in the solution. The "free" chloride ions, shown outside the brackets, are also separate entities in solution. Werner had thus shown that in these complexes Pt(IV) has a coordination number of 6. By his reasoning complex ions with the formulas $[Pt(NH_3)Cl_5]^-$ and $[PtCl_6]^{2-}$ should also exist, and these are indeed now known. Werner extended his analysis to show correctly that Cr(III) and Co(III) also have coordination numbers of 6, whereas Pt(II) and Pd(II) have coordination numbers of 4.

The geometries of the complexes were then inferred from the numbers of isomers. For example, two distinct species were known to share the formula $[Co(NH_3)_4Cl_2]^+$; one is green, the other lavender. Werner supposed correctly that the difference lies in the geometric arrangement of the ligands. If the structure were highly irregular, there ought to be more than two isomers; indeed, if all six sites were geometrically different, the number of possible isomers would be $\frac{1}{2}(6 \times 5) = 15$. Only if all the sites are essentially equivalent can the number of isomers be as low as two. A polyhedron with six equivalent vertices is an octahedron, and an octahedral arrangement of the ligands does give rise to only two isomers, the *cis* and *trans* structures of Fig. 9.17*a* (*cis* and *trans* refer to the relative positions of the Cl^- ligands). Similar pairs of isomers are found for most other 6-coordinate complexes with ligands in 4:2 ratio. If Co(III) complexes are octahedral, then $[Co(NH_3)_3Cl_3]$ should also have two isomers, with the structures shown in Fig. 9.17*b;* this is indeed the case. Similarly, the 4-coordinate complex $[Pt(NH_3)_2Cl_2]$ has two isomers and is thus assumed to be square-planar, Fig. 9.17*c*. Evidence of this sort has been used to determine the structures of a great number of complexes.

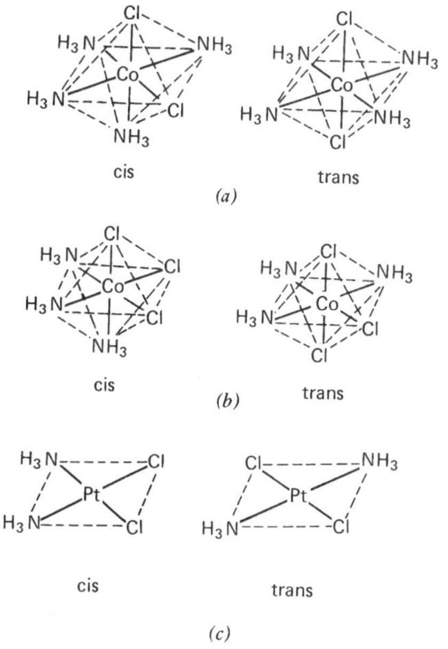

Figure 9.17 Geometric isomerism in complexes: *cis*- and *trans*-isomers of (*a*) the octahedral complex $[Co(NH_3)_4Cl_2]^+$; (*b*) the octahedral complex $[Co(NH_3)_3Cl_3]$; (*c*) the square-planar complex $[Pt(NH_3)_2Cl_2]$. Note that in each case all the vertices are equivalent.

[14] Here we use the notation now common in inorganic chemistry for the *formal* positive charge on the ion; thus Pt(IV) indicates a formal charge or oxidation state of +4 on platinum. The electric fields around a real Pt^{4+} ion are far too strong for such a highly charged ion to exist as such near other atoms or molecules. Although highly charged ions are found in sparks and plasmas, they are not found in condensed matter. Note, however, that we continue to write the charge in the usual way for ions such as $[Pt(NH_3)_6]^{4+}$ because the complex ion is an identifiable species even with its charge of 4+.

Thus far we have discussed only *monodentate* ("one-toothed") ligands, those that can occupy only one site in a complex's coordination shell. An example of a *bidentate* ligand is ethylenediamine, $H_2NCH_2CH_2NH_2$, which behaves almost like two ammonia molecules linked by a short, flexible chain. Both the —NH_2 groups are able to act as ligands, so that the ethylenediamine molecule occupies two sites in the coordination shell, not necessarily around the same central atom. Other common bidentate ligands include propylenediamine, $H_2N(CH_2)_3NH_2$, and acetylacetonate, $(CH_3COCHCOCH_3)^-$. Still other species exist that can act as ligands of three, four, or more sites. When a polydentate ligand can occupy two (or more) sites around the *same* central atom, thus forming a ring, it is called a *chelate* (from the Greek *chele,* "claw"), and the central atom is said to be chelated. Chelation is a good method of stabilizing ions in solution, since the bulky chelate rings make it difficult for any reactive species to get at the central atom.

9.6 Chirality and Optical Rotation

The existence of bidentate ligands makes a different kind of isomerism possible. Consider the complex $[CoCl_2en_2]^+$ ("en" is the standard abbreviation for ethylenediamine), which is found to have three isomers. One of the three is chemically, and in most ways physically, different from the other two. The second and third isomers are identical in almost every way—spectroscopically, chemically, and with regard to most physical measurements. They differ only in properties that distinguish the left- or right-handedness, the *chirality* of the molecule. Before seeing what this means for the structure, we must explain what chirality is.

Left- and right-handed gloves of a pair are a commonplace example of two objects that differ only in their chirality. In other words, they are mirror images of each other. To detect such a difference on the microscopic level, one must use a physical phenomenon that is itself inherently chiral. The most straightforward such phenomenon is interaction with circularly polarized light. "Circular polarization" means that the electric field vector **E** at a point in space rotates in the plane perpendicular to the direction of propagation of the light. This is in contrast to plane polarization (Fig. 3.4), in which **E** at a point in space merely oscillates back and forth along a single axis. At a given instant, the field vectors at successive points along a ray of circularly polarized light describe a helix, either right-handed or left-handed. This is illustrated in Fig. 9.18.

How does one use circularly polarized light to study chirality? The physically significant fact is that the combination of a left-handed thing together with a right-handed thing is physically different from a pair of things of the same handedness—where "physically different" means a difference in more than just handedness. An example may make this clearer: A left-handed glove glued palm-first to a right-handed glove makes a quite different-shaped object than the result of the same operation on two left-handed gloves. Similarly, a right-handed molecule interacts with right-polarized light in just the same way that a left-handed molecule interacts with left-polarized light; neither of these interactions is physically equivalent to the way a right-handed molecule interacts with left-polarized light, but the latter is the same as the interaction of a left-handed molecule with right-polarized light.

The optical properties most commonly used to study chirality are the differences, either in absorption or refraction,

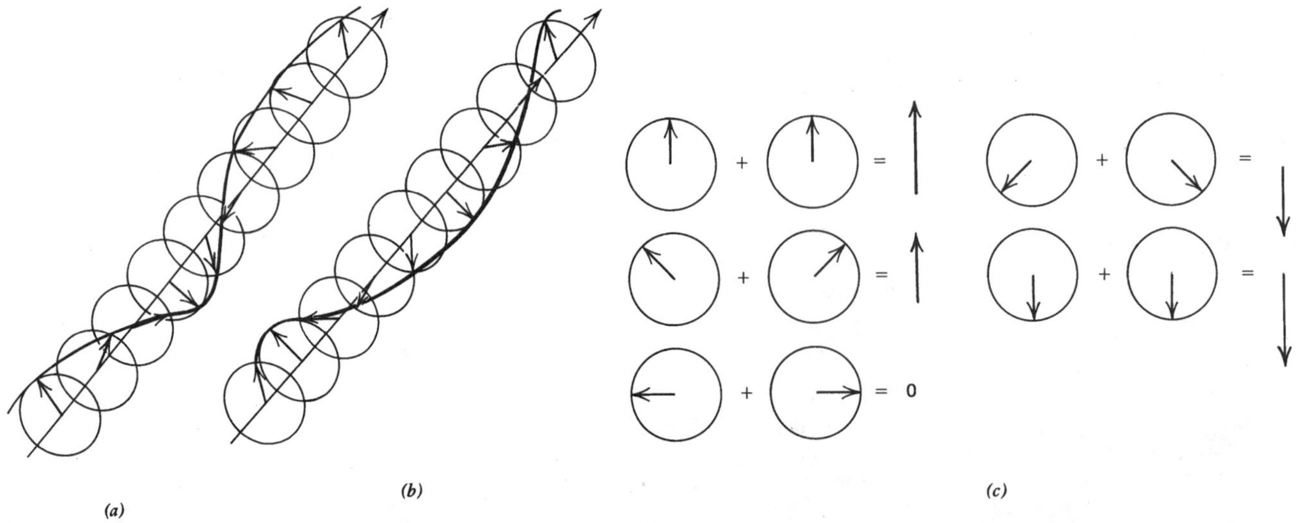

Figure 9.18 Right- (*a*) and left- (*b*) circularly polarized light: The field vectors **E** at points along the light ray are indicated by the small arrows, the points of which are joined to form a helix. Part (*c*) of the figure shows how left- and right-circularly polarized light in phase can add to give plane-polarized light; the **E** vectors are again shown from the viewpoint of one looking down the direction of propagation.

between left- and right-circularly polarized light interacting with molecules of a specific chirality. The difference between the optical absorption (extinction) coefficients for left- and right-polarized light is called the *circular dichroism* (CD); it is a function of the frequency of the light, and may have either sign. The other property, the difference between the refractive indices for left-circular and right-circular polarized light has a much longer history of use in physical chemistry. It is still sometimes called *optical activity,* but a more meaningful and precise name is *optical rotatory dispersion* (ORD).[15] In the most common form of this method, plane-polarized light is passed through a chiral sample. Plane-polarized light can be described as the resultant of two circularly polarized beams with equal amplitudes and opposite directions of rotation. If the two components are exactly in phase, they will add constructively when both **E** vectors point in the same direction, but exactly cancel when both are at right angles to that direction (Fig. 9.18c). But if the refractive index of the sample differs for the two components, then one will have a higher velocity of propagation than the other, and the phase relationship between the components will constantly change with distance along the propagation axis.[16] The resultant light remains linearly polarized, since there is always some axis along which the two rotating **E** vectors are parallel; but this axis of polarization rotates as the light passes through the sample. The angle of rotation α is given by

$$\alpha = \alpha_n lc, \qquad (9.9)$$

where α_n is a constant (the *optical rotatory power*) characteristic of the sample, l is the path length through the sample, and c is the molar concentration of chiral molecules in the sample; α_n depends on wavelength, and is most commonly determined with light of the sodium D lines, at 589.0 and 589.6 nm.

---

[15] ORD and CD together are referred to as the *Cotton effect.*

[16] One can also use the difference in refractive index to separate the two components of polarization. This is just how circularly polarized light is obtained. Both plane and circular polarization are ordinarily produced by passing light through suitably cut prisms of an anisotropic crystal such as quartz (SiO_2), in which the crystal structure rather than the individual molecule is chiral.

9.7 Chiral and Other Complex Ions

Now that we know what chirality is, let us return to the complex $[CoCl_2en_2]^+$. The three isomers must be those shown in Fig. 9.18. Note that the *trans* species has no chiral character, whereas the two *cis* species are identical except for their chirality: Like the left and right gloves, they are mirror images of each other. The trans species is a *geometric isomer* (or *diastereoisomer*) of either *cis* species; the two *cis* species are *optical isomers* of each other (such pairs are called enantiomorphic—"opposite-shaped"—isomers or *enantiomers*). Any molecule that has a plane of symmetry is identical with its mirror image, and thus can have no optical isomers. Conversely, any molecule that cannot be superimposed on its mirror image must have an optical isomer of opposite chirality. Optical isomers are also commonly found among organic molecules; they are usually, but not always, associated with an "asymmetric carbon" (or chiral center), a tetrahedral carbon atom bound to four different substituents, as in

$$
\begin{array}{ccc}
\text{F} & & \text{CH}_3 \\
| & & | \\
\text{H—C—Cl} & \text{or} \quad \text{CH}_3\text{CH}_2\text{—C—COOH,} \\
| & & | \\
\text{Br} & & \text{H}
\end{array}
$$

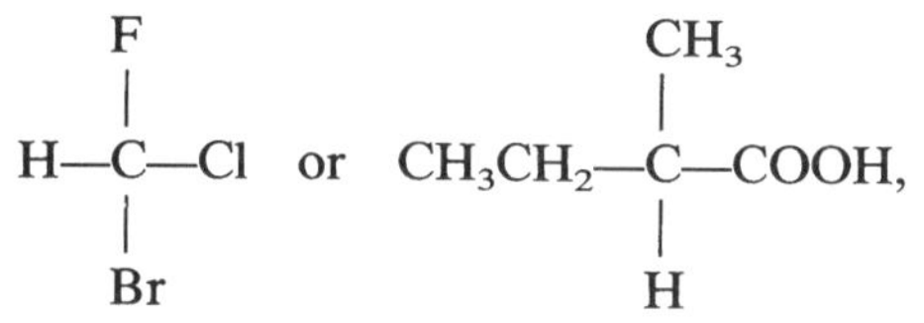

since a tetrahedron with four differently labeled corners is not identical to its mirror image.

The existence of the set of isomers shown in Fig. 9.19 is one of the clearest demonstrations of the octahedral structure of this complex and, by extension, of other Co(III) complexes. A great number of complexes with chelate ligands are optically active, and this is nearly always useful in determining their structures. Optical isomers can often be separated or *resolved,* usually by precipitating the ion in combination with another optically active species; again, the combination of two chiral properties results in a nonchiral physical difference.

Thus far we have discussed isomerism almost exclusively in octahedral complexes, but similar analyses can be applied to determine the structure of complexes with other geometries. We mentioned earlier that 4-, 5-, and 6-coordinate complexes are the most common; Fig. 9.20 illustrates the

(a)   (b)   (c)

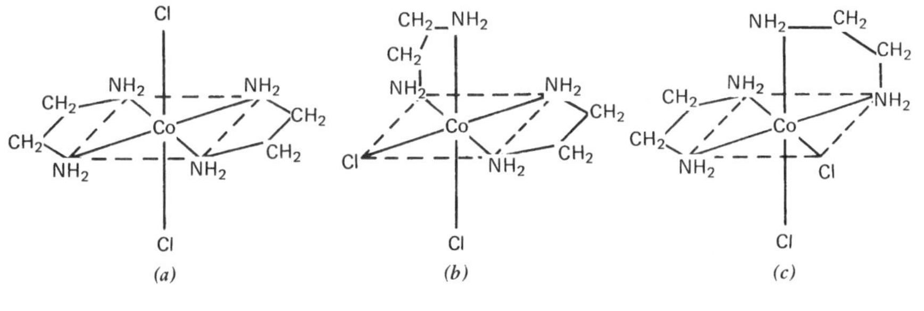

Figure 9.19 The three isomers of $[CoCl_2en_2]^+$: (*a*) *trans;* (*b*), (*c*) the two *cis* optical isomers.

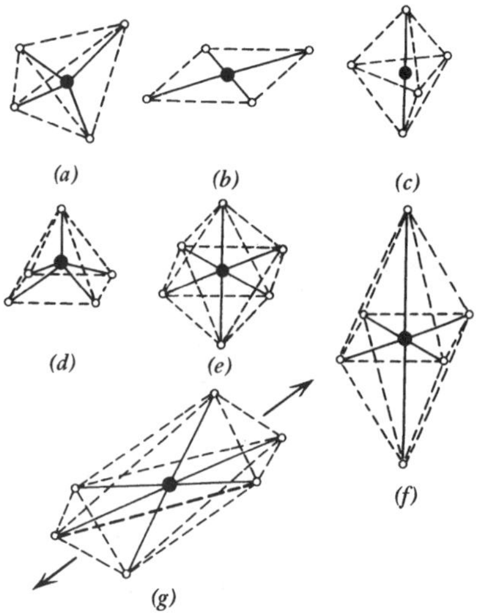

Figure 9.20 Common geometries for complexes: 4-coordinate (*a*) tetrahedron, (*b*) square-planar; 5-coordinate, (c) trigonal bipyramid, (*d*) square pyramid; 6-coordinate, (*e*) octahedron, (*f*) tetragonally distorted octahedron, (*g*) trigonally distorted octahedron; (axis of elongation indicated by arrows).

usual structures found for each of these. (As one might expect, the symmetries of these structures are often the same as those observed in small covalent molecules: cf. Fig. 9.2.) We shall briefly survey here some types of complexes with each of these structures.

The principal structures known for 4-coordination are the tetrahedral and the square-planar. Many of the tetrahedral complexes are tetrahalides, such as $[ZnCl_4]^{2-}$, $[CdBr_4]^{2-}$, $[FeCl_4]^{-}$, $[CoI_4]^{2-}$, and similar species such as $[Hg(CN)_4]^{2-}$. Square-planar complexes are more common for noble metals, as in $[PtCl_4]^{2-}$ and $[Pd(NH_3)_4]^{2+}$, as well as for Ni(II) and Cu(II), as in $[Ni(CN)_4]^{2-}$ and $[CuCl_4]^{2-}$. (Cu(II) is also found as a distorted tetrahedron.) The square-planar geometry is sometimes interpreted as an extreme case of 6-coordination in which two *trans* vertices of a tetragonally distorted octahedron have extremely weak bonding capacity.

Five-coordinate complexes are relatively uncommon. They usually have trigonal-bipyramidal form, like PCl_5; the simplest instances are pentacarbonyls such as $[Fe(CO_5)]$ or $[Mn(CO)_5]^{-}$. Rare examples, such as $[Ni(CN_5)]^{3-}$, are square-pyramidal, with four of the cyano groups at the base of the figure, the fifth at its unique apex, and the nickel ion at the center.

Six-coordinate complexes, probably the most common of all (at least the most studied), are almost always either octahedral or distorted-octahedral. We have already given many examples of the normal octahedron, in which all six vertices are equivalent. The most common distortions are either *tetragonal,* with a lengthened (or shortened) axis perpendicular to the square equatorial plane of the distorted octahedron; or *trigonal,* with the octahedron stretched (or compressed) along an axis perpendicular to two opposite triangular faces. A noteworthy example of tetragonal distortion is the $[Cu(H_2O)_6]^{2+}$ complex (which produces the bright blue color of aqueous Cu(II) solutions): two of the Cu—O bonds are longer than the other four. Ferrous salts, such as ferrous ammonium sulfate and $FeSiF_6$, display trigonal distortions, particularly in the solid state.

Now let us summarize the more exotic coordination numbers. The coordination number 2 occurs in a few linear complexes of singly charged cations, such as $[CuCl_2]^{-}$ and $[Ag(NH_3)_2]^{+}$. Most species once thought to have coordination number 3 have turned out to be polymeric;[17] one real example appears to be the triangular $[HgI_3]^{-}$. Many species with apparent coordination numbers of 2 and 3 may actually have higher values, with very weakly bound solvent molecules occupying additional sites in their coordination shells. Coordination numbers greater than 6 are fairly rare and limited almost entirely to large cations; some examples include $[UF_7]^{3-}$ (pentagonal bipyramid), $[Mo(CN)_8]^{4-}$ (dodecahedron), $[TaF_8]^{3-}$ (square antiprism—a cube with one face rotated 45° relative to the opposite face), and $[ReH_9]^{2-}$.

We note in passing that there also exist *polynuclear* complexes, which have two or more central atoms connected by bridges. These bridges may be ordinary monodentate ligands shared by two coordination shells, or bidentate ligands with one end in each coordination shell. Some examples of the first kind are

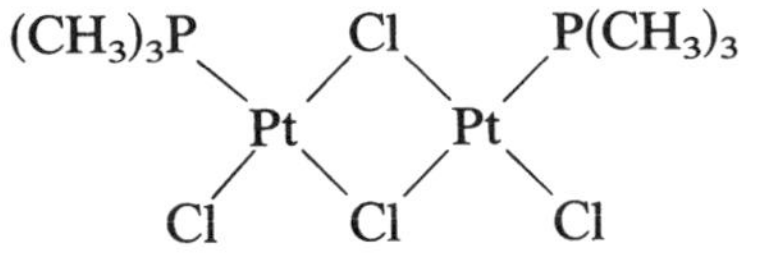

and

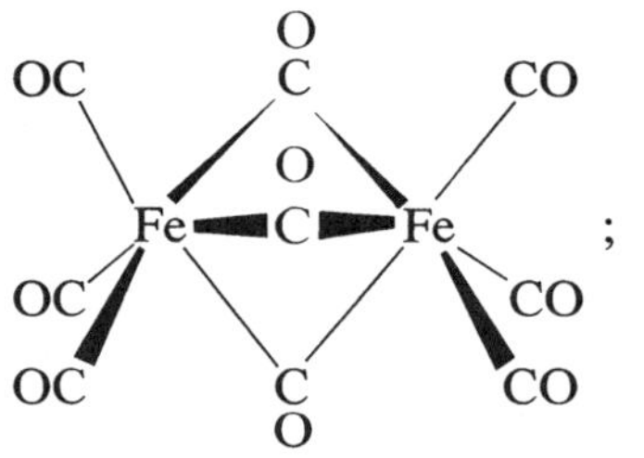

;

the platinum compound is planar and has *cis* and *trans* isomers. There are also polynuclear complexes with direct

---

[17] For example "$CuCl_3$" is really

$$\begin{array}{ccc} Cl & Cl & Cl \\ | & | & | \\ -Cl-Cu-Cl-Cu-Cl-Cu-Cl-, \\ | & | & | \\ Cl & Cl & Cl \end{array}$$

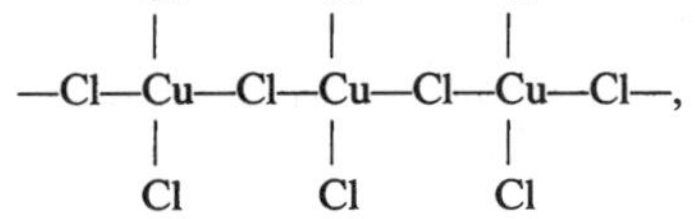

with coordination number 4.

metal-to-metal bonds, some with quite elaborate structures; for example, in $[Mo_6Cl_8]^{4-}$ and $[Rh_6(Co)_{16}]$ the metal atoms form an octahedron!

This completes our preliminary survey of complexes. In the next two sections we shall see how their properties are related to their electronic structure.

9.8 Magnetic Properties of Complexes

Among the most telling diagnostic properties of complexes are their magnetic and optical characteristics, from which one can infer much about their electronic structures. It is true that we can learn little in this way about tightly bound closed-shell molecules such as PF_5, or about the complexes of closed-shell ions such as Zn(II), which are colorless and essentially nonmagnetic. However, the complexes of ions with partially empty *d* shells, such as Cr(III) or Fe(III), are often strongly colored and paramagnetic; this class includes most complex ions of the transition metals. One of the attractions of this field has always been the range of hues and intensities that the transition metal complexes exhibit. For example, copper forms the light blue $[Cu(H_2O)_6]^{2+}$ complex in water, and the more strongly bound, intensely deep blue $[Cu(NH_3)_6]^{2+}$ complex in liquid ammonia, plus all the intermediate species; and we have already mentioned the lavender and green isomers of $[Co(NH_3)_4Cl_2]^+$. The colors of these species are generally associated with low-lying energy levels to which electrons can be excited by visible light; in the next section we shall look into the nature of these levels. Let us turn first, however, to the magnetic properties.

The magnetic properties of complexes have been studied by two rather different methods. One of these measures the bulk property called magnetic susceptibility; the other probes the magnetic moments of individual molecules by the spectroscopic technique generally called electron spin resonance. We mentioned both of these in our study of atomic magnetism (Section 5. 1), but we shall go into more detail here. In both cases the quantity one seeks to determine is $\boldsymbol{\mu}_m$, the magnetic dipole moment of the complex.

Magnetic susceptibility is a quantity that measures the capacity of a substance to develop magnetization when subjected to a magnetic field. In a vacuum the magnetic induction (magnetic flux density) **B** is related to the magnetic field strength **H** by the equation $\mathbf{B} = \mu_0\mathbf{H}$, where μ_0 is the permeability of free space ($\mu_0 \equiv 4\pi \times 10^{-7}$ H/m, where H means henrys). Within a piece of matter **B** is in general not equal to $\mu_0\mathbf{H}$ (where **H** is the external field), but may be greater or less. We thus write the equation

$$\mathbf{B} = \mu_0(1+\chi)\mathbf{H}, \qquad (9.10)$$

which defines the magnetic susceptibility χ. The *magnetization* is $\mathbf{M} = \chi\mathbf{H}$. If χ is negative ($\mathbf{B} < \mu_0\mathbf{H}$), the substance is said to be *diamagnetic;* if χ is positive ($\mathbf{B} > \mu_0\mathbf{H}$), it is *paramagnetic.*

All materials have the capacity for diamagnetism, that is, for responding to an external magnetic field by setting up a weak field that tends to cancel the applied field. Diamagnetism is due in large part to the precession of the electronic orbital angular momentum vectors induced by the applied field, and in lesser part to actual distortion of the orbital wave functions. Diamagnetism is generally a small effect, and is by far outweighed by paramagnetism when the latter is present. In paramagnetic species the diamagnetic effects can be more than adequately taken into account by approximation methods in reducing experimental data.

Paramagnetism, on the other hand, results from the existence of permanent magnetic moments within the molecular units of a substance. An external magnetic field tends to align these microscopic magnets parallel to itself, and thus produces a net positive magnetization on the macroscopic level. We have outlined the theory of the magnetic moment in Section 5.1. When a molecule has both orbital angular momentum **L** and spin angular momentum **S** (and the two are separable), its magnetic dipole moment is given by

$$\boldsymbol{\mu}_m = \boldsymbol{\mu}_L + \boldsymbol{\mu}_S = -\frac{e}{2m_e}(\mathbf{L} + g_s\mathbf{S}). \qquad (9.11)$$

The Lande *g*-factor g_s is found experimentally to have the value 2.002319 for the free electron; the value of g_s varies slightly with environment, but for many purposes can be approximated by 2. The magnitude of $\boldsymbol{\mu}_m$ depends on the quantum state of the molecule. Some special cases can be evaluated immediately. For a state with only orbital angular momentum ($S = 0$) we have

$$\begin{aligned} |\boldsymbol{\mu}_m|_{\text{orb}} &= |\boldsymbol{\mu}_L| = \frac{e\hbar}{2m_e}[L(L+1)]^{1/2} \\ &= \mu_B[L(L+1)]^{1/2}, \qquad \mu_B \equiv e\hbar/2m_e, \end{aligned} \qquad (9.12)$$

and for a spin-only state ($L = 0$) we have

$$|\boldsymbol{\mu}_m|_{\text{spin}} = |\boldsymbol{\mu}_S| = g_s\mu_B[S(S+1)]^{1/2}, \qquad (9.13)$$

where *L* and *S* are the orbital and spin quantum numbers and μ_B is the Bohr magneton. The situation in states with both orbital and spin angular momentum is more complicated, depending on both the coupling conditions and the distribution of molecules over the term levels (which in a complex, as we shall see in the next section, are split by the presence of the ligands). For *LS* coupling, with spin–orbit interaction negligible relative to the ligand field splitting, one can show that the magnetic moment should be

$$|\boldsymbol{\mu}_m| = \mu_B[4S(S+1) + L(L+1)]^{1/2}, \qquad (9.14)$$

assuming that $g_s = 2$. This equation is in principle applicable to most transition metal complex ions, but we shall see that there are complications. In general, the effective g-factor g_{eff}, defined by

$$|\boldsymbol{\mu}_m| = g_{\text{eff}}\,\mu_B, \tag{9.15}$$

may or may not be given by Eq. 9.14.

Before seeing how $|\boldsymbol{\mu}_m|$ can be related to the macroscopic susceptibility, let us digress to consider *electron spin resonance* (esr)—more properly called *electron paramagnetic resonance* (epr) for complex ions, because more than electron spin is usually involved. This method utilizes the Zeeman-effect splitting of energy levels in a magnetic field, described by Eq. 5.8 for a free atom. Complications can again occur; to encompass these, one can write the Zeeman shift as

$$E - E_0 = -\boldsymbol{\mu}_m \cdot \mathbf{B} = -g_{\text{eff}} M_J \mu_B B$$
$$(M_J = M_L + M_S), \tag{9.16}$$

where M_J is the quantum number for the component of total angular momentum in the direction of the field **B**. The separation of adjacent Zeeman levels is then

$$\Delta E = g_{\text{eff}}\,\mu_B B. \tag{9.17}$$

To measure this quantity, one normally applies a magnetic field of variable strength and an oscillatory radiation field of precisely known frequency ν in the microwave region. The variable field is adjusted until the Zeeman splitting $g_{\text{eff}}B$ coincides with the energy $h\nu$ of the microwave quantum, at which point one observes resonant absorption of energy. This sort of spectroscopy, in which one tunes the energy-level spacing by varying the magnetic field, rather than the frequency of the radiation for a fixed magnetic field, is usually more convenient in the radiofrequency and microwave regions. The reason is that it is easier to maintain a very precisely known, fixed frequency and then measure the magnetic field accurately than to fix the magnetic field and then make accurate measurements of a variable frequency.

In modern practice, it is common to use quite another technique to determine magnetic resonance spectra. In this approach, one holds the system in a fixed magnetic field and applies either a fast pulse of "white" radiation—radiation consisting of many frequencies in the appropriate microwave or radiofrequency region, analogous to white visible light—or a fast, carefully determined sequence of pulses of such light and then measures the temporal response of the system to the pulsed radiation. The frequency response—i.e., the spectrum—of the system is revealed in the decomposition of that temporal response into component sine and cosine waves of all the frequencies contained in the pulse. Such a decomposition can be done with any continuous function $f(x)$; if one considers the function only within a fixed, finite interval—e.g., from 0 to π or from −1 to 1—then that function can be represented exactly by a sum of sines and cosines,

$$f(x) = \sum_{n=0}^{\infty} (A_n \sin \alpha n x + B_n \cos \alpha n x), \tag{9.18a}$$

where the constant α scales the distance x so that sin αx and cos αx each have just one half-cycle in the given interval. Such a series is called a *Fourier series,* after the French mathematician and Egyptologist Joseph Fourier, 1768–1830. If the interval is not finite but infinite, we must take the sum to its limit of an integral and, instead of summing over discrete integers n, we integrate over a continuous variable k,

$$f(x) = \int_{-\infty}^{\infty} (A(k)\sin kx + B(k)\cos kx)\,dk, \tag{9.18b}$$

where the scale factor α is no longer needed because we do not have to deal with the fixed scale of a given interval and, because k is continuous, we write it as an argument rather than as a discrete index. The continuous variables x and k are complementary; their product must be dimensionless. If x has the dimension of length, then k has the dimension of $(\text{length})^{-1}$, which we already recognize is, apart from a factor of h, a momentum. If x has the dimension of time, then k has the dimension of frequency! Thus, Eq. 9.18*b* is the expression in which we can find the spectral information we want. All the amplitudes in the response signal from the sample, $A(k)$ and $B(k)$, are just those of the input pulse signal, except those at frequencies at which the system has a resonant response. If we measure the output signal as an absorption, then the output amplitudes at the resonant frequencies of the sample are diminished from those of the input signal. The process of inverting the output signal that we have identified with our $f(x)$ (with x taken as time and k as the spectral frequency) is called *Fourier transformation* or *Fourier inversion.* The process is very much analogous to expressing a wave function as a sum of contributions from a set of basis functions, which we have already seen; the basis functions here are simply sines and cosines. The mathematical process of extracting the amplitudes is also like the way one determines the contribution of a basis function to a full wave function: here, one multiplies $f(x)$ by sin kx or cos kx and integrates over all x. Since a product of the form (sin kx sin $k'x$) has as many negative parts as positive over the entire range of x, the integral of this product from $-\infty$ to $+\infty$ is zero—unless $k = k'$. In that case, the integrand is $\sin^2 kx$, whose average value is $\frac{1}{2}$, so the integral does not vanish—that is, the sines and cosines form an orthogonal set, either the discrete, indexed sines and cosines of Eq. 9.18*a,* over a finite interval, or the continuous set of Eq. 9.18b, over an infinite interval (which may be from zero to infinity instead of from $-\infty$ to $+\infty$). Writing out the multiplication by sin kx and integration of Eq. 3.18b over x yields $A(k)$. (In fact one must be careful to assure that the result does not become infinite. However, exploring that point would carry us further than this text can go.) In practice, fast Fourier transformations are now carried out extremely efficiently by numerical computation so that one can extract graphs of the desired amplitudes $A(k)$ and $B(k)$ virtually as fast as a printer can produce them or a screen can display them.

We can now return to the magnetic susceptibility χ, which is defined by Eq. 9.10. This quantity is usually measured with a *magnetic balance:* A sample is suspended half in and half out of a strong magnetic field, and the force exerted on the sample is measured directly. This force is given by

$$F = \tfrac{1}{2}\chi A\mathbf{B}\cdot\mathbf{B}, \tag{9.19}$$

where A is the cross-sectional area of the sample and $\mathbf{B}$ is the magnetic field. A paramagnetic material ($\chi > 0$) is pulled into the field; a diamagnetic material ($\chi < 0$) is pushed out. One usually expresses results in terms of the chemically more significant *molar susceptibility,*

$$\chi_M \equiv \frac{M\chi}{\rho}, \tag{9.20}$$

where M is the molecular weight and ρ is the density. Although χ_M for a paramagnetic sample contains both paramagnetic and diamagnetic contributions, the diamagnetic part is very small and can be estimated with sufficient accuracy; subtracting this out, one obtains the "corrected" molar susceptibility χ_M^{corr}. This quantity is usually found to be inversely proportional to the absolute temperature T,

$$\chi_M^{\text{corr}} = \frac{C}{T}, \tag{9.21}$$

where C is a constant: Eq. 9.21 is known as the *Curie law.*

One can derive the Curie law directly from the microscopic theory. The magnetization $\mathbf{M}$ ($= \chi\mathbf{H}$) of a bulk sample is simply the average magnetic moment per unit volume. For paramagnetic materials it is zero in the absence of an external field, since the microscopic magnets are then randomly oriented. The external field tends to align the individual magnetic moments with itself, but because of molecular motion this alignment is only partial; what results is a distribution over a range of angles. If a given molecule's moment $\boldsymbol{\mu}_i$ is directed at an angle θ_i away from the direction of $\mathbf{H}$, its contribution to the net moment is $|\boldsymbol{\mu}_i|\cos\theta_i$. One obtains the average moment of the bulk material (and thus $\mathbf{M}$ and χ) by averaging over all possible angles; Part II deals with the way in which one obtains such averages. As a result of this calculation, one finds that at thermal equilibrium the corrected molar susceptibility should be

$$\chi_M^{\text{corr}} = \frac{N_A|\boldsymbol{\mu}_m|^2}{3k_BT}, \tag{9.22}$$

where N_A is Avogadro's number, k_B is Boltzmann's constant, and $\boldsymbol{\mu}_m$ is as usual the magnetic moment of an individual molecule. If the sample contains more than one kind of molecule or ion, $|\boldsymbol{\mu}_m|^2$ is replaced by an average over the various species present. Equation 9.22 is of the same form as the Curie law, with $C = N_A|\boldsymbol{\mu}_m|^2/3k_B$; this equation is used to evaluate $|\boldsymbol{\mu}_m|$ from a measurement of χ_M^{corr}.

Whether one measures the spin resonance or the susceptibility, one obtains a value of $|\boldsymbol{\mu}_m|$. This observed moment is expressible, as in Eq. 9.15, in terms of the Bohr magneton multiplied by an effective g-factor g_{eff}. For gaseous atoms g_{eff} has the values predicted by Eq. 9.14. But for complex molecules and ions, and even for "free" atoms and monatomic ions in liquid and solid media, the magnetic moments are generally not as large as Eq. 9.14 would indicate. Very often the observed moments are much closer to the values predicted for the electron spins alone by Eq. 9.13.

Let us first consider some species that have only spin angular momentum. The ions Mn^{2+} and Fe^{3+} in their ground-state configurations have five $3d$ electrons, corresponding to a half-filled $3d$ subshell; all other occupied shells in these ions are filled. According to Hund's first rule, the lowest-energy state of this configuration should be that with the maximum value of S; here this means that all the $3d$ spins should be parallel (have the same value of m_s), giving $S = \frac{5}{2}$ for the ion. But if all five $3d$ electrons have the same m_s, they must all be assigned different values of m_l (from +2 to −2). This means that all five orbitals of the $3d$ subshell are equally populated with one electron each: Thus the $3d$ charge distribution, like that of the filled shells, is spherically symmetric, and the orbital angular momentum of the ion is zero. With $L = 0$ and $S = \frac{5}{2}$, the ground states of these ions must be ^6S. Since they have no orbital angular momentum, the magnetic moments of the Mn^{2+} and Fe^{3+} should be given by Eq. 9.13:

$$|\boldsymbol{\mu}_m| = 2\left[\tfrac{5}{2}\left(\tfrac{5}{2}+1\right)\right]^{1/2}\mu_B = 5.92\,\mu_B\,.$$

The observed values are, in fact, approximately 5.9 Bohr magnetons for these species.

Most ions with permanent magnetic moments have both spin and orbital angular momenta. Even in most of those cases the observed moments are strikingly close to the spin-only values. For example, the Cu^{2+} ion has the ground-state configuration $3d^9$, with a single unpaired d electron, and thus is ^2D ($L = 2$, $S = \frac{1}{2}$). The spin-plus-orbital magnetic moment predicted by Eq. 9.14 is $3.00\mu_B$, whereas the spin-only value predicted by Eq. 9.13 is $1.73\mu_B$. The experimental values for Cu(II) in various complexes with ligands that are not themselves paramagnetic range between 1.7 and 2.2 Bohr magnetons. Similarly, for Ni(II) ($3d^8$) the ground state of the free ion is ^3F ($L = 3$, $S = \frac{3}{2}$), the spin-plus-orbital and spin-only calculations give $4.47\mu_B$ and $2.83\mu_B$, respectively, and the observed moments for various Ni(II) complexes in solution span the range 2.8–$4.0\mu_B$.

In cases in which the observed magnetic moment is less than the full amount expected from the combination of spin and orbital angular momenta, we say that the orbital angular momentum is quenched. Quenching occurs because the electrons are in a potential energy field that is not spherically symmetric; this is the polyhedral field of the ligands in a complex, or that of the surrounding ions in a crystal. As we pointed out in Chapter 5, every departure from symmetry removes some degeneracy and causes the "spoiling" of some conservation law. In complexes, the deviations from spherical symmetry ("quenching potential") stimulate interchange of orbital angular momentum between an ion and its surroundings. Classically, one might say that torques are exerted on the orbiting electrons. Thus **L** is no longer a well-conserved quantity. In particular, its component L_z in the direction of an external field is not conserved, and it is this component that interacts with the field to produce magnetization: The orbital part of the interaction energy is $\boldsymbol{\mu}_L \cdot \mathbf{B} = -(e/2m_e)L_zB$. In some cases the interchange of angular momentum is so rapid that L_z averages to zero, and quenching is complete. This corresponds to a quenching potential much greater than the Zeeman splitting due to the external field. In the next section we shall inquire into the nature of the quenching potential in complexes.

Thus far we have tacitly assumed that the magnetic behavior of a complex can be interpreted in terms of the magnetic properties of the central atom or ion in isolation. This is indeed true in many instances and provides us with yet another example of the separability of properties. For example, our analysis of the Fe^{3+} ion, with its ^6S ground state, corresponds well to the condition of Fe(III) in the colorless $[FeF_6]^-$ complex, in the brown $[Fe(H_2O)_6]^{3+}$ complex, in alums,[18] in beryl, and as an impurity in rutile (TiO_2). But the red $[Fe(CN)_6]^{3-}$ (ferricyanide) complex has a magnetic moment of only about $2.3\mu_B$ corresponding to a single unpaired electron with some unquenched orbital angular momentum (since for $S = \frac{1}{2}$ the spin-only moment is $1.73\mu_B$). The isoelectronic Mn(II) complexes exhibit the same pattern, with $[Mn(CN)_6]^{4-}$ an anomalous "low-spin complex" like $[Fe(CN)_6]^{3-}$. For another example, consider the Co^{3+} ion, which has six $3d$ electrons and a ^5D ($L = 2$, $S = 2$) ground state. The blue $[CoF_6]^{3-}$ complex has a magnetic moment of $5.2\mu_B$, corresponding to the full spin-plus-orbital value of Eq. 9.14, but the yellow $[Co(NH_3)_6]^{3+}$ complex is diamagnetic, corresponding to $L = 0$, $S = 0$. Clearly such large deviations from the isolated-ion model cannot be accounted for merely by quenching, especially when spin as well as orbital angular momentum is involved. We are dealing here with a phenomenon that can be interpreted only by examining the electronic structures of complexes.

[18] A typical alum is $KFe(SO_4)_2 \cdot 12H_2O$, which contains $[K(H_2O)_6]^+$ and $[Fe(H_2O)_6]^{3+}$ ions in its crystal lattice.

9.9 Electronic Structure of Complexes

Some complexes have electronic structures and chemical bonds that are well described by a classical electrostatic model, that is, a central ion to which the ionic or dipolar ligands are attracted by Coulomb forces only. For example, the hydrated ions of the heavier alkali metals approach this limit rather closely. The classical electrostatic model, similar to that we introduced for ionic molecules in Section 7.4, works best with closed-shell central ions and simple ligands. At the other extreme are covalent polyhedral molecules (Section 9.1), which are best described in terms of delocalized molecular orbitals. Most of the species ordinarily considered as complexes are intermediate between these limits. To describe their properties, especially those arising from partially filled electron shells of the central atoms, we must use an approach involving at least some of the quantum mechanical apparatus.

One of the simplest methods is basically electrostatic, in that the ligands are treated as point charges or dipoles located at the vertices of a polyhedron; one then considers the effect of this electrostatic field on the orbitals of the central atom. This simple but often effective approach was first used to interpret the spectra of ions in crystals, and is thus called *crystal field theory* (CFT); most of this section is written on the level of crystal field theory. For more exact results one must introduce at least some aspects of the molecular orbital approach, with orbitals that may encompass both the central atom and the ligands. The term *ligand field theory* (LFT) is generally applied to the range of models intermediate between the CFT and the full molecular orbital treatment.

Because of the high symmetries of polyhedral complex ions, the analysis of their properties becomes particularly elegant when it is done with the techniques of group theory, that part of mathematics concerned specifically with symmetry. Space is too limited here to permit us to develop the elements of group theory; several of the references at the end of this chapter present and apply its methods. Here, we concentrate on the physical properties of complexes and their origins in electronic structure, without drawing extensively on the symmetries of the ions.

What is the role of the ligands in determining the properties of a complex? First, they shield the central atom or ion from the environment; more important, they create a potential field within which the valence electrons of that central atom move. If this field is weak, the central atom behaves much as it would in isolation. But if the field is strong, the energy levels and orbitals of the central atom are no longer like their counterparts in the free atom or ion. Fortunately, one can analyze the properties of complexes, treating both weak-field and strong-field limits, without performing elaborate calculations. The problem lends itself to an empirical representation in terms of simple parameters. Our discussion here will be qualitative, but indicates how the quantitative treatment proceeds.

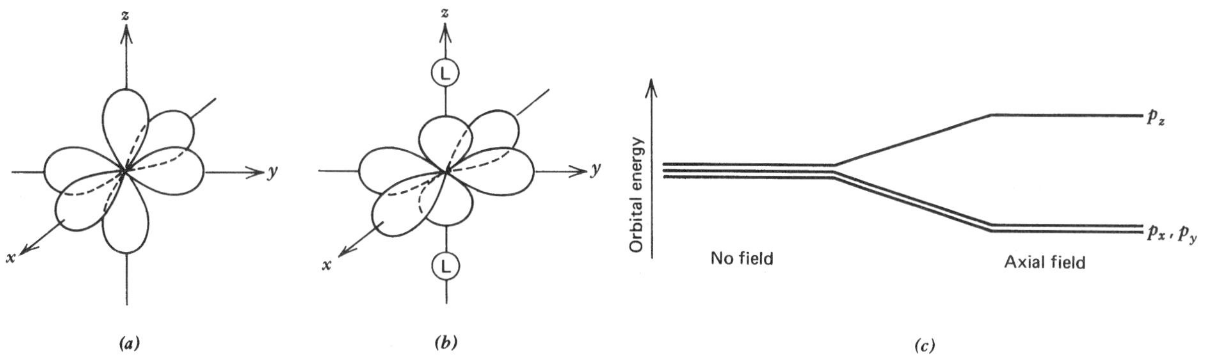

Figure 9.21 Behavior of p orbitals in an axially symmetric field. (*a*) The three degenerate orbitals with no external field. (*b*) The effect of ligands (L) along the z axis, splitting the orbitals into a twofold-degenerate pair (p_x, p_y) and a nondegenerate (p_z) orbital. (*c*) Schematic illustration of the splitting of orbital energies (the variation of which is presumably a continuous function of the strength of the axial field).

One usually assumes that the equilibrium geometries of the complexes of interest are known. It is possible but costly to find the geometry from direct computations of the potential energy surface; in practice, it is frequently more convenient to infer the structure from experiments or, for the more elaborate complexes, to assume the geometry from chemical analogs. Fortunately, the experimental knowledge of the geometry of complexes is usually sound. Structural information can be obtained from chemical methods such as Werner's, from x-ray diffraction (Chapter 11), from spectroscopy, or from magnetic measurements. We shall see why the spectroscopic and magnetic properties depend sensitively on the shape of the complex.

Given the geometry, then, what can we say about the electronic structure? Consider the outer electrons of the central atom or ion. In the free atom these electrons would be subject to a potential field that is spherically symmetric, except for the interactions between the electrons themselves, because of the isotropy of free space. There is no physical distinction between different orientations of the electrons and those of the atom itself. Remember that a symmetric Hamiltonian leads to degeneracy. In this case, for example, an atomic state[19] with a given value of J corresponds to $2J + 1$ degenerate wave functions with the same energy. But in a complex the situation is quite different. The polyhedron of ligands produces a potential field that is less than spherically symmetric, and the usual result is some removal of degeneracy. For a specific polyhedral geometry, it may be possible to observe a physical distinction between different orientations.

The latter statement does not apply to an orbital or state with zero angular momentum, an s orbital or ^1S state: Since such states are nondegenerate and have spherically symmetric wave functions, they have no capacity to distinguish orientations. The presence of a set of ligands may shift the energy of such a state and distort its wave function from spherical symmetry, but the directional character must still be the most symmetric allowed by the ligand field. Thus an s orbital in a pure octahedral field may distort, but only into octahedral symmetry; it will still have the same amplitude for all the equivalent directions of the octahedral environment.

Orbitals and states with nonzero angular momentum do not respond to the presence of ligands in so simple a manner. For the most elementary example of the directionality induced by ligands, let us consider what happens to the central atom's p orbitals in a linear "complex" such as XeF_2 (coordination number 2). We call the internuclear axis the z axis, with the origin of coordinates at the central atom (Fig. 9.21). In the free atom one could define the usual p_x, p_y, p_z orbitals extending along the three coordinate axes, and all three of these orbitals would be equivalent in every physically distinguishable way. But if we put a ligand on each side of the central atom along the z axis, the fields acting on the different p orbitals are no longer equivalent. The p_z orbital becomes physically different from the p_x and p_y orbitals, which remain equivalent to each other. By "physically different" we mean that the shape and energy of the p_z orbital must be different from those of the p_x and p_y. We say that the ligand field has split the threefold degeneracy of the p orbitals into a twofold-degenerate pair (p_x,p_y) and a single nondegenerate orbital (p_z). One would expect the central atom's p_z orbital to distort somewhat so as to move electron density away from the filled orbitals of the ligands because of electron–electron repulsion. The result is an increase in energy of the p_z orbital relative to the p_x and p_y orbitals. These effects are illustrated in Fig. 9.21.

If we considered XeF_2 as a complex made up of Xe^{2+} and F^- ions, the four $5p$ electrons of Xe^{2+} would probably be best described as occupying the lower-energy degenerate orbitals. However, this is a poor model for XeF_2, in which the bonding is strongly covalent and better treated in terms of three-center MOs. We have gone through this analysis with p orbitals only for illustrative purposes: As we shall see, CFT and LFT are ordinarily most powerful when they are applied to d and f orbitals, which partake less in covalent bond formation than do p orbitals.

[19] By "state" here we mean, say, a particular component of a multiplet term, such as $^2P_{1/2}$: cf. Section 5.7.

Here is another way to look at the splitting. In a field with only axial symmetry (with the z axis taken as the symmetry axis), the space-quantized orbitals with $m_l = \pm 1$ (or their real combinations, p_x and p_y) differ physically from the orbital with $m_l = 0$ (p_z). Clearly a similar splitting must occur for $d, f, \ldots$ orbitals in an axially symmetric field. The splitting always takes the same form: From a set of orbitals that are degenerate in the isolated atom, the one with $m_l = 0$ becomes nondegenerate, whereas those with any given value of $|m_l|$ greater than zero form a degenerate pair with its own energy. The twofold degeneracy that remains is the result of the equivalence of the x and y directions.

In an octahedral ligand field, on the other hand, the x, y, and z directions are all physically equivalent just as in the isolated atom. To see that this is so, we need merely choose the coordinate axes to coincide with the bonds to the ligands, that is, along the lines connecting opposite vertices of the octahedron, as shown in Fig. 9.22*a*. Because of this equivalence the three components of a p state remain equivalent and unsplit in an octahedral environment. It may be slightly less obvious that a tetrahedral ligand field is also incapable of splitting p levels. To demonstrate this, we inscribe the tetrahedron in a cube and choose the axes as shown in Fig. 9.22*b:* One can see that the x, y, and z axes are again equivalent. Note that in these cases, as with s orbitals, the energy level may be shifted and the wave function distorted, but the degeneracy remains.

For a degenerate set of atomic orbitals to display splitting in an octahedral or tetrahedral field, the orbitals must be numerous enough, and have enough angular dependence, to distinguish physically different directions in space. Although the p orbitals do not have this capacity, we shall now see that the five components of a set of d orbitals do satisfy this requirement. The same is true *a fortiori* of $f, g, \ldots$ orbitals. A great many complexes have octahedral or tetrahedral symmetry or can be treated in terms of distortions of these structures, and we expect splitting to occur in such complexes when there is a partially filled d or f subshell on the central atom—that is, in complexes of transition metal or rare earth ions. The d and f electrons of these ions are largely responsible for the bonding, optical, and magnetic properties of the complexes. It is to complexes of this type that CFT and LFI are mainly applied.

The most useful way to define five d orbitals is to pick the real functions we described in Section 4.4, five independent functions that vary respectively as xy, xz, yz, $x^2 - y^2$, and $z^2 - \frac{1}{2}(x^2 + y^2)$. (The last is called d_{z^2} for convenience.) These are suitable analogs of the real p functions that vary as x, y, z. We are concerned here only with their shapes, that is, their angular dependence,[20] which has been illustrated in Fig. 4.7.

In an octahedral ligand field the d_{xy}, d_{xz}, and d_{yz} orbitals are physically equivalent. Each has two equivalent axes of maximum amplitude that bisect the angles between pairs of coordinate axes. In the coordinate system of Fig. 9.23*a*, they are directed toward the midpoints of the octahedron's edges, that is, midway between pairs of ligand sites. The $d_{x^2-y^2}$ and d_{z^2} orbitals, in contrast, have their maxima along the coordinate axes—toward the vertices of the octahedron, and thus toward the ligand sites, as Fig. 9.23*b* shows. If the valence orbitals of the ligands are filled, as is usually the case, then Coulomb and exclusion repulsion will tend to drive electrons of the central atom away from the regions near the ligands. This is especially true when the ligand is negatively charged, or is a dipole with its negative end inward, both common cases. This effect is strongest in the $d_{x^2-y^2}$ and d_{z^2} orbitals, in which the electrons are on the average closer to the ligands to begin with. As a result, the $d_{x^2-y^2}$ and d_{z^2} orbitals generally have higher energies than the d_{xy}, d_{xz}, and d_{yz} orbitals in an octahedral field. The five d orbitals that are degenerate in the free atom split into three degenerate low-energy orbitals and two degenerate high-energy orbitals.[21] The splitting is illustrated in Fig. 9.23c.

We have thus far dealt only with one-electron orbitals. However, the same splitting rules apply to the many-electron atomic wave functions. Thus in an octahedral field an atomic D term (a set of states with $L = 2$) splits into a triply degenerate low-energy set and a doubly degenerate high-energy set, whereas a P term ($L = 1$) remains unsplit.

A tetrahedral ligand field also splits a set of d orbitals, but not in quite the same way. In the tetrahedral field, as shown in Fig. 9.24, it is the d_{xy}, d_{xz}, and d_{yz} orbitals that have their

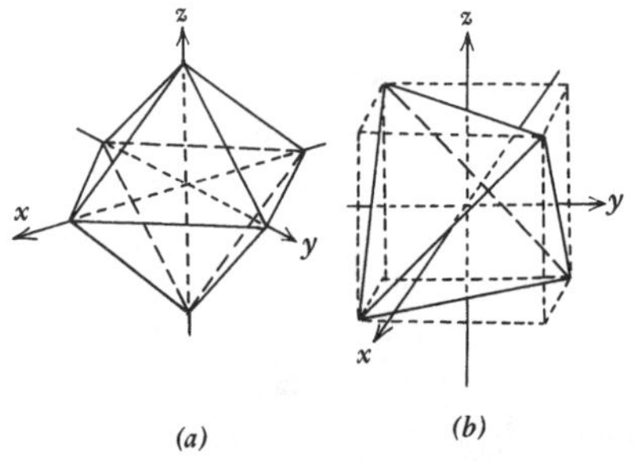

Figure 9.22 Equivalence of x, y, and z axes in (a) octahedral and (b) tetrahedral ligand fields. In (b) the coordinate axes pass through the centers of opposite faces of the circumscribed cube, and thus through the midpoints of opposite edges of the tetrahedron.

[20] The angular parts of these functions are given by the following table:

d_{z^2}	d_{xz}	d_{yz}	$d_{x^2-y^2}$	d_{xy}
$Y_{2,0}$	$Y_{2,\cos\phi}$	$Y_{2,\sin\phi}$	$Y_{2,\cos 2\phi}$	$Y_{2,\sin 2\phi}$

where the Y's are defined by Eq. 4.24 and Table 3.1. The magnitude of the angular function $Y_{2,\sin 2\phi}$ varies as xy/r^2 for d_{xy}, etc. (division by r^2 removes the radial dependence).

[21] It may not be obvious why the $d_{x^2-y^2}$ and d_{z^2} orbitals remain degenerate, since their shapes are quite different (Fig. 4.7). However, since

$$z^2 - \tfrac{1}{2}(x^2 + y^2) = \tfrac{1}{2}[(z^2 - x^2) + (z^2 - y^2)],$$

the d_{z^2} orbital is formally equivalent to a sum $d_{z^2-x^2}$ and $d_{z^2-y^2}$ orbitals, each with the same shape as the $d_{x^2-y^2}$ and thus degenerate with it when the x, y, and z directions are equivalent (as is the case in both octahedral and tetrahedral fields).

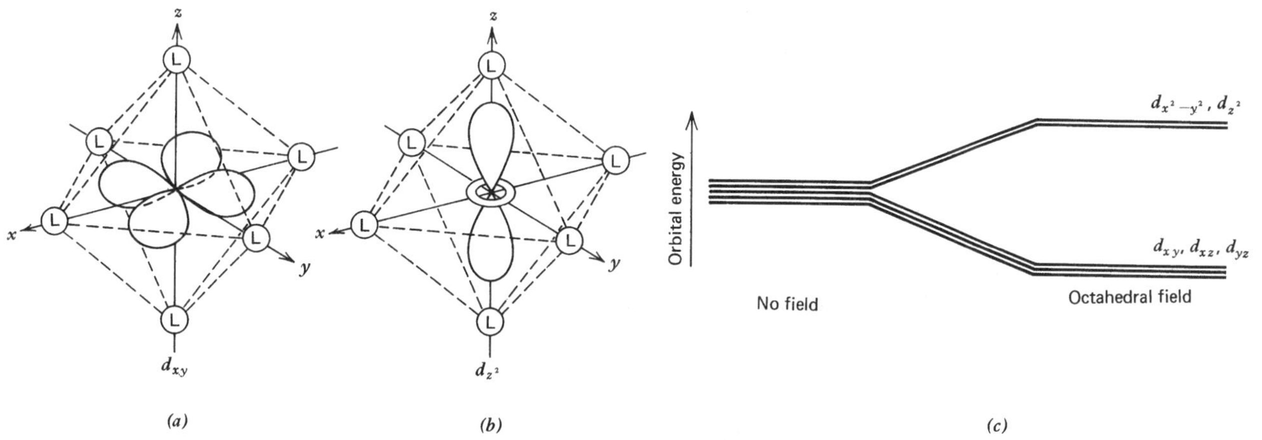

Figure 9.23 Splitting of d orbitals in an octahedral field. (*a*) The d_{xy} orbital (d_{xz} and d_{yz} have the same shape and also extend between the coordinate axes). (*b*) The d_{z^2} orbital ($d_{x^2-y^2}$ resembles d_{xy}, but extends along the x and y axes). (*c*) Splitting of orbital energies; in no field, all five orbitals are degenerate. In an octahedral field, these split into a set of three degenerate levels and a degenerate pair of levels.

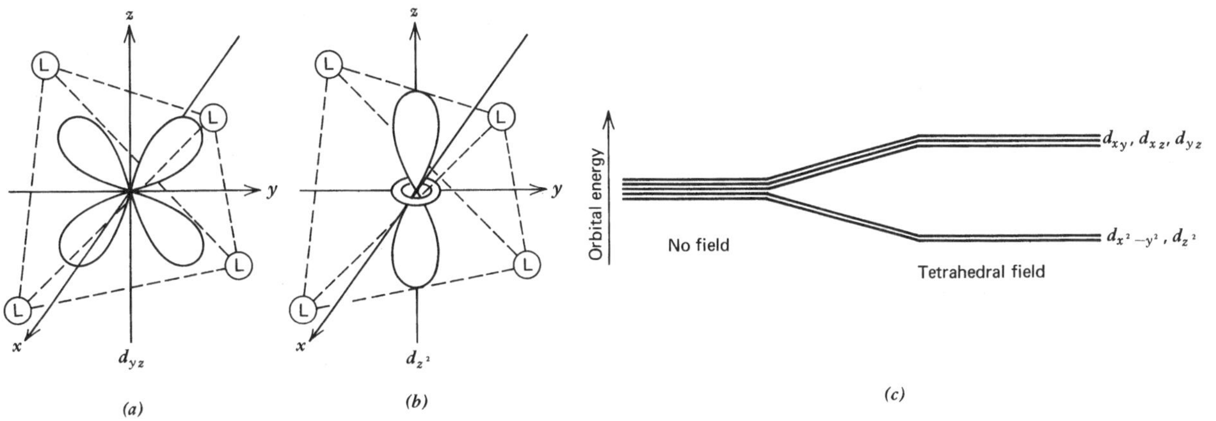

Figure 9.24 Splitting of d orbitals in a tetrahedral field. (*a*) The d_{yz} orbital (the lobes extend close to but not directly toward the ligand sites, none of which are in the yz plane). (*b*) The d_{z^2} orbital (the lobes point midway between ligand sites). (*c*) Splitting of orbital energies. The five degenerate d orbitals again split into a triply degenerate set and a degenerate pair, but, in contrast to the octahedral case, the pair usually has lower energy.

maxima relatively close to the directions of the ligand sites, whereas the $d_{x^2-y^2}$ and d_{z^2} orbitals point between the ligand sites; Fig. 9.24*a* and 9.24*b* illustrate this. Thus the d orbitals again split into sets of two and three, but the energy relationships are reversed from those in an octahedral field. Most commonly the interaction between d electrons and ligands is predominantly repulsive, so the doubly degenerate pair of $d_{x^2-y^2}$ and d_{z^2} orbitals are the low-energy set and the other three are the high-energy set. This splitting is shown in Fig. 9.24*c*; for similar ion-ligand interactions, the energy splitting of the d orbitals in the tetrahedral complexes is only about half as great as in the octahedral cases.

Distortions, if they occur, lower the symmetry of a tetrahedral or octahedral complex. We mentioned earlier the tetragonal and trigonal distortions of the octahedron. The result of such distortions is that the energy levels split even further, as the orbitals that are equivalent in octahedral or tetrahedral symmetry become nonequivalent. In a tetragonal distortion along the z axis, for example, the energy of the orbitals that extend in the z direction (d_{xz}, d_{yz}, and especially d_{z^2}) are shifted relative to those in the xy plane (d_{xy} and $d_{x^2-y^2}$). The 4-coordinate square-planar complex, as we have pointed out, is an extreme case of this distortion. In tetragonal complexes the d_{z^2} orbital has a binding energy considerably greater than the d_{xy} and $d_{x^2-y^2}$ pair; sometimes the d_{z^2} orbital lies lower than the d_{xz} and d_{yz} levels, which in this case remain degenerate. Other distortions can be similarly analyzed (cf. Problem 20 at the end of the chapter). Distortions are relatively common, though often small. One can show that any complex in which an orbital degeneracy occurs can lower its total energy by undergoing a geometric distortion that splits the degeneracy (the *Jahn–Teller effect*).

Our analysis has now gone far enough to allow us to give some interpretation of the relationships between the structures of complexes and their spectroscopic and magnetic properties.

As we mentioned earlier, the characteristic colors of transition metal complexes must be due to the absorption of light in the visible region of the spectrum. What is the nature of the transitions involved? Most transitions involving the valence electrons of free atoms—and especially of positive ions—are associated with absorption or emission of ultraviolet radiation. Relatively few such transitions have energies low enough to give rise to spectral lines in the visible region. But in complexes the ligand field is relatively weak compared to the strong central field of the atom. The bonding between the central atom and the ligands is relatively weak for the same reason. Thus the ligand field splitting of *d* levels is very often small enough to put the absorption bands due to *d–d* transitions into the visible region, and thus make the complexes colored. The spectra of complexes can usually be explained in considerable detail with the full apparatus of ligand field theory. The spectra may be obtained under many different conditions, and their appearances change accordingly. In solutions at room temperature broad bands rather than discrete spectral lines are usually observed, primarily because of the vibrational motion of the ligands. In crystals at very low temperatures very sharp spectra often appear, in which vibrational levels can be assigned with confidence.

As for the magnetic properties, it is easy to see how they can be interpreted in terms of energy-level diagrams like those in Figs. 9.23 and 9.24. Let us examine the magnetic and optical properties for several cases of increasing complexity.

The simplest transition metal ion in which *d* orbitals are occupied is Ti^{3+}, with a single 3*d* electron outside an argon closed shell. The magnetic moments of TI(III) complexes average $1.7\mu_B$, corresponding to the spin-only value with quenching of orbital angular momentum. We can now see that the "quenching potential" we spoke of in the last section is the ligand field, which has a much greater effect on the electron's motion than does the external magnetic field; detailed calculations of this effect are possible, but will not be discussed here. For an example of the spectroscopic properties, consider the hexaquo ion $[Ti(H_2O)_6]^{3+}$. It has a broad absorption band whose peak lies at about 4930 Å, giving the complex a violet color. (The transmitted light on both sides of the band is mainly red and blue, combining to produce violet.) The complex is presumably octahedral, with an energy-level diagram like that shown in Fig. 9.23*c*. We assume that the single *d* electron occupies the lower, triply degenerate level in the ground state, and that the visible absorption corresponds to a transition to the upper, doubly degenerate level.

The V^{4+} ion is isoelectronic with Ti^{3+}, and forms some octahedral complexes with similar properties. It also forms some distorted tetrahedral compounds, of which VCl_4 is the best known. As we noted above, the ligand field splittings in tetrahedral complexes are typically about half as great as those in octahedral complexes for ligands of the same sort. Thus VCl_4 is brown because of an absorption band centered near 9000 Å in the near infrared, presumably due to a transition from the lower, doubly degenerate to the upper, triply degenerate level in Fig. 9.24*c*. Both types of complexes have magnetic moments of $1.7–1.8\mu_B$, again the spin-only value for one electron.

Ions with more *d* electrons offer richer examples for interpretation. The electron–electron interactions play a role that, in effect, competes with the ligand field effects. A good and rather simple example is the Fe^{3+} ion, which has the configuration $[Ar]3d^5$, with a half-filled *d* shell. As we indicated in the last section, the ground state of the isolated ion is ^6S, with maximum spin and zero orbital angular momentum. In a weak ligand field—one whose effects are small compared with the electron–electron interactions responsible for Hund's rules—the designation ^6S remains characteristic of the ground state of an Fe(III) complex. That is, the total spin is $\frac{5}{2}$, the magnetic moment is $5.9\mu_B$, and we can assume that each of the five 3*d* orbitals still contains one electron despite the splitting of the orbital levels by the ligand field. We have mentioned a number of complexes, such as $[Fe(H_2O)_6]^{3+}$, in which this behavior is observed. The electron configuration in these complexes is presumably that shown in Fig. 9.25*a*. The best-known *d–d* absorption bands[22] in these systems involve either transfer of an electron from the doubly degenerate level to the formally lower-energy triply degenerate level, at the expense of increased electron–electron repulsion, of course, or transitions between different terms of the ground-state configuration. In either case the energy difference is relatively small and the bands weak.

In a strong ligand field, on the other hand, the splitting between the two sets of *d* levels can become so large that the difference in orbital energies becomes greater than the separation of the ^6S and higher terms arising from the electron–electron repulsion. In these circumstances all five 3*d* electrons of the Fe^{3+} ion can be assigned to the lower-energy triply degenerate level, as shown in Fig. 9.25*b*. In this configuration there should be only one unpaired electron—and, as we mentioned earlier, the $[Fe(CN)_6]^{3-}$ complex, in which the CN^- ligand exerts a very strong field, does have a magnetic moment characteristic of only one unpaired spin, instead of five.

Many other ions exhibit a similar distinction between weak-field (also called *high-spin*) and strong-field (also called *low-spin*) complexes. In general these have the maximum and minimum possible numbers of unpaired spins, respectively; it is exceedingly rare to find an intermediate case. Of the examples we cited at the end of Section 9.8, the Mn(II) complexes have the same configuration as those of Fe(III), whereas for Co(III) (. . . $3d^6$) the low-spin com-

[22] Often there are much stronger charge-transfer bands, involving the transfer of an electron from a metal orbital to a ligand orbital or vice versa. These bands have their peaks in the ultraviolet but are often strong enough in the visible region to have a dominant effect on the color and other optical properties.

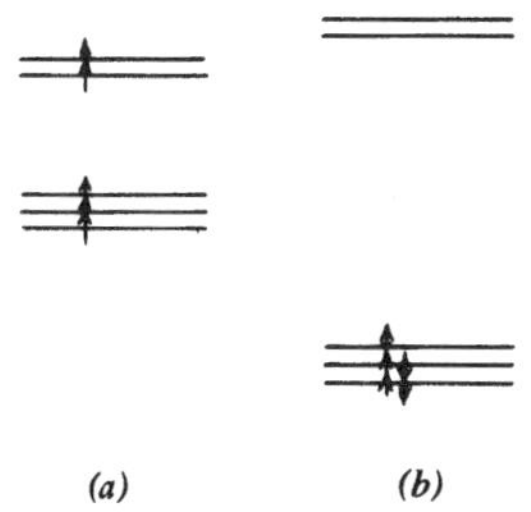

Figure 9.25 Configurations of the five 3*d* electrons in the ground states of Fe(III) complexes: (*a*) weak field (high-spin); (*b*) strong field (low-spin).

plexes have all six *d* electrons in the triply degenerate lower level, leaving no electrons unpaired and making the complex diamagnetic. One can easily see from Fig. 9.22 that high-spin and low-spin configurations should be possible for all octahedral complexes with four to seven *d* electrons in the outermost shell. Similarly, for tetrahedral complexes (Fig. 9.24) the phenomenon appears with three to six *d* electrons. But which complexes are high-spin and which low-spin? This depends on the strength of the ligand field, and thus on the nature of the ligands. For a number of common ligands the ability to split *d* orbitals is found to fall in the order:

$$\begin{aligned} CN^- > NO_2^- > \text{en} > NH_3 > H_2O \\ > OH^- > F^- > Cl^- > Br^- > I^-, \end{aligned}$$

(the *spectrochemical series*).

Thus far we have not faced the question of how bonding occurs between the central atom and the ligands. In the original crystal field theory the bonding was interpreted in terms of classical electrostatics. As useful as the simple ionic CFT is in giving a qualitative interpretation of magnetic and spectral properties, it is inadequate to account for bonding. When it became possible to interpret spectra in terms of microscopic interactions, it was clear that the CFT predictions were not in satisfactory quantitative agreement with the observations. The reason, of course, is that the metal–ligand bonds have a significant covalent contribution. To some extent this can be accounted for by inserting empirical parameters into the theory, but this procedure is not altogether satisfactory either.

It was a major step forward when the concept of molecular orbitals was extended to the bonding in complexes. The MOs used, of course, must be symmetry orbitals with the symmetry of the complex as a whole. In general, all the valence-shell orbitals of the central atom (*s* and *p* as well as *d*) are used in constructing the MOs, together with at least one orbital from each ligand. Central-atom orbitals that extend toward the ligands (Figs. 9.23 and 9.24) usually form σ bonds, whereas those that extend between the ligands can form π bonds. This approach has made it possible to bring the experiments and the theory together, at least as well as contemporary measuring and computing techniques can justify. We already know that complexes span a range from highly "electrostatic" and weakly bound to highly covalent, with strong short-range attractive forces. Experiment shows, and the MO calculations confirm, that the more covalent complexes tend to be low-spin (strong-field); this is to be expected, since the ligands whose orbitals overlap the central-atom orbitals to the greatest extent are naturally those that produce the strongest ligand fields. Thus, the pure *d* character of the central atom's orbitals is maintained best in the high-spin complexes.

Before we close this section, a few words are in order about complexes of the rare earth (lanthanide) elements. The partially filled 4*f* shell of these elements is buried deeper within the atom than are the outer *d* shells of the ordinary transition metals. Hence the ligand field splittings of the *f* levels are generally quite small so there are no low-spin states, and the bands associated with transitions between these split levels often lie in the infrared. The *f–f* bands are also relatively narrow, because of the shielding effect of the outer electrons. Since spin–orbit interaction is strong in these atoms, the magnetic moments should be given by $g_J\mu_B[J(J+1)]^{1/2}$ rather than Eq. 9.13 or 9.14. Quenching is not significant, again because of shielding, and the experimental values agree well with these equations. In the actinide elements, by contrast, the 5*f* shell is relatively poorly shielded, so the complexes and other compounds of these elements have properties intermediate between those of the lanthanides and the transition metals.

APPENDIX 9A

Schmidt Orthogonalization

The most convenient way to generate a set of n orthogonal functions $\psi_i(i = 1, \ldots, n)$ from a set of n arbitrary but independent functions $\varphi_j(j = 1, \ldots, n)$ is called *Schmidt orthogonalization*. Suppose that the functions are listed serially. In essence, one picks the first function φ_1 arbitrarily. Then one makes the second function φ_2 orthogonal to the first by removing any component of φ_1 from it; next, one makes the third function φ_3 orthogonal to the first two by removing any component of φ_1 and the now-orthogonalized version of φ_2. One continues this way through the list until every function has been made orthogonal to all the preceding functions in the list. Finally, it is convenient to renormalize the functions so they all give unity for the integrals of their (absolute) squares. The set that one obtains in the end obviously depends on the sequence in which the initial functions φ_j were chosen.

Formally, the procedure is this. Assume that the set φ_j is normalized; otherwise normalize each function to unity. Let

$$\psi_1 = \varphi_1 \tag{9A.1}$$

and construct

$$\psi_2' = \varphi_2 - \psi_1 \int \psi_1^* \varphi_2 \, d\tau, \tag{9A.2}$$

which is φ_2 without its ψ_1 component. Hence ψ_2' is orthogonal to ψ_2, thus

$$\begin{aligned}\int \psi_1^* \psi_2' \, d\tau &= \int \psi_1^* \varphi_2 \, d\tau - \left(\int \psi_1^* \psi_1 \, d\tau \right) \left(\int \psi_1^* \varphi_2 \, d\tau \right) \\ &= \int \psi_1^* \varphi_2 \, d\tau - 1 \cdot \int \psi_1^* \varphi_2 \, d\tau \\ &= 0. \end{aligned} \tag{9A.3}$$

Now normalize ψ_2',

$$\psi_2 = \frac{\psi_2'}{\left(\int \psi_2'^* \psi_2' \, d\tau \right)^{1/2}}, \tag{9A.4}$$

so that

$$\int \psi_2^* \psi_2 \, d\tau = 1. \tag{9A.5}$$

Next, construct

$$\psi_3' = \varphi_3 - \psi_1 \int \psi_1^* \varphi_3 \, d\tau - \psi_2 \int \psi_2^* \varphi_3 \, d\tau, \tag{9A.6}$$

which is orthogonal to ψ_1 and ψ_2, and normalize to get

$$\psi_3 = \frac{\psi_3'}{\left(\int \psi_3'^* \psi_3' \, d\tau \right)^{1/2}}. \tag{9A.7}$$

Continue the procedure through ψ_n.

• FURTHER READING

Ballhausen, C. J., *Introduction to Ligand Field Theory* (McGraw-Hill Book Co., Inc., New York, 1962).

Burdett, J. K., *Molecular Shapes—Theoretical Models of Inorganic Stereochemistry* (John Wiley and Sons, New York, 1980).

Cotton, F. A., and Wilkinson, G., *Advanced Inorganic Chemistry*, Second Ed. (Interscience Publishers, New York, 1966).

Dewar, M. J. S., *The Molecular Orbital Theory of Organic Chemistry* (McGraw-Hill Book Co., Inc., New York, 1969).

Doggett, G., *The Electronic Structure of Molecules: Theory and Application to Inorganic Molecules* (Pergamon Press, Oxford and New York, 1972).

Dunn, T. M., McClure, D. S., and Pearson, R. G., *Some Aspects of Crystal Field Theory* (Harper and Row, New York, 1965).

Herzberg, G., *Molecular Spectra and Molecular Structure. III. Electronic Spectra and Electronic Structure of Polyatomic Molecules* (D. van Nostrand Co., Inc., New York, 1966).

Jørgensen, C. K., *Absorption Spectra and Chemical Bonding in Complexes* (Pergamon Press, Oxford, England, 1962).

Langhoff, S. R., ed., *Quantum Mechanical Electronic Structure Calculations with Chemical Accuracy* (Kluwer Academic Publishers, Dordrecht, 1995).

Mulliken, R. S., and Ermler, W. C., *Polyatomic Molecules—Results of ab initio Calculations* (Academic Press, New York, 1981).

Orville-Thomas, W. J., editor, *Internal Rotation in Molecules* (John Wiley & Sons, Inc., New York, 1974).

Shubnikov, A. V., and Koptsik, V. A., *Symmetry in Science and Art*, translated by G. D. Archart, edited by D. Harker (Plenum Press, New York, 1974), esp. Chapter 3.

Siegbahn, K., et al., *ESCA Applied to Free Molecules* (North-Holland, Amsterdam, 1971).

Streitweiser, A., *Molecular Orbital Theory for Organic Chemists* (John Wiley & Sons, Inc., New York, 1961).
Tables of Interatomic Distances, Special Publications Nos. 11 and 18, The Chemical Society, London, 1958 and 1965. (This is a standard reference for data on molecular structures.)
Wells, A. F., *Structural Inorganic Chemistry,* 4th Ed. (Oxford University Press, Oxford, 1975).
Yarkony, D. R., ed., *Modern Electronic Structure Theory, Parts 1 and 2* (World Scientific, Singapore, 1995).

PROBLEMS

1. The claim was made that the orbitals of Eq. 9.1 are equivalent, but with different spatial orientations. Show that they are equivalent, and determine the angles θ, ϕ along which they have maximum amplitude.

2. What structure do you predict for UF_5 monomer?

3. The ammonia molecule has a bond angle (HNH) of 106.7°. Suppose the N—H bond orbitals are symmetric about their respective N—H axes and are expressible as sums of hydrogenic 1*s* orbitals and nitrogen hybrids of the form $\alpha\varphi(2s) + \beta\varphi(2p)$. Find α and β from the value of the bond angle. If the lone pair orbital contains the rest of the 2*s* and 2*p* orbitals, what is its form?

4. Use single-lobed Slater-type orbitals with $Z = 3.18$ to compute the charge density in a radial direction along the direction of maximum charge for (a) a 2*p* orbital; (b) an *sp* hybrid; (c) an sp^2 hybrid; (d) an sp^3 hybrid. These are traditional approximations for the atomic orbitals of carbon. Their explicit forms are

$$\varphi(2s) = (Z^3/32\pi)^{1/2}(2Zr)\exp(-Zr/2)$$
$$\varphi(2p_z) = (Z^3/96\pi)^{1/2} Zr\exp(-Zr/2)\cos\theta$$
$$\varphi(2p_x) = (Z^3/96\pi)^{1/2} Zr\exp(-Zr/2)\sin\theta\cos\varphi$$
$$\varphi(2p_y) = (Z^3/96\pi)^{1/2} Zr\exp(-Zr/2)\sin\theta\sin\varphi.$$

Suppose the length of a carbon–carbon bond can be determined approximately by superposing the points of maximum charge density for the appropriate hybrid orbitals from the two separate atoms. Use this supposition to make a table of C—C bond lengths for the various hybrid orbital combinations. Compare your results with the values given in Table 9.3.

5. Hydrogen peroxide, HO—OH, has a structure in which the hydrogens lie off the O—O axis, making a dihedral angle of roughly 90°, as shown. In ethane, H_3C—CH_3, the hydrogens at one end of the molecule lie between those at the other end, when the molecule is viewed along the C—C axis.

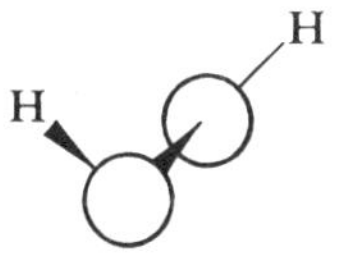

H_2O_2

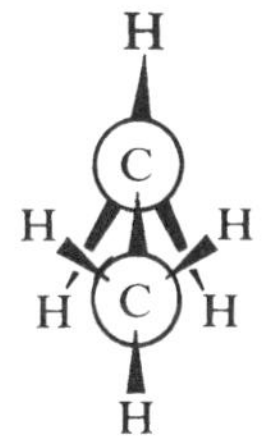

C_2H_6

What physical effects could be responsible for these structures?

6. When two identical groups such as those in Table 9.1 are near each other, their oscillations—strictly, their oscillating charge distributions—induce a coupling between the two oscillators. The result is that the pair of vibrating groups oscillate synchronously in their stationary states. Show that the first two excited states would be degenerate in the absence of any coupling, and that with the nonzero coupling the stationary states correspond to motion exactly in phase and exactly out of phase. If two carbonyl (> C = O) groups are coupled by an interaction of 100 cm^{-1}, what is the resulting splitting in the infrared spectrum? (The form of the calculation should be suggested by footnote 10; justify the use of such a form.)

7. Of the binding energies derived from photoelectron spectroscopy that are given in Table 9.2, which could be determined by photoionization with a lamp in which the light must pass through a lithium fluoride window passing light with $\lambda > 120$ nm? a sapphire window passing $\lambda > 150$ nm? Which could be studied by excitation with the light from the first allowed transition ($2p \rightarrow 1s$) of helium? from the first allowed transition ($3s \rightarrow 2p$) of neon? from the first allowed transition ($4s \rightarrow 3p$) of argon?

8. Predict the ultraviolet (58.5 nm) photoelectron spectra of
(a) allyl alcohol, H_2C=CH—CH_2OH;
(b) 1,5-hexadiene, H_2C=CH—CH_2—CH_2—CH=CH_2;
(c) benzaldehyde, C_6H_5CHO.

Compare your predictions with the data given in the reference of Table 9.2.

9. Set up and solve the secular equations for the Hückel molecular orbitals of
(a) the allyl radical, $H_2\dot{C}$—$\dot{C}H$—$\dot{C}H_2$;
(b) the cyclopropenyl radical;
(c) cyclobutadiene.
(*Hint:* use the molecular symmetry to factor the 3 × 3 and 4 × 4 matrices so you have nothing larger than a quadratic to solve.)

10. Find an *approximate* value for the lowest value of the unknown ϵ in Eq. 9.8 by expanding the determinant in minors. Terminate the expansion successively after one, two, and three steps.

11. Consider a ring of $2n$ carbon atoms, in a molecule $C_{2n}H_{2n}$. Show that the Hückel molecular orbital model implies that the energies ϵ_j of the molecule's π-orbitals have the form

$$\epsilon_j = \alpha + 2\beta\cos(j\theta), \quad \text{where} \quad \theta = \pi / n$$
$$\text{and} \quad j = 0, \pm 1, \ldots, \pm(n-1), n.$$

Be careful to note how the range of j is defined. Show this implies that if n is odd, so that the molecule contains $4m + 2$ carbons (with m an integer), then all the orbitals occupied in the ground state are fully occupied, but that if n is even, so the number of carbons is divisible by four, the highest normally occupied orbitals are degenerate and only two electrons occupy these two degenerate orbitals.

12. Construct a set of molecular orbitals for B_2H_6 from the $1s$ atomic orbitals of the hydrogens and the $2s$ and $2p$ orbitals of the boron atoms. Assume that the outer B—H bonds are localized but that the bonds in the central boron-hydrogen ring are delocalized over the four atoms. What is the number of electrons in each orbital in the molecular ground electronic state? Indicate the bonding and antibonding character of each orbital; be specific with respect to which atoms the orbital is bonding or antibonding.

13. Estimate the torsional oscillation frequency for ethane on the basis of the 0.126 eV barrier shown in Fig. 9.12. Use a quadratic approximation for the minimum of the sinusoidal well and neglect any tunneling. Approximately how many vibrational levels lie below the maximum of the barrier to internal rotation?

14. The barrier to internal rotation in nitromethane, CH_3NO_2, is roughly a hundredfold smaller than the barrier in ethane. Give a physical explanation in terms of the structures of these molecules why this is so. The NO_2 group defines a plane that contains the C—N axis.

15. How many isomers are there for each of the following complexes? Assume the six-coordinate species are regular octahedra and the five-coordinate species are trigonal bipyramids. (The abbreviation "en" is used for ethylene diamine, $H_2NCH_2CH_2NH_2$.) $[Co(NH_3)_2(H_2O)_2Cl_2]^+$; $[Co(NH_3)_3(H_2O)Cl_2]^+$; PF_2Cl_3; PF_2ClH; $[Co\ en_2ClF]^+$; $Co\ enCl_3$.

16. The argument leading to Eq. 9.9 implies that an optically active material rotates the direction of polarization for plane-polarized light by having different velocities of propagation for the left- and right-circularly polarized components of the light. Show that this phenomenon can be explained by supposing that the material has different absorption cross sections for left and right circularly polarized light.

17. If every atom in an ideal gas had a magnetic moment of 0.1 Bohr magneton, what would be the magnetic susceptibility of the gas at room temperature, when the density is 10^{19} atoms/cm^3?

18. Prove equation 9.14.

19. Potassium chromium alum, containing Cr^{3+}, has a magnetization of 1.00 Bohr magneton when H/T is 3.5×10^3 oersted/K or $3.5 \times 10^6/4\pi$ A/mk. What is its effective magnetic moment and how does this value compare with the pure spin-only value for Cr^{3+}?

20. What distortions of an octahedron leave it with threefold and twofold symmetry but no fourfold symmetry? On the basis of physical equivalence or inequivalence, indicate what further splittings one could expect for the $3d$ orbitals beyond those shown in Fig. 9.22, if a threefold (trigonal) distortion took place in an octahedral complex? Indicate which functions—d_{z^2}, d_{xy}, d_{xz}, d_{yz}, $d_{x^2-y^2}$—remain degenerate. (Be careful in your choice of axes.)

21. Assume a charge $-q$ on each ligand L of the octahedron in Fig. 9.22. Compute the electrostatic energy of a charge $-\epsilon$ interacting with the six ligand charges when (a) the negative charge is midway between the center of the octahedron and one of the ligands (i.e., one of the vertices) and (b) when the negative charge is midway between the center of the octahedron and the center of one of its edges. This is a very rough model for estimating the difference in energy between the upper two and lower three orbitals of Fig. 9.22*c*.

22. Carry out an electrostatic energy calculation analogous to that of Problem 21 for the tetrahedron of Fig. 9.21. Compare the splitting with that of the octahedral complex.

23. Consider a hypothetical octahedral complex of a transition metal ion and six hydrogen atoms. From the six hydrogen $1s$ orbitals, construct six molecular orbitals that have symmetries corresponding to s-, p-, and d-like orbitals in an octahedral structure. Show which (if any) of these will mix with each of the $3d$-orbitals of the central ion.

24. Estimate the strength of an external magnetic field large enough to decouple the electrons in an Fe^{3+} complex from a low-spin configuration (Fig. 9.24b) and align them as in the high-spin configuration of Fig. 9.24a, if the ion absorbs light of 900 nm in the absence of any field. Assume the absorption is due to the d–d transition.

25. Determine the equilibrium geometry (bond lengths and bond angles) of NH_3 using (a) the HF–SCF, (b) the CISD, and (c) the MP2 methods along with a 6–31G (*d, p*) basis set. Further, compute the value of the NH_3 dipole moment and compare with the experimental value of 1.471 Debye. (See Problems 5.26 and 7.27 for basis set decriptions and software packages.)

26. Use (a) the HF–SCF, (b) the CISD, and (c) the MP2 methods along with a 6–31G (*d, p*) basis set to determine the equilibrium geometry (bond lengths and angles) as well as the fundamental harmonic vibrational frequencies of the ground electronic state of the CH_4 molecule.

27. Use Hückel molecular orbital theory to estimate the delocalization energy of (a) benzene C_6H_6, (b) $C_5H_5^+$, and (c) $C_5H_5^-$.

28. $TiCl_6^{2-}$ is an octahedral complex ion. Compute the equilibrium bond lengths of the ground electronic state of the complex ion using the HF–SCF method while constraining the geometry of the ion to be octahedral. Employ a 3–21G basis and the TZV basis (a triple zeta basis set) found in GAMESS.

29. CrO_4^{2-} is a tetrahedral complex ion. Compute the equilibrium bond lengths of the ground electronic state of the complex ion using the HF–SCF method while constraining the geometry of the ion to be tetrahedral. Employ a 3–21G basis and the TZV basis (a triple zeta basis set) found in GAMESS.

CHAPTER 10

Intermolecular Forces

In the last four chapters we have examined the bonding within isolated molecules. Now we shall consider the interaction between one molecule and another. By a "molecule" we mean a group of atoms linked by chemical bonds strong enough for the entity to retain its identity over a significant period of time at ordinary temperatures. Except in very dilute gases, this means that the energy required to break the bonds must be large compared with the average energy available in a molecular collision. For example, we have shown that the system of two hydrogen atoms has a ground state with a binding energy of 4.5 eV (432 kJ/mol) relative to the isolated atoms. Since an average gas molecule at 25°C has a kinetic energy of only 0.04 eV, the H_2 molecule is not likely to be dissociated by collisions. In contrast, although the system of two hydrogen molecules also has bound states, the binding energy is only about 0.004 eV, and the system cannot remain bound as $(H_2)_2$ at ordinary temperatures. Interactions of the latter type are too weak to produce stable molecules, but they nevertheless play important roles in the physics and chemistry of materials. In Parts II and III we shall have much to say about the effects of these interactions; in this chapter we consider their origin.

That there are forces between molecules can be deduced from two simple macroscopic observations: (1) All substances form condensed phases at sufficiently low temperatures; this indicates the existence of intermolecular attractions that are strong enough to bind together molecules with low kinetic energy. (2) All condensed phases strongly resist further compression; this indicates the existence of short-range repulsive forces that keep the molecules from coalescing once they are in "contact." The ranges of the repulsive forces should be, roughly, the average intermolecular distances in a liquid. Since attraction and repulsion are dominant at long and short ranges respectively, the intermolecular potential energy must look something like the interatomic energy in a diatomic molecule (cf. Fig. 6.2), but the scale and exact shape of the curve are different, and these are what we now investigate.

Ultimately all the forces we need to consider between atoms are either Coulombic or exchange forces—that is, due either to the direct interaction of charged particles or to the quantum effects that give rise to the Pauli exclusion principle. At long range the Coulomb forces are dominant, and we examine these first, looking at the hierarchy of interactions from the electrostatic force between two ions to the "dispersion forces" that exist between any two atoms. These long-range forces are for the most part attractive, but the overlap of electron clouds causes molecules to repel one another, primarily through the unshielding of the nuclei as the Pauli Exclusion Principle forces the electrons away from each other, at short distances. The repulsive interaction is more difficult to calculate than the attractive, and to a large extent has been treated in terms of empirical models, a number of which we shall describe.

The distinction between intramolecular and intermolecular forces is somewhat arbitrary, since the entire range of possible interaction energies can be found in one system or another. We have already discussed one intermediate case, the hydrogen bond. This and other types of "weakly associated" species are surveyed in the final section of the chapter.

10.1 Long-Range Forces: Interactions between Charge Distributions

Any atom, molecule, or ion can be considered as a spatial distribution of electric charge. At long range—that is, distances great enough that the overlap of electron distributions can be neglected—the interaction of two such species can be treated largely in terms of classical electrostatics, although one major component, the "dispersion force," requires quantum mechanics for its description. All atoms and molecules exert attractive forces on one another at sufficiently great distances; generally the attractive region extends down to distances at which the atoms "touch," that is, at which the electron clouds begin to overlap. The most important exception is the repulsive Coulomb force between two ions of like sign. However in a medium that maintains electrical neutrality, because of the presence of other ions of opposite sign that shield the Coulomb force, a weak attraction becomes dominant at *very* long distances. It is the purpose of this section to examine the origins and strengths of these attractive forces.

First, let us introduce the notion of the *moments* of a charge distribution. The electric dipole moment is one example that we have already considered; now we shall generalize the concept. The nth "moment" of a distribution is the average, over that distribution, of the nth power of the coordinates of the charges, normally taken from the center of the distribution. Consider a set of point charges Q_i at various points in space. The total charge,

$$Q = \sum_i Q_i , \tag{10.1}$$

is called the *zeroth moment* or *monopole moment* of the charge distribution. If our charge distribution is continuous, with a total charge density (charge per unit volume) of $\rho(x, y, z)$ at each point in space, the sum in Eq. 10.1 must of course be replaced by an integral,

$$Q = \iiint \rho(x,y,z)\; dx\; dy\; dz, \tag{10.2}$$

with the integration taken over all space. In the vicinity of an atom or molecule it is usually appropriate to consider the nuclei as point charges, and the electrons as continuous charge distributions. In that case, we must sum over the point charges and integrate over the continuous distributions. (If we are using the square of a molecular orbital wave function to represent the electronic charge distribution, the electronic density ρ may be written as $\sum_i \rho_i$, where ρ_i is the charge density in the ith orbital; but remember that this is only an approximation.)

The (electric) *dipole moment* or *first moment* has been defined in Section 4.5 for a system of two point charges. The general definition for a point-charge distribution is

$$\boldsymbol{\mu} = \sum_i Q_i \mathbf{r}_i, \tag{10.3}$$

where $\mathbf{r}_i$ is the radius vector of the ith particle from some origin, in a molecule usually taken to be the center of mass. For a continuous distribution the dipole moment becomes

$$\boldsymbol{\mu} = \iiint \mathbf{r}(x,y,z)\rho(x,y,z)\; dx\; dy\; dz; \tag{10.4}$$

$\mathbf{r}(x, y, z)$ is the vector whose Cartesian components are x, y, z. Thus the electric dipole moment is the set of averages of x, y, and z over the charge distribution $\rho(x, y, z)$. Note that the dipole moment of a distribution with zero total charge is simply the absolute charge of either the positive or the negative part, multiplied by the vector difference between the centers of positive and negative charge:

$$\boldsymbol{\mu} = Q^+(\mathbf{r}^+ - \mathbf{r}^-) = |Q^-|(\mathbf{r}^+ - \mathbf{r}^-). \tag{10.5}$$

In this case the dipole moment is independent of the origin of coordinates. Thus a permanent dipole moment exists in all molecules in which the centers of positive and negative charge do not coincide. We have already shown that such *polar molecules* include all heteronuclear diatomics (HCl, CO) and many more complicated species (H_2O, NH_3); their dipole moments are usually of the order of 1 debye (1 D = 3.33564×10^{-30} C m, which is $0.2e \times 1$ Å) but can be considerably larger if the molecule is ionic—e.g., NaCl.

The higher moments are called the second, third, . . . , nth moments of the charge distribution, or more commonly the *electric quadrupole, octupole*, . . . , 2^n-pole moments. They are not quite as easily described as the dipole moment, but fortunately we shall not need to evaluate them. The quadrupole moment is the set of averages of all pairs of the form x^2, xy, xz, y^2, yz, and z^2 over the distribution ρ. The quadrupole moment is what is called a second-rank tensor, which has six distinct components, five of which are in general independent. It may be written in the form of a matrix,

$$\mathrm{Q} \equiv \begin{vmatrix} x^2 & xy & xz \\ yx & y^2 & yz \\ zx & zy & z^2 \end{vmatrix}$$

but it has the properties of two vectors, $\mathbf{rr}$, taken together, each of which may perform vector operations, the left vector to the left, and the right vector to the right. We shall encounter second-rank tensors again in Part III in the context of transport properties.

For nonpolar molecules with cylindrical symmetry, such as in H_2 or CO_2, the quadrupole moment consists of only one independent quantity, which can then be called *the* quadrupole moment. The quantity usually reported is this moment divided by the electronic charge,

$$|\mathbf{q}| \equiv \frac{\sum_i Q_i(3z_i^2 - r_i^2)}{e}, \tag{10.6}$$

with the dimensions of an area. (The z axis is taken as the symmetry axis.) For homonuclear diatomic molecules $|\mathbf{q}|$ is typically about 3×10^{-21} m² (0.3 Å²).

It is the structure of a molecule that determines whether it has a nonzero dipole, quadrupole, or higher moment when it stands in isolation—that is, whether it has permanent moments. A spherically symmetric neutral charge distribution, such as that of any isolated atom, has no permanent multipole moments at all.[1] Any heteronuclear diatomic molecule can be expected to have an electric dipole moment: A

[1] In saying that an atom's charge distribution is spherically symmetric, we neglect the internal structure of the nucleus. Nuclei do not have dipole moments, but many do have nonzero quadrupole moments, with $|\mathbf{q}|$ of the order of 10^{-30} m². The quadrupole energy levels split in an electric field, and the splittings can be measured by the technique called *nuclear quadrupole resonance*.

difference in nuclear charge implies differences in electronegativity, which in turn places the center of electronic charge toward one or the other nucleus, on average, relative to the center of nuclear charge. A homonuclear diatomic molecule has no permanent dipole moment, but does have a permanent quadrupole moment. So does a symmetrical planar molecule such as cyclopropane. And in a tetrahedral molecule such as CH_4 both dipole and quadrupole moments are zero, but the permanent octupole moment is nonzero. All nonspherical species must exhibit some nonvanishing permanent electric moments. The first nonzero moment is independent of the choice of origin, but the values of any subsequent moments do depend on where the origin lies. This means that a molecule's lowest nonvanishing moment is always by far the most important to us.

In any electric field an atom or molecule suffers some distortion of its charge cloud. As we noted in Section 6.2, the effect is to separate the centers of positive and negative charge, thus creating an induced dipole moment. In a uniform field **E** the induced moment can be written as

$$\boldsymbol{\mu} = \alpha \mathbf{E}, \tag{10.7}$$

an equation that defines the *polarizability* α. In fields of ordinary strength, such as those of common electric lights, sunlight or even low-power lasers, α can be treated as a constant, but in the more intense fields of powerful lasers, we must recognize that the induced moment need not increase linearly with field strength. In very strong fields, such as those of focused laser beams, of the order of 10^8 V/m, can one observe significant variation in α. Fields of even greater magnitude may be involved when two molecules collide, but α can safely be taken as constant[2] when considering long-range interactions. Polarizability is a measure of the "softness" of a charge cloud, that is, the ease with which it can distort. In classical electrostatics, a perfectly conducting sphere of volume V has a polarizability of exactly $4\pi\epsilon_0 V$; similarly, for molecules the quantity $\alpha/4\pi\epsilon_0$ is usually of the order of the molecular volume, a few cubic angstroms. The polarizabilities of the elements in fact correlate well with their atomic sizes (Section 5.5). The large alkali atoms are highly polarizable, whereas the relatively small inert gas atoms have low polarizabilities; a graph of α versus atomic number thus looks much like Figure 1.7. Typical polarizabilities are given in Table 10.1.

Positive ions generally have polarizabilities much smaller than the corresponding neutral atoms—both because of their smaller sizes and because the decreased shielding of the nuclei makes it harder to overcome the electron–nuclear attractions. Negative ions are much more polarizable than neutral atoms for the converse reasons: They are large, and the outer electrons are not strongly bound to their nuclei.

A nonuniform electric field induces not only a dipole moment, but higher moments as well. Just such a nonuniform field is exerted on one charge distribution by another

Table 10.1a Polarizabilities of Neutral Atoms in Their Electronic Ground States[a]

Atom	α (Å^3)
H	0.666793
He	0.204956
Li	24.3
Be	5.6
B	3.03
C	1.76
N	1.10
O	0.802
F	0.557
Ne	0.3946
Na	23.6
Cl	2.18
Ar	1.64
K	43.4
Kr	2.48
Rb	47.3
Sr	27.6
I	3.9
Xe	4.04
Cs	59.6
Hg	5.1

[a] From T. M. Miller and B. Bederson, *Adv. At. Mol. Phys.* **13,** 1 (1977).

Table 10.1b Polarizables and Anisotropies of Polarizabilities of Neutral Molecules in Their Ground States[a,b]

Molecule	α (Å^3)	$\alpha_\parallel - \alpha_\perp$ (Å^3)
H_2	0.819	0.314
D_2	0.809	0.299
N_2	1.77	0.70
CO	1.98	0.53
O_2	1.60	1.10
HCl	2.60	0.311
CO_2	2.63	2.10
NH_3	2.22	0.288
Cyclopropane	5.64	−0.81
Benzene	10.4	−5.6
CH_3Cl	4.53	1.55
$CHCl_3$	8.50	−2.68

[a] From N. J. Bridge and A. D. Buckingham, *Proc. Roy. Soc.* (London) **A295,** 334 (1966).

[b] Note that $\alpha = \frac{1}{3}(\alpha_\parallel + 2\alpha_\perp)$, the average of parallel and perpendicular polarizability components, for a linear molecule or a molecule with at least a threefold rotational symmetry axis.

[2] For a nonspherical molecule α is in general a tensor with nine components,

$$\mu_i = \sum_{j=1}^{3} \alpha_{ij} E_j \quad (i, j = x, y, z),$$

so that $\boldsymbol{\mu}$ and **E** need not point in the same direction.

nearby—that is, on one molecule by another—unless at least one charge distribution is spherically symmetric and neutral. Hence we can expect all atoms and molecules to exhibit induced moments when they are either in an external field or interacting with other nonspherical molecules. The induced moments are added to whatever permanent moments the species may have.

When molecules are free to orient themselves in the presence of an electric field, they of course tend to occupy the lowest-energy orientations. In a uniform field, for example, a dipolar molecule will tend to align itself so that $\boldsymbol{\mu}$ and $\mathbf{E}$ are parallel. The same is true when the field is due to another molecule, and at moderately large distances (beyond the region where the repulsive forces are significant) the lowest-energy orientations are those in which the forces acting on the electric moments are attractive.

What now concerns us is finding the dependence of these forces on the distance between the molecules. In general, the strength of the interaction between two moments is directly proportional to the product of their magnitudes. The moments are all defined so that an nth moment (2^n-pole moment) has the dimensions of charge × (length)^n, whereas the interaction energy must have the dimensions of ϵ_0^{-1} × (charge)^2 × (length)$^{-1}$ as in Coulomb's law. Combining these relationships, we surmise that the long-range interaction energy between two moments M_1 and M_2, of orders n_1 and n_2, respectively, should have the form

$$V(R) \propto \epsilon_0^{-1} |M_1||M_2| R^{-n_1-n_2-1}, \tag{10.8}$$

where R is the distance between molecules, the length we expect to be most significant for long-range interactions. More detailed calculations confirm this law, as we shall shortly illustrate.

The longest-range interactions (by which we mean those falling off most slowly with distance) are those between two monopoles ($n_1 = n_2 = 0$), for which Eq. 10.8 predicts an energy proportional to R^{-1}. For point charges this is simply a restatement of Coulomb's law in the form of Eq. 2.55, and we have shown in Section 7.4 that two ions interact in the same way at large R. Monopole–monopole interactions of course exist only when both partners are charged. If one molecule is charged and the other is neutral but dipolar, the monopole–dipole interaction is longest-range, varying as R^{-2}. Similarly, dipole–dipole interactions fall off as R^{-3}, dipole–quadrupole as R^{-4}, quadrupole–quadrupole as R^{-5}, and so forth. For most systems the longest-range interaction is also the strongest at all distances greater than that at which the repulsive forces become dominant.[3]

Let us now derive some of the exact laws corresponding to Eq. 10.8, beginning with the direct interaction between an ion and a molecule with a permanent dipole moment. At long range the field of the ion will be essentially that of a point charge, whereas the field of the polar molecule will be essentially that of an ideal dipole: two point charges $+q$ and $-q$ a distance δ apart, such that

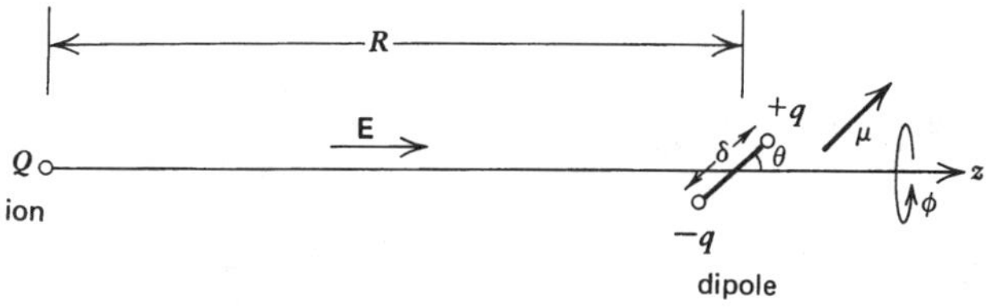

Figure 10.1 Ion–dipole interaction. The ion has a radial electric field **E**, which near the dipole is parallel to the z axis; since **E** is the force per unit positive charge, it points away from the ion (as shown) for $Q > 0$, toward the ion for $Q < 0$.

$$\boldsymbol{\mu} = q\boldsymbol{\delta}; \tag{10.9}$$

the vectors $\boldsymbol{\mu}$ and $\boldsymbol{\delta}$ both point from $-q$ toward $+q$. Now suppose that the center of the dipole lies a distance R from the ion of charge Q, in a direction that forms an angle θ with $\boldsymbol{\mu}$ (Fig. 10.1). If $R >> \delta$, we can take the distances from Q to $\pm q$ to be simply $R \mp \frac{1}{2}\delta \cos\theta$, and the Coulomb force on the dipole then has the z component[4]

$$\begin{aligned} F_z = F_+ + F_- &= \frac{qQ}{4\pi\epsilon_0 (R+\frac{1}{2}\delta\cos\theta)^2} - \frac{qQ}{4\pi\epsilon_0 (R-\frac{1}{2}\delta\cos\theta)^2} \\ &= \frac{qQ}{4\pi\epsilon_0 R^2}\left[\left(1-\frac{\delta\cos\theta}{R}+\cdots\right)-\left(1+\frac{\delta\cos\theta}{R}+\cdots\right)\right] \\ &\approx -\frac{qQ\delta\cos\theta}{2\pi\epsilon_0 R^3} = \boldsymbol{\mu}\cdot\frac{\partial \mathbf{E}}{\partial R}, \end{aligned} \tag{10.10}$$

where **E** is the electric field of the ion, of magnitude

$$|\mathbf{E}| = \frac{Q}{4\pi\epsilon_0 R^2}. \tag{10.11}$$

For a molecule with permanent dipole $\boldsymbol{\mu}$, this force corresponds to an interaction energy

$$V(R,\theta) = -\boldsymbol{\mu}\cdot\mathbf{E} = -\frac{\mu Q\cos\theta}{4\pi\epsilon_0 R^2} \tag{10.12}$$

since $F_z = -\partial V/\partial R$, in agreement with Eqs. 4.30 and 10.8.

Now consider the interaction between an ion and a nonpolar molecule. The ion's field will induce a dipole moment

---

[3] There are exceptions such as HD, which has a tiny dipole moment (about 10^{-3} D) but an appreciable quadrupole moment.

[4] The second line uses the binomial expansion

$$(1 \pm x)^{-2} = 1 \mp 2x + 3x^2 \mp \cdots.$$

in the molecule as given by Eq. 10.7, in which we assume α constant and scalar; in general, α must depend on the molecule's orientation to the field, but one can take an average value (which is what is usually measured anyway). The result of Eq. 10.10, giving the instantaneous force on the dipole, is valid whether the dipole is permanent or induced. We thus have, for $R >> \delta$,

$$F_z = \boldsymbol{\mu} \cdot \frac{\partial \mathbf{E}}{\partial R} = \alpha \mathbf{E} \cdot \frac{\partial \mathbf{E}}{\partial R} = \frac{\alpha}{2} \frac{\partial (\mathbf{E} \cdot \mathbf{E})}{\partial R}$$
$$= \frac{\alpha}{2} \frac{\partial |\mathbf{E}|^2}{\partial R} = -\frac{\alpha Q^2}{8\pi^2 \epsilon_0^2 R^5}. \tag{10.13}$$

In this case $\boldsymbol{\mu}$ varies with R, so to find the interaction energy at a given R we must integrate over the distance from infinity (where V is zero by definition) to R. Applying Eq. 2.46, we have

$$V_{\text{ind}}(R) = -\int_{\infty}^{R} \mathbf{F} \cdot d\mathbf{s} = -\int_{\infty}^{R} F_z dR$$
$$= \frac{\alpha Q^2}{8\pi^2 \epsilon_0^2} \int_{\infty}^{R} \frac{dR}{R^5} = -\frac{\alpha Q^2}{32\pi^2 \epsilon_0^2 R^4}. \tag{10.14}$$

Note that no angle θ appears, since the induced dipole is taken to be parallel to the field. Here $V(R)$ corresponds to the Born–Oppenheimer $E(R)$ of the ion-molecule system; this result confirms our statement in Section 6.2 that $E(R) \propto -1/R^4$ for $H + H^+$ at long range. Note that an induction effect of this kind is also present when the molecule has a permanent dipole moment, but at long range is much smaller than the direct interaction given by Eq. 10.12.

One can carry out similar calculations for the direct and induced interactions between dipoles, quadrupoles, and so on—calculations necessarily more complicated, since the orientations of both molecules must be taken into account. We shall mention only the result for two dipolar molecules. For two permanent dipoles A and B, each oriented at angles θ, ϕ relative to the A–B axis (cf. Fig. 10.1), the long-range interaction energy is

$$V(R, \theta_A, \theta_B, \phi_A, \phi_B)$$
$$= \frac{\mu_A \mu_B}{4\pi\epsilon_0 R^3} [-2\cos\theta_A \cos\theta_B + \sin\theta_A \sin\theta_B \cos(\phi_B - \phi_A)] \tag{10.15}$$

in agreement with Eq. 10.8. A dipolar molecule, as well as an ion, can induce a dipole moment in another molecule, whether or not the latter has one to begin with. The magnitude of this interaction should again be proportional to the product of the two dipoles divided by R^3, but the induced dipole is as usual $\alpha\mathbf{E}$, where the field is that of the first dipole, varying as μ/R^3. Thus the induced dipole has a magnitude proportional to $\alpha\mu/R^3$, and the interaction energy between the permanent and induced dipoles must vary as $\alpha\mu^2/R^6$. A detailed calculation gives

$$V_{\text{ind}}(R, \theta_A) = -\frac{\alpha_B \mu_A^2 (3\cos^2\theta_A + 1)}{2(4\pi\epsilon_0)^2 R^6}. \tag{10.16}$$

In the interaction of two permanent dipoles there are two such induction terms to be added to Eq. 10.15.

Let us estimate the magnitudes of the interactions we have been talking about, for typical values of molecular quantities and a reasonable distance R. Thus if we take $Q = \pm e$ (a singly charged ion), $\mu = 1.5$ D (about that of NH_3), $\alpha/4\pi\epsilon_0 = 3$ Å^3 (typical for small molecules), and set all the θ's equal to zero (dipoles aligned along the A–B axis), we obtain the following interaction energies:

	$R = 5$ Å	**$R = 10$ Å**
Ion–ion (Coulomb's law), Eq. 2.55	2.88 eV	1.44 eV
Ion–dipole, Eq. 10.12	0.180 eV	0.045 eV
Ion-induced dipole, Eq. 10.14	0.0346 eV	0.0022 eV
Dipole–dipole, Eq. 10.15	0.0225 eV	0.0028 eV
Dipole–induced dipole, Eq. 10.16	5.4×10^{-4} eV	8.4×10^{-6} eV

The dipole–dipole term is the largest that can appear in the interaction of two neutral molecules, and even at $R = 3$ Å (a typical "touching" distance for small molecules) this term only reaches about 0.10 eV. The interaction energy of neutral molecules is thus at best comparable to ordinary thermal energies ($\frac{3}{2} k_B T$, or 0.039 eV for a gas molecule at 300 K), and much smaller than the energies of chemical bonds.

Note that Eqs. 10.12, 10.15, and 10.16 give the instantaneous interaction energy at a particular orientation. The average interaction energy among a large number of such molecules (as in a gas) will fall off with R more rapidly, since the tendency for dipoles to align with the field also decreases with distance. Alignment also decreases with increasing temperature, that is, with increasing random motion of the molecules. These average energies are important, since many of the methods for investigating intermolecular forces involve measurements on bulk samples. For example, it is possible to show, by methods such as those to be introduced in Part II, that the average direct interaction energy between two dipoles at distance R is

$$\langle V(R) \rangle = -\frac{2}{3k_B T} \left(\frac{\mu_A \mu_B}{4\pi\epsilon_0} \right)^2 \frac{1}{R^6}, \tag{10.17}$$

where k_B is Boltzmann's constant and T is the absolute temperature. This is equivalent to Eq. 10.15 averaged over all possible values of the angles θ_A, θ_B, ϕ_A, ϕ_B. If we again take $\mu = 1.5$ D and set $T = 300$ K, Eq. 10.17 gives us average direct dipole–dipole energies of 3.3×10^{-3} eV at 5 Å and 5.1×10^{-5} eV at 10 Å, considerably smaller than the values for perfectly aligned dipoles. The corresponding average for the induction energy of two dipoles, from Eq. 10.16, is

$$\langle V_{\text{ind}}(R)\rangle = -\frac{\alpha_A \mu_B^2 + \alpha_B \mu_A^2}{(4\pi\epsilon_0)^2 R^6}; \tag{10.18}$$

here the average is just half the value for $\theta = 0$, and for weakly polar molecules such as CO or HI may be larger than the average direct interaction. Note that no temperature factor appears here, since the induced moment is always parallel to the field. It is interesting that both average dipole–dipole energies vary as R^{-6}, since there is an additional interaction that varies as R^{-6} for *all* molecules; we shall now look at this interaction.

The final kind of long-range attractive interaction is that arising from the so-called *dispersion forces* (or *London forces*). These forces are always present between any two molecules whether or not there are permanent multipoles of any order. The dispersion forces are often said to be quantum mechanical in origin, and they are indeed, in the sense that they result from the correlated motion of the electrons in the two molecules, or, more specifically, from correlation between the amplitudes of the electronic wave functions. Still, this correlation can be thought of crudely in a semiclassical way: Although the positions of the electrons in an isolated nonpolar molecule average over time to give a dipole moment of zero, at any instant their fluctuating positions give rise to a transient dipole moment. This moment generates a transient field that tends to polarize the electrons in a nearby molecule, producing (or rather modifying) a transient dipole moment there. The fluctuations of the charge distributions in the two molecules tend to be correlated, since the field of each transient dipole must influence the other. This argument is a convenient way of visualizing the interaction, but it should not be taken too literally; the actual effect is not time-dependent in the sense of classical fluctuations occurring. It is an effect of the correlation of the electronic wave functions of two neighboring molecules that produces a fluctuating dipole–dipole interaction that, in turn, gives rise to a small net attractive force.

The accurate calculation of the dispersion forces is extremely complicated and will not be discussed here. One relatively simple model gives an approximate result which can furnish some insight into how the forces arise:

$$V(R) = -\frac{3}{2}\left(\frac{h\nu_A \nu_B}{\nu_A + \nu_B}\right)\frac{\alpha_A \alpha_B}{(4\pi\epsilon_0)^4}\frac{1}{R^6}. \tag{10.19}$$

That is, the interaction energy for the dispersion force between two molecules is proportional to the product of their polarizabilities and varies as the inverse sixth power of their separation[5] independent of orientation. The frequencies ν_A and ν_B are parameters characteristic of the two molecules, and are approximately equal to the frequencies of their first allowed electronic transitions. Thus Eq. 10.19 suggests that the distortion of each molecule in the fluctuating field of the other is expressed as a tendency to undergo an optical transition to the first excited state. A real transition of this sort is impossible, because energy would not be conserved; but one can say that the effect of the mutual polarization is to mix a little of the excited state into each molecule's wave function. More detailed calculations also give an R^{-6} dependence, with additional terms varying as R^{-8}, R^{-10}, . . . ; if one writes the leading term as $-C/R^6$, a typical value of C would be 5×10^{-78} J m^6 (for molecules such as Ar, N_2, O_2). For comparison with our earlier calculations, this value gives energies of 2×10^{-3} eV at 5 Å and 3×10^{-5} eV at 10 Å. Thus the dispersion energy is comparable to the dipole–dipole energy for polar molecules, and is the dominant contribution to the long-range interaction energy for nonpolar molecules.

The term *van der Waals* forces is often used to include all the attractive forces between neutral molecules—the dispersion forces in all cases, and the dipole–dipole and other multipole interactions whenever these are present. The name refers to J. D. van der Waals, who was concerned with intermolecular forces in his studies of the equation of state (Chapter 21).

10.2 Empirical Intermolecular Potentials

The discussion in the previous section applies only to the interactions of molecules at long range, that is, at distances appreciably greater than the dimensions of the molecules themselves. At shorter distances things become much more complicated, as the electron distributions of the molecules begin to overlap. Under these conditions one can no longer use simple electrostatic models for the molecules, just as within a single molecule quantum mechanics must be used to calculate the interactions. In an exact calculation, indeed, one cannot meaningfully separate such a system into two molecules, but must treat all the electrons and nuclei as a single system. In this sense, for example, "the interaction energy of two (monatomic) Ar molecules" and "the energy of the Ar_2 molecule" are but two ways to describe the same quantity, the ground-state energy of the system of two argon nuclei and 36 electrons as a function of internuclear distance. Although theoretical calculations can be and have

[5] At very long distances—a few hundred angstroms or more—this dependence changes from R^{-6} to R^{-7} because of the finite velocity of light: It takes a significant time for the transient fields to propagate from one molecule to the other and back again.

been made for many two-atom systems such as He—He, Ne—Ne, Ar—Ar, and H_2—H_2, it will always be important to determine intermolecular potentials from experiment. Such determinations were being made long before *ab intitio* theoretical calculations were possible, so it is not surprising that many empirical formulas have been devised to correlate and summarize the inferences from experimental observation.

Without attempting to examine the calculations, what can we say about the nature of the potential energy curve? We take the interaction energy to be zero when the particles are infinitely far apart. At long range the potential must be attractive ($E < 0$) because of the effects described in the previous section. At short range, however, two stable molecules ordinarily repel each other strongly. (The exceptions are those pairs of molecules that readily undergo chemical reaction, and even there a "hard core" eventually takes control.) This is easy to understand in terms of the crude MO theory that we outlined for atom–atom interactions in Section 7.6. Like the inert gas atoms, most individual stable molecules have only filled orbitals. Thus the "supermolecule" made of two such molecules would have equal numbers of bonding and antibonding orbitals filled, giving a net antibonding effect. In short, the H_2—H_2 and N_2—N_2 systems have repulsive potential energy curves for much the same reasons as the He—He and Ne—Ne systems. In all these cases, however, the repulsive energy, which falls off roughly exponentially with distance (cf. Section 7.4), is eventually overtaken at large distances by the long-range attraction. And since we are interested primarily in the region where attractive and repulsive forces are comparable, let us look at the experimental evidence.

For simplicity let us consider only the interaction of two monatomic molecules, so that the internuclear distance R is the only parameter of significance. One then finds that the interaction energy $V(R)$ is of a form familiar to us: steep repulsion at small R, attraction trailing off gradually at large R, with a minimum between—in short, quite similar to the $E(R)$ curve for the ground state of a strongly bound diatomic molecule. The difference is principally one of scale; we illustrate this with Fig. 10.2, which compares the energy curves for the H—H and Ne—Ne systems, with well depths of about 5 eV and 0.003 eV, respectively. That we do not ordinarily consider Ne_2 as a molecule is because the average room temperature kinetic energy of two neon atoms is far more than enough to dissociate such a molecule. Thermal energies of molecules are usually of the order of k_BT ($k_B = 1.3807 \times 10^{-23}$ J/K $= 8.617 \times 10^{-5}$ eV/K). It is thus customary to describe intermolecular energies for systems having no "chemical bond" in terms of the parameter ϵ/k_B, where ϵ is the well depth, which has the units of temperature. All the inert gas diatomic systems have extremely shallow potential wells: ϵ is 9.5×10^{-4} eV for He—He, 0.0027 eV for Ne—Ne, and 0.012 eV for Ar—Ar; the corresponding ϵ/k_B values are 11 K, 31 K, and 144 K, and at

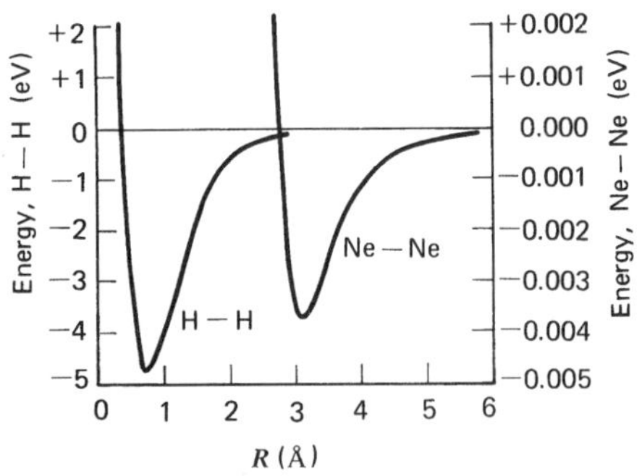

Figure 10.2 Potential energy curves for the H—H and Ne—Ne systems; note the difference in energy scales. The first curve is usually called "the binding energy of the H_2 molecule," the second "the interaction energy of two Ne molecules," yet they are essentially the same except for their scale.

room temperature these systems are almost totally monatomic. For comparison, the rather weakly bound alkali metal diatomic molecules have ϵ's of 0.5–1 eV, yet alkali vapors are only about 1% diatomic, largely because of entropy effects; see Chapter 21.

How does one obtain intermolecular potential energy curves? In some cases spectroscopy can be used, as with strongly bound molecules. The He—He well is so shallow that it barely contains a single, fragile vibrational state, but the Ne—Ne well contains three vibrational levels and the Ar—Ar well six; the shapes of the curves can be obtained by the usual methods from vibration-rotation or electronic spectra. But since so few two-molecule systems are likely to occupy bound states, other methods are more common. Scattering of molecular beams off one another gives a direct measurement of the interaction energy, especially its repulsive part (see Chapter 27). Such bulk properties of gases as the second virial coefficient and the viscosity are related to the intermolecular potential, as we shall discuss in Chapters 21 and 28, respectively; these measurements used to be the most important source of intermolecular energy data. When individual molecules retain their identity in the solid state (molecular crystals), the cohesive energy and compressibility of the crystal give much information on the shape of the curve near its minimum. And x-ray scattering can give the average distances between molecules, and thus tell something about their interactions, in both solid and liquid states. In a few cases, as we have already noted, theoretical calculations are also available. It is a sort of triumph that all these approaches usually give consistent results, in the sense that they agree within their expected errors, which are small relative to ϵ in cases where several data are available.

The relationships between the intermolecular potential and the various measurable properties are quite complex, and the bulk properties are usually quite insensitive to the exact form of the potential. That is, a large variation in the shape of the potential well may make a difference too small

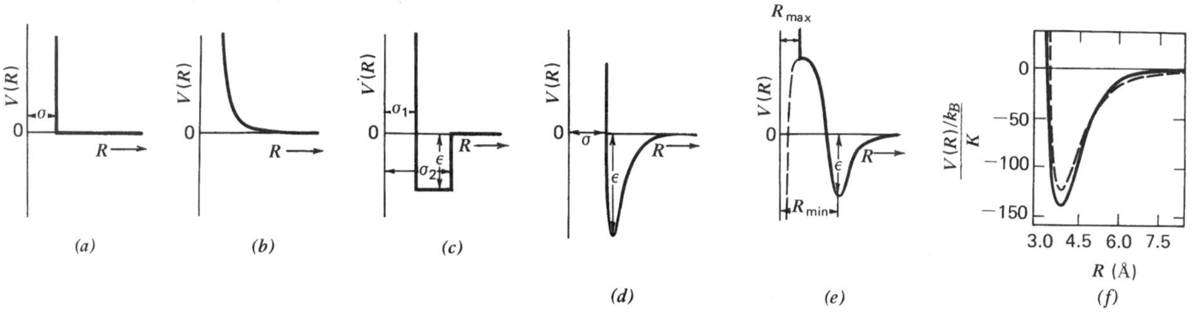

Figure 10.3 Some empirical intermolecular potential functions: (*a*) hard spheres; (*b*) point centers of repulsion ($\delta = 4$ shown); (*c*) square-well potential; (*d*) Lennard-Jones potential; (*e*) exp-6 potential (schematic); (*f*) comparison of the Dymond–Alder numerical potential (——) with the Lennard-Jones potential (- - - -) for the Ar—Ar interaction.

to measure in the value of a given macroscopic property.[6] For this reason it is customary to approximate the potential by some simple and mathematically tractable function with only a few adjustable parameters, which are then fitted as well as one can manage to the experimental data. There are many such empirical potential functions in use, some of which are important enough to deserve discussion here; these are illustrated in Fig. 10.3.

(a) ***Hard spheres.*** This is the simplest possible model, representing molecules as rigid spheres of diameter σ: Two such molecules will bounce off each other elastically if their centers approach to a distance σ, but otherwise do not interact at all. The potential energy can thus be written in the form

$$V(R) = \infty \quad (R < \sigma), \qquad V(R) = 0 \quad (R \geq \sigma), \tag{10.20}$$

as shown in Fig. 10.3*a*. This model takes account of the short-range repulsion only, and even that in idealized form, since the real repulsive force cannot be infinitely large. But it has the great advantage of mathematical simplicity, so that calculations with this model can be carried out with relative ease. In many cases such calculations give a good qualitative picture of the effects of molecular collisions. This is especially true at relatively high temperatures, where molecules have enough kinetic energy that they encounter mainly the upper portion of the repulsive curve. Most of the principles of kinetic theory can be developed with only this model, which we thus use almost exclusively in Part III of this book.

(b) ***Point centers of repulsion.*** A somewhat more realistic representation of the repulsive energy is given by the potential

$$V(R) = dR^{-\delta}, \tag{10.21}$$

Fig. 10.3*b*, where d is a constant and the index of repulsion δ must be larger than 3 (values between 9 and 15 are usually found).[7] This model recognizes the fact that more energetic molecules can approach each other more closely before being repelled, and thus can represent some of the temperature dependence of properties. But it again neglects the attractive part of the molecular interaction, and has mainly mathematical convenience to recommend it.

(c) ***Square well.*** The simplest potential model that includes both attractive and repulsive contributions is that shown in Fig. 10.3*c*, a hard-sphere core surrounded by an attractive well of constant depth. The square-well potential is defined by the equations

$$V(R) = \infty \quad (R \leq \sigma_1), \qquad V(R) = -\epsilon \quad (\sigma_1 < R < \sigma_2), \qquad V(R) = 0 \quad (R \geq \sigma_2), \tag{10.22}$$

with the three adjustable parameters σ_1, σ_2, and ϵ. This model is a good compromise between mathematical simplicity and realism.

(d) ***Lennard-Jones potential.*** The most widely used empirical potential is that named for J. E. Lennard-Jones, the first to use it extensively in analyzing the properties of gases. It is the simplest of the models considered here that gives a potential energy curve qualitatively similar to the correct one. The Lennard-Jones (12–6) potential, shown in Fig. 10.3*d*, is

$$V(R) = 4\epsilon\left[\left(\frac{\sigma}{R}\right)^{12} - \left(\frac{\sigma}{R}\right)^{6}\right], \tag{10.23}$$

where ϵ is the well depth and σ the distance at which $V(R) = 0$. One can easily show that the minimum energy,

[6] Quantum effects in scattering experiments, by contrast, are sensitive to the potentials so their prediction offers a severe test of any model. These typically appear as oscillations in angular and energy distributions of scattered particles.

[7] If $\delta \leq 3$, the expression for the second virial coefficient diverges [cf. Sec. 21.7 and J. O. Hirschfelder, C. F. Curtiss, and R. B. Bird, *Molecular Theory of Gases and Liquids* (Wiley, New York, 1954; rev. ed., 1964), especially chapters 3 and 6].

$V(R) = -\epsilon$, occurs at the distance $R = 2^{1/6}\,\sigma$. The attractive term is proportional to R^{-6}, thus giving the correct long-range dependence of the dispersion energy.[8] The repulsive term is more arbitrary, in that a wide range of exponents would fit the available data about as well; 12 is chosen for convenience (since R^{-12} is the square of R^{-6} and this makes it possible to find analytic representations of some properties). The Lennard-Jones parameters ϵ and σ have been determined for a wide range of molecules; some of the most accurate of these values are listed in Table 10.2. Calculations of many macroscopic properties have been made on the basis of the Lennard-Jones model, partly because of its mathematical convenience.

(e) *Exp-6 potentials.* Since the repulsive part of the potential energy is best described by an exponential term (cf. Section 7.4), a number of potential functions have been devised to make use of this. One problem is that any simple function of the form $Ae^{-\alpha R} - BR^{-6}$ has a maximum at small R and plunges to $-\infty$ as $R \to 0$; this is usually overcome by introducing a hard-sphere core. One of the most commonly used potentials of this type is

$$V(R)=\frac{\epsilon}{1-6/\alpha}\left\{\frac{6}{\alpha}\exp\left[\alpha\left(1-\frac{R}{R_{\min}}\right)\right]-\left(\frac{R_{\min}}{R}\right)^{6}\right\} \quad (R>R_{\max}),$$
$$V(R)=\infty \quad (R\le R_{\max}), \tag{10.24}$$

where $R_{\min}$ is the usual potential minimum, $R_{\max}$ is the distance at which the first function has its maximum, ϵ is the well depth, and α is an adjustable parameter (usually in the range 12–15). This exponential-six (exp-6) potential is shown schematically in Fig. 10.3*e*, with the rejected function for $R \le R_{\max}$ indicated by a dashed line. The maximum would actually be a much higher energy than shown here, a typical value being $V_{\max} = 3 \times 10^4\,\epsilon$ at $R_{\max} = 0.2R_{\min}$ (for $\alpha \approx 14$). This potential has three adjustable parameters, ϵ, $R_{\min}$, and α.

Many other potential models exist besides those we have described here. The double-exponential Morse potential, Eq. 7.11, gives a good fit near the minimum but falls off too rapidly at large R to represent intermolecular van der Waals interactions, where the Lennard-Jones potential falls off too slowly. Nonspherical molecules are sometimes represented by hard ellipsoids or cylinders rather than hard spheres; however, the Lennard-Jones and other spherically symmetric models work reasonably well for small molecules such as N_2 or CO_2. For polar molecules the dipole–dipole interaction must be taken into account; the most commonly used

Table 10.2 Lennard-Jones Potential Parameters for Selected Pairs of Atoms

Atom Pair	ϵ (10^{-21} J)	ϵ/k_B (K)	σ (Å)
LiAr	0.92	67	3.7
LiKr	1.5	110	3.0
LiXe	2.4	175	3.8
NaAr	1.0	72	3.8
NaKr	1.4	100	4.5
NaXe	2.1	150	4.4
Ar_2	2.0	145	3.8
Kr_2	2.7	201	3.6
Xe_2	3.9	288	3.9

From W. R. Hindmarsh and J. M. Farr, *Progress in Quantum Electronics* **2,** 141 (1972); J. A. Barker, R. O. Watts, J. K. Lee, T. P. Schafer, and Y. T. Lee, *J. Chem. Phys.* **61,** 3081 (1974); J. M. Parson, P. E. Siska, and Y. T. Lee, *J. Chem. Phys.* **56,** 1511 (1972).

model is the *Stockmayer potential,* which is simply Eq. 10.15 added to the Lennard-Jones potential.

However, it is unlikely that any simple function with a limited number of adjustable parameters can represent the true intermolecular potential accurately over all distances. In recent years attempts have been made to compute the true potential directly from a wide range of experimental data. One such potential for the Ar—Ar interaction, obtained in numerical form by Dymond and Alder,[9] is shown in Fig. 10.3*f* in comparison with the Lennard-Jones potential; the difference in the shape of the attractive potential is quite apparent. And even the Dymond–Alder potential, which was based on all the data available for the thermodynamic and transport properties of gaseous argon, gives a poor representation of the Ar_2 vibrational spectrum. To reproduce the results of the most sensitive and demanding experiments, such as high-resolution rotation-vibration spectroscopy or measurement of the angular distributions of particles scattered from colliding beams of atoms one must work with very flexible, elaborate mathematical representations of potential curves. These are now accurate enough to reproduce the properties of species such as Ar_2.[10]

Most determinations of intermolecular potentials derive from measurements on one-component systems (which are relatively easy to analyze), and thus give information only on the interaction between two molecules of the same kind. The entry under Ar_2 in Table 10.2, for example, gives the parameters for the Ar—Ar interaction. Some potential parameters have been determined for unlike molecules, but

[8] However, when the dispersion energy $-CR^{-6}$ is calculated directly from molecular polarizability—by expressions such as Eq. 10.19—the value of C obtained is usually only half the Lennard-Jones value ($C = 4\epsilon\sigma^6$) for the same molecule. This demonstrates that the Lennard-Jones potential is not quite adequate to represent the true shape of the intermolecular potential.

[9] J. H. Dymond and B. J. Alder, *J. Chem. Phys.* **51,** 309 (1969).

[10] The most elaborate and accurate of these have been constructed by Aziz and collaborators; R. A. Aziz and H. H. Chen, *J. Chem. Phys.* **67,** 5719 (1977); R. A. Aziz, *Molec. Phys.* **38,** 177 (1979); and R. A. Aziz and M. Slaman, *Mol. Phys.* **58,** 679 (1986) are some of the presentations of this work.

in only a limited number of cases. In the absence of such information, one often estimates the Lennard-Jones parameters by the empirical combining laws

$$\sigma_{AB} = \tfrac{1}{2}(\sigma_{AA} + \sigma_{BB}), \qquad (10.25)$$

which would be exactly correct for hard-sphere molecules, and

$$\epsilon_{AB} = (\epsilon_{AA}\epsilon_{BB})^{1/2}. \qquad (10.26)$$

The latter equation would follow from Eq. 10.19 for the dispersion energy if the parameters ν_A and ν_B had the same value; for many pairs of molecules this is approximately the case. But these combining rules and similar ones for other potentials are only approximate, and it is unlikely that any universal relationship of this type actually exists.

10.3 Weakly Associated Molecules

In this section we survey a number of species in which atoms or groups of atoms are attracted to one another by bonding forces intermediate in strength between the strong interactions we call "chemical bonds" and the weak "intermolecular" interactions just discussed. The "molecules" thus formed tend to be loose, fluctuating aggregations more like liquids than the rigid geometric structures characteristic of strongly bound molecules. We usually refer to such loosely bound species as "associated."

The most weakly bound "compounds" we know are those formed by two inert gas atoms. The only contribution to the bonding is that from dispersion forces, without even any permanent multipoles to help hold the atoms together. Yet as we noted in the last section, a molecule such as Ar_2 is stable enough for its vibrational spectrum to be measured. Even the molecule He_2 has a single bound vibrational state, albeit with a binding energy measured in milli-degrees Kelvin and an average internuclear distance of many tens of Ångström units. The fact that the inert gases do condense to form liquids implies that they must also be capable of forming triatomic and larger associations. These species exist in measurable concentrations in special conditions, usually transient, such as in the efflux of gas from a high-pressure jet or trapped in a matrix of other, nonreactive material. They are difficult to observe under equilibrium conditions, but their structures and spectra have been studied, and they have been subjects of very extensive computer simulation—usually with potentials such as the Lennard-Jones or Morse as the basis of the computer model. These polyatomic associations are called "atomic clusters" (and there are of course molecular clusters as well); in Chapter 24 we shall examine why the transition from a gas of isolated molecules to the indefinitely large cluster we call a liquid is so abrupt.

The next species we consider are those in which both dispersion and multipole forces contribute to the attraction, giving molecules somewhat more tightly bound than the inert gas dimers. The study of these species has a very short history because they can exist for extended periods of time only under conditions of low temperature and low density—low temperature, to keep the kinetic and vibrational energy low enough to prevent molecules from breaking up with every collision, and low density so that collisions are infrequent. It is difficult to find these conditions on or near the earth, but they do occur in interstellar space. Nonetheless, many such species have been studied in the laboratory where they exist as transient but readily observable "compounds." A few examples of the many weakly bound molecules that have been observed and characterized are XeHCl, HeI_2, He_2I_2, NeI_2, ArI_2, Ar_2I_2, $(C_6H_6)_2$, $(HF)_2$ (cf. Section 7.9), $(HCl)_2$, and $(HCN)_2$. The last three of these are best described in terms of hydrogen bonds, of which we shall have more to say.

In some cases of "cluster formation" the participating molecules may have one or more unpaired electrons, which are available to form weak chemical bonds. (Cf. our discussion of electron-deficient bonds in Section 9.3.) A classic example is the dimer of nitrogen dioxide, which appears to have the structure

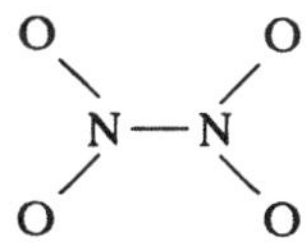

;

the weakness of the N—N bond is indicated by its length of 1.78 Å, compared with 1.47 Å for the normal single bond in hydrazine (H_2N—NH_2). The NO_2 monomer is a brown gas; the color is associated with the presence of an odd or unpaired electron, and thus of a partially occupied orbital. The N_2O_4 dimer, by contrast, has no unpaired electrons and is a colorless gas.[11] The color difference makes it easy to measure the relative amounts of NO_2 and N_2O_4 in a mixture of the two: The solid and liquid are almost pure N_2O_4, but the gas consists mostly of NO_2 molecules.

An even weaker association of odd-electron molecules occurs in the solid form of nitric oxide, made up of dimers with the structure

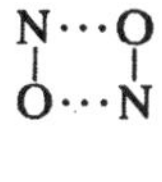

;

the short and long N—O bonds have lengths of 1.10 Å and 2.38 Å, respectively. Yet unpaired electrons do not always

[11] A large number of odd-electron compounds (including many of those called "free radicals") are colored. The usual rationalization is that the unpaired electron is somewhat weakly bound, and thus undergoes electronic transitions in the visible rather than the ultraviolet. But the phenomenon is not universal; one obvious exception is NO, which is colorless in the gas phase.

lead to association: The oxygen molecule, with two unpaired electrons in its ground state, shows little tendency to form O_4 dimers.

Hydrogen-bonded molecules are a special case whose attractive forces and association can be interpreted in terms of multipole interactions. However these are usually considered as a separate category of associated molecules, both because of their ubiquity and the relative strength of the forces that hold them together. The poor shielding of the hydrogen nucleus allows relatively strong interactions to take place, in some cases leading to quite well-defined structures. One such example is the dimer of acetic acid, with the structure

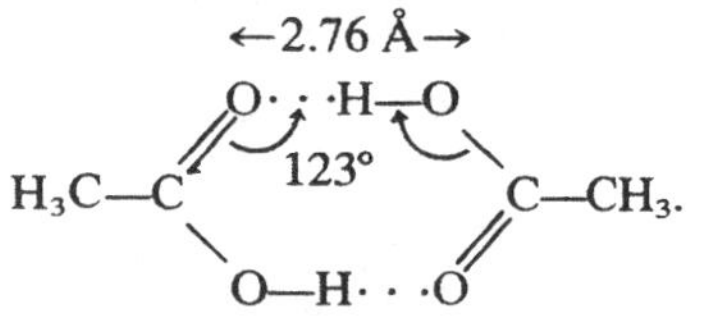

Among other examples is the dimer of HF (Section 7.9), with its F—H · · · F bond; the F—F distance is fixed at about 2.79 Å, but the hydrogen atoms oscillate and wobble easily. And in Section 9.4 we have described how internal hydrogen bonding stabilizes the structures of DNA and other biological macromolecules.

Thus far we have discussed only association between neutral molecules, but a stronger electrostatic interaction is obviously possible when one component is an ion. In some cases such an "association" actually leads to a quite strongly bonded molecule. For example, the H_3O^+ (hydronium) ion might be thought of as an association between the H^+ ion (proton) and a water molecule, but the energy required to split H_3O^+ into $H_2O + H^+$ in the gas phase is 7.34 eV, so one can hardly think of this as a case of weak bonding. This is why bare protons are virtually nonexistent in aqueous solutions of acids. The properties these solutions share in common are due to the presence of H_3O^+ or even larger ions. The NH_4^+ (ammonium) ion is also strongly bonded, with about 9.0 eV needed to split it into $NH_3 + H^+$. The NH_4^+ ion plays a role in liquid-ammonia solutions like that of H_3O^+ in water.[12] Both H_3O^+ and NH_4^+ ions are stabilized by delocalization, since all the hydrogen atoms in each are equivalent.

But whereas the H^+ ion has so intense a field that its "associations" are strongly bonded, other ions do form weakly associated species in combination with neutral molecules. These are usually called *ion-molecule clusters* in the gas phase, *solvated ions* in solution. They are normally composed of a single ion with one or more neutral molecules in a shell around it. The relationship to the complex ions of the last chapter is obvious; the difference is simply the strength of the bonding. An ion-molecule cluster or solvated ion can be thought of as a coordination complex in which the ligands are loosely held—"loosely" being arbitrarily defined by wherever one wants to draw the line between the two classes.

Definitely on the weakly bound side of such a line are the species formed by the alkali metal ions in association with water molecules. These hydrated ions are held together mainly by the attraction between the positive charge of the central alkali ion and the permanent dipoles of the water molecules, augmented by the additional dipole moments that the central ion induces in the water molecules. The bonding in these clusters may be so weak that, unlike the more tightly bound complexes, they do not maintain definite shapes for long intervals of time. Even the number of associated water molecules may fluctuate. The fluctuations result from the balance between the attractive force of the ion and the random motion of the solvent molecules. The association is nevertheless real on the average. Rather than evidence for a definite structure, one finds the weaker evidence that at certain distances from the central ion there is a higher than random concentration of water molecules. That is, there are strong peaks in the radial distribution function, which we discuss in Chapters 23 and 26.

The attachment of water molecules does not stop with a coordination shell of nearest neighbors; a second shell, and perhaps even more, can be bound somewhat less strongly. Some of the properties of the hydrated alkali ions are given in Table 10.3. Because Li^+ is the smallest alkali ion, it can exert the strongest binding forces on surrounding water molecules which can approach more closely than with the larger ions; the result is that Li^+ has the largest effective radius as a hydrated ion, the greatest hydration energy (energy released by the formation of the hydrated ion), and the highest average number of associated water molecules. In the Li^+ hydrated ion the inner shell of water molecules appears to be

Table 10.3 Properties of Hydrated Alkali Ions

Property	**Li^+**	**Na^+**	**K^+**	**Rb^+**	**Cs^+**
Radii of central ions (in crystals) (Å)	0.68	0.98	1.33	1.48	1.67
Radii of hydrated ions (Å)	3.40	2.76	2.32	2.28	2.28
Hydration energy[a] (eV)	5.3	4.2	3.3	3.0	2.7
Average number of associated water molecules[b]	25.3	16.6	10.5	—	9.9

[a] These energies are slightly lower than the *heats of* hydration in Table 26.4.

[b] These numbers differ from those derived by computer simulation and quoted in Chapter 26. The difference is a measure of the ambiguity inherent in the concept of hydration.

[12] There are many solid hydronium salts ("acid hydrates") completely analogous to the corresponding ammonium salts, except that most of the hydronium salts melt at much lower temperatures. For example hydronium chloride ($H_3O^+Cl^-$, but usually written $HCl \cdot H_2O$) melts at −15.4°C, whereas NH_4Cl remains solid up to 340°C.

bound tightly enough to form a well-defined tetrahedron; Na^+ and K^+ also have four molecules in the first hydration shell, but Rb^+ and Cs^+ may have six. Similar solvated ions are formed in polar solvents other than water: in liquid ammonia, for example, the $[Na(NH_3)_4]^+$ ion is a well-defined tetrahedral complex.

Wherever ions and neutral molecules are both abundant, it is usual to find associations between the two. This is true not only in liquid solutions but in ionized gases, where the resulting ion-molecule clusters can be readily detected by mass spectrometry. The binding energies of these clusters are usually strongly dependent on the charge and size of the ion, but not on the detailed structure of the neutral molecule(s). Both $K^+(CO_2)$ and $NO^+(CO_2)$ have binding energies of about 0.5 eV, but the corresponding values for $NO^+(N_2)$ and $NO^+(O_2)$ are only about 0.2 eV. The binding energy of Ar_2^+ is considerably greater, about 1.5 eV; like such other homonuclear ions as H_2^+, He_2^+, Li_2^+ (cf. Chapter 7), it should be considered a strongly bound molecule-ion rather than a cluster. For most clusters of one ion and one nonpolar molecule, the binding energies are consistent with the values predicted by Eq. 10.14 for the equilibrium separation. Clusters with more than two entities also occur: The Ar_3^+ ion has been observed, as have such aggregates of nitrogen molecules as N_4^+, N_6^+, and N_8^+.

Especially large clusters can be formed with molecules capable of forming hydrogen bonds. Additional water molecules readily attach themselves to the H_3O^+ ion to form $H_5O_2^+$, $H_7O_3^+$, and $H_9O_4^+$; even higher numbers of water molecules can be bound in the gas phase, apparently going into an outer shell of the cluster. The $H_9O_4^+$ cluster is believed to have the polyhedral structure shown in Fig. 10.4; the smaller clusters are assumed to be fragments of this polyhedron, with additional H_2O molecules going on its exterior. However, the binding energies for successive addition of one, two, three, four, and five H_2O molecules to H_3O^+ are approximately 1.5, 1.0, 0.75, 0.65, and 0.55 eV, respectively, with no sharp break at the completion of the assumed first shell. The limiting value is simply the energy of vaporization of water, which is about 0.46 eV per molecule. By extending the hydration energy measurements in the gas phase to larger and larger clusters, it is possible to estimate the total energy of hydration of the proton in aqueous solution. The value obtained in this way is 12.0 ± 0.3 eV, in rather good agreement with the accepted value of 11.35 ± 0.15 eV, based on a variety of other measurements.

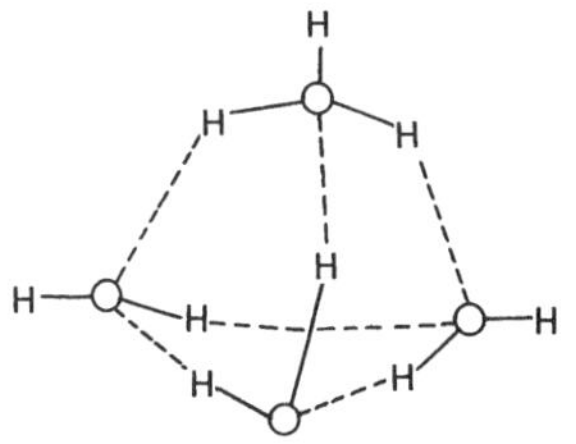

Figure 10.4 Presumed structure of the $H_9O_4^+$ ion in the gas phase.

The $H_9O_4^+$ ion in Fig. 10.4 has been drawn as made up of one H_3O^+ ion (at the top of the figure) and three H_2O molecules. But the protons involved in hydrogen bonds are extremely mobile, so that either of the bonding protons in the H_3O^+ ion could move readily to the oxygen atom in the adjacent H_2O molecule. A rotation of the entire $H_9O_4^+$ complex would give a result indistinguishable from this permutation, but would be far slower. The exterior protons can also exchange places with bridging protons, but again more slowly than the latter can "jump across" the hydrogen bonds. As a result of this mobility, the lifetime of an individual H_3O^+ ion in aqueous solution is estimated to be only about 10^{-13} s.

Ammonia also forms clusters with the ammonium ion, and species of the type $NH_4^+(NH_3)_n$ with n as large as 20 have been observed. The notion of a shell structure is somewhat better supported by experiment here than in the case of the $H_3O^+(H_2O)_n$ clusters. The energies for attachment of NH_3 molecules to NH_4^+ drop smoothly from about 1 eV for the first NH_3 to about 0.75 eV for the fourth, and then abruptly to about 0.4 eV for the fifth. The limit for $n \rightarrow \infty$ is in this case the energy of vaporization of NH_3, which is roughly 0.2 eV. There is no evidence for a well-defined second shell.

Negative-ion clusters can also be formed. For example, the hydroxyl (OH^-) ion can be hydrated to form $H_3O_2^-$, $H_5O_3^-$, and $H_7O_4^-$; the energies associated with these steps are very similar to those for hydration of the H_3O^+ ion. Although neutral $(O_2)_2$ associations seem not to exist, the O_4^- cluster is observed; the energy required to dissociate it into $O_2 + O_2^-$ is 0.54 eV. The hydrate $O_2^-(H_2O)$ is slightly more tightly bound, with a dissociation energy of about 0.8 eV; the energies for adding successive water molecules are 0.6–0.8 eV each through the pentahydrate.

The tendency for ions to form large clusters—especially those involving H_2O molecules—is quite in line with the accepted notion that ions are good nuclei for initiating condensation. The formation of such clusters, growing into aerosol droplets, is naturally important as a major mechanism in the production of clouds and rain. (But it is not the only mechanism: Solid particles—which are often themselves electrostatically charged—can also act as condensation nuclei.) The ions that initiate the process are formed in the upper atmosphere by the action of cosmic rays, natural radioactivity, and ultraviolet light. Ions may be formed from any of the neutral species present in the atmosphere, but a series of exchange reactions leads primarily to the formation of $H_3O^+(H_2O)_n$ clusters.

All the ion-molecule clusters we have mentioned thus far are singly charged. Solvated ions with charges greater than 1 are well known in solution, where solvent molecules can bind to and thus stabilize the ion. In gases, however, no such clusters are known. There is invariably some mechanism that allows the splitting of the charge between separate fragments, which, by escaping, lower the repulsive energy of the system. For example, a hypothetical $[Fe(H_2O_2)_2]^{2+}$ ion would presumably decompose at once to $FeOH^+$ and H_3O^+.

FURTHER READING

Blaney, B. L., and Ewing, G. E., "Van der Waals Molecules," *Ann. Rev. Phys. Chem.* **27,** 553 (1976).

Hirschfelder, J. O., Curtiss, C. F., and Bird, R. B., *Molecular Theory of Gases and Liquids* (Wiley, New York, 1954; rev. ed., 1964).

Kauzmann, W., *Quantum Chemistry* (Academic, New York, 1957), Chapter 13.

Kihara, T., *Intermolecular Forces* (Wiley, New York, 1978).

Kondratyev, V., *The Structure of Atoms and Molecules* (Noordhoff, Groningen, The Netherlands, 1964), Chapter 9.

Margenau, H., and Kestner, N. R., *Intermolecular Forces* (Pergamon Press, Oxford, 1969).

Maitland, G. C., Rigby, M., Smith, E. B., and Wakeham, W. A. *Intermolecular Forces* (Oxford University Press, Oxford, 1981).

Rigby, M., Smith, E. B., Wakeham, W. A., and Maitland, G. C., *The Forces Between Molecules* (Oxford University Press, Oxford, 1986).

Amdur, I., and Jordan, J. E., "Elastic Scattering of High Energy Beams: Repulsive Forces," in *Molecular Beams,* J. Ross, ed., Vol. X of *Advances in Chemical Physics* (Wiley, New York, 1966), Chapter 2.

Buck, U., "Elastic Scattering," in *Molecular Scattering: Physical and Chemical Applications,* K. P. Lawley, ed., Vol. XXX of *Advances in Chemical Physics* (Wiley, New York, 1975).

Dalgarno, A., and Davison, W. D., "The Calculation of van der Waals Interactions," *Advances in Atomic and Molecular Physics* **2,** 1 (1966).

Pauly, H., and Toennies, J. P., "The Study of Intermolecular Potentials with Molecular Beams at Thermal Energies," *Advances in Atomic and Molecular Physics* **1,** 195 (1965).

PROBLEMS

1. Prove that the value of the dipole moment μ of a charge distribution is independent of the choice of origin of coordinates if the total charge is zero, but depends on the choice of origin if the net charge is nonzero.

2. Estimate the equilibrium distance R_e and the polarizability of the neutral partner, assuming that atomic polarizabilities are approximately $4\pi\epsilon_0 V$ and that molecular polarizabilities are approximately the sums of the polarizabilities of the atoms. Then evaluate the dissociation energy for Ar_2^+, K^+CO^2, and N_4^+.

3. Carry the approximation of Eq. 10.13 one step further by deriving the term involving the induction of a dipole in the ion by the field of the induced dipole of the neutral. Show by examples based on data of Chapter 5 that this effect is far more important in negative ion clusters than in positive ion clusters.

4. Polarizabilities, the derivatives of induced moments with respect to increases in an applied electric field, are usually treated as constants. In reality, polarizabilities are not constant; for example in sufficiently strong uniform fields, polarizabilities change linearly with the field strength, implying that the electric dipole moment is changing quadratically with the field strength. Which way does the polarizability change with increasing field and why?

5. (a) Evaluate the quadrupole moment of a charge distribution in which the charges are as follows: $+2e$ at (0, 0, 1) and (0, 0, −1), and $-1e$ at (1, 0, 0), (−1, 0, 0), (0, 1, 0), and (0, −1, 0), with all distances in Å. (b) Evaluate the quadrupole moment of a distribution in which the positive charges of $+2e$ lie at (0, 0, 1) and (0, 0, −1) and the negative charge of $-4e$ is distributed uniformly throughout a sphere of radius 4 Å.

6. Estimate the magnitudes of electric dipole–dipole, dipole–quadrupole, and quadrupole–quadrupole interaction energies for separation distances of 4 Å and 8 Å, basing your values on magnitudes of the molecular moments. Estimate the magnitude of the dispersion energy at the same distance on a similar basis.

7. Compute the electric dipole–dipole interaction energy for two HCl molecules 1 nm (10 Å) apart when the two molecules are aligned as shown: (see Table 7.4).

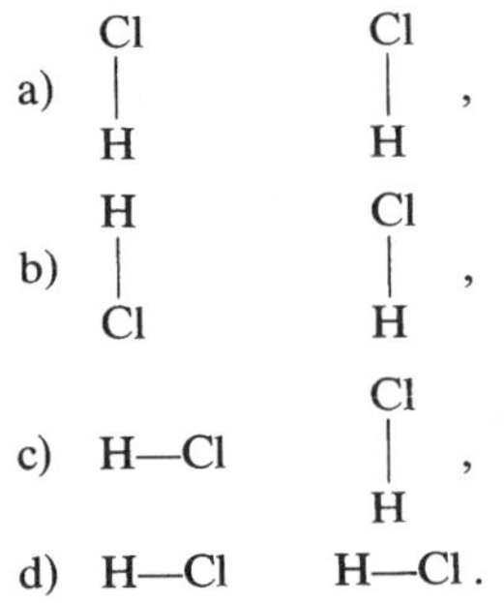

8. Compute the interaction energy between a polar diatomic molecule with electric dipole μ and an ion with charge Q when the molecule is rotating in a state with $J = 1$ quantized along the ion-molecule axis with (a) $M_J = 0$; (b) $M_J = \pm 1$. [The wave functions for rotational motion can be taken here to be the normalized spherical harmonics $Y_L^M(\theta, \phi)$.]

9. When two alkali halide molecules are far from each other, their interaction potential is that of two permanent dipoles. When the diatomic molecules are close together, their interaction is the sum of the separate ion–ion potentials that eventually give rise to a stable compound, for example for 2NaCl, as shown:

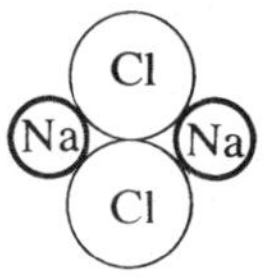

From the ionic radii, estimate the geometry of Na_2Cl_2. Then sketch the form of the effective potential as the dimer is separated into two diatomic molecules. Indicate the analytic form of the potential for the short-range and long-range regions, and show how the former transforms to the latter as the intermolecular distance grows.

10. Evaluate and graph the dipole moment induced in an atom of argon by the field of a proton approaching from 20 Å to 5 Å. Assume the polarizability of Ar is 1.64 Å^3, as given in Table 10.1.

11. The dipole moment of every diatomic molecule in its ground state is zero when the internuclear distance $R = 0$ and when $R \to \infty$. Sketch the behavior of $|\mu(R)|$ and explain why it goes to zero at both limits. Be sure to indicate an approximate scale of distance.

12. Compute the force constants, vibrational frequencies, and equilibrium internuclear distances for NaAr, NaXe, and Ar_2, based on the parameters of their Lennard-Jones potentials.

13. What are the coefficients of the first nonvanishing terms beyond the quadratic if one expands the Lennard-Jones and exp-6 potentials around their minima at $R = R_{min}$? Express these in terms of the parameters of the Lennard-Jones and exp-6 potentials, respectively. Evaluate the former for Ar_2 and compare it with the observed value.

14. The polarizability of CO_2 has two independent components, one parallel to the O—C—O axis and the other perpendicular. These have values of 4.03 and 1.93 Å, respectively. The first allowed optical transition of CO_2 falls at 46,000 cm^{-1} or 5.70 eV. Estimate the potential energy of interaction of the two CO_2 molecules in a linear configuration, a T-shape and side by side, when the C—C distances are 5 Å and 10 Å, if the dispersion forces are the dominant component of the internuclear force.

15. Compare the magnitudes of the three kinds of attractions behaving as R^{-6} for the following pairs:
(a) Ar—Ar;
(b) Ar—HCl;
(c) He—HCl;
(d) HCl—HCl.

16. The quadrupole moment of CO_2 is 4.32×10^{-26} esu cm^2 or 4.32×10^{-30} esu m^2. For the same configurations and distances considered in Problem 7, compute the interaction energy of two CO_2 molecules, assuming the interactions are pure quadrupole–quadrupole. Show that the T-shaped geometry has the lowest energy. Compare the energies based on quadrupole–quadrupole interactions with those based on dispersion forces as derived in Problem 7.

17. Construct a square-well potential for Ar_2 that has the same depth (0.012 eV) as the actual potential well of Ar_2 and contains the same number (6) of bound vibrational states with no rotation. Are there any remaining free parameters in this square-well potential?

18. Rationalize the observation that He_2 has only one bound state but He_3 has several.

19. Compare the bonding of Ar_2^+ with that of Cl_2 and Ar_2 in terms of the occupied molecular orbitals and their bonding or antibonding character. On this basis, interpret the differences between the dissociation energies of these three molecules.

20. The hydrogen fluoride molecule forms hydrogen-bonded dimers. Referring to Table 7.2, we find that the equilibrium H—F distance in the monomer is 0.9168 Å and the vibrational frequency $\tilde{\nu}_e$ is 4139 cm^{-1}. The $(HF)_2$ dimer has a linear F—H · · · F arrangement with the F atoms 2.79 Å apart at equilibrium. The other H is off at an angle of about 108° from the F—H—F axis. Suppose the bonding H atom can be attached to either F, so that the configurations F—H · · · F and F · · · H—F are equivalent. Construct the potentials for these two structures assuming that only the H moves, and find the energy of their intersection, thereby setting an approximate upper limit for the potential barrier between the two structures.

21. Estimate the total binding energy of a complex $[Na(H_2O)_4]^+$, taking into account the interaction of the charge of the Na^+ ion with the dipoles of the water molecules and the interaction of the dipoles with each other. What distances and geometry do you choose? Explain your choices. Compare your estimates with the values given in Table 10.3.

22. Rationalize the difference between the "hydration energy" of the proton, 7.34 eV, that is, its binding energy to a single water molecule, and the total hydration energy of a lithium ion of 5.3 eV, as given in Table 10.3.

23. Alkali halide molecules, particularly those of the lighter alkalis and halogens, have a strong tendency to form dimers and higher polymers in the gas phase. What geometries do you expect to correspond to the configurations of minimum energy for Li_2F_2 and Li_3F_3. Estimate the energy required to dissociate these into separated molecules. For Li_2F_2 compare this with the energy to dissociate into $Li_2F^+ + F^-$. Use the data from Tables 7.3 and 7.4.

24. Construct molecular orbital representations of O_4 and O_4^-, assuming that they are plane, rectangular molecules.

25. The potential energy surfaces of argon dimer, Ar_2, and krypton dimer, Kr_2, possess shallow van der Waals minima. Use (a) the HF–SCF, (b) the CISD, and (c) the MP2 methods along with a 6–311G (*d, p*) basis set for Ar_2 and a 3–21G basis set for Kr_2 to determine values of D_e and R_e for the ground electronic state of the dimers; compare the calculated values to the Lennard-Jones potential parameters given in Table 10.2. A 6–311G (*d, p*) basis set is a triple-zeta basis set in which each inner shell atomic orbital is represented as one linear combination of 6 Gaussians and each valence shell atomic orbital is represented by three basis functions, one a linear combination of three Gaussians, the other two by one Gaussian function each. Further, a set of *d*-type polarization functions to nonhydrogen atoms and a set of *p*-type polarization functions to hydrogen have been added. See problems 5.26 and 7.27 for basis set descriptions and software packages.

26. Use (a) the HF–SCF methods along with a 6–311G (*d, p*) basis set to determine values of D_e and R_e for the ground electronic state of the LiAr and NaAr molecules, which also exhibit shallow van der Waals minima on their potential energy surfaces. Compare the calculated values to the Lennard-Jones potential parameters given in Table 10.2. A 6–311G (*d, p*) basis set is a triple-zeta basis set in which each inner shell atomic orbital is represented as one linear combination of six Gaussians and each valence shell atomic orbital is represented by three basis functions, one a linear combination of three Gaussians, the other two by one Gaussian function each. Further, a set of *d*-type polarization functions to nonhydrogen atoms and a set of *p*-type polarization functions to hydrogen have been added. See Problems 5.26 and 7.27 for basis set descriptions and software packages.

27. The HF dimer is an example of hydrogen bonding. Find the optimized equilibrium geometry of the dimer and the value of D_e using (a) the HF–SCF, (b) the CISD, and (c) the MP2 methods along with a 6–31G++ (*d, p*) basis set. Compare your results to the experimental values given in Problem 10.20 as well as to the experimental values for the HF monomer given in Table 7.2. A 6–31G++ (*d, p*) basis set is constructed by adding highly diffuse *s*-type and *p*-type functions to nonhydrogen and hydrogen atoms of the 6–31G (*d, p*) basis set. See Problems 5.26 and 7.27 for basis set descriptions and software packages.

CHAPTER 11

The Structure of Solids

Solids and liquids are the states of matter called "condensed phases." On the microscopic scale they are characterized by the proximity of each atom or molecule to its neighbors. In a solid or a liquid every particle is always moving in the force fields due to several neighbors, usually near the minimum of the net potential energy surface produced by all these interactions. The equilibrium intermolecular distance is in the region in which both attractive and repulsive forces are significant, and nearly balance each other. Gases, by contrast, are so rarefied that each molecule spends most of its time outside the range of strong interaction with any neighbor—when molecular interactions do occur in gases under ordinary conditions away from conditions of condensation, they occur almost invariably only through two-body collisions dominated by long-range attractive forces.

In this chapter we shall examine some bulk properties of solids from a phenomenological viewpoint, study the crystalline and electronic structure of solids from a microscopic perspective, and begin to interpret the dramatic differences between the properties of solids and gases. This will take us into the most important method for determining structures, into the connections among bonding, structure, and stability, and finally into the electronic structure of periodic solids.

In a solid or a liquid, the simultaneous attractions of each particle for its several neighbors are the cause of *cohesion.* The *cohesive energy* of a solid is the energy that would be required to separate it into its constituent atoms, ions, or molecules at distances so large that they effectively no longer interact with one another. The cohesive energy is the analog for solids of the dissociation energy of a diatomic molecule, and is often of similar magnitude, per constituent atom or molecule. Indeed, in many cases a solid can be thought of as an infinitely large molecule.

11.1 Some General Properties of Solids

To begin with, what is the difference between a solid and a liquid? Both are condensed phases with comparable values of intermolecular distances (although for a given substance the liquid is usually slightly less dense), and the cohesive energy of a liquid is not much less than that of a solid. The great difference is that a solid has a fixed structure: Two molecules in a given piece of solid can maintain their relative positions, except for vibrations, over a long period of time, and the solid as a whole thus maintains a fixed shape. In a liquid the molecular positions are not fixed, and the liquid readily changes its shape under even a weak external force such as gravity; we say that the liquid assumes the shape of its container. A liquid is distinguished not by any structure or structural property but by its dynamical character: A liquid, subjected to any but the tiniest or fastest stress, responds by moving to a new, stable structure. In short, a liquid flows whenever it responds to a force; a solid, by contrast, responds elastically to a force, unless it is stronger than some minimum (which is characteristic of each kind of solid). A liquid responds to the microscopic forces of thermal motion of its constituent particles by exhibiting *diffusion* on a time scale not very different from the time scale of vibrations; diffusion occurs in solids but far, far slower than vibrations.

Solids fall into two classes according to the degree of order in their structure. Those in which the constituent atoms or molecules are arranged in a *repeating* pattern that may be continued arbitrarily far are called *crystalline.* An example is the gold crystal whose surface is shown in Fig. 11.1. Those in which no regular pattern of repetition can be found are called *amorphous.* The distinction between amorphous solids and very viscous liquids is not a sharp one, naturally. A crystalline material may occur in a single crystal or as a collection of many small crystals; in the latter case the material is called *polycrystalline* or, if the crystals are very small, *microcrystalline.* (An unusual form of solid is that of a *quasicrystal,* in which the particles pack in an orderly but nonperiodic [nonrepeating] fashion. We shall say just a bit more about such systems when we study crystal structures in the following discussion.)

When we speak of a repeating pattern of atoms or molecules in a crystal, we of course mean that their *equilibrium positions* generate a repeating pattern. Just as the nuclei of a molecule vibrate about their equilibrium positions, the

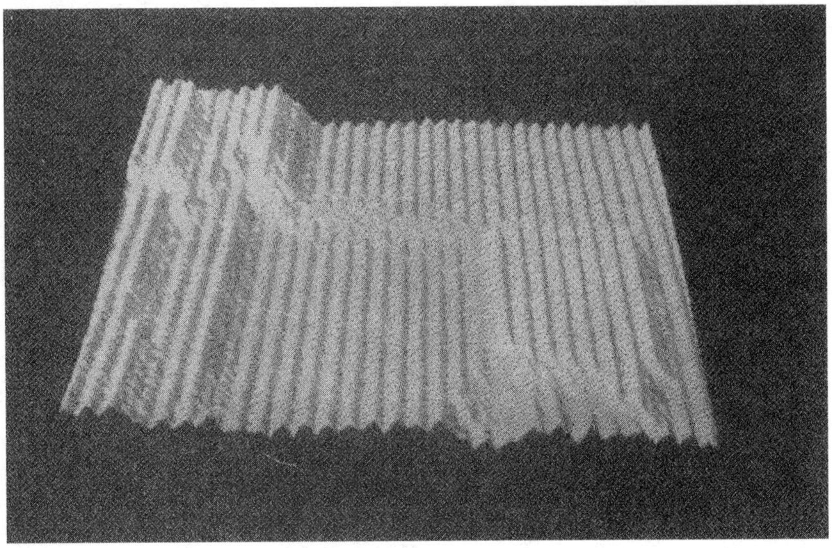

Figure 11.1 An image from atomic-level microscopy of the surface of a gold crystal. The rows are truly rows of atoms. (Taken, with permission, from I. Amato, *Science* **276,** 1983 (1997)).

constituents of a crystal vibrate about their equilibrium positions. And also as in a molecule, analysis of these vibrations can give us the shape of the potential energy surface—in simplest terms, the variation of the cohesive energy with interparticle distance. Information about the interparticle forces and potentials can be obtained from such bulk properties as the isothermal (constant-temperature) compressibility and the coefficient of thermal expansion. But these topics are better treated as part of the thermodynamics of solids, and have thus been postponed to Chapter 22. In the remainder of this chapter we shall treat solids as if the nuclei were at rest at their equilibrium positions.

Let us return to the distinction between crystalline and amorphous solids. A substance is crystalline if its microscopic structure consists of units that are replicated along three independent directions in space (although the term "two-dimensional crystal" is sometimes used for regular arrays on surfaces and their theoretical models); the replication must be capable of extending arbitrarily far. In the next section we shall examine the nature of the replicated units. An amorphous material contains no replicating unit. Liquids are amorphous; so are glasses and some other solids. It is not sufficient for a material to have local regions of ordered structure to be crystalline. Liquids have some degree of local order at any instant, particularly if we examine the structures about which their constituent particles are vibrating, but because the structure of a liquid changes fairly rapidly it exhibits no reproducible, replicating pattern. Amorphous solids often have regions of local order but no such order persists over macroscopic distances. We shall devote most of this chapter to crystalline solids; the structure of liquids and other amorphous materials will be considered in Chapter 23.

Four types of crystals can be distinguished readily on the basis of the forces that hold them together. One consists of molecules bound together by forces that are so weak, compared with the intramolecular forces, that the molecules retain their identity. In such a crystal, called a *molecular crystal,* the constituents are recognizable entities, and, at least in their lowest electronic states, exhibit properties very similar to the properties of the free gaseous molecules. The infrared spectra, bond lengths, and magnetic properties of the constituents all change only slightly when the molecules condense into molecular crystals. The forces responsible for the existence of molecular crystals are the weak attractions we examined in Chapter 10: dipole–dipole and other multipole interactions and van der Waals forces, especially the latter. Some examples of molecular crystals are those of argon, benzene, and naphthalene (often used for moth balls), typically weakly bound and relatively volatile.

A second type is an *ionic crystal,* such as that of sodium chloride. The best simple model of such a crystal supposes that it is composed of a three-dimensional array of alternating positive and negative ions. The cohesive energy comes almost entirely from the strong Coulomb attraction of the oppositely charged ions, so that ionic crystals are tightly bound. Because the binding forces are strong, ionic crystals are hard and involatile. The ions are packed together as tightly as possible, so as to minimize the total energy, which is composed of three kinds of interactions: the attractions of oppositely-charged ions, the repulsions of ions of the same charge, and the hard-core repulsions exhibited by all pairs of nearest neighbors.

Covalent crystals such as diamond constitute the third category. These crystals are also hard, tightly bound, and fairly close-packed. The forces that bind them are virtually identical with the forces that bind large molecules together (cf. Chapter 9).

Metallic crystals constitute the last of our categories. They are often soft, ductile, and malleable, but nevertheless have large cohesive energies. Metals as we usually find them are composed of very small crystals pressed together into polycrystalline bulk materials. However, large single crystals of metals can be grown and studied. In metals we normally cannot distinguish molecules or whole neutral atoms,

Table 11.1 Comparison of Properties of Typical Crystals of Various Sorts

Material	Type of Crystal	Cohesive Energy (eV)	Electrical Conductivity (MS/m)[a]	Near-Neighbor Distance or Molecular Dimension (Å)
Argon	Molecular	0.08	(insulator)	3.83
Methane (CH_4)	Molecular	0.10	(insulator)	4.49 (nearest neighbor)
Naphthalene ($C_{10}H_8$)	Molecular	0.39	(insulator)	5.10 (nearest neighbor)
Rock salt (NaCl)	Ionic	7.9	(insulator)	5.628 (Na–Na)
Cesium fluoride	Ionic	7.48	(insulator)	6.008 (Cs–Cs)
Diamond (C)	Covalent	7.32	(insulator)	1.5445
Graphite (C)	Covalent	7.34	7.27×10^{-2}	1.421
Quartz (SiO_2)	Covalent	6.17	(insulator)	1.50 (Si–O)
Silicon	Valence-metallic	4.34	10	2.352
Sodium	Metallic	1.1	23.8	3.716
Copper	Metallic	1.0	59.8	2.556

[a] MS = megasiemens; 1 siemens is 1 ohm^{-1}.

but the cores—nuclei and inner shells of electrons—retain their identities.

An important intermediate case is that of *hydrogen-bonded molecular crystals,* of which ice is no doubt the most obvious example. As in the context of intermolecular forces, the hydrogen bonds make intermolecular binding considerably stronger than in systems bound by typical, weak, intermolecular attractions.

Examples of some of the properties that distinguish the types of crystals are collected in Table 11.1. In addition to the cohesive energy, we have cited the electrical conductivity and the nearest-neighbor distance or some other typical molecular dimension. Molecular solids and covalent crystals are usually insulators for both electricity and heat. (The two conduction processes are sometimes related, as we shall see.) Ionic and metallic crystals are normally electrical conductors, but of different kinds. Conduction in ionic crystals occurs by motion of the ions themselves under the influence of an applied electric field, as in electrolytic solutions. In metallic conduction, only the electrons move under the influence of the electric field, and the conductivity is thus much greater than in ionic conductors. We shall return to these properties when we discuss particular kinds of solids in more detail.

11.2 Space Lattices and Crystal Symmetry

We have said that the structure of a crystal is made up of replicated (duplicated and repeated) units. Let us examine just what this means. Suppose that we wish to construct a model of a crystal, and have a basic physical unit with which to build; the crystal consists of copies of this unit. We can create a three-dimensional structure as follows. Through the basic unit we lay out three axes, not coplanar and not necessarily orthogonal. Then we copy the basic unit along one of the axes, say the x axis, at regular intervals, maintaining its orientation. This gives us a row of physical units extending from $-\infty$ to $+\infty$. Then we lay out an infinite set of identical x axes at regular intervals along the y axis, and replicate the row of physical units at each new x axis; this gives us a two-dimensional structure, an infinite planar array of identical physical units. Finally we repeat this plane at regular intervals along the z axis and replicate the physical units throughout to create a three-dimensional array, our crystal. We define three vectors, one along each axis, that define a parallelepiped enclosing one physical unit. Then we let these vectors have the shortest lengths consistent with the condition that periodic repetition of their parallelepiped at each physical unit covers the whole crystal. These vectors, **a**, **b**, and **c**, are called the *primitive vectors* of the crystal and their lengths, a, b, and c, are the *primitive distances* of the crystal. The primitive lengths may be equal but they need not.

The parallelepiped defined by the primitive vectors **a**, **b**, and **c** is the *primitive cell.* Note that the primitive cell is a parallelepiped in space, not the physical unit that the cell encloses. If we repeat the primitive vectors throughout space, we define an infinite abstract periodic structure, a set of points called a *lattice,* that has the same translational symmetry as the physical crystal. The directions and distances of translation that superimpose the abstract lattice on itself are the same as the directions and distances of translation that leave the physical crystal looking unchanged. The special lattice that has the same periodicity as the physical crystal is the *Bravais lattice* corresponding to that crystal. We may think of the Bravais lattice as the set of vertices that define the space-filling but non-overlapping periodic structures that occupy the primitive elementary parallelepipeds.

A primitive elementary parallelepiped is the simplest set of vertices of the Bravais lattice incorporating all the information necessary to construct the entire Bravais lattice by translations. The primitive elementary parallelepiped can contain no points of the Bravais lattice in its body or on its faces; its only Bravais lattice points are its eight vertices.

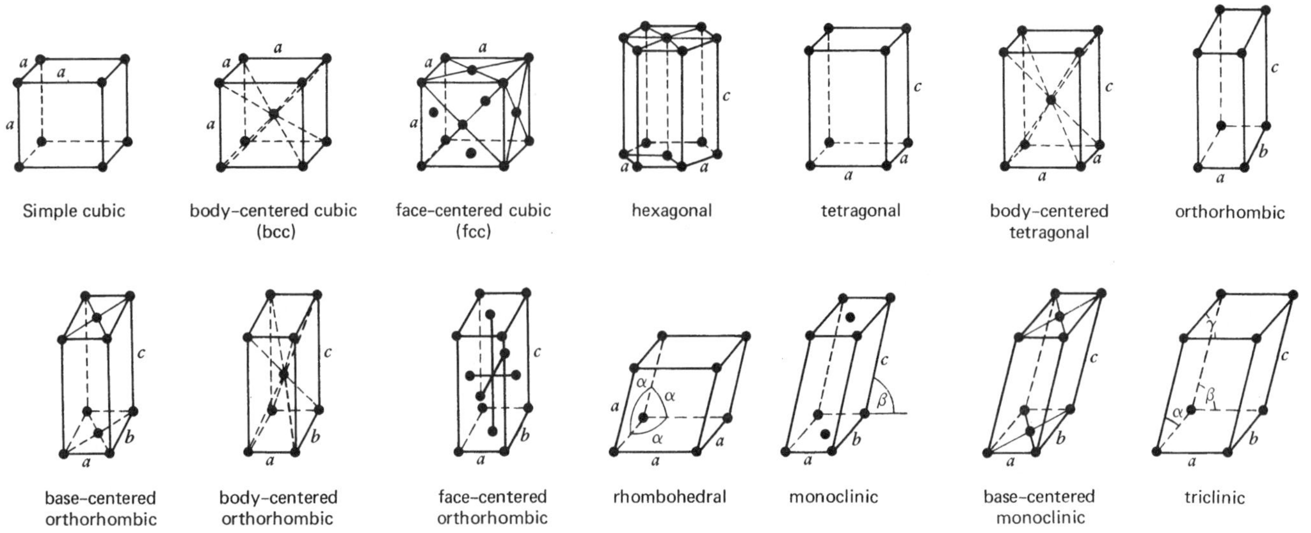

Figure 11.2. The 14 Bravais lattices. The first three belong to the cubic system; the next, to the hexagonal; the next two, to the tetragonal. The orthorhombic system has three possible lattices; the rhombohedral, one; the monoclinic, two; and the trielinic, only one.

In three dimensions, there are 14 Bravais lattices, in the sense that there are exactly 14 sets of relations of equality or inequality among the primitive distances and the angles between the primitive vectors. These are illustrated in Fig. 11.2, as the parallelepipeds defined by the primitive vectors. The figures shown are not in general the primitive elementary parallelepipeds, because several of the units shown in Fig. 11.2 have vertex points of the Bravais lattice at the centers of their bodies or faces. In this presentation it is easier to see the spatial relations among the lattice vertices than it is if only the eight lattice points of the primitive elementary parallelepiped are shown. The hexagonal right prism of Fig. 11.2, for example, is composed of three primitive elementary parallelepipeds; each of these is a right prism with rhombuses as top and bottom faces whose vertices are the central atoms of the hexagonal faces, as shown, and three successive vertices of those hexagonal faces.

The Bravais lattices fall into seven subsets, called the seven *crystal systems*. These are classified as follows in terms of relations among the primitive distances a, b, and c, and the angles α (between **b** and **c**), β (between **a** and **c**), and γ (between **a** and **b**):

Cubic: $a = b = c;\ \alpha = \beta = \gamma = 90°$.
Hexagonal: $a = b \neq c;\ \gamma = \beta = 90°,\ \gamma = 120°$ (Note that this corresponds to three equal coplanar axes in the ab plane intersecting at 60°, and a fourth, the c axis perpendicular to the ab plane.)
Tetragonal: $a = b \neq c;\ \alpha = \beta = \gamma = 90°$.
Orthorhombic: $a \neq b \neq c;\ \alpha = \beta = \gamma = 90°$.
Rhombohedral: $a = b = c;\ \alpha = \beta = \gamma \neq 90°$.
Monoclinic: $a \neq b \neq c;\ \alpha = \beta = 90°;\ \gamma \neq \alpha$.
Triclinic: $a \neq b \neq c;\ \alpha \neq \beta \neq \gamma$.

Crystals themselves have *structures*. Any ideal crystal structure can be associated with—and therefore considered as a realization of—a Bravais lattice belonging to one or another crystal system. In some crystals each atom or molecule lies on a lattice point. The alkali metals, for example, crystallize with a structure that lies on a body-centered cubic lattice. The Bravais lattice of this structure can be constructed of four points defining one square face of a cube of side length a (see Fig. 11.2b) and four points defining another square with sides and face parallel to those of the first, one of which points corresponds to the one shown in the center of the cube of Fig. 11.2b. In other crystals we must consider pairs or larger groups of atoms, ions, or molecules as the basic units that define the Bravais lattice. The smallest arrangement of basic constituents that both represents the chemical composition of the material and can generate the Bravais lattice by replication at integral multiples of the displacement vectors **a**, **b**, **c** is called the *unit cell* of the physical crystal. The unit cell is not necessarily the same as the primitive elementary parallelepiped; a unit cell may contain one or more primitive cells.

Let us consider some examples of real crystals. All of the alkali metals have body-centered cubic (bcc) lattices; one way of constructing the unit cell of sodium locates one sodium atom in the center and eight in the corners. Since the corner atoms are shared among eight adjoining cells, the unit cell consists of $1 + 8(\frac{1}{8}) = 2$ sodium atoms. If we were to neglect the differences between different kinds of atoms, cesium chloride would also be bcc, with a Cs^+ ion in the center and Cl^- ions at the corners, or vice versa (see Section 11.8), and would have one Cs^+Cl^- unit per unit cell. Because Cs^+ and Cl^- are different, we must consider the unit cell as something containing at least one atom of each. This

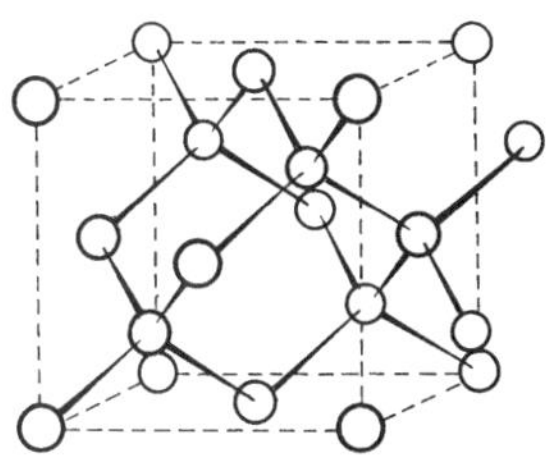

Figure 11.3 The crystal structure of diamond. The figure shows how the structure consists of a tetrahedron inside a face-centered cubic frame. The unit cell consists of only two carbon atoms. From A. F. Wells, *Structural Inorganic Chemistry,* 2nd ed. (Oxford University Press, Oxford, 1950).

implies that the lattice of CsCl is simple cubic. However, the *crystal structure* of CsCl is more important for many purposes than its Bravais lattice, and gives its name to a general class of like structures. Other crystals are more complicated, sometimes involving two interpenetrating lattices. For example, if the difference between Na^+ and Cl^- were negligible, the NaCl crystal would be face-centered cubic (fcc) (see Section 11.8); its unit cell contains four Na^+Cl^- units. Again, because of the difference, the crystal structure of NaCl is not fcc, but, similar to CsCl, has given its name, the "rock salt" structure, to a class of crystals. The diamond crystal (Section 11.10) has a structure built on the tetrahedral bonds of carbon (Fig. 11.3), but its unit cell contains only two atoms. The naphthalene crystal is monoclinic, with two *molecules* per unit cell, and so forth. Once the structure is known, whatever it is, the unit cell dimensions can be easily related to the macroscopic density of a single crystal.

Crystals exhibit symmetry. The simplest symmetry of a crystal is always translational: If the crystal, supposed infinite in extent in all directions, is translated through an integral multiple of any lattice vector, the displaced crystal is physically indistinguishable from the initial crystal. Thus, given any integers n_a, n_b, and n_c, we can translate the crystal by the vector $n_a\mathbf{a} + n_b\mathbf{b} + n_c\mathbf{c}$ and find it apparently unchanged. Most crystals also have rotational symmetry. If a cubic crystal is rotated through 90° about any of the three mutually perpendicular axes **a**, **b**, **c**, it is left unchanged. Similarly, a cubic crystal is invariant under a rotation of 120° about any of the body-diagonal axes of the cube. Because a fourfold repetition of rotation through 90° returns the crystal to its original orientation, the symmetry of the cubic structures with respect to rotation about the **a**, **b**, or **c** axes is said to be fourfold; similarly, rotation about the body diagonals of the cube exhibits threefold symmetry.

The tetragonal lattice retains fourfold symmetry with respect to rotation about the **c** axis, but has only twofold symmetry with respect to rotation about the **a** and **b** axes, and has no threefold symmetry. The rhombohedral lattice has the same threefold symmetry as the cubic lattice, but exhibits only twofold symmetry with respect to rotation about the **a**, **b**, **c** axes. The orthorhombic crystal has twofold symmetry about the **a**, **b**, **c** axes but no threefold axes. Monoclinic crystals have only twofold symmetry with respect to rotation about the **b** axis (the axis perpendicular to the plane containing the angle not equal to 90°), and the triclinic lattice has no symmetry other than translation.

What we have just described is the symmetry of the Bravais lattice, or of any of the solid models having the forms of the drawings of Fig. 11.2. Only when there is a single molecule per unit cell is the symmetry of the crystal simply the symmetry of the lattice. Otherwise the unit cell has its own symmetry. (The molecules or atoms comprising such a crystal may have symmetries of their own, but normally such internal symmetries do not correspond to transformations that leave the crystal in an unchanged condition, and thus are not part of the symmetry of the crystal.) In a crystal containing two or more identical constituents per unit cell, however, the symmetry of the crystal is greater than that of the lattice. By this we mean that there are operations that cannot be generated by combinations of the symmetry operations of the lattice, but that nevertheless leave the crystal unchanged. These additional operations are the symmetry operations of the unit cell. Often the unit cell has some symmetries that do correspond to operations already included in the lattice symmetry; if we find that all the symmetries of the unit cell are also lattice symmetries, then we must go back and redefine the unit cell. It is the presence of extra symmetry operations that are not among those of the lattice that gives rise to the concept of unit cell. The extra kinds of operations may include reflections, specific combinations of translations, specific translations plus rotations, or other combinations of these operations. The lattice symmetry and the unit cell symmetry together define the symmetry of the entire crystal.

11.3 X-ray Diffraction from Crystals: The Bragg Model

The most powerful tool for determining the structures of crystals (and of the molecules that constitute them) is the analysis of the patterns we record when their regular structures diffract waves. Waves are scattered by any particles they pass. If the wavelength λ is comparable to the distances between the scattering particles, then the scattering can be explained in terms of Huygens' classical theory of diffraction: Each scatterer produces a spherical wave going outward, and the amplitudes of all these spherical waves add together to produce a scattered wavefront. Each kind of atom has its own characteristic capacity to scatter waves of any particular kind, defined by a quantity called a *scattering factor,* to which the scattered amplitude is proportional. The kinds of waves of principal interest to us are x-rays, electrons, and neutrons, because their wavelengths are comparable to atomic dimensions. In general, x-rays are by far the most widely used for structural work, and neutrons are the most recently exploited for problems of chemical interest.

One carries out diffraction experiments by measuring the intensity of scattered radiation as a function of the scattering angle θ, the angle of deviation from the propagation axis of the incident beam (cf. Section 2.5). If the scattering material is amorphous, such as a glass, a liquid, or a gas, then θ is the only significant angle. In such a case the intensity of scattered radiation $I(\theta)$ is symmetrical about the propagation axis and tends to fall off as θ increases. From the behavior of $I(\theta)$ one can calculate the probability of finding a particle a given distance R away from any other selected particle: If $\mathcal{P}(R)\,dR$ is the probability of finding a particle between R and $R + dR$, given that one particle is at the origin, then one can show that (for an amorphous medium)

$$I(\theta) \propto \int_0^\infty \mathcal{P}(R)\frac{\sin kR}{kR}\,dR, \quad \text{where} \quad k \equiv \frac{4\pi}{\gamma}\sin\left(\frac{\theta}{2}\right). \tag{11.1}$$

Since $\mathcal{P}(R)$ would be proportional to $4\pi R^2$ in a completely uniform medium, it is convenient to introduce a function called the *pair correlation function* $g_2(R)$ defined by

$$\mathcal{P}(R) = 4\pi R^2 n g_2(R), \tag{11.2}$$

where n is the material's number density. The more nearly random the distribution of particles, the less $g_2(R)$ varies with R; if the medium were continuous and homogeneous, $g_2(R)$ would be constant. For liquids $g_2(R)$ exhibits oscillations such as those in Fig. 11.4*a:* There is a most probable nearest-neighbor distance corresponding to the first peak, but all other distances in this vicinity also occur; the next maximum in $g_2(R)$ corresponds to the most probable second-nearest-neighbor distance, and so forth. As the distance R grows larger, it is increasingly difficult to identify maxima; as the temperature increases, fewer maxima can be found. We shall have more to say about the pair correlation functions of liquids in Chapter 23.

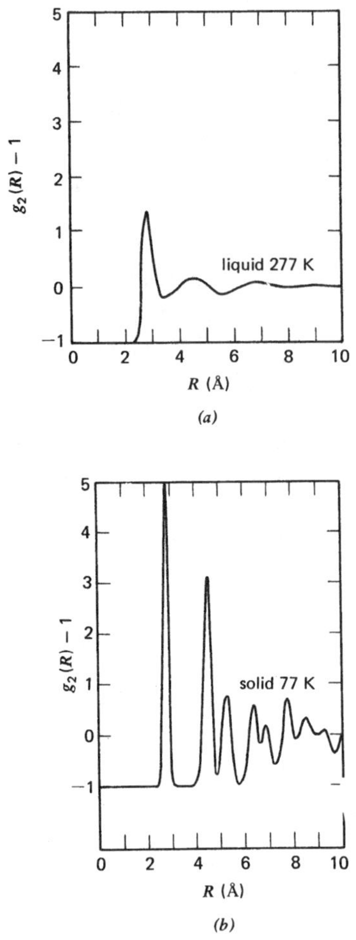

Figure 11.4 Pair correlation functions $g_2(R)$ for (*a*) liquid water, (*b*) solid ice. From A. H. Narten, C. G. Venkatesh, and S. A. Rice, *J. Chem. Phys.* **64,** 1106 (1976).

Pair correlation functions can be drawn for crystals, too, as in Fig. 11.4*b*. If there were no vibrational motion in the crystal, its pair correlation function would be a series of lines, one at each of the regular interparticle distances. The height of each bar would be proportional to the number of occupied lattice sites at the corresponding distance from a selected reference site. But real atoms in crystals do vibrate, so crystals must also have pair correlation functions that are smooth curves. Equation 11.1 applies only to an amorphous material, so how does one obtain the pair correlation function for a crystal? One way is to use a *powder* sample, made up of many small and randomly oriented crystals: Averaging over the random orientations again gives Eq. 11.1. When a beam of x-rays (or electrons or neutrons) is scattered from such a powder sample, the scattered x-rays come out in cones as shown in Fig. 11.5. The intensity $I(\theta)$ is independent of the angle ϕ about the propagation axis of the x-ray beam but varies with the angle θ of the cone from the propagation axis. In the apparatus originally used for such measurements, $I(\theta)$ was recorded on a strip of photographic film

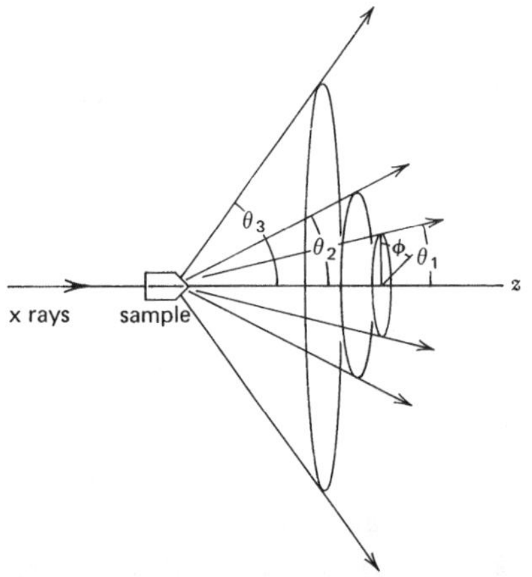

Figure 11.5 Diffraction of x-rays from a powder sample. The x-ray beam propagates from the left along the z axis. Three cones of diffracted x-rays are shown, corresponding to angles θ_1, θ_2, and θ_3; the intensity is independent of the angle ϕ about the z axis for a powder or any other sample that is, on average, isotropic, meaning a sample having no distinguishable internal directions.

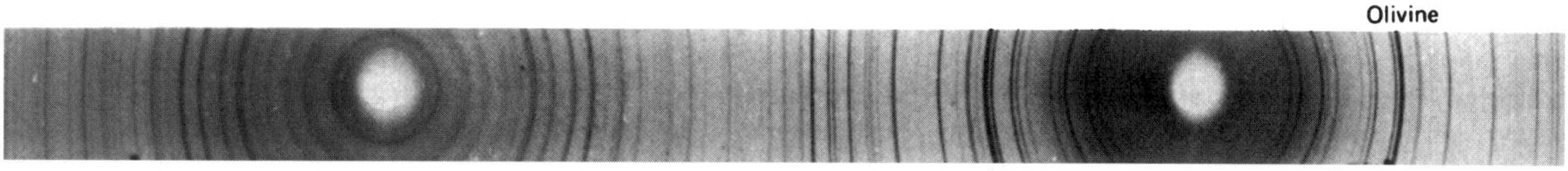

Figure 11.6 A typical x-ray powder pattern: olivine. (Supplied by Prof. Joseph Smith, The University of Chicago.)

or by a counter sweeping across a line or arc over a range of θ. The $I(\theta)$ so obtained has a series of narrow peaks, which appear as circular arcs on a photographic film. These *powder patterns* are characteristic of the crystal structure, and can thus be used to identify a powdered substance by comparison with the powder patterns of known substances. An example is shown in Fig. 11.6. Now it is common to use arrays consisting of many contiguous detectors, in a linear arrangement or distributed over an area. These must all have the same sensitivity or be calibrated to account for any differences in their response. With this in hand, the modern detectors enable researchers to collect the data from diffraction studies at rates vastly higher than in the earlier studies.

Diffraction and scattering from crystals yield much more information than just the pair correlation function $g_2(R)$. This is suggested by the fact that a crystal has an ordered structure with symmetry axes of its own. We should expect the intensity of scattered radiation to depend on the angles of orientation of the crystal to the axis of the incident beam, as well as on the angle *about* the latter axis. There are two ways to see how this occurs; before we can introduce them, however, we must develop some nomenclature.

The structure of a space lattice can be described with reference to the various planes of atoms that can be constructed in the lattice. We define such planes by a set of indices, customarily designated as h, k, l, which in turn are defined with respect to the primitive distances a, b, c. Imagine a plane that intersects at least one of the principal axes of the crystal inside (or at the boundary of) a unit cell which we take as a reference. If this plane cuts the principal axes of the crystal at the intercepts pa, qb, rc, measured from a chosen corner of the unit cell, then the *Miller indices* of that plane are the smallest integers h, k, l such that

$$h:k:l = \frac{1}{p}:\frac{1}{q}:\frac{1}{r}. \qquad (11.3)$$

The plane with indices h, k, l is conventionally designated as an (hkl) plane. A negative index is indicated by a bar over the corresponding number; for example, a plane with indices 1, −2, 3 is a $(1\bar{2}3)$ plane. Note that p, q, and r are all between 0 and 1. A plane parallel to one of the principal axes is thought of as intercepting that axis at infinity, so the corresponding Miller index is $1/\infty$ or 0. For example, in a simple cubic lattice the (100), (010), and (001) planes correspond to faces of the cubic primitive cell. A set of planes equivalent (because of the symmetry of the crystal) to a given plane (hkl) is designated with braces: $\{hkl\}$. Thus we say that all

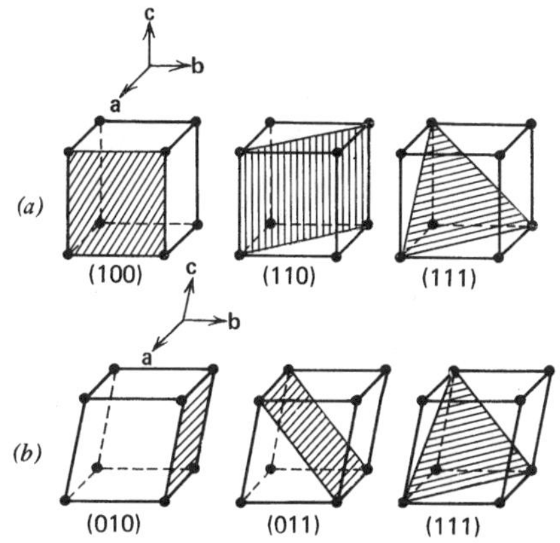

Figure 11.7 Miller indices of some crystal planes. (*a*) Cubic lattice. (*b*) Monoclinic lattice.

the planes of the square faces in a simple cubic lattice are $\{100\}$ planes. In Fig. 11.7 we illustrate some crystal planes and their corresponding Miller indices. The distance d_{hkl} between successive planes of a set $\{hkl\}$ can be expressed in terms of the lattice parameters and the Miller indices. In a cubic lattice, for example, we have

$$d_{hkl} = a(h^2 + k^2 + l^2)^{-1/2}, \qquad (11.4)$$

where a, as usual, is the lattice parameter.

We can now resume our discussion of diffraction in crystals. The simplest picture of x-ray diffraction, conceptually, is that introduced in 1913 by W. H. and W. L. Bragg. The physical picture underlying the usual derivation of this model (which we shall use) is rather crude because it expresses the ideas in terms of rays, the elements of geometric optics, rather than in terms of the scattered waves, the elements of "physical optics"—i.e., wave optics. This might leave one feeling that the conclusions rest on insecure ground. In fact, the results are equivalent to those obtained by more painstaking methods that follow the history of the scattered waves, which we examine in the next Section. The Braggs' concept supposes that x-rays are reflected by the planes of atoms on which they impinge. This reflection is supposed to follow *Snell's law* for the optics of specular (mirrorlike) reflection: The propagation vectors of the incident and reflected beams and the normal to the surface of the crystal all are coplanar, and the angles of incidence and reflection are equal. This behavior is illustrated in Fig. 11.8.

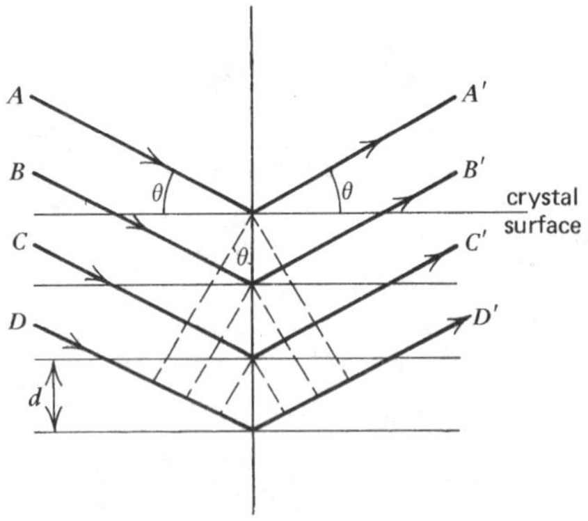

Figure 11.8 Bragg reflection of x-rays from parallel planes of atoms. The incident beam A is reflected as the beam A', B as B', and so on; in each case the angles of incidence and reflection have the same value θ.

In general, the beams reflected from successive planes parallel to the crystal surface interfere with each other so that no scattered radiation emerges for most angles of incidence. At certain angles, however, constructive interference of all the scattered waves produces net outgoing rays with intensities well above zero. Let us analyze how this phenomenon comes about.

For constructive interference to occur, the emerging beams A', B', C', D', . . . of Fig. 11.8 must all come out of the crystal with a common phase; the emerging beams then regenerate a plane wavefront corresponding to the incident wavefront $ABCD$. . . . But the ray BB' that strikes the second plane travels farther than the ray AA', and the ray CC' goes farther still. If the reflected rays are to emerge in phase, then the extra distances must all be integral multiples of λ, the wavelength of the radiation. Because we are dealing with a crystal, where planes of any given sort are precisely equally spaced, we know that if the first and second planes generate rays in phase with each other, so will all the others; the third plane thus has exactly the same relation to the second as the second has to the first, and so forth. So we need consider only the first two planes. The extra distance traveled by BB' relative to AA' is $2d \sin \theta$, where d is the interplanar distance and θ is the angle of incidence. Hence we require that

$$2d\sin\theta = n\lambda, \tag{11.5}$$

where n is an integer.

What Eq. 11.5 tells us is that x-rays reflected from a given series of crystal planes with separation d will give a diffraction pattern with maxima at the angles

$$\theta = \sin^{-1}\left(\frac{n\lambda}{2d}\right), \qquad (n = 1, 2, \ldots). \tag{11.6}$$

The integer n tells us the number of "extra" wavelengths traveled by BB', or the number of cycles that BB' is retarded relative to AA'. We call n the *order* of the spectrum. Because of Eq. 11.4, we can write for a cubic lattice

$$\begin{aligned}\frac{2a\sin\theta}{\lambda} &= n(h^2 + k^2 + l^2)^{1/2} \\ &= [(nh)^2 + (nk)^2 + (nl)^2]^{1/2},\end{aligned} \tag{11.7}$$

where θ is the angle at which we obtain nth-order reflection from the (hkl) plane. This expression is the form in which we shall derive the diffraction conditions from another viewpoint.

11.4 The Laue Model

A real lattice is not composed of reflecting planes, but of discrete atoms; thus the Bragg model is useful and conceptually simple but fundamentally unsatisfying. Max von Laue (1912) used the more realistic model that crystals are composed of small scattering centers that act on wavefronts passing by. In the tradition of the Huygens construction, Laue supposed that each scattering center generates a spherical outgoing wave whenever it responds to the periodic force of an applied electromagnetic wave. The strength or intensity of the scattered wave is a property of each atomic scatterer, represented by a quantity called the *atomic scattering factor* f_s. When an electromagnetic wave passes through a crystal it induces a small forced oscillation in each atom, which then becomes the center of an outgoing spherical wave, the scattered wave. The amplitudes of the waves coming from all the atoms add together algebraically. In almost every direction coming out of the crystal these individual amplitudes add to give exceedingly small total amplitudes, because the positive contributions at any instant are almost always nearly equal to the negative contributions. However, Laue showed that in a few special directions the amplitudes of the individual scattered waves add constructively, producing narrow pencils of scattered radiation. More important still, these directions depend intimately on the structure of the crystal.

To understand the scattering process in a quantitative way, we can proceed stepwise from a one-dimensional array to two- and then three-dimensional periodic structures. A little trigonometry will suffice to describe the process. Suppose that, as shown in Fig. 11.9*a*, a wavefront is incident on a line of atoms spaced regularly a distance a apart; let $\mathbf{a}$ be a vector of length a directed along the line of atoms. Let α_0 be the angle of incidence, and α one of the infinitely many possible angles of refraction; $\mathbf{s}_0$ and $\mathbf{s}$ are unit vectors in the respective directions of propagation. Then the refracted waves from two adjacent atoms X and Y will interfere constructively only for those values of α such that the path lengths $\mathbf{a} \cdot \mathbf{s}_0 = a \cos \alpha_0$ and $\mathbf{a} \cdot \mathbf{s} = a \cos \alpha$ differ by an integral number of wavelengths:

$$a(\cos\alpha - \cos\alpha_0) = \mathbf{a}\cdot(\mathbf{s} - \mathbf{s}_0) = n\lambda. \tag{11.8}$$

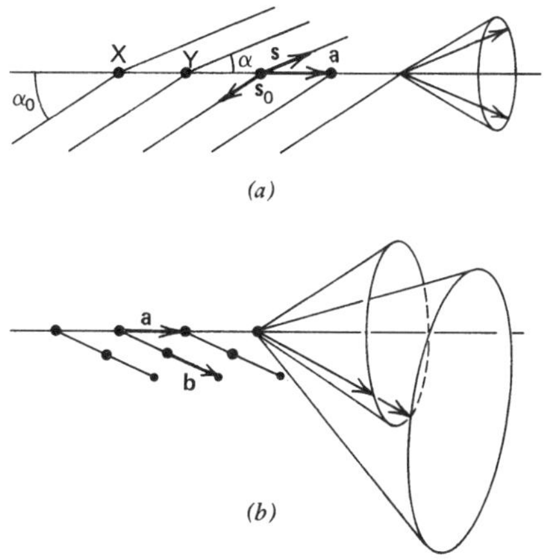

Figure 11.9 Diffraction from linear and planar lattices. (*a*) For a line of atoms, reinforcement occurs only for the cones of radiation defined by $\mathbf{a} \cdot (\mathbf{s} - \mathbf{s}_0) = n\lambda$. (*b*) Reinforcement occurs for a two-dimensional array only along the lines of intersection of the two sets of cones from the simultaneous one-dimensional diffraction. The doubleheaded arrows indicate these directions.

For a fixed wavelength λ, Eq. 11.8 defines a cone of outgoing rays at a particular angle α for each integer n. If these cones are to be far enough apart for useful measurements, the wavelength λ must be of the same order as (or smaller than) the lattice parameter a, which is of the order of 10^{-8} cm. This is why x-rays are used; this is just the range of their wavelengths.

To apply the same analysis to both x and y directions of a two-dimensional lattice, we need two conditions. Assume that the lattice parameters are a in the x direction, b in the y direction. Let α_0 and α be defined as before relative to the x axis, with β_0 and β the corresponding angles relative to the y axis. Then we require, for constructive interference, that

$$\mathbf{a}\cdot(\mathbf{s}-\mathbf{s}_0)=a(\cos\alpha-\cos\alpha_0)=n_1\lambda \qquad (11.9a)$$

and

$$\mathbf{b}\cdot(\mathbf{s}-\mathbf{s}_0)=b(\cos\beta-\cos\beta_0)=n_2\lambda \qquad (11.9b)$$

be fulfilled *simultaneously.* For each n_1, n_2 these conditions define the intersection of two cones, one for the solution of each equation, as in Fig. 11.9*b*. But the intersection of two cones with the same vertex is a pair of lines. Thus for the two-dimensional lattice we obtain constructive interference along a set of rays corresponding to all the pairs of integers n_1, n_2 for any incident direction of radiation $\mathbf{s}_0$. Incidentally, this two-dimensional portrayal of diffraction is the one used to represent diffraction of low-energy, long-wavelength *electrons* by the surface layers of solids, the process known as *LEED* (*low-energy electron diffraction*).

If the lattice is three-dimensional, we must have a third condition for constructive interference, associated with periodicity in the z direction. Let the lattice parameter along the z axis be c, with γ_0 and γ the angles of incident and refracted beams relative to the z axis. Then constructive interference requires that

$$\mathbf{c}\cdot(\mathbf{s}-\mathbf{s}_0)=c(\cos\gamma-\cos\gamma_0)=n_3\lambda \qquad (11.9c)$$

be satisfied simultaneously with Eqs. 11.9*a* and 11.9*b*. With three conditions to be satisfied simultaneously, the diffracted radiation can no longer exhibit constructive interference for almost all directions of incident radiation. Rather, $\mathbf{s}_0$ itself must be in one of a set of special directions if Eqs. 11.9*a*, 11.9*b*, 11.9*c* are all to apply. To see how this follows, we square and add the three equations, obtaining

$$\begin{aligned}&a^2(\cos^2\alpha+\cos^2\alpha_0-2\cos\alpha\cos\alpha_0)\\&+b^2(\cos^2\beta+\cos^2\beta_0-2\cos\beta\cos\beta_0)\\&+c^2(\cos^2\gamma+\cos^2\gamma_0-2\cos\gamma\cos\gamma_0)\\&=(n_1^2+n_2^2+n_3^2)\lambda^2.\end{aligned} \qquad (11.10)$$

For simplicity let us restrict our discussion to cubic crystals, for which $a^2 = b^2 = c^2$; applying the law of cosines to both incident and refracted angles,

$$\begin{aligned}\cos^2\alpha+\cos^2\beta+\cos^2\gamma&=1,\\ \cos^2\alpha_0+\cos^2\beta_0+\cos^2\gamma_0&=1,\end{aligned} \qquad (11.11)$$

we obtain

$$\begin{aligned}&2a^2(1-\cos\alpha\cos\alpha_0-\cos\beta\cos\beta_0-\cos\gamma\cos\gamma_0)\\&=(n_1^2+n_2^2+n_3^2)\lambda^2.\end{aligned} \qquad (11.12)$$

We now define the angle between the unit vectors $\mathbf{s}$ and $\mathbf{s}_0$ as ϕ,

$$\mathbf{s}\cdot\mathbf{s}_0=\cos\phi, \qquad (11.13)$$

so that

$$\begin{aligned}\cos\phi&=\cos\alpha\cos\alpha_0+\cos\beta\cos\beta_0+\cos\gamma\cos\gamma_0\\&=1-2\sin^2\left(\frac{\phi}{2}\right).\end{aligned} \qquad (11.14)$$

Hence, Eq. 11.12 becomes

$$\begin{aligned}2a^2(1-\cos\phi)&=4a^2\sin^2\left(\frac{\phi}{2}\right)\\&=(n_1^2+n_2^2+n_3^2)\lambda^2\end{aligned} \qquad (11.15)$$

or

$$2a\sin\left(\frac{\phi}{2}\right)=(n_1^2+n_2^2+n_3^2)^{1/2}\lambda. \qquad (11.16)$$

Since $\phi/2$ is half the angle between **s** and $\mathbf{s}_0$, it can be used to define the direction with respect to which **s** and $\mathbf{s}_0$ are symmetrical. This is the direction relative to which $\mathbf{s}_0$ and **s** correspond to *incidence* and *specular reflection,* respectively, with a reflecting plane perpendicular to the direction of $\mathbf{s} - \mathbf{s}_0$. Hence, $\phi/2$ is equivalent to the θ of Eq. 11.5. By identifying n_1, n_2, n_3, respectively, with the *nh, nk, nl* of Eq. 11.7, we establish the equivalence of Eqs. 11.7 and 11.16. This means that *n,* the order of the reflection, is the largest common factor of n_1, n_2, and n_3.

11.5 Determination of Crystal Structures

Now that we have derived the basic formula for x-ray diffraction in crystals by two different routes, let us see how this can be used to obtain crystal structures from diffraction patterns.

Two structural characteristics enter into the formation of diffracted beams: the phase differences among the waves scattered in a given direction by the various atoms in the unit cell, and the amplitude of the wave scattered by each atom. We shall now see how these characteristics determine the directions (scattering angles) and intensities of the beams one detects, and how these directions and intensities in turn enable us to infer the crystal structure. Our discussion is necessarily rudimentary, but will illustrate the general character of the diffraction process.

The preceding discussions of Bragg and Laue diffraction were phrased in terms of waves that emerge from a crystal exactly in phase, in the sense that all such waves have either the same phase or phases differing by an integral multiple of 2π. Yet the general case of scattering in arbitrary directions involves a phase difference that need not be an integral multiple of 2π. The phase difference φ between two rays is related to *l,* the difference in path length, by the equation

$$l = \frac{\varphi\lambda}{2\pi}. \tag{11.17}$$

This path difference is $2d \sin \theta$ in the Bragg picture (Fig. 11.8), where *d* is the spacing between Bragg planes. Hence, we have

$$\varphi = \frac{4\pi d}{\lambda}\sin\theta. \tag{11.18}$$

In the Laue picture we must express *l* in terms of spacings between *points,* rather than spacings between planes of atoms. We therefore write the position vector of the *j*th atom of the unit cell relative to the origin (at an atom of the cell) in terms of the lattice vectors **a**, **b**, **c** and the coordinate values u_j, v_j, and w_j of the nucleus of atom *j:*

$$\mathbf{r}_j = u_j\mathbf{a} + v_j\mathbf{b} + w_j\mathbf{c}. \tag{11.19}$$

We suppose that the atoms defining the coordinate origins of unit cells already satisfy the Laue or Bragg conditions. Then the phase difference between waves diffracted from the origin and atom *j* of the same cell is

$$\varphi_j = \frac{2\pi}{\lambda}\mathbf{r}_j \cdot (\mathbf{s} - \mathbf{s}_0), \tag{11.20a}$$

where $\mathbf{r}_j \cdot (\mathbf{s} - \mathbf{s}_0)$ is the difference in path lengths, or

$$\varphi_j = 2\pi(u_j n_1 + v_j n_2 + w_j n_3), \tag{11.20b}$$

because we require that Eq. 11.9 be satisfied.

Consider once more a cubic lattice, with atoms only at the corners of the unit cell. Let us choose **s** and $\mathbf{s}_0$ such that $n_1 = 1$, $n_2 = n_3 = 0$, that is, such that these corner atoms give a first-order diffracted beam in the direction corresponding to reflection from the (100) plane; the phase of this beam is by definition zero. Suppose now that additional atoms are placed in the unit cell, in particular at the positions $(a/2, 0, 0)$, where $u_j = \frac{1}{2}$, $v_j = w_j = 0$. For the same **s** and $\mathbf{s}_0$, these atoms produce waves with the phase $\varphi_j = 2\pi(\frac{1}{2} \cdot 1) = \pi$, which are exactly out of phase with, and thus cause destructive interference with, the waves scattered by the corner atoms. If the two sets of atoms scatter with the same strength, then the first-order beam in the given direction simply disappears. The first nonvanishing beam from the (100) plane is the one for which $n_1 = 2$, rather than $n_1 = 1$, for then $\varphi_j = 2\pi$, and the two sets of waves add constructively. Hence, the angle at which we see the lowest-order beam from the (100) plane is

$$\theta = \sin^{-1}\left(\frac{2\lambda}{2a}\right) = \sin^{-1}\left(\frac{\lambda}{a}\right), \tag{11.21}$$

from Eq. 11.7 or 11.16 (with $\theta = \phi/2$). This agrees with our previous conclusion, in connection with Eq. 11.8, that to observe diffraction we must have $\lambda \lesssim a$.

The intensity of scattered radiation from an individual atom with *Z* electrons is most conveniently expressed in terms of the ratio of the *amplitude* of a scattered wave to the amplitude that would be produced by one free electron at the position of the nucleus. This ratio is roughly proportional to the total electronic charge. More specifically, it is the sum of contributions of all parts of the electronic charge cloud, each with its own appropriate phase φ relative to the nucleus. At a point *P* defined by the radius vector **r**, the charge density $\rho(\mathbf{r})$ is, as usual, proportional to the square of the wave function $\psi(\mathbf{r})$, whereas the phase is given by Eq. 11.18 as

$$\varphi(\mathbf{r}) = \frac{4\pi d(\mathbf{r})}{\lambda}\sin\theta, \tag{11.22}$$

where now $d(\mathbf{r})$ refers to the distance between the zero-phase reflecting plane and the point *P;* here θ is again half the angle between the incident and diffracted beams. Figure 11.10 shows these relations and defines the angle ζ; we have

$$d(\mathbf{r}) = r\cos\zeta, \tag{11.23}$$

and thus,

$$\varphi(\mathbf{r}) = \frac{4\pi}{\lambda} r\cos\zeta \sin\theta = kr\cos\zeta. \tag{11.24}$$

We have here introduced the quantity

$$k \equiv \frac{4\pi}{\lambda}\sin\theta, \tag{11.25}$$

which has the dimensions of a wavenumber. It is not hard to show that $\hbar k$ is the magnitude of the momentum transferred to the scattered photon.[1]

The amplitude of the waves scattered from the point P is proportional to the electron density, and thus to $|\psi(\mathbf{r})|^2$, weighted by the phase factor $e^{i\varphi(\mathbf{r})}$. The atomic scattering factor f_s is thus defined as the integral of $e^{i\varphi(\mathbf{r})}|\psi(\mathbf{r})|^2$ over the whole volume of the atom:

$$f_s = \int_{\phi=0}^{2\pi}\int_{\zeta=0}^{\pi}\int_{r=0}^{\infty} e^{i\varphi(\mathbf{r})}|\psi(\mathbf{r})|^2 r^2 \sin\zeta \, dr\, d\zeta\, d\phi. \tag{11.26}$$

We now substitute the value of $\varphi(\mathbf{r})$ from Eq. 11.24 and assume that the atomic wave function $\psi(\mathbf{r})$ is spherically symmetric, so that

$$\begin{aligned} f_s &= 2\pi\int_{r=0}^{\infty}\left[\int_{\zeta=\pi}^{0} e^{ikr\cos\zeta}\, d(\cos\zeta)\right]|\psi(\mathbf{r})|^2 r^2\, dr \\ &= 2\pi\int_0^{\infty}\frac{e^{-ikr}-e^{ikr}}{-ikr}|\psi(\mathbf{r})|^2 r^2\, dr \\ &= 4\pi\int_0^{\infty}|\psi(\mathbf{r})|^2\frac{\sin kr}{kr} r^2\, dr. \end{aligned} \tag{11.27}$$

Note that a similar derivation leads to Eq. 11.1, in which θ corresponds to the angle we now call ϕ. The scattering factor f_s is a function of k, and thus of both θ and λ; it becomes equal to the total number of electrons in the atom when $(\sin kr)/kr \approx 1$, that is, when $\sin\theta \approx 0$. It is a decreasing function of k, approaching zero as $k \to \infty$ or $\lambda \to 0$; that is, all atoms become weak scatterers when the wavelength of the incident radiation becomes small enough. Typical x-ray scattering factors for several atoms are given in Table 11.2.

Atomic wave functions give rather good representations of the scattering factors f_s. Strictly, if one is dealing with

[1] A photon of wavelength λ has momentum $h\nu/c = 2\pi\hbar/\lambda$; since the wavelength is not changed by scattering, the momentum change is that required to rotate the momentum vector through an angle 2θ without changing its magnitude.

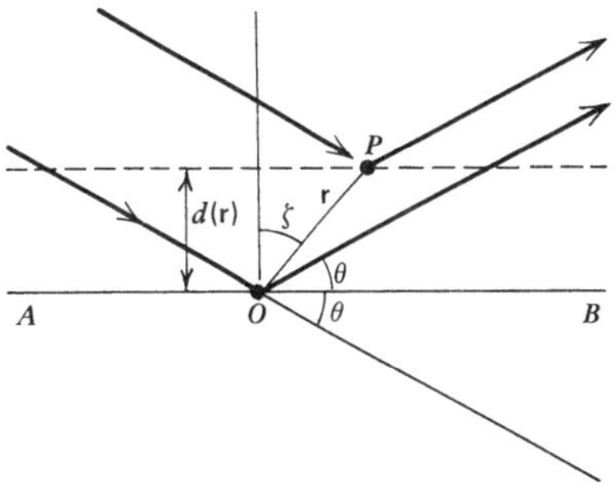

Figure 11.10 Scattering within an atom, from a point P defined by the radius vector $\mathbf{r}$ from the nucleus O. Here AOB is the base plane for Bragg reflection, ζ is the angle between $\mathbf{r}$ and the normal to AOB, $d(\mathbf{r}) = r\cos\zeta$ is the distance from P to AOB, and θ is the scattering angle.

Table 11.2 Typical Atomic Scattering Factors for X-rays[a]

	$(\sin\theta)/\lambda$			
Element	**0**	**0.2**	**0.5**	**1.0**
H	1	0.481	0.071	0.007
Li	3	1.741	1.032	0.320
C	6	3.581	1.685	1.114
N	7	4.600	1.944	1.263
O	8	5.634	2.338	1.374
Na	11	8.34	4.29	1.78
Na^+	10	8.390	4.328	1.785
Cl	17	12.00	7.29	4.00
Cl^-	18	12.20	7.28	4.00
K	19	13.73	7.87	4.84
K^+	18	13.76	7.86	4.84
Fe	26	20.09	11.47	6.51

[a] The units of the atomic scattering factor f_s are electrons per atom; f_s is a function of the wavelength λ of the x-ray and the scattering angle θ, through the variable $(\sin\theta)/\lambda$. When $(\sin\theta)/\lambda$ is zero, f_s is the number of electrons in the atom. (From *International Tables for Crystallography, Vol. III, Physical and Chemical Tables,* published for the International Union of Crystallography by the Kynoch Press, Birmingham, England, 1962.)

scattering from atoms in a molecule, the appropriate wave function $\psi(\mathbf{r})$ for use in Eq. 11. 27 is the part of the *molecular* wave function around the atom in question. Then, when the total scattering of the crystal is calculated, one should integrate over all the charge density in the unit cell. In practice, however, one frequently uses atomic wave functions to find f_s and sums over atomic scattering factors to obtain the total *structure factor F:*

$$\begin{aligned} F &= \sum_j f_{sj} e^{i\varphi_j} \\ &= \sum_j f_{sj} e^{2\pi i(u_j n_1 + v_j n_2 + w_j n_3)}, \end{aligned} \tag{11.28}$$

where f_{sj} and φ_j are the scattering factor and phase (at the nucleus) of the jth atom. It has been possible in a few cases to distinguish between the scattering predicted by Eq. 11.28 and the more accurate value one would obtain if one were to use

$$F = \int_{\phi=0}^{2\pi}\int_{\zeta=0}^{\pi}\int_{r=0}^{\infty} e^{i\varphi(\mathbf{r})}\left|\psi_{\text{crystal}}(\mathbf{r})\right|^2 r^2 \sin\zeta \, dr\, d\zeta\, d\phi. \tag{11.29}$$

However, one usually does not know the wave function ψ_{crystal}, and the simple atomic $\psi(\mathbf{r})$ serves quite well for almost all purposes. Moreover, the atomic $\psi(\mathbf{r})$'s have the computational advantage of being spherically symmetric. The most precise measurements, particularly for small molecules, most of whose electrons participate in chemical bonding, can distinguish between charge distributions of real molecules and the hypothetical but calculable charge distributions that correspond to separate, noninteracting atoms.

Finally, one can calculate the total relative scattered intensity, defined in terms of the structure factor F. The relative scattered intensity (the absolute magnitude is of no concern to us) is directly proportional to the quantity

$$|F|^2 = \left|\sum_j f_{sj} e^{i\varphi_i}\right|^2 = \left(\sum_j f_{sj}\cos\varphi_j\right)^2 + \left(\sum_j f_{sj}\sin\varphi_j\right)^2. \tag{11.30}$$

Here $|F|^2$ is the absolute square of the sum of the amplitudes of the waves scattered by individual atoms, and thus displays the interferences among these waves.

To see how the structure factor depends on the indices n_1, n_2, n_3 and how structures can be inferred from diffraction patterns, we consider some examples. First, in a simple cubic structure (of which there are very few known examples in nature) there is only one atom per unit cell, so we have

$$|F|^2 = |f_s|^2, \tag{11.31}$$

and all rays appear for any integers n_1, n_2, n_3.

Next we consider a body-centered-cubic structure. The unit cell (Fig. 11.1) has atoms at (0, 0, 0) and at (a/2, a/2, a/2); thus, in Eq. 11.28 we have $u_1 = v_1 = w_1 = 0$ and $u_2 = v_2 = w_2 = \frac{1}{2}$. Suppose that the atoms at (0, 0, 0) and equivalent points have scattering factors f_1 and the other atoms have scattering factors f_2. Then we obtain

$$F = f_1 + f_2 e^{\pi i(n_1+n_2+n_3)} \tag{11.32}$$

and

$$\begin{aligned}|F|^2 &= f_1^2 + 2f_1 f_2\cos[\pi(n_1+n_2+n_3)] \\ &\quad + f_2^2\{\cos^2[\pi(n_1+n_2+n_3)] \\ &\quad + \sin^2[\pi(n_1+n_2+n_3)]\} \\ &= f_1^2 + f_2^2 + 2f_1 f_2\cos[\pi(n_1+n_2+n_3)].\end{aligned} \tag{11.33}$$

If f_1 and f_2 are equal, then $|F|^2$ vanishes whenever $\cos[\pi(n_1 + n_2 + n_3)]$ is -1, or whenever $n_1 + n_2 + n_3$ is odd. Hence, the "odd" reflections vanish, but the even reflections appear. If f_1 and f_2 are not equal, then the odd reflections appear. If f_1 and f_2 are similar in magnitude but not identical, then the odd reflections are weak, but not absent.

Face-centered-cubic crystals have atoms of type 1 at the eight corners, and atoms of type 2 at the centers of the six sides. We leave as an exercise (see Problem 11) the determination of the diffracted beam intensities in terms of f_{s1} and f_{s2}, the scattering factors for these two sorts of atoms. When $f_{s1} = f_{s2}$, two kinds of interference are possible: If n_1, n_2, n_3 are all even or all odd, constructive reinforcement occurs. Otherwise (two even, one odd, or vice versa) the reflections vanish.

11.6 Techniques of Diffraction

We have spoken of two types of x-ray diffraction by crystals. One, considered only briefly, is powder diffraction by monochromatic x-rays; the other, which we have examined in some detail, is diffraction of monochromatic x-rays by single crystals. We implicitly assumed in discussing the single-crystal method that the crystal could be oriented at any desired angle relative to the direction of propagation of the x-rays, so as to meet the diffraction conditions. Both powder and single-crystal techniques are used, the first for classifying and identifying solids, the second for determining structures of molecules by finding the locations of atoms within unit cells. A third method, also used for classification, uses polychromatic ("white") x-rays for determining crystal symmetry. It is useful here to inject a point of realism: While it is easy to produce crystals of rock salt and sucrose, it sometimes requires considerable skill (and perhaps good fortune) to produce crystals of very complex molecules such as proteins. It is not uncommon now for the most difficult step in the determination of the structure of a protein molecule to be making a crystal large enough and of high enough quality to yield clear x-ray diffraction patterns.

The sources for x-rays were, for many years, tubes in which high-energy electron beams bombarded metal targets, exciting and ionizing inner-shell electrons of the metal atoms, in turn creating inner-shell vacançies into which outer electrons could fall. This last process generated the short-wavelength, high-energy radiation we call x-rays. Now another kind of source has become very important, the synchrotron radiation from free electrons accelerated in large machines called storage rings. In these, bunches of electrons cycle in circular orbits, radiating as they go because they are subject to forces that keep them in those orbits. The radiation they generate propagates in a narrow cone in the direction tangential to the orbit, so the radiation coming to any one detector appears as a brief pulse each time a bunch of electrons passes the point where the tangent to the orbit lies on a line from the ring to the detector. The energy distribution of the radiation depends on the energy of the orbiting electrons. If the electrons have energies of approximately 6 GeV (6×10^9 eV/electron), the peak of the power distribution of their synchrotron radiation is at approximately 0.2 Å. This makes it possible to attain spatial resolution from x-ray diffraction comparable to that from electron diffraction.

In addition to x-rays, both electrons and neutrons are used for diffraction experiments. *Electron diffraction* is also used for studying gaseous molecules and clusters. Scattering of electrons by atoms differs from scattering of x-rays in three ways. First, nuclei contribute significantly to electron scattering; often they furnish the dominant part of the atomic scattering factor, especially at large angles. Second, electrons are scattered far more effectively than x-rays of the same energy. Third, the wavelengths are different. According to the de Broglie relationship, Eq. 3.1, the wavelength of an electron is

$$\lambda = \frac{h}{p} = \frac{h}{(2mE)^{1/2}}, \tag{11.34}$$

where p and E are the electron's momentum and kinetic energy, respectively. Substituting the values of h and m_e, and manipulating units, we find that

$$\lambda(\text{Å}) \approx [150 / E(\text{eV})]^{1/2},$$

so that a 15-keV electron has a wavelength of about 0.1 Å. This was well below the wavelengths one could use for x-ray diffraction prior to the advent of synchrotron radiation as a source for x-rays. It used to be that one could, at least in principle, obtain much better resolution with electrons than with x-rays. In practice, however, the ease with which electrons are scattered restricted this technique for many years to gaseous (or easily volatilized) samples. In crystals it is difficult for electrons to penetrate more than a few tens of atomic layers, and electron diffraction is used primarily to study surface phenomena. Electrons, more than x-rays, may be scattered more than once when they penetrate a sample; multiple scattering, if it is extensive, can cause considerable difficulty by generating "noise" in the diffraction pattern.

The atomic scattering factor for electrons contains contributions from both nuclei and bound electrons because electrons, unlike x-rays, can transfer momentum moderately effectively to nuclei as well as to other electrons. The presence of the two contributions with opposite signs causes the electron scattering factor to have a minimum in the forward direction. The electronic contribution falls off with angle faster than the nuclear part, so that the high-angle scattering is due primarily to the nuclei.

Neutron diffraction is the third diffraction technique used for structure determination. Neutrons are scattered in two ways; through close encounters with nuclei, or through interaction with *magnetic* scatterers. Because neutrons are essentially as massive as protons, their de Broglie wavelengths are 1800 times smaller than those of electrons with the same kinetic energy. Their scattering cannot be described by a simple systematic function of the Z of the scatterer because the process depends in detail on the nuclear structure. Neutron scattering, in contrast to x-ray or electron diffraction, also involves a significant amount of recoil because the similarity of the masses of atomic nuclei and of neutrons leads to non-negligible exchanges of momentum between these when they collide. However, even isolated protons are capable of scattering neutrons effectively; hence, neutron diffraction offers one way to locate hydrogen atoms. Also, because neutrons scatter from nuclei, they are particularly suited to the determination of the amplitudes of vibrational motions. X-rays can also be used to locate hydrogen atoms, and to determine, or at least to estimate, vibration amplitudes, but only with considerable refinement of the data beyond what is necessary to find molecular structures.

Neutrons are unique in their capability to detect magnetic order superposed on the chemical order of the crystal. If, for example, a crystal is antiferromagnetic, then the spins of the component atoms or molecules are paired to generate an order whose unit cell is typically twice the size of the "chemical" unit cell. This is because the unit cell in the magnetically paired crystal must have equal numbers of subunits with spin α and spin β. Doubling the size of the unit cell increases the number of terms entering into F, and hence the number of reflections. One can actually detect the onset of antiferromagnetic order by watching new diffracted beams appear when a sample is cooled below the temperature of its normal-to-antiferromagnetic transition. An example is MnO, whose transition temperature is 122 K.

11.7 Molecular Crystals

Thus far we have surveyed the types of symmetry found in crystals and described methods for the experimental determination of crystal structure. In the remainder of this chapter we shall study the types of bonding and electronic structure that exist in crystals. We begin with those solids in which the bonding is most like what we have described in earlier chapters.

Some solids have properties that can be described easily in terms of their separate component molecules. Each molecule retains its identity, and is held to the aggregate structure by the weak dispersion forces we described in Chapter 10. The simplest examples of these *van der Waals crystals* or *molecular crystals* are solid phases of the inert gas elements, in which the molecules are single atoms. The dispersion forces, it will be recalled, are associated with the fluctuating, transient moments of individual atoms, moving in an only partially correlated way. They are thus weak and only instantaneously directional. Because of this lack of directionality, the atoms crystallize in *close-packed* structures.

There are two close-packed structures, the face-centered cubic (fcc) and the hexagonal close-packed (hcp). Both can be constructed by making three-dimensional arrays of hard spheres: Around one central atom, we can make a planar hexagon of nearest neighbors. In the interstices below this plane, we can place three more hard spheres of the same radius, by using every other one of the six interstices next to the central atom. Then we can place three more atoms above the original plane, again over alternating interstices. There are two options at this stage: We can put the last three spheres above the three covered interstices, or above the open interstices. If we choose the first option, we get the hexagonal close-packed structure; the second gives us the face-centered cubic structure. These are illustrated in Figs. 11.11*a* and 11.11*b*.

The actual structures of the solids of neon, argon, krypton, and xenon are face-centered cubic. Understanding why these are close-packed structures is simple: Hard spheres pack this way, and the inert gas atoms behave almost like hard spheres when they condense. The reasons for the choice of the fcc structure are more subtle, involving the balance between the effective interactions at the level of next-nearest and third-nearest neighbors. Model calculations are just about accurate enough to indicate that the preference for the fcc structure is consistent with all one can include in the best theoretical descriptions. However, predicting the structures of these crystals is actually a far more difficult problem than predicting the structures of most other solids, simply because the inert gas interactions are weak, and the two alternative structures differ only very slightly in energy. It is interesting to note that no species of any kind has been found that exhibits both close-packed structures. Each atomic or molecular system that forms a close-packed crystal chooses one or the other.

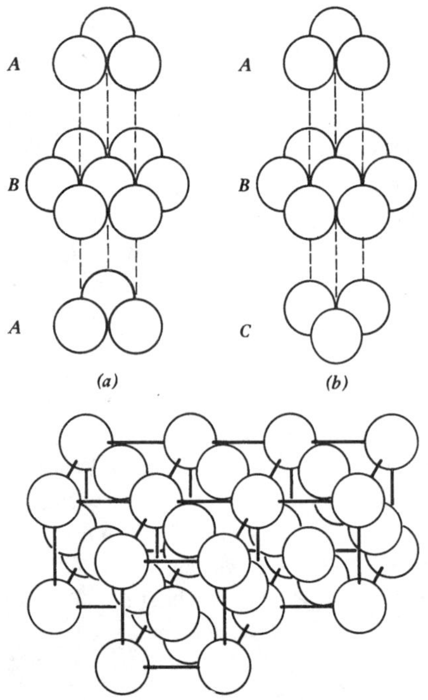

Figure 11.11 Close-packed crystal structures: (*a*) hexagonal close-packed (hcp); (*b*) face-centered cubic (fcc); (*c*) fcc drawn to show the hexagonal structure in a "cubic picture."

Molecular crystals of diatomic or larger molecules are rarely found to be close-packed. The structures are varied, and simple molecules do not necessarily form crystals with simple structures. Structures of some of the crystals of simple linear molecules are shown in Fig. 11.12, and of some hydrocarbons in Fig. 11.13. In all these examples the individual molecules retain their identity, and the forces that hold the molecules together in the crystal are far smaller than the forces that hold the atoms together in the molecules.

As we saw in the previous chapter, the forces between molecules nearly always include some contributions from permanent electric moments associated with the nonspherical shapes and permanent local bond moments of the molecules. Even in a species as symmetrical as the tetrahedral methane molecule, the deviations of the charge distribution from spherical symmetry generate short-range forces on neighboring methane molecules. The existence of these forces is reflected in the cohesive energies listed in Table 11.1 and in the way the molecules pack together in the crystal.

There is a useful (but not universal) rule of thumb for the crystal forms of large molecules. Rodlike molecules such as polyenes, $CH_2{=}CH{-}(CH{=}CH)_n{-}CH{=}CH_2$, tend to lie in close, parallel rows, forming sheets, so that their crystals are frequently flat plates. Flat molecules sometimes tend to stack into columns, which give rise to needlelike crystals, but this is not true for aromatic hydrocarbons. Large saturated molecules tend to arrange themselves so as to put large areas of adjacent molecules near one another, maximizing the near-neighbor attractions. Aromatics tend to arrange themselves so the hydrogens of one molecule are near the π-electron cloud of its neighbors. These generalizations rationalize the primary structure of many crystals; we find rodlike arrays of saturated flat molecules or flat arrays of rodlike molecules, and structures like those of Fig. 11.13 for aromatic molecules. These arrays then combine into secondary structures that reflect the primary crystal structure.

The dynamical properties of molecular crystals reflect those of their component molecules. The vibrational modes of these crystals separate into two quite distinct groups, particularly for large, stiff molecules such as the aromatic hydrocarbons. One set consists of the intramolecular vibra-

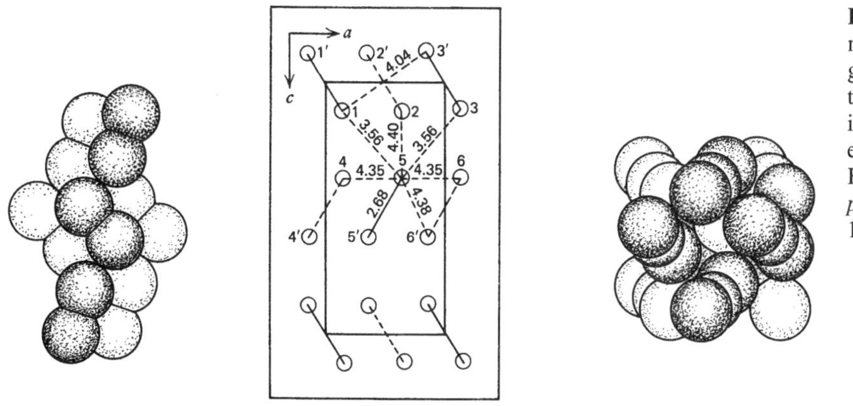

(a) (b) (c)

Figure 11.12 Crystal structures of typical linear molecules. (*a*) The array of molecules in the halogens. (*b*) The projection of the iodine structure on the *ac* plane of the I_2 crystal, showing the shortest intermolecular distances. (*c*) Packing of CO_2 molecules in the carbon dioxide crystal. From A. I. Kitiagorodskii, *Organic Chemical Crystallography,* translated by Consultants Bureau, New York, 1961.

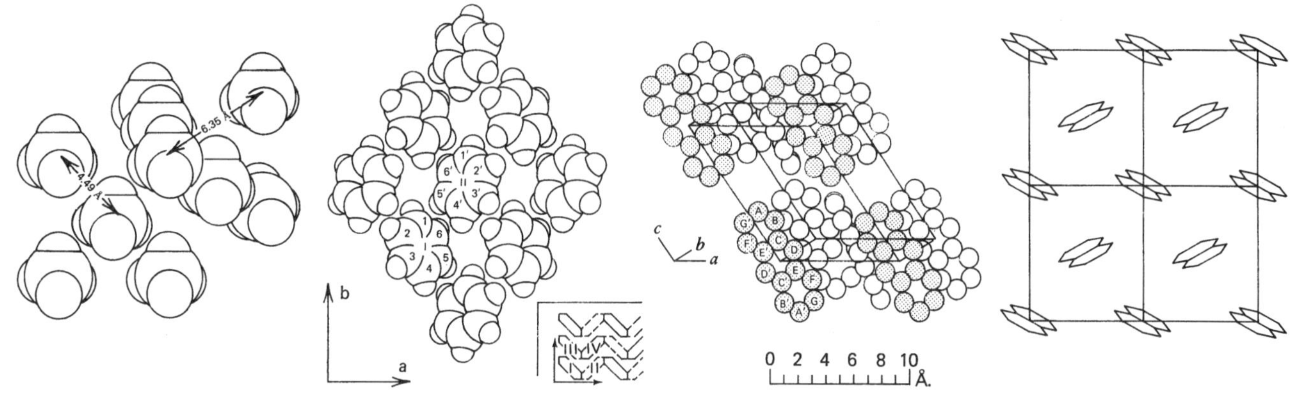

(a) (b) (c) (d)

Figure 11.13 Crystal structures of hydrocarbons. (*a*) Methane. (*b*) Benzene. (*c*) Anthracene, projected along [010]. (*d*) Naphthalene, projected along [001]. Figures are from D. Fox, M. M. Labes, and A. Weissberger, *Physics and Chemistry of the Organic Solid State* (Interscience, New York, 1963); (*a*), (*b*) from Chapter 11, by D. A. Dows; (*c*), (*d*) from Chapter 6, by P. Hartman.

tions, which are very much like the vibrations of the atoms in the free molecules. The other set consists essentially of the vibrations of the molecular centers of mass about their equilibrium positions; these include the *librations,* the rocking motions that would be rotations if the intermolecular forces produced no barrier to molecular rotation in the crystal. The intramolecular vibrations, as one might expect, have the high frequencies characteristic of chemical bonds and atomic masses, with $\tilde{\nu}$ of the order of 1000 cm^{-1}. The vibrations of the molecules as a whole, however, have lower frequencies characteristic of the weaker intermolecular forces and the larger molecular masses. If we think of the entire crystal as a single weakly bound molecule, these vibrations are its normal modes. Since they are characteristic of the crystal lattice as a whole, they are called *lattice modes.* The highest frequencies of lattice modes typically correspond to wavenumbers well below 100 cm^{-1}. The frequencies of the very lowest modes, in principle, extend almost to zero, because they consist of one-half of the crystal oscillating against the other half. Lattice modes occur even in crystals having single-atom unit cells. Lattice modes are the modes of oscillation responsible for the ordinary temperature dependence of the heat capacity of solids. The discussion in Section 2.9 used an oversimplified representation, in which all the lattice modes were taken to be degenerate, but they were nonetheless lattice modes. In Chapter 22 we shall give a full explanation of this relationship, with a more complete discussion of the vibrational spectra of solids.

Electronic excitation in molecular crystals is, in some respects, like that in the free molecules, at least in a dense gas. The electronic spectral band systems of many molecular crystals can be identified with similar transitions in the gaseous molecules. However, the bands in spectra of crystals are usually broader than those of the free molecules (at least at room temperature), partly because of the varying amounts of lattice vibrational energy that may be superposed on the molecular electronic and vibrational energy, and partly because of electronic perturbations of one molecule by another. These perturbations can be systematized, and even estimated rather accurately, because their form is strongly dominated by the symmetry of the crystal. The general phenomenon involves two kinds of symmetries: the translational symmetry of the entire crystal and the local symmetry of the unit cell. We shall discuss the effects of translational symmetry in a later section.

Molecular interactions cause each electronic transition of the free molecule to split into closely spaced multiplets, with as many components as there are distinguishable molecules in the unit cell of the crystal. Thus in the anthracene crystal there are two molecules per unit cell, and the spectral bands responsible for molecular electronic transitions are split into doublets. The reason for this, first elucidated by Davydov, is as follows. Let Φ_A be the wave function for the ground state of molecule A, and Φ_B the corresponding wave function for molecule B. Let $\Phi_A^\star$ and $\Phi_B^\star$ be the wave functions for A and B in their first excited states. The ground state of the unit cell has a wave function that we can approximate well as

$$\Psi_0 = \Phi_A \Phi_B, \tag{11.35}$$

because A and B retain most of their molecular identity. In lowest approximation we could write two degenerate excited-state wave functions, $\Phi_A^\star \Phi_B$ and $\Phi_A \Phi_B^\star$. We can reestablish this equivalence in the usual way by writing the functions

$$\Psi_1 = \frac{1}{\sqrt{2}}(\Phi_A^\star \Phi_B + \Phi_A \Phi_B^\star) \tag{11.36}$$

and

$$\Psi_2 = \frac{1}{\sqrt{2}}(\Phi_A^\star \Phi_B - \Phi_A \Phi_B^\star), \tag{11.37}$$

in each of which A and B are interchangeable. The functions Ψ_1 and Ψ_2 are wave functions for the *excited unit cell* that reflect the equivalence of A and B. But now the functions Ψ_1 and Ψ_2 themselves *are* physically distinguishable because of the difference in sign; hence, Ψ_1 and Ψ_2 describe states with different energies. The splitting between these two excited states is just twice the "interaction energy" between $\Phi_A \Phi_B^\star$ and $\Phi_A^\star \Phi_B$, that is, the Hamiltonian term $\int(\Phi_A \Phi_B^\star) \mathsf{H}(\Phi_A^\star \Phi_B)\, d\tau$. It is known that such crystals retain their molecular character, even in their excited states, when this splitting is small relative to the energy intervals between excited electronic states of the free molecules. To describe the entire crystal of N cells, we multiply Ψ_1 and Ψ_2 by the ground-state wave function for all the other unit cells of the crystal, and then add to this all the other $N - 1$ functions in which the excitation is localized in another unit cell. Then, to normalize, we multiply the sum by $N^{-1/2}$. This wave function represents the state that is usually the first electronically excited state of the crystal.

There are certain exceptions to this simple picture of electronic excitation in molecular crystals. Sometimes an excited molecule can react with a neighboring molecule to form a dimeric molecule. These excited dimers may persist as chemical entities called *photodimers.* Alternatively the excited dimers can radiate their energy and return the system of two molecules to the ground state. Such transient species are called *excimers.* When excimers form, they give a little vibrational energy to the lattice around them in order to gain stability. When they radiate, they generally emit light of considerably longer wavelength than that of the exciting light. This shift occurs because the downward, emissive transition usually takes the dimer to a very repulsive region high on the potential curve of the ground state of the pair. Hence excimer formation and emission shift the light to the red, and transfer large amounts of the remaining excitation energy to the lattice when the ground-state pair flies apart. This is shown schematically in Fig. 11.14 for the diatomic Xe_2 system. This particular pair occurs as an excimer in liquid or high-pressure gaseous xenon when excited by the light of the first allowed electronic transitions of the xenon atom, at 147.0 and 129.5 nm in the vacuum ultraviolet region. The emission occurs throughout the region between about 160.0 and 185.0 nm, also in the vacuum ultraviolet region. Other species that are known to form excimers include small molecules such as argon and krypton (which give rise to bound excited states of Ar_2 and Kr_2, even when their ground states exhibit only van der Waals interactions), as well as larger molecules such as benzene, anthracene, and pyrene. There are also examples of mixed excimers, in which the two partners need not be of the same species.

Photodimers are formed by species such as anthracene and many of its derivatives. The tendency for neighboring molecules to combine when one is photoexcited can be exploited very nicely, particularly when the crystal structure holds the molecules in an orientation especially suited to the formation of one desired product.

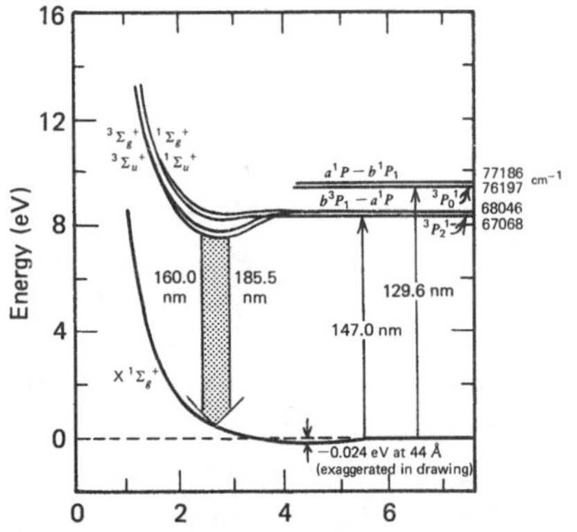

Figure 11.14 Potential curves and absorption and emission frequencies for formation and decay of the excimer Xe_2 in condensed xenon. From R. S. Mulliken, *J. Chem. Phys.* **52,** 5170 (1970).

11.8 Structures of Ionic Crystals

After molecular crystals, the easiest to understand are the *ionic crystals.* The alkali halides are the simplest examples, but many other materials form crystals that are essentially ionic. These include species as CaCN, CaF_2, Na_2O, and most other substances composed of something strongly electropositive combined with something strongly electronegative. In contrast to molecular crystals, it is quite easy to calculate the cohesive energy of ionic crystals because the forces can be approximated well by a simple classical model. For the simplest crystals, even *ab initio* quantum mechanical calculations are feasible and accurate. We begin by describing some of the common structural types found in ionic crystals.

The basic structures of all ionic crystals ensure that each ion is surrounded by a shell of ions of opposite charge. The *rock salt* structure, Fig. 11.15*a,* puts each positive ion in the center of a regular octahedron of negative ions and each negative ion in the center of an octahedron of positive ions. This is *not* the same as a simple cubic lattice; the ions of each kind do not define a simple cubic lattice. (What kind of lattice do they define?) The rock salt structure occurs in all the lithium, sodium, potassium, and rubidium halides, in CsF, in the silver halides (except AgI), and in the alkaline earth oxides and sulfides (except BeO and BeS). This is the most common structure for ionic crystals with the empirical formula MX.

The other cesium halides exhibit a structure called simply the *cesium chloride* or *CsCl* structure, Fig. 11.15*b;* the thallium salts TlCl, TlBr, and TlI also have this structure. The CsCl structure is based on interpenetrating simple cubic lattices of positive and negative ions, with each ion surrounded by a cubic shell of oppositely charged nearest neighbors. We refer to the number of nearest neighbors as the *coordination number,* as in complex ions (see Chapter 9). The rock salt lattice has a coordination number of 6, and the cesium chloride lattice has a coordination number of 8.

Two common ionic structures combine the empirical formula MX and the coordination number 4 associated with tetrahedral coordination. These are the *zincblende* and *wurtzite* structures shown in Figs. 11.15*c* and 11.15*d,* respectively, and named after the two crystalline forms of ZnS. A number of similar compounds, such as CdS and ZnTe, also exhibit both forms. Some other II–VI compounds, for example, BeO and MgTe, exhibit only the wurtzite structure; still others, such as BeS, BeSe, BeTe, only the zincblende structure; and some, as already noted, crystallize in the rock salt structure.

The rock salt, cesium chloride, zincblende, and wurtzite structures are by no means the only structures exhibited by MX compounds. In the tenorite structure of CuO, for example, the oxygens are tetrahedrally 4-coordinated, but the coppers are square planar. There are still other structures, but those illustrated in Fig. 11.15 are the most important.

Two other examples will illustrate the kinds of structures encountered when the number of ions of one kind is unequal to the number of the other. We have chosen two of the most common: Figure 11.16*a* shows the *rutile* structure, named for a form of TiO_2. Figure 11.16*b* shows the *fluorite* structure, named for the crystalline form of CaF_2. In rutile the more highly charged ions are octahedrally 6-coordinated, whereas the ions with lesser charge are at the centers of triangles, and hence 3-coordinated. In fluorite, on the other hand, the ions of lesser charge are tetrahedrally 4-coordinated, whereas the ions of greater charge are 8-coordinated at the centers of cubes. The names "fluorite" and "rutile" are reserved for compounds of the type MX_2, in which the positive ions are more highly charged; the corresponding structures for M_2X compounds are called *antifluorite* and *antirutile.*

A great variety of structures are exhibited by compounds with more complex formulas, but we have given enough

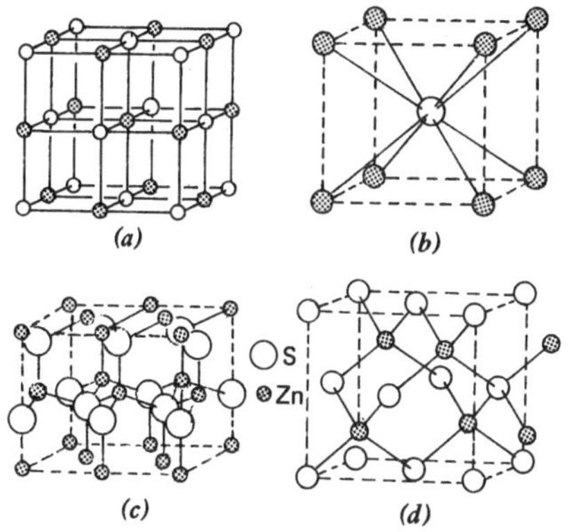

Figure 11.15 Structures of MX-type ionic crystals. (*a*) Rock salt. (*b*) Cesium chloride. (*c*) Wurtzite. (*d*) Zincblende.

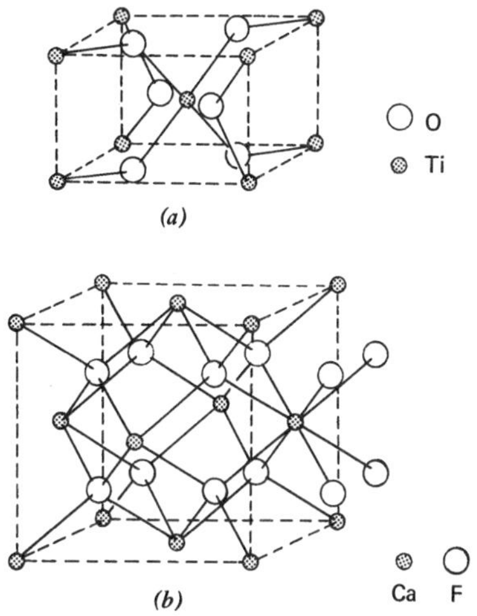

Figure 11.16 Two common structures for MX_2 crystals. (*a*) Rutile. (*b*) Fluorite.

examples for illustrative purposes. The next question we must ask ourselves is, why does a given ionic compound crystallize in one structure rather than another? As with the structures of molecules, the complete answer to this question requires a comparison of the binding energies of the various possible structures. But in ionic solids the simplicity of the forces between the ions makes it possible to guess the correct structure by an elementary but usually accurate method.

In our considerations of ionic diatomic molecules (Section 7.4) we saw that the interaction between two ions is represented well by a point-charge Coulomb force combined with a short-range repulsion. The same model is useful in describing ionic crystals. Since the short-range repulsion falls off quite steeply, to a good approximation we can represent each ion by a charged sphere of definite radius. In other words, just as one can assign approximate "sizes" to neutral atoms (Section 5.5), it is similarly possible to define ionic sizes. One determines the ionic radii in much the same way as covalent radii, by measuring the nearest-neighbor distances in a large number of crystals and selecting a set of radii that most nearly reproduce these distances. Table 11.3 gives such sets of ionic radii calculated by Zachariasen[2] and by Pauling and Sherman,[3] with some atomic radii for comparison.

Given this model, how can we determine the structure that maximizes the binding energy in an ionic crystal? Clearly we want ions to be as close as their radii allow to ions of opposite charge (the radii are of course defined to satisfy this condition) and be as far as possible from ions of the same charge. If we pack ions as though they were charged hard spheres, each ion must necessarily be in contact with its nearest neighbors (of opposite charge). But to minimize the repulsive energy it is best that these nearest neighbors (which have a common charge) not be in contact with one another. The situation in which an ion is in contact both with nearest neighbors (of opposite charge) and with its next-nearest neighbors (of the same charge) is a limit of stability.[4] We illustrate this in Fig. 11.17, with which we introduce the concept of the *radius-ratio effect.*

The stability principle (sometimes called *Pauling's first stability criterion*) that we are now proposing is this: Consider a "central" ion in an ionic crystal; the most stable structure is that which maximizes the number of its nearest-neighbor ions of opposite charge, subject to the condition that the central ion be "in contact" with all these nearest neighbors. Here "in contact" means that the average distance between the nucleus of any nearest neighbor and the nucleus of the central ion is equal to the sum of their ionic radii. Thus, simple geometry shows that a square of A ions surrounding a smaller B ion satisfies the limiting condition

$$\sqrt{2}R_A \leq R_A + R_B, \quad \text{or} \quad \frac{R_B}{R_A} \geq \sqrt{2} - 1 = 0.414.$$

The equality applies to the case shown in Fig. 11.18*a*, in which all the possible contacts between ions are made. By

Table 11.3 Ionic and Atomic Radii of Selected Elements (Å)

	Pauling–Sherman	Zachariasen	Corresponding Neutral Atom[a]
Li^+	0.60	0.68	1.45
Na^+	0.95	0.98	1.80
K^+	1.33	1.33	2.20
Rb^+	1.48	1.48	2.35
Cs^+	1.69	1.67	2.60
Be^{2+}	0.31	0.39	1.05
Mg^{2+}	0.65	0.71	1.50
Ca^{2+}	0.99	0.98	1.80
Sr^{2+}	1.13	1.15	2.00
Ba^{2+}	1.35	1.31	2.15
F^-	1.36	1.33	0.50
Cl^-	1.81	1.81	1.00
Br^-	1.95	1.96	1.15
I^-	2.16	2.19	1.40
O^{2-}	1.40	1.40	0.60
S^{2-}	1.84	1.85	1.00
Se^{2-}	1.98	1.96	1.15
Te^{2-}	2.21	2.18	1.40

[a] From J. C. Slater, *J. Chem. Phys.* **41,** 3199 (1964); these are slightly different from those of Table 5.2, by Pauling, and are not as widely used, even though they are more recent. The differences reflect the arbitrariness and ambiguities associated with defining atomic and ionic radii.

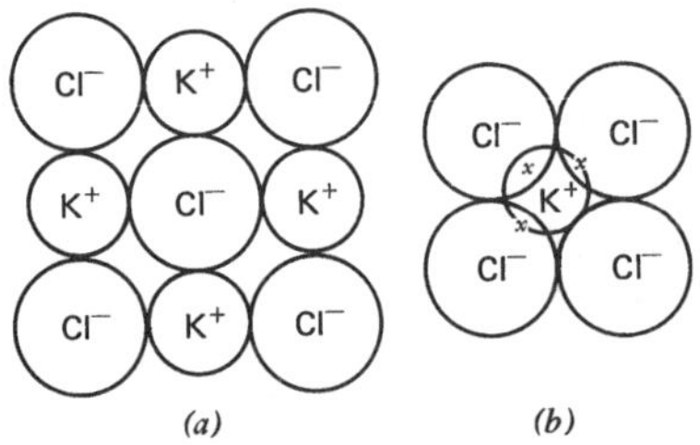

Figure 11.17 The radius-ratio problem. (*a*) Packing of KCl in one layer of a rock salt structure, stable for this salt; $R(K^+) = 1.33$ Å, $R(Cl^-) = 1.81$ Å, so the Cl^- ions around each K^+ ion can all touch the K^+ ion without touching one another. (*b*) Packing of KCl in a CsCl structure, unstable for this species. Here, even with the Cl^- ions touching one another, only three of the four required by the structure can simultaneously be in contact (at the points marked *x*) with a given K^+ ion in the next layer; thus Cl^-—Cl^- repulsion would be high, K^+—Cl^- attraction low, and the structure is unstable.

[2] W. H. Zachariasen, *Z. Krist.* **80,** 137 (1931).

[3] L. Pauling, *Proc. Roy. Soc.* (London) **A114,** 181 (1927); L. Pauling and J. Sherman, *Z. Krist.* **81,** 1 (1932).

[4] This condition implies that the truly close-packed structures of Fig. 11.9—hcp and fcc—*cannot exist* in ionic crystals, because the next-nearest neighbors touch and the repulsive Coulomb forces are thus too strong for stability.

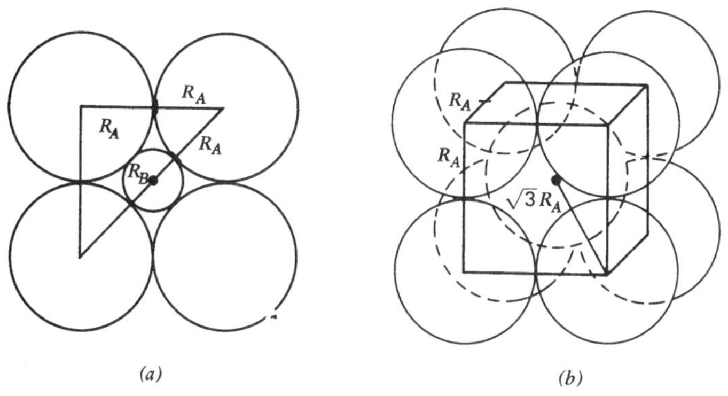

Figure 11.18 Radius-ratio limits for (*a*) square planar and (*b*) cubical nearest-neighbor geometries.

convention we write the radius ratio with the smaller radius, here R_B, in the numerator. The square geometry of this condition is that of a rock salt crystal, so that salts of equally charged ions with radius ratios greater than 0.414 can be expected to have rock salt structures—unless they can also satisfy the criterion for 8-coordination, as indicated in Fig. 11.18*b* for the CsCl structure. The condition for stability of this structure is

$$\sqrt{3}R_A \le R_A + R_B, \quad \text{or} \quad \frac{R_B}{R_A} \ge \sqrt{3} - 1 = 0.732.$$

Where both criteria are satisfied, the CsCl structure is more stable than the rock salt structure, since each ion has eight nearest neighbors rather than six. Thus, for radius ratios between 0.732 and 1, we expect CsCl structures; for radius ratios between 0.414 and 0.732, we expect rock salt structures. If the radius ratio is below 0.414, we expect coordination numbers of 4 (as in the ZnS structures) or 3.

The minimum radius ratios for several structures are given in Table 11.4, where "coordination polyhedron" refers to the arrangement of nearest neighbors about a given central ion. Note that the radius-ratio limit for the most stable 5-coordinate arrangement, a trigonal bipyramid, is that of the equilateral triangle, 0.155. The tetrahedron has a higher limit of 0.225 (see Problem 12), so that any truly ionic species that might be expected to 5-coordinate would really exist as a 4-coordinate tetrahedron (unless it had a ratio as large as 0.414, in which case it would occur as an octahedron).

Similar limiting conditions can be derived for crystals of other types. For example in $(B^+)_2\,A^{2-}$ crystals the fluorite structure is the more stable when $R_B/R_A > 0.73$, but the rulite structure is predicted to be more stable for lower ratios.

The radius-ratio rule must be used with great caution, and generally more in the spirit of a first guess than as a predictive tool. The radius-ratio conditions do not even apply to all the $A^+\,B^-$ crystals that we might consider ionic. The lithium halides all have rock salt structures, but only the fluoride has a radius ratio $R(Li^+)/R(A^-) > 0.414$. This is true for both the Pauling-Sherman and Zachariasen values of ionic radii. Similarly, KF, KCl, RbF, RbCl, RbBr, and CsF all have rock

Table 11.4 Minimum Radius Ratios for Various Coordination Polyhedra[a]

Polyhedron	Coordination Number	Minimum Radius Ratio
Cubo-octahedron	12	1.000
Cube	8	0.732
Square antiprism	8	0.645
Octahedron	6	0.414
Tetrahedron	4	0.225
Triangle	3	0.155

[a] From A. F. Wells, *Structural Inorganic Chemistry,* 2nd ed. (Oxford University Press, Oxford, 1950).

salt structures but radius ratios $R_B/R_A > 0.732$. (See for example, J. K. B. Burdett, *Chemical Bonding in Solids,* cited in the references at the end of this chapter.) Attempts have been made to define other sets of ionic radii that satisfy radius-ratio rules, but these radii tend to improve agreement near one ratio boundary at the expense of agreement at the other boundary. We shall find other examples later (Section 11.10) where the radius-ratio rules are not met by presumably ionic crystals.

11.9 Binding Energy of Ionic Crystals

The binding energy of an ionic crystal, which is commonly called the *cohesive energy* or *lattice energy,* is a readily calculable quantity. One can write it, to an excellent approximation, in terms of a sum of pair interactions; and if one is clever about it, this sum can quite easily be evaluated accurately.

Let us suppose, as in Section 7.4, that the pair interaction between ions consists of a point-charge Coulomb interaction and a short-range repulsion. For simplicity we shall neglect polarization (which, because of the symmetric environment around each ion, is much less in crystals than in ionic molecules), van der Waals forces, and other weak interactions, but these could readily be included in the model. The repulsive part of the potential energy can be approximated as an exponential function.[5] The potential energy between two ions (pair potential) is then given by the *Born–Mayer* potential

$$V(R_{ij}) = \frac{Z_i Z_j e^2}{4\pi\epsilon_0 R_{ij}} + \lambda_{ij} e^{-R_{ij}/\rho}, \tag{11.38}$$

where R_{ij} is the distance between ions i and j, Z_i and Z_j are the ionic charges (in units of e), λ_{ij} is in general different for each kind of ionic pair, and ρ is the same for all pairs.

The total potential energy of the crystal is obtained by summing the pair potential $V(R_{ij})$ over all i, j and then dividing by 2 (to avoid counting each pair twice):

$$U = \frac{1}{2}\sum_i \sum_j \left(\frac{Z_i Z_j e^2}{4\pi\epsilon_0 R_{ij}} + \lambda_{ij} e^{-R_{ij}/\rho} \right). \tag{11.39}$$

The evaluation of this sum is then easily carried out for any particular lattice, provided that one knows the values of the λ_{ij} and ρ. To take a concrete example, let us consider the rock salt lattice, Fig. 11.15*a*. We can scale the interaction in terms of the equilibrium nearest-neighbor distance R_0 (which in this case is half the unit cell length). If we start with a cation as origin, there are six anions at a distance R_0, twelve cations at $\sqrt{2}R_0$, eight anions at $\sqrt{3}R_0$, and so forth; with an anion as origin, we obtain the same distances with cations and anions exchanged.

The Coulomb potential sum in Eq. 11.39 is easily evaluated, since R_0 can be factored out to leave only a numerical sum. Consider a crystal with the rock salt lattice and containing $2N$ ions (or N ion pairs). Since each ion (whether cation or anion) has exactly the same Coulomb interaction energies with its nearest neighbors, next-nearest neighbors, and so on, we can replace the double sum over i, j with $2N$ identical sums over j:

$$U_{\text{Coul}} = \frac{1}{2}\sum_i \sum_j \frac{Z_i Z_j e^2}{4\pi\epsilon_0 R_{ij}} = \frac{1}{2}\left(2N \sum_j \frac{Z_i Z_j e^2}{4\pi\epsilon_0 R_{ij}} \right)$$
$$= \frac{Ne^2}{4\pi\epsilon_0} \sum_j \frac{Z_i Z_j}{R_{ij}}. \tag{11.40}$$

Since Z_+ and Z_- have the same magnitude Z in the rock salt lattice, we can write

$$U_{\text{Coul}} = \frac{NZ^2e^2}{4\pi\epsilon_0}\left(-\frac{6}{R_0} + \frac{12}{\sqrt{2}R_0} - \frac{8}{\sqrt{3}R_0} + \cdots \right)$$
$$= -\frac{NZ^2e^2}{4\pi\epsilon_0 R_0}\left(6 - \frac{12}{\sqrt{2}} + \frac{8}{\sqrt{3}} - \cdots \right) \equiv -\frac{NMZ^2e^2}{4\pi\epsilon_0 R_0}, \tag{11.41}$$

[5] As we noted in Section 7.4, the screened Coulomb potential

$$V_{\text{rep}}(R_{12}) = \frac{Be^{-R_{12}/\rho}}{R_{12}}$$

is more accurate.

Table 11.5 Madelung Constants for Several Lattice Types[a]

Lattice	Example	M	M/ν	Average Coordination No.
Cesium chloride	CsCl	1.7627	0.8813	8
Rock salt	NaCl	1.7476	0.8738	6
Fluorite	CaF_2	2.5194	0.8398	$5\frac{1}{3}$
Wurtzite	ZnS	1.641	0.820	4
Zincblende	ZnS	1.6381	0.8190	4
Rutile	TiO_2	2.408	0.803	4
Cuprite	Cu_2O	2.0578	0.6859	$2\frac{2}{3}$

[a] The Madelung constant is the numerical factor M obtained when one evaluates the Coulomb energy in the form

$$U_{\text{Coul}} = -\frac{NMZ_+Z_-e^2}{4\pi\epsilon_0 R_0},$$

where R_0 is the nearest-neighbor distance. In the fourth column, ν is the number of ions per "molecule" (formula unit).

where the constant M, equal to the numerical sum in parentheses, is known as the *Madelung constant.* This constant has a value of 1.747558 for the rock salt lattice. For any other lattice one obtains an expression of the same form as Eq. 11.41, but with a different sum and value of M. The evaluation of Madelung constants is tedious, since these sums converge very slowly, but the calculation only has to be done once for a given lattice. Values of the Madelung constant[6] for a number of common lattices are given in Table 11.5. (The last two columns of Table 11.5 will be considered later.)

The calculation of the repulsive energy is somewhat more complicated. Since there are only two kinds of ions in the rock salt lattice, we have three λ_{ij}'s: λ_{+-}, λ_{++}, and λ_{--}; we cannot in general make the assumption of Eq. 11.40, that cations and anions have the same total energy. There are N cations and N anions in our crystal, so we write the repulsive part of Eq. 11.39 as

$$U_{\text{rep}} = \frac{1}{2}\sum_i \sum_j \lambda_{ij} e^{-R_{ij}/\rho}$$
$$= \frac{N}{2}\left[\sum_j (\lambda_{ij} e^{-R_{ij}/\rho})_{i=\text{cation}} + \sum_j (\lambda_{ij} e^{-R_{ij}/\rho})_{i=\text{anion}} \right]$$
$$= \frac{N}{2}[6\lambda_{+-}e^{-R_0/\rho} + 12\lambda_{++}e^{-\sqrt{2}R_0/\rho}$$
$$+ 8\lambda_{+-}e^{-\sqrt{3}R_0/\rho} + \cdots + 6\lambda_{+-}e^{-R_0/\rho}$$
$$+ 12\lambda_{--}e^{-\sqrt{2}R_0/\rho} + 8\lambda_{+-}e^{-\sqrt{3}R_0/\rho} + \cdots]$$
$$= 6N\left[\lambda_{+-}e^{-R_0/\rho} + (\lambda_{++} + \lambda_{--})e^{-\sqrt{2}R_0/\rho} \right.$$
$$\left. + \frac{4}{3}\lambda_{+-}e^{-\sqrt{3}R_0/\rho} + \cdots \right]. \tag{11.42}$$

[6] The actual value of M depends on how it is defined: some authors subsume Z_+Z_- into M, and others use a different unit length than R_0 (the unit cell length, for example).

Bear in mind that the repulsive energy falls off very rapidly with distance: In a typical ionic crystal ρ is only about 0.1 R_0. If λ_{+-}, λ_{++}, and λ_{--} are all approximately the same, the ratio between second-nearest-neighbor and nearest-neighbor terms in Eq. 11.42 should be

$$\frac{\lambda_{++}+\lambda_{--}}{\lambda_{+-}}e^{-(R_0/\rho)(\sqrt{2}-1)} \approx 2e^{-10(\sqrt{2}-1)} \approx 0.032;$$

for the next term the ratio is only 0.0009, and the series converges rapidly. To a good approximation (better than 4%), then, we can neglect all but nearest-neighbor repulsions in calculating the lattice energy. If z is the number of nearest neighbors (6 for the rock salt lattice), we can write

$$U_{\text{rep}} \approx zN\lambda_{+-}e^{-R_0/\rho}. \qquad (11.43)$$

We shall use this approximation to simplify certain later results, but as with the Madelung constant the complete sum can be used in exact calculations.

Combining Eqs. 11.41 and 11.43, we have, for the total potential energy of an ionic crystal,

$$U = -\frac{NMZ_+Z_-e^2}{4\pi\epsilon_0 R_0} + zN\lambda_{+-}e^{-R_0/\rho}, \qquad (11.44)$$

where Z_+ and Z_- are the magnitudes of the positive and negative charges. Since R_0 is the equilibrium value of the nearest-neighbor distance, the energy U must be a minimum with respect to R; we therefore have

$$\left(\frac{\partial U}{\partial R}\right)_{R=R_0} = \frac{NMZ_+Z_-e^2}{4\pi\epsilon_0 {R_0}^2} - \frac{zN\lambda_{+-}e^{-R_0/\rho}}{\rho} = 0. \qquad (11.45)$$

Combining these two equations to eliminate the exponential, we obtain

$$U = -\frac{NMZ_+Z_-e^2}{4\pi\epsilon_0 R_0}\left(1-\frac{\rho}{R_0}\right), \qquad (11.46)$$

Note that this result assumes only that the total repulsive energy is of the form 11.43, varying exponentially with R_0; the actual value of z drops out of the calculation. To calculate U, we still need to know the value of ρ, and for this we require thermodynamics; such a derivation will be found in Section 14.9, including numerical results for the NaCl lattice. (The latter section also contains a description of how lattice energies are obtained experimentally.) The values obtained for ρ in various crystals cluster around 0.33 Å.

Let us now look again at the Madelung constant M, whose calculation we passed over rather perfunctorily above. The problem of finding M is merely one of adding the positive and negative contributions to the Coulomb energy due to interactions with more and more distant neighbors, until the sum has converged to the desired accuracy. The most obvious, naive way to do this sum would be first to add the contributions from the nearest neighbors, then the contributions of the next-nearest neighbors, and so on. Unfortunately this is a very inefficient method, because the terms of the series alternate in sign, and although the distances increase, so do the numbers of contributions to each successive term. The most straightforward way of getting round the convergence problem was presented by H. M. Evjen in 1932. Rather than taking the discrete charges of successive shells of neighbors as the terms of the sum, one breaks the sum into groups of ions (or parts of ions) with zero total charge. To evaluate the Madelung constant for the rock salt crystal, one makes a first approximation by combining all the charges within the unit cell cube of Fig. 11.15*a:* The six nearest-neighbor ions (on the cube faces) are half within the cube, the 12 next-nearest-neighbors (cube edges) one-fourth within, and the eight ions in the next shell (cube corners) one-eighth within. Thus the first estimate of the Coulomb energy for the rock salt lattice is, by analogy with Eq. 11.41,

$$\begin{aligned}(U_{\text{Coul}})_0 &= \frac{NZ^2e^2}{4\pi\epsilon_0}\left[-\frac{6\left(\frac{1}{2}\right)}{R_0}+\frac{12\left(\frac{1}{4}\right)}{\sqrt{2}R_0}+\frac{8\left(\frac{1}{8}\right)}{\sqrt{3}R_0}\right] \\ &= -\frac{NZ^2e^2}{4\pi\epsilon_0 R_0}\left(3-\frac{3}{\sqrt{2}}+\frac{1}{\sqrt{3}}\right) \\ &= -\frac{NZ^2e^2}{4\pi\epsilon_0 R_0}(1.4560), \end{aligned} \qquad (11.47)$$

corresponding to a Madelung constant of 1.4560. If one carries out the same process for a cube with twice the edge length, one obtains a Madelung constant of 1.750; successive calculations of this type converge on the limiting value of 1.747558. There are other methods that converge even more rapidly, and these have been used in calculating the values of M in Table 11.5. Note that M/ν, the Madelung constant per ion, increases with the number of nearest neighbors; this is consistent with our earlier discussion of Pauling's first stability criterion.

The cohesive energies calculated by this simple model from measurements of lattice constants, structures, and compressibilities are remarkably close to the experimental values for the alkali halides, the silver halides, and the chlorides, bromides, and iodides of thallium and copper. The largest discrepancies among these compounds are about 4–6% in the copper halides, for which one might question the model because of covalent contributions.

Calculations of the cohesive energy of alkali halides from the Hartree–Fock self-consistent-field model have also been quite successful; the results for NaCl and KCl are given in Table 11.6. These are *ab initio* calculations, not based on fits to empirical parameters. The results indicate that the Hartree–Fock independent-particle model is a good one for alkali halide crystals, in which one might expect the electrons to be rather well localized around individual atoms. This conclusion justifies the use of independent-particle models for covalent and metallic crystals, where localization and therefore correlation should be even less important than in ionic crystals.

Table 11.6 Comparison of Theoretical and Experimental Values for Cohesive Energies and Compressibilities of Alkali Halides[a]

	NaCl		KCl	
	Theoretical	**Experimental**	**Theoretical**	**Experimental**
Lattice constant (Å)	5.50	5.58	6.17	6.23
Cohesive energy (eV)	7.89	7.87	7.19	7.08
Compressibility (10^{-10} Pa^{-1})	4.6	3.3	6.0	4.8

[a] Based on the calculations of P.-O. Löwdin, *Ark. Mat. Astron. Fysik* **35A,** No. 9, 30 (1947), and cited by C. Kittel, *Introduction to Solid State Physics* (Wiley, New York, 1953).

11.10 Covalent Solids

We now turn to the solids in which bonding between the atoms is essentially the same as bonding in covalent molecules. At one extreme are the homopolar or homonuclear solids such as diamond and germanium, in which the atoms are all of the same kind. From these there is a regular gradation of properties in crystals made up of increasingly dissimilar atoms. An example is the progression from Ge to the binary compound gallium arsenide (GaAs), to zinc selenide (ZnSe), to copper bromide (CuBr). Qualitatively, this takes us through the range from near one extreme of covalency to near the other extreme of ionic character. Naturally, this parallels the progression we saw in Section 7.10 for series such as C_2, BN, BeO, and LiF, which first illustrated the regular transition from covalent to ionic bonding.

Let us examine the diamond crystal, Fig. 11.3. Each carbon atom in the lattice occupies the center of a regular tetrahedron, with another carbon atom at each vertex. The crystal is therefore a three-dimensional molecule, but with no limits on its extension; in other words, diamond is a three-dimensional polymer of carbon atoms. The molecule neopentane, $C(CH_3)_4$, might be thought of as a precursor, with "ends" or bond termini stabilized by hydrogen atoms.

The bonds in diamond are very similar to the carbon–carbon bonds of saturated hydrocarbons (cf. Section 9.2). We can think of them as sp^3 hybrids rather localized along the internuclear axes. However, just as the electrons in extended molecules are actually free to travel throughout the molecule, the electrons in diamond are free to travel throughout the entire crystal. Again, as with molecules, we can think of the electrons as occupying extended molecular orbitals. With the possible exception of a (relatively) minute number of localized sites at the surface, all the bonding orbitals of diamond are fully occupied; this statement must be true whether we conceive of the crystal in terms of localized bonds or delocalized bonding orbitals. Hence, although an electron is free to move from one localized orbital φ_a to another φ_b, an electron must simultaneously move out of φ_b and into φ_a for the Pauli exclusion principle to be obeyed. The same statement can be put in terms of delocalized orbitals: The filled condition of the bonding orbitals assures that for any electron moving in the direction **r** there is a complementary electron moving in the direction −**r**, with the same average rate $\langle | d\mathbf{r}/dt | \rangle$. (We neglect transient fluctuations, which average to give no net contribution to the motion.)

The implication of the Pauli principle is stringent: It says that electrons in covalent solids move only in a correlated way, so that there is no way for current to flow when a small voltage is applied to a crystal such as diamond. In other words, covalent crystals are *insulators* because their bonding orbitals are all fully occupied and there are no states readily available to allow electrons to be accelerated along an electric field. Electrons can be free to move from one site to another without a compensating "back-flow" only if they are excited up to unoccupied orbitals. However, this requires considerable energy, approximately 7 eV in diamond, which is far, far more than either a typical thermal energy of an electron or an amount of energy an electron in a solid can acquire from even a large external field. Consequently, we can expect covalent crystals to become conducting only when they are excited by radiation energetic enough to promote electrons to normally unoccupied orbitals—that is, to be *photoconductors.* Some photoconductors can be excited by visible light; many more require ultraviolet radiation.

Now compare diamond and germanium (Fig. 11.19*a*) with their "III–V" counterparts, the structurally similar forms of BN and GaAs. Both borazon and gallium arsenide (Fig. 11.19*b*) are tetrahedral like diamond: Each boron has four nitrogen nearest neighbors; each gallium has four arsenic nearest neighbors. The chemical difference between BN or GaAs and diamond, silicon, or germanium is that we have replaced a uniform group IV solid with group III and group V elements from the same row of the periodic table in equal numbers. This produces what we call a III–V compound. They are true compounds, in the sense that their stoichiometry is 1:1 and their structures are perfectly regular; there are no III–III or V–V nearest neighbors in a perfect crystal of a III–V compound. The difference between diamond and borazon, or between Ge and GaAs, in terms of electronic structure, is precisely analogous to the difference between the C_2 and BN diatomic molecules.

When we come to zinc sulfide, we seem to have reached a paradox. Here we are calling ZnS an example of a covalent solid with the tetrahedral structure of the diamond lattice, since it is isoelectronic with Ge. But in Section 11.8 we cited ZnS as meeting the criteria of the ionic model because it assumes a tetrahedral structure with a radius ratio of 0.40

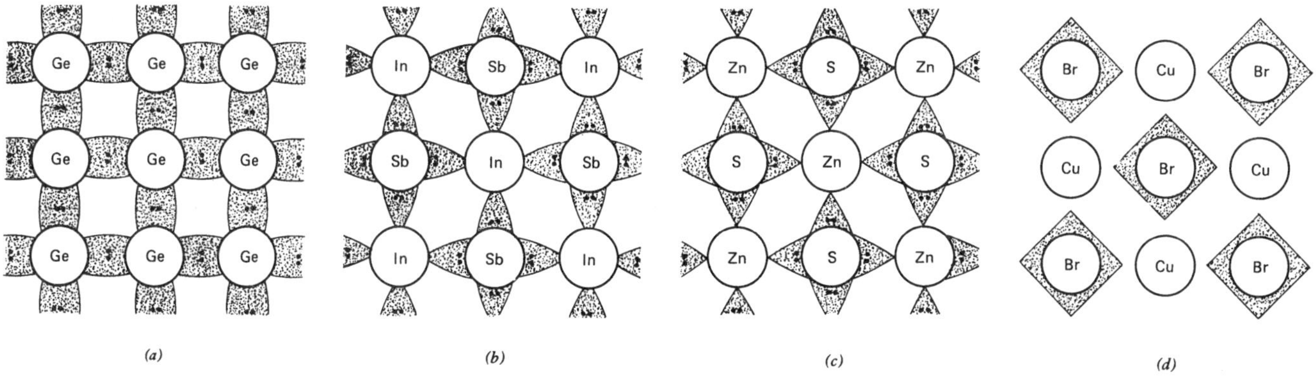

Figure 11.19 Projections of the (schematic) electron distributions in (*a*) germanium, (*b*) indium antimonide (which has the same structure as gallium arsenide), (*c*) zinc sulfide, and (*d*) copper(I) bromide. All four are tetrahedral structures. Think of the two nearest neighbors above and below the central atom as projecting up out of the page, and the two neighbors at the sides as projecting down below the page. From J. M. Ziman, *Theory of Solids* (Cambridge University Press, London, 1964).

(see Table 11.4). Is one of these interpretations more nearly correct than the other? Probably not; ZnS does have a closed-shell electronic structure, and a zincblende or wurtzite lattice, in either model, Fig.11.19*c*. If we were to use an ionic model to solve the Schrödinger equation for the ZnS crystal, the model would take as its starting point the wave functions for Zn^{2+} and S^{2-} ions, and then allow these functions to be distorted by the presence of the other ions. This corresponds to allowing the charge on the negative ions to be polarized in the directions of the Zn^{2+} ions—in other words, to be localized along the tetrahedral bond directions of the diamond structure. If we were to solve the Schrödinger equation for ZnS as a covalent crystal, we would start by generating some form of localized bond orbital between Zn and S—which would lie primarily in the neighborhood of the sulfur atom, with its maximum density along the Zn–S axis. As with other approximations we have described, the two pictures become quite equivalent as one refines them. In truth, ZnS is almost an ideal example of the fortunate kind of intermediate case: In this instance the models for both extremes describe the intermediate case correctly.

Even copper(I) bromide, CuBr, the next member of the sequence from Ge, GaAs, and ZnS, has the zincblende structure, as shown in Fig. 11.19*d*. Here, however, the covalent model and the ionic model do not give the same results. The radius ratio $R(Cu^+)/R(Br^-)$ is approximately 0.5, which implies that CuBr should have the octahedral arrangement of the rock salt crystal. This case shows the limitations, interestingly, of the *ionic* model, rather than of the covalent model—the opposite to what one might expect for an A^+B^- compound. One might think of rationalizing this disparity by invoking the ideas of the ligand field model (Sections 9.5–9.9) to account for stabilization of a tetrahedral rather than an octahedral geometry around Cu^+. Unfortunately, this approach offers no help: Since Cu^+ has a filled 3*d* shell, neither a tetrahedral nor an octahedral field causes a net energy change of the 3*d* shell, at least so long as one neglects mixing with higher, unoccupied orbitals. As much energy is lost by the electrons in orbitals destabilized by the ligands as is gained by the orbitals whose energy is lowered. If one admits higher orbitals and electron correlation into the ionic description, one naturally moves this model closer to reality, but at the expense of simplicity. Most self-consistent models become more and more accurate if they are refined indefinitely, but at some point they lose the utility of the simple physical concept inherent in the simplest form of the model.

Our discussion of covalent and ionic solids has had molecular orbitals, localized or delocalized, at its conceptual basis. Such a picture is useful for qualitative considerations, and for quantitative calculations as well. However, some of the properties of solids, especially the relationship between insulators and metallic solids and the origins of the electronic properties of metals, are clearer in terms of another set of models, the free-electron and band models. We now turn to this subject; once we have developed it for metals, we shall return to add a little reinterpretation to our picture of molecular, ionic, and covalent solids.

11.11 The Free-Electron Theory of Metals

The most significant characteristics of metals are the properties that we associate with mobile electrons: high electrical conductivity, high reflectance, and strong bonding coupled with rather easy deformability. The conductivity comes directly from the ease with which electrons in metals can move under the influence of a static or slowly oscillating electric field. The reflectance involves the capability of nearly free electrons to be accelerated by a high-frequency electromagnetic field and then reradiate the energy absorbed; since the electrons very near the surface of a metal absorb and reradiate virtually all the energy of an incident light beam, the metal is opaque and (because of relations governing the directions of reradiation, which we shall not derive) gives rise to *specular* or mirror-like reflection, with an angle of reflection equal to the angle of incidence. The strong bonding coupled with deformability is the consequence of the relatively nondirectional character of bonds in metals, by comparison with the bonds in covalent solids. That is, when one changes a "bond angle" (or on a macroscopic scale, when one introduces a shear deformation), the resulting change in crystal energy is much smaller than the corresponding effect in a covalent solid.

Ease of motion and lack of localized character are two properties suggesting that electrons in metals might be considered as particles free to move everywhere inside the metal. This picture is the starting point for the *free-electron model,* which we shall quickly refine into the more general band model of electrons in solids.

The starting point for the free-electron model is the notion that each valence electron moves independently of all the others, and that the positive atomic cores—composed of nuclei and tightly bound core electrons—can be treated approximately as a continuous background of massive positive charge density. Thus, each electron is treated as a particle in a three-dimensional box the size of the metal crystal. From Eq. 3.119 we know that the energy levels of a particle of mass m in a cubical box of side L (within which the potential energy is constant) are given by

$$E = \frac{\hbar^2}{2m}\left(\frac{\pi}{L}\right)^2 (n_x^2 + n_y^2 + n_z^2) = \frac{\hbar^2 \pi^2 n^2}{2mL^2} = \frac{\hbar^2 k^2}{2m}, \tag{11.48}$$

where n_x, n_y, n_z (= 1, 2, . . .) are the quantum numbers for motion in the x, y, z directions, with

$$n^2 \equiv n_x^2 + n_y^2 + n_z^2, \tag{11.49}$$

and k^2 is defined by

$$k^2 \equiv \frac{\pi^2 n^2}{L^2}. \tag{11.50}$$

We can think of the quantity k^2 as the squared magnitude of a vector **k**, with components k_x, k_y, k_z ($k_x \equiv \pi n_x/L$, etc.); the magnitude of **k** has the dimensions of a wavenumber, whereas $\hbar\mathbf{k}$ is the momentum of the free particle. The possible values of **k** define points in what we call a *k space,* which is equivalent to a momentum space. Each point in such a "space" represents a particle's quantized magnitude of momentum rather than its position. (Strictly, the momentum itself is not a constant of motion because if it were, the electron would not be confined; only the magnitude is constant and hence has a corresponding quantum number.)

Now let us determine the *density of states* of the free-electron gas, that is, the number of states per unit volume (in real space) with momentum magnitude between $\hbar k$ and $\hbar k + \hbar\, dk$, or with wavenumber between k and $k + dk$. This is the number of states lying in k space in the spherical shell with radius k and thickness dk. The volume of this spherical shell in k-space is

$$dV_k = 4\pi k^2\, dk. \tag{11.51}$$

However, since k_x, k_y, k_z must all be positive,[7] only one-eighth of this volume actually contains possible states of the system. Equations 11.48–11.50 show that the energy eigenstates correspond to points evenly spaced π/L apart (in a simple cubic lattice) in k space, with a k-space volume of $(\pi/L)^3$ per eigenstate. For each of these points, however, there are actually two possible states of the electron, corresponding to the two values of the spin. We take the real-space volume of our crystal to be L^3. Combining all these results, we find that the number of one-electron states with momentum magnitude between $\hbar k$ and $\hbar k + \hbar\, dk$ per unit volume of real space is

$$G(k)\, dk = \frac{2}{L^3}\cdot\frac{1}{8}\cdot\frac{4\pi k^2\, dk}{(\pi/L)^3} = \frac{k^2}{\pi^2}\, dk. \tag{11.52}$$

The function $G(k)$ is known as the density-of-states function for free electrons; strictly, it is a density of momentum states. We can similarly define a density-of-states function $g(\epsilon)$, where $g(\epsilon)\, d\epsilon$ is the number of one-electron states with energy between ϵ and $\epsilon + d\epsilon$. We differentiate Eq. 11.48,

$$d\epsilon = \frac{\hbar^2}{m} k\, dk, \tag{11.53}$$

and substitute to obtain

$$g(\epsilon) = G(k)\frac{dk}{d\epsilon} = \frac{k^2}{\pi^2}\cdot\frac{m}{\hbar^2 k} = \frac{mk}{\pi^2\hbar^2} = \frac{m^{3/2}\epsilon^{1/2}}{\sqrt{2}\pi^2\hbar^3}. \tag{11.54}$$

If we let $C \equiv m^{3/2}/\sqrt{2}\pi^2\hbar^3$, we can write

$$g(\epsilon) = C\epsilon^{1/2}. \tag{11.55}$$

Thus, the density of momentum states $G(k)$ of a free-electron gas rises as the square of the wavenumber, or the density of energy states $g(\epsilon)$ as the square root of the energy. This behavior is shown in Fig. 11.20, in parts (*a*) and (*b*) of which the area under the curve represents the number of states.

Now consider the metal at 0 K, where the electrons should have no excitation at all. The Pauli exclusion principle requires that no more than two electrons have the same

[7] Recall that we are dealing with standing waves only, and thus with only positive n_x, n_y, n_z.

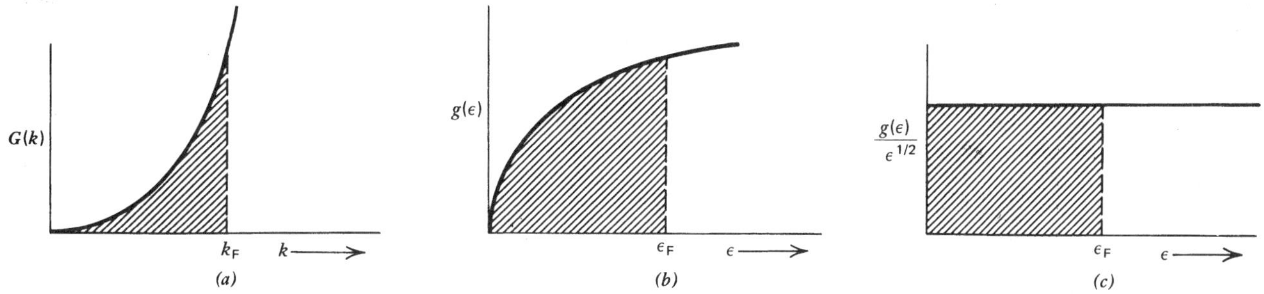

Figure 11.20 The density of states of a free-electron gas (*a*) in momentum or wavenumber space; (*b*) in energy space; (*c*) in energy space, but scaled by $\epsilon^{1/2}$. Hatched areas indicate the occupied levels at $T = 0$ K.

k, and the temperature condition requires that each electron be in as low an energy level as possible. Together these imply that all the levels are doubly occupied, from the lowest upward, until all the electrons have found a state. This means that there is a maximum energy of the occupied states; all states with less energy than this maximum are occupied; states with more energy are empty. This maximum energy of the occupied levels in a very cold metal (or insulator) is called the *Fermi energy,* ϵ_F, and the set of filled levels below ϵ_F is often called the "Fermi sea."

In Part II we shall show how knowledge of the density of states and the fundamental nature of the particles—classical, Fermi–Dirac, or Bose–Einstein—is all we need to know, in principle, to determine the average occupation of every quantum state for a system in thermal equilibrium at a given temperature. At this point we shall merely quote the result, which we can express in the form

$$d\left(\frac{N}{V}\right) = f(\epsilon, T) g(\epsilon)\, d\epsilon, \tag{11.56}$$

where $d(N/V)$ is the number of *occupied* states in the energy range between ϵ and $\epsilon + d\epsilon$, $g(\epsilon)$ is still the density of states (whether or not occupied), and $f(\epsilon, T)$ is called the *distribution function.* In general $f(\epsilon, T)$ is a function of both the energy and the absolute temperature T; it is defined in such a way as to approach unity as $T \to 0$. For the class of particles that includes electrons (called Fermi–Dirac particles or fermions), we shall show in Chapter 21 that

$$f(\epsilon, T) = \frac{1}{e^{(\epsilon-\mu)/k_BT} + 1}, \tag{11.57}$$

where μ is a constant at any given temperature. The value of μ can be evaluated by integration of Eq. 11.56 over all energies, which must give the total number of effectively free electrons per unit volume (for example, one electron per atom for alkali metals). Since $f(\epsilon, T) = \frac{1}{2}$ when $\epsilon = \mu$, we conclude that μ equals the energy at which $f(\epsilon, T)$ has half its maximum value. At absolute zero we have

$$\begin{aligned} f(\epsilon, 0) &= 1 \qquad (\epsilon < \epsilon_F), \\ f(\epsilon, 0) &= 0 \qquad (\epsilon > \epsilon_F), \end{aligned} \tag{11.58}$$

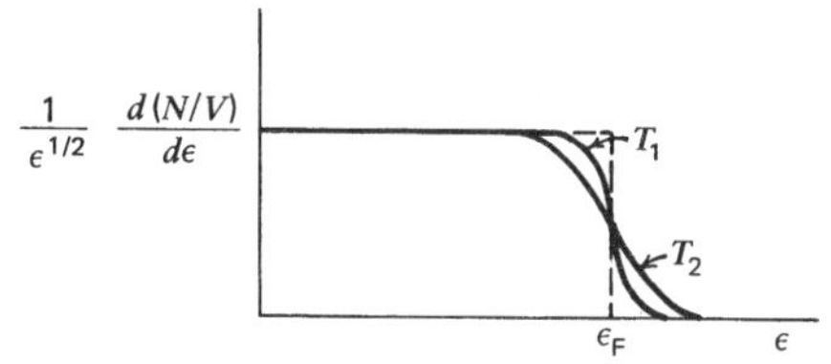

Figure 11.21 The electron distribution in a metal at temperatures above 0 K; $0 < T_1 < T_2$. The dashed line indicates the distribution at 0 K.

which is consistent with Eq. 11.57 only if $\lim_{T\to 0} \mu = \epsilon_F$; at other temperatures it is a good approximation to set $\mu \approx \epsilon_F$, as long as $k_BT << \epsilon_F$ (for most metals ϵ_F is of the order of 5 eV, equivalent to a temperature of about 6×10^4 K). Making this approximation, we can write the distribution of electrons over energy levels as

$$d\left(\frac{N}{V}\right) = \frac{C\epsilon^{1/2}\, d\epsilon}{e^{(\epsilon-\epsilon_F)/k_BT} + 1}, \tag{11.59}$$

using the value of $g(\epsilon)$ from Eq. 11.55.

At low temperatures $(\epsilon - \epsilon_F)/k_BT$ is large and negative for $\epsilon < \epsilon_F$ and large and positive for $\epsilon > \epsilon_F$. Hence $f(\epsilon, T)$ is slightly less than 1 for levels below ϵ_F and goes rapidly to zero as the energy goes above ϵ_F. In Fig. 11.21 we show how the distribution of electrons over energy levels varies with temperature. In a normal metal at room temperature, the distribution differs only slightly from that at 0 K; only a very small fraction of the electrons are normally out of their states in the "Fermi sea," occupying levels above ϵ_F. But it is just these electrons that have left the Fermi sea, and the "holes" they leave behind, that are responsible for ordinary metallic electrical conductivity and the electronic contribution to the heat capacity of metals.

Because electrons excited out of the Fermi sea are sparse relative to the number of levels available for their occupancy, the constraints of the Pauli exclusion principle hardly affect them at all. They can move rather freely under the influence of an applied voltage, and thereby give rise to a current. So also can the holes left in the Fermi sea; they can move freely, just as their electron counterparts can. Electrons and holes

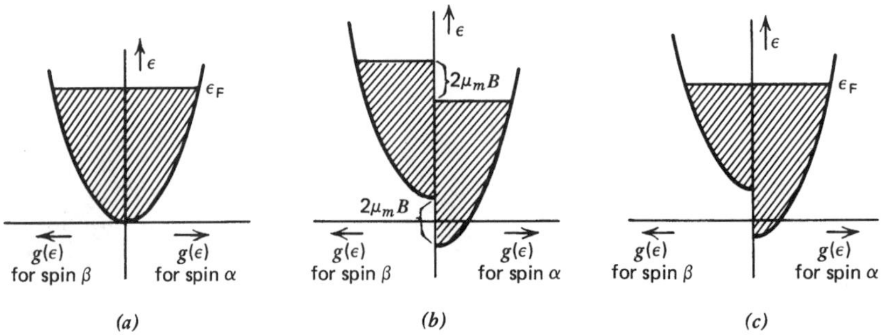

Figure 11.22 The paramagnetism of free electrons in a metal, for $T \approx 0$ K. (*a*) The population distribution for spins α (right side of *y* axis) and β (left side of *y* axis) in the absence of a field: $g(\epsilon) = C(\hbar^2k^2/2m)^{1/2}$, $N_\alpha = N_\beta$. (*b*) The transient population after a magnetic field **B** is applied but before the spins relax to equilibrium,

$$g(\epsilon) = C\left(\frac{\hbar^2 k^2}{2m} \pm \mu_m B\right)^{1/2},$$

$$N_\alpha = N_\beta.$$

(*c*) The population of the spin system in the presence of **B** after equilibrium: $g(\epsilon)$ as in (*b*), $N_\alpha > N_\beta$.

both contribute to conductivity. The relative importance of electrons and holes to the conductivity of a metal depends on detailed properties of the metal, particularly on whether electrons or holes are more easily scattered by the atomic nuclei and core electrons.

When we introduced the heat capacity of solids in Section 2.9, we described the empirical law of Dulong and Petit, which states that many solids have a temperature-independent heat capacity of about 25 J/mol K. We shall see in Chapter 22 how this result can be derived from models (those of Einstein and Debye) in which the heat capacity arises from the vibrational motions of the nuclei in the lattice. In these models the contribution of the electrons to the heat capacity is entirely neglected, and we can now see why. Only those electrons that can absorb energy contribute to the heat capacity, but at ordinary temperatures (as we see in Fig. 11.21) only a small fraction of the electrons, far less than one per atom, have their energies significantly affected by a temperature change. Actually, the Einstein and Debye models predict a constant heat capacity of $3R$ per mole, where $R \equiv N_A k_B = 8.314$ J/mol K because these models neglect electrons as contributors to the storage of energy. Real heat capacities of metals are, on average, slightly higher—Dulong and Petit's own value was equivalent to $3.1R$—because the electronic contribution is not completely negligible.

The model of the nearly filled Fermi sea with a few excited electrons also accounts for the slight paramagnetism of most metals. The overwhelming majority of electrons in the Fermi sea are in states with their spins paired. However, some of the electrons with energies near ϵ_F are unpaired and thus free to orient, so that their spins can orient themselves in the low-energy direction if an external magnetic field is applied. The application of the magnetic field thus causes a redistribution of electrons among α and β spin states until the system has achieved a new equilibrium in which a majority of electrons have spins parallel to the field. One way to look at the effect on the electrons of a uniform external magnetic field **B** is as though it shifts the energy of the entire set of electrons of one spin upward by an amount $\mu_m B$ (where μ_m is the effective electronic magnetic moment), and the energy of the electrons with the opposite spin a like amount downward. The system then comes to a new equilibrium, with both sets of spin states filled *up to the same energy level.* Figure 11.22 shows this process as if it occurred in the two steps just described: first a separation due to the Zeeman effect, and then a redistribution to achieve a common highest filled level. Since there is an excess of one spin over the other, the metal has a net magnetization as a result of this process. As with the heat capacity, however, it is only the small fraction of the electrons with energies near ϵ_F that actually change spin; for $k_B T << \epsilon_F$ this fraction can be shown to be approximately equal to $k_B T/\epsilon_F$. Thus the molar magnetic susceptibility of a metal in the free-electron model is found to be

$$\chi_M = \frac{N_A \mu_m^2}{k_B T_F} \qquad (T_F \equiv \epsilon_F / k_B), \tag{11.60}$$

which, unlike Eq. 9.21, is temperature-independent; most metals do follow Eq. 11.60 approximately. (The quantitative evaluation of the magnetic susceptibilities of metals is considerably more involved than the evaluation of the spin contribution; the orbital motion must also be included, and is often quite complicated.) Some magnetic susceptibilities are given in Table 11.7, to show the range of magnitudes they span.

We shall not treat the problem of the electrical conductivity of metals here, except in outline. The free-electron model would imply that the conductivity is infinite for a perfectly free-electron gas because free electrons could attain

Table 11.7 Typical Molar Magnetic Susceptibilities of Some Elements and Compounds[a]

Element	Susceptibility
Aluminum	+16.5 (× 10^{-6}/mol)
Argon	−19.6
Bismuth	−280.1
Carbon (diamond)	−5.9
Cerium (α phase, 80 K)	+5,160.0
Chromium	+180.0
Chromium(II) chloride, $CrCl_2$	+7,230.0
Copper	−5.46
Copper(II) chloride, $CuCl_2$	+1,030.0
Gadolinium	+755,000.
Helium	−1.88
Hydrogen (*g*)	−3.98
Nitrogen	−12.0
Nitric oxide, NO (147 K)	+2,324.0
Oxygen (*g*)	+3,449.0
Potassium	+20.8
Sodium	+16.0
Water (273 K)	−12.65
(373 K)	−13.09

[a] Positive values correspond to paramagnetic substances and negative values, to diamagnetic substances. In SI units, the magnetic susceptibility is in mol^{-1}; in cgs units, the values are the same but χ_M has units of cm^3/mol.

arbitrarily high velocities under the influence of a constant field. To obtain a finite conductivity the model must be extended to include a mechanism for electrons to lose energy more and more effectively as they are accelerated. In effect, the existence of electrical resistance demands that there be some kind of friction to retard the electrons. This friction can be put into the free-electron theory if one supposes that electrons scatter occasionally from vibrating atoms of the lattice, and that on average the electrons have no velocity component parallel to the electric field after collision. With this addition, the theory can give a good representation of the electrical conductivity of a metal.

The free-electron theory has several major limitations. First, the electrons in a real metal see not the perfectly uniform potential of a particle in a box, but rather the potential of a regular assembly of ion cores and other electrons. Second, the free-electron theory gives no clue about the relationship between metals and other solids. Third, the theory gives a totally inadequate picture of all the properties of metals except those that depend on the excited electrons and corresponding holes that exist near the surface of the Fermi sea at temperatures above 0 K. The stimulus of these limitations led to the development of the *band theory* of solids. We shall not try to pursue the band theory of metals to the point of actually deriving properties, but we shall develop the theory enough to make the model consistent with our previous ideas about the structure of solids, and to show how metals and insulators are related.

11.12 The Band Theory of Solids

The mathematical basis of the band model is the periodicity of the crystal lattice. Periodicity is important not because it is responsible for the properties of solids, but because it furnishes an easy way to solve the mathematical problem of the electronic structure. In fact, the physical processes in an amorphous solid do not differ very much from those in a chemically similar material with a periodic structure. Even properties of liquid metals are much like those of solid metals, properties such as the reflectivities and electrical and thermal conductivities. Low-melting metals such as mercury, gallium, and tin are about as shiny as liquids as they are as solids. (It is difficult to judge this by eye with high-melting metals because they typically emit fairly intense visible light when they are molten.) Nevertheless, the practical and conceptual value of the periodic model is extremely high, and we certainly should not slight the model merely because it is not general enough to describe all condensed materials.

The band theory is an extension of the Hartree–Fock model, and assumes that one can write one-electron wave functions for the electrons in a solid. In essence, we first suppose that we can find *local* solutions to the Hartree–Fock equations—solutions that describe the electrons well when they are in the vicinity of atoms—and then we build linear combinations of these functions that have the periodicity of the lattice. The process is similar to that we introduced in Chapters 6 and 7 for diatomic molecules, but now the number of atomic wave functions combined is essentially infinite.

The most striking consequence of the periodic potential is the appearance of *energy bands.* When we combine two identical atoms into a diatomic molecule, each pair of identical (and therefore degenerate) energy levels of the separated atoms splits into two distinct levels, as illustrated in Figs. 6.9 or 7.14. Much the same thing happens when we combine a large number of identical atoms into a periodic solid; however, since the number of atoms is effectively infinite, each level of the separated atoms now becomes a continuous band of energy levels over some range of energy. More precisely, as the number of states becomes infinite the spacing of the levels becomes indefinitely small, that is, the spectrum becomes more and more nearly continuous. But these bands of energy levels never cover the entire energy spectrum; each kind of atomic orbital defines a sharply-bounded range of energy within which lies a continuum of levels made from those localized orbitals. There are intervals of energy (*gaps*) within which no physically allowable states exist. The formation of energy bands from separated-atom energy levels is illustrated schematically in Fig. 11.23. We shall see how the distribution of allowed and forbidden

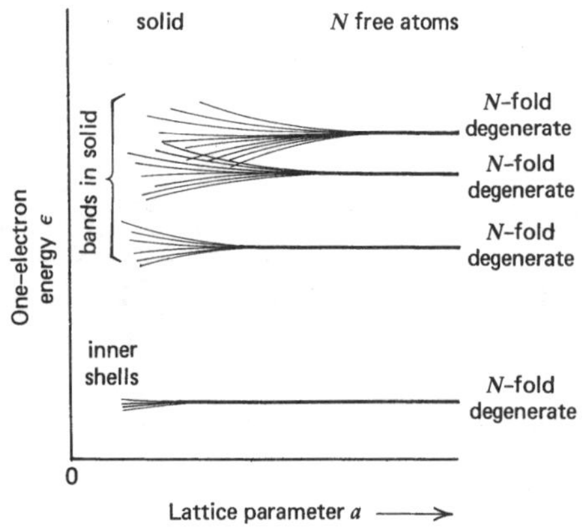

Figure 11.23 Schematic representation of the formation of bands of energy levels as the component atoms approach one another.

energy intervals determines the nature of a solid. Different bands may overlap; for example, the band of metallic Ca composed of atomic 4*s* orbitals overlaps with the band at slightly higher energy composed of atomic 4*p* orbitals.

Some features of the free-electron theory are retained in the band model, in particular the meaning of the wavenumber *k*. The energy spectrum of the band model differs from that of the free-electron model by being discontinuous. Nevertheless, in both models, all values of *k* from zero up to the inverse of the lattice parameter remain physically allowed. Thus it is clear that the electron energy can no longer be such a simple function of *k* as that given for free electrons by Eq. 11.48, and in what follows we shall direct our attention to the derivation of the function $\epsilon(k)$ for a periodic potential. But these remarks are premature; let us proceed to the actual development of the band model in a relatively simple form.

We treat the crystal as made up of mobile but not completely free electrons and a periodic lattice of fixed atomic cores. Let $V(\mathbf{r})$ be the total potential acting on an electron at the point $\mathbf{r}$. Then the periodicity condition requires that the potential must be invariant under translation through a lattice vector $\mathbf{a}$, so that

$$V(\mathbf{r}) = V(\mathbf{r}+\mathbf{a}). \qquad (11.61)$$

Consider now the simple case of a one-dimensional circular lattice, a very large ring with N members. Thus $2\pi/N$ is the angle subtended by the arc between neighbors. The periodicity condition requires that

$$V(\theta) = V\left(\theta + \frac{2\pi}{N}\right), \qquad (11.62)$$

where the angle θ defines the location in the ring. The Schrödinger equation for an electron in this ring is

$$V(\mathbf{r})\left[\frac{\mathrm{p}^2}{2m} + V(\theta)\right]\psi(\theta) = \epsilon\psi(\theta). \qquad (11.63)$$

Since two positions $2\pi/N$ apart are physically equivalent, we must have

$$|\psi(\theta)|^2 = \left|\psi\left(\theta + \frac{2\pi}{N}\right)\right|^2; \qquad (11.64)$$

in other words, rotation of the ring through an angle $2\pi/N$ can at most change the wave function by a phase factor:

$$\psi\left(\theta + \frac{2\pi}{N}\right) = C\psi(\theta), \qquad (11.65)$$

where $C^*C = 1$. There is an additional condition that $C^N = 1$, since N-fold repetition of the rotation through $2\pi/N$ must restore the original state. These conditions are satisfied if C has the form $e^{2\pi i/N}$.

Let us write the wave function in terms of distance; also, let j be an indexing integer that we shall relate to the location of the atoms, and then to a sort of momentum. Call the distance along the ring x and the spacing between neighbors a. The function $\psi(x)$ could therefore have a form as simple as

$$\psi_j(x) = e^{2\pi ijx/Na} \qquad (j = 0, 1, \ldots, N-1), \qquad (11.66)$$

or

$$\psi_k(x) = e^{ikx}, \qquad (11.67)$$

where we have anticipated the following discussion by defining

$$k \equiv \frac{2\pi j}{Na}. \qquad (11.68)$$

But these are merely the wave functions for an electron on a uniform ring. We can make our model much more general, taking into account local structure, because we can multiply $\psi_k(x)$ by any other function $u_k(x)$ that has the periodicity of the system, that is, for which $u_k(x+a) = u_k(x)$. This gives us our general solution

$$\psi_k(x) = e^{ikx}u_k(x). \qquad (11.69)$$

If the ring consisted of a set of N atoms, the natural first approximation to a solution would take for u_k a sum of atomic functions, $\varphi_j(x)$, one from each atom j in our ring model:

$$u_k(x) = \sum_{j=1}^{N}\varphi_j(x). \qquad (11.70)$$

This would make $\psi_k(x)$ linear combination of atomic orbitals, modulated by the sinusoidal wave e^{ikx}. For $k = 0$ there is no modulation. For the function with $j = N/2$, k becomes π/a so that e^{ikx} changes sign, and the wavelength of the modulation is $2a$.

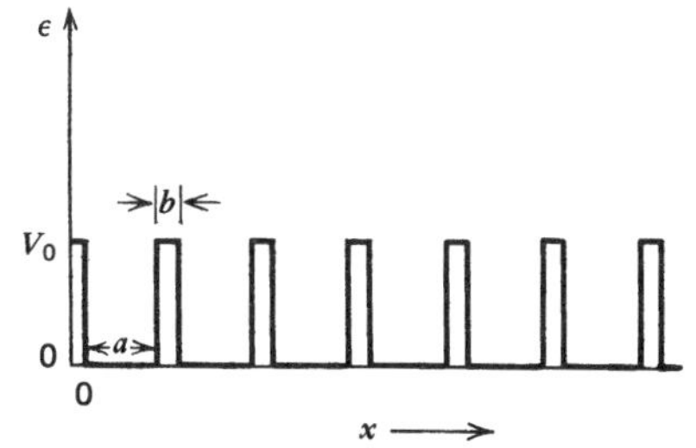

Figure 11.24 The Kronig–Penney model potential for a one-dimensional solid, prior to the limiting process $b \to 0$ subject to $V_0 b = \text{const}$.

Now let us move on from this simple periodic ring structure to see how energy bands arise. We carry out a calculation for a particle in a special sort of one-dimensional box. We suppose that the box is the limit of the model shown in Fig. 11.24; we obtain that limit by letting the higher value V_0 of the potential grow arbitrarily higher, and the width of the regions under the teeth diminish, so that the area under each tooth remains constant. This potential, the *Kronig–Penney potential,* is given by

$$\lim_{\substack{b\to 0\\ V_0 b=\text{const.}}} \begin{cases} V(x)=0, & n(a+b)\le x\le n(a+b)+a,\\ V(x)=V_0, & n(a+b)-b<x<n(a+b). \end{cases} \tag{11.71}$$

This form makes it particularly easy to develop the banded character of the energy spectrum, because in the limit $b \to 0$ the wave functions can be required to have no change in wavelength at the boundaries of the teeth of $V(x)$. For convenience let us call each periodic unit a cell.

We set up the Schrödinger equation for the general case $b > 0$ and take the limit in which the teeth are so narrow and high that we can require the wave function $\psi(x)$ that solves the equation to have the same wavelength everywhere. We assume, that is, that the width of each barrier is negligible. The Schrödinger equation has its usual form:

$$\frac{d^2\psi(x)}{dx^2}+\frac{2m}{\hbar^2}(\epsilon-V)\psi(x)=0. \tag{11.72}$$

On the basis of the insight we just obtained concerning periodicity, we substitute

$$\psi(x)=e^{ikx}u_k(x) \tag{11.73}$$

to get equations for each local atomic function $u_k(x)$:

$$\frac{d^2u(x)}{dx^2}+2ik\frac{du(x)}{dx}+\frac{2m}{\hbar^2}\left(\epsilon-V-\frac{\hbar^2k^2}{2m}\right)u(x)=0. \tag{11.74}$$

The quantity k is a kind of momentum for motion from cell to cell, which appears explicitly in the effective potential $(\epsilon - V - \hbar^2k^2/2m)$ because we have made the separation of Eq. 11.73.

Let us define

$$\alpha \equiv \left(\frac{2m\epsilon}{\hbar^2}\right)^{1/2}, \tag{11.75}$$

the momentum in the region where $V = 0$ for a particle with energy ϵ, and

$$\beta \equiv \left(\frac{2m(V_0-\epsilon)}{\hbar^2}\right)^{1/2}, \tag{11.76}$$

the momentumlike quantity in the region where $V(x) = V_0$. Then in the region $0 < x < a$, where $V(x)$ is zero, the solution to Eq. 11.74 is

$$u(x)=Ae^{i(\alpha-k)x}+Be^{-i(\alpha+k)x}, \tag{11.77}$$

and in the region $a < x < a + b$, where $V(x) = V_0$, the solution is

$$u(x)=Ce^{(\beta-ik)x}+De^{-(\beta+ik)x}. \tag{11.78}$$

(Note that these solutions are exactly those of Eqs. 4.2 and 4.3.) We set the conditions that $u(x)$ and its derivative be continuous at a and b. When we let $b \to 0$, the condition of periodicity on the function is

$$u(0)=A+B=u(a)=Ae^{i(\alpha-k)a}+Be^{-i(\alpha+k)a}, \tag{11.79}$$

and the corresponding condition on the derivative is

$$\left(\frac{du}{dx}\right)_0=\left(\frac{du}{dx}\right)_{a+b}. \tag{11.80}$$

We may now approximate the derivative at $a + b$ by the leading terms of its Taylor expansion at a:

$$\begin{aligned}\left(\frac{du}{dx}\right)_{a+b}&=\left(\frac{du}{dx}\right)_a+\left(\frac{d^2u}{dx^2}\right)_a b\\ &=\left(\frac{du}{dx}\right)_a+b\beta^2u(0),\end{aligned} \tag{11.81}$$

because $u(0) = u(a)$. Now let us state the limiting process more precisely. We define the area under a tooth but above the particle energy to be

$$P=\frac{ab\beta^2}{2}, \tag{11.82}$$

and require that

$$\lim_{\substack{V_0\to\infty\\ b\to 0}}\frac{ab\beta^2}{2}=\frac{2mab}{\hbar^2}(V_0-\epsilon)=P. \tag{11.83}$$

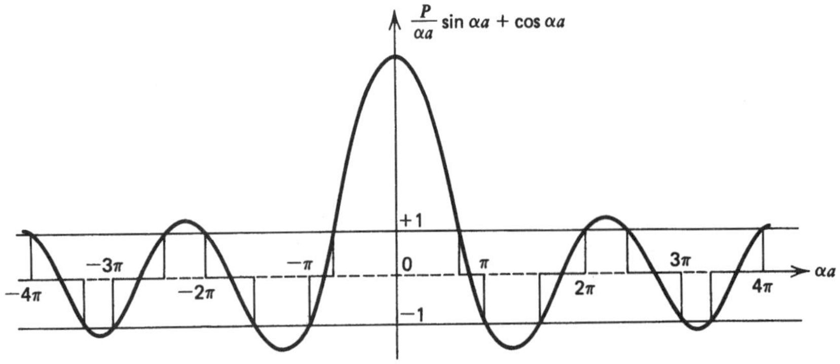

Figure 11.25 The function $Q = P(\sin \alpha a)/\alpha a + \cos \alpha a$, as a function of αa, and for a value of P that generates forbidden ranges of this function. Energy values are allowed when αa takes on values for which $-1 \le Q\ (\alpha a) \le 1$. From C. Kittel, *Introduction to Solid State Physics* (Wiley, New York, 1953).

If we now take the limit of $b \to 0$, the derivative equation can be transformed into an expression for the A and B of Eq. 11.78, the relative amplitudes of waves moving to the right and left:

$$\left[i(\alpha - k) - \left(\frac{2P}{a}\right)\right]A - \left[i(\alpha + k) + \left(\frac{2P}{a}\right)\right]B$$
$$= i(\alpha - k)Ae^{i(\alpha - k)a} - i(\alpha + k)Be^{-i(\alpha + k)a}. \qquad (11.84)$$

We can solve the continuity equations 11.78 and 11.82 simultaneously for A and B by requiring the determinant of the coefficients of A and B to be zero. That is, when we write these equations as

$$a_1 A + b_1 B = 0, \qquad (11.85a)$$
$$a_2 A + b_2 B = 0, \qquad (11.85b)$$

the solution is given by the condition

$$\begin{Vmatrix} a_1 & b_1 \\ a_2 & b_2 \end{Vmatrix} = 0, \qquad (11.86)$$

where

$$a_1 = 1 - e^{i(\alpha - k)a}, \qquad (11.87a)$$
$$b_1 = 1 - e^{-i(\alpha + k)a}, \qquad (11.87b)$$
$$a_2 = i(\alpha - k)(1 - e^{i(\alpha - k)a}) - \frac{2P}{a}, \qquad (11.87c)$$
$$b_2 = -i(\alpha + k)(1 - e^{-i(\alpha + k)a}) - \frac{2P}{a}. \qquad (11.87d)$$

Equation 11.84 leads to the condition on P and the momentum

$$P\frac{\sin \alpha a}{\alpha a} + \cos \alpha a = \cos ka. \qquad (11.88)$$

This equation has solutions only when the left-hand side has an absolute value less than or equal to unity, because $|\cos ka| \le 1$. There are regions in which this is true, but in general, for large enough P (which is always positive), there are also values of αa for which the absolute value of the left-hand side is greater than 1. Figure 11.25 illustrates this point; here we show $P(\sin \alpha a)/\alpha a + \cos \alpha a$ versus αa.

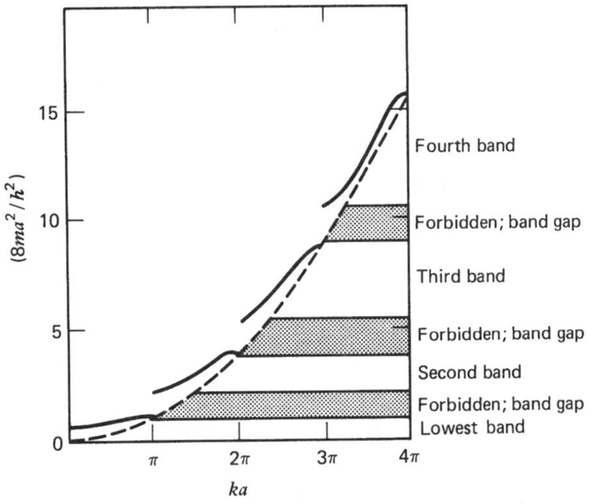

Figure 11.26 The energy as a function of the wave number k, for the Kronig–Penney potential with $P = 3\pi/2$. Adapted from C. Kittel, *Introduction to Solid State Physics* (Wiley, New York, 1953).

When $abm/\hbar^2$ is small or ϵ approaches V_0, then P is small and there may be no forbidden regions. However, in general there are, and the energy as a function of k takes on the form shown in Fig. 11.26 for the Kronig–Penney model. There are regions for which the energy is a continuous function of the wave vector **k**; these are the allowed *energy bands*. There are also values of k at which the energy is a discontinuous function of **k**, leaving ranges of ϵ for which there is no physical solution. These ranges of ϵ are called *band gaps*.

We saw how sets of degenerate levels split when isolated atoms or groups are brought into proximity and interaction. We can think of each band of the Kronig–Penney model as originating with one discrete level of a square potential well, repeated many times. The special character of infinite systems or infinitely long chains, which makes them differ from ordinary molecules, is that their level structure becomes continuous. Like the free particle, the electron in a Kronig–Penney potential can go on out to infinity. Its wave function is not constrained to go to zero at large distances from the origin; hence, its spectrum is continuous. However, it is constrained by the boundary conditions imposed by the potential, which continue to appear at all distances. This constraint is responsible for closing off some values of energy, but the continuous bands remain.

11.13 Conductors, Insulators, and Semiconductors

The band spectrum is not unique to the Kronig–Penney potential or to one-dimensional potentials. The phenomenon occurs in two-dimensional and three-dimensional situations as well. In one form or another, we find it in all solids, whether they be insulators or conductors, ionic, molecular, covalent, or metallic. The differences among these are (1) the widths of the bands relative to the excitation energies of the separated atoms or molecules that comprise the crystal, and (2) the way the electrons fill the bands.

We stated that covalent solids are insulators because their energy levels consist of filled bands of bonding orbitals, which we call *valence bands* with band gaps above the highest filled level of the uppermost occupied band. Metals, by contrast, are *conductors* because they have only partly filled bands called *conduction bands.* The highest occupied band of a metal is in almost every way like the highest occupied set of orbitals of a large molecule. In the metal there are not enough electrons to fill that band. The difference between the bands of metals and the orbital levels of finite molecules is the continuous range of allowed energy of the band, in contrast to the sharp discrete energies of orbital levels. In any solid at 0 K, electrons fill the bands up to the Fermi level ϵ_F. In an insulator, the electrons are like electrons in molecules, and cannot move freely; the Fermi level of an insulator is the top of a band. In a metal, the occupied states at ϵ_F are only infinitesimally below empty levels just above ϵ_F. Hence any extra energy, however small, can promote an electron from a state locked by the exclusion principle into immobility in the Fermi sea, into a state above ϵ_F where there are essentially no constraints on the motion of the electron. The availability of states at ϵ_F (or infinitesimally above it) makes it possible for electrons of a metal to move under the influence of any external electric field, as weak as we wish.

Each band of a metallic crystal can be associated with one particular level of the individual atoms. Thus, the conduction band of sodium is the 3*s* band; each atom contributes its 3*s* orbital toward the formation of the band, and one electron to occupy a state in the band. The result, therefore, is a 3*s* band exactly half-filled.

This argument suggests that magnesium would have a filled 3*s* band and an empty 3*p* band, and hence would be likely to be an insulator. In fact, the low-energy limit of the 3*p* band lies below the upper limit of energy of the 3*s* band,

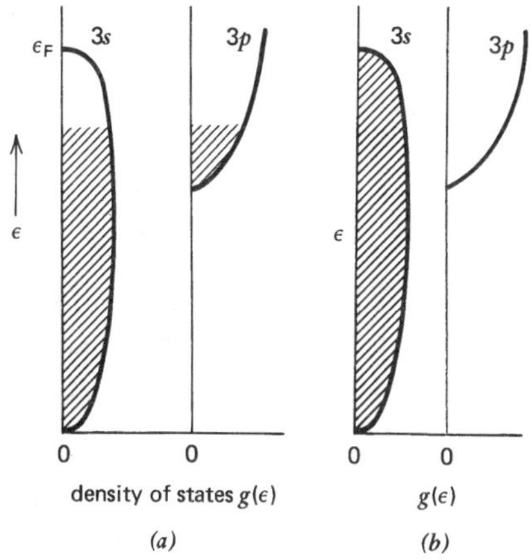

Figure 11.27 Bands of magnesium metal (schematic): (*a*) as actually occupied; (*b*) as they could be occupied if the 3*s* levels were filled and the 3*p* band were empty.

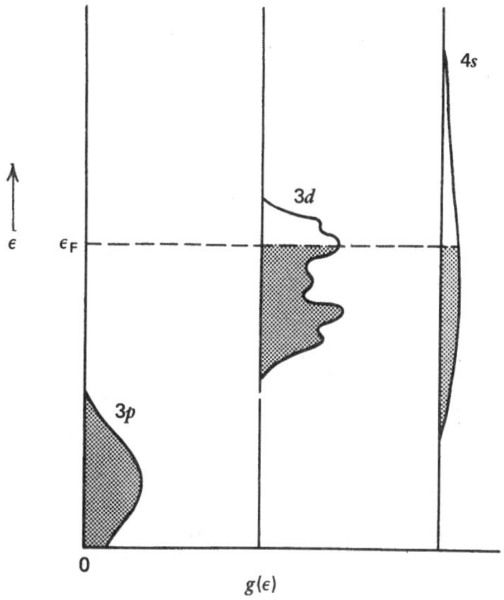

Figure 11.28 Schematic representation of the density of states $g(\epsilon)$ for a transition metal such as cobalt. The 3*d* band has multiple peaks and covers a range of energies much narrower than the 4*s* band. Only a portion of the 3*p* band is shown, and lower bands are omitted.

as indicated schematically in Fig. 11.27. Hence the electron population is as shown in Fig. 11.26*a*, not as in Fig. 11.26*b*, so that the empty parts of both the 3*s* and 3*p* bands contribute to the conductivity of magnesium. It is not at all rare for bands to overlap in energy in this way; the 4*s* band is very broad, and overlaps about half the 3*d* band in most of the transition metals, as indicated in Fig. 11.28.

In some cases the levels of a band all crowd near a common energy E_0; in others, they spread over a broad range of energy. In the former case, the density of states of the band is high near E_0, but necessarily low at other energies. In the latter case the density of states is a rather smooth function of energy or momentum.

We have now seen qualitatively how a partially filled band is responsible for metallic conduction. We have also seen how the bands have their origins in the interactions between equivalent states on the separate atoms of the crystal. Previously, we recognized how the filled orbitals of covalent crystals give rise to the insulating properties of these crystals. Now it only remains to identify the sets of delocalized orbitals of our earlier picture with the bands of the section just preceding. The band picture puts emphasis on the set of levels, rather than on the individual atomic parentage of each, but the concept of individual atomic components lurks behind all band models except the pure free-electron picture. Recall how we wrote the one-electron function in Eqs. 11.69 and 11.70 as a modulated periodic function, described locally at each site by an atomic orbital $u_k(\mathbf{r})$ there. The picture of solids based on covalent bonds emphasizes the microscopic, local structure of the orbitals that go to make up the band, rather more than the band structure itself. The split degeneracy of the local pairs or small sets of orbitals occurs because of the delocalization and interaction between sites. We saw a similar situation previously with the splitting of degenerate orbital energy levels in the formation of bonds in molecules.

The most successful modern treatments of energy levels and electronic structures of metals and covalent solids have converged, from both covalent and band model directions, to a form in which the electronic wave functions are superpositions of free-electron waves and localized atomic orbitals. The result is a consistent and rather accurate description of the electronic properties—conductivity, magnetic susceptibility, electronic spectra—of both conductors and insulators.

The particular electrical character of insulators is the immobility of their electrons, because their highest occupied bands are normally filled. This is much like the closed-shell structure of inert gases. For an electron to be free, in the band picture of an insulator, it must gain an energy of excitation large enough to carry it to the next band. The spacing between bands of an insulator is large compared with mean thermal energies. Typically, the radiation required to promote an electron from a filled valence band to an empty conduction band of an insulator lies in the ultraviolet region. Hence the population of electrons in the conduction band is normally almost nil.

However, there are insulators whose filled valence bands are close to the next empty conduction bands, close enough that a few electrons occupy conduction levels at room temperature, or perhaps a few hundred degrees higher. Pure elements such as silicon and germanium, and oxides such as ZrO_2, have this character. So do typical III–V compounds such as GaAs and InSb. These materials are called *semiconductors.* They differ from conductors in that their highest occupied level coincides with the top of a band at 0 K; they differ from insulators in that their conduction bands can be populated thermally. (Insulators may melt or vaporize at temperatures lower than those required to populate their conduction bands.) The three types of level structures are shown schematically in Fig. 11.29.

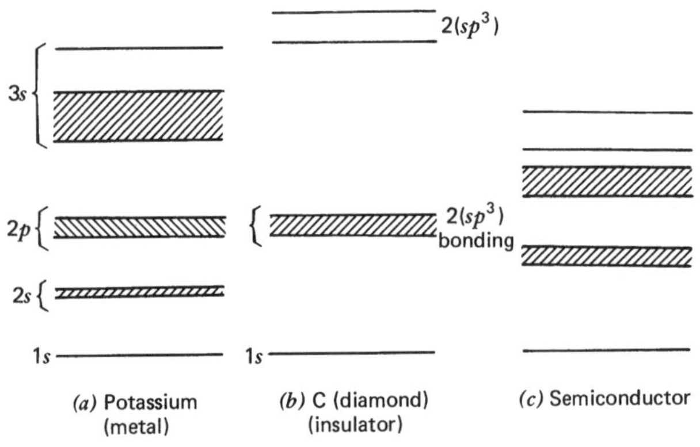

Figure 11.29 Schematic diagram of levels of (*a*) a metal; (*b*) an insulator; (*c*) a semiconductor. Hatched areas indicate filled levels.

There are also impurity semiconductors—in which foreign atoms in low concentration provide occupied levels that contribute electrons to conduction bands (*n*-type semiconductors), or empty levels that trap electrons from the valence band (highest filled insulator band) leaving mobile holes (*p*-type semiconductors). These are the basis for transistors and other solid-state electronic devices.

The conductivity of semiconductors increases with temperature because the number of conduction electrons increases rapidly with temperature, roughly as $\exp(-\Delta E_{gap}/k_BT)$. (The Boltzmann constant k_B is not to be confused with the k of Eqs. 11.68 and 11.73.) The number of conducting electrons increases with temperature in a metal also, but not so rapidly. Moreover, the scattering of electrons by vibrating nuclei, which is responsible for positive electrical resistance in all conductors, also increases with temperature. In a semiconductor the increase in resistance is less important to the conduction than the increase in the number of mobile electrons, and the resistance drops with increasing temperature. In a metal, where electrons can be set free with any amount of energy, the increased scattering dominates the temperature-dependent aspects of conductance and the resistance increases with increasing temperature.

We close this chapter with a few comments about the band structure of a *molecular* crystal, and use this to introduce the concepts of *excitons* and *effective mass.* Here the basic building blocks are molecules, and the interactions between these building blocks are weak. Hence we expect, and find, that (1) the electronic and vibrational structures of molecular crystals are characteristic of the separated molecules, only slightly perturbed, and (2) the *bands* of electronic states are very narrow, because of the weak intermolecular forces. Often the widths of bands in molecular crystals are much less than the vibration frequencies of the free molecules. When this is the situation, the bands can be characterized not only by their parent electronic states, but by parent vibrational states as well. In cases of this sort we envision a crystal in which a quantum of vibrational-plus-electronic excitation is delocalized, traveling through the crystal with a local wave function that we call $\varphi_j^{\star}(\mathbf{r})$, an excited-state function (hence the star) analogous to a component of Eq. 11.70. The state function for this case is best constructed as $\Psi(\mathbf{r}_1, \ldots, \mathbf{r}_N)$, an N-electron function in which excitation may appear on any molecule:

$$\Psi_k^\star(\mathbf{r}_1,\ldots,\mathbf{r}_N) = \text{const.}\sum_i A_j \Phi_j^\star(\mathbf{r}_1,\ldots,\mathbf{r}_N). \tag{11.89}$$

From our examination of wave functions in periodic structures, we now expect a phase term $\exp(i\mathbf{k}\cdot\mathbf{r}_j)$ in the constant A_j. (We avoid explicit sorting out of the antisymmetrization, for simplicity.) Each term $\Psi_k^\star$ has one localized excitation and $N-1$ unexcited sites; thus, one such term is

$$\Phi_j^\star = \left\|\varphi_1(\mathbf{r}_1)\varphi_2(\mathbf{r}_2)\cdots\varphi_j^\star(\mathbf{r}_j)\cdots\varphi_N(\mathbf{r}_N)\right\|, \tag{11.90}$$

so that, altogether, Ψ_k has in it one component with the excitation on each atom, and the excitation propagates with the modulation $\exp(i\mathbf{k}\cdot\mathbf{r})$. As in the case of one-electron functions, $\mathbf{k}$ acts like a momentum. The quantum of excitation behaves much like a particle, but the part associated with $\mathbf{k}$ travels from electron to electron. The propagating excitation is called a *quasi-particle;* in particular, a quasi-particle of electronic excitation in a crystal is called an *exciton.* This leads us to the ingenious concept of *effective mass* for a quasi-particle.

To associate excitons and electrons in bands with velocities and effective mass, we must know the relation between the energy ϵ and the wave vector $\mathbf{k}$, the so-called *dispersion relations* of the bands. This is a generalization of the dispersion relation discussed in Chapter 2. We turn to classical mechanics to generate our definitions. For a free particle, $\epsilon = h\nu = 2\pi\hbar\nu$ and $p = \hbar k$, so,

$$\epsilon = \frac{\hbar^2 k^2}{2m}. \tag{11.91}$$

Hence the energy and momentum are related:

$$\frac{1}{\hbar}\frac{d\epsilon}{dk} = \frac{\hbar k}{m}, \tag{11.92}$$

which is just the velocity, and

$$\left(\frac{1}{\hbar^2}\frac{d^2\epsilon}{dk^2}\right)^{-1} = m, \tag{11.93}$$

the mass.

Now we apply Eqs. 11.92 and 11.93 to any system for which we know $\epsilon(k)$. We can define an effective velocity

$$v_{\text{eff}}(k) = \frac{1}{\hbar}\frac{d\epsilon(k)}{dk}, \tag{11.94}$$

and an effective mass

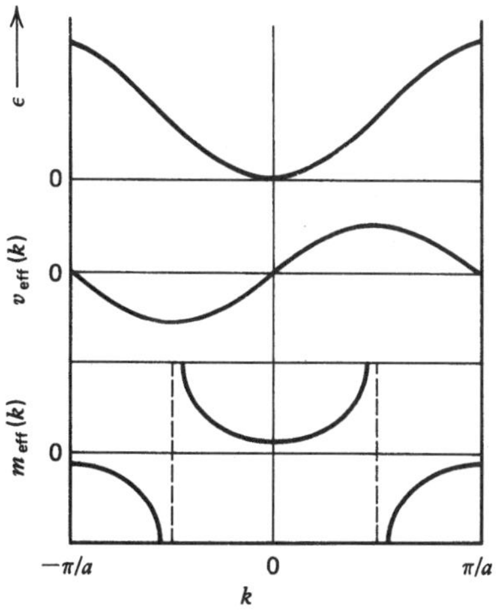

Figure 11.30 Energy, velocity, and effective mass for a particle in the first band of a one-dimensional solid. (*a*) Energy ϵ. (*b*) Effective velocity $v_{\text{eff}}(k)$. (*c*) Effective mass $m_{\text{eff}}(k)$.

$$m_{\text{eff}}(k) = \frac{1}{\hbar^2}\frac{d^2\epsilon(k)}{dk^2}. \tag{11.95}$$

The behavior of $v_{\text{eff}}(k)$ and $m_{\text{eff}}(k)$ are displayed in Fig. 11.30 for a typical $\epsilon(k)$ such as one obtains from the first band of the Kronig–Penney model. The velocity is greatest in the middle of the band and is zero at the edges. The effective mass, on the other hand, goes off to $\pm\infty$ at these points. Moreover m_{eff} is positive for $0 \le |k| < \pi/2a$, and negative for $\pi/2a < |k| \le \pi/a$, where π/a is the upper limit of the band. Hence, the direction of acceleration of an electron is *toward* a negative electrode when $|k| \le \pi/2a$, because the sign is reversed from the usual one in the expression $\mathbf{F} = m\mathbf{a}$. This means that an electron in the high-k half of the band acts like a positive particle, not a negative one.

11.14 Other Forms of Condensed Matter

Crystalline solids are the best known among all the forms of condensed matter, but several others are extremely important in our daily lives, in the solar system and the universe, and in the context of our understanding of the properties of matter. Some forms, which we shall not discuss here, occur only under conditions in which atoms or even nuclei lose their identities; these are the extremely dense conditions of many of the objects of which galaxies consist. Some examples are normal stars such as our Sun, which are very hot and consist of nuclei in a sea of electrons; neutron stars, made as their name suggests of very densely packed neutrons; and black holes, which are exceedingly dense, so dense that their gravitational fields trap light and whose interiors are mysteries to us. Here we shall briefly survey forms of condensed matter in which atoms do retain their identities and in which reactions are those of atoms, in contrast to those of nuclei or of subnuclear particles. This means that we shall summarize characteristics of *amorphous materials* (meaning liquids and glasses), *polymers, colloids and other suspensions, micelles, surfaces* (or interfaces), and *clusters.* Several of these topics will appear in Part II, in the context of thermodynamics and statistical mechanics.

Liquids and Glasses

Crystals have periodic symmetry. Ideal, infinite crystals have infinite translational symmetry and in some cases, other periodic symmetry as well, all corresponding to *discrete* changes either in the spatial positions of the constituent particles or in the choice of coordinates. (The two are complementary ways to represent the same symmetry operation.) A crystal displaced along any of its principal symmetry axes by one or any integral number of unit cell dimensions is indistinguishable from the initial crystal. However, if the crystal is shifted by a fraction of a cell dimension, it is distinguishable from the unshifted crystal. An *amorphous* material has no such regularity. A thoroughly amorphous material is so irregular that it can be shifted along any axis by any displacement whatsoever, and we cannot distinguish the shifted material from the unshifted. This is because a truly disordered material has, somewhere, essentially every possible local arrangement of its particles, consistent with the energy or temperature of the material. Examined afar, or in terms of properties averaged over a sample of sufficient size, a conventional *liquid* is amorphous at almost any instant and so is a *glass.* We shall return shortly to refine this point. The most obvious way the glass and the liquid differ is that a glass is solid; its component atoms, molecules, or ions have no overall order, but they do have definite positions of local equilibrium around which they vibrate. A liquid is characterized by the ease of motion of its component atoms, molecules, or ions. This ease of motion expresses itself in many ways, of which two may be particularly apparent. The first is a macroscopic property of *compliance.* Liquids respond readily to forces, whether the gravitational force that makes them take the shape of their containers or the force of an object that penetrates into the liquid—a finger into syrup or a diver's body into water. The only exceptions occur when the force is applied too fast for the molecules of the liquid to comply and move; this is what every beginning diver learns with the first belly-flop. The second property is the microscopic property of *diffusion.* Diffusion occurs in solids, liquids, and gases; it is slow in all three forms of matter, as one can see by letting a drop of ink fall into a glass of unstirred water and watching how long the drop, however complex a filamented shape it assumes, retains most of its identity, or by letting a drop of gasoline evaporate and seeing how many seconds or minutes it takes for the odor of the gasoline to reach an observer a few feet away. However, in solids, the process by which atoms or molecules wander is extremely slow; in effect, for an atom to leave its "home base," it must find a vacancy very nearby and have sufficient energy to move to that vacancy. (Atoms can, in some solids, also diffuse by moving interstitially, but vacancy diffusion is far more common.) In not-very-dense gases, diffusion is faster than in solids because the molecules have mean free paths longer than the nearest-neighbor distance. In effect, there are almost always nearby "vacancies" to which the particles can move before they collide with another particle. Liquids fall between the two extremes of the solid and the liquid. We observe the diffusion of one liquid into another whenever we pour milk carefully into coffee and watch the two mix without stirring. Such diffusion is, of course, so slow that we usually help things along by stirring, if we are more interested in having coffee than studying diffusion.

Viscosity is the rate of transfer of momentum in directions perpendicular to the momentum itself. Because momentum may be transferred as readily by collisions as it is by mass transport, viscosity is not simply related to the rate of diffusion. A high viscosity implies a low rate of diffusion because if momentum transfer is slow, then neither momentum nor mass transfer can be effective. However, a substance may transport momentum very readily—that is, it may have a low viscosity—and yet exhibit only very slow diffusion. Glycerine is virtually a stereotypical viscous liquid; water is rather inviscid but more viscous than, for example, liquid nitrogen.

Most liquids are less dense than their corresponding solids. This is because most liquids have empty spaces, some literally voids where atoms or molecules might otherwise be, and others, simply more space than the corresponding crystal because "defects" occur, corresponding to fewer nearest neighbors around some molecules than occur in a perfect crystal. These empty spaces allow the molecules in a liquid to move more easily than they do in a solid. Water is an exception to this picture; liquid water, between 0 and 4 C, is more dense than ice at 0°C. The reason is that the H_2O molecules in ice are held in a rigid, hydrogen-bonded, diamond-like structure, which is both open and stiff. When ice melts, many of the hydrogen bonds break, the structure becomes irregular, and the molecules can move more freely than in the ice crystal structure.

One issue regarding liquids that remains an open challenge is the question of how much structure we should attribute to any given liquid. As stated previously, the instantaneous configuration of a liquid is almost always disordered, amorphous, and apparently structureless. However, this lack of positional order need not be an indication that the liquid has no order at all. In fact, some computer simulations of liquids, notably by F. Stillinger and T. Weber, have probed the question, "Around what geometry is a liquid vibrating?" The answer is that some model liquids do vibrate around amorphous, locally stable structures most of the time, but other model liquids appear to vibrate, a large fraction of the time, around very ordered structures—not perfectly crystalline but with only a few defects in the crystal structure. Their vibrations have such large amplitudes, and are so anharmonic, that the instantaneous structures of the latter type are indistinguishable from those of the former. Only an examination of the local-equilibrium structures reveals the difference. The open problems here are finding what characteristics are responsible for the difference between the two kinds of model liquids and then determining the extent to which one or both models may apply to real liquids.

Colloids and Micelles

Finely divided matter has many characteristics that differentiate it from aggregated, bulk material. The most notable is probably the visible buffeting the particles of such matter undergo, just from the random collisions of the molecules of the medium that surrounds them. These particles may be in a gas, as fine liquid drops of an aerosol such as we produce when we sneeze, or as solid particles such as smoke or the motes of dust in a sunbeam or interstellar dust. They may be solid particles of a colloid in a liquid, such as the particles of colloidal gold that may give a solution of any of many colors depending on the size of the particles, or they may be particles of liquid in a liquid, such as the tiny fatty droplets in homogenized milk. Or they may be particles of liquid or solid dispersed in a solid.

In gases and liquids, such tiny particles exhibit at least two important kinds of behavior. First, although they may be subject to the gravitational force of the earth, the collisions they undergo with the molecules of their surrounding medium keep them suspended; the particles do not *sediment.* They remain suspended, and the stable distribution of their concentration with height can be used to evaluate the Boltzmann constant, the scale factor that relates kinetic energy per particle to the absolute temperature. Second, they undergo random motion, called *Brownian motion,* as a result of those collisions. Brownian motion was first reported in 1827 by Robert Brown, a botanist who observed the random motions of small particles of pollen in solution. This motion can be described by a phenomenological force law—i.e., a law based on physical assumptions and observations, but not on first principles. The law supposes that the force, that is, the mass x acceleration, $m d\mathbf{v}/dt$, is equal to the sum of a frictional term like that used in the Millikan oil drop experiment, just a constant γ times the velocity $\mathbf{v}$, and a term representing a very rapidly fluctuating *random* force, $\mathbf{A}'(t)$. The equation of motion for this behavior becomes

$$m\frac{d\mathbf{v}}{dt} = -\gamma'\mathbf{v} - \mathbf{A}'(t), \tag{11.96a}$$

or, if we divide through by the mass to arrive at the more commonly used form,

$$\frac{d\mathbf{v}}{dt} = -\gamma\mathbf{v} - \mathbf{A}(t). \tag{11.96b}$$

This is known generally as a *Langevin equation,* notable because of the last term representing random forces. The solution of this equation, a probability distribution of velocities at a later time resulting from a given initial velocity, is discussed in Chapter 29. Brownian motion is an example of diffusion, a special case in which we may follow the path of an individual particle.

Micelles are a special sort of small particle. They have organized structures, originating in the chemical properties of the molecules that constitute them, or at least their outer layer. The molecules of soaps are examples that form micelles. Soaps are typically made of sodium or potassium salts of fatty acids, long hydrocarbon chains with carboxylate groups, —COO—, at one end, with their charges balanced by Na^+ or K^+ ions nearby. The long hydrocarbon chains are readily soluble in oils, greases, and other nonpolar, organic materials, but not in water; the carboxylate parts are soluble in water but do not dissolve readily in oils. Small amounts of soap in a large quantity of water form tiny, approximately spherical globules, or *micelles,* with the hydrocarbon tails of the molecules pointed inward and the carboxylate groups in contact with the surrounding water. If we use the soap to clean grease from our hands, the grease gets carried into the middle of the micelles and eventually gets washed away when we rinse off the soap.

Clusters

Colloidal particles may have mean diameters of microns, 10^{-3} mm; smaller still than the micron or "micro" scale are nanoparticles, whose mean diameters are typically tens or hundreds of nanometers. Below that scale are particles consisting of only thousands, hundreds, tens, or still smaller numbers of atoms or molecules. Such particles are called *clusters* because many of their properties are at least as much like those of molecules as they are like bulk matter. Like bulk matter and the microscale and nanoscale particles but unlike molecules, clusters have no definite composition and generally may have many stable structures. Like bulk matter, many kinds of clusters may be solid or liquid, but they do not have the sharp melting points that bulk solids exhibit. However, unlike bulk matter or microscale particles, most of the atoms or molecules comprising a cluster are on the surface of the cluster, exposed to the environment. Solid clusters of many substances, particularly clusters composed of atoms, take on regular structures that are not truly crystalline. The best studied of these are the rare gas clusters, which form icosahedra or portions of icosahedra. The clusters composed of precisely the numbers of atoms required to complete a regular icosahedron are particularly stable toward evaporation or dissociation. These numbers begin with 13, 55, 147, 339, ... The icosahedron cannot be the basis of a periodic structure, so, unlike the cube or the octahedron, it cannot generate a true crystal. (Of course, it is possible to make a crystal from icosahedral balls, but such a crystal must have one of the allowed symmetries of a true lattice, as discussed earlier, and not icosahedral symmetry.) Clusters of more than a few thousand rare gas atoms are more stable in face-centered cubic (fcc) crystalline structures than in icosahedral geometries. The reason is that the atoms in the outer layers of an icosahedron are a little further apart than the atoms in a crystal lattice.

Clusters of other substances have other magic numbers. For example carbon clusters of 7, 11, 15, 19, 23, . . . atoms show unusually large binding energies. Clusters of alkali atoms have magic numbers determined not by the packing of the atoms but by the electronic shell structure, the shell structure of the valence electrons of the metallic cluster. The magic numbers in these clusters arise when an electronic shell is precisely filled, rather than when a geometric shell is precisely complete. The magic numbers of this kind begin with 2, 8, 20, 40, ... The extra stability of alkali metal clusters with these numbers of atoms—and hence of valence electrons—reflects itself not only in the binding energy but also in the high ionization energies of the clusters consisting of these numbers of atoms.

Surfaces

The last of the special forms of matter we examine here is that of a surface. In typical bulk solids and liquids, only a very small fraction of the component atoms or molecules are at the surface, so many of their properties are uninfluenced by the particles on the surface. There are obvious exceptions: The compliance of a liquid to the shape of its container and the shape that any open face of the liquid takes on are strongly influenced by the special interactions at the surface. The facets of crystals and the amorphous surfaces of glasses have many properties different from those of the particles in the interior of the same materials. Cavities in porous materials such as the zeolite clays provide surfaces where interesting chemical and physical processes occur. The chemistry of particles on surfaces is especially rich and useful: Catalytic reactions at surfaces are responsible, for example, for the conversion of nitric oxide to molecular nitrogen and oxygen in the catalytic converters used in automobiles.

Liquid surfaces follow the simple rules that they take the shape that produces the most stable system. What shape that is depends on the nature of the liquid, the shape of the container and the interactions between the molecules of the liquid and those of the container, often phrased in terms of the tendency of the liquid to wet the walls of the container. The general rules governing such stability fall into the realm of thermodynamics, particularly of free energies, and are the subject of much discussion in Part II of this book. The most important bulk property arising just from the liquid's molecules is the *surface tension,* the work, per unit area, necessary to increase the area of the surface. One can think of the surface tension as the attractive force that pulls a bit of free liquid into a spherical drop; when the drop is at equilibrium, that inward pull balances the short-range repulsive forces that keep all molecules apart.

The surfaces of crystals, by contrast, retain the order imposed by their lattices. Hence, there are many possible kinds of surfaces even for a very simple crystal. We categorize these kinds of surfaces, naming them according to what kind of crystal face is exposed. For example a face-centered, close-packed crystal, such as a crystal of argon atoms, may be cut to expose a (100) face in which each atom has four nearest neighbors in that face. Alternatively, we may expose a (111) face, in which each atom on the surface has six nearest neighbors in the open face. The (111) surface is of course smoother than the (100) face; in fact, (111) faces are the smoothest we know.

Because of the differences in numbers of neighbors in the surface, these different faces have somewhat different properties. It is easier to remove atoms from less-dense faces than from close-packed faces, in general. The energy required to remove an electron from a surface depends on what kind of face it is; that is, the work functions of different kinds of faces are different. Even the chemical properties of different faces may differ; some faces of a material are more reactive than others.

The stability and reactivity of a face depend very much on the smoothness of the face. If we try to cut a crystal face to be at a low angle; e.g., 10° to 20° from a (100) face in a cubic crystal, the atoms of this "sloping" face may well rearrange into steps, whose treads are (100) faces and whose risers are another such stable face; e.g., a (010) or (111) face—although they may be only one atom high. The new, more stable face is said to be "stepped" and will have an average angle equal to the angle at which the crystal was cut. In a metal crystal, for example, the atoms comprising the steps of such a crystal are often especially reactive sites, in contrast to the atoms in the surfaces of the flat faces. Such steps are apparent in the surface of the gold crystal shown in Fig. 11.1. This is easy to rationalize insofar as the atoms at the top edges of the steps are more exposed, less surrounded by neighbors to which they are bound and hence less tied into the crystal than those in the flat faces. The kinetics of reactions on surfaces is a substantial area within chemical kinetics and hence is appropriately considered in Part III.

At this point, we have just completed Part I by abstracting from a complicated many-body problem the simplifying generalizations of quasiparticles, effective velocities, and effective masses. In Part II we shall treat the properties of complex systems at equilibrium by creating far more general and powerful variables with the science of thermodynamics and, by relating these variables to the microscopic structural information of Part I, with statistical mechanics.

• FURTHER READING

Blakemore, J. S., *Solid State Physics,* 2nd Ed. (W. B. Saunders Company, Philadelphia, 1974).

Bunn, C. W., *Chemical Crystallography* (Oxford University Press, London, 1961).

Burdett, J. K., *Chemical Bonding in Solids* (Oxford University Press, New York, 1995).

Catlow, R., and Cheetham, A., *New Trends in Materials Chemistry* (Kluwer Academic Publishers, Boston, 1997).

Hannay, B. N., *Treatise on Solid State Chemistry,* especially Vol. 1 (Plenum Press, New York, 1975).

Kittell, C., *Introduction to Solid State Physics* (John Wiley & Sons, Inc., New York, 1953).

Kittell, C., *Quantum Theory of Solids* (John Wiley & Sons, Inc., New York, 1967).

Koerber, G. G., *Properties of Solids* (Prentice-Hall, Inc., Englewood Cliffs, N.J., 1962).

Morrison, M. A., Estle, T. L., and Lane, N. F., *Quantum States of Atoms, Molecules, and Solids* (Prentice-Hall Inc., Englewood Cliffs, N.J., 1976), Chapters 18–24.

Peierls, R. E., *Quantum Theory of Solids* (Oxford University Press, London, 1965).

Seitz, F., *The Modern Theory of Solids* (McGraw-Hill Book Co., Inc., New York, 1940).

Wells, A. F., *Structural Inorganic Chemistry,* 2nd Ed. (Oxford University Press, London, 1950).

Wheatley, P. J., *The Determination of Molecular Structure* (Oxford University Press, London, 1968).

Ziman, J. M., *Theory of Solids* (Cambridge University Press, London, 1964).

• PROBLEMS

1. Show that the base-centered tetragonal, face-centered hexagonal, and body-centered monoclinic structures do *not* define additional space lattices beyond the 14 of Fig. 11.2.

2. Show that the diamond crystal defines a Bravais lattice in which there are two carbon atoms per unit cell.

3. Sketch the following planes of a cubic lattice and indicate which of them are equivalent:
(a) (102);
(b) ($\bar{1}$1);
(c) ($\bar{1}\bar{1}$1);
(d) (021);
(e) (003).

4. How do the translational symmetries of the bcc and fcc structures differ from that of the simple cubic structure?

5. A plane parallel to the x axis of a simple cubic lattice, with lattice vector a, intersects the y and z axes at $3a$ and $4a$. What are its Miller indices? Sketch the plane (222) of a simple cubic lattice. Do the same for a base-centered monoclinic lattice.

6. Show that for a cubic lattice the spacing between adjacent (hkl) planes is

$$d = \frac{a}{(h^2 + k^2 + l^2)^{1/2}}.$$

7. Calculate the x-ray scattering pattern (i.e., the angles of the diffracted beams) for a powder of body-centered cubic crystals of plutonium. The lattice parameter is 3.638 Å; assume that the x-ray line is the Cu K_α line, whose wavelength is 1.54 Å. Identify each maximum in the powder pattern by the Miller indices of the reflecting plane, up to at least the eighth reflection. Compare this pattern to the pattern that would be generated by a simple cubic lattice with the same lattice parameter.

8. The derivation for x-ray diffraction based on Fig. 11.5 and the Bragg model as given in Eqs. 11.5 and 11.6 implies that constructive interference occurs at discrete angles θ, so that the diffracted x-rays come out in sheets or planes. Extend the derivation to show that a three-dimensional crystal produces pencils or rays of diffracted x-rays. (Note that the derivation from the Laue model, Section 11.4, gives this result.)

9. Calculate the scattering pattern from a single body-centered cubic crystal of Ag, with the lattice parameter 4.086 Å, and with Cu K_α x-rays of wavelength 1.54 Å. Suppose that the crystal is oriented to give Bragg reflection from the (100) plane; from the (111) plane. Give the value of a suitable orientation angle. Identify reflections by their order and, if more than one plane can give reflections at this angle, identify the plane associated with each reflection.

10. Iron has three modifications with body-centered cubic structures, called a α-Fe, β-Fe, and δ-Fe. Their lattice parameters are, respectively, 2.8665 Å, 2.91 Å, and 2.94 Å. What are the quantitative differences of the diffraction patterns of single crystals of these materials, when the Mo K_α radiation, 0.75 Å, is used to study the diffraction?

11. Show that a face-centered cubic crystal composed of only one kind of atom has reflections for n_1, n_2, n_3 all even or all odd, but not if one integer differs in oddness from the other two.

12. Show that salts of equally charged ions should exhibit zincblende or wurtzite structures when the ratio of radii lies between 0.225 and 0.414. Devise a structure that could exist for a ratio below 0.225 and find its limits. Check the radius ratios of zinc sulfide and other species known to exhibit 4-coordination against the limits 0.225 and 0.414.

13. The Born–Mayer model of the ionic crystal, Section 11.9, gives the energy $U(R_0)$. The compressibility of an isotropic crystal can be obtained from this expression, because the compressibility κ_T is the fractional decrease in volume per unit increment of pressure, or of force unit area:

$$-\kappa_T = \frac{1}{V}\left(\frac{\partial V}{\partial p}\right)_T = \left(\frac{\partial \ln V}{\partial p}\right)_T$$

(Strictly, as indicated, the derivative is taken at constant temperature.) Noting that $\partial U(R_0)/\partial R = -F(R_0)$, an effective force, derive an expression for κ_T from the expression 11.44 for $U(R_0)$. Evaluate κ_T for a rock salt crystal.

14. The volume of a crystal always has the form $V = cNR^3$, where R is a characteristic interparticle distance. If R is taken to be the metal–metal distance, what is the value of c for the rock salt structure? For the CsCl structure? For the zincblende structure?

15. Find the equilibrium nearest-neighbor distance for a linear chain or particles, alternately positive and negative, with pairwise interactions of the form $V_{ij} = \pm e^2/4\pi\epsilon R_{ij} + Ae^{-\beta R_{ij}}$.

16. Evaluate the Fermi energy ϵ_F for a free-electron metal with the lattice spacing of sodium. Estimate the fraction of valence electrons with energies above ϵ_F when the temperature is 300 K.

17. Consider a ring with the potential of Eq. 11.71, with $a = b$ rather than with $b \to 0$. Suppose the ring has length $2Na$. Write the wavefunction in the nth valley (with $l = a + b$) as

$$\psi(x) = A_n e^{ik(x-nl)} + B_n e^{-ik(x-nl)}.$$

Show that the coefficients A_{n+1} and B_{n+1} of the exponential in the next valley can be expressed as

$$A_{n+1} = (a_1 - i\beta_1)e^{ikl}A_n - i\beta_2 e^{ikl}B_n,$$
$$B_{n+1} = i\beta_2 e^{-ikl}A_n + (\alpha_1 + i\beta_1)e^{-ikl}B_n.$$

18. Compute the effective mass as a function of k for an electron in the first band of a line of "atoms" with a Kronig–Penney potential and a lattice spacing of 1.5 Å. Compute the density of states for the first band of the same system.

19. Compare the eigenfunctions and eigenvalues for an electron on a one-dimensional ring of length L with those of an electron in a one-dimensional box of the same length. In what ways do these two examples differ and why?

20. Show that Eq. 11.88 follows from Eq. 11.84 and find ϵ at the lower and upper limits of the first three bands. At what values of ϵ (eV) do these *band edges* occur for an electron in a lattice with $a = b = 1.5$ Å and $V_0 = 3$ eV?

Appendices

APPENDIX I

Systems of Units

The measurement of physical and chemical properties requires the definition of units. Frequently the chosen units in a given investigation are first the commonly used ones, such as inches, pounds, and so on. The complexity and multiplicity of common units for weights and measures is analogous to that of any other aspect of culture, such as language. The need for easy comparison of measurements requires a common "language," an agreed-upon set of units. Just as an agreed-upon language evolves in time, so do agreed-upon units. The currently recommended units are called the *International System of Units* (abbreviated *SI,* for *Système International d'Unités*). This system is the result of much discussion and compromise. For a fuller description of the process of development of units, see the NIST (National Institute of Science and Technology) Web page "Fundamental Physical Constants" http://physics.nist.gov/cuu/index.html. M. L. McGlashan, *Physicochemical Quantities and Units,* 2nd ed. (Royal Institute of Chemistry, London, 1971), and *Ann. Rev. Phys. Chem.* **24,** 51 (1973); F. D. Rossini, *Fundamental Measures and Constants for Science and Technology* (CRC Press, Cleveland, 1974); *The International System of Units* (*SI*), National Bureau of Standards (U.S.) Special Publication 330 (Washington, D.C., 1977); M. A. Paul, *J. Chem. Soc.* **11,** 3 (1971). The following tables list the essentials of the SI units, and give their relationship to some other commonly used units (including the older cgs and mks systems).

The International System of Units (SI)

			BASE UNITS
Quantity	**Unit**	**Symbol**	**Definition (abridged)**
Length	meter	m	The length of the path travelled by light in a vacuum during a time interval of 1/299 792 458 of a second.
Mass	kilogram	kg	The mass of the international prototype of the kilogram (a piece of platinum-iridium kept in Sèvres, France).
Time	second	s	9 192 631 770 periods of the radiation corresponding to the transition between the hyperfine levels of the ^{133}Cs ground state.
Electric current	ampere	A	The current that produces a force of 2×10^{-7} newton per meter of length between two infinitely long, negligibly thick parallel conductors 1 meter apart in a vacuum.
Thermodynamic temperature	kelvin	K	The fraction 1/273.16 of the thermodynamic temperature of the triple point of water.
Amount of substance	mole	mol	The amount of substance of a system containing as many specified elementary entities (atoms, molecules, ions, etc.) as there are atoms in 0.012 kilograms of ^{12}C.
Luminous intensity	candela	cd	The perpendicular luminous intensity of a surface that emits monochromatic radiation of 540×10^{12} hertz with a radiant intensity of 1/683 watt per steradian.

		PREFIXES			
Fraction	**Prefix**	**Symbol**	**Multiple**	**Prefix**	**Symbol**
10^{-1}	deci	d	10	deca	da
10^{-2}	centi	c	10^2	hecto	h
10^{-3}	milli	m	10^3	kilo	k
10^{-6}	micro	μ	10^6	mega	M
10^{-9}	nano	n	10^9	giga	G
10^{-12}	pico	p	10^{12}	tera	T
10^{-15}	femto	f			
10^{-18}	atto	a			

DERIVED UNITS

Quantity	Unit	Symbol	Definition
Frequency	hertz	Hz	s^{-1}
Energy	joule	J	kg m^2/s^2
Force	newton	N	kg m/s^2 = J/m
Power	watt	W	kg m^2/s^3 = J/s
Pressure	pascal	Pa	kg/m s^2 = N/m^2 = J/m^3
Electric charge	coulomb	C	A s
Electric potential difference	volt	V	kg m^2/s^3 A = J/A s = J/C
Electric resistance	ohm	Ω	kg m^2/s^3A^2 = V/A
Electric conductance	siemens	S	s^3A^2/kg m^2 = Ω^{-1}
Electric capacitance	farad	F	A_2s^4/kg m^2 = A s/V = C/V
Magnetic flux	weber	Wb	kg m^2/s^2A = V s
Inductance	henry	H	kg m^2/s^2A^2 = V s/A = Wb/A
Magnetic flux density (magnetic induction)	tesla	T	kg/s^2A = V s/m^2

The SI is an extension of the *meter-kilogram-second* (*mks*) *system.* A parallel system of units formerly predominant in scientific work is the *centimeter-gram-second* (*cgs*) *system,* in which the indicated three units (cm, g, s) are used as base units. The cgs mechanical units with special names include:

Quantity	Unit	Symbol	Definition	Relation to SI Unit
Energy	erg	erg	g cm^2/s^2	10^{-7} J
Force	dyne	dyn	g cm/s^2	10^{-5} N
Viscosity	poise	P	g/cm s	10^{-1} Pa s
Kinematic viscosity	stokes	St	cm^2/s	10^{-4} m^2/s

The following named units are decimal multiples of SI or cgs units:

Quantity	Unit	Symbol	SI Definition	cgs Definition
Length	angstrom	Å	10^{-10} m	10^{-8} cm
Length	micron	μ	10^{-6} m	10^{-4} cm
Volume	liter	l	1 dm^3 (=10^{-3} m^3)	10^3 cm^3
Mass	tonne	t	10^3 kg	10^6 g
Pressure	bar	bar	10^5 Pa	10^6 dyn/cm^2
Concentration	(molarity)	*M*	mol/dm^3	10^{-3} mol/cm^3 (=mol/liter)

The degree Celsius (°C) is equal in magnitude to the kelvin, and may be used in either system for temperature *differences* or for the *Celsius temperature,* defined as $T - 273.15$ K, where T is the thermodynamic temperature (cf. Section 12.5).

Electric and magnetic units present a problem, since several inconsistent systems have been used, with even the basic equations (and thus the unit dimensions) differing between them. The SI is a so-called four-quantity system, with the ampere (on which other electromagnetic units are based) defined independently of the mechanical units; a four-quantity cgs system is also possible, for example, with the franklin (Fr), equal to the esu of charge, as the fourth unit. However, the usual practice was to use a three-quantity cgs system, defining electric and magnetic units in terms of mechanical units by suppressing the constants in either Coulomb's law (*electrostatic system*) or Ampere's law (*electromagnetic system*); the *Gaussian system* uses electrostatic units for electric quantities and electromagnetic units for magnetic quantities. Some of the units in these systems include:

Quantity	Unit	Definition	Relation to SI Unit
Electrostatic Units			
Electric charge	statcoulomb ("esu of charge")	$g^{1/2}\ cm^{3/2}/s$	3.33564×10^{-10} C
Electric current	statampere	statcoulomb/s	3.33564×10^{-10} A
Electric potential	statvolt	erg/statcoulomb	299.793 V
Electric resistance	statohm	statvolt/statampere	$8.98758 \times 10^{11}\ \Omega$
Dipole moment	debye (D)	10^{-18} statcoulomb cm	3.33564×10^{-30} C m
Electromagnetic Units			
Electric current	abampere ("emu of current")	$g^{1/2}\ cm^{1/2}/s$	2.99793×10^{9} A
Magnetic field strength	oersted (Oe)	abampere/cm	79.5775 A/m
Magnetic flux density	gauss (G)	abampere/cm	10^{-4} T
Magnetic flux	maxwell (Mx)	abampere cm	10^{-8} Wb

The following units belong to neither the SI nor any variation of the cgs system, but have been in sufficiently common use to require definition (a few of the major English-system units are included):

Quantity	Unit	Symbol	Value	Other Equivalents
Length	inch	in.	2.54 cm	
Length	bohr	(a_0)	5.291772×10^{-11} m	0.5292 Å $(\equiv \epsilon_0 h^2/\pi m_e e^2)$
Mass	pound	lb	0.45359237 kg	
Mass	atomic mass unit	amu	1.660539×10^{-27} kg	$(\equiv 1\ g/N_A)$
Time	minute	min	60 s	
Time	hour	h	3600 s	60 min
Energy	calorie[a]	cal	4.184 J	
Energy	electron volt	eV	1.602176×10^{-19} J	$(\equiv e \times 1\ V)$
Energy	hartree	$(R_\infty hc)$	4.35974×10^{-18} J	27.2114 eV $(\equiv m_e e^4/8\epsilon_0^2 h^2)$
Energy/*hc* (wavenumber)	—	cm^{-1}	1.98648×10^{-23} J/*hc*	1.23985×10^{-4} eV/*hc*
Pressure	torr	Torr	133.3224 Pa	(1/760) atm ≈ 1 mmHg
Pressure	atmosphere	atm	1.01325×10^{5} Pa	1.01325 bar = 760 Torr
Celsius temperature	degree Fahrenheit	°F	(5/9)°C	$(\theta/°F = \frac{9}{5}\ \theta/°C + 32)$
Thermodynamic temperature	degree Rankine	°R	(5/9) K	$(T/°R = \theta/°F + 459.688)$

[a] This is the defined thermochemical calorie (cf. Section 13.7); note that several other "calories," with slightly different values, are also in use.

APPENDIX II

Partial Derivatives

Suppose that a given function $u(x, y)$ depends on the values of two independent variables, x and y, and that we are interested in just how u varies with x and y. Such a problem can be conveniently analyzed by separating the effects of the variables, that is, varying only one at a time. By a generalization of the process used to define the derivative in elementary calculus, we can define the *partial derivative* of u with respect to x at constant y:

$$\left(\frac{\partial u}{\partial x}\right)_y = \lim_{\Delta x \to 0} \frac{u(x+\Delta x, y) - u(x, y)}{\Delta x}. \qquad \text{(II.1)}$$

The symbol ∂ (rather than d) is used for partial differentiation; a subscript (here y) is used to designate a variable held constant. Similarly, we have

$$\left(\frac{\partial u}{\partial y}\right)_x = \lim_{\Delta y \to 0} \frac{u(x, y+\Delta y) - u(x, y)}{\Delta y}. \qquad \text{(II.2)}$$

The derivatives of various functions have the same form as in ordinary differentiation; for example, if $u(x, y) = x^2y$, the partial derivatives are

$$\left(\frac{\partial u}{\partial x}\right)_y = y\frac{d}{dx}(x^2) = 2xy \text{ and } \left(\frac{\partial u}{\partial y}\right)_x = x^2\frac{d}{dy}(y) = x^2.$$

The extension to functions of three or more variables is obvious: The function $v(x, y, z)$ has the partial derivatives $(\partial v/\partial x)_{y,z}$, $(\partial v/\partial y)_{x,z}$, $(\partial v/\partial z)_{x,y}$. Second and higher derivatives can be taken in the usual way; one can easily show that

$$\left[\frac{\partial}{\partial x}\left(\frac{\partial u}{\partial y}\right)_x\right]_y = \left[\frac{\partial}{\partial y}\left(\frac{\partial u}{\partial x}\right)_y\right]_x = \frac{\partial^2 u}{\partial x \partial y}, \qquad \text{(II.3)}$$

that is, that the order of differentiation is immaterial, as long as u and its derivatives are continuous.

As with ordinary differentiation, we can obtain a simple geometric interpretation of the partial derivative. A function of only one variable can be represented by a simple curve in a plane; this curve, so long as it is smooth, has at any point a single well-defined slope, equivalent to the derivative of the function. To plot all the possible values of a function like $u(x, y)$ we need a three-dimensional surface, as shown in Fig. II.1. Such a surface has an infinite number of slopes at a given point, corresponding to the infinite number of directions in which the slope can be taken. As can be seen from the figure, the derivatives defined in Eqs. II.1 and II.2 correspond to the slopes parallel to the x and y axes. More generally, for a function of n variables, the partial derivatives are equivalent to slopes on an $(n + 1)$-dimensional surface.

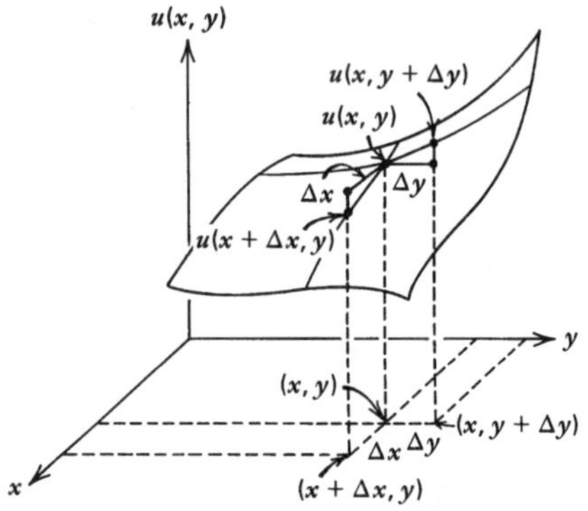

Figure II.1 Geometric interpretation of partial differentiation. The figure shows a portion of the surface $u(x, y)$. The partial derivatives $(\partial u/\partial x)_y$ and $(\partial u/\partial y)_x$ for given values of x and y are the slopes of this surface parallel to the x and y axes, respectively, at the point $u(x, y)$.

The partial derivatives we have defined correspond to slopes parallel to the principal axes representing the independent variables. However, it is possible and sometimes important to take slopes in other directions. For example, one might want to know how $u(x, y)$ varies when we impose the condition that $x = \alpha y^2$ at all times; this will be given by the slope along the curve $x = \alpha y^2$. The derivative of u in the direction corresponding to the steepest slope is called the *gradient* of u; it equals the magnitude of a vector (in the xy plane) whose x and y components are $(\partial u/\partial x)_y$ and $(\partial u/\partial y)_x$, respectively.[1]

How does $u(x, y)$ change when x and y vary simultaneously? For infinitesimal changes in x and y, the resulting infinitesimal change in u is clearly

$$du = \left(\frac{\partial u}{\partial x}\right)_y dx + \left(\frac{\partial u}{\partial y}\right)_x dy; \qquad \text{(II.4)}$$

[1] In three dimensions the gradient corresponds to the vector **grad** u or ∇u defined in footnote 9 on page 65.

this expression is called the *total differential* of u. Suppose we vary x and y together in such a way that u remains constant; then $du = 0$ and we have

$$\left(\frac{\partial u}{\partial x}\right)_y dx + \left(\frac{\partial u}{\partial y}\right)_x dy = 0 \quad (u = \text{const.}). \tag{II.5}$$

The ratio of dx to dy at constant u is simply $(\partial x/\partial y)_u$, so we can rearrange Eq. II.5 in the forms

$$\left(\frac{\partial x}{\partial y}\right)_u = -\frac{(\partial u/\partial y)_x}{(\partial u/\partial x)_y} \quad \text{or}$$

$$\left(\frac{\partial u}{\partial x}\right)_y \left(\frac{\partial x}{\partial y}\right)_u \left(\frac{\partial y}{\partial u}\right)_x = -1. \tag{II.6}$$

(The second form is easy to remember because the variables appear in cyclic order.) Note that u is as legitimate an independent variable as x or y: If we have an equation $f(u, x, y) = 0$ connecting the values of u, x, and y, we can solve for any one in terms of the other two.

Suppose that there is a set of related variables x, y, z, u, . . . of which only two—*any* two—are independent. (An example would be the pressure, temperature, density, energy, . . . of a fluid.) We can then, for example, express the variable u as a function of x and y, of x and z, of y and z, and so forth. By differentiating Eq. II.4 we can obtain

$$\left(\frac{\partial u}{\partial x}\right)_z = \left(\frac{\partial u}{\partial x}\right)_y + \left(\frac{\partial u}{\partial y}\right)_x \left(\frac{\partial y}{\partial x}\right)_z \tag{II.7}$$

(since $\partial x/\partial x = 1$, no matter what is held constant),

$$\left(\frac{\partial u}{\partial z}\right)_v = \left(\frac{\partial u}{\partial x}\right)_y \left(\frac{\partial x}{\partial z}\right)_v + \left(\frac{\partial u}{\partial y}\right)_x \left(\frac{\partial y}{\partial z}\right)_v, \tag{II.8}$$

and many other such relationships. Generalization of all this to more than two independent variables is straightforward.

Two other equations that hold for any combination of variables are the "chain rule,"

$$\left(\frac{\partial u}{\partial y}\right)_z = \left(\frac{\partial u}{\partial x}\right)_z \left(\frac{\partial x}{\partial y}\right)_z, \tag{II.9}$$

and the inverse relationship,

$$\left(\frac{\partial x}{\partial y}\right)_z = \frac{1}{(\partial y/\partial x)_z}; \tag{II.10}$$

we used the latter in deriving Eq. II.6.

Consider a general differential expression

$$df = M(x,y)\ dx + N(x,y)\ dy, \tag{II.11}$$

where M and N are some functions of x and y. If this expression is actually the differential of some function f, it is called an *exact differential*. According to Eqs. II.3 and II.4, a differential of the form II.1 is exact if

$$\left(\frac{\partial M}{\partial y}\right)_x = \left(\frac{\partial N}{\partial x}\right)_y. \tag{II.12}$$

For the two-variable case, an inexact differential can always be converted to an exact differential by multiplying by some *integrating factor*. For example, $2y\ dx + x\ dy$ is inexact; multiplication by x converts it to $2xy\ dx + x^2\ dy$, which is the exact differential of x^2y.

We often have occasion to transform from one set of variables to another—to introduce new coordinates in mechanics, or new independent variables in thermodynamics. Given the equations relating the new to the old variables, it is a straightforward process to transform an algebraic expression; for an example, see the derivation of Eq. 3.125. Things are more difficult when one has to differentiate or integrate with respect to the variables in question.

Suppose that we want to transform from the variables x, y, z, . . . to the variables u, v, w, ... Derivatives with respect to one of the old variables can be transformed by an equation of the form

$$\left(\frac{\partial}{\partial x}\right)_{y,z,\ldots} = \left(\frac{\partial u}{\partial x}\right)_{y,z,\ldots} \left(\frac{\partial}{\partial u}\right)_{v,w,\ldots} + \left(\frac{\partial v}{\partial x}\right)_{y,z,\ldots} \left(\frac{\partial}{\partial v}\right)_{u,w,\ldots} + \cdots, \tag{II.13}$$

which is simply the generalization of Eq. II.8 in operator form (cf. Section 3.4). For an example, consider the transformation from Cartesian coordinates x, y to circular polar coordinates r, ϕ. We must evaluate the operators

$$\left(\frac{\partial}{\partial x}\right)_y = \left(\frac{\partial r}{\partial x}\right)_y \left(\frac{\partial}{\partial r}\right)_\phi + \left(\frac{\partial \phi}{\partial x}\right)_y \left(\frac{\partial}{\partial \phi}\right)_r,$$
$$\left(\frac{\partial}{\partial y}\right)_x = \left(\frac{\partial r}{\partial y}\right)_x \left(\frac{\partial}{\partial r}\right)_\phi + \left(\frac{\partial \phi}{\partial y}\right)_x \left(\frac{\partial}{\partial \phi}\right)_r. \tag{II.14}$$

From the transformation equations given with Fig. 3.9, we obtain

$$\left(\frac{\partial r}{\partial x}\right)_y = \frac{x}{(x^2+y^2)^{1/2}} = \frac{x}{r} = \cos\phi,$$
$$\left(\frac{\partial r}{\partial y}\right)_x = \frac{y}{(x^2+y^2)^{1/2}} = \frac{y}{r} = \sin\phi,$$
$$\left(\frac{\partial \phi}{\partial x}\right)_y = \frac{-y}{x^2+y^2} = \frac{y}{r^2} = -\frac{\sin\phi}{r},$$
$$\left(\frac{\partial \phi}{\partial y}\right)_x = \frac{x}{x^2+y^2} = \frac{x}{r^2} = \frac{\cos\phi}{r}. \tag{II.15}$$

Substitution in Eqs. II.14 then yields

$$\left(\frac{\partial}{\partial x}\right)_y = \cos\phi\left(\frac{\partial}{\partial r}\right)_\phi - \frac{\sin\phi}{r}\left(\frac{\partial}{\partial \phi}\right)_r,$$

$$\left(\frac{\partial}{\partial y}\right)_x = \sin\phi\left(\frac{\partial}{\partial r}\right)_\phi + \frac{\cos\phi}{r}\left(\frac{\partial}{\partial \phi}\right)_r; \qquad \text{(II.16)}$$

these expressions are used in Appendix 3B to obtain the Hamiltonian operator in circular coordinates.

When transforming integrals, one needs to know the "volume element" in the new variables. That is, for any quantity f that depends on the variables, one must know the factor J for which

$$\int_V f(x,y,\ldots)\,dx\,dy\ldots = \int_V f(u,v,\ldots)J\,du\,dv\ldots \qquad \text{(II.17)}$$

when the two integrals are taken over corresponding ranges of coordinates. The expression $J\,du\,dv$ . . . has the same dimensions as $dx\,dy$ . . . , which for Cartesian coordinates is a volume element. We state without proof the rule giving the value of J:

$$J\left(\frac{x,y,\ldots}{u,v,\ldots}\right) = \begin{vmatrix} \left(\frac{\partial x}{\partial u}\right)_{v,w,\ldots} & \left(\frac{\partial x}{\partial v}\right)_{u,w,\ldots} & \cdots \\ \left(\frac{\partial y}{\partial u}\right)_{v,w,\ldots} & \left(\frac{\partial y}{\partial v}\right)_{u,w,\ldots} & \cdots \\ \cdots & \cdots & \cdots \end{vmatrix}; \qquad \text{(II.18)}$$

the determinant J is called the *Jacobian* of the transformation. For two-variable systems the Jacobian has the simple form

$$J(x,y/u,v) = \left(\frac{\partial x}{\partial u}\right)_v\left(\frac{\partial y}{\partial v}\right)_u - \left(\frac{\partial x}{\partial v}\right)_u\left(\frac{\partial y}{\partial u}\right)_v. \qquad \text{(II.19)}$$

For the transformation from Cartesian to circular coordinates, we have (again using Fig. 3.9)

$$\begin{aligned} J(x,y/r,\phi) &= \left(\frac{\partial x}{\partial r}\right)_\phi\left(\frac{\partial y}{\partial \phi}\right)_r - \left(\frac{\partial x}{\partial \phi}\right)_r\left(\frac{\partial y}{\partial r}\right)_\phi \\ &= (\cos\phi)(r\cos\phi) - (-r\sin\phi)(\sin\phi) \\ &= r(\cos^2\phi + \sin^2\phi) = r, \end{aligned} \qquad \text{(3A.20)}$$

and the area element corresponding to $dx\,dy$ is $r\,dr\,d\theta$. One can also deduce this result directly from a diagram, as shown in Fig. II.2. On the other hand, the volume element of Fig. 2A.2 could have been obtained by calculating

$$J(x,y,z/r,\theta,\phi) = r^2\sin\theta \qquad \text{(II.21)}$$

from the transformation relations.

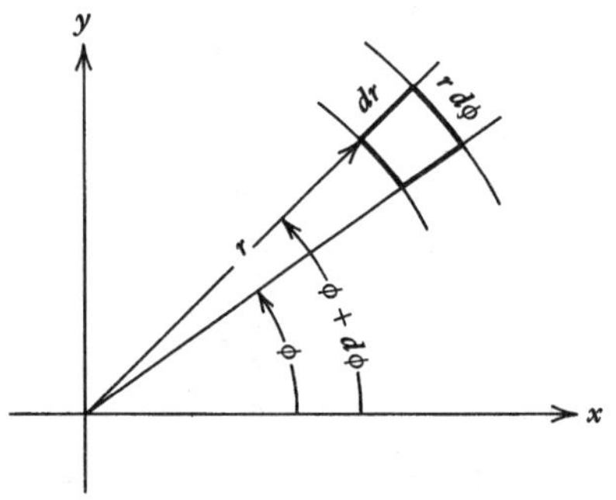

Figure II.2 Area element in circular polar coordinates: $dA = (dr)(r\,d\phi) = r\,dr\,d\phi$.

APPENDIX III

Glossary of Symbols

Symbol	Meaning	Where used
A	Helmholtz free energy	general
	action	Part I; Eq. 2.25
	area	general
	frequency factor in rate coefficient	Part III; Eq. 30.42
A	operator for a classical property A	Part I; Section 3.4
$A(X)$	electron affinity of X	Eq. 14.46
A_∞	high-pressure frequency factor for unimolecular reactions	Part III; Eq. 30.174
$\mathscr{A}$	amplitude of perturbation	Part III; Eq. 30.127
a	molar Helmholtz free energy	Part II; general
	ionic radius	Eq. 26.69
	radius of Brownian particle	Eq. 29.53
$\mathbf{a}$	acceleration	general
a_i	activity of component i	Parts II and III; general
a_0	Bohr radius	general
$a_\pm$	mean ionic activity	Eq. 26.8
$\mathbf{a}, \mathbf{b}, \mathbf{c}$	primitive vectors of a crystal	p. 315 ff
a, b, c	primitive distances in a crystal	p. 315 ff
a_x, a_y, a_z	x-, y-, z-components of acceleration	general
B	second virial coefficient of a gas	Part II; Eq. 21.17
	second virial coefficient in the concentration expansion of the osmotic pressure of a solution	Part II; Eq. 25.127
	magnitude of the magnetic field strength **B**	general
$\mathbf{B}$	magnetic field strength	general
B_e	rotational constant for a rigid rotor	Eq. 7.21
B_υ	rotational constant for a molecule in the vibrational state designated by quantum number υ	Eq. 7.21
B_2^*	second virial coefficient in liquid solution	Eq. 25.196
$B(R)$	bridge function in theory of liquids	Eq. 23.53
b	impact parameter	general
	excluded volume term for the van der Waals equation of state	Table 21.2
	excluded volume in a lattice model of a mixture	Eq. 25.193
C_p; C_υ	heat capacity at constant pressure; heat capacity at constant volume	general
C_σ	heat capacity along the liquid-vapor coexistence line	Eq. 24.19
C	the Coulomb; SI unit of charge	Table 1.1
C_2	symmetry operator for 180° rotation	Section 31.6
$C(R)$	direct correlation function of a liquid	Eq. 23.24
C	third virial coefficient of a gas	Eq. 21.17
	third virial coefficient in the concentration expansion of the osmotic pressure of a solution	Eq. 25.127
C_d	differential capacitance of electrolyte-electrode double layer	Eq. 26.132
c	speed of light	general
	speed of sound	Section 28.10
c^*	"equivalent" concentration	Eq. 29.103
c_i	concentration of component i	general

Symbol	Meaning	Where used
c_{ij}	coefficient of the jth basis orbital in the expression for the ith molecular orbital	Part I; Eq. 8.17
C^S_{ijkl}	second order adiabatic elastic constant of a crystal	Eq. 22.18
C^T_{ijkl}	second order isothermal elastic constant of a crystal	Eq. 22.18
c_p; c_v	molar heat capacity at constant pressure; molar heat capacity at constant volume	general
c^x_v	contribution of x (for example, vibration) to the constant volume molar heat capacity	Eq. 22.44
D	dielectric constant ($\mathcal{D}$ in Section 30.8)	Eq. 26.58
	Debye unit	Eq. 4.29 ff
	(self-) diffusion coefficient	Table 28.1
D_k	(mutual) diffusion coefficient of component k	Eq. 20.15; App. 16A
D_0	dissociation energy measured from the zero point energy level	general
D_{0i}	energy required to dissociate a molecule of species i into atoms, measured from the zero point level	general
D_e	dissociation energy measured from the minimum in the potential well (well depth)	general
$D(w, w^*)$	distance between points w and w^* in phase space	p. 423
d	spacing between lattice planes	Fig. 11.8
	hard-sphere diameter	Part III; Fig. 27.7
$d(\mathbf{r}_2, \theta_{12} \mid \mathbf{r}_1)$	conditional probability density for finding electron 2 at $\mathbf{r}_2$, θ_{12} when electron 1 is at $\mathbf{r}_1$	Part I; Eq. 5.17
E	energy	general
E	operator for energy	Part I; Eq. 3.32
$\mathbf{E}$	electric field strength	general; Eq. 1.8
$E(R)$	eigenvalue of the Schrödinger equation based on the Hamiltonian $H_{\text{elec}}(r, R)$ for fixed nuclei	Part I; Eq. 6.3
E_{rot}	rotational energy; an eigenvalue of the Schrödinger equation for rotational energy of a molecule	Eq. 7.19
E_{vib}	vibrational energy; an eigenvalue of the Schrödinger equation for vibrational energy of a molecule	Eq. 7.6
E_a	activation energy	Part III; Eq. 30.42
E_v	energy for formation of a vacancy	Section 29.5
E_∞	high-pressure activation energy for a unimolecular reaction	Part III; Eq. 30.174
ΔE_{reac}	energy change in a reaction	general
ΔE^0_0	net change in energy in conversion of reactants to products at $T = 0$	Eq. 21.152
ΔE^*_f, ΔE^*_r	barrier height in forward and reverse directions of reaction	Part III; Fig. 30.6
$\Delta E^\ddagger$	energy of activation (energy difference between activated complex and reactants)	Part III; Section 30.6
$\epsilon_i(R)$	energy of ith molecular orbital	Part I, esp. Chapter 7, ff
e	elementary electronic charge	general; Table 1.1
$\mathcal{E}$	electromotive force	Eq. 26.38
	electric potential	Table 13.1
$\mathcal{E}^0$	standard emf	Eq. 26.37
$\mathcal{E}_{1/2}$	half-cell potential	Part II; Eq. 25.566
F	structure factor	Eq. 11.28
	magnitude of the force **F**	general
	tension	Part II; Table 13.1
$\mathbf{F}$	force	general
F_{bonding}	magnitude of the force bonding two nuclei	Part I; Eq. 6.8
$\mathcal{F}$	The Faraday constant	general
f	number of degrees of freedom of the macroscopic variables in a multicomponent multiphase system	Part II; Chapter 24
	flux density of a beam	Eq. 2.13; Eq. 30.86
$f(\mathbf{r}, \mathbf{v}, t)$	velocity and space distribution function; number of molecules per unit "volume" in phase space at time t	Part III; Section 28.1
$f(\theta, E)$	quantum mechanical scattering amplitude	Eq. 27.73
$f(\mathbf{r}, t)$	density of molecules in configuration space	Part III; Eq. 28.14
$f(\mathbf{v}, t)$	density of molecules in velocity space	Part III; Eq. 28.2

Symbol	Meaning	Where used
$\mathscr{f}(\mathbf{r}, \mathbf{v}, t)$, $\mathscr{f}(\mathbf{r}, t)$, $\mathscr{f}(\mathbf{v}, t)$	probability densities corresponding to $\mathscr{f}(\mathbf{r}, \mathbf{v}, t)$, $\mathscr{f}(\mathbf{r}, t)$, and $\mathscr{f}(\mathbf{v}, t)$, $\mathscr{f}(\mathbf{v}, t) \equiv 1/N\, \mathscr{f}(\mathbf{r}, \mathbf{v}, t)$, etc. Note that $\mathscr{f}(\mathbf{v}, t)$ corresponds to the $\mathscr{f}(\mathbf{v})$ of Section 19.7	Part III; Eqs. 28.4 ff
$f(\mathbf{v})$	equilibrium velocity distribution function	Part II; Eq. 12.8
$f(v)$	equilibrium speed distribution function	Part II; Eq. 12.10
$f(\omega)$	distribution of orientations of a molecule	Eq. 24.96
$f(t)$	randomly fluctuating force on a Brownian particle	Eq. 29.38
f_i^0	fugacity of pure component i	Eq. 21.14
f_i	fugacity of component i in a mixture	Eq. 21.37
f_s	atomic scattering factor	Part I; Eq. 11.26
$f(\epsilon)$	one-particle energy distribution function	general; Eq. 11.56
$f(j, t)$	number of oscillators in quantum state j at time t	Part III; Eq. 28.15
G	Gibbs free energy	general
G_{ij}	Integrated correlation function in Kirkwood–Buff Theory of mixtures	Eq. 25.189
$G(k)$	density of momentum states	Part I; Eq. 11.52
g	molar Gibbs free energy	Part II; general
	acceleration due to gravity	general; Eq. 1.22
g_m	mean molar Gibbs free energy of a mixture	Chapter 25
g_k	degeneracy of the level with energy E_k; the number of states with energy in the range $E_k \le E \le E_k + dE$	Eq. 19.115
g_1	osmotic coefficient	Part II; Eq. 25.125
g_s	Landé magnetic g-factor	Part I; Eq. 5.7
g_{eff}	effective magnetic g-factor	Part I; Eq. 9.15
$g(\epsilon)$	density of one-electron energy states	Eq. 11.54
$g_2(\mathbf{R}_1, \mathbf{R}_2)$, $g_2(R)$	pair correlation function of a liquid	Eq. 11.2; Eq. 23.15
$g_3(\mathbf{R}_1, \mathbf{R}_2, \mathbf{R}_3)$	triplet correlation function of a liquid	Eq. 23.23
$g(t)$	autocorrelation function	Part III; Eq. 29.62
$g_v(t)$	velocity autocorrelation function	Part III; Eq. 29.74
H	enthalpy	general
H	Hamiltonian operator	general; Eq. 3.62
H_{AB}	integral of the Hamiltonian over the orbital functions φ_A^* and φ_B	Part I; Eq. 6.37
H_{HF}	Hartree–Fock Hamiltonian operator	Part I; Eq. 8.16
H_{nuc}	Hamiltonian operator for motion of the nuclei of a molecule in the effective potential of the electrons $E(R)$	Eq. 6.1
H_{elec}	Hamiltonian operator for electronic energy and nuclear potential energy	Eq. 6.1
$H_n(z)$	nth Hermite polynomial	Eq. 4.9
$\mathbf{H}$	magnetic field strength	Table 13.1
$\hat{\mathsf{H}}$	Hamiltonian operator	p. 448
h	Planck's constant $\hbar \equiv h/2\pi$	general
	molar enthalpy	general
h, k, l	Miller indices	Part I; Eq. 11.3
$h(R)$	correlation function in theory of liquids	Eq. 26.89
h_1	1s hydrogenic orbital on atom 1	Part I; p. 360
h_m	mean molar enthalpy of a mixture	Chapter 25
Δh_{fus}	molar heat of fusion	general
Δh_{vap}	molar heat of vaporization	general
Δh_{subl}	molar heat of sublimation	general
$\bar{h}_i$	partial molar enthalpy of component i	general
$\Delta\bar{h}_i \equiv \bar{h}_i$	partial molar enthalpy of component i	Section 14.6
ΔH	heat of formation	general
ΔH^0	standard heat of formation per mole of product	general
$\Delta H^0_{\text{atom'n}}$	enthalpy change on atomization of one mole of a compound under standard state conditions	Section 14.8

Symbol	Meaning	Where used
$\Delta H_{\text{sol'n}}$	integral heat of solution	Section 14.6
ΔH_{∞}	enthalpy change when one mole of solute is added to an effectively infinite amount of solvent	Section 14.6
ΔH_{dil}	integral heat of dilution	Section 14.6
I	intensity of a beam	Eq. 2.8
	electric current	general
	ionic strength	Eq. 26.49
	moment of inertia of diatomic molecule	Eq. 3.129
	nuclear spin quantum number	Eq. 18.15
I_x, I_y, I_z	principal moments of inertia of a polyatomic	Eq. 3.173
I_{IR}	moment of inertia for internal rotation	Eq. 21.129
I_0	intensity of incident light	Eq. 31.22
$I(\text{M})$	ionization potential of M	Eq. 14.46
$\mathbf{i}, \mathbf{j}, \mathbf{k}$	unit vectors in the x-, y-, z-directions	Eq. 3.54
i	net current in an electrolytic cell	Section 31.13
$\mathcal{I}(x, t)$	intensity of a wave at the point x at time t	Eq. 3.47
$\mathbf{J}$	total angular momentum	general; Eq. 5.18
J	rotational quantum number	general; Eq. 7.19
	cell partition function	Part II; Eq. 25.207
J	Joules	general
J	Coulomb integral over atoms A and B	Eq. 6.40
$J(x, y, \ldots/u, v, \ldots)$	Jacobian of the transformation from variables $x, y, \ldots$ to variables $u, v, \ldots$	Eq. 11.18
$\mathbf{J}^{Q}(\mathbf{r}, t)$	vector flux of Q at $\mathbf{r}$, t	Part II; Eq. 20.2
$\mathbf{J}^{m}(\mathbf{r}, t)$	mass flux at $\mathbf{r}$, t	Part II; Eq. 20.6
$\mathrm{J}_k^{m}(\mathbf{r}, t)$	mass flux of component k, relative to the center of mass, at $\mathbf{r}$, t	Part II; Eq. 20.13
$\mathbf{J}^{p}$	momentum flux	Part II; Eq. 20.21
$\mathbf{J}_{ij}^{p}$	component ij of the momentum flux tensor	Part II; Eq. 20.23
j	rotational quantum number (usually J in Parts I and II)	Part III; general
$\mathbf{j}^{r}$	electric current density	Part II; Eq. 29.87
K	Kolmogorov entropy	p. 460
$K(j \mid j')$	Transition probability for $j' \to j$	Eq. 15C.2
$K(T)$	equilibrium constant for a gas phase reaction (independent of pressure)	Eq. 21.64
K_{p}	equilibrium constant in terms of component pressures	Eq. 21.75
K_V	equilibrium constant in terms of component concentrations	Eq. 21.88
K_x	equilibrium constant in terms of component mole fractions	Eq. 21.86
$K(T, p, \text{solvent})$	equilibrium constant for a reaction in a condensed phase	Eq. 25.131
K_a	equilibrium constant in terms of component activities	Eq. 30.82
K_w	ion product of water ($\equiv [H_3O^+][OH^-]$)	Eq. 31.104
$K_V^{\ddagger}$	value of K_V for formation of an activated complex ($K^{\ddagger}$, Eq. 30.79, is the corresponding equilibrium constant in terms of activities)	Part III; Eq. 30.44
K_{AB}	exchange integral over atoms A and B (sometimes K)	Eq. 6.40
K_b	boiling point constant	Part II; Eq. 25.111
K_f	freezing point constant	Part II; Eq. 25.114
k	force constant of a harmonic oscillator	Eq. 2.32
	wave number ($\equiv 2\pi/\lambda$)	Eq. 3.3
	rate coefficient of a chemical reaction	Part III; general
$\mathbf{k}$	wave vector	general
	unit normal to a surface	general
k_B	Boltzmann's constant	general
$k_2(T, p, \text{solvent})$	Henry's law constant	Part II; Eq. 25.60
k_{α}	force constant for the αth normal mode of vibration of a molecule	Part I; Eq. 8.38

Symbol	Meaning	Where used
k_a, k_b	rate coefficients for acid and base catalysis	Part II; Eq. 31.92
k_D	diffusion-controlled limit of rate coefficient	Part III; Eq. 30.103
k_f, k_r	rate coefficients for forward and reverse reactions	Part III; general
k_{uni}	effective unimolecular rate coefficient	Part III; Eq. 30.172
k_∞	high-pressure limit of k_{uni}	Part III; Eq. 30.174
k_1, k_2, . . .	rate coefficients for steps 1, 2, . . . in a reaction mechanism (k_{-1} is the coefficient for the step inverse to step 1)	Part III; general
$\mathbf{L}$	orbital angular momentum	Eq. 3.151
L	magnitude of angular momentum $\mathbf{L}$	Eq. 2.31
L^2	operator for the square of the angular momentum $\mathbf{L}$	Eq. 3.176
L_z	operator for the z-component of the angular momentum $\mathbf{L}$	Eq. 3.176
$\mathbf{L}_i$	orbital angular momentum of the ith electron	Eq. 5.20
L_{ik}	Onsager linear transport coefficient	Eq. 20.91
l	length	general
	mean free path	Part III; Eq. 28.65
	orbital angular momentum quantum number	Eq. 3.162
M	Madelung constant	Eq. 11.41
	total mass (in a system of two or more particles)	general
$M(T)$	chemical potential along the critical isochore	Part II; Eq. 24.67
M_i	molecular weight of component i	Eq. 14.14
$\mathbf{M}$	magnetic moment	Table 13.1
M_L	quantum number for the z-component of the total orbital angular momentum $\mathbf{L}$	Eq. 5.22
M_S	quantum number for the z-component of the total spin angular momentum $\mathbf{S}$	Eq. 5.22
m	mass	general
m_e, m_n, m_p	mass of the electron, neutron, proton	Table 1.1
m_{eff}	effective mass	Eq. 11.95
m_i	molality of component i	general
$m_\pm$	mean ionic molality	Eq. 26.8
m_1	molality of pure solvent	Part II; Eq. 25.137
m	quantum number of angular momentum for a particle in a circular box, or for the z-component of angular momentum of an electron	Eq. 3.142; Eq. 3.161
	z-component of any angular momentum	
m_s	spin quantum number; the z-component of the spin angular momentum of an electron	Eq. 5.2
m_{li}, m_{si}	quantum number for the z-components of the orbital and spin angular momentum of electron i	Eq. 5.22
N	number of molecules in a system	general
N_A	Avogadro's number	general
$\mathcal{N}$	number of replicas in an ensemble	Part II; Chapter 15
$N(E_n)$	number of states with energy E_n	Part I; Eq. 4.17
N_v	number of vacancies in a lattice	Eq. 29.86
n	number density	general
	number of moles	general
	integer	Eq. 2.53
	a quantum number	general
n_i	number of moles of component i	general
	principal quantum number for the ith one-particle quantum state	Part I; Eq. 2.10
n_i^*	effective principal quantum number for the ith one electron quantum state	Part I; Eq. 2.12
$n(v)$	number density of molecules with speed v	Part III; Section 27.5
$P(x)$	probability distribution of x	general
$\mathcal{P}(x)$	probability density of x	general
$\mathcal{P}$	excess pressure in a perturbation	Part III; Eq. 28.125

Symbol	Meaning	Where used
P_{12}	operator indicating permutation of particles 1 and 2	Eq. 6.53
p	pressure	general
	magnitude of momentum **p**	general
	probability aftereffect factor	Part II; Eq. 16.A.12
	molecular orbital bond order	Part I; Eq. 9.4
	reaction probability factor	Part III; Eq. 30.27
$\mathfrak{p}$	unit of pressure	Part II; Eq. 21.12
$\mathbf{p}$	momentum	general
p	momentum operator	general; Eq. 3.36
p_θ, p_ϕ	angular momentum	general; Eq. 3.128
p_ϕ	angular momentum operator	general; Eq. 3.138
p_{app}	applied pressure	general
p_σ	vapor pressure of a liquid along the liquid-vapor coexistence line	Eq. 23.1
p^*	reduced pressure (with respect to pair potential parameters)	Eq. 21.178
$\tilde{p}$	reduced pressure (with respect to the critical pressure)	Chapter 21
p, q	probabilities of positive and negative displacements in Brownian-motion model	Part III; Section 29.2
	numbers of protons on acid and proton-acceptor sites on conjugate base	Part III; Eq. 31.106
Q	electric charge	general
	ratio of activities in an electrochemical-cell reaction	Part III; Section 31.13
Q_α	αth normal coordinate	Eq. 8.34
Q_N	canonical partition function for N particles in a volume V at temperature T	general; Eq. 19.131
q	heat absorbed by a system	general
	spatial coordinate	Part I; Eq. 3.98
	electric charge	Eq. 1.5
$\mathbf{q}$	quadrupole moment	Part I; Eq. 10.6
	heat flux vector	Part II; Eq. 20.64
$q_i(R)$	charge density around ion i	Part II; Eq. 26.74
q_1	single-molecule partition function per unit volume	Parts II and III; Eq. 21.148; Eq. 30.54
R	gas constant	general
	interatomic or intermolecular distance	general
	electrical resistance	Eq. 29.101
R_e	internuclear distance for which $E(R)$ is a minimum	general
R_H	Rydberg constant for hydrogen	Eq. 2.7
R_∞	Rydberg constant for a hydrogenic atom with infinite nuclear mass	Eq. 2.69
$R_{nl}(r)$	solution of the radial wave equation for the hydrogen atom	Eq. 4.18
R^*	reduced distance (with respect to pair potential parameters)	Eq. 26.90
$\mathcal{R}$	excess density in a perturbation	Part III; Eq. 28.126
r	radial distance; radial polar coordinate	general
$\mathbf{r}$	position vector	general
r^*	critical distance for a diffusion-controlled reaction	Part III; Eq. 30.94
r_{kl}	distance between electrons k and l	Eq. 8.14
r_{Kk}	distance between nucleus K and electron k	Eq. 8.14
r_{max}	distance at which the electronic probability density has its maximum	Eq. 4.23
S	entropy	general
	screening constant	Part I; Eq. 2.6
$\mathbf{S}$	spin angular momentum	general; Eq. 5.1
$\mathbf{S}_i$	spin angular momentum of the ith electron	general; Eq. 5.20
S_{AB}	overlap integral over atoms A and B	general; Eq. 6.30
s	molar entropy	general

Symbol	Meaning	Where used
	number of internal oscillators in a molecule	Part III; Section 31.1
s_m	mean molar entropy of a mixture	Chapter 25
$\bar{s}_i$	partial molar entropy of component i	
$\mathbf{s}$	scattering vector	Eq. 23.16
T	absolute temperature	general
T	kinetic energy operator	general
T_e	kinetic energy operator of a molecule when $R = R_e$	Part I; p. 151
T_{vap}, T_{fus}	temperature of vaporization, fusion	general
T^*	reduced temperature (with respect to pair potential parameters)	Eq. 21.178
$\tilde{T}$	reduced temperature (with respect to the critical temperature)	Chapter 21
$\mathcal{T}$	excess temperature in a perturbation	Part III; Eq. 28.127
t	time	general
	centigrade (Celsius) temperature	general
t^*	limiting zero-pressure temperature on the centigrade scale	Eq. 12.34
$t_{1/2}$	half-life for a first-order reaction	Part III; Eq. 30.8
U	internal energy	general
	potential energy of an ionic crystal	Part I; Eq. 11.39
u	molar internal energy	general
	magnitude of the mass velocity (local average velocity)	Part III; Section 28.10
	mobility of a charged particle	Part III; Eq. 29.57
$\mathbf{u}$	velocity	App. 2A
u_m	mean molar internal energy of a mixture	general
$u_2(R)$	pair interaction potential	Part II; general
$u_3(\mathbf{R}_1, \mathbf{R}_2, \mathbf{R}_3)$	potential energy of three molecules	Part II; Eq. 21.159
V	volume	general
	electric potential difference	general
$V(r)$, $V(R)$	potential energy as a function of r, R	general
V	potential energy operator	general, Eq. 3.40b
V^*	reduced volume (with respect to pair potential parameters)	Eq. 21.178
$\tilde{V}$	reduced volume (with respect to the critical volume)	Chapter 21
$\mathbf{V}$	center-of-mass velocity	Eq. 27.14
$\mathcal{V}$	volume of subregion in phase space	Eq. 15A.2
$\mathbf{v}(\mathbf{r}, t)$	local fluid velocity of $\mathbf{r}$, t	Eq. 20.6
$\mathcal{V}_N$	total potential energy of N molecules	Eq. 21.157
$V_{\text{ind}}(R)$	induction energy of interaction between charge distributions	Eq. 10.16
V_f	free volume in solution	Part III; Eq. 30.76
v	molar volume	general
	quantum number for vibrational energy	general
	volume per unit mass	Chapter 20
	speed (magnitude of velocity $\mathbf{v}$)	general
$\bar{v}_i$	partial molar volume of component i	Eq. 25.24
v_m	mean molar volume of a mixture	Eq. 25.17
v_m^E	excess volume per mean mole of mixture	Eq. 25.18
v_x, v_y, v_z	x-, y-, z-components of velocity $\mathbf{v}$	general
$v_{\text{eff}}(k)$	effective velocity	Part I; Eq. 11.94
$\mathcal{W}$	difference between reversible and irreversible heat transfers	Eq. 20.116
w	work done on a system (sometimes W)	general
	parameter in regular solution model	Part II; Eq. 25.29
w_{ad}	adiabatic work	general
w_2	pair potential of average force in a liquid	Eq. 23.18

Symbol	Meaning	Where used
X	intensive generalized force	Eq. 13.13
$\mathbf{X}_i$	force associated with rate of variation of entropy	Eq. 20.89
x, y, z	Cartesian coordinates	general
x_i	mole fraction of component i	general
Y	extensive generalized displacement	Eq. 13.13
$Y_{lm}(\theta, \phi)$	spherical harmonic	Eq. 3.166
y_i	mole fraction of component i in the vapor over a liquid (two-phase, two-component system)	Eq. 25.20
Z	atomic number	general
Z_i	number of elementary charges on ion i	general
Z_N	configurational partition function for N molecules in a volume V at temperature T	Eq. 21.C.5
Z_{12}	collision frequency between species 1 and 2	Eq. 28.17
z	number of electrons transferred in an electrochemical reaction	Chapters 26, 31
α	coefficient of thermal expansion	general
	polarizability	general
	diagonal element of the energy matrix of the Hückel Hamiltonian	Part I; Eq. 9.6
	critical exponent for the temperature dependence of the heat capacity near the critical temperature	Part II; Eq. 24.36
	attenuation constant for sound waves	Part III; Eq. 28.149
	proportionality constant in Hinshelwood-Lindemann mechanism	Part III; Eq. 31.9
	net number of chain carriers produced in a branching step	Part III; Eq. 31.50
	coefficient of thermoelectric potential	Probs. 20.15, 20.16; p. 545
$\alpha(1), \beta(1)$	spin functions for electron 1, with $m_s = \frac{1}{2}$ and $-\frac{1}{2}$, respectively	general; Eq. 6.47
β	$1/k_BT$	general
	nearest-neighbor off-diagonal element of the energy matrix of the Hückel Hamiltonian	Part I; Eq. 9.7
	critical exponent for the temperature dependence of the density near the critical temperature along the coexistence line	Part II; Eq. 24.33
γ	gyromagnetic ratio, magnetic moment per unit of angular momentum	Part I; Eq. 8.52 ff
	ratio of heat capacities, c_p/c_v	general
	critical exponent for the temperature dependence of the compressibility near the critical temperature along the isochore $\rho = \rho_c$, $T > T_c$	Part II; Eq. 24.37
$\gamma_i, \gamma_i', \gamma_i''$	activity coefficient of component i, mole fraction scale, molality scale, molarity scale, respectively	general; Eq. 25.1; Eq. 25.136; Eq. 25.139
$\gamma_\pm$	mean ionic activity coefficient	general; Eq. 26.8
γ_V	$(\partial p/\partial T)_V$	Eq. 23.6
γ_G	Grüneisen constant	Eq. 22.26
$\Gamma(E, V, N)$	total number of quantum states with energy less than E for given N, V	Eq. 15.12
$\Gamma(x)$	Gamma function of x	Eq. 21.170
$\Gamma_n, \Gamma_m, \Gamma_E$	flux densities for numbers of molecules, mass, and energy crossing a plane	Part III; Eqs. 28.65 ff
$\mathbf{\Gamma}_n, \mathbf{\Gamma}_m, \mathbf{\Gamma}_E$	flux density vectors corresponding to Γ_n, Γ_m, and Γ_E ($\Gamma_n = \mathbf{k} \cdot \mathbf{\Gamma}_n$, etc.)	Eqs. 28.67 ff
Γ_{mv}	flux density for momentum: pressure tensor (or stress tensor)	Part III; Eq. 28.75
δ	quantum defect	Part I; Eq. 2.12
	critical exponent for the density dependence of the pressure near the critical density along the critical isotherm	Part II; Eq. 24.35
	range of concentration gradient near an electrode	Part III; Section 31.13
δ_{ij}	Kronecker delta	general
$\delta(x)$	delta function	general
Δ	volume of a lattice cell	Eq. 25.191
Δ_{r-1}	determinant of the derivatives $(\partial^2 g_m/\partial x_i \partial x_j)$, $I, j = 2, \ldots, r$	Eq. 25.169
$\Delta_3(1, 2, 3)$	excess of the energy of three molecules over the sum of the pair interaction energies	Eq. 21.159
ϵ	energy parameter (well depth) in the intramolecular potential	general; Eq. 10.23
ε	energy per molecule	general

Symbol	Meaning	Where used
ϵ_0	permittivity of free space	general; Eq. 1.5
ϵ_F	Fermi energy	Eq. 11.58
ϵ_{n_1,n_2,n_3}	energy of a molecule in a state with quantum numbers n_1, n_2, n_3	general
ε_{ijk}	Levi–Civita symbol	Eq. 20.A.1
ζ	spin-orbit coupling constant	Eq. 5.24
	bulk viscosity	Eq. 20.46
	friction coefficient	Eq. 29.38
η	shear viscosity	general; Eq. 1.21
	ground state degeneracy	Part II; Eq. 18.14
	critical exponent for the distance dependence of the pair correlation function near the critical temperature	Part II; Eq. 24.102
	overpotential	Part III; Section 31.13
	efficiency of a heat engine	Eq. 120.132
	information entropy	p. 460
	packing fraction	Eq. 123.40
η_{ij}	rotation independent strain parameter	Eq. 22.5
θ	gas scale temperature	Eq. 12.29
	empirical temperature	Chapter 12
	Lindemann ratio	Eq. 24.98
	angular polar coordinate	general
	fraction of surface sites covered by adsorbed molecules	Part III; Eq. 31.111
θ^*	gas scale temperature in the limit of zero gas pressure	Eq. 12.31
$\Theta(\theta)$	factor of the wave function, for a particle in a spherically symmetric potential which contains the θ dependence of that function	Eq. 3.155
Θ_{rot}, Θ_{vib}, Θ_{elec}	characteristic temperature for rotational, vibrational, and electronic excitation, respectively	Eq. 21.134; Eq. 21.138; Eq. 21.141
Θ_E, Θ_D	characteristic temperature of the Einstein and Debye models of a solid, respectively	Chapter 22
κ	Debye screening length	Part II; Eq. 26.67
	transmission coefficient in activated-complex theory	Part II; Eq. 30.49
κ_T	isothermal compressibility	general
κ_S	adiabatic compressibility	general
κ_f, κ_r	theoretical rate coefficients for forward and reverse reactions	Part III; Eq. 30.34
λ	wavelength	general
	quantum number for orbital angular momentum of an electron along the axis of a linear molecule	general; Eq. 6.43
	absolute activity	Part II; Eq. 21.A.5
	thermal conductivity	general
	progress variable	Eq. 19.95
	length parameter in theory of diffusion-controlled reactions	Part III; 30.101
	reorganization energy	Section 31.13; Eq. 30
	Lyapunov exponent	Eq. 15A.8
λ_{ij}	parameter in the Born–Mayer repulsion between ions i and j	Eq. 14.33
	strain in a crystal	Eq. 22.2
Λ	quantum number for total orbital angular momentum along the axis of a linear molecule	general
	molar conductance	Part III; Eq. 29.108
μ	reduced mass	general; Eq. 2.74
$\boldsymbol{\mu}$, $\boldsymbol{\mu}_e$	electric dipole moment	Eq. 4.29
μ_0	magnetic permittivity of free space	general; Chapter 18
$\boldsymbol{\mu}_l$	orbital magnetic moment	Eq. 5.4
μ_N	nuclear magneton	Eq. 5.9
μ_B	Bohr magneton	Eq. 5.6

Symbol	Meaning	Where used
$\boldsymbol{\mu}_m$	magnetic dipole moment	Eq. 5.3
$\boldsymbol{\mu}_s$	spin magnetic dipole moment	Eq. 5.7
$\boldsymbol{\mu}_{n'n}$	transition dipole moment between states characterized by quantum numbers n' and n	Eq. 4.33
μ_i	chemical potential of component i	general
μ_i^0	chemical potential of the reference state of component i (the reference state is the pure liquid or solid)	general; Eq. 25.5
$\mu_i^{\ominus}$	chemical potential of the reference state of component i (the reference state is a hypothetical substance having the same properties as does the solute at infinite dilution)	general; Eq. 25.23
μ_i^E	excess chemical potential of component i	Eq. 25.13
μ_{JT}	Joule–Thomson coefficient	Prob. 21.7
ν	frequency in cycles per second (Hz)	general
	number of degrees of freedom of a system	Part II; Chapter 15
$\tilde{\nu}$	wavenumber	general
$\tilde{\nu}_e$	frequency factor in the power series expansion of E_{vib} as a function of the quantum number υ (approximately equal to the fundamental vibration frequency in cm^{-1})	Eq. 7.10
$\tilde{\nu}_e x_e$	second coefficient in the power series expansion of E_{vib} as a function of the quantum number υ	Eq. 7.10
ν_i	stoichiometric coefficient of the ith species in a chemical reaction	general
ν_+, ν_-	numbers of positive and negative ions in a salt	Eq. 26.6
ξ	distance scaled to dimensionless units	Part I; Eq. 3.65
	extent of reaction ($\dot{\xi}$, rate of reaction)	Part III; Eq. 30.2
	scale factor in the geometric mean combining law for mixed species pair interaction	Part II; Eq. 25.211
	correlation length in the critical region of a fluid	Part II; Eq. 24.102
ξ_i	ith normal coordinate	Part II; Eq. 24.84
Ξ	grand partition function	general; Eq. 19.126
π_g, π_u	designations of orbitals of a linear molecule with quantum numbers $\lambda = 1$ and with even and odd parity, respectively	general; Fig. 6.8
π	osmotic pressure	general; Eq. 25.121
Π	designation of an electronic state of a linear molecule for which $\Lambda = 1$	general; p. 203
	Peltier coefficient	Eq. 20.95
ρ	mass density	general
	Hammett reaction parameter	Part III; Eq. 30.112
	scale length in Born–Mayer repulsion between ions	Eq. 14.34
$\tilde{\rho}$	reduced density (with respect to the critical density)	Chapter 21
$\rho(\mathbf{r}, t)$	local mass density	Part III; Eq. 28.114
$\rho_Q(\mathbf{r}, t)$	local density of Q at point $\mathbf{r}$ and time t	Eq. 20.1
$\rho_k(\mathbf{r}, t)$	mass density of component k at $\mathbf{r}$, t	Eq. 20.12
ρ_ε	local density of potential energy in a fluid	Eq. 20.64
$\hat{\rho}$	von Neumann density operator	Eq. 15B.6
ρ_{mn}	element of von Neumann density operator	Eq. 15B.4
σ	total scattering cross section	general; Eq. 2.9
	size parameter in the pair interaction between molecules	Eq. 10.23
	electrical conductivity	Eq. 29.86
	time interval	Part III; Eq. 29.74
	Hammett substitution parameter	Part III; Eq. 30.112
	symmetry number	Eq. 21.111
	surface tension	Table 13.1
$\boldsymbol{\sigma}$	stress tensor	Part II; Eq. 20.43
σ_g, σ_u	designations of orbitals of a linear molecule with quantum number $\lambda = 0$ and with even and odd parity, respectively	general, Fig. 6.8

Symbol	Meaning	Where used
σ_v	symmetry operator for reflection through a plane	Section 31.5
$\sigma(v)$	total scattering cross section	Part III; Eq. 27.62
$\sigma(v, \alpha, \beta)$	differential scattering cross section in laboratory coordinates	Eq. 27.53
$\sigma(v, \chi)$	differential scattering cross section, in relative (center-of-mass) coordinates	Eq. 23.54
$\sigma_R(v)$	cross section for reactive scattering	Eq. 30.32 ff
Σ	designation for an electronic state of a linear molecule for which $\Lambda = 0$	general; p. 165
τ	period	general; Eq. 2.38
	relaxation time	general; Eq. 28.111
	effective lifetime of photoexcited state	Part III; Eq. 31.27
	scattering volume	Section 27.5
τ_f	mean free time	Fig. 28.3
ϕ	practical osmotic coefficient	Fig. 26.20
	angle; azimuthal polar coordinate	general
	work function for electron emission	Part II; Eq. 26.55
	phase shift	Part III; Eq. 28.158
	electric potential	Eq. 30.89
ϕ_A	volume fraction of component A	Eq. 25.20
φ_A, φ_B	atomic orbitals centered on nuclei A, B	Eq. 6.19
Φ	quantum yield for a photochemical reaction	Eq. 31.26
Φ_i	electrostatic potential acting on ion i	Eq. 26.61
Φ_A, Φ_B	wave function for molecules A, B	Eq. 11.35
$\Phi(\phi)$	factor of the wave function for a particle in a spherically symmetric potential which contains the dependence of that function on the angle ϕ	Eq. 3.155
χ	magnetic susceptibility	Footnote 4, p. 118
	relative scattering angle	Fig. 27.3, 27.4
$\boldsymbol{\chi}$	susceptibility tensor of an anisotropic crystal	p. 598
χ_i	ith product orbital	Section 31.6
ψ	wave function	general
	electrostatic potential	Part II; general
	general transport property	Part III
$\psi(x)$	spatial part of the wave amplitude along the x-direction	Eq. 3.59
$\psi(1, 2)$	molecular orbital wave function for electrons 1 and 2	Eq. 6.46
$\psi_{\text{elec}}(r, R)$	solution of the electronic wave equation for fixed nuclei (separated by R)	Eq. 6.3
$\psi_{\text{nucl}}(R)$	factor of the molecular wave function describing rotation and vibration	Eq. 6.2
ψ_{MO}	molecular orbital wave function	Section 6.6; Eq. 6.46
ψ_{VB}	valence bond wave function	Section 6.8; Eq. 6.72
Ψ	wave amplitude	Eq. 3.14
Ψ^*	complex conjugate of Ψ	general
ω	angular frequency ($\equiv 2\pi\nu$)	general
	small element of volume	Part II; Chapter 15
$\boldsymbol{\omega}$	angular velocity	general
	set of angles defining the orientation of a molecule	Part II; Eq. 24.96
$\omega(E)$	density of states at energy E	general
$\bar{\omega}_{sl}$	degeneracy of the sth state of molecule 1	Eq. 21.100
Ω	solid angle	general; Eq. 2.15
$\Omega(E, V, N)$	number of quantum states in the energy range between E and $E + dE$ f or given V, N	general; Eq. 15.12
Z	rate of entropy creation	Eq. 20.85

Symbol	Meaning	Where used
Commonly Used Superscripts and Subscripts		
0	Generally used to denote a standard state, or a change in which both initial and final states are standard states.	Part II
0	Denotes pure component.	
$\ominus$	Denotes the hypothetical reference state defined to be a pure substance which has the same properties as does the solute at infinite dilution.	
*	Denotes the perfect gas reference state, or the limiting value of the gas property when the pressure approaches zero.	
E	Denotes, for a mixture, the excess value of a thermodynamic function over that expected for an ideal mixture, and, for a pure component, the residual value of a thermodynamic function over that expected for a perfect gas at the same temperature and density (the latter is used only in Chapter 23).	
$\bar{z}_i$	Denotes partial molar "zee" of component i.	
i	component i	
V,p,T, . . .	constant volume, pressure, temperature, . . .	
σ	along the coexistence curve	
0	ground state	
m	mean molar value for a mixture	
±	mean ionic quantity	
reac,fus vap,subl inv	value of the thermodynamic function for the process indicated	
′	value (of velocity, impact parameter, etc.) after collision	Part III; Eq. 27.1
′ ″	Designates quantum numbers in the upper and lower states of a transition	Eq. 7.28
*	complex conjugate	general
	excited state	Section 5.3
	unusual configuration	Section 5.3
	value at the maximum of probability distribution	Eq. 15.49
	value corresponding to top of potential energy barrier	Eq. 30.22
‡	value for activated complex	Part III; Eq. 30.44

APPENDIX IV

Searching the Scientific Literature

by Andrea Twiss-Brooks and Fritz Whitcomb

Introduction

Finding chemical and physical data is one of the most difficult tasks in searching the literature, at times frustrating even experienced researchers and information professionals. Chemical and physical data appear in the journal articles, data handbooks and compilations, technical reports, commercial databases available online or on CD-ROM and diskette, and more recently in sites on the World Wide Web. Navigating this vast landscape of resources requires at least a basic understanding of these various resources and how to use them. This appendix is intended to start the physical chemistry student on the search for chemical and physical data by providing references to information guides and printed data resources, general guidelines for effective searching, specific examples of data searches and results, and a selected list of on-line and Web resources that report chemical and physical data.

The print literature remains a valuable source of chemical and physical data, even in this era of desktop computers and widespread access to the Internet. Effective searching of the printed literature requires access to a research library and a significant investment of time. Guidance for searching the printed literature may be found from a research library's science reference staff as well as from a number of guides devoted to the chemical and physical literature, a few of which are listed later.

In addition to the print literature, there are numerous physical and chemical databases available for searching through commercial database sources. Need for familiarity with the search interface and the often high cost involved means that these resources are often available to students only through a science library's search service or other supervised access. However, some providers offer special academic discounts, and many universities have taken advantage of these discounts to offer online searching to their students and research staff. Despite the costs involved, the sophisticated search interfaces offer faster retrieval of high-quality data and sometimes may be the only source of needed data.

In recent years, the growth of the World Wide Web has provided some new and exciting resources for chemical and physical data. In some cases, the Web has provided a simplified search interface and delivery vehicle for commercial sources. In others, it has provided a more effective distribution mechanism for data sources at little or no cost to the user. These factors have allowed many universities and their libraries to offer the student direct access to chemical and physical data resources. The examples in this appendix are taken from both commercial and noncommercial applications on the Web and are intended to provide some idea of the wealth of data available as well as a few starting points in a search for chemical and physical data. There are many excellent guides to searching the chemical and physical print, commercial on-line and Internet-based literature. The selected list of titles below provide overviews of data producing agencies, descriptions and guides to use of print format standard data reference sets, discussions of scope and currency of resources, lists of on-line databases for physical and chemical data, and in some cases, specific search strategies and examples. Many of these titles may be found at any research or university library.

Selected Guides to the Literature of Chemical and Physical Data

Arny, Linda Ray, *The Search for Data in the Physical and Chemical Sciences* (Special Libraries Association, New York, 1984).

Information Sources in Chemistry, 4th ed., R. T. Bottle and J. F. B. Rowland, eds. (Bowker-Saur, London, 1993).

Information Sources in Physics, 3rd ed., Dennis F. Shaw, ed. (Bowker-Saur, London, 1994).

The Internet: A Guide for Chemists, Steven M. Bachrach, ed. (American Chemical Society, Washington, D.C., 1996).

Maizell, Robert E. (Robert Edward), *How to Find Chemical Information: A Guide for Practicing Chemists, Educators, and Students,* 3rd ed. (Wiley, New York, 1998).

Wiggins, Gary, *Chemical Information Sources* (McGraw-Hill, New York, 1991).

www.nist.gov/srd/#begin.htm

General Guidelines for Chemical and Physical Data Searching

Perhaps the most important part of the search for chemical and physical data is the formulation of the question. Clearly, one first must know what is the question before it is certain that the answer has been found. Always review what information is known already that might be helpful with a search. Some of the questions to ask before going to the library or the computer workstation are:

- Will the data be recognized when found? What are the names or labels of the data? Are there synonyms (in English and/or other languages), alternate spellings, standard symbols that are used by different sources?
- What units will the desired data have? Standard SI or other system? What conversions might be necessary?
- What conditions of data measurement are desired? What temperature or pressure? Are the data sought independent of some conditions but not others? Are actual laboratory measurements required, or will extrapolated or calculated values suffice?
- How many data are likely to be found? Will the data be for a common substance with thousands of literature references? Is a Chemical Abstracts Registry Number (CAS RN) available? Can a systematic name be found, or a list of common synonyms for each compound? Is the molecular formula or structure known?
- When the data are found, how are they known to be accurate? What are the approximate values expected? What institutions and/or research groups are reliable sources of the data? Will older values be accurate, or have the methods of measurement changed significantly over time?
- Where should the search for data begin? Would the data most likely be found in the older printed literature? Are there guides or references that might point to a likely resource? Have some of the standard handbooks been consulted to see what kind of data is already compiled in those resources?

Examples of Searches for Chemical and Physical Data

Since there are many excellent guides to searching the printed literature for chemical and physical data, the examples given here will focus on data searches in Web-based resources. However, it is important to remember to search the printed literature thoroughly for data suspected to have been measured some years ago. Many values that have been previously reported in journals, handbooks and other resources will not be repeated in online resources. This is just a small sample of the many printed resources available that contain collections of physical and chemical data:

Beilstein, Friedrich Konrad, *Beilstein Handbook of Organic Chemistry,* 4th ed., 5th suppl. series (Springer-Verlag, Berlin, 1984–).

Chemical Rubber Company, *CRC Handbook of Chemistry and Physics,* 79th ed., David R. Lide, ed. (CRC Press, Cleveland, Ohio, 1998).

Journal of Physical and Chemical Reference Data, Malcolm W. Chase, ed. (American Chemical Society, Washington, D.C., 1972–).

Landolt, Hans, Richard Börnstein, and K. H. Hellwege, *Zahlenwerte und Funktionen aus Naturwissenschaften und Technik. Neue Serie (Numerical Data and Functional Relationships in Science and Technology. New Series)* (Springer, Berlin, 1961–).

TRC Thermodynamic Tables. Hydrocarbons (Thermodynamic Research Center, Texas A & M University, College Station, TX, 1985–).

The three examples shown next are not intended as a substitute for more formal search instruction, database documentation, or consultation with an information specialist trained in chemical and physical literature searching. The examples are intended rather to illustrate a few of the possibilities available to students and researchers, and to demonstrate some general principles of data searching. They are also intended to point out some pitfalls for searchers to avoid. Consulting a local expert (instructor, librarian or other experienced person) for help in planning searchers for coursework or research is strongly encouraged. The URLs (uniform resource locators) for Web resources mentioned in the examples are included in the list at the end of the appendix.

Example 1: Electronic Spectra of Aromatic Hydrocarbons

An area of significant research is the investigation of theoretical models of aromatic hydrocarbons and the measurement of molecular spectra to test these models. Studies of spectra obtained at extremely low temperatures are an important contribution to this research. Identification of an article reporting a high-resolution, low-temperature spectrum of naphthalene using a bibliographic database is presented. There are many excellent chemical and physical databases to choose from, including INSPEC (Physics Abstracts) and SciFinder Scholar (Chemical Abstracts), to name two. The search shown here was performed using a Web-based version of the INSPEC database. The search terms are entered via a Web form (Fig. IV.1) to construct a search strategy.

All citations, abstracts and subject keywords for articles published from 1969 to the present are searchable. In the search shown, the words *naphthalene* and *high resolution* are combined using a logical operator *and* to give a set of articles containing the word in the title, abstract, or subject fields in the record (Fig. IV.2).

An example of a citation retrieved by the above search strategy is shown in Fig. IV.3. The search resulted in about 45 citations. More citations may be retrieved by broadening the scope of the search by using synonyms, truncation to allow different endings, or using alternate spellings. If too many articles are retrieved by a particular search, it may be necessary to limit the search by specifying a span of years,

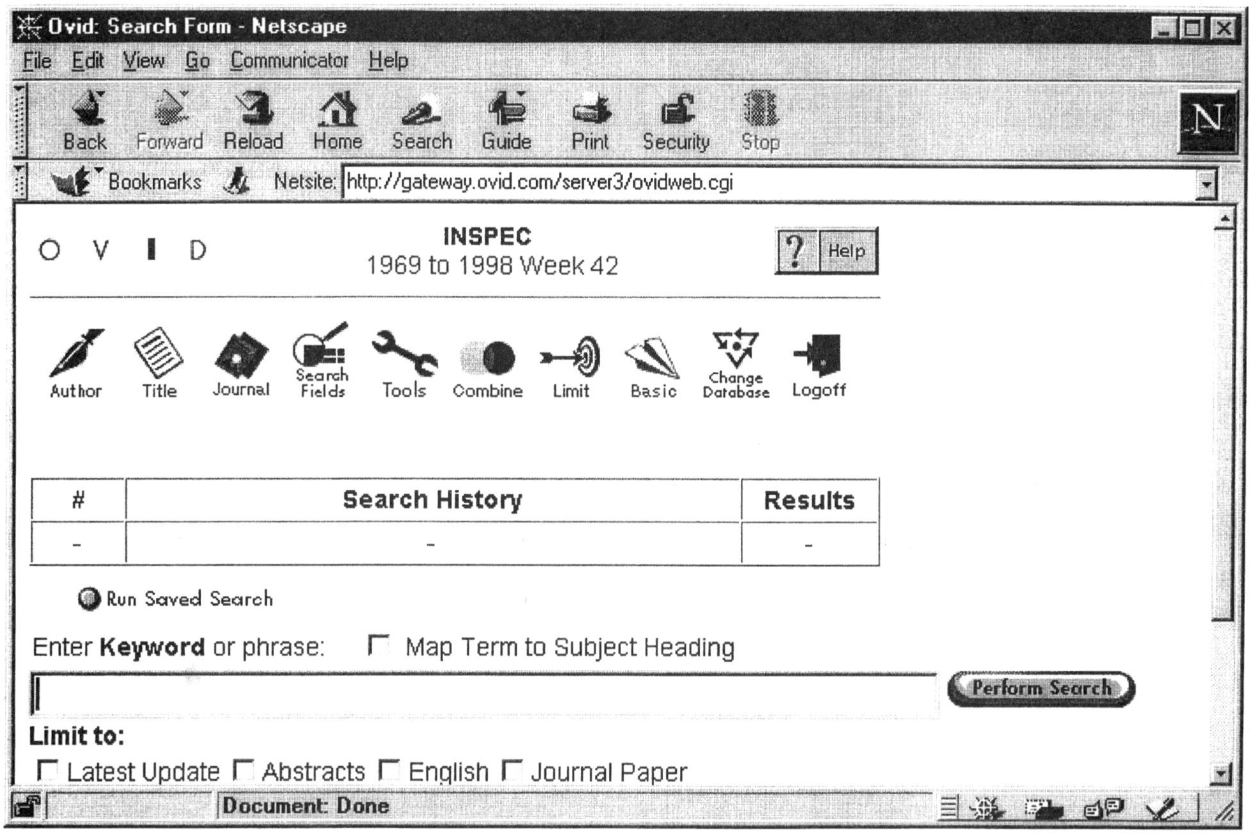

Figure IV.1 Search screen for *INSPEC via OVID* database.

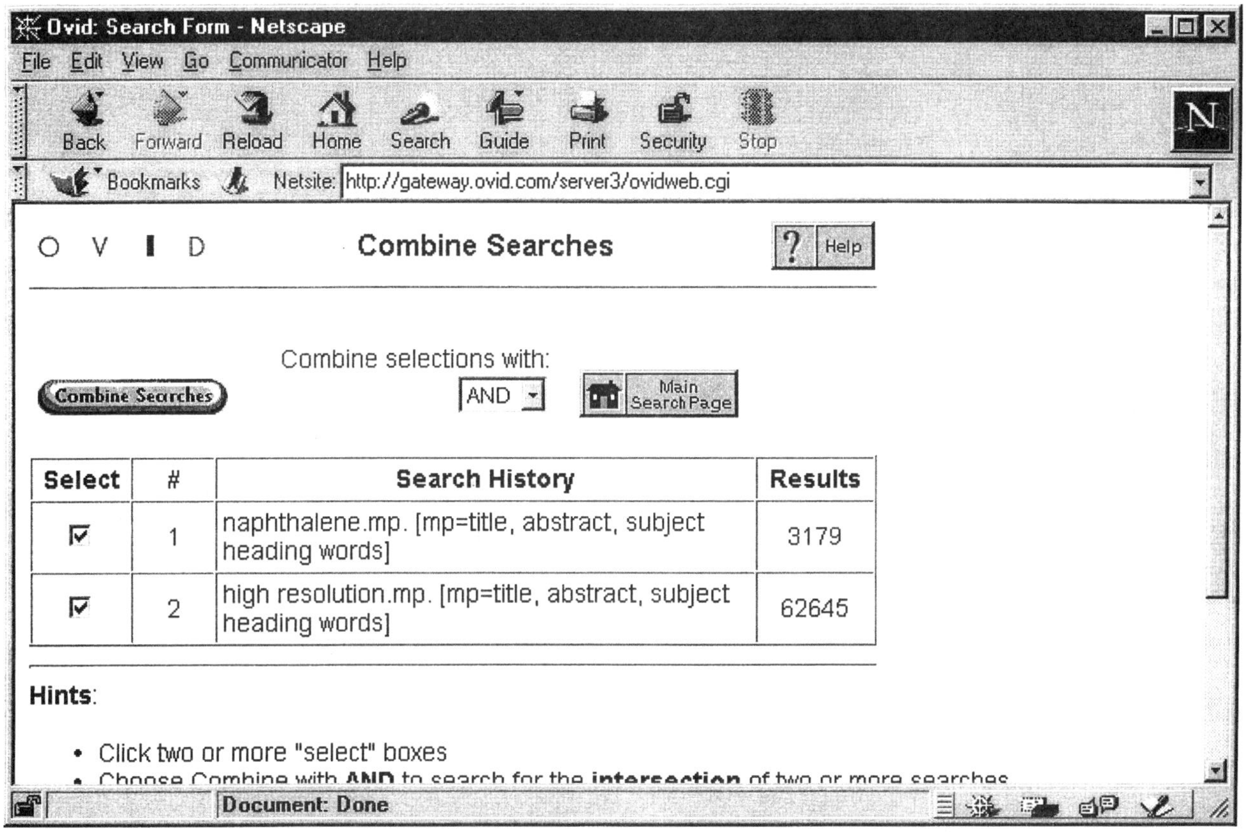

Figure IV.2 Combining search terms in *INSPEC via OVID* database.

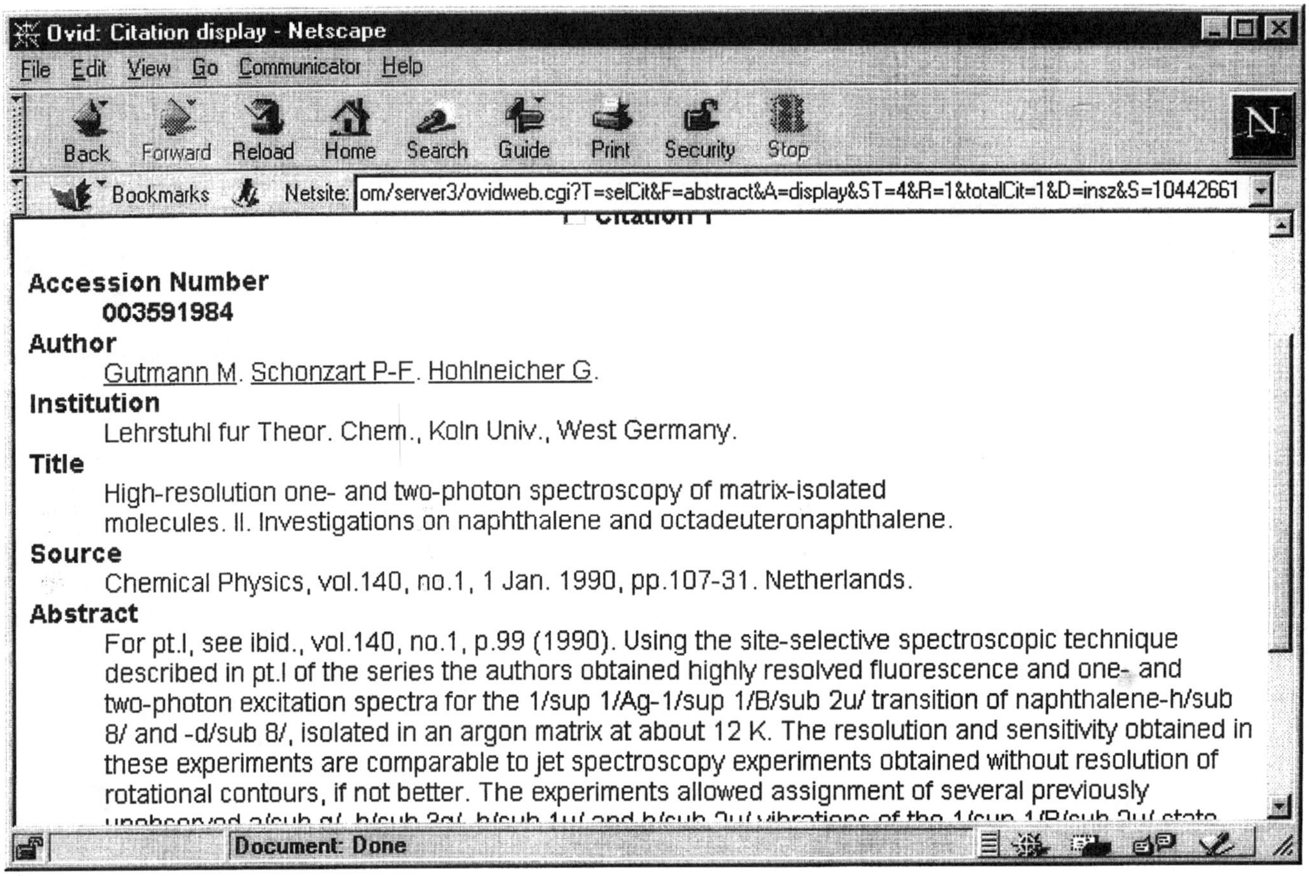

Figure IV.3 Sample citation and abstract retrieved from *INSPEC via OVID* database.

combining the subject words with particular author names, or including additional search terms.

Often, an article abstract will contain important data values from the original article, although it is usually advisable to consult the article itself for complete experimental details and conditions. Full-text on-line journals are currently available (including titles produced by the American Institute of Physics, the Institute of Physics, and several commercial publishers), and it may be possible to retrieve the article from the Web or other networked sites. However, if this option is not available, users may still go to the library to obtain the original report for full details of the complete data available.

Example 2: Thermodynamic Properties for NH_3 (Ammonia)

There are very simple substances of economic and research importance that have a wealth of chemical and physical data available. Chemical manufacturing and purifying process industries rely on these data in their plant designs. Researchers doing precise theoretical calculations or interpreting experimental measurements also need reliable data. In many cases, this means identifying sources of evaluated data from highly reliable sources. The next example demonstrates the use of one source of evaluated thermophysical and thermochemical data for ammonia over a range of temperatures.

It is possible to examine a number of reliable print sources for data on ammonia. Electronic data sources are also available. One organization that produces a number of databases is the National Institute of Standards Technology (NIST) through their Standard Reference Data program. The *Standard Reference Data Series* includes about 50 databases, including materials properties, surface data, atomic and molecular physics, and chemical kinetics titles. NIST also produces a Web resource called the *NIST Chemistry WebBook*. The *NIST Chemistry WebBook* contains thermochemical data for 5000 organic and small inorganic compounds, thermophysical property data for 16 fluids and a variety of other data on a number of compounds, including spectra, ion energetics, and reaction thermochemistry. After connecting to the *NIST Chemistry WebBook,* it is possible to search for a particular substance by chemical name, molecular formula, partial formula, or other identifier (e.g., Chemical Abstracts Service Registry Number). In this example, the molecular formula, NH_3, is used to retrieve the appropriate record from the database (Fig. IV.4). In some search interfaces, it is necessary to enter the molecular formula in a particular order (e.g., Hill order), or with special syntax for the numeric subscripts, so always consult the help files, search examples, or other available documentation for guidance if zero answers in a search are found.

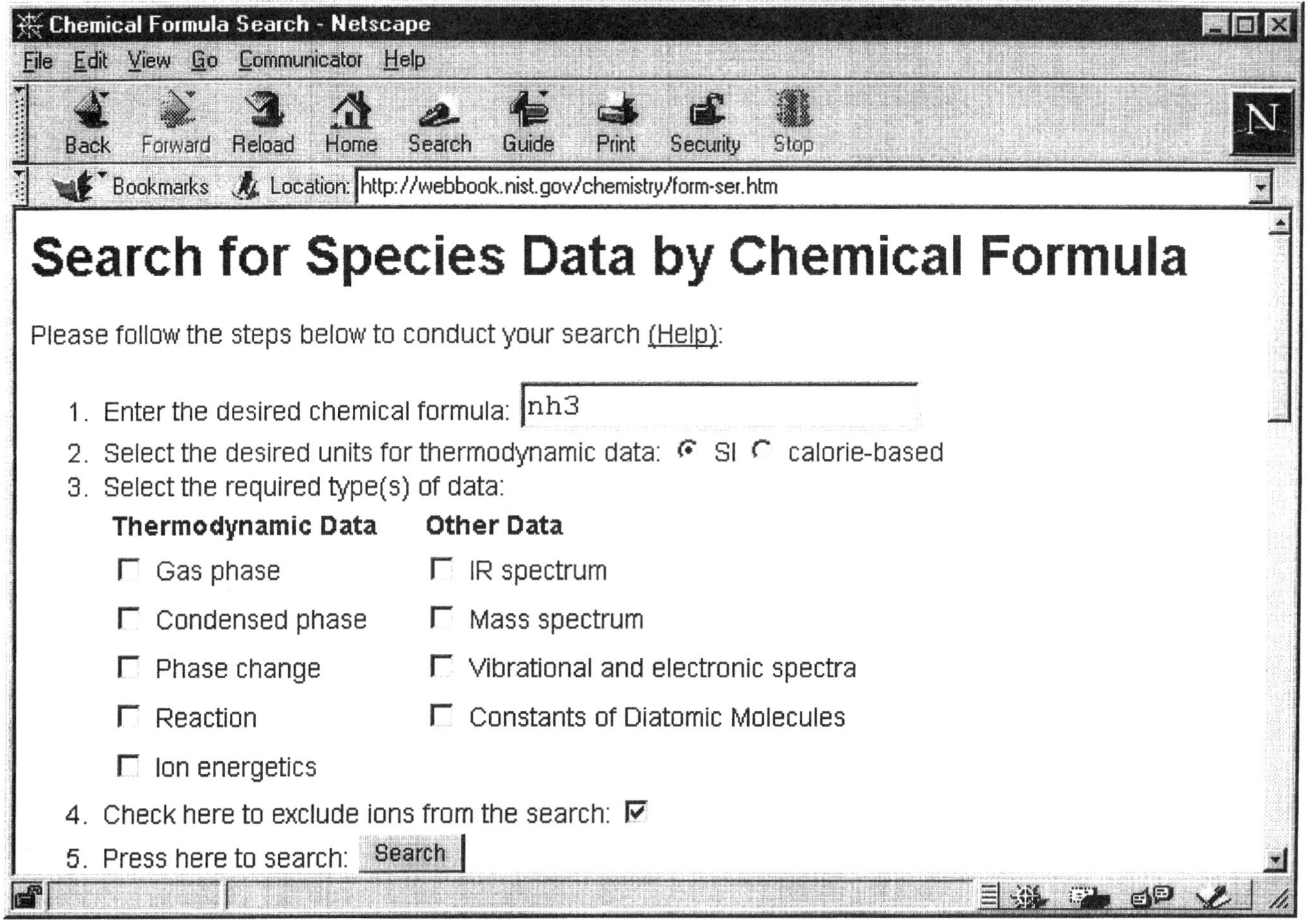

Figure IV.4 Search form for *NIST Chemistry WebBook,* molecular formula option.

After completion of the search, the results screen indicates structure, alternate names, and available data for ammonia. The Chemical Abstracts Registry number (CAS RN) and molecular weight are also given. Note the data available include thermodynamic data, reaction data, fluid properties, ion energetics and spectral data. Choosing fluid properties will lead to another input screen (Fig. IV.5) for specifying the parameters of the fluid properties data table.

Let's examine the isobaric fluid properties data for ammonia at 200 MPa and over a range of temperatures of 200–500 °K. When searching for chemical and physical data, it may be necessary to make conversions in units for temperature, pressure or other criteria to obtain the desired results. The isobaric data retrieved may be viewed in tabular or graphical format. The isobaric fluid data for ammonia is shown in chart format in Fig. IV.6.

References for the source(s) of the data in the *NIST Chemistry WebBook* are given for all available data. The availability of references for any data compilation is an important feature that should be consulted to judge the quality of the data. Another indication of the quality is acceptance of the data by other scientific or research organizations. For example, the Committee on Data for Science and Technology (CODATA) is an organization dedicated to the production and distribution of critically evaluated numeric data. Many of the *NIST Chemistry WebBook* data values are marked as CODATA accepted values. CODATA also has a list of approved fundamental physical constants available on the Web (see the list of Web sites listed at the end of this appendix for the URL).

Example 3: Kinetic Data for Ozone Reactions with Chlorinated Atmospheric Pollutants

Chlorofluorocarbons, or CFCs, have been implicated in the destruction of the ozone layer surrounding the Earth. A hole in the ozone layer discovered over Antarctica has spurred an increase in scientific and public interest in the reaction of various chemical pollutants with atmospheric gases. This example asks for articles reporting recent studies on the kinetics of atmospheric reactions involving chlorinated species and ozone. Since this is a topic familiar even to the nonscientist, there is a lot of information to be found on the World Wide Web. However, starting a review of ozone reaction chemistry and kinetics by doing a simple text word search on *ozone* and *CFCs* using a Web search engine would retrieve hundreds or even thousands of hits, many of which are of dubious value. Adding the word *kinetics* to the search

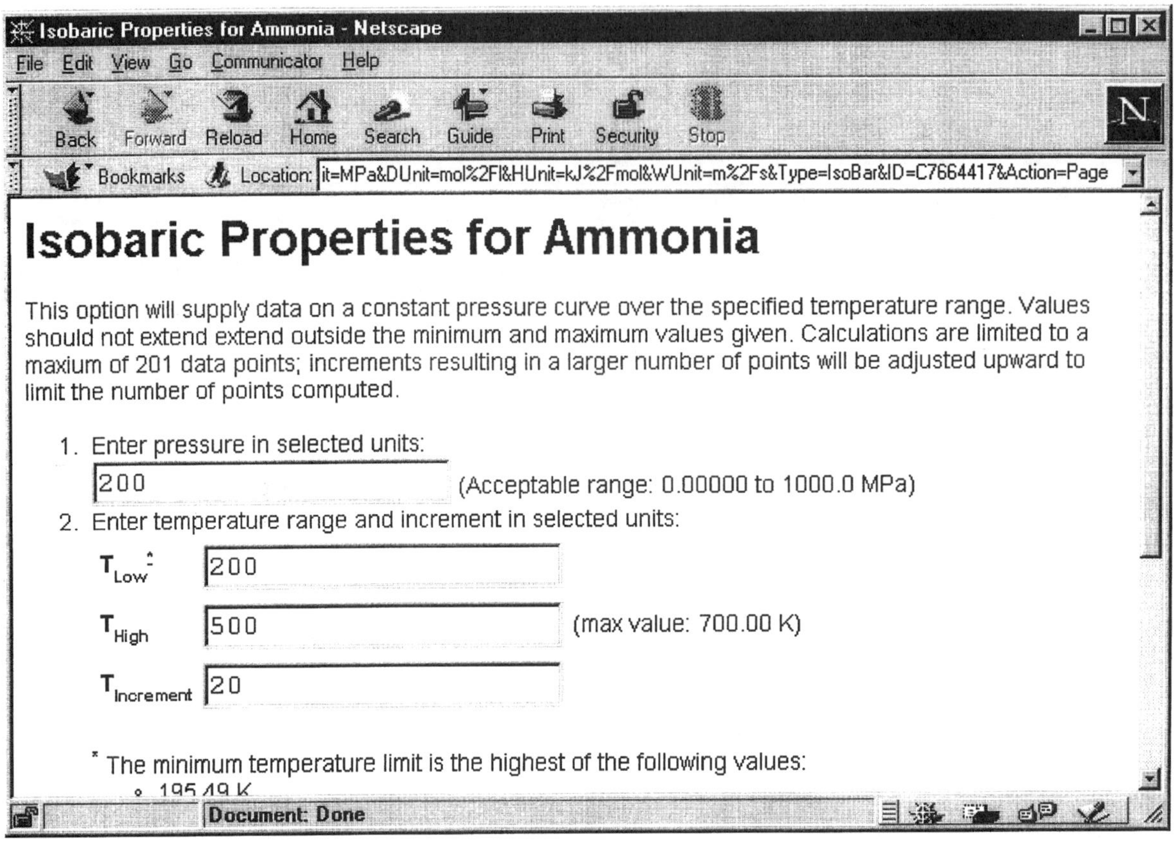
Isobaric Properties for Ammonia - Netscape

File Edit View Go Communicator Help

Back Forward Reload Home Search Guide Print Security Stop

Bookmarks Location: it=MPa&DUnit=mol%2Fl&HUnit=kJ%2Fmol&WUnit=m%2Fs&Type=IsoBar&ID=C7664417&Action=Page

Isobaric Properties for Ammonia

This option will supply data on a constant pressure curve over the specified temperature range. Values should not extend extend outside the minimum and maximum values given. Calculations are limited to a maxium of 201 data points; increments resulting in a larger number of points will be adjusted upward to limit the number of points computed.

1. Enter pressure in selected units:
   200 (Acceptable range: 0.00000 to 1000.0 MPa)
2. Enter temperature range and increment in selected units:
   T_{Low}*: 200
   T_{High}: 500 (max value: 700.00 K)
   $T_{Increment}$: 20

* The minimum temperature limit is the highest of the following values:
   ∘ 195.49 K

Document: Done

Figure IV.5 Form for defining parameters of isobaric properties of ammonia data display in *NIST Chemistry WebBook.*

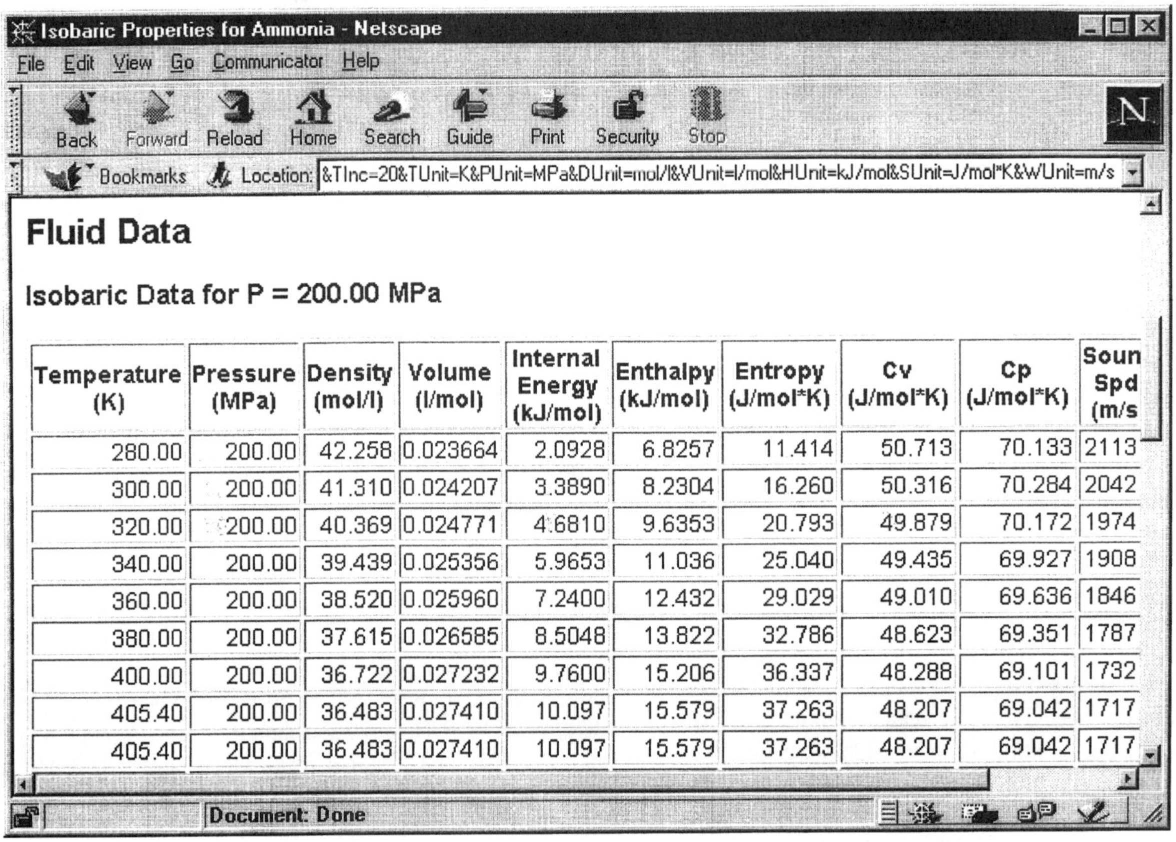
Isobaric Properties for Ammonia - Netscape

File Edit View Go Communicator Help

Back Forward Reload Home Search Guide Print Security Stop

Bookmarks Location: &TInc=20&TUnit=K&PUnit=MPa&DUnit=mol/l&VUnit=l/mol&HUnit=kJ/mol&SUnit=J/mol*K&WUnit=m/s

Fluid Data

Isobaric Data for P = 200.00 MPa

Temperature (K)	Pressure (MPa)	Density (mol/l)	Volume (l/mol)	Internal Energy (kJ/mol)	Enthalpy (kJ/mol)	Entropy (J/mol*K)	Cv (J/mol*K)	Cp (J/mol*K)	Soun Spd (m/s
280.00	200.00	42.258	0.023664	2.0928	6.8257	11.414	50.713	70.133	2113
300.00	200.00	41.310	0.024207	3.3890	8.2304	16.260	50.316	70.284	2042
320.00	200.00	40.369	0.024771	4.6810	9.6353	20.793	49.879	70.172	1974
340.00	200.00	39.439	0.025356	5.9653	11.036	25.040	49.435	69.927	1908
360.00	200.00	38.520	0.025960	7.2400	12.432	29.029	49.010	69.636	1846
380.00	200.00	37.615	0.026585	8.5048	13.822	32.786	48.623	69.351	1787
400.00	200.00	36.722	0.027232	9.7600	15.206	36.337	48.288	69.101	1732
405.40	200.00	36.483	0.027410	10.097	15.579	37.263	48.207	69.042	1717
405.40	200.00	36.483	0.027410	10.097	15.579	37.263	48.207	69.042	1717

Document: Done

Figure IV.6 Isobaric fluid data for ammonia (displayed as HTML table) found in *NIST Chemistry WebBook.*

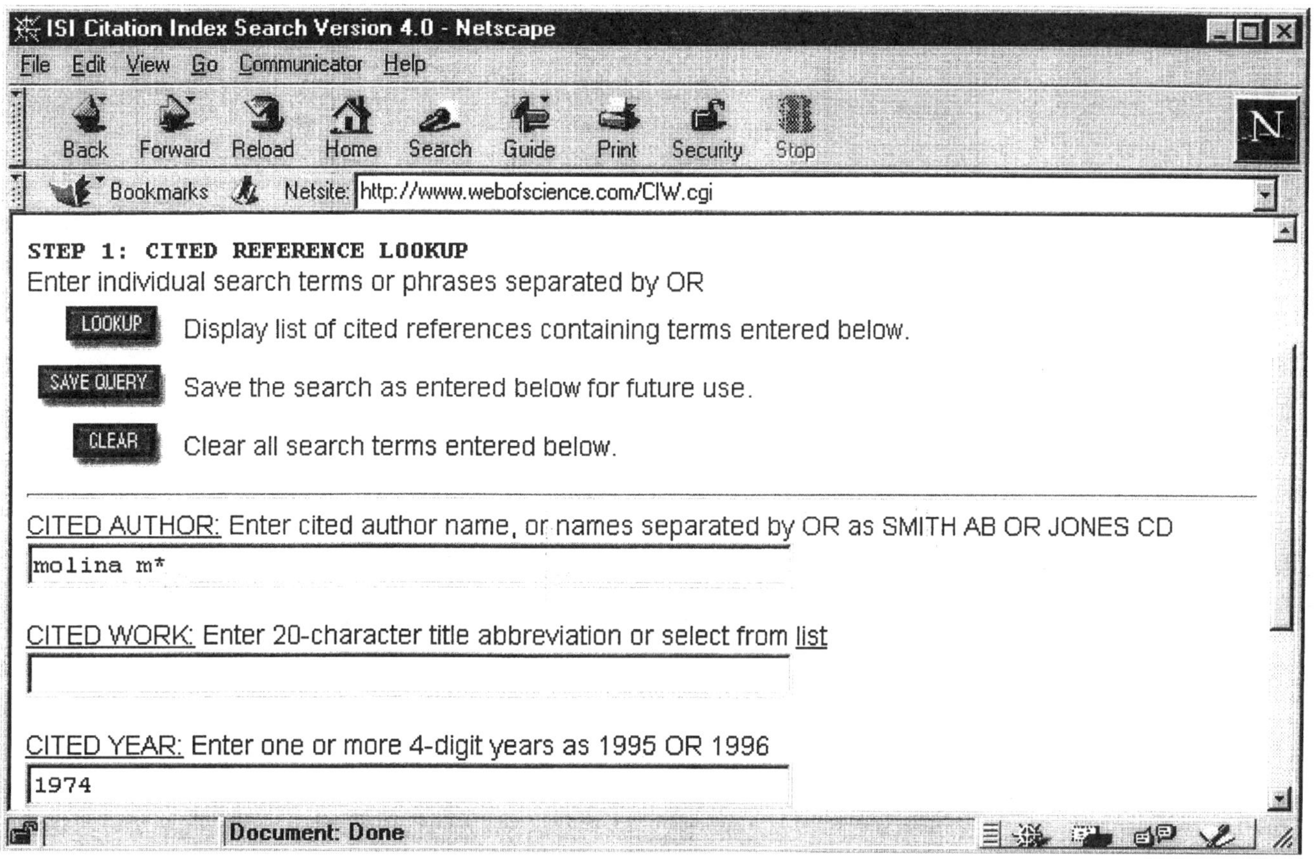

Figure IV.7 Cited reference search form for *Web of Science (Science Citation Index).*

might help decrease the number of answers, but there are other, more direct ways of focusing a search.

One approach which might be more effective and cover the most appropriate range of the available scientific literature is a citation search. This approach relies on identifying a "classic" paper in a field and searching for subsequent articles that list the "classic" paper in their references. In this case, the classic paper is fairly easy to identify, once one remembers that there was a Nobel Prize awarded to some scientists working in the area of atmospheric ozone chemistry. Mario Molina and F. Sherwood Rowland were two of the corecipients of the Nobel Prize for their work on ozone and CFCs. By tracing the citation history of a particular paper they published, later research efforts that published kinetic and other reaction studies on ozone and various species can be identified. To trace this history, use the Web version of the *Science Citation Index* called *Web of Science.* This database is available only at sites that have licensed it; its on-line counterpart is available from a number of database providers. There is also a printed version available.

In *Web of Science,* search terms are entered into a Web form. Molina's name is truncated after the first initial, since his middle initial may not be known, or he may not use it in publications. Since there are many, many M. Molina's listed in the cited references in this database, narrowing the lookup down to the year in which the paper was published will retrieve manageable results (Fig. IV.7). By looking up some biographical information on Molina (on the Nobel Prize web site) before starting a citation search, it can be found that Molina and Rowland published their landmark paper in 1974.

Once the lookup is performed, see that there is one particular paper published by Molina in 1974 that has been very heavily cited since then (Fig. IV.8). This paper appeared in volume 249 of the journal *Nature.* To verify that it is the correct paper, you probably could go to the library and look at this reference. However, performing the cited reference search at this point should give evidence from the results as to whether or not the papers deal with atmospheric ozone reactions.

The results retrieved are for research articles on atmospheric ozone chemistry. In fact, a number of articles retrieved might have been missed if the search had involved using species or compound names or formula, since many of the articles discuss compounds related to ozone and CFC's. For the particular reference shown below, some of the rate constants for various reactions of interest are given in the abstract (Fig. IV.9).

The cited references approach is an excellent way to bring together related papers without constructing a complex search statement including many formula or name search terms. This type of approach is also useful in research areas where compilations of data have not been assembled, or in areas where publication of data is in a broad range of

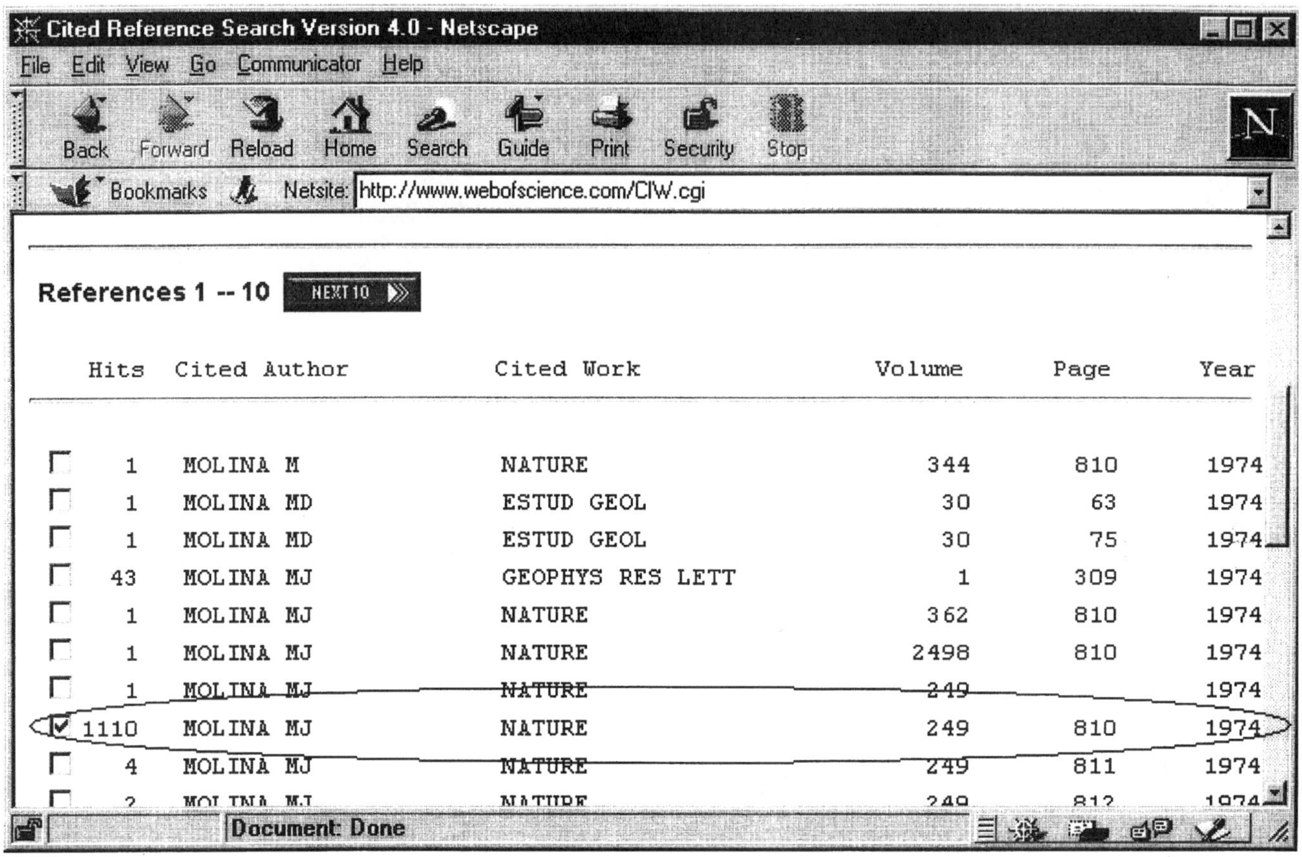

Figure IV.8 List of papers written by M. Molina in 1974 that are cited by later papers as reported in *Web of Science (Science Citation Index)*.

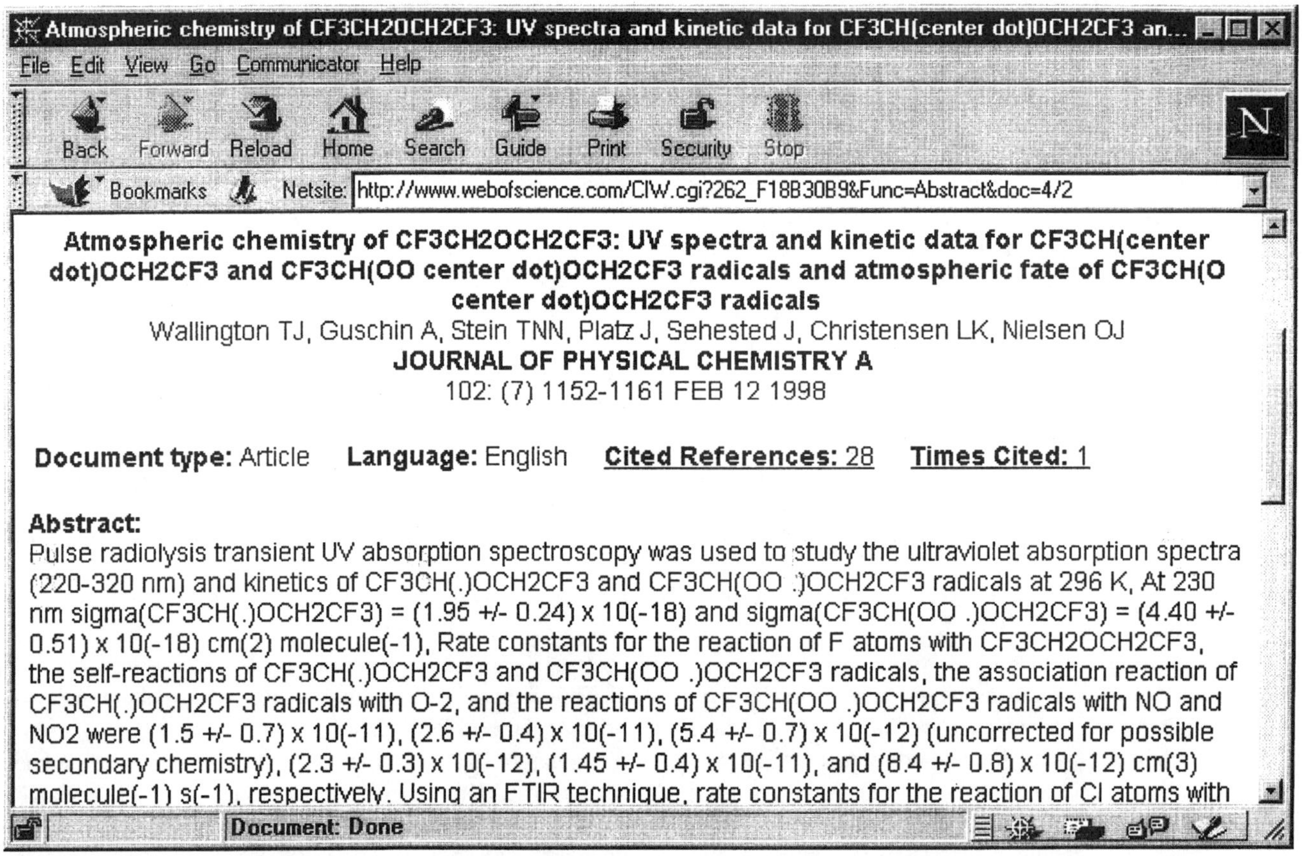

Figure IV.9 Sample citation and abstract for article citing Molina, M. J., and Rowland, F. S., *Nature* 249:810–814, 1974.

journals in a variety of fields like chemistry, physics, atmospheric science, and environmental studies.

Conclusion

The range of resources available to the searcher for data in the chemical and physical sciences should now be known. Successful searches require planning and preparation, as well as practice. The following list of resources on the World Wide Web are just a few of the many places to begin the search for chemical and physical data:

Andrea Twiss-Brooks, Chemistry Librarian/Chemical Information Specialist
Fritz Whitcomb, Bibliographer for Physics, Astronomy, and Technology
The John Crerar Library, The University of Chicago
http://www.lib.uchicago.edu/LibInfo/Libraries/Science/

Some Web Sites of Interest

http://www.lib.uchicago.edu/LibInfo/SourcesBySubject/Chemistry/data_bibliography.html
An annotated bibliography of print and Web resources for data collections maintained at the University of Chicago.

http://www.nist.gov/srd/jpcrd.htm
Cumulative index to the *Journal of Physical and Chemical Reference Data.*

http://physics.nist.gov/cuu/Constants/
CODATA recommended values of fundamental physical constants.

http://ojps.aip.org/journals/doc/JCPSA6-home/top.html
Journal of Chemical Physics home page.

http://webbook.nist.gov/
NIST Chemistry WebBook.

http://www.isinet.com/
Institute for Scientific Information, producer of *Web of Science.*

http://www.nobel.se/prize/index.html
The Electronic Nobel Museum Project, including a searchable index of Nobel laureates.

Index

Page references of tables and figures are in italics

Table of Atomic Weights

Element	Symbol	Atomic Number	Atomic Weight
Actinium	Ac	89	(227)[a]
Aluminum	Al	13	26.98154
Americium	Am	95	(243)
Antimony	Sb	51	121.75
Argon	Ar	18	39.948
Arsenic	As	33	74.9216
Astatine	At	85	(210)
Barium	Ba	56	137.34
Berkelium	Bk	97	(247)
Beryllium	Be	4	9.01218
Bismuth	Bi	83	208.9804
Boron	B	5	10.81
Bromine	Br	35	79.904
Cadmium	Cd	48	112.40
Calcium	Ca	20	40.08
Californium	Cf	98	(251)
Carbon	C	6	12.01115
Cerium	Ce	58	140.12
Cesium	Cs	55	132.9054
Chlorine	Cl	17	35.453
Chromium	Cr	24	51.996
Cobalt	Co	27	58.9332
Copper	Cu	29	63.546
Curium	Cm	96	(247)
Dysprosium	Dy	66	162.50
Einsteinium	Es	99	(254)
Erbium	Er	68	167.26
Europium	Eu	63	151.96
Fermium	Fm	100	(257)
Fluorine	F	9	18.99840
Francium	Fr	87	(223)
Gadolinium	Gd	64	157.25
Gallium	Ga	31	69.72
Germanium	Ge	32	72.59
Gold	Au	79	196.9665
Hafnium	Hf	72	178.49
Hahnium[b]	Ha	105	(260)
Helium	He	2	4.00260
Holmium	Ho	67	164.9304
Hydrogen	H	1	1.00797
Indium	In	49	114.82
Iodine	I	53	126.9045
Iridium	Ir	77	192.22
Iron	Fe	26	55.847
Krypton	Kr	36	83.80
Kurchatovium[c]	Ku	104	(260)
Lanthanum	La	57	138.9055
Lawrencium	Lr	103	(256)
Lead	Pb	82	207.19
Lithium	Li	3	6.941
Lutetium	Lu	71	174.97
Magnesium	Mg	12	24.305
Manganese	Mn	25	54.9380
Mendelevium	Md	101	(258)
Mercury	Hg	80	200.59
Molybdenum	Mo	42	95.94
Neodymium	Nd	60	144.24
Neon	Ne	10	20.179
Neptunium	Np	93	237.0482
Nickel	Ni	28	58.71
Niobium	Nb	41	92.9064
Nitrogen	N	7	14.0067
Nobelium	No	102	(255)
Osmium	Os	76	190.2
Oxygen	O	8	15.9994
Palladium	Pd	46	106.4
Phosphorus	P	15	30.97376
Platinum	Pt	78	195.09
Plutonium	Pu	94	(244)
Polonium	Po	84	(210)
Potassium	K	19	39.098
Praseodymium	Pr	59	140.9077
Promethium	Pm	61	(147)
Protactinium	Pa	91	231.0359
Radium	Ra	88	226.0254
Radon	Rn	86	(222)
Rhenium	Re	75	186.2
Rhodium	Rh	45	102.9055
Rubidium	Rb	37	85.4678
Ruthenium	Ru	44	101.07
Samarium	Sm	62	150.4
Scandium	Sc	21	44.9559
Selenium	Se	34	78.96
Silicon	Si	14	28.086
Silver	Ag	47	107.868
Sodium	Na	11	22.98977
Strontium	Sr	38	87.62
Sulfur	S	16	32.06
Tantalum	Ta	73	180.9479
Technetium	Tc	43	98.9062
Tellurium	Te	52	127.60
Terbium	Tb	65	158.9254
Thallium	Tl	81	204.37
Thorium	Th	90	232.0281
Thulium	Tm	69	168.9342
Tin	Sn	50	118.69
Titanium	Ti	22	47.90
Tungsten	W	74	183.85
Uranium	U	92	238.029
Vanadium	V	23	50.9414
Xenon	Xe	54	131.30
Ytterbium	Yb	70	173.04
Yttrium	Y	39	88.9059
Zinc	Zn	30	65.38
Zirconium	Zr	40	91.22

[a] Value in parentheses is the mass number of the most stable or best-known isotope.
[b] Suggested by American workers but not yet accepted internationally.
[c] Suggested by Russian workers. American workers have suggested the name Rutherfordium.